ENGINEERING THERMODYNAMICS

McGRAW-HILL BOOK COMPANY

New York St. Louis San Francisco Auckland Bogotá Düsseldorf
Johannesburg London Madrid Mexico Montreal New Delhi
Panama Paris São Paulo Singapore Sydney Tokyo Toronto

ENGINEERING THERMODYNAMICS

SECOND EDITION

WILLIAM C. REYNOLDS
Department of Mechanical Engineering
Stanford University

HENRY C. PERKINS
Department of Mechanical Engineering
University of Arizona

ENGINEERING THERMODYNAMICS

34567890 KPKP 78321098

This book was set in Times New Roman.
The editors were B. J. Clark and Douglas J. Marshall;
the designer was Anne Canevari Green;
The production supervisor was Leroy A. Young.
New drawings were done by J & R Services, Inc.
Kingsport Press, Inc., was printer and binder.

Library of Congress Cataloging in Publication Data

Reynolds, William Craig, date
Engineering thermodynamics.

Includes index.
1. Thermodynamics. I. Perkins, Henry Crawford, joint author. II. Title.
TJ265.R38 1977 621.4'021 76-43284
ISBN 0-07-052046-1

CONTENTS

PREFACE

This text was developed as an alternative version of the senior author's book, *Thermodynamics*. The primary spirit of these two books is the same; microscopic arguments are used to provide insight into the basic macroscopic postulates. The parent text treats a broad range of applications in engineering, physical chemistry, and includes introductory chapters in statistical thermodynamics, kinetic theory, and irreversible thermodynamics. In contrast, this book concentrates along the lines of more traditional engineering courses. The applications possess a strong engineering flavor, and introductory chapters on applied one-dimensional gas dynamics and heat transfer are included.

Throughout the text, the value of a systematic methodology in analyses is emphasized. Such an approach is absolutely essential and should be required in the student's problem assignments. A lack of understanding of the fundamentals of engineering frequently is caused by students consistently starting problems "in the middle." Overly easy homework problems can often be successfully solved in this manner, and we have purposely provided longer and more difficult problems, particularly in the later chapters where several of the thermodynamic principles can be brought to bear in a single analysis. We have found that getting into the analysis of simple thermodynamic systems as soon as possible provides good motivation for futher developments in theory. For this reason, energy-balance applications are taken up before the introduction of second-law concepts. This arrangement also provides a period for digestion of state and first-law concepts and helps spread the introduction of new ideas more evenly over the course.

Our objective has been to develop the subject matter in a way that retains the generality and simplicity of purely macroscopic thermodynamics and yet draws upon the student's insight into microscopic matters. To this end the microscopic arguments are used to provide an intuitive basis for macroscopic postulates; the laws of thermodynamics are *not* derived from microscopic postulates. This approach preserves the generality of macroscopic thermodynamics and at the same time places the roots for energy, entropy, and temperature firmly in the microscopic world.

Some modest changes in the ordering of second-law material have been made for this edition. While the microscopic insights that are characteristic of the first edition are retained, the exemplary microscopic calculations have been placed in a position where they may be omitted by the instructor who prefers to follow a more tranditional macroscopic approach. More emphasis has been placed on the concept of entropy production for we have found in our own teaching that this assists the students in understanding and working with the second law. The format for applied second-law analysis is more strongly stated, and a number of new examples of second-law analysis have been added. In these examples emphasis is placed on analysis of the best possible performance, and availability concepts are introduced. Motivation for engineering applications of the second law is provided by the obvious importance of such analysis to the national program of energy conservation.

This second edition is very strongly "dimensionally bilingual," with equal emphasis given to metric (SI) and English units. We believe that it is important that today's engineers think metric but also be fully comfortable with the English units. There are many reasons for this, including the need to be able to communicate with persons who are not comfortable with metric units and the need to use data in English units (today very little of the information actually used by engineers is available in SI). Examples, problems, and thermodynamic data in both SI and English units are included.

Thermodynamics is often characterized as a difficult subject. Indeed, if one's approach is to memorize every equation developed in the course, the subject will be very difficult. However, we urge the student to adopt a fundamental approach; work to understand the concepts and develop the ability to apply the basic principles in a systematic way. The student who takes this approach will find that the subject is really quite easy, and that it provides a tremendously useful set of tools for engineering analysis.

This book could not have been written without the continued encouragement and suggestions of faculty colleagues and students at our two institutions. In particular, we both obtained a real appreciation for the value of systematic methodology in engineering thermodynamic analysis from Professor A. L. London. We are also indebted to colleagues at other institutions who, through use of our earlier edition, have helped us improve the current version. And to our wives and families, who patiently endured our discussions and long hours over manuscripts, we express our special appreciation.

William C. Reynolds
Henry C. Perkins

ENGINEERING THERMODYNAMICS

CHAPTER ONE

SOME INTRODUCTORY CONSIDERATIONS

1·1 THE NATURE OF THERMODYNAMICS

Thermodynamics is one of the most important areas of engineering science. It is the science used to explain how most things work, why some things do not work the way that they were intended, and why other things just cannot possibly work at all. It is a key part of the science engineers use to design automotive engines, heat pumps, rocket motors, power stations, life support systems, gas turbines, air conditioners, firefighting equipment, artificial kidneys, superconducting transmission lines, chemical refineries, high-power lasers, and solar heating systems. Students who have an interest in any of these need to understand and use thermodynamics.

Thermodynamics centers about the notions of *energy*; the idea that energy is conserved is the *first law of thermodynamics*. It is the starting point for the science of thermodynamics and for engineering analysis. A second concept in thermodynamics is *entropy*; entropy provides a means for determining if a process is possible. Processes which produce entropy are possible; those which destroy entropy are impossible. This idea is the basis for the *second law of thermodynamics*, about which you may have heard. It also provides the basis for an engineering analysis in which one calculates the maximum amount of useful power that can be obtained from a given energy source, or the minimum amount of power input required to do a certain task. These calculations can be made without any specific notions about the nature of the systems involved, and it is this great generality that gives thermodynamics its tremendous power.

A clear understanding of the ideas of energy and entropy are essential for one who needs to use thermodynamics in engineering analysis. Traditionally thermodynamics has been regarded as a difficult subject, and hundreds of textbooks on the subject have been written in an effort to develop better clarity and rigor. Most of these books ignore the fact that matter is inherently molecular, and try to develop the ideas by forgetting that molecules exist. This *classical* or *macroscopic* approach to thermodynamics, which is mathematically rather simple, is of considerable intellectual interest, but because it is so abstract it is hard to grasp for someone who does not already know the subject. Another approach is to start at the molecular level and develop equations from this base. But this *microscopic* approach is mathematically rather involved and difficult for someone who does not already know the basic ideas. We shall follow an intermediate approach, drawing upon microscopic ideas where these provide understanding of the macroscopic mathematics one actually uses in engineering analysis. We think this approach makes thermodynamics easily understood, and this is highly desirable for such an important subject.

There is another important difference between thermodynamics as taught, say, by physicists and chemists, and as taught and used by engineers. The pure scientist is interested primarily in analyzing specific pieces of matter that interact with one another, while engineers are usually interested in analyzing complex systems through which matter is flowing. Scientists are interested in using thermodynamics to predict and relate the properties of matter; engineers are interested in using this data, together with the basic ideas of energy conservation and entropy production, to analyze the behavior of complex technological systems. In this book we deal with both points of view, but our emphasis is clearly on the engineer's needs and approach to thermodynamics.

Figure 1·1 is an example of the sort of system of interest to engineers, a large central power station. In this particular plant the energy *source* is petroleum in one of several forms, or sometimes natural gas, and the function of the plant is to convert as much of this energy as possible to electric energy and to send this energy down the transmission line. Simply expressed, the plant does this by boiling water and using the steam to turn a turbine which turns an electric generator. The simplest such power plants are able to convert only about 25 percent of the fuel energy to electric energy. The plant shown in Fig. 1·1 converts approximately 40 percent; it has been ingeniously designed through careful application of the basic principles of thermodynamics to the hundreds of components in the system. The design engineers who made these calculations used data on the properties of steam developed by physical chemists who in turn used experimental measurements in concert with thermodynamic theory to develop the property data. Plants presently being studied could convert as much as 55 percent of the fuel energy to electric energy, if they indeed perform as predicted by thermodynamic analysis. Improvement in power plant efficiency is one of the main objectives of national energy conservation programs.

Figure 1·1. The Moss Landing power station. (*Courtesy of Pacific Gas and Electric Co.*)

With the advent of the energy crisis, thermodynamics has enjoyed renewed popularity in engineering schools. As oil prices rise, other energy forms become competitive, and a great deal of new and exciting engineering will have to be done to develop these new energy resources: solar, geothermal, wind, coal gasification, biomass energy plantations, and eventually fusion. In each case thermodynamics will play a crucial role in the engineering analysis. In addition, there is great need to develop ways to do things using less energy. An engineer with a solid background in thermodynamics and the ability to use it accurately in engineering analysis will be able to participate in these important technological developments.

An important attribute of a good engineer is the ability to work accurately in a careful and organized manner. We cannot emphasize enough the importance of a systematic methodology, without which easy problems become hard and much time can be wasted stewing over wrong answers or building devices that will not work. In parallel with the developments of thermodynamic theory, we shall present a methodology we have found to be quite effective in engineering analysis. Understanding of the basic thermodynamic concepts and principles and the ability to apply them in engineering are the primary objectives of our study.

In this chapter we shall attempt to establish a point of view through discussion of ideas already familiar to the student. The fundamental approach and philosophy adopted in this review of the basic concepts, models, and laws of related branches of physics will be carried over to the new thermodynamic ideas in subsequent chapters.

1·2 CONCEPTS, MODELS, AND LAWS

Concepts form the basis for any science. These are ideas, usually somewhat vague (especially when first encountered), which often defy really adequate definition. The meaning of a new concept can seldom be grasped from reading a one-paragraph discussion. There must be time to become accustomed to the concept, to integrate it with prior knowledge, and to associate it with personal experience. Inability to work with the details of a new subject can often be traced to inadequate understanding of its basic concepts.

The physical world is very complicated, and to include every minute detail in a theoretical analysis would be impracticable. Science has made big steps forward by use of *models* which, although always representing some simplification over reality, reduce the mathematics to a tractable level. The range of validity and utility of the resulting theory is consequently restricted by the idealizations made in formulating the model. Newtonian mechanics is quite adequate for analysis of the great majority of everyday mechanical processes, and inclusion of relativistic effects in such mechanical analysis is an unnecessary complication. However, in many instances such effects are important, and it is the responsibility of the user of any theory to know both its bases and its limitations.

Concepts and models are not enough in themselves for a physical theory. These notions must be expressed in appropriate mathematical terms through basic equations, or *laws*. We choose to look upon a physical law as a contrivance of man that allows him to explain and predict phenomena of nature. Such predictions will be only as accurate and encompassing as the models on which the laws are based, and as new information is gathered and new understanding is developed, man may find it convenient, or perhaps necessary, to alter the basic laws. For example, mechanics is a direct outgrowth of Kepler's astronomical studies and his laws relating to the motion of planets about the sun. Newton generalized these observations and formed new, more basic laws, from which Kepler's rules could be deduced as special consequences. Later Newton's mechanics became merely a special case of Einstein's relativistic mechanics. In general, laws are replaced not because they are incorrect, but because their range of validity is restricted. Such was the case in the early development of thermodynamics, where at one time heat was thought of as something contained within matter. A useful but extremely limited caloric theory of heat, built upon this concept, was discarded more than a century ago; unfortunately, carryover of this misconception inhibits understanding of contemporary thermodynamics.

In many fields of science the concepts are very close to everyday experience, and the difficulties are primarily mathematical in nature. In most of thermodynamics the converse is true; the mathematics is not complicated but the concepts are sometimes difficult to grasp at the beginning, and most of the errors in thermodynamic analysis arise because of lack of clarity in either concepts or methodology. For this reason we shall spend a good deal of time on these matters; they should not be taken lightly, even though it may not be evident why so much attention is paid to apparently small details.

1·3 CONCEPTS FROM MECHANICS

Let's review some basic ideas from mechanics, the science that describes the motion of objects. Practically all problems in thermodynamics can be handled with newtonian mechanics, i.e., without reference to relativistic effects; exceptions include analyses of the interior of stars and the flow of gases at speeds approaching that of light. The mechanics we shall use is entirely newtonian.

The ideas from mechanics we shall find most important are *force*, *mass*, *velocity*, *acceleration*, *work*, *torque*, *kinetic energy*, and *potential energy*. Some of these are concepts and some are definitions. Newton's laws of motion are also important, though some are really parts of other concepts and not laws at all. Mechanics is important because measurement of energy is ultimately tied to measurement of work, in both newtonian and relativistic thermodynamics. Work is in turn defined in terms of forces and motions. Force is a basic concept that must be very well understood by the student of thermodynamics. We expect that you are familiar with the concept; you should find the following presentation different and, hopefully, refreshing.

Force. Force is a concept, and so it defies definition. We can use words like "a push" or "a pull" to help explain what we mean, but these are not definitions of force. However, these words do suggest that a force has a point of application and a direction. And, since something has to do the pushing or pulling, a force must somehow be connected with the interaction between two things. Pushes and pulls tend to produce acceleration, bending, denting, or other behavioral changes in the object pushed or pulled, and so force is an essential idea in describing the reasons for these sorts of changes. These changes can be produced in a given system by interaction with an infinite variety of other systems. For example, people could push a car, a truck could push the car, or a strong wind could blow the car along a road. If the pushes of the people, truck, and wind were all identical, the car would experience the same motion, and blindfolded riders in the car could not tell why they were moving. In analyzing the car we would mentally replace whatever it was that was doing the pushing by a force (Fig. 1·2), and we could calculate the motion of the car in terms of the magnitude of the force without any reference to what was producing it. So, force is a convenient abstraction we use to represent mentally the pushing or pulling interaction between things.

When we use arrows passing through particular points to identify forces, we imply that forces are vectorlike quantities having magnitude, direction, and point of application. So, the mathematics of forces is that of vectors, if due regard

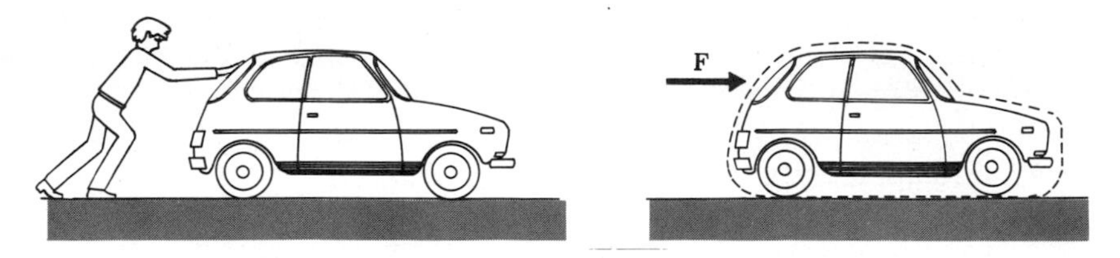

Figure 1·2. Forces replace external pushes or pulls

is given to the point of application. Thus the net effect of two forces acting on the same point is exactly as would be produced by the vector sum of the forces. This vector nature of force is a key part of the force concept.

Force cannot be made into a useful quantitative idea without the addition of two other conceptual aspects. The first is usually billed as *Newton's third law:* If one object exerts a force **F** on another, then the second object exerts an equal but opposite force on the first. This is really part of the force concept; without it forces cannot be measured, hence the law cannot be tested. The second additional part of the force concept is the notion that it takes a net unbalance of force to produce acceleration of the object on which the forces act; if the net forces are balanced (i.e., if the vector sum is zero), the object will not accelerate. This is called *Newton's first law*, but it is really part of the concept of force; it is impossible to establish a means for measuring forces without invoking the law, hence the law cannot be tested.

No conceptual quantity becomes operationally useful until some way for its measurement has been established. One possible way of setting up a scale for force is to select some standard spring and say that the force it exerts is some selected constant times its deflection. This scheme has the distinct disadvantage of making the force scale dependent on the choice of material in the spring, among other factors. Suppose someone else set up a similar scale, based on a different kind of spring; the two scales could be adjusted to agree at one point but could not be expected to agree elsewhere. To each one the other would be nonlinear. It is always more desirable to devise scales of measure that are completely independent of the nature of any substance. In principle it is possible to do this for force, taking advantage of the notion that the resultant force on a stationary body is zero. Imagine selecting any reproducible force, such as that produced by a selected spring compressed some selected amount, and designating this as a unit force. Let this force act on a body in sole opposition to two identical forces selected so as to keep the spring at its standard deflection when the body is motionless (the two identical forces could be obtained from any two identical springs, for example). The two identical forces must each be half the unit force, and either can be used to measure such a force (see Fig. 1·3). This process can be continued, and we can collect a set of springs, each measuring some rational fraction or multiple of the unit force. We can therefore, in principle, measure any unknown force to any desired degree of accuracy. The force scale is unique in that it is independent of the

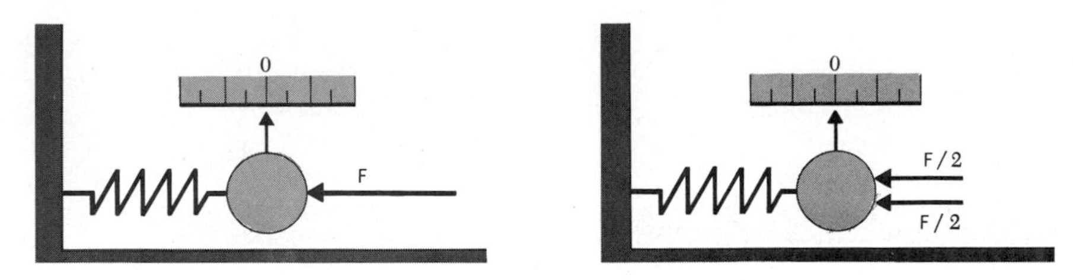

Figure 1·3. A unique force scale can be established using symmetry and the concepts of force

nature of any substance. It will be the same regardless of the material of which the springs are made.

Torque. Torque is another aspect of force that is important in engineering. Recall that the torque, or moment, of a force about a particular point is defined as the product of the force and the distance from the point to the force vector.

The *law of the lever* is a torque balance on a stationary object. It is not a law at all, but is instead something that one can prove using the ideas of force discussed above and one of the most basic ideas for scientific reasoning: *symmetry*. The thought processes involved are shown in Fig. 1·4. The symmetry of the coplanar force set of Fig. 1·4*a* requires that it be a *zero set*, meaning that the forces would produce neither translational nor rotational accelerations if they acted on the bar. Figure 1·4*b* also shows a zero set, and the sum of these two zero sets (Fig. 1·4*c*) forms a third *nonsymmetric zero set* which is equivalent to the force set in Fig. 1·4*d*. In Fig. 1·4*d* the ratio of the forces at A and C is the inverse of the ratio of their distance from point O. This argument can be continued, and one can construct an infinite number of zero sets with one force at O, another at C, and the third at some point A along the bar. In every case the ratio of the force at A to that at C is found to be exactly the inverse of the ratio of their distances from the potential pivot point O; hence, the law of the lever:

$$\frac{\mathsf{F}_A}{\mathsf{F}_C} = \frac{L_C}{L_A}$$

This may be recast as

$$\mathsf{F}_A L_A = \mathsf{F}_C L_C \qquad (1\cdot 1a)$$

or

$$\mathsf{T}_A = \mathsf{T}_C \qquad (1\cdot 1b)$$

Equation (1·1*b*) states that the torque of force F_A, about the pivot point O, is exactly balanced by the torque of force F_B.

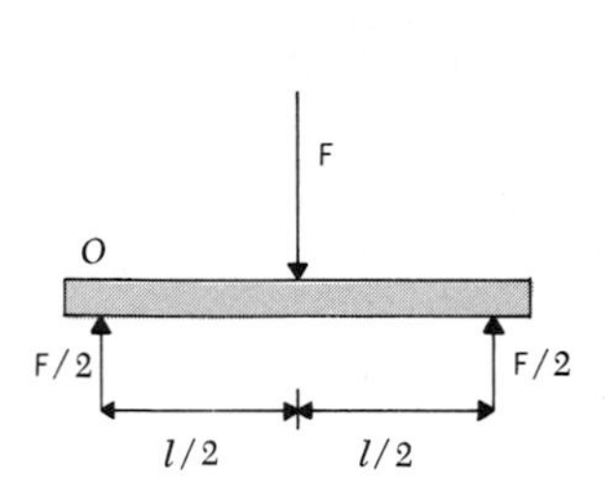

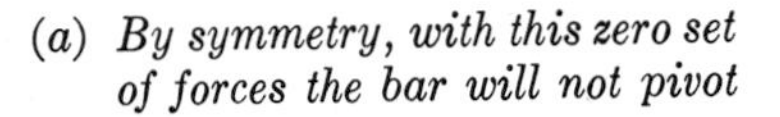
(a) *By symmetry, with this zero set of forces the bar will not pivot*

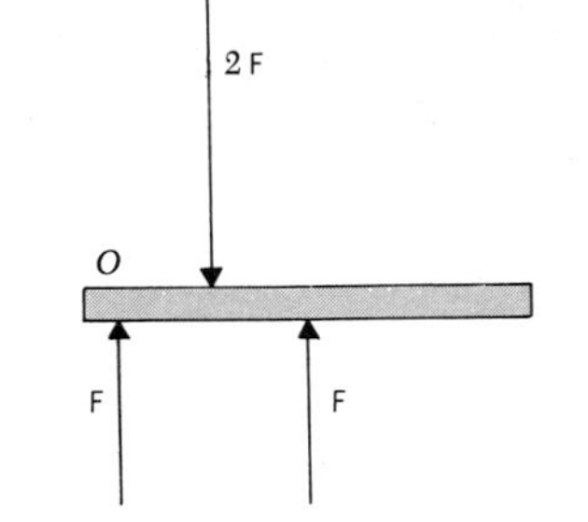

(b) *By symmetry, this is also a zero set*

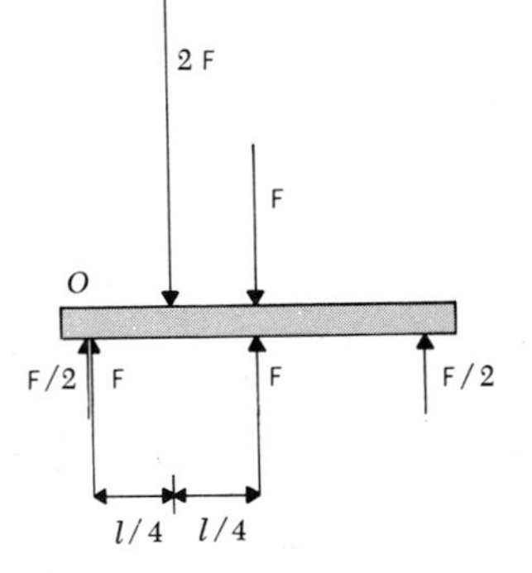

(c) *Therefore this is also a zero set*

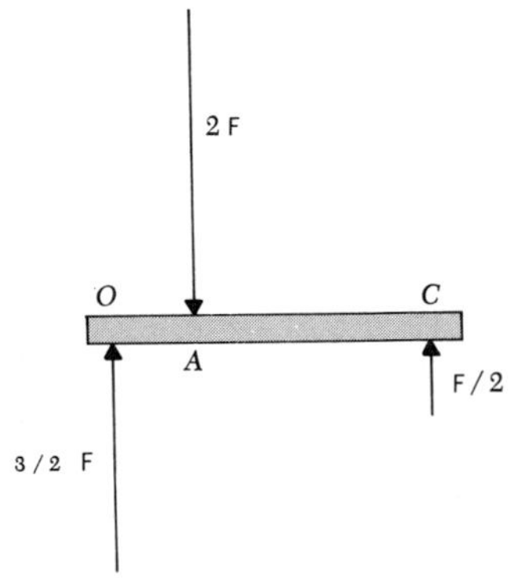

(d) *So this must be a zero set*

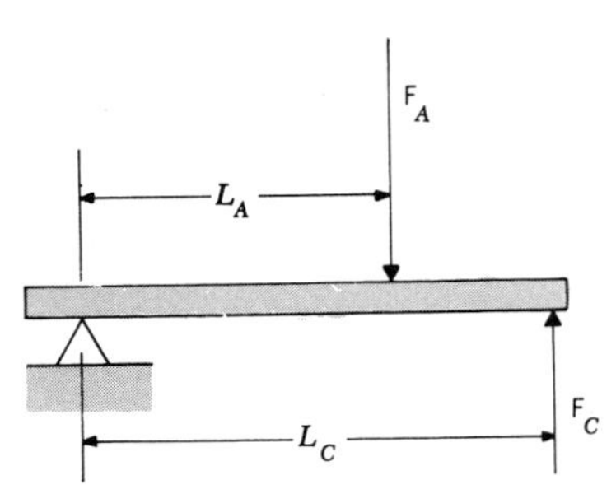

Figure 1·4. The law of the lever can be deduced by symmetry

Mass. Quantitative prediction of the acceleration of bodies requires the introduction of another familiar concept, *mass.* A body acted on by an unbalanced force does not suddenly increase its speed, but rather accelerates gradually at a rate dependent on the magnitude of the force. The *mass* of an object is conceived as a property characteristic of its resistance to velocity change. Any two objects

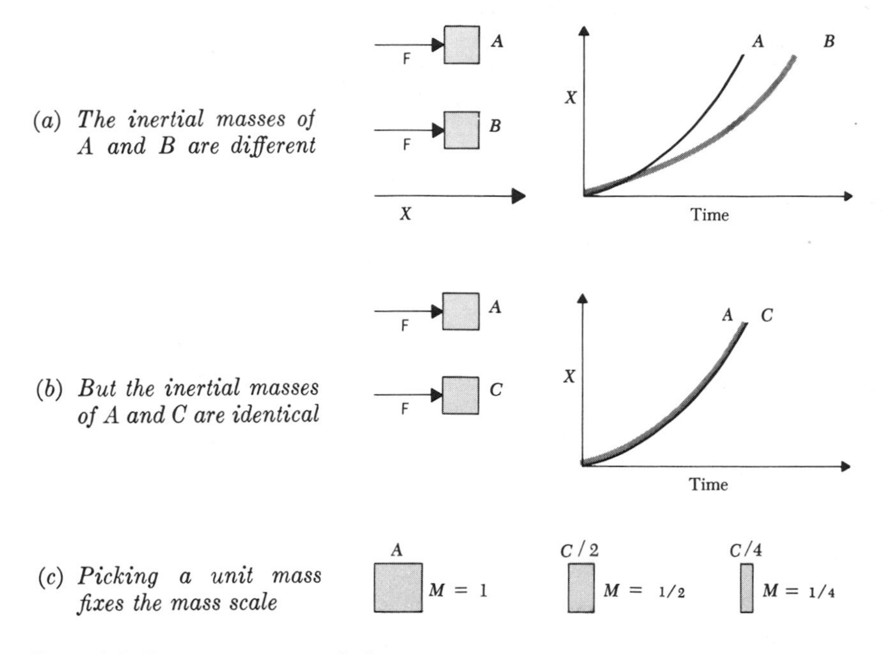

Figure 1·5. Devising a unique scale for mass using symmetry arguments

that undergo identical translational accelerations when acted upon by the same force have identical masses.

An idea inherent in the newtonian concept of mass is that the mass of two objects taken together is the sum of their individual inertial masses; cutting a homogeneous body into two identical parts produces two identical masses, each with half the original mass. This idea is equivalent to the law of *conservation of mass*. In principle, a unique scale for mass may be constructed using the same sort of arguments used to construct a force scale. The reasoning involved is indicated in Fig. 1·5.

Gravitation. Any object that is dropped will accelerate as it falls, even though it is not in physical contact with any other body. To explain this behavior humans had to conceive *gravitational forces*, which one body, such as the earth, can exert on another, even though they are far apart. The gravitational attraction of two objects depends upon the mass of each and the distance between them. The equation expressing this relationship is called the *gravitational law*, and its form can be deduced using mental experiments. The idea is outlined in Fig. 1·6. Let F represent the force exerted by an object having mass M_1 on a second object with mass M_2. Then, place an object identical to M_1 next to M_1. It too will pull with force F on M_2. Think of the two identical objects as a single object having mass $M_1' = 2M_1$; we see that the force exerted on M_2 by this object is doubled, i.e., the force is proportional to M_1. By similar reasoning, the force exerted on M_1 is proportional to M_2. But the forces on the right- and left-hand objects are equal

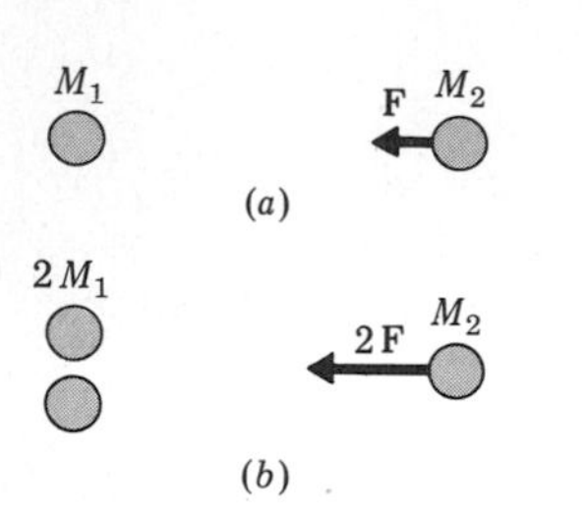

Figure 1·6. Gravitational attraction is proportional to mass

and opposite, and so this force must be proportional to both masses, i.e., to the product of the masses.

The gravitational attraction also depends upon the distance R between the two objects. To deduce the form of this dependence, imagine a large body with small objects distributed over a sphere around it (Fig. 1·7). The forces exerted on the surrounding objects by the central object take the form of rays pulling the small objects towards the center. Suppose we regard these as the *force field* of the central body. Then the number of rays per unit of area passing through the sphere's surface is a measure of the strength of this force field. The rays diverge as the radius of the sphere increases, and the number of rays per square meter varies as $1/R^2$. Thus the attractive force between the central object and one of the surrounding objects should decrease like $1/R^2$.

We have just reasoned out the form of the *universal gravitational law:*

$$\mathbf{F} = k_G \frac{M_1 M_2}{R^2} \tag{1·2}$$

Here k_G is a physical constant, the value of which depends upon the particular unit system used. Note that the key ideas used were (*a*) symmetry and (*b*) the concept of a gravitational field. This law is confirmed by experiments.

Newton's law. This same line of reasoning can be used to deduce the form of the equation relating the acceleration of a body to the forces that act upon it. Consider a mass M acted on by a force **F**. It will undergo an acceleration **a**. Now, place an identical mass right next to it, acted on by the identical force. It too will

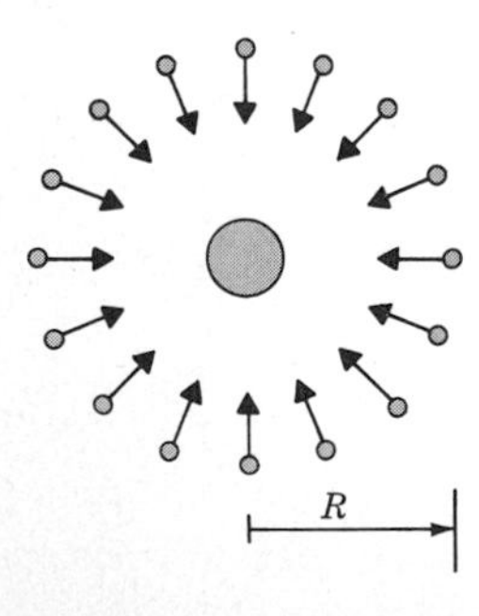

Figure 1·7. Deducing the gravitational law

undergo acceleration **a**. Now think of the combined mass as a single system which has mass $2M$ and is acted on by force $2\mathbf{F}$. Note that the force required to produce a given acceleration is proportional to the mass. What about the dependence on acceleration **a**? By symmetry the force must be an odd function of **a** (otherwise, if the force were reversed, the reverse acceleration would not be found). The simplest odd function is **a** itself, and so we might be tempted to try

$$\mathbf{F} = k_N M\mathbf{a} \tag{1·3}$$

where k_N is another physical constant that, like k_G, has a value that depends upon the particular unit system used. As you know, this works, and is called *Newton's second law*. You have usually seen it written as $\mathbf{F} = M\mathbf{a}$, which is correct only if one uses a unit system in which $k_N = 1$. However, k_N is a basic physical constant connecting the quite different concepts of force, mass, and acceleration, and its hidden presence must be remembered when one writes $\mathbf{F} = M\mathbf{a}$. This point will be discussed in greater detail later in this chapter.

The other mechanical concepts of primary interest to us are work, kinetic energy, and potential energy. These will be discussed shortly.

1·4 CONCEPTS FROM ELECTROMAGNETICS

Since thermodynamics applies to any situation involving energy, it is useful in electromagnetics. We shall not dwell on these aspects in this book, but a brief review of the concepts of charge and current and Coulomb's law will be useful.

The fundamental concept of electromagnetics is that of *charge*. Its invention arose out of the need for an explanation of forces which can exist between physically separated objects, forces which could not be accounted for in terms of gravitation. Charge is conceived as a property of electric particles and can give rise to either attractive or repulsive forces. The concept of charge is intimately tied to the idea that charge is a conserved quantity. Two objects that have been rubbed together will attract one another; we explain this by saying that one object has been positively charged and the other negatively charged (in order to conserve charge). We observe that bodies having charges of equal sign repel one another and that objects with opposite charge attract, and we adopt these conventions as part of the charge concept.

With the concept that electrostatic forces are somehow associated with charge, it becomes possible to tell when two objects have the same charge. This occurs when they each exert the same force on a third charged body. When a charged object is touched to an identical uncharged object, the objects assume identical charges. Symmetry considerations, together with the concept of charge as a conserved quantity, indicate that each object then has half the original charge. The notion that charge is conserved allows us to establish a unique scale of charge which is independent of the nature of any substance (see Fig. 1·8).

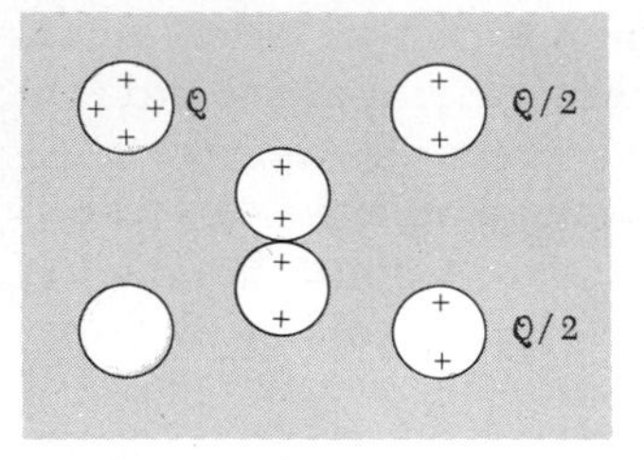

Figure 1·8. Establishing a unique charge scale with symmetry arguments

Experimentally it is found that the repulsive force between two stationary point charges is proportional to the product of their charges and inversely proportional to the square of the distance between them,

$$\mathsf{F} = k_C \frac{Q_1 Q_2}{R^2} \tag{1·4}$$

This is *Coulomb's law*, the form of which can be deduced using arguments of the sort made previously. k_C is a physical constant that depends upon the unit system. It is possible to construct systems in which k_C comes out to be unity, and we shall discuss this later in the chapter.

Moving charges exert additional forces upon one another. If the charges are not moving too rapidly, the additional forces are given by the *Biot-Savart law*.† The constant of proportionality in this expression similarly depends upon the particular choices of force and charge measures.

Current is the rate of charge flow, i.e., the rate at which charge moves across a given area. Although it is conceptually less basic than charge, it is used as the basis for defining electrical units in the international system of units, or Systeme International, which we shall now discuss.

1·5 DIMENSIONAL AND UNIT SYSTEMS

You probably were not surprised to see the physical constant k_G in the gravitational law, or the constant k_C in Coulomb's law, but may have been quite mystified by the fact that we put k_N in Newton's law. This is undoubtedly because you are most familiar with dimensional systems in which k_N is *by choice* equal to unity but k_C and k_G have weird values and units. However, there is absolutely nothing sacred about any particular dimensional system. We could just as well pick $k_G = 1$, and have a funny value for k_N; or we could set up a system in which k_N, k_G, and k_C are *all* unity. Some more advanced intelligence on another planet might well have decided to work in just such a system.‡

Here on earth we are slowly switching to the metric system, more properly called the *Systeme International* (SI) solely to place the entire world on a common

† See Appendix A, Table A·2.

‡ There are those who argue that this would be clear evidence of inferior intelligence.

basis, not because the system is "better" or "simpler" than any other system. It happens to be a base-10 system, which makes it consistent with the number base in most common use on earth. However, there is nothing especially convenient or better about base-10 systems, which stem solely from the number of fingers on human hands. Perhaps on a planet where creatures have three fingers on each of three hands they work in base-9 systems, with k_N, k_G, and k_C all 9 for mystical reasons.

Eventually the entire world may be fully converted to the SI, but today a significant segment of the world still works in other units systems, of which the *old English system*,† is the most important. Much of the data engineers need for analysis is only available in the old English system. Many engineers still think in this system, and a great deal of our manufacturing equipment is based on this system. For many years engineers will have to refer to books and papers written in the old English system. For this reason it is important that engineers understand and be able to work in *both* SI and old English units, for essentially the same reasons that astronauts and cosmonauts must speak both English and Russian. Therefore this book is "multilingual" dimensionally. The student should work hard to develop a "feel" for the size of numbers in both the SI and the old English system, as well as the ability to execute engineering calculations correctly in either system. Shorthand abbreviations for the various units of these and other systems are given in Tables A·1 and A·2.

In this section we will develop the conceptual basis for dimensional systems. Once these concepts are understood, you should find it very easy to work in any system you might encounter, including the IN‡ system.

Vocabulary. Let's first establish some vocabulary that we can use to discuss the basic ideas of different dimensional systems. We shall use the word "dimension" to mean a name given to any measurable quantity. Length, time, mass, area, and velocity are *dimensions*. The *primary quantities* of a particular dimensional system are those for which we decide to set up arbitrary scales of measure. *Secondary quantities* are those whose dimensions are expressed in terms of the dimensions of primary quantities. In the dimensional systems we customarily employ, length and time are primary quantities, and velocity and area are secondary quantities. The primary scales of measure are expressed in terms of *units*. For example, feet, inches, meters, yards, miles, and light-years are all units for the dimension length. The basic difference between dimensional systems is not in the *units* employed but in the primary *dimensions*. The SI uses length, time, and *mass* as primary quantities, and all other quantities are secondary. The cgs system is similar; these are called *MLT* (mass-length-time) systems. In contrast, the *absolute* old English system, still used in some sectors of the U.S. aerospace and civil engineering fields today, is an *FLT* system; it uses force, length, and time as primary quantities, and

† In deference to England, which has converted to the SI, the English system will be called the old English system in this text.

‡ Intergallactic Nine-based system.

everything else, including mass, is secondary. The old English system uses *four* primary quantities: length, time, mass, *and* force. It is an *FMLT* system. It is possible to devise dimensional systems that use only two primary quantities, say, length and time. It is even possible to devise a system that uses only *one* primary quantity (time). The fact that the SI has three primary quantities is a reflection of human choice and not of basic nature.

Role of basic laws. Earlier we introduced two laws relating force, mass, and motion: Newton's second law and the law of gravitation. In the SI and cgs systems, force is a secondary quantity whose dimensions and units are defined by the *arbitrary* choice of $k_N = 1$ in Newton's law. The dimensions of force *in these systems* (but not in others) are therefore those of mass times acceleration, which we can express symbolically as

$$F \overset{d}{=} \frac{ML}{T^2}$$

Here the symbol $\overset{d}{=}$ means "has the dimensions of," and M, L, and T represent the dimensions mass, length, and time. In contrast, in the absolute old English system *mass* is a secondary quantity whose dimensions and units are set by the *arbitrary* choice of $k_N = 1$; in this system (see Table A·1),

$$M \overset{d}{=} \frac{FT^2}{L}$$

In neither of these systems is the gravitational constant k_G equal to unity. However, one could just as easily, and just as rationally, choose $k_G = 1$ in an *MLT* or *FLT* dimensional system, in which case the constant in *Newton's* law would not be unity. The fact that we often write $\mathbf{F} = M\mathbf{a}$ and we never write $F = M_1 M_2/R^2$ is due to historical practice, not fundamental science. If Newton had been more interested in planets than apples, it might have happened the other way around.

If we wanted, we could develop an *LT* system in which *both* k_N and k_G were unity. If we choose to define the speed of light as unity, we could work with a system in which the only primary quantity is time, and discard the standards of mass and length required by the SI. We could go further, making the constant k_C in Coulomb's law equal to unity, and dispose of the charge standard. Then all the equations would be quite simple. The trouble is that mass would have the dimensions of time, force would be dimensionless, and we would have to be very careful in our thinking (see Prob. 1·10).

In an *FMLT* system, force, mass, length, and time are all primary quantities; this is the essence of the old English system. The units of length and time are the foot and second. Mass is measured in *pounds mass*, written lbm, and force is measured in *pounds force*, written lbf. A lbf is not equal to a lbm any more than a chicken is equal to an apple; mass and force are separate concepts. Therefore, in the old English system one thinks of using separate standards for force and mass. These happen to be such that the numerical values for the mass of a body in lbm and its weight in lbf at sea level on earth are numerically identical (but the units and dimensions are of course quite different). This makes it very easy to figure the mass of an object; if it weighs 1000 lbf, it has a mass of 1000 lbm. The value of k_N in the old English system is

$$k_N = \frac{1}{32.1739} \text{ lbf}\cdot\text{s}^2/(\text{ft}\cdot\text{lbm})$$

It is customary to denote the reciprocal of k_N by g_c,

$$g_c = 32.1739 \text{ ft}\cdot\text{lbm}/(\text{lbf}\cdot\text{s}^2) \qquad (1\cdot5)$$

To facilitate work in a variety of systems, we shall keep the Newton constant explicitly in Newton's law, which henceforth we shall write as

$$\mathbf{F} = \frac{1}{g_c} M\mathbf{a} \qquad (1\cdot6)$$

We emphasize that g_c is *not* a conversion factor. It is a fundamental physical constant that expresses the proportionality between force and momentum change, a constant similar to k_G and k_C, and whose presence is hidden when one writes simply $\mathbf{F} = M\mathbf{a}$. In the SI and cgs systems, $g_c = 1$ *by choice;* in the old English system it has the *experimental* value given by Eq. (1·5).

Aliases. The units of secondary quantities become complicated in all the unit systems. For example, the units of force in the SI are kg·m/s². Rather than carry lengthy expressions, it is convenient to give names to complex combinations of units. So, the term *newton* (N) is used in place of kg·m/s²; the N is the *alias* for kg·m/s².

Table 1·1 gives the SI units, aliases, and symbols for mechanical quantities of most interest in this text. Table 1·2 gives corresponding information for the old English system. Appendix A contains other information on various unit systems. Table A·1 gives the aliases for secondary quantities in the various mechanical unit systems. Table A·2 gives additional information on electrical unit systems, and may be helpful to electrical engineers struggling to read an old paper. Table 1·3 gives the names and abbreviations of common prefixes.

Standards. By the international agreements that established the SI, standards for length, time, mass, electric current, temperature, and luminous intensity have been established. For example, the meter is defined as 1,650,763.73 times the wavelength of a prominent orange-red line in the spectrum of krypton 86, and the second is defined in terms of a cesium radiation frequency. The older units are now defined in terms of these basic SI units.

Table 1·1 SI mechanical units

Quantity	Dimensions	Units	Alias	Symbol
Length	L	meter	—	m
Time	T	second	—	s
Mass	M	kilogram	—	kg
Velocity	L/T	m/s	—	—
Acceleration	L/T^2	m/s^2	—	—
Frequency	$1/T$	1/s	hertz	Hz
Force	ML/T^2	$kg \cdot m/s^2$	newton	N
Pressure	M/T^2L	$kg/(s^2 \cdot m) = N/m^2$	pascal	Pa
Energy	ML^2/T^2	$kg \cdot m^2/s^2 = N \cdot m$	joule	J
Power	ML^2/T^3	$kg \cdot m^2/s^3 = J/s$	watt	W

Table 1·2 Mechanical quantities in the old English system

Quantity	Dimensions	Units	Alias	Symbol
Length	L	foot	—	ft
Time	T	second	—	s
Mass	M	pound mass	—	lbm
Force	F	pound force	—	lbf
Velocity	L/T	ft/s	—	—
Acceleration	L/T^2	ft/s^2	—	—
Frequency	$1/T$	1/s	hertz	Hz
Pressure	F/L^2	lbf/ft^2	—	psf†
Energy	LF	ft·lbf	—	—
Power	LF/T	ft·lbf/s	—	—

† psi and psf are frequently used for lbf/in^2 and lbf/ft^2.

Table 1·3 Scale factors

Number	Prefix	Symbol	Example
10^{12}	tera	T	terahertz (THz)
10^{9}	giga	G	gigajoule (GJ)
10^{6}	mega	M	megawatt (MW)
10^{3}	kilo	k	kilometer (km)
10^{-2}	centi	c	centimeter (cm)
10^{-3}	milli	m	milliwatt (mW)
10^{-6}	micro	μ	microsecond (μs)
10^{-9}	nano	n	nanometer (nm)
10^{-12}	pico	p	picofarad (pF)

A caution on the SI. When using the SI, the student must be careful to realize that the basic mass unit is the kg, not the g. A k in front of m, i.e., km, means 1000 of the basic length units. But kg means only *one* of the basic mass units; kg must be regarded as a pair, never to be separated. Perhaps someday this confusion will be eliminated by an international agreement that shortens kg simply to k, for kilo, and abolishes the g. Then g would simply be a mk.

Another place to be careful is in the treatment of prefixes. For example, volume measured in mm^3 means $(mm)^3 = 10^{-9}\ m^3$, not $10^{-3} \times m^3$. The prefixes in front of a unit become part of that unit, and, if the unit is raised to a power, it is understood that the prefix is raised to the power as well.

A fun example. To illustrate the ideas of dimensional systems, let us set up a system in which only length and time are chosen as primary quantities. Both force and mass will be secondary quantities, and we shall determine their dimensions and units by arbitrarily selecting the constant in Newton's law to be 2 and the constant in the gravitational law to be unity. In our new system these laws will be

$$\mathsf{F} = 2Ma \tag{1·7a}$$

$$\mathsf{F} = \frac{M_1 M_2}{R^2} \tag{1·7b}$$

Considering two identical mutually gravitating masses and the accelerations they undergo, we equate the above two equations and solve for M, finding

$$M = 2aR^2 \tag{1·8}$$

Mass will therefore have the dimensions of $\text{length}^3/\text{time}^2$ in this new system. If we adopt the meter and the second as primary standards, mass will have the units m^3/s^2. We might choose the alias "chunk" for this combination of primary units. The dimensions of force would then be $\text{length}^4/\text{time}^4$, and we could give the alias "push" to 1 m^4/s^4.

Conversion factors from the SI to our new system will not be found in standard tables, but we can work out our own. Consider two bodies, each having a mass of 1 kg in the SI, freely gravitating toward one another. Suppose they are 1 m apart. The acceleration they undergo can be computed first from the SI equations

$$\mathsf{F} = Ma \tag{1·9a}$$

$$\mathsf{F} = \frac{k_G M^2}{R^2} \tag{1·9b}$$

Then
$$a = \frac{k_G M}{R^2} = \frac{6.67 \times 10^{-11}\ m^3/(kg \cdot s^2) \times 1\ kg}{1\ m^2}$$
$$= 6.67 \times 10^{-11}\ m/s^2$$

The mass in the new unit system may now be computed from Eq. (1·8) as

$$M = 2 \times (6.67 \times 10^{-11}\ \text{m/s}^2) \times 1\ \text{m}^2 = 13.34 \times 10^{-11}\ \text{m}^3/\text{s}^2$$

We therefore have the conversion equivalence

$$1\ \text{kg} = 13.34 \times 10^{-11}\ \text{m}^3/\text{s}^2 = 13.34 \times 10^{-11}\ \text{chunks}$$

The conversion between newtons and pushes could be obtained in the same way.

1·6 MECHANICAL CONCEPTS OF ENERGY

Let's now review the concepts of kinetic and potential energy as developed in mechanics from the laws of motion and gravitation. For purposes of generality we will use the more general form of Newton's law for a particle, Eq. (1·6). Since acceleration is the time rate of change in velocity, Newton's law can be written as

$$\mathbf{F} = \frac{1}{g_c} M \frac{d\mathbf{V}}{dt} \tag{1·10}$$

Multiplying by dt and integrating from t_1 to t_2,

$$\int_{t_1}^{t_2} \mathbf{F}\, dt = \frac{1}{g_c} M(\mathbf{V}_2 - \mathbf{V}_1) \tag{1·11}$$

It is customary to call the integral the *impulse* and to call the quantity $M\mathbf{V}$ the *momentum.* The basic equation of mechanics therefore tells us that the impulse provided is proportional to the increase in momentum of the particle.

Recognizing that by definition $\mathbf{V} = d\mathbf{X}/dt$, we can multiply Eq. (1·10) by $d\mathbf{X}$ (vector dot product) and integrate between any two points in space, obtaining†

$$\int_{\mathbf{X}_1}^{\mathbf{X}_2} \mathbf{F} \cdot d\mathbf{X} = \frac{M}{2g_c} (\mathsf{V}_2^2 - \mathsf{V}_1^2) \tag{1·12}$$

It is customary to call the integral the *work* done on the particle and to refer to the quantity $M\mathsf{V}^2/2g_c$ as the *kinetic energy* of the particle. The basic mechanical law then tells us that the work done by the force on the particle is equal to the increase in its kinetic energy.

For example, let's compute the kinetic energy of a 1000-kg (2200-lbm) car hurtling down the road at 60 mph (88 ft/s, 26.8 m/s).

† Here V^2 means $\mathbf{V} \cdot \mathbf{V}$.

In the SI,

$$KE = \tfrac{1}{2}MV^2 = \tfrac{1}{2} \times 1000 \text{ kg} \times (26.8 \text{ m/s})^2 = 3.59 \times 10^5 \text{ kg}\cdot\text{m}^2/\text{s}^2$$
$$= 0.359 \text{ MJ}$$

In the old English system,

$$KE = \frac{MV^2}{2g_c} = \frac{2200 \text{ lbm}}{2 \times 32.2 \text{ ft}\cdot\text{lbm}/(\text{lbf}\cdot\text{s}^2)} \times (88 \text{ ft/s})^2$$
$$= 2.64 \times 10^5 \text{ ft}\cdot\text{lbf}$$

To develop the relationship between weight and mass, let's consider an object lifted by a rope in the earth's gravitational field (Fig. 1·9). The weight is the force exerted by the earth on the object. If the object is motionless, the force exerted by the rope on the object has the same value, but opposite direction. We can relate the weight to the mass by noting that if the rope is cut the object will fall at the acceleration of gravity g. An average value at the surface of the earth is

$$g = 9.80 \text{ m/s}^2 = 32.17 \text{ ft/s}^2$$

Hence, while falling

$$\mathsf{F} = \mathsf{w} = \frac{1}{g_c} Ma = \frac{1}{g_c} Mg \tag{1·13}$$

so the weight is related to the mass by

$$\mathsf{w} = \frac{g}{g_c} M \tag{1·14}$$

For example in the SI, if the mass is 1000 kg, the weight is

$$\mathsf{w} = (9.80 \text{ m/s}^2) \times 1000 \text{ kg} = 9.8 \times 10^3 \text{ kg}\cdot\text{m/s}^2 = 9.8 \times 10^3 \text{ N}$$

In the old English system, for an object of mass 2000 lbm the weight is simply 2000 lbf.

Now, suppose we raise the object by pulling on it with the rope. During this process the rope force is

$$\mathsf{F} = \mathsf{w} + Ma$$

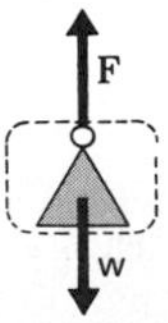

Figure 1·9. The system

acting upward on the object. If we start from rest and lift the object a distance z above its original position, where again we stop, the work done on the object by the rope will be

$$\text{Work} = \int_1^2 \mathsf{F}\,dx = \int_1^2 \left(\mathsf{w} + \frac{M}{g_c}\frac{d\mathsf{V}}{dt}\right) dx$$

$$= \mathsf{w}z + \int_1^2 \frac{M}{g_c}\,\mathsf{V}\,d\mathsf{V} = \mathsf{w}z \tag{1·15}$$

Note that the last integral, which represents net work done in accelerating and decelerating the object, is zero, since the object is motionless at the start and finish of the process. The work represents an expenditure of energy by whatever pulled on the other end of the rope. This energy now resides in the raised object, and it is customary to call the term $\mathsf{w}z$ the *gravitational potential energy* of the object.

Note that we used the idea that *energy is conserved* when we imagine energy flowing from the lifting agent into the raised body through a *transfer of energy as work*. Similarly, when an object is accelerated, energy is transferred as work from the agent producing the force to the accelerating body, where it then resides in the form of kinetic energy.

The concept of energy as a conserved quantity has been central in science since the late seventeenth century. At first the idea was applied only to a freely falling body; as additional systems were considered, new kinds of energy had to be introduced in order that the conservation aspect be retained. Considerations of charge led to coulomb energy and eventually to the concept of energy in an electromagnetic field. With the modifications of mechanics introduced by relativity theory came rest-mass energy. Physicists have conceived new particles solely in order to retain the conservation of energy at the nuclear and subnuclear levels. Man has become fond of the conservation-of-energy concept and, being unwilling to discard it, has been forced to visualize new kinds of energy. One wonders if simpler explanations of nature might be obtained by forgetting about the concepts of force and energy and starting over with new concepts. While this may some day happen, we engineers of today must use present-day concepts, models, and laws to devise our systems, and the notions of force, mass, work, and in particular, the conservation of energy, are quite adequate for our purposes.

1·7 ENERGY UNITS

In any unit system energy has the dimensions of force times distance; in the SI this corresponds to *newton meters*, abbreviated N·m. Now N is the alias for kg·m/s^2, so a N·m is really a kg·m^2/s^2. As indicated in Table 1·1, this group has the alias *joule* (J); a J is a N·m.† In the *old English* system, energy is measured in ft·lbf.

† If you do not yet have your personal calibration in the SI, remember that a N is about $\frac{1}{4}$ lbf, i.e., about the weight of an apple, and a m is about how far the apple that hit Newton fell. So, a N·m, i.e., a J, is about the amount of energy the apple delivered to Newton.

There are other energy units of historical origin. The *Btu* (British thermal unit) is now *defined* to be 1.05506 kJ; it is approximately the amount of energy required to raise 1 lbm of water 1°F. The *calorie* (cal) is approximately the amount of energy required to elevate 1 g of water 1°C. The calorie used by dietitians is really a *kilocalorie* (kcal), or 1000 cal. A daily diet might total 1500 kcal.

In the United States, reports and discussions of energy usually involve the Btu, and sometimes the *quad* (Q); 1 Q = 10^{15} Btu. In 1975 the United States used about 70 Q. Energy costs are often quoted per million Btu; since 1 Btu is very nearly 1 kJ, 1 million Btu is approximately 10^9 J, or 1 GJ. A barrel of oil contains approximately 6 million Btu; so at $12/barrel, oil energy costs about $2/MBtu, i.e., about $2/GJ.

The *electronvolt*, abbreviated eV, is another common energy unit. It is defined to be 1.602×10^{-19} J, and is approximately the amount of energy required to move an electron up a potential of 1 V. The eV is used primarily by physicists and chemists in discussing atomic and molecular processes.

Power is defined as the *rate* of energy transfer or energy usage, and so has the dimensions of energy/time. In the SI the J/s carries the alias *watt* (W). The *horsepower* (hp) is defined to be 550 ft·lbf/s.

Other units and conversions for energy and power are given in Appendix A.

1·8 A PRACTICAL EXAMPLE

To illustrate the practical utility of what we have been and will be discussing, let's consider a windmill (Fig. 1·10) and estimate its power output. To do this we will simply compute the kinetic energy in the column of air that passes through the windwill rotor disc at velocity V in a time period t; the best that the windmill could possibly do would be to convert all this energy to electric energy. The volume of this air is $LA = \mathsf{V}t\pi D^2/4$. Denoting the density of air by ρ (kg/m^3 or lbm/ft^3), the mass of the air column is

$$M = \frac{\rho \mathsf{V} t \pi D^2}{4}$$

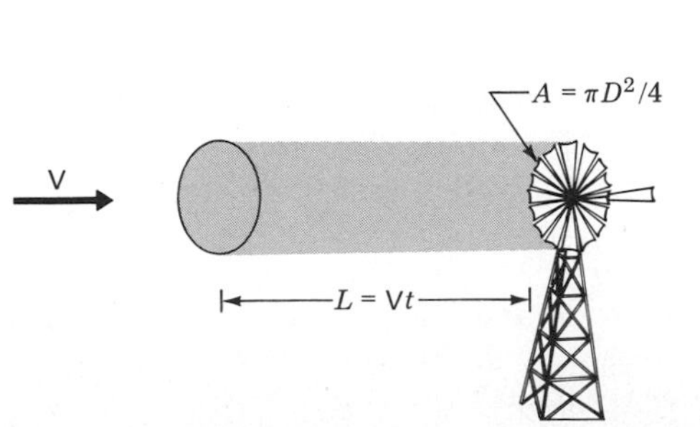

Figure 1·10. Windmill analysis

Hence the kinetic energy of the air column is

$$KE = \frac{M\mathsf{V}^2}{2g_c} = \frac{\pi}{8g_c}\rho D^2\mathsf{V}^3 t$$

The kinetic energy per unit time, KE/t, is the *power* of the air flow

$$\mathsf{P} = \frac{\pi}{8g_c}\rho D^2\mathsf{V}^3 \tag{1·16}$$

Note that our simple analysis tells us that the power will vary as the *square* of the rotor diameter and the *cube* of the wind velocity.

The density of atmospheric air is about 1.2 kg/m^3. Suppose the rotor diameter is 10 m and the wind velocity is 8 m/s. Then, the wind power is

$$\mathsf{P} = \frac{\pi}{8 \times 1} \times 1.2 \times 10^2 \times 8^3 \frac{\text{kg}}{\text{m}^3} \times \text{m}^2 \times \frac{\text{m}^3}{\text{s}^3} = 2.41 \times 10^4 \text{ W} = 24.1 \text{ kW}$$

This is the maximum possible power. A real windmill might capture 30 percent of this, or about 7.2 kW. A city of 1 million people requires about 1000 MW of power. Under the assumed conditions, this would require 139,000 of such windmills. If the wind velocity dropped by a factor of 2, down to 4 m/s, it would take 1,112,000 windmills to supply this power. This calculation illustrates why the utilities employ coal, oil, and nuclear energy rather than wind power.

SELECTED READING

Kestin, J., *A Course in Thermodynamics*, chap. 1, Blaisdell Publishing Co., Inc., Waltham, Mass., 1966.

Resnick, R., and D. Halliday, *Physics, Part 1*, chap. 1, chap. 5, John Wiley & Sons, Inc., New York, 1966.

Zemansky, M. W., M. M. Abbott, and H. C. Van Ness, *Basic Engineering Thermodynamics*, 2d ed., secs. 1.1–1.4, McGraw-Hill Book Company, New York, 1975.

QUESTIONS

1·1 Explain your concepts of length and time.

1·2 Why is it not true that $g/g_c = 1$ in the old English system, even though the numerical values of g and g_c may be the same?

1·3 What is the difference between work and power?

1·4 Consider $\mathsf{F}\,dX$, $d(\mathsf{F}X)$, and $X\,d\mathsf{F}$. Which is work, and why? Which can be integrated without knowing how F varies with X?

1·5 What are microscopic and macroscopic viewpoints?

1·6 How might you explain the concepts of energy, force, and charge to a high school freshman?

1·7 How do concepts differ from definitions?

1·8 At what age did you first hear about energy? At what age did you first understand the concept? Have you ever heard about entropy?

1·9 What is your mass in lbm? What is your height in statfarads?

1·10 Push very hard on a stone wall. How much work did you do on the wall? Why did you get tired?

1·11 Why is it important to make a distinction between lbm and lbf? Can they be canceled in an equation?

1·12 What do the following mean: SI, MW, GJ, mW?

PROBLEMS

1·1 Discuss the differences between concepts and definitions. What are the basic concepts of geometry? What are some geometric definitions?

1·2 In the Middle Ages it was thought that length up was somehow not the same as length horizontally (after all, things fall down but never fall sideways). Discuss how the concepts of these two lengths have changed since that time.

1·3 Discuss the possible means for devising fundamental measures of length and time.

1·4 Consider an observer in a frame which is accelerating linearly at acceleration a_0. What should he write for Newton's law in his coordinate frame? What would be his energy accounting for the work done by a force on a particle?

1·5 A body having a mass of 1 kg falls from rest from a height of 30 ft. Express its kinetic energy in Btu, ft·lbf, and eV.

1·6 Using the inverse-square gravitational law, calculate the work required to lift a 1000-kg satellite from the earth's surface to an altitude of 100 mi and inject it into orbit at 17,500 mph. How long would a 300-hp engine have to be run to produce this much energy? How many automotive batteries would be required if the orbiting could be done electrically (600 Wh is the total available energy from a typical automotive wet cell)?

1·7 Estimate your kinetic energy while running 10 mph. State the result in J, Btu, and eV. Did you do any work on the ground in achieving this speed?

1·8 A 100-megaton hydrogen bomb releases 10^{18} J of energy. Compare this to the kinetic energy of an iron asteroid 10 km in diameter approaching the earth at 50,000 m/s.

1·9 Set up a dimension system in which length and time are the only primary quantities. Pick the constants in Newton's law, Coulomb's law, and the universal gravitational law to be unity. Use the meter and the second as the primary standards and determine the equivalence of a 1-kg mass and a 1-C charge in the new unit system.

1·10 Set up a dimension system in which time is the only primary quantity. Pick the constants in Newton's second law, Coulomb's law, and the gravitational law to be unity and, in addition, pick the speed of light to be unity (dimensionless). Length will then have the dimensions of time. Use the s as the unit of time, and determine the equivalents of 1 kg, 1 m, and 1 C.

1·11 Set up a dimension system in which force and charge are the primary quantities. Pick the constants in Coulomb's law, Newton's law, and the universal gravitational law to be $\frac{1}{2}$, and state the dimensions of length, time, and mass in terms of the primary dimensions. Determine the equivalents of 1 m, 1 kg, and 1 s in the new unit system, taking the N and the C as the fundamental units of force and charge.

1·12 Set up a dimension system in which the primary quantities are mass, length, time, and charge. Pick the constant in Coulomb's law to be unity. Find the dimensions of force and of the constant in Newton's law in this system (in terms of M, L, T, and $\mathcal{Q}$). Choosing kg, m, s, and C as the basic units for the primary quantities, calculate the value of the constant in Newton's law for the new dimensional system and determine the equivalence of 1 N in the new system. Is this the same as 1 N in the SI?

1·13 Discuss the point of the previous four problems.

1·14 Discuss the implications of Prob. 1·10; do you think there might be any unknown basic laws relating mass, force, length, time, and charge?

1·15 Large central power stations typically have a power output capacity of 1000 MW and convert about 35 percent of the input energy to electricity. Consider such a plant that burns coal containing 10,000 Btu/lbm. How many tons of coal are required by this plant per year, assuming it operates continuously? If 10 percent of the coal mass remains as residual ash, how many tons will develop in the ash pile over 30 years? If 1 ton of SO_2 is produced for every 50 tons of coal consumed, how much SO_2 is produced in a year?

1·16 The total electrical generating capacity of the United States was about 400,000 MW in 1974. If trends continue, this will double in about 10 years time. How many new plants of the size of the system in Prob. 1·15 will be required? If half of these are coal-fired, how many tons of coal will be required per year for the new plants, and how many tons/yr of ash and SO_2 will have to be handled in 1984?

1·17 Suppose that coal costs \$7/ton at the power plant, and oil costs \$12/bbl (1 bbl is approximately 6×10^6 Btu). Using the data from Prob. 1·15, calculate the annual fuel bill for coal-fired and oil-fired plants at these prices, and express the fuel bill in cents/kW·h of output energy.

1·18 Suppose you eat two 100-kcal cookies per hour. What is your power intake in W and hp?

1·19 A strong person could put out about 0.05 hp on a continuous basis. Suppose this person works an 8-h day and sells this energy for \$1.50/million Btu to be competitive with \$9/bbl oil. How much will be earned in a day?

1·20 In the United States the per capita energy consumption is of the order of 200 kW·h/day. Suppose you are in the United States purchasing your energy from people working as in Prob. 1·19, but they are being paid \$5/h. How much would you have to pay per day for your energy?

1·21 This problem deals with the energy utilization and power requirements for an urban vehicle. The problem will let you review some basic ideas in mechanics, practice your calculus on some real problems, and give you the opportunity to take an independent approach on an interesting and important engineering problem. Write your analysis up briefly as you might if asked to do it as an engineer at General Motors.

The vehicle has the following specifications:

Laden weight (for design purposes), 3200 lbf
Hill-climbing capability, +3 percent grade at 55 mph
Uniform acceleration capability, 0 to 55 mph in 15 s
Cruising speed, 55 mph
Frontal area, 20 ft^2

(*a*) *Hill-climbing power*—Calculate the engine power required to satisfy the hill-climbing requirement, neglecting wind friction and rolling losses.

(*b*) *Acceleration power*—Calculate the power necessary to produce the design acceleration, neglecting wind friction and rolling losses. Plot the instantaneous engine power output (hp) and speed versus time for the 15-s acceleration period.

(*c*) *Wind power*—A consultant has suggested that the drag can be estimated using simple aerodynamic concepts as

$$\mathsf{F} = C_D \frac{\rho \mathsf{V}^2}{2g_c} A$$

where F = drag force, lbf

C_D = drag coefficient, dimensionless

ρ = air mass density, lbm/ft^3

V = vehicle velocity, ft/s

g_c = Newton's law constant, $g_c = 32.2\ \text{ft}\cdot\text{lbm/(lbf}\cdot\text{sec}^2)$

A = projected frontal area of the vehicle, ft^2

The consultant says that the density of air is about $0.074\ \text{lbm/ft}^3$, and that for a vehicle with moderate streamlining $C_D \simeq 0.6$ is appropriate.

Calculate the wind power as a function of speed using this information. (Plot hp versus mph.)

(*d*) *Rolling frtction*—Based on experience, the consultant thinks that the power necessary to overcome rolling friction can be estimated from

$$P_r = \frac{2V}{50} + \frac{4\mathsf{V}^2}{2500}$$

where P_r is the power in hp, and V is the velocity in mph. Calculate the rolling power as a function of speed. (Plot hp versus mph.)

(*e*) *Maximum engine power*—Specify the maximum engine power output required to meet the specifications, all things considered.

(*f*) *Energy consumption*—Consider the following driving cycles:

(i) Highway: 55 mph for 1 h.

(ii) Suburban: Accelerate at $7\ \text{ft/s}^2$ to 30 mph, maintain for 30 s, decelerate to rest at $4\ \text{ft/s}^2$, idle for 10 s. Repeat for 1 h.

Neglecting energy consumption by the engine during the deceleration phases, calculate the total energy expenditure (kJ, Btu, and ft·lbf) for each driving cycle.

Suppose that the fuel consumption of your vehicle's engine is 0.64 lbm/(hp·h) when loaded, and 2 lbm/h when unloaded. In other words, it uses 0.64 lbm of fuel to produce 1 hp·h of energy, and 2 lbm/h when decelerating or idling. Suppose the fuel used weighs $50\ \text{lbf/ft}^3$. Calculate the mi/gal used for the driving cycles.

(*g*) *Energy conservation*—In order to promote better fuel economy, several design changes have been suggested:

(i) Lighten vehicle to laden weight of 2600 lbf.

(ii) Reduce frontal area to $15\ \text{ft}^2$.

(iii) Through research on vehicle aerodynamics, reduce the drag coefficient to 0.4.
(iv) Install a 500-lbf flywheel to store energy on deceleration and release this energy on acceleration.
(v) Shut off fuel when decelerating (above 5 mph only).

Examine the impact of each of these design changes.

CHAPTER TWO

ENERGY AND THE FIRST LAW

2·1 SYSTEMS

In any scientific or engineering analysis it is very important to identify clearly whatever it is that is under consideration. We shall use the term *system* in a very broad sense to identify the subject of discussion or analysis. The system is something defined by the analyst for the particular problem at hand. A system might be a particular collection of matter, such as the gas in a bottle. Or it might be a region in space, such as the bottle and whatever happens to be in it at the moment. Sometimes we include fields in our definition of the system; for example, the gas in the bottle and the electric field in the bottle might be defined as the system. At other times fields are defined to be outside the system; thus the gas may be the system, but the fields that occupy the same space are considered external to the system. Another situation in which two systems share the same space occurs in the analysis of ionized gases; the ions are often treated as one system, and the electrons as another. Interacting systems are often of quite different types; for example, in the study of liquid droplets the liquid interior to the surface is sometimes treated as one system, and the surface molecules as another. A system might be very simple, such as a piece of matter, or very complex, such as a nuclear power generation plant. Matter may flow through a system, such as a jet engine, or the system may be completely devoid of matter, such as the system of radiation in an enclosed volume. The system is whatever we wish to discuss, and we must be very careful each time to describe precisely just what it is that we are talking about. Hence we shall place considerable emphasis on the definition of systems in this text.

When motions are involved, the definition of the system must include a *reference frame* in which the motions will be measured. An *inertial frame* is one in which any free particle moves at constant velocity. The laws of mechanics as normally written apply only in inertial frames, hence one must be careful in choosing the reference frame if these laws are to be invoked in the analysis.

It is often helpful to indicate the system under consideration by enclosing it with dotted lines in a sketch. Of course, one cannot tell from the sketch whether the system is the *matter* inside the lines or the *space* inside the lines (which may contain different material at different times). In order to differentiate, the term *control mass* is used to indicate a system consisting of specified matter, and the term *control volume* is used to indicate a system specified by space. In working with the properties of materials one usually uses control masses; however, much engineering analysis involves some sort of flow process, and then control volumes are used instead. Often the sketch is not enough to convey the system definition to the reader, in which case a few words of definition are desirable. The student should learn to give adequate system definitions, for much confusion can result when an analyst and his reader are talking about different systems (Do you mean the space or the matter in it?). Many examples of system definition can be found in this and later chapters, particularly in Chaps. 5 and 9; inspection of these systems at this time will also give the student some idea of where we are going in our study of thermodynamics.

Having carefully defined the system, everything else is automatically its *environment*. The interactions between a system and its environment are the main interest in thermodynamics. It is frequently convenient to make the conceptual idealization that the system is *isolated* from any interaction with its environment; we say conceptual because isolation requires walls around the system that are impermeable to matter, rigid, nonconducting to electric charge, and so on. All conserved quantities are "trapped" within an isolated system. For example, since no mass can escape, the mass of an isolated system is constant. So is its charge, and so is its energy.

In the rest of this chapter we shall confine our attention to systems consisting of specified matter, that is, to control masses. Energy, but *not* matter, can cross the boundary of such a system; the various mechanisms for this energy transfer are the main topic of this chapter.

2·2 MICROSCOPIC ENERGY MODES

What is energy? In Chap. 1 we reviewed the mechanics of a particle and found that integration of the equations of motion leads to a term $M\mathsf{V}^2/2g_c$, which is called the kinetic energy of the particle. This term is a function of the velocity and mass of the particle and can be computed without knowledge of how this velocity was attained. In this sense it is something that belongs to the particle, that is, it is a *property* of the particle.† If the particle collides with another particle, the second particle can be put into motion, that is, its kinetic energy will be increased, and the laws of mechanics tell us that the kinetic energy of the first particle will decrease. It is clear that this particular type of energy, kinetic energy, is somehow associated with the motions of the particle; it is "what makes the particle go." In thermodynamics we want to generalize the concept of energy; all matter and all systems have energy, and energy plays a dominant role in the explanation of interactions between systems.

We can view each piece of matter as being composed of many fundamental particles, each whizzing about in accordance with the laws of mechanics. We know from physics that each particle can have energy in several forms, and it will be helpful to review these forms now. We shall call them *microscopic energy modes* (Fig. 2·1).

Molecules possess energy by virtue of their translation through space. This kinetic energy is termed *translational energy*. For polyatomic molecules *rotational kinetic energy* is also important. If the atoms of the molecule are vibrating back and forth about their common center of mass, we say that the molecule also has *vibrational energy*. The energy of gas molecules at low temperatures is largely associated with the translational and rotational modes, while at higher temperatures the vibrational modes begin to contribute significantly to the total energy.

Electrons whirling about their nucleus have kinetic energy, the amount of which depends on their orbit. Usually the electrons tend to be in the inner low-energy orbits; atoms with electrons in more distant orbits are said to be in *excited states* and have more energy than normal atoms of that species. It is also necessary to think of electrons as spinning and thereby possessing some *spin energy*. Many of the other fundamental particles also seem to have spin energy.

Molecules are held together by molecular binding forces, including coulomb and gravitational forces. All these forces seem to be conservative, and thus we think of potential energy in association with these intermolecular forces. Forces between the electrons and the nucleus are responsible for keeping the electrons in orbit, and there is potential energy associated with these forces. The nucleus is held together by forces much stronger than molecular binding forces, and consequently the nuclear binding energy is considerably larger than the binding energy of a molecule.

The orbiting electrons constitute tiny electric currents, producing little magnetic dipoles. In the presence of an external magnetic field these dipoles can be twisted, and there is energy associated with these dipole moments. An electrically neutral atom does not have all its charges in the same place, and consequently constitutes an electric-dipole moment, which can have energy in the presence of an external electric field. These dipole-moment energies are important in dielectrics and paramagnetic substances.

† The concept of property will be developed more fully in the next chapters.

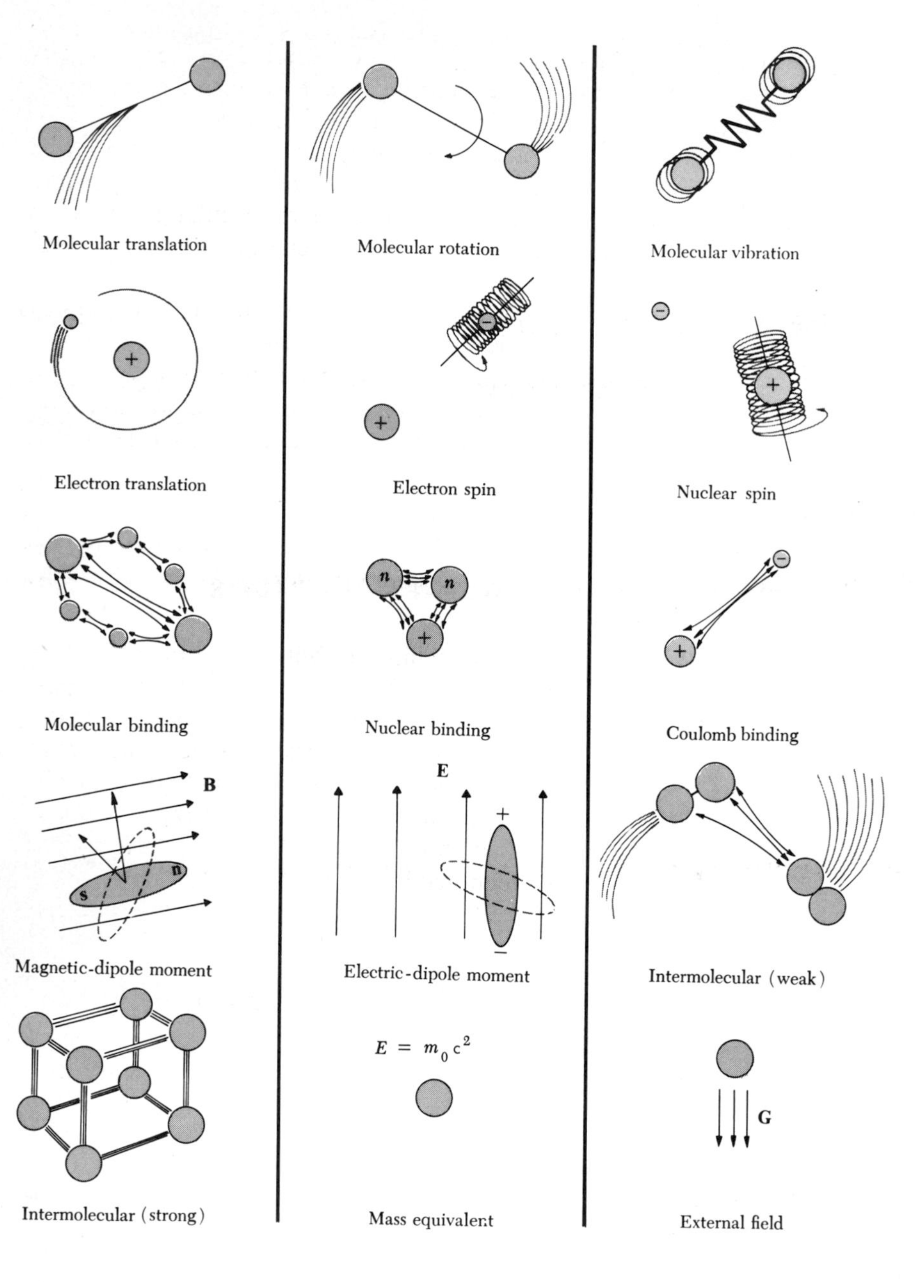

Figure 2·1. Microscopic energy modes

Ionized molecules and free electrons have forces exerted on them by external electric and magnetic fields and consequently can have associated energies. All particles with mass possess potential energy in gravitational fields.

Collections of particles have additional energy associated with the forces between molecules. In liquids and solids these energies are especially important, but in low-density gases the amount of energy due to intermolecular potentials is quite small. Evaporation is a process requiring enough energy to free molecules from these strong bonds.

In relativistic mechanics it becomes necessary to include the energy equivalence of matter; we think of a particle having rest mass M_0 as possessing rest-mass energy $M_0 c^2$. This energy constitutes a large percentage of the energy of a molecule, but in nonnuclear reactions the change in rest mass is so small that changes in rest-mass energy are negligible compared to changes in the other forms of energy.

2·3 MACROSCOPIC ENERGY REPRESENTATIONS

Consider now a control mass consisting of billions of billions of molecules. One way to discuss the energy of this system would be to give values for the translational, vibrational, rotational, etc., energy of each molecule. Clearly this is impractical. Another way would be to give values for the total translational, vibrational, rotational, etc., energy, but this is usually much more information than one really needs. What is practical is simply to give a value for the total energy possessed by all the molecules in all the ways discussed above; this energy is called the *internal energy* of the system, and it is usually denoted by the symbol U. Evaluation of U as a function of the condition or state of the system is one of the central problems in thermodynamics.

When dealing with systems containing many, many molecules, it is convenient to think of the material as being continuous, for this greatly simplifies the mathematical description. In continuum analysis, the kinetic energy of a piece of material having mass M moving at velocity V is $M\mathsf{V}^2/2g_c$, just as it would be for a particle. If different pieces of the continuum are moving at different speeds, the total energy can be found by summing up (integrating) over little pieces of the continuum; you have no doubt done this in a mechanics course. Likewise, the potential energy is computed from the object's weight and height above some arbitrary datum, as for a particle. These two calculations use macroscopically observable properties of the continuum—velocity, mass, weight, and height—for computation of the kinetic and potential energies. Clearly the kinetic energy will not include that portion due to *random* molecular motions, which are "hidden" from macroscopic view. The internal energy must be brought in to account for this energy. So, in continuum analysis one expresses the energy of a chunk of material as

$$E = KE + PE + U \tag{2·1}$$

Here KE and PE represent the macroscopically observable kinetic and potential energy of the object, computed as described above. Note that these are organized forms of energy, associated, for example, with a coherent net translation of all the molecules in one direction, or coherent revolution around an axis, or coherent placement above the ground. In contrast, the internal energy U is associated with the random or disorganized aspects of the molecules. You might correctly guess that organized energy is much more useful than disorganized energy, which is quite the case. One key job of the engineer is to find clever ways to convert disorganized energy into organized molecular motions capable of being used macroscopically.

An aspect of the energy concept which we have already used is that the energy of two systems taken together is the sum of their individual energies. Hence the energy of the whole is the sum of the energy of the parts. This idea is very useful when we want to evaluate the energy of a complex system, for we can evaluate the energy of individual pieces, and then add to obtain the total system energy. In the language of thermodynamics, this additive feature is indicated by saying that energy is *extensive*.

2·4 CONSERVATION OF ENERGY

A fundamental aspect of the energy concept is that energy is conserved, that is, the energy of an isolated system is constant. Equations stating this law provide the basis for quantitative analysis of the changes that take place between interacting systems.

For example, suppose that two moving bodies hit head on, and then come to rest. One of the laws of matter, the momentum principle, says that, since the bodies come to rest, the momenta of the two bodies must have been exactly equal in magnitude and opposite in sign at the moment of collision. What happens to their energy? Two frequently heard *incorrect* statements are

1. The energy was lost.
2. The energy was dissipated into heat.

The first statement is correct only if applied to the organized mechanical energy of motion. As a result of the impact, the bulk kinetic energy of the bodies is converted into internal energy. The energy is no longer macroscopically evident as kinetic energy, but the increase in internal energy is clearly evidenced by an increase in the temperature of each body. No energy was lost; the energy was just rearranged. The second statement could be made correct by replacing "heat" by "internal energy." As we shall shortly see, heat and internal energy are two quite different things; the energy associated with microscopic motions and forces is internal energy, not heat. The use of the term "dissipation" in the second statement suggests that energy in the form of bulk kinetic energy is somehow more desirable than internal energy. This is indeed true, for the perfect organization of bulk motion makes it fully useful (say, to raise a weight), but the random, disorganized microscopic motions reflected by the internal energy make energy in this form less available for use in a practical macroscopic device.

The conservation of energy can be expressed algebraically for the two colliding bodies a and b; the *system* we consider consists of the two bodies together (Fig. 2·2), and we consider them over the *time period* from an instant before collision, denoted by 1, to just after collision, when both are at rest. The initial energy of the system is

$$E_1 = E_{a1} + E_{b1} = \frac{1}{2g_c} M_a \mathsf{V}_{a1}^2 + \frac{1}{2g_c} M_b \mathsf{V}_{b1}^2 + U_{a1} + U_{b1} \qquad (2\cdot 2a)$$

The final energy of the system is

$$E_2 = U_{a2} + U_{b2} \qquad (2\cdot 2b)$$

Since the two bodies form an isolated system, the energy of the system must not change, hence

$$\underset{\substack{\text{initial}\\ \text{energy}}}{E_1} = \underset{\substack{\text{final}\\ \text{energy}}}{E_2} \qquad (2\cdot 3)$$

Substituting Eqs. (2·2) into Eq. (2·3), we can solve for the total increase in internal energy U of the system,

$$U_2 - U_1 = (U_{a2} - U_{a1}) + (U_{b2} - U_{b1}) = \frac{1}{2g_c}(M_a \mathsf{V}_{a1}^2 + M_b \mathsf{V}_{b1}^2) \qquad (2\cdot 4)$$

Knowing the masses and velocities, we could calculate $U_2 - U_1$. However, we do not have enough information to determine the increase in internal energy of each body individually.

The example above illustrates two important things. First, the notion of conservation of energy, as applied to an *isolated system*, says that the energy of the isolated system is constant. Second, the energy balance provides a means for determining the *change* in internal energy of a body from *macroscopic* data (in this

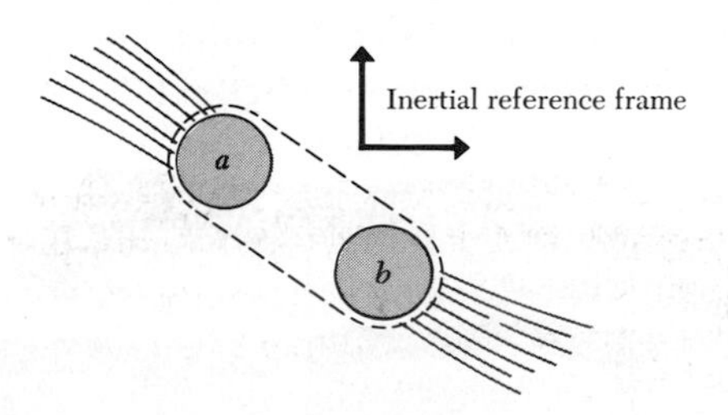

Figure 2·2. The system includes both bodies; it is an isolated system

case the mass and velocities of the bodies). Thus the very idea that energy is conserved provides a means for inferring changes in the internal energy.

Where does the principle of conservation of energy come from? Can it be proved? These are frequent questions, the answers for which lie in the understanding of the *concept* of energy. We want to be able to predict nature, and so we conceive of energy. In a very real sense, energy is an invention of *humans*, not of nature. Rather than being "that which makes things go," perhaps we should view it as something that we use to predict and explain how things go. Whenever conservation of energy is apparently violated, the physicist "discovers" a new form of energy. That is, a new form of energy is *defined* in order to keep the principle of energy conservation unviolated. The thing that makes energy "real" is that it is very useful. It is not very often that we have had to "invent" new energy forms, hence we can proceed with our scientific and engineering analysis with considerable confidence that our energy balances contain the relevant energy terms. Asking for proof of the conservation-of-energy principle is like asking for proof that addition is commutative; the fact that $A + B = B + A$ is an inherent part of the concept of addition. What *is* a relevant question is: What can I do with addition? In thermodynamics, the relevant counterpart is: What can I do with the conservation-of-energy principle? The answer to this is that practically all scientific and engineering analysis involves energy considerations.

2·5 ENERGY TRANSFER AS WORK

We come now to considering the mechanisms by which the energy of a non-isolated system can be altered. Since the system and its environment form an isolated system, if the energy of the system increases the energy of the environment must decrease a corresponding amount in order to maintain conservation of energy. We can therefore view the interaction as a process of energy transfer, and *work* is one of the mechanisms for such energy transfer.

Following the definition used in mechanics, the amount of energy transfer *to* a system as work (the "work done *on* the system") associated with some infinitesimal change in the position of matter inside is

$$\blacktriangleright \quad đW = \mathbf{F} \cdot d\mathbf{X} = \mathsf{F}\, dX \tag{2·5}$$

Here F is a force exerted by the environment on matter within the system, and dX is the infinitesimal motion of that matter in the direction of F which occurs during the period of observation. $\mathbf{F} \cdot d\mathbf{X}$ is the equivalent vector definition. Both F and X must be macroscopic measurables; they represent the net visible effect of billions of molecules, and not forces or motions of individual molecules.

We emphasize that dX must be motion observed with respect to a chosen coordinate frame. If the frame is attached to the matter, then there is no energy transfer as work in that particular analysis. Also, the force must be exerted by the environment on the matter within the system; forces exerted between matter in different parts of the system may cause internal rearrangements of energy, but lead to no energy transfer across the system boundary. One must therefore focus on the boundary of the system in order to identify and evaluate energy transfer as work.

If the motion extends from point 1 to point 2 in space, then the total amount of energy transfer as work is obtained by summing over all infinitesimal displacements, that is, by integrating Eq. (2·5). If we represent the total amount of energy transfer as work by W_{12}, then we have

$$▶ \quad W_{12} \equiv \int_1^2 đW = \int_1^2 \mathbf{F} \cdot d\mathbf{X} \tag{2·6}$$

There are important physical differences between the quantities standing behind the symbols $đ$ and d. The term dX may be immediately integrated, and the change in X thereby found,

$$\int_{X_1}^{X_2} dX = X_2 - X_1 \tag{2·7}$$

The quantity X is something which exists at each point in space, and the symbols X_1 and X_2 have definite meanings. In contrast, the integral of $đW$ cannot be computed unless we know how the force F varies with X. The value of the integral of $đW$ between any two points X_1 and X_2 therefore depends on how the force varies with position, that is, on the particular process involved. The work at a given point has absolutely no meaning. Whereas the integral of dX represents the difference in the values of X at two points, the integral of $đW$ depends upon the particular manner in which the system is taken from one configuration to another. The ratio dX/dt would mean the derivative of X with respect to t, or the rate of change of X. In contrast, the ratio $\dot{W} \equiv đW/dt$ would be the *amount* of work done per unit of time, or the *rate of energy transfer as work*. We have adopted this special symbology to emphasize the conceptual physical differences between things that are associated with the conditions of a system and those that are associated with the processes the system undergoes.

2·6 EVALUATION OF ENERGY TRANSFER AS WORK

In order to carry out a successful system energy analysis one must be able to evaluate the magnitudes of energy transfers as work, that is, the work done on or by the system. Having defined the system, the next step in this evaluation is to define the direction of positive energy transfer.† We shall *always* indicate our choice for positive energy transfers by showing the energy transfers on the system sketch. For example, consider the system of Fig. 2·3*a*. The system is defined by

† Some texts always treat work input as positive, and work output as negative. Others do just the opposite. We prefer to make an appropriate choice for each problem, showing the choice in a sketch.

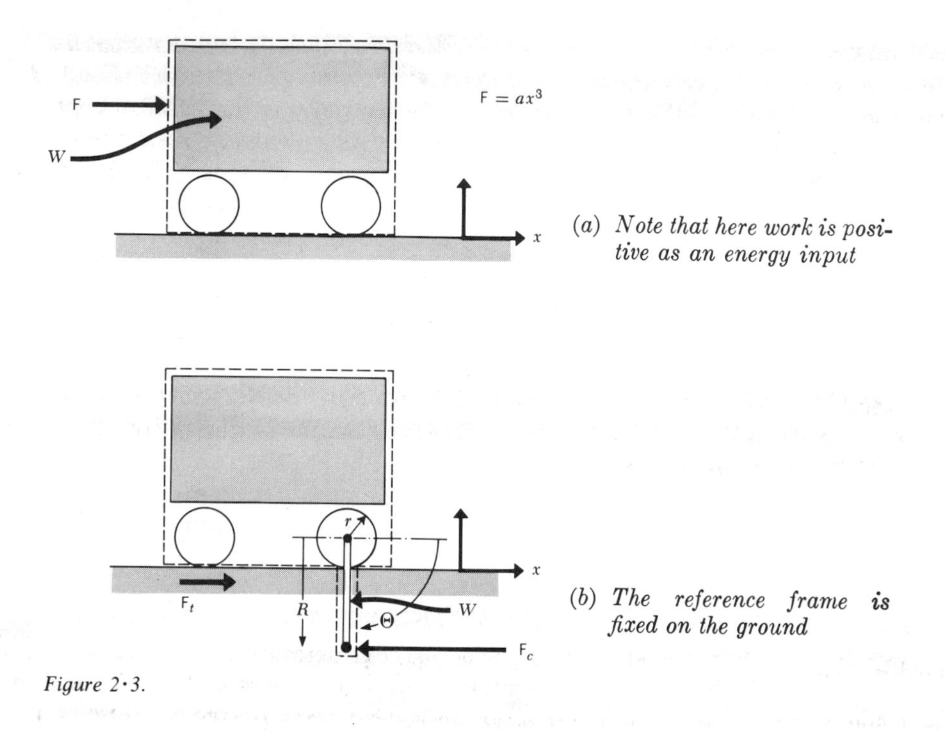

Figure 2·3.

the dotted lines, and is the little cart. The reference coordinate system is attached to the ground, and the force F is exerted as a push on the back of the cart by a metal rod. Note that we have elected to consider the work done by F positive if it leads to an energy transfer *into* the system. (If the work was subsequently calculated to be −35 ft·lbf we would then know that the energy transfer actually took place in the direction opposite our initial assumption.)

Let us analyze the system of Fig. 2·3*a*. The amount of energy transfer as work to the cart due to its motion under the influence of F is, from the basic definition,

$$W_{12} = \int_1^2 \mathsf{F}\, dX = \int_1^2 ax^3\, dx = \frac{a}{4}(x_2^4 - x_1^4) \tag{2·8}$$

If there were no other energy transfers involved, this work would show up as an increase in the energy stored within the system. We could neglect any changes in the internal energy or in the rotational energy of the wheels, in which case the only significant energy change would be in the bulk translational kinetic energy of the cart as a whole. An energy balance on the cart could then be used to calculate the final cart velocity.

A somewhat more subtle example is provided by the system of Fig. 2·3*b*. Now we have added a crank to one of the wheels, and we exert a force F_c to the left on the crank handle, causing the cart to move to the right. There is also a traction force F_t on the wheels; under ideal rolling conditions, the motion of the material in the wheel at the contact point is vertically up and down, hence this matter has *no* component of motion in the direction of F_t. Consequently there is no energy transfer as work associated with the force F_t. There is some work associated with F_c, however, and a common mistake is to say that this work is $\int F_c\, dx$, where x is the position of the cart. In computing work dX must be the displacement of the matter on which the force acts, which in this case is the handle and not the cart body. If R is the length of the handle, then $R\, d\theta$ is the horizontal displacement of the handle relative to the wheel axis associated with an infinitesimal clockwise rotation $d\theta$ (when the crankshaft is in the position shown). Is the work associated with F_c equal to $F_c R\, d\theta$? *No*, not for the system as defined, for the displacement $R\, d\theta$ is relative to a reference frame on the cart, while we chose a frame on the ground! Since the axle moves to the right a distance $r\, d\theta$, the displacement of the handle relative to the ground frame is $r\, d\theta - R\, d\theta$. But this is a displacement to the *right*; while the force F_c acts to the *left*. Hence the amount of energy transfer as work to the system as defined, relative to the indicated reference frame, is

$$đW = -F_c(r - R)\, d\theta = F_c(R - r)\, d\theta \tag{2·9}$$

This example illustrates the need for very careful consideration of just what the system is *before* one begins to evaluate work.

Suppose in the example of Fig. 2·3*b* we instead choose a reference frame attached to the cart. In this frame the displacement of the handle in the direction of F_c is $R\, d\theta$, hence the amount of energy transfer as work in this frame is

$$đW' = F_c R\, d\theta \tag{2·9b}$$

The product $F_c R$ is defined as the *torque* T on the wheel shaft, hence $đW'$ could be written as†

$$đW' = T\, d\theta \tag{2·9c}$$

Equation (2·9*c*) is a special case of the general expression for the work done on a system by a torque. If $\mathbf{T}$ is the torque vector, and $d\boldsymbol{\theta}$ is the vector angular displacement of the matter within the system on which the torque acts, then the amount of energy transfer as work *to* the system relative to a frame attached to the torque axis, is

$$▶ \qquad đW = \mathbf{T}\cdot d\boldsymbol{\theta} \tag{2·10}$$

Since mechanical systems frequently involve rotating shafts, the expression is rather important.

† Note that the cart would not have any kinetic energy in this reference frame (see Prob. 1·4).

The forces of Fig. 2·3 are examples of *surface forces*. They are exerted on the matter at the surface of the system, and the appropriate displacement is the displacement of the matter at the surface. Another type of force is a *body force*, which acts on the material in the system interior. For such forces the appropriate displacement must be that of the matter on which the body force acts. This may or may not be the same as the displacement of a system boundary. For example, consider the system of Fig. 2·4*a*. The system does not include the electric field. The system boundaries are held stationary, but the charged plate within the system moves when the electric field strength is increased. If the charge on the plate is $\mathcal{Q}$, and the electric field strength is **E**, the force on the plate is $\mathbf{F} = \mathcal{Q}\mathbf{E}$ to the right. The amount of energy transfer as work to the system associated with motion of the plate an infinitesimal amount $d\mathbf{x}$ to the right is then

$$đW = \mathcal{Q}(\mathbf{E} \cdot d\mathbf{x}) \tag{2·11a}$$

Integrating between two positions x_1 and x_2, the total energy transfer as work for the process 1-2 is

$$W_{12} = \int_1^2 \mathcal{Q}\mathbf{E} \cdot d\mathbf{x} \tag{2·11b}$$

To complete the integration we would have to know $\mathbf{E}(\mathbf{x})$ during the process. The energy transferred into the system as work would result in an increase in the system energy, and an energy balance would be the central tool for analysis of this system.

In the example of Fig. 2·4*a* the system was defined to include the plate, the sponge, and the gas, but to exclude the electric field. The energy transfer as work between the plate and the gas was internal to the system, hence would not appear as an energy transfer term in the energy balance. However, if the system is instead

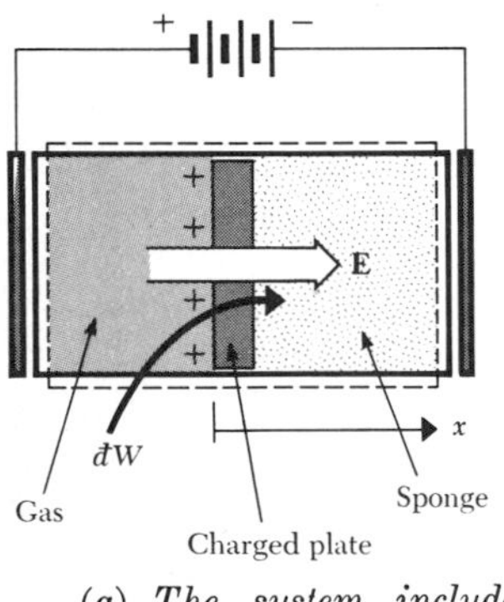

(*a*) *The system includes the material but not the electric field*

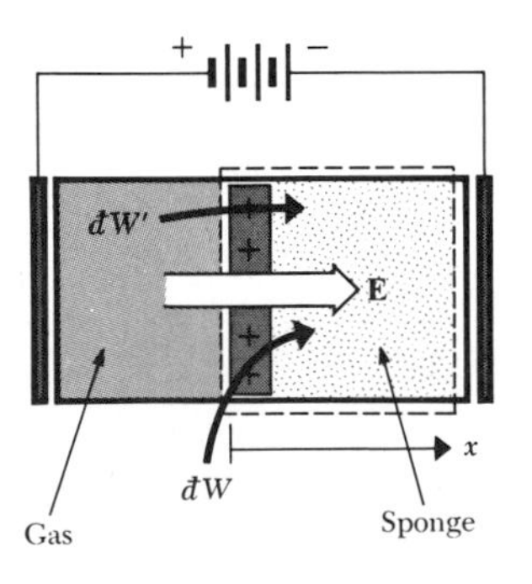

(*b*) *Note that the work terms depend on the system definition*

Figure 2·4.

defined as in Fig. 2·4*b*, the energy transfer as work from the gas to the plate must be considered. The energy transfer as work from the environment (that is, the electric field, which is external to the system) is again given by Eqs. (2·11). If the pressure exerted by the gas on the plate is *P*, and the plate area is *A*, the force on the plate is *PA*, and the amount of energy transfer as work at the left-hand boundary of the system is

$$đW' = PA\,dx \tag{2·12a}$$

Integrating, the total amount of energy transfer as work from the gas to the plate is

$$W'_{12} = \int_1^2 PA\,dx \tag{2·12b}$$

We would have to know how the gas pressure varied with x in order to complete the integration. This example illustrates that the work terms included in an energy analysis depend very much on the choice of the system. A clear system definition is therefore essential to a clear analysis.

Procedure. Correct evaluation of the energy transfer as work is a necessary part in any thermodynamic analysis. The steps in this evaluation are summarized below.

1. Define the system and reference frame.
2. Define the forces acting on the system and the motions of the material on which these forces act.
3. Define the sign of positive energy transfer for each different work mode by an arrow on the system sketch.
4. Apply the basic definition of work [Eq. (2·5)] consistently with the above definitions over the time period under examination.
5. Bring in the additional information needed to complete the integrations.

One must be particularly consistent in evaluating work done by a conservative force field in order to avoid "double accounting" in the subsequent energy balance. The potential-energy change is by definition the work done by the conservative force, hence both the potential energy and the work done by the conservative force must never appear in the same energy analysis. Two alternative approaches are demonstrated in Fig. 2·5. In Fig. 2·5*a* we account for the work done by the rope force F_H and the body force *w*, but do not include a potential-energy term in the system energy. In Fig. 2·5*b* we instead use the potential energy and include only the work done by the rope force. The net energy input as work, $đW$, is equal to the increase in energy, dE, in either case.†

† Another approach is to assign the potential energy to the field. If the field is part of the system, Fig. 2·5*b* is appropriate. If the system is defined to exclude the field, Fig. 2·5*a* is proper.

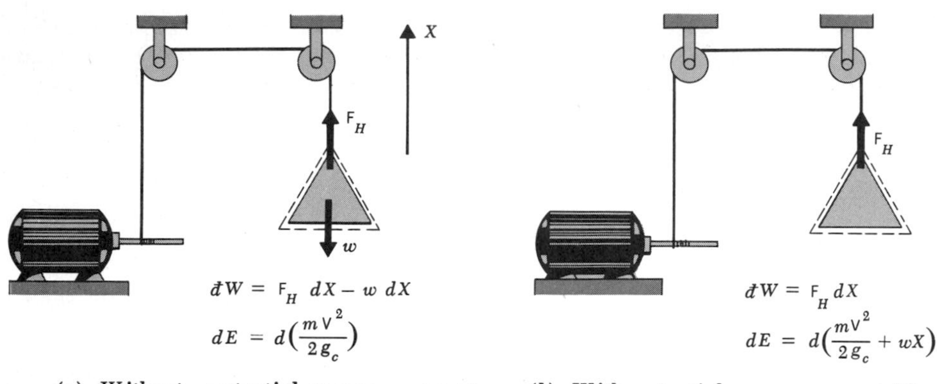

(*a*) *Without potential-energy concept (the system excludes the gravitational field)*

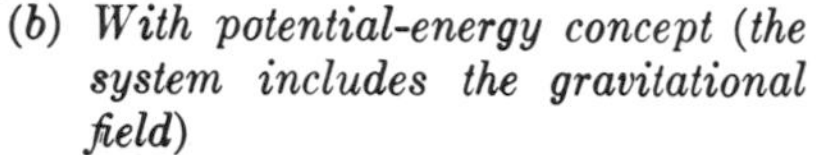

(*b*) *With potential-energy concept (the system includes the gravitational field)*

Figure 2·5. One must be consistent when making an energy analysis where a conservative force is involved. Note đW = dE in either case

2·7 SOME PARTICULAR WORK MODES

There are a number of work modes that occur frequently in thermodynamic analysis, and they deserve special mention here.

Expansion or compression of a fluid. A great many thermodynamic systems involve fluids (that is, a liquid or a gas), hence the work associated with expansion or compression of a fluid is very important. The piston-cylinder system of Fig. 2·6*a* provides an easy way to visualize and compute the amount of energy transfer as work associated with a change in the fluid volume. The force per unit area exerted by the fluid on the piston is the fluid pressure P. Denoting the piston area by A, the force on the piston is PA, and the amount of energy transfer as work *from* the fluid *to* the piston (see the sign convention of the sketch) is, for an elemental expansion dx,

$$đW = PA\,dx$$

But since $A\,dx = dV$ is the increase in fluid volume, the amount of energy transfer as work from the fluid may be written as

$$đW = P\,dV \tag{2·13a}$$

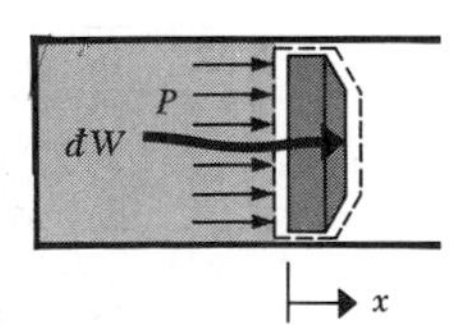

(*a*) *The fluid is expanding*

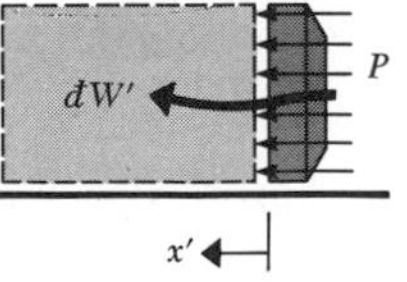

(*b*) *The fluid is being compressed*

Figure 2·6.

By considering a fluid volume of *arbitrary* shape it may easily be shown that this same expression applies for the energy transfer as work *from* the fluid *to* its environment, hence Eq. (2·13*a*) has considerable utility in thermodynamic analysis.

Suppose we consider the compression of a fluid. Now it is easier to consider the work positive if energy is transferred *to* the fluid (Fig. 2·6*b*). The force that the piston exerts on the fluid at the system boundaries is PA, and displacement of the fluid at the piston surface is dx'. Hence the amount of energy transfer as work *from* the piston *to* the fluid is

$$đW' = PA\,dx'$$

We can again relate this work to the change in fluid volume. Denoting the *increase* in fluid volume by dV (in the conventions of calculus dx always represents an increase in x), we write $dV = -A\,dx'$ (dx' is positive, so dV will be negative). Hence

$$đW' = -P\,dV \tag{2·13b}$$

is the amount of work done *by* the piston *on* the fluid. Comparing the energy-transfer sign conventions of Figs. 2·6*a* and 2·6*b*, we see that $đW' = -đW$. Hence Eqs. (2·13*a*) and (2·13*b*) are consistent. This will always be the case; for a particular choice of the direction of positive energy transfer as work, the expression for the work will be independent of whether the energy is actually added or removed (dx will be positive in one case and negative in the other).

Extension of a solid. We consider extension of a solid rod as shown in Fig. 2·7. Denoting the normal stress in the x direction by σ_x, the "pulling" force exerted by the holder on the solid is $\sigma_x A$. The amount of energy transfer as work *from* the holder *to* the solid when the solid is stretched is†

$$đW = \sigma_x A\,dx$$

Now, the *strain* ε_x in the x direction is defined as the deformation per unit length, so

$$d\varepsilon_x = \frac{dx}{L} = \frac{A\,dx}{V}$$

† Note if we replace σ_x by $-P$ (P is a "push," σ_x is a "pull") and take proper account of the differing directions of positive energy transfer in this and the fluid expansion case, equivalent results are obtained.

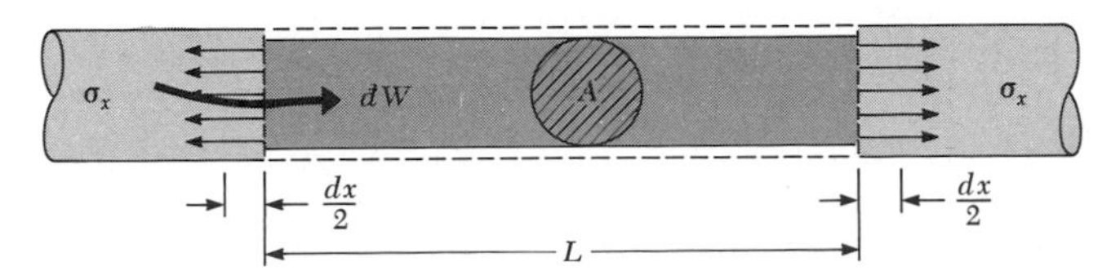

Figure 2·7. Stretching a thin rod requires work

hence the amount of energy transfer as work *to* the solid can be written as†

$$\blacktriangleright \qquad đW = V\sigma_x \, d\varepsilon_x \qquad (2\cdot 14)$$

In the microscopic view we see individual molecules of the holder pulling on molecules of the solid, thereby straining the crystal bonds in the solid structure. The net macroscopic effect of these microscopic interactions is represented by Eq. (2·14).

Stretching of a liquid surface. Consider a liquid sheet suspended between two plates, as shown in Fig. 2·8. The molecules near the liquid surface are attracted more strongly by internal liquid molecules than by the very distant gas molecules, and this gives rise to a macroscopically measurable force in the liquid-gas surface. The *interfacial tension* σ is defined as the force *per unit of length* acting normal to a line in the surface. The total interfacial tension force exerted on each plate is therefore $2\sigma b$, where b is the depth of the liquid sheet normal to the plane of the paper. The energy transfer as work *from* the plates *to* the liquid when the plate separation is increased an amount dx is

$$đW = 4\sigma b \frac{dx}{2} = 2\sigma b \, dx$$

Now, the change in surface area is simply $dA = 2b \, dx$, hence the energy transfer as work *to* the liquid can be written as

$$\blacktriangleright \qquad đW = \sigma \, dA \qquad (2\cdot 15)$$

In the microscopic view, work is required to pull individual molecules into the new surface, and Eq. (2·15) represents the macroscopic evaluation of this microscopic effect.

† Here we considered only one-dimensional deformation. It may be shown that the amount of energy transfer as work per unit volume *to* a solid undergoing arbitrary three-dimensional deformation is

$$đW = \sum_{i=1}^{3} \sum_{j=1}^{3} \sigma_{ij} \, d\varepsilon_{ij}$$

where σ_{ij} and ε_{ij} are the stress and deformation tensors.

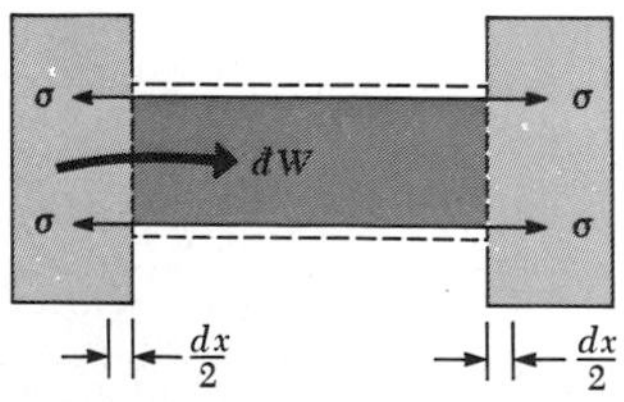

Figure 2·8. The liquid extends a large distance b into the paper

Work due to magnetization and polarization. Microscopic electric dipoles within dielectrics resist turning, and work is therefore done on such substances when they are polarized. For a system through which the electric and polarization fields are uniform, it may be shown that the energy transfer as work to the dielectric material† from the electric field when the polarization is infinitesimally increased is

$$\blacktriangleright \qquad đW = \mathbf{E}\cdot d(V\mathbf{P}) \qquad (2\cdot16)$$

Here **E** represents the strength of the electric field within the dielectric, **P** the polarization field (electric-dipole moment per unit of volume), and V the system volume.

Similarly, work can be done on magnetic materials when the magnetization is changed, for the microscopic magnetic-dipole moments resist turning. For a system through which the magnetic and magnetization fields are uniform, it may be shown that the energy transfer as work to the material† from the magnetic field when the magnetization is increased an infinitesimal amount is

$$\blacktriangleright \qquad đW = \mu_0\, \mathbf{H}\cdot d(V\mathbf{M}) \qquad (2\cdot17)$$

Here **H** is the strength of the applied magnetic field, **M** is the magnetization vector (magnetic-dipole moment per unit of volume), and μ_0 is a constant (the "permeability" of free space). Unlike work due to compression or extension, polarization and magnetization do not result in macroscopic motions. However, they do result in alignment of the particles, and this ordered alignment is macroscopically detectable as an increase in the total dipole moment.

The expressions for work in the particular cases discussed here are all of the form

$$đW = \mathsf{F}\, dX$$

For example, for liquid extension $\mathsf{F} = \sigma$ and $X = A$. F and dX do not necessarily have dimensions of force and displacement, but they are frequently called *generalized force* and *generalized displacement*, respectively. They may be scalars (P and V), vectors (**E** and **P**), or tensors (σ_{ij} and ε_{ij}). These F's and X's are very important in the general thermodynamic theory of materials. The amounts of energy transfer as work for these modes are summarized in Table 2·1.

2·8 WORK DEPENDS UPON THE PROCESSES

The amount of energy transfer as work for a given process can be computed if the variation of F with X during the process is known. For example, consider a system

† The electric and magnetic fields are excluded from the systems. These expressions are for the rationalized mksc unit system. See, for example, David Halliday and Robert Resnick, *Physics*, combined ed., secs. 27-26 and 33-34, John Wiley & Sons, Inc., New York, 1962.

Table 2·1 Some particular work modes

Mode	$đW$, energy transfer to substance as work	Restrictions
Fluid compression	$-P\,dV$	P uniform over surface
Solid extension	$V\sigma_x\,d\varepsilon_x$	one-dimensional strain only
Liquid surface extension	$\sigma\,dA$	σ uniform over area
Polarization	$\mathbf{E}\cdot d(V\mathbf{P})$	**P** and **E** uniform through volume
Magnetization	$\mu_0\,\mathbf{H}\cdot d(V\mathbf{M})$	**H** and **M** uniform through volume

consisting of mass M of a certain gas, which we compress from some state 1 to another state 2. The work done *on* the gas is determined by integrating Eq. (2·13*b*), and is

$$W_{12} = -\int_1^2 P\,dV$$

We must know how P varies with V during the process in order to complete the integration; Fig. 2·9 demonstrates two different cases.

For example, suppose we idealize that the pressure exerted by the gas is related to the volume and temperature by†

$$PV = MRT$$

where M is the gas mass, and R is a constant for the gas. The energy transfer as work to the gas is then

$$W_{12} = -M\int_1^2 \frac{RT}{V}\,dV$$

† This is valid for an *ideal*, or *perfect*, gas. The concept of temperature will be developed in the next chapter.

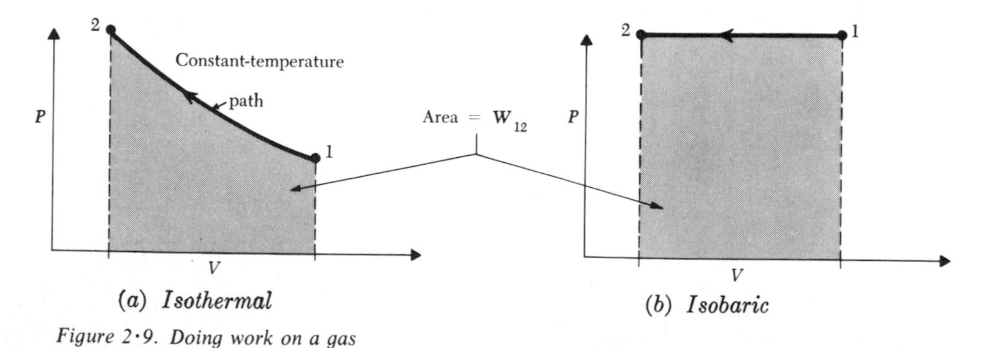

(*a*) *Isothermal* (*b*) *Isobaric*

Figure 2·9. Doing work on a gas

We would have to know how the temperature varies with the volume during this process in order to perform the integration. In particular, for an *isothermal* (constant-temperature) *process*

$$W_{12} = -MRT \ln \frac{V_2}{V_1}$$

Another case of interest is the *isobaric* (constant-pressure) *process*, where the energy transfer to the gas as work is

$$W_{12} = \int_1^2 -P\, dV = P(V_1 - V_2)$$

As a second example, consider the amount of energy transfer as work to a paramagnetic Curie salt,† for which the magnetic field strength and magnetization are related to the temperature by

$$\mathsf{M} = \frac{C\mathsf{H}}{T}$$

Here C is the *Curie constant* of the salt. From Eq. (2·17),

$$W_{12} = \int_1^2 \frac{\mu_0 VT}{C} \mathsf{M}\, d\mathsf{M}$$

The manner in which the temperature varies during this process would be an important factor in determining the amount of work done for a given increase in magnetization (see Fig. 2·10). In particular, for an *isothermal* magnetization process, the energy transfer to the salt as work is

$$W_{12} = \frac{\mu_0 VT}{2C} (\mathsf{M}_2^2 - \mathsf{M}_1^2)$$

† We assume constant volume.

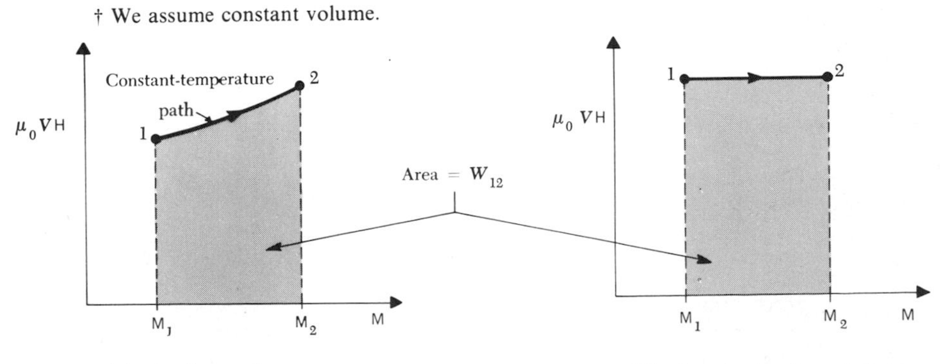

(a) *Isothermal* (b) *Constant external field*

Figure 2·10. Doing work on a magnetic substance

Alternatively, if the substance were magnetized in a constant external field (H constant), the amount of energy transfer to the material as work would be

$$W_{12} = \mu_0 V \mathsf{H}(\mathsf{M}_2 - \mathsf{M}_1)$$

In these examples simple algebraic expressions could be found. In many cases of interest it will be necessary to integrate numerically or graphically to obtain the amount of energy transfer as work for a given process.

2·9 ENERGY TRANSFER AS HEAT

We have discussed several means by which macroscopically observable work can be done on or by a system, causing its energy to change. However, it is possible to transfer energy to a system in ways which are not observable as macroscopic work. Consider the system shown in Fig. 2·11. We might picture the atoms in the walls surrounding the system as little vibrating masses. Some will be heading toward the boundary when some are heading away and, as a result, there might be no macroscopically observable motion. However, the interaction of these atoms with the molecules in the system can result in changes in the energy of individual particles, hence a change in the internal energy of the system. The macroscopically observable work is zero, so we must find some other way to account for the energy change. This second mechanism of energy transfer depicted in Fig. 2·11 is called *energy transfer as heat*. *Heat* is energy transfer as work on the microscopic scale which fails to be accounted for by macroscopic evaluations of work. In the microscopic view there is no such thing as heat; heat is required in the macroscopic view as a means for accounting for microscopically disorganized energy transfer which, through its disorganization, is "hidden" from direct macroscopic view. Heat is a central *concept* in thermodynamics.

Heat, like work, is energy in the process of being transferred. Heat and work are not stored within matter; they are "done on" or "done by" matter. Energy is what is stored, and work and heat are two ways of transferring energy across the boundaries of a system. Once energy has entered the system, it is impossible to tell whether the energy was transferred in as heat or as work. The term "heat of a substance"† is thus as meaningless as "work of a substance."

† The literature contains some remnants of the caloric theory of heat, in which heat was pictured as a conserved substance. In present thermodynamic theory heat is not conserved and is not a property of matter. Still, some handbooks tabulate "heats of the liquid," "sensible heats," and "latent heats." These terms will be related to modern terminology in subsequent chapters.

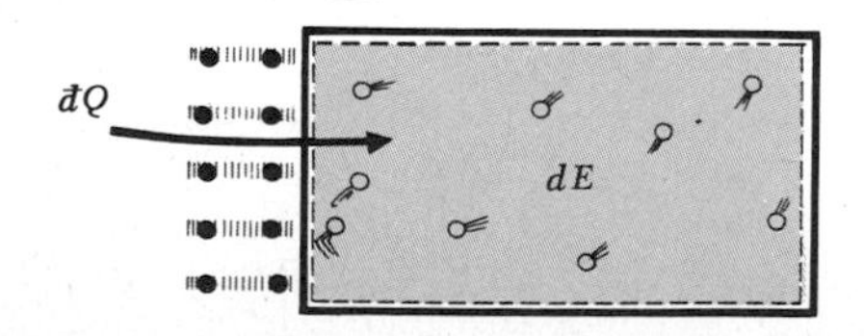

Figure 2·11. Energy transfer not observable as macroscopic work is called energy transfer as heat

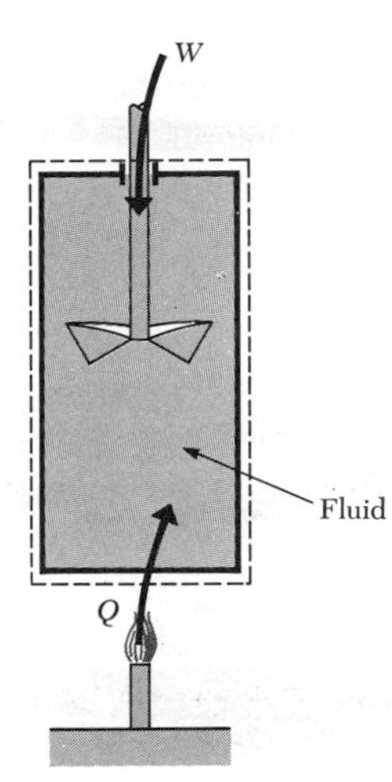

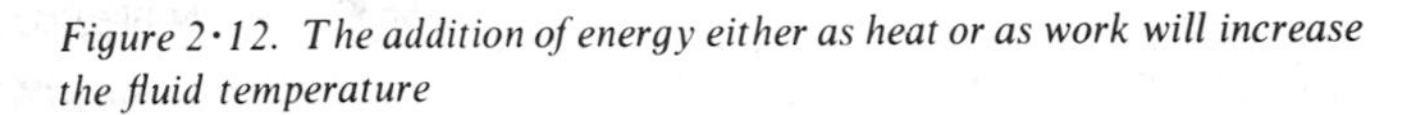

Figure 2·12. The addition of energy either as heat or as work will increase the fluid temperature

The symbol Q is usually employed to represent an amount of energy transfer as heat; $đQ$ therefore represents an infinitesimal amount of energy transfer as heat. As with work, we shall always indicate the direction of positive energy transfer as heat by an arrow on the system sketch. The value of Q depends on the details of the process and not on the end conditions of the system. For example, consider the system in Fig. 2·12. The fluid temperature can be changed by transferring energy in as work through the stirrer, or as heat via the flame. In the first case $Q = 0$, and in the second case $Q > 0$. The initial and final system states, and the change in the system internal energy, could be the same in both cases.†

Temperature. In the next chapter we shall carefully treat a concept closely related to energy transfer as heat, namely, *temperature*. We mention it now in the

† It is very important to distinguish clearly between the concepts of internal energy and heat, and an account of the experiences of Professor J. Yule will assist in this regard. Professor J. Yule was experimenting with the apparatus of Fig. 2·12, which involved a fluid in a container with a mechanical stirrer. One Friday afternoon in December 1843, Professor Yule left the device in his laboratory under the care of his assistant, Dr. B. T. Ewe, who was known for his clever practical jokes. At that time the fluid temperature was 25°C. On returning to the laboratory the following Monday, Professor Yule found that the fluid temperature was 40°C, indicating a greater internal energy. His student, Cal Orick, immediately presumed that Ewe had heated the device with the bunsen flame during the chilly weekend. "It *must* be so," said Orick, "for see, the temperature has clearly risen!" But Yule, being a properly cautious scientist, was not so sure. He had previously discovered that it was possible to increase the fluid temperature (and the internal energy) by driving the stirrer, a process that clearly fits in the category of *work*. Even after much argument Yule could not convince Orick that they just could not tell whether the clever Ewe had added the extra energy to the fluid as work using the stirrer or as heat using the bunsen flame. Ewe finally confessed to a very neat trick: he had first taken energy *out* of the fluid as heat by placing the device on a block of ice, and then had raised the temperature using the stirrer. Hence, in the actual process carried out by Ewe, there had been a *removal* of energy as heat and an *addition* of energy as work. The *net* effect of this *cooling-work* process had been an *increase* in the *internal energy* (and temperature) of the fluid. After this experience Cal Orick went mad, Professor J. Yule became a fellow of the Royal Society, and Dr. B. T. Ewe became a laboratory instructor at an obscure western university.

hope of preventing any confusion between the concepts of heat, temperature, and internal energy. Energy transfer as heat always takes place *from* a body at the higher temperature *to* one at a lower temperature. The words "hot" and "cold" describe the relative temperatures of the two bodies, respectively. Temperature can be viewed as a driving potential for energy transfer as heat.

Adiabatic. If we wish to isolate a system, we must prevent all flows of energy to or from the system. A rigid wall will prevent any $P\,dV$ work; a wall impermeable to electric fields will prevent any polarization work. In order to prevent any energy transfer as heat we conceive of an *adiabatic wall* which cannot transmit any energy as heat. The adiabatic wall is a useful fiction (like the rigid wall). The hollow space in a vacuum flask forms adequate approximation to an adiabatic wall for many laboratory experiments. The term "adiabatic process" is used to denote any process in which no energy transfer as heat occurs across the boundaries of the system under study. The concepts of an adiabatic wall and an adiabatic process are very important in thermodynamics.

2·10 ENERGY BALANCE FOR A CONTROL MASS

We are now able to express the conservation-of-energy notion analytically for an important special kind of system, a control mass. A *control mass* is a system of specified matter. As our previous discussion indicates, we can change the energy of a control mass by transfer of energy either as heat or as work, and *these are the only ways*. The control mass and its environment together form an isolated system; their total energy must remain constant. If the energy of one increases, the energy of the other must decrease by precisely the same amount. Work and heat are the sole mechanisms by which such energy transfers take place.

The total energy input to the control mass must account precisely for the increase in the control-mass energy; the algebraic statement of this accounting is called the *energy balance*. In terms of the system and symbols of Fig. 2·13, the energy balance is

$$\underbrace{W + Q}_{\text{energy input}} = \underbrace{\Delta E}_{\text{increase in energy storage}} \qquad (2\cdot18a)$$

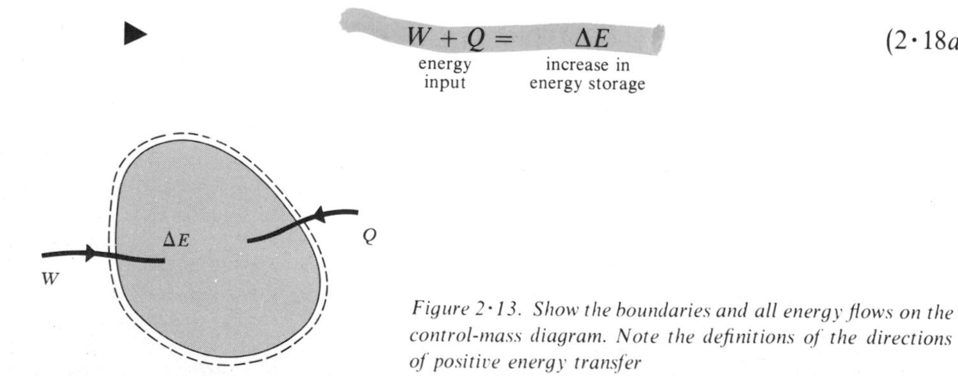

Figure 2·13. Show the boundaries and all energy flows on the control-mass diagram. Note the definitions of the directions of positive energy transfer

Here W and Q represent the amounts of energy transfer *to* the control mass as work and heat, respectively, and ΔE is the *increase* in the energy of the control mass,

$$\blacktriangleright \qquad \Delta E \equiv E_{\text{final}} - E_{\text{initial}}$$

(The symbol Δ will *always* be used to mean "final-minus-initial," that is, "the increase in . . . .") Alternatively, we could analyze the control mass over an infinitesimal part of the process, in which case the energy balance would be

$$\blacktriangleright \qquad \underset{\substack{\text{energy}\\ \text{input}}}{đW + đQ} = \underset{\substack{\text{increase in}\\ \text{energy storage}}}{dE} \tag{2·18b}$$

Here $đW$ and $đQ$ represent infinitesimal amounts of energy transfer to the control mass, and dE represents the infinitesimal increase in the control-mass energy. (The symbol d always means "an infinitesimal increase in . . ."; the symbol $đ$ means "an infinitesimal amount of . . . .")

First law. The first basic principle of thermodynamics is that matter has energy, and energy is conserved; this is called the *first law of thermodynamics*. Equations (2·18) are particular forms of the first law of thermodynamics and apply only to a control mass; later we shall extend the mathematical representations of the first law to control volumes.

The analysis of a thermodynamic system invariably begins with energy-balance considerations, which are called *first-law analyses;* some simple examples of such analyses follow.

2·11 EXAMPLES OF CONTROL-MASS ENERGY BALANCES

The primary aim of this chapter is the development of the concepts of energy, internal energy, work, and heat. We need more tools before we can begin to do much productive thermodynamic analysis, but some examples at this stage should assist in clarifying the basic ideas. The examples that follow are intended to point up some important ideas from the preceding discussions and to indicate a general approach to the analysis of thermodynamic systems. A more complete energy-balance methodology will be given in Chap. 5.

Gas compression. Two kg of a gas are squeezed in a device from a volume of 1.4 m^3 to a volume of 0.9 m^3. During this time the pressure remains constant at 100,000 N/m^2, and it is known from other considerations that the internal energy decreases by 12,000 J. How much energy was transferred as heat to or from the gas for this process?

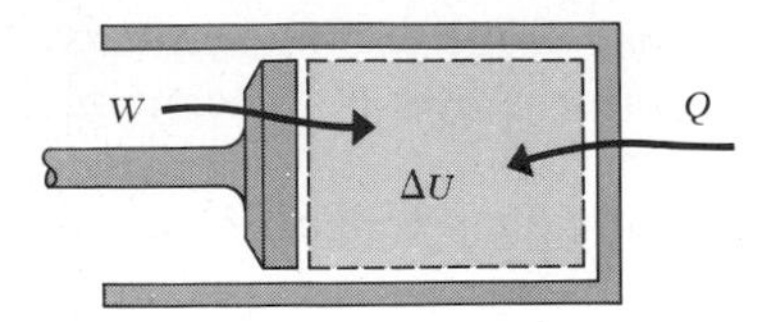

Figure 2·14. The control mass and energy flows

We first define the control mass to include only the gas. It is clear from the problem statement that energy will be put into the system as work, and so we choose the sign of positive work as indicated in Fig. 2·14. It may not be immediately clear which way the energy transfer as heat takes place; let's define Q to be positive as an energy input, and we indicate this on Fig. 2·14. We also assume that the only system energy is its internal energy U. In terms of the system and symbols of Fig. 2·14, the energy balance is then

$$\underbrace{W + Q}_{\text{energy input}} = \underbrace{\Delta U}_{\text{increase in energy storage}} \tag{2·19}$$

where $\Delta U = U_2 - U_1$ and the subscripts 1 and 2 denote the initial and final states. We know the internal energy change and can compute Q from the energy balance if we can evaluate W. We have already derived an expression for the work done *on* a gas during compression, namely,

$$W = -\int_1^2 P\,dV$$

Since the pressure P is constant, we integrate and have

$$W = P(V_1 - V_2) = 10^5 \text{ N/m}^2 \times (1.4 - 0.9) \text{ m}^3 = 50{,}000 \text{ J}$$

Solving Eq. (2·19) for Q,

$$Q = \Delta U - W = -12{,}000 - 50{,}000 = -62{,}000 \text{ J}$$

The minus number means that we guessed the actual direction of energy transfer as heat incorrectly. The answer is that 62,000 J of energy was transferred as heat *from* the gas.

If we had correctly guessed the actual direction of energy transfer as heat, and pointed the Q arrow the other way on Fig. 2·14, the energy balance would read

$$\underbrace{W}_{\text{energy input}} = \underbrace{Q}_{\text{energy output}} + \underbrace{\Delta U}_{\text{increase in energy storage}}$$

and the calculation would lead to

$$Q = W - \Delta U = 62{,}000 \text{ J}$$

Note that the correct problem solution can be obtained with either analysis; the main requirement is consistency between the sketch and the energy balance.

A pneumatic catapult. A light airplane catapult uses high-pressure steam to launch aircraft with the system shown schematically in Fig. 2·15. Initially the cylinder volume is 10 ft^3, and finally it is 35 ft^3. The launch velocity is 200 ft/s, and the combined mass of the piston, linkage, and aircraft is 6000 lbm. The process occurs very rapidly, so there is not much time for energy transfer as heat between the steam and the cylinder walls, hence $Q = 0$ is a good idealization. We wish to determine the change in internal energy of the steam for this process.

We take the control mass shown in Fig. 2·15. It includes the steam, the launch gear, and the aircraft, and we consider it over the launch period. To simplify the analysis we shall neglect the interaction with the air during the launch, and we shall neglect friction. These idealizations permit us to treat the control mass as an isolated system. There is no transfer of either energy or matter across its boundaries. The energy balance is therefore simply

$$\underset{\substack{\text{increase in}\\ \text{energy storage}}}{\Delta E} = 0 \qquad (2\cdot 20)$$

Now, the energy consists of the internal energy U_s of the steam, the internal energy of the mechanical parts, and bulk kinetic energy KE. We assume that the internal energy of the metal does not change during the process, and that the bulk kinetic energy of the steam is negligible. Then, denoting the initial and final conditions by subscripts 1 and 2, respectively, the energy change ΔE is

$$\Delta E = \Delta U_s + \Delta KE_m$$

The increase in the kinetic energy of the mechanical parts is

$$\Delta KE_m = \Delta\left(\frac{1}{2}\frac{M}{g_c}\mathsf{V}^2\right) = \frac{1}{2}\frac{M}{g_c}(\mathsf{V}_2^2 - \mathsf{V}_1^2)$$

$$= \frac{6000 \text{ lbm}}{2 \times 32.2 \text{ ft}\cdot\text{lbm}/(\text{lbf}\cdot\text{s}^2)}[(200)^2 - 0] \text{ ft}^2/\text{s}^2$$

$$= 3.73 \times 10^6 \text{ ft}\cdot\text{lbf}$$

The energy balance, Eq. (2·20), then says

$$\Delta U_s = -\Delta KE_m = -3.73 \times 10^6 \text{ ft}\cdot\text{lbf}$$

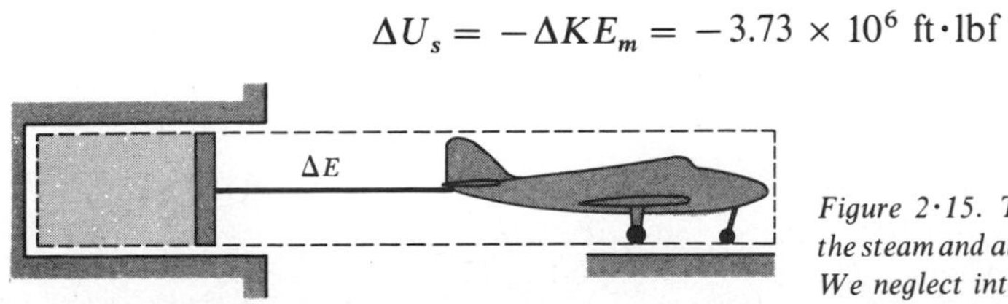

Figure 2·15. The control mass includes the steam and all moving mechanical parts. We neglect interaction with the air

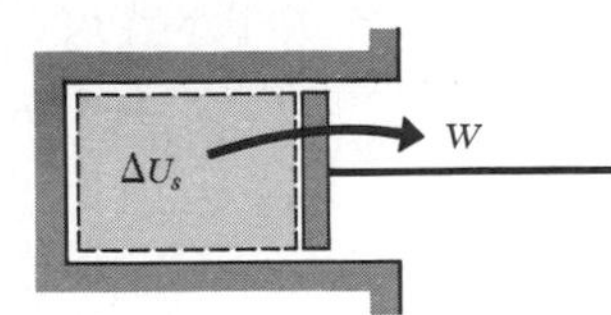

(*a*) *The control mass is just the gas*

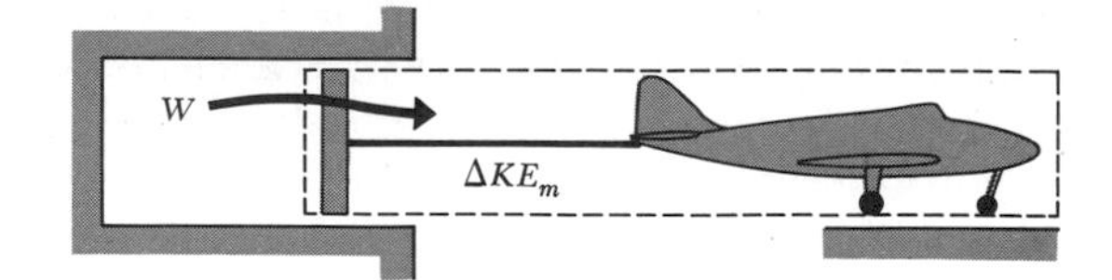

(*b*) *The control mass is just the moving mechanical parts*

Figure 2·16.

Hence the internal energy of the steam *decreases* by 3.73×10^6 ft·lbf during the launch. The catapult extracts some of the disorganized molecular kinetic energy and transforms it into the organized kinetic energy of the aircraft.

Suppose we wish to calculate the work done by the steam on the piston. We could analyze either of the control masses of Fig. 2·16. If we take the steam as the system, the energy balance is

$$\underset{\text{energy input}}{0} = \underset{\text{energy output}}{W} + \underset{\text{increase in energy storage}}{\Delta U_s}$$

or

$$W = 3.73 \times 10^6 \text{ ft·lbf}$$

If instead we take the launch gear and aircraft as the system, the energy balance is

$$\underset{\text{energy input}}{W} = \underset{\text{increase in energy storage}}{\Delta KE_m} = 3.73 \times 10^6 \text{ ft·lbf}$$

Neglect of interaction with the air in the analysis above may introduce significant error if the steam pressure is not well above atmospheric pressure. The next approximation should include work done by the system in pushing air out of the way as the control mass expands. Calculation of this work would require more detailed knowledge of the geometry of the cylinder system. We see that the analyst must sometimes consider more than one system to complete a calculation. There are often several ways to do the analysis, and all correct ways will give the same answer.

The energetics of a dielectric medium. Consider a dielectric in which the dipole moment per unit volume $\mathbf{P}$ is related to the applied electric field $\mathbf{E}$ by

$$\mathbf{P} = \varepsilon_0(\kappa - 1)\mathbf{E} \tag{2·21}$$

where κ is the dielectric constant. Derive an expression for the work done by the electric field when the electric field is changed from $\mathbf{E}_1$ to $\mathbf{E}_2$.

The control mass consists of the dielectric, but does not include the external electric field which occupies the same space. The amount of energy transfer as work from the electric field to the dielectric material is [Eq. (2·16)]

$$W = \int_1^2 \mathbf{E}V \cdot d\mathbf{P} \tag{2·22}$$

where V is the fixed system volume. Using Eq. (2·21),

$$W = \varepsilon_0 V(\kappa - 1) \int_1^2 \mathbf{E} \cdot d\mathbf{E} = \tfrac{1}{2}\varepsilon_0 V(\kappa - 1)(\mathsf{E}_2^2 - \mathsf{E}_1^2) \tag{2·23}$$

Now, the energy balance on the control mass is (Fig. 2·17)

$$\underset{\text{energy input}}{Q + W} = \underset{\text{increase in energy storage}}{\Delta U} \tag{2·24}$$

where U is the internal energy of the dielectric. Combining Eqs. (2·23) and (2·24),

$$\Delta U = Q + \tfrac{1}{2}\varepsilon_0 V(\kappa - 1)(\mathsf{E}_2^2 - \mathsf{E}_1^2) \tag{2·25}$$

In particular, for the special case of $Q = 0$, which is called an *adiabatic process*, the increase in the energy of the dielectric is equal to the work done by the electric field, which may be represented as the change in the quantity $\frac{1}{2}\varepsilon_0 V(\kappa - 1)\mathsf{E}^2$. It is sometimes stated in books on electrostatics that this quantity is the energy of the dielectric. We see that this is the case only if it is assumed that the dielectric always undergoes adiabatic polarization. This is normally an excellent assumption in electronic circuits where the frequencies of polarization are so rapid that there is no time for any significant amount of energy transfer as heat to take place.

For example, if the dielectric constant is 2.4, the volume is 10^{-6} m^3, and the electric field strength is initially zero and finally 10^4 V/m, the change in energy of the dielectric is

$$\Delta U = 0.5 \times 8.854 \times 10^{-12} \frac{\mathrm{C^2 \cdot s^2}}{\mathrm{kg \cdot m^3}} \times 10^{-6}\ \mathrm{m^3} \times 1.4 \times \left(10^4 \frac{\mathrm{kg \cdot m}}{\mathrm{C \cdot s^2}}\right)^2$$

$$= 6.2 \times 10^{-10}\ \mathrm{J}$$

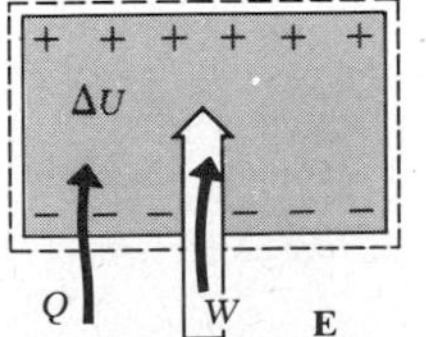

Figure 2·17. The system includes the dielectric but not the electric field

A complex power system. A nuclear power station consists of the hardware shown in Fig. 2·18. In the primary circulation loop, liquid NaK (sodium-potassium eutectic) is circulated by a pump through the reactor, which provides the energy source, and then through a boiler in which water flowing in the secondary loop becomes steam. This steam is fed to a turbine, which drives the electric power generation equipment. In order to close the secondary loop, the steam is condensed and then pumped up to a high pressure. This complicated system is the sort we shall be able to analyze in detail before very long; a simple analysis of certain aspects is possible now with only the tools at hand.

Suppose the desired turbine shaft power output is 100 MW, and it is expected that a system having an energy-conversion efficiency of 33 percent can be built. The energy-conversion efficiency η is the ratio of the power output $\dot{W}$ to the reactor power input $\dot{E}$. A simple energy balance will permit us to determine the required reactor power and the rate at which energy is discarded to the atmosphere via the condenser.

From the definition of the efficiency,

$$\eta = \frac{\dot{W}}{\dot{E}}$$

so

$$\dot{E} = \frac{\dot{W}}{\eta} = \frac{100 \text{ MW}}{0.33} = 300 \text{ MW}$$

Now, consider an energy balance on the control mass shown. It includes all the complicated hardware, but does not include the fissionable material. The rate of energy transfer to the control mass from the fission system is $\dot{E}$. The rate of energy

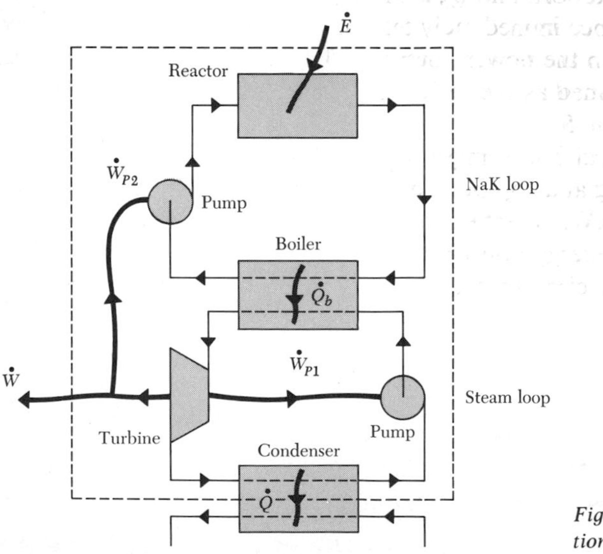

Figure 2·18. A nuclear power generation system

transfer from the control mass as heat is $\dot{Q}$, and the rate of energy transfer from the control mass as work is $\dot{W}$. We assume that the system is operating steadily, and consequently as time passes there is no change in the energy stored inside the control mass. The energy balance, made over any time period t, therefore gives

$$\underset{\text{energy input}}{\dot{E}t} = \underset{\text{energy output}}{\dot{W}t + \dot{Q}t} + \underset{\text{energy storage}}{0}$$

Hence
$$\dot{Q} = \dot{E} - \dot{W} = 300 - 100 = 200 \text{ MW}$$

This large energy-rejection rate may seem outlandish; we discard more energy than we use. Yet the efficiency of 33 percent is typical of modern power stations. We shall see why these efficiencies are so low later in our study.

A heat pump. An attractive device for household heating is the heat pump, and a simple energy analysis quickly shows us why. We shall discuss the inner workings of heat pumps later, and for now all we need to know is that they take in electric power and energy as heat from the cool outdoors, deliver energy as heat to the inside of the house, and operate continuously. Figure 2·19 shows the basic energy flows. An energy balance on the control mass shown, made over a time interval t, gives

$$\underset{\text{energy input}}{(\dot{E}_e + \dot{Q}_c)t} = \underset{\text{energy output}}{\dot{Q}_h t} + \underset{\text{energy storage}}{0}$$

where $\dot{E}_e$ is the electric power input, $\dot{Q}_c$ is the rate of energy transfer as heat into the heat pump from the outdoors, and $\dot{Q}_h$ is the rate of energy transfer as heat to the house. The energy balance immediately tells us that we get more energy into the house than we buy from the power company. The *coefficient of performance* (cop) of a heat pump is defined as the ratio $\dot{Q}_h/\dot{E}_e$, and under the proper conditions can be as much as 4 or 5.

For example, a typical house might require 100,000 Btu/h of heat input. With a heat pump operating at a cop of 4, the electric power input would only be 25,000 Btu/h, or about 7.3 kW. Direct electric heating would require 29.3 kW, so there clearly would be an energy and operating cost saving over direct electric heating. However, it is not clear that there would be a saving over direct gas

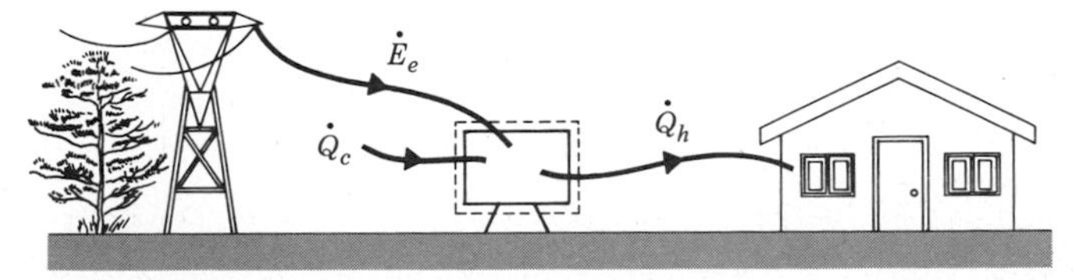

Figure 2·19. Heat pump system

heating; to study this question one would have to make a careful energy analysis of the electric energy production and transmission systems and a similar analysis of the gas system, and then compare the two from the point of view of energy and cost. This is typical of the simpler but extremely important problems tackled by engineering thermodynamicists.

2·12 SUMMARY

To summarize, there are four ideas inherent in the concept of energy:

1. Every system has energy (E).
2. Energy is extensive ($E_{a+b} = E_a + E_b$).
3. Energy is conserved.
4. Work provides the fundamental measure of energy.

A number of concepts and definitions have been introduced in this chapter; the key ideas are summarized below.

System Whatever we define as the thing being studied.
Control mass A system defined to be a specific piece of matter.
Environment Everything except the system.
Isolated system A system which does not interact with its environment.
Internal energy Energy of matter associated with the randomly oriented motions of the molecules and the forces between them (the energy of the "hidden" microscopic modes).
Work Energy transfer by the action of a macroscopically measurable force on matter within the system (organized microscopic work).
Heat Energy transfer which is not recognized macroscopically as work (disorganized microscopic work).
Temperature A characteristic of matter which serves as a driving potential for energy transfer as heat. Energy is transferred as heat from the body at the greater temperature to the one at the lower temperature.
Hot, cold Adjectives describing bodies of high and low temperature.
Adiabatic wall One which prevents energy transfer as heat.

SELECTED READING

Lee, J. F., and F. W. Sears, *Thermodynamics*, secs. 3-1 through 3-10, Addison-Wesley Publishing Co., Inc., Reading, Mass., 1963.

Reif, F., *Fundamentals of Statistical and Thermal Physics*, secs. 2.6–2.11, McGraw-Hill Book Company, New York, 1965.

Resnick, R., and D. Halliday, *Physics, Part 1*, chap. 21, secs. 22-1, 22-6, 22-7, John Wiley & Sons, Inc., New York, 1966.

Van Wylen, G. J., and R. E. Sonntag, *Fundamentals of Classical Thermodynamics*, 2d ed., chap. 4, John Wiley & Sons, Inc., New York, 1973.

Zemansky, M. W., M. M. Abbott, and H. C. Van Ness, *Basic Engineering Thermodynamics*, 2d. ed., secs. 3.1–4.7, McGraw-Hill Book Company, New York, 1975.

QUESTIONS

2·1 What concepts were introduced in this chapter? What basic postulates? What definitions?

2·2 A control volume is any defined region in space; under what conditions is a control volume a control mass? An isolated system?

2·3 How might you explain the concepts of energy, heat, work, and internal energy to a 6-year-old?

2·4 The oven is at 300 degrees of heat. What is wrong with this statement?

2·5 The heat within a gas is evidenced by the random motion of its molecules. What is wrong with this statement?

2·6 Can energy be transferred to one molecule as work? As heat?

2·7 When two molecules of a gas collide, is there any friction?

2·8 Stir a pail of water. What happens to the energy you transfer to the water as work?

2·9 If you say that a rock has potential energy in the earth's gravitational field, what must you say about the work done by the weight force when the rock is dropped?

2·10 Explain the difference between heat and internal energy.

2·11 Can you conceive of matter in a zero energy state?

2·12 Under what circumstances is work equal to $F(X_2 - X_1)$?

2·13 A house is receiving energy as heat from resistance heating elements, but the temperature of the house is not changing. What do you think is happening to the energy transferred as heat?

2·14 It is possible to increase the temperature of a substance, such as a gas, without any energy being transferred as heat. Think of an example, define the system, and identify the energy transfers across the boundaries.

PROBLEMS

2·1 Consider the rectangular free-surface water piston-cylinder system shown. The width of the chamber is b (normal to the page), and the atmospheric pressure is P_0. (*a*) From basic considerations, show that the force exerted by the water on the piston is $F = (P_0 + \gamma h/2)hb$, where γ is the weight density (weight per unit of volume) of water.

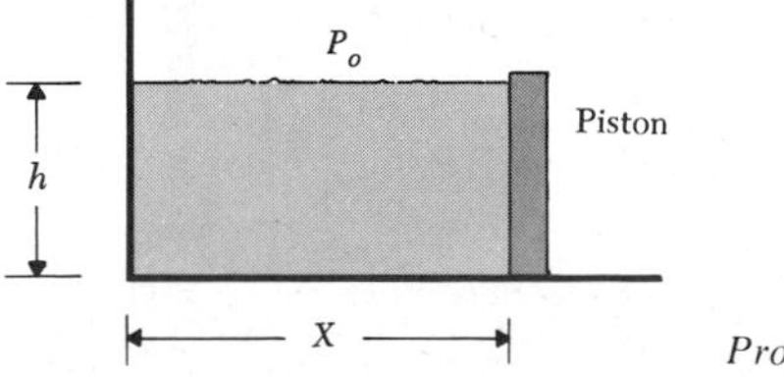

Problem 2·1.

(*b*) Calculate the work done by the water on the piston and the work done by the atmosphere on the water when the chamber length is increased slowly from X_1 to X_2. Express in terms of P_0, b, h_1, h_2, X_1, X_2, and γ.

2·2 A hollow steel sphere of mass M and volume V is hung from a cable in a fluid having weight density (weight per unit of volume) γ. Compute the work an external agent must do on the cable in order to lift the sphere a distance h through the fluid. The motion is very slow, and frictional forces may be neglected. Is any work done on the fluid in this process? What happens to the energy transferred as work?

2·3 Consider the pulley-belt system shown. The tension force in the top belt is T, and the tension in the lower part may be neglected. The pulleys rotate at angular velocity ω. Which is the driving pulley? The power being transmitted is $T\,D\omega/2$. Define a system, show the forces acting on it and the direction through which they move, and derive this result from the definition of work.

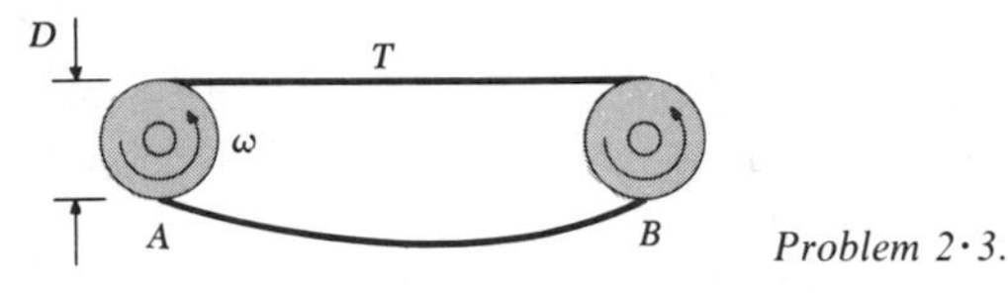

Problem 2·3.

2·4 Repeat the development of Eq. (2·13) for a spherical system where the boundaries move radially.

2·5 Assuming that the atmosphere is locally isothermal, the density variation with pressure is given by

$$\rho = \rho_0 \frac{P}{P_0}$$

where ρ is the density, P the pressure, and the subscript zero denotes the earth's surface. Show that the pressure variation through the isothermal atmosphere is

$$\frac{P}{P_0} = \exp\left(-\frac{\rho_0 g z}{g_c P_0}\right)$$

where z is the height above the surface, and g is the (constant) acceleration of gravity.

2·6 Using the atmospheric pressure distribution of Prob. 2·5, derive an expression for the work done by the skin of a balloon on the atmosphere in rising slowly to a height h above the surface. Assume that the initial balloon volume is V_0 and that the volume varies inversely with pressure during the ascent. Neglect local variations in pressure around the balloon.

2·7 Using the atmospheric pressure distribution of Prob. 2·5, derive an expression for the work done by the skin of a balloon in rising slowly to a height h above the surface. Assume that the initial balloon volume is V_0, and that the volume varies linearly with height, attaining a value V_1 at h.

2·8 Work Prob. 2·7 assuming that the volume varies exponentially with height.

2·9 Using the atmospheric pressure distribution of Prob. 2·5, derive an expression for the work done by the skin of a balloon on the atmosphere in rising slowly from the earth's surface to a height h. Assume that the balloon is spherical and does not change size. (Some approximations relating to the small size of the balloon may be useful in carrying out necessary integrations.)

2·10 A person on a moving truck lifts a 100-kg mass up 2 m vertically. If the truck is moving horizontally, how much work did the person do on the weight relative to a person on the opposite side of the earth? How much relative to a pixie on the weight?

How much relative to the truck driver? In which reference frame did the person get the most tired?

2·11 A person in an elevator moving upward at a velocity of 10 ft/s lifts a 50-lbf weight 3 ft off the elevator floor. This all takes 30 s. How much work did the person do on the weight relative to the elevator operator? How much relative to an observer on the ground floor? How much relative to an observer in an elevator going down at the same speed? Explain why the work is different and what happens to the energy transferred to the weight as work in the three cases. In which reference frame does the person get the most tired?

2·12 A 3-lbm quantity of a substance is made to undergo a process within a piston-cylinder system, starting from an initial volume of 2 ft^3 and an initial pressure of 100 lbf/in^2. The final volume is 4 ft^3. Compute the work done by the substance for the following processes: (*a*) pressure remains constant, (*b*) pressure times volume remains constant, (*c*) pressure is directly proportional to volume, and (*d*) pressure is proportional to the square of the volume.

2·13 Liquid water at 32°F weighs 62.42 lbf/ft^3, and ice at the same temperature weighs 57.2 lbf/ft^3. Consider a system which is initially an ice cube 1 in on a side. How much work does this system do on the atmosphere when it melts?

2·14 Six kg of a substance is compressed in a piston-cylinder system from an initial volume of 0.4 m^3 to a final volume of 0.2 m^3. The initial pressure is 0.7 MN/m^2. Compute the amount of energy transfer as work to the substance if (*a*) the pressure is constant during compression, and (*b*) the pressure varies inversely with volume. If the process of case (*b*) is adiabatic, what is the change in internal energy of the substance?

2·15 Consider the cylinder of liquid suspended between two cones as shown in the sketch. Show that the amount of work done on the cones when the cone separation is increased an infinitesimal amount can be written as

$$đW = \sigma(dA - \cos\theta \, dA_w)$$

where σ is the interfacial tension between the liquid and the gas, A is the liquid-gas interfacial area, A_w is the "wetted area," and θ is the contact angle. *Hint:* What is the liquid pressure?

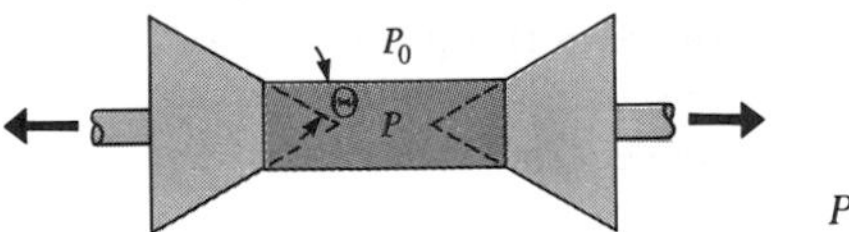

Problem 2·15.

2·16 Most dielectrics are such that the polarization is related to the electric field by $\mathbf{P} = (\kappa - 1)\varepsilon_0 \mathbf{E}$, where κ is the dielectric constant, and ε_0 is a physical constant (Table A·2). The dielectric constant is typically a function of temperature. (*a*) Derive an expression for the work done on a dielectric when it is polarized at constant temperature. (*b*) The dielectric constant of acetone as a function of temperature is given in the table. Compute the work done (J) on 1 cm^3 of acetone when the electric field **E** is increased from zero to 10^4 V/m while the temperature varies from 0 to 50°C, if the temperature varies linearly with **E** during this process. *Hint:* Do not be afraid of graphical integration.

T, °C	κ
−80	31.0
20	21.4
50	18.7

2·17 A water cannon for rock tunneling operates by using compressed gas for accelerating a 65-kg high-strength titanium alloy piston to 225 km/h over a distance of 0.3 m. The energy of the piston is transferred to 0.95 kg of water, which is fired out of a small nozzle. Calculate the energy of the piston as it contacts the water. Assuming all this energy is transferred to the water, calculate the velocity the water achieves.

2·18 Five kg of a material is heated at constant volume from a state where its internal energy is 40 MJ/kg to a state where its internal energy is 60 MJ/kg. Compute the amount of energy transfer as heat for this process (J).

2·19 A substance expands from $V_1 = 1$ ft^3 to $V_2 = 6$ ft^3 in a constant-pressure process at 100 lbf/in.^{2} The initial and final internal energies are $U_1 = 40$ Btu and $U_2 = 20$ Btu, respectively. Find the direction and magnitude of the energy transfer as heat for this process (Btu).

2·20 A sample of gas (not ideal) is made to undergo an expansion process during which its pressure and volume are related as shown in the table below. The energy of the gas at the start and finish of the process is measured and found to be 5 and 3.2 Btu, respectively. Determine the amounts of energy transferred as heat and work during this process. Express in Btu and MJ.

V, ft^3	P, lbf/in^2
1	40
2	27
3	21
4	18

2·21 A 200-kg chunk of lead falls from a height of 30 m and smashes into a rigid concrete floor. Calculate the increase in the internal energy ΔU (MJ), assuming that no energy is transferred as heat from the lead.

2·22 An ideal gas ($PV = MRT$) is heated at constant volume until its temperature is doubled and then cooled at constant pressure until it is returned to its initial temperature. Derive an expression for the work done on the gas. It is a peculiarity of an ideal gas that its internal energy is a function only of temperature. What is the net energy transfer as heat for the process described?

2·23 A paramagnetic salt obeying Curie's law ($\mathbf{M} = C\mathbf{H}/T$) is made to undergo a process at constant **M** in which the applied field **H** is doubled (and the temperature doubled), followed by a magnetization process during which the applied field is held constant. This process returns the salt to its initial temperature. Derive an expression for the energy transfer as work for this process in terms of the initial values of **H** and **M**. It may be shown that the internal energy U of such a salt is a function only of temperature. What is the net energy transfer as heat for the process described?

2·24 A paramagnetic salt obeying Curie's law ($\mathbf{M} = C\mathbf{H}/T$) is made to undergo a process at constant temperature T, during which its magnetization **M** is doubled. This is followed by a heating process in which its temperature is doubled, while the applied magnetic field **H** is held fixed. It may be shown that the internal energy of such a *Curie substance* depends only on temperature. Derive expressions for the total amounts of energy transfer as heat and work to the substance for the two-step process; express the results in terms of the initial values of T, **H**, and **M**, and the initial and final values of the internal energy U.

2·25 Assuming that the internal energy of iron is $0.2T$ kJ/kg, where T is the iron temperature in kelvins, estimate the maximum increase in temperature of the iron asteroid of Prob. 1·8 when it crashes to earth.

CHAPTER THREE

PROPERTIES AND STATE

3·1 CONCEPTS OF PROPERTY AND STATE

Any engineering system is made up of various amounts and kinds of matter. Description of such a system and prediction of its performance require knowledge of the properties of the various materials. Thus understanding the concepts of state and property is essential to any engineer.

A *property* is any characteristic or attribute which can be quantitatively evaluated. Volume, mass, energy, temperature, pressure, magnetization, polarization, and color are all properties of matter.

Properties are things that matter "has." *Work and heat are not properties, for they are things that are "done" on a system in order to produce changes in the properties.* Energy transfer as work or heat will be evidenced by changes in properties, but the amounts of energy transfer depend on the manner in which a given change takes place. For example, consider the problem of compressing a gas from one condition where the pressure is P_1 and the volume is V_1 to some other state P_2, V_2. This can be accomplished in an infinite variety of ways, two of which are shown in Fig. 3·1. The amount of work done on the gas in these two cases is obviously different, yet the changes in properties are identical. The integral $\int đW = \int P\,dV$ between two states depends on the path of integration $P(V)$; in contrast, the integral of any property, such as $\int dV$ or $\int dP$, depends only on the initial and final properties.

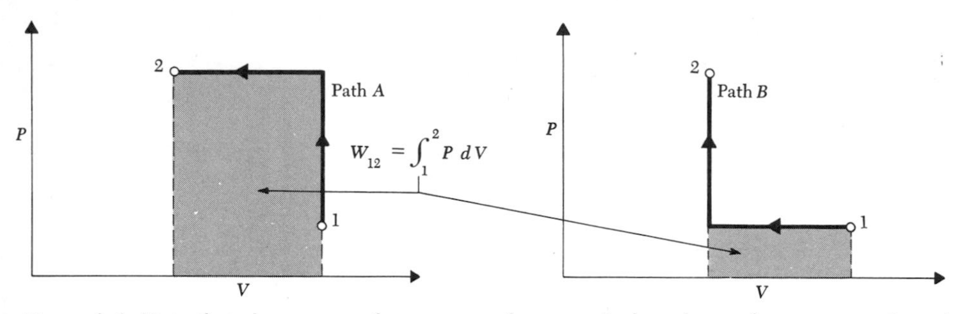

Figure 3·1. Note that the amount of energy transfer as work depends on the process path, and hence the work done is not a property of the end states

The *state* of something is its condition, as described by a list of the values of its properties. For example, the position coordinates and velocity components can completely describe the state of a single particle. On the other hand, in some instances it is sufficient merely to know the mass and energy of the particle, and its state is therefore satisfactorily described by fewer properties. The complete description of a system of many particles, such as 1 cm^3 of a gas, would require specification of the position coordinates and velocity components for each particle within the boundaries. The list of these properties would be very long—indeed, so long that it probably could never be written down, certainly not before some of the properties had changed. More economical descriptions of state are certainly desirable; we must find some way to reduce the number of relevant properties from something of the order of 10^{23} to a few. We might settle for knowing merely the volume of the system and the energy of each particle; still, this list would also be very long. A less complete way to describe the system would be to give merely its energy and volume, but these may fluctuate rapidly in time as the particles within the system interact with the particles outside the boundary (see Fig. 3·2). We might hope that the *average* properties would be sufficiently informative to allow adequate solution to our engineering problems; indeed, we normally work with properties representing the average of some feature of all the particles comprising the system. These averages are called *macroscopic properties*.

We can further reduce the list of relevant properties by noting that not all macroscopic properties are relevant in a particular analysis. For example, the color of a jet airplane is certainly not relevant in an analysis of its lift-drag characteristics. On the other hand, the color may be relevant in an analysis of its sales potential. It is helpful to group properties into classes that are relevant in

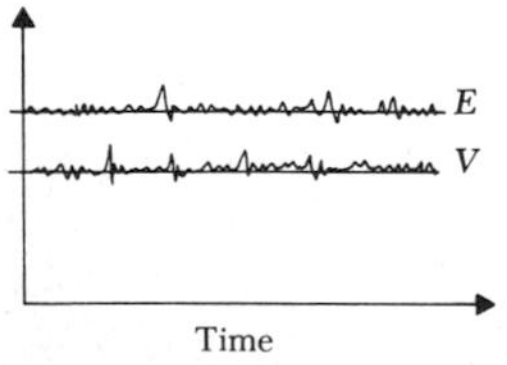

E
V
Time

Figure 3·2. The average properties are used in thermodynamics

Table 3·1 Some different types of state

State	Properties
Geometric	length, width, breadth, moment of inertia, volume, etc.
Kinematic	position, speed, acceleration, etc.
Hydrodynamic	pressure, shearing stress, strain rate, etc.
Electromagnetic	electric field strength, magnetic-dipole moment, charge, etc.
Chemical	chemical composition, free charge, energy, entropy, etc.
Esthetic	smell, color, eye-catchingness, etc.
Thermodynamic	energy, temperature, volume, pressure, stress, magnetic-dipole moment, entropy, etc.
Quantum-mechanical	momentum and energy of each particle, total volume, etc.

different kinds of analysis. We do this by thinking about different sorts of states. For example, the lift-drag characteristics of the airplane would be a property of its geometry, that is, of its *geometric state*. Its speed and altitude would be properties of its *kinetic state*, and the temperature and humidity in the cabin would be properties of its *thermodynamic state*. The color would be a property of its *esthetic state*. Some different types of state are listed in Table 3·1.

Thermodynamics centers around energy; thermodynamic properties are related to energy. For example, the amount of energy transfer as work *to* a fluid when it is compressed is $đW = -P\,dV$, where P is the pressure and V is the volume. The *pressure* is therefore a thermodynamic property. This equation holds regardless of how the volume changes, that is, irrespective of the *shape* of the fluid. The shape is therefore irrelevant in the thermodynamic analysis of fluids; the length, width, and depth of a fluid piece are properties of its *geometric* state, but are *irrelevant* to its thermodynamic state. But the *volume* is relevant to the energy transfer as work, hence volume is a thermodynamic property.

In order to complete a thermodynamic analysis it may be necessary to consider other kinds of state. For example, consideration of the geometric state might allow one to calculate the volume needed in the thermodynamic analysis. Consideration of the kinematic state of an accelerating gas stream may be necessary in order to calculate the gas temperature. A thermodynamic analysis often involves a companion geometric, dynamic, electrodynamic, or chemical analysis.

3·2 EQUILIBRIUM AND THERMODYNAMIC PROPERTIES

The state of a system is determined by the molecules within the system boundaries. These molecules undergo continual changes in their individual states as

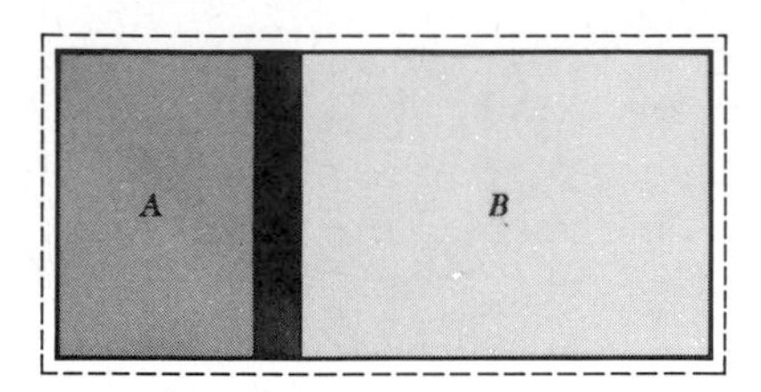

Figure 3·3. The pressures of A and B must be equal for mechanical equilibrium

they interact with one another. If we isolate a system and allow its molecules to freely interact with one another, the state of the system will undergo visible macroscopic change. But after some time the changes that can be detected with macroscopic instruments will cease; the microscopic activity continues, but somehow the macroscopic state has reached equilibrium. The macroscopic measurables have definite constant values; they are properties of the system in this equilibrium configuration.

Consider the system of Fig. 3·3. The cylinder contains a floating piston, with different amounts of two different gases A and B on opposite sides. Suppose the piston is locked in a certain position, and the pressures of the two gases are unequal. If we unlock the piston, the unbalance of pressure will accelerate the piston in one direction. We would say that the two gas systems A and B were not in *mechanical equilibrium* with one another. If, on the other hand, the two pressures were identical, then the piston would not move upon release; A and B would be in mechanical equilibrium with one another. We see that *pressure* is a property that two systems have in common when they are in *mechanical equilibrium*. This, in fact, is the *thermodynamic concept of pressure*. In Chap. 7 we will use this idea to construct a fundamental *thermodynamic definition of pressure*.

Suppose now we lock the piston and allow the two gases to exchange energy across the piston. Any energy transfer that takes place must be as heat, for locking the piston prevents energy transfer as work. When energy transfer as heat is possible, but none occurs, we say that the systems A and B are in *thermal equilibrium*, and that they have the same *temperature*. Temperature is the property that two systems have in common when they are in thermal equilibrium; this is the *thermodynamic concept of temperature*. In Chap. 7 we will use this idea to construct a fundamental *thermodynamic definition of temperature*.

There are other kinds of equilibrium, three of which deserve brief comment. Two systems are said to be in *electrostatic equilibrium* if there is no tendency for a net charge flow between them when they are brought into contact. The electrostatic potential is a property that two systems have in common when they are in electrostatic equilibrium. Two phases of a substance (for example, solid and liquid) are said to be in *phase equilibrium* if there is no tendency for phase transformation (for example, melting) when they are brought into contact. A mixture of gases is said to be in *chemical equilibrium* if there is no tendency for a net reaction to take place when they are allowed to interact. For each kind of equilibrium there is one thermodynamic property that two systems must have in common, and in thermodynamic theory this idea is used to define the property.

It is important to appreciate that these definitions only apply to systems in equilibrium.

The term *thermodynamic equilibrium* is used to indicate a condition of equilibrium with respect to *all* possible macroscopic changes in a system where the molecules are free to interact with one another in any way. There is no macroscopic energy, matter, or charge flow within a system in thermodynamic equilibrium, though the molecules are free to produce such flows. In order to test a piece of matter to see if it is in thermodynamic equilibrium, we imagine isolating the matter and watching for macroscopically observable changes. If none occur, the matter was in thermodynamic equilibrium at the moment of isolation. We emphasize that the molecules must be free to move about in order to rearrange the energy, mass, or charge when we make this test.

For a system in thermodynamic equilibrium, those properties that are in some way relevant to energy are called *thermodynamic properties*. Any property defined in terms of other thermodynamic properties must be a thermodynamic property. It is important to emphasize that the term *thermodynamic properties* refers only to a system in thermodynamic equilibrium.

The *thermodynamic state* is the condition of the matter as described by all the thermodynamic properties. The thermodynamic properties are all fixed when the thermodynamic state is fixed. But they are not all independently variable, hence the thermodynamic state can be fixed by giving values for just a few thermodynamic properties. We shall develop a rule for how many properties must be known to fix the thermodynamic state. In later chapters we shall use thermodynamic theory to discover equations relating various thermodynamic properties. Again, we must emphasize that these relationships apply only for states of thermodynamic equilibrium.

When we treat a complex system not in equilibrium, the usual procedure is to dissect the system mentally into small pieces which can treat *individually* as being in thermodynamic equilibrium. For example, in studying the gas flow in a rocket nozzle we might treat each small piece of fluid as being in thermodynamic equilibrium and then consider the changes in its state as it interacts with the other fluid pieces. The electrons and ions in a plasma are often treated as separate systems, each in thermodynamic equilibrium but not in equilibrium with each other. The analysis of the interaction between these two systems allows one to predict the approach to equilibrium for the entire plasma. By dissecting the systems in this manner their states can be evaluated by giving the thermodynamic properties of each part.

3·3 PRESSURE

The concept. Previously we have used the concept of pressure as the force per unit area exerted by a fluid on the face of a piston, the surface of a submarine, or the base of a barometer column. This is a *mechanical concept of pressure*. We can think of pressure as arising from the billions of collisions between the molecules of

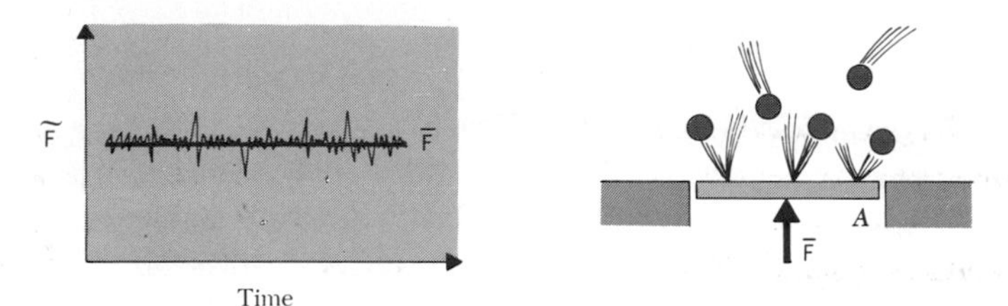

Figure 3·4. Pressure; $P = \bar{F}/A$

the fluid and the solid wall that take place every second. Figure 3·4 depicts this idea.

Although it is convenient to measure pressure at a wall, we often think of the pressure *within* the fluid. In so doing we imagine cutting out a little fluid box, replacing it by a solid box, and measuring the pressure on the walls. In liquids and gases it is usually an excellent assumption that the force per unit area is independent of the orientation of the little box (Fig. 3·5), i.e., that the pressure is independent of direction. But if the fluid is moving rapidly in some nonuniform way, viscous forces may also be important, and in this case a more careful definition of pressure must be made. For the most part, the simpler definition of pressure will be quite adequate for all systems analyzed in this book.

In a fluid at rest, the pressure throughout a connected fluid region will be the same at all points of the same height; you can convince yourself of this by considering the forces acting on tubes of fluid cut out of the fluid region, noting that horizontal pressure differences would cause the fluid to accelerate horizontally.

Measurement. Pressures can be measured by a variety of devices. All these really measure the difference in two pressures; only if one of these is a perfect vacuum will the device measure the true pressure. For example, the manometer of Fig. 3·6 measures the pressure difference $P_B - P_A$.

Analysis of the pressure in a fluid at rest is called *hydrostatics* and will be familiar to you if you have had a course in fluid mechanics. Hydrostatics is used to determine the relationships between pressure differences and heights in manometers. For example, a force balance on a short column of fluid (Fig. 3·7) tells us that the pressure at the bottom exceeds the pressure at the top by the product of the *weight density* of the fluid γ (N/m^3 or lbf/ft^3) and the height h (m or ft). If this idea

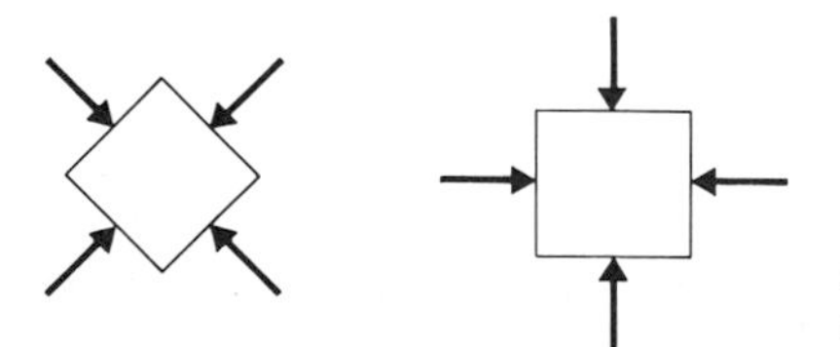

Figure 3·5. In a fluid, the pressure on each cube face is independent of the cube orientation

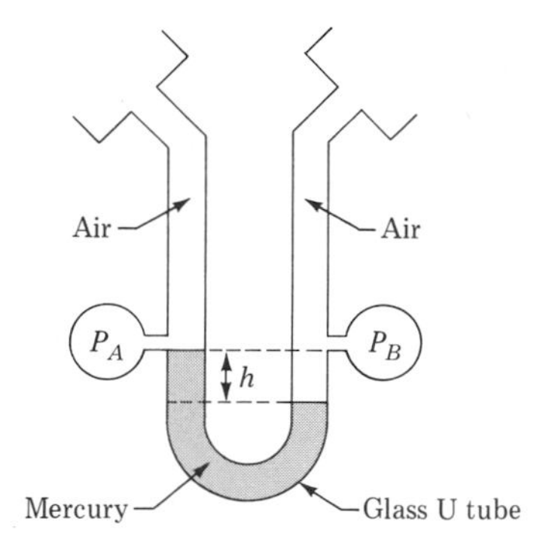

Figure 3·6. Manometers are used to measure pressure differences

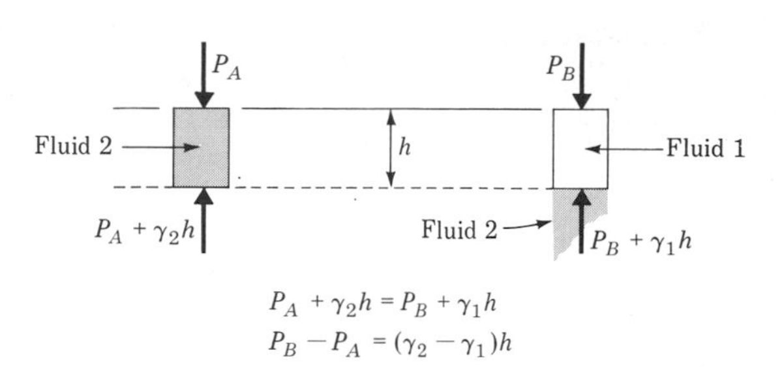

Figure 3·7. Deriving the equation for a U tube manometer (see Fig. 3·6)

is applied to the manometer of Fig. 3·6, along with the idea that the pressure in fluid 2 at the bottoms of the two boxes is equal (see Fig. 3·7), one finds

$$P_B - P_A = h(\gamma_2 - \gamma_1)$$

where the subscripts 1 and 2 denote the two fluids. If fluid 1 is a gas, $\gamma_2 \gg \gamma_1$, and so γ_1 is often neglected in practice. With manometers as a basic way to measure fluid pressures, other types of gauges can be calibrated.

Most pressure gauges read the excess of the "test" pressure over the atmospheric pressure; this is called the *gauge pressure*, often denoted by Pag (Pascals gauge) or psig (lbf/in^2 gauge). True or absolute pressures are sometimes indicated by psia (lbf/in^2 absolute). In the literature Pa, N/m^2, and psi *usually* (but not always) refer to absolute pressures. In this book the symbol P will always stand for absolute pressure.

We mentioned earlier the thermodynamic notion that pressure is something two systems have in common when they are in equilibrium. In Chap. 7 we will introduce a thermodynamic definition of pressure and show that it is equivalent to the mechanical definition of pressure discussed here.

Another common unit of pressure is the *bar*, defined as 10^5 N/m^2. This is approximately the average atmospheric pressure. One atm is usually taken as

$$1 \text{ atm} = 1.013 \times 10^5 \text{ N/m}^2 = 14.696 \text{ lbf/in}^2$$

Example. A gauge shows a pressure of 100 psig. What is the absolute pressure?

$$P = 100 + 14.7 = 114.7 \text{ psia} = 7.908 \times 10^5 \text{ N/m}^2 = 0.7908 \text{ MPa}$$

Example. A mercury-water manometer shows a height difference of 0.8 m. What is the pressure difference? For water $\gamma = 9.87 \times 10^3$ N/m^3, while for mercury $\gamma = 133.7 \times 10^3$ N/m^3 so

$$P_B - P_A = 0.8 \text{ m} \times (133.7 - 9.87) \times 10^3 \text{ N/m}^3 = 9.91 \times 10^4 \text{ Pa}$$

3·4 TEMPERATURE

The concept. One of the important properties is temperature. What is temperature? A primitive view is that temperature is the reading on a mercury-in-glass thermometer. But who put the markings there, and what do they really mean? What is so special about the mercury-glass combination? These are deeply probing questions, questions which seek some more fundamental idea of the *concept* of temperature.

The housekeeper thinks of temperature as how hot the oven must be set to cook the roast. The computer engineer thinks of temperature as how cool the transistors must be kept to make them work reliably. The plasma physicist views temperature as a measure of the kinetic energy of molecules or electrons. And the astronomer views temperature as a measure of the radiant energy emission from stars. These rather diverging concepts of temperature have one thing in common: They all relate to energy or energy transfer, and hence they clearly mark temperature as a *thermodynamic* property.

Another key idea about temperature is that it is a "pointer" for energy transfer as heat. Energy will tend to be transferred as heat from regions of high temperature to regions of low temperature. Molecular motions tend to be more violent at high temperatures, and energy tends to pass from the molecules that form a region of higher temperature to the more sluggish molecules that comprise a region of lower temperature.

Now, if two systems are in thermal equilibrium, then they must have the same temperature. If each is in equilibrium with a third, then all three have the same temperature, hence any two or all three are in thermal equilibrium. This notion is sometimes called the *zeroth law of thermodynamics*. It is actually an implicit part of the concept of temperature. We shall use it in Chap. 7 to develop a very fundamental thermodynamic definition of temperature.

It is important to make a clear distinction in one's mind between the concepts of heat, temperature, and internal energy, hence we now review these again. Internal energy is the energy possessed by molecules that is "hidden" from direct macroscopic view by the disorganized character of the microscopic state. That energy may or may not have been put into the matter through transfer of energy as heat. Heat is energy transfer that is not accounted for in macroscopic evaluation of energy transfer as work; heat is microscopic work that is "hidden" from direct macroscopic view by the disorganized nature of the energy transfer process. Temperature is a property of matter; if the temperature of one body is greater than the temperature of a second, then any energy transfer as heat will take place from the first body to the second. The internal energy of a substance will depend in part, but not exclusively, on the temperature. Hence the temperature is *not* in general a complete measure of the internal energy. As we shall see, temperature can only be defined when a body is in equilibrium; the body has energy regardless of whether it is in equilibrium or not. A body does not "have" any heat; it may have received some energy as heat, but now that energy appears as internal energy, and one cannot tell from the body's state whether that energy entered the body as heat or as work. Internal energy and temperatures are *properties* of matter; heat is not a property. Temperature and internal energy are also inherently different kinds of properties. The temperature of a small piece of a large body is the same as the temperature of the body as a whole; the energy of the small piece is but a small fraction of the energy of the entire body.

Measurement. To make the temperature concept operational, we need some sort of temperature *scale*. The uniformly spaced markings on a mercury-in-glass thermometer provide one such scale over a limited range. If instead a uniformly marked alcohol-in-glass thermometer is used, the two temperature scales can be made to agree at two points, but not in between. It seems odd that if temperature is such a fundamental concept its scales should be subject to the whims of a thermometer maker. It would be much more desirable if some scale of temperature could be established which is independent of the choice of thermometer. This can indeed be done; means can be devised for measuring the *ratio* of two temperatures without reference to any thermometric material. A conceptually unique scale for temperature can thereby be established in a manner similar to the method outlined in Chap. 1 for establishing unique scales for force, mass, and charge. Being able to measure ratios, one is free to select the value for *one* point on the temperature scale, and the measured ratios then give all other temperatures.

In order to develop the thermodynamic definition and scales for temperature we need to understand both the first and second laws of thermodynamics, and we are not yet ready to make these developments. Consequently we shall for the moment work with an empirical temperature scale, that based on the gas thermometer. This scale appears to be identical with the thermodynamic temperature scale, and so we can consider the perfect-gas thermometer a device for measuring thermodynamic temperature. In Chap. 7 we shall develop the thermodynamic definition of temperature and show the equivalence with the empirical perfect-gas temperature scale.

The empirical gas temperature is based on the observation that the temperature of a gas confined at constant volume is a monotonically increasing function of the gas pressure. The scale is established by arbitrarily selecting a value for the temperature of a mixture of water, water vapor, and ice in thermal equilibrium. Such a mixture can exist at only one temperature (called the *triple point*), and the temperature of this mixture provides an excellent easily reproducible standard.

The thermometer consists of a glass bulb into which various amounts of gas can be placed. The pressure may be measured in several ways, but the volume change so introduced must be zero. Suppose we wish to determine a particular temperature, say, the boiling point of water at 1 atm pressure. A given amount of gas is placed in the bulb, which is then immersed in the triple-point mixture. The thermometer is allowed to reach thermal equilibrium with the mixture, and the gas pressure is then measured. We denote this pressure by P_s. The thermometer is then inserted into the boiling water, and the procedure is repeated; the new pressure we call P_m. The ratio P_m/P_s might be used to define the ratio of the two temperatures T_m/T_s; however, if this is done, one obtains different temperature

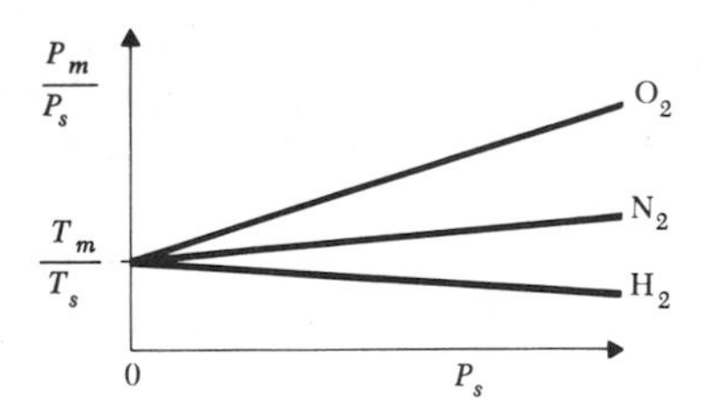

Figure 3·8. Readings from a constant-volume gas thermometer

scales depending on the amount of gas in the thermometer. The procedure followed in precision thermometry is to repeat the above measurements with less and less gas in the bulb and to plot the ratio P_m/P_s versus P_s, extrapolating to zero pressure, as shown in Fig. 3·8. The surprising result is that many gases seem to yield the same limiting ratio, which is therefore, in a limited way, independent of the nature of the substance chosen. The temperature scale is then set by defining

$$\frac{T_m}{T_s} = \lim_{P_s \to 0} \frac{P_m}{P_s}$$

By an international agreement, the temperature at the H_2O triple point is set at 273.16 K (273.16 kelvins). The number may seem quite arbitrary, but with this particular choice the newer *one-point scale* corresponds closely with earlier temperature scales defined on the basis of assumed linear behavior between two arbitrarily selected points. It is important to emphasize that the scales of force, mass, length, time, charge, and temperature are all one-point scales which are set by selecting values and units for only *one* easily reproducible situation. The kelvin is one of the six basic SI units. The ° symbol has officially disappeared from the SI, so current literature uses simply K. In general we will follow this practice when using the SI.

A second absolute temperature scale, used with the old English system, is the Rankine scale, defined by

$$T(\text{rankines}) = \tfrac{9}{5}T(\text{kelvins})$$

This makes the H_2O triple-point temperature correspond to 491.69 R.

Associated with the Kelvin and Rankine scales are the Celsius (formerly centigrade) and Fahrenheit *relative temperature scales*. Temperature *differences* are identical in degrees Celsius (°C) and kelvins, but the Celsius scale has its zero at 273.15 K; this makes 0 K correspond to −273.15°C. Similarly, temperature *differences* are identical in degrees Fahrenheit (°F) and rankines, but a *level* of 0°F corresponds to 459.67 R. These four temperature scales are compared in Fig. 3·9.

It should be pointed out that the empirical gas scale can be used only at temperatures above the boiling point of the gas and below the melting point of the container. Outside this range other empirical temperatures must be used. The

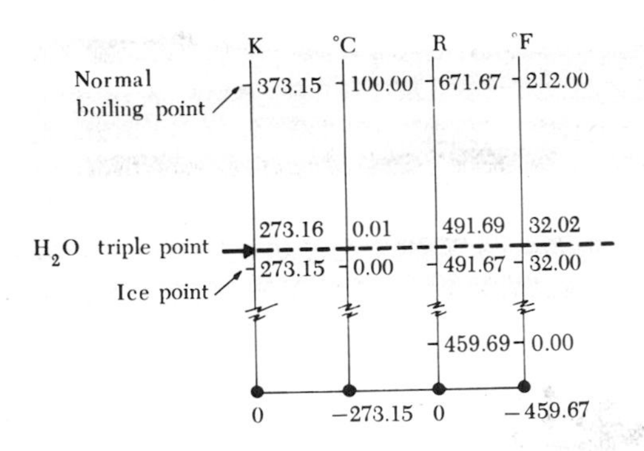

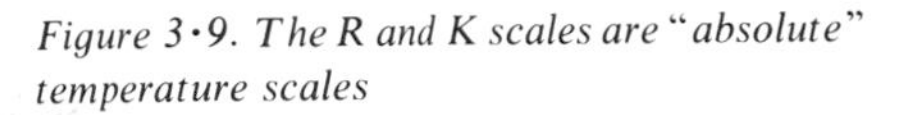
Figure 3·9. The R and K scales are "absolute" temperature scales

thermodynamic definition of temperature given in Chap. 7 is a continuous definition of temperature valid over all ranges and therefore provides an essential link between the several empirical measures of temperature.

Example. A thermometer reads 240°C. Calculate the temperature *level* in °F, K, and R.

To find °F: $$240°\text{C} \times \frac{9°\text{F}}{5°\text{C}} + 32.00°\text{F} = 464°\text{F}$$

To find R: $$464°\text{F} + 459.69\ \text{R} = 923.69\ \text{R}$$

To find K: $$240°\text{C} + 273.15\ \text{K} = 513.15\ \text{K}$$

Example. One object is 30°C hotter than another. What is the temperature *difference* in °F, R, and K?

$$T_1 - T_2 = 30°\text{C} = 30\ \text{K} = 30°\text{C} \times \frac{9°\text{F}}{5°\text{C}} = 54°\text{F} = 54\ \text{R}$$

3·5 INTENSIVE AND EXTENSIVE STATE

It is convenient to distinguish between two types of properties. Suppose we have two pieces of the same substance in equilibrium with one another. If we bring them together and consider them as one system, the energy and volume of the new system will be the *sum* of the energies and volumes of the two parts. But the temperature and pressure of the new system will be the same as the temperature and pressure of each part. Properties that depend on the *size* or *extent* of the system are called *extensive* properties; volume, mass, energy, surface area, and electric-dipole moment are all extensive properties. Extensive properties have values regardless of whether the system is in an equilibrium state or not. In contrast, properties that are independent of the size of the system are called *intensive properties*; temperature, pressure, and electric field intensity are intensive properties. As a rule, these properties only have meaning for systems in equilibrium states.

It is customary to define some additional intensive properties associated with extensive properties. For example, the volume per unit mass is called the *specific volume*,

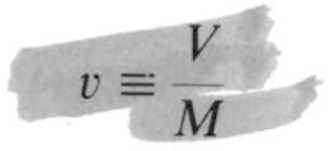

$$v \equiv \frac{V}{M}$$

and the internal energy per unit mass is called the *specific internal energy*,

$$u \equiv \frac{U}{M}$$

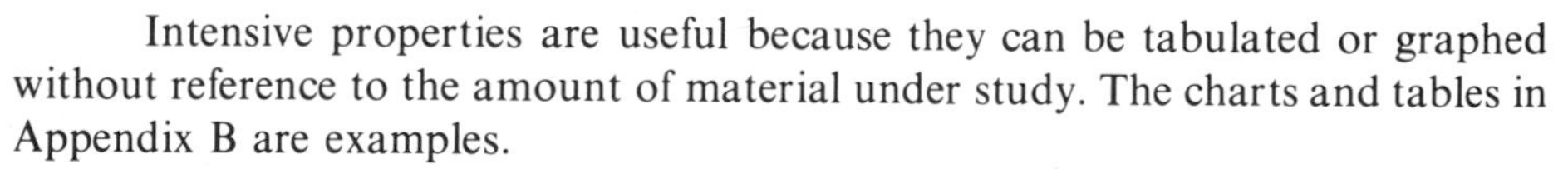

Intensive properties are useful because they can be tabulated or graphed without reference to the amount of material under study. The charts and tables in Appendix B are examples.

Consider a system consisting of a single substance in thermodynamic equilibrium. When the intensive properties of the substance are specified, the thermodynamic state of the system is completely known, apart from a single measure of size (say, the mass). The state of the system as specified by intensive thermodynamic properties is called the *intensive thermodynamic state*. The addition of a size measure (mass) completes the description of the *extensive thermodynamic state*.

3·6 INDEPENDENT VARIATIONS OF THE THERMODYNAMIC STATE

The thermodynamic properties of a substance are not independently variable, for various relationships exist between the intensive thermodynamic properties. For example, the pressure, temperature, and specific volume of a gas are frequently idealized to be related by

$$Pv = RT$$

where R is a constant for the gas. Only two of P, v, and T can be independently varied. In this section we shall develop a rule for determining the number of independently variable thermodynamic properties for any substance.

Thermodynamics deals with energy, and thermodynamic properties are those which in some way are related to energy. The number of ways in which we can independently vary the energy of a given substance tells us the number of independent thermodynamic properties.

Let's first consider the various ways in which we can transfer energy as work to the substance under study. If the substance is compressible, we can increase its energy through $-P\,dV$ work. If it is magnetic, we can increase its energy through magnetization work $\mu_0\mathbf{H}\cdot d(\mathbf{M}V)$. For a particular substance we consider all the relevant work modes shown in Table 2·1; there will be at least one independently variable property for each mode (volume, magnetic moment, and so on). In addition, we can hold these properties fixed and vary the energy through transfer of energy as heat (which varies the temperature). This gives us one more free variable. The count is in fact complete; for each of the independent ways of varying the energy of a given substance there is one independently variable thermodynamic property. We shall formalize this idea following some additional arguments on its behalf.

Let us first consider the nature of the work modes. As we noted in Chap. 2, each is of the form $\mathsf{F}\,dX$, where F is some sort of generalized force and dX is a generalized displacement. If F is independent of the direction and rate of change of the process, then the amount of energy transfer to the system when X is increased by dX will be exactly the same as the amount of energy transfer from the system when X is decreased by this same amount. This means that *the work mode is reversible;* the amount of energy added in a forward process can be removed by the reverse process. It is clear that any work mode for which F is a *property* of the thermodynamic state of the substance will be reversible in this sense. Each of the work modes listed in Table 2·1 is of this category. If, on the other hand, F depends on the direction or rate of the process and not just on the thermodynamic state, the process can exhibit hysteresis, hence is irreversible. Viscous work on a fluid falls into this category; in contrast, $-P\,dV$ work for the same fluid can be reversible. The concept of reversible work developed here will suffice for the moment. In Chap. 6 we shall give a much deeper meaning to reversible processes.

Let us now consider a piece of fluid, and see how we might change its thermodynamic state. We take a certain mass of a given fluid, so the composition of the system is fixed. Now, we can clearly change the state by compressing the fluid, thereby changing both the volume and energy. We can simultaneously cool the fluid in order to keep its energy fixed; this gives us a change in state at fixed energy, where the volume is the adjustable variable. We can also hold the volume fixed and adjust the energy independently through energy transfer as heat. So, the volume and energy are clearly two independently variable properties. Can we hold the volume and energy fixed and vary other thermodynamic properties? We might think of trying to vary the pressure, but this is impossible if we keep the volume and energy fixed. Squeezing the fluid would indeed raise the pressure, but would change the volume; heating would also raise the pressure, but would change the energy. Suppose we stir the fluid; this will increase the pressure, but it will also increase the energy. In fact, the same change in state could alternatively be produced through energy transfer as heat. We might try other irreversible work modes, but we always see that the effects of the irreversible work modes can be accomplished by a combination of reversible work and energy transfer as heat. We are inescapably led to the conclusion that there is but one freely variable property for each *reversible* work mode; this free variable is the generalized displacement X. But we can also hold all the X's fixed and vary the energy through energy transfer as heat. Thus, if there are n relevant reversible work modes for a given substance, there are only $n + 1$ independently variable thermodynamic properties.

We can view the set of properties $(X_1, X_2, \ldots, X_n, U)$ to be the independently variable set. However, we can give up control over one of these and thereby obtain the freedom to vary some other property. For example, we can vary the pressure and volume of a fluid system independently, if we do not insist on simultaneous control over the energy, simply by heating the substance at the desired volume until the desired pressure is achieved.

3·7 THE STATE POSTULATE

We formalize the ideas discussed above in the *state postulate:*

> *The number of independently variable thermodynamic properties for a specified system is equal to the number of relevant reversible work modes plus one.*

There are several ideas implicit in the words used. "Specified system" implies a specified amount of some specified matter; "thermodynamic properties" implies that we refer to those characteristics relevant to energy and to *thermodynamic equilibrium states.* The reference "relevant reversible work modes" means that we count only important work modes for the system in question and do not count irreversible work modes. The "plus one" is for the independent control of energy through heating or irreversible work.

Note that the rule gives the number of independent properties but does *not* say that *any* $n + 1$ properties are independent. The n X's and the energy always constitute an independent set, however.

The postulate as stated above deals with any amount of a specified substance. It is often convenient to work with a unit mass of the substance, and a form of the state postulate relating to intensive thermodynamic properties is therefore useful. Imagine that the system consists of a unit mass of the substance. The state postulate then can be interpreted as a rule for the number of independent thermodynamic properties of a unit mass, that is, for the number of independent *intensive* thermodynamic properties of the substance:

> *The number of independent intensive thermodynamic properties of a specified substance is equal to the number of relevant reversible work modes plus one.*

Again we interpret the implications of the wording: "specified substance" implies the percentages of each kind of molecule; "thermodynamic properties" implies those properties of the substance relevant to energy and to *thermodynamic equilibrium states.* Again only the relevant reversible work modes are to be counted. The rule gives the number of independent intensive thermodynamic properties, but does *not* imply that *any* set of $n + 1$ intensive properties will always be independently variable. For example, the internal energy and temperature of an ideal gas are not independently variable, as we shall see.

To illustrate the use of the state postulate, consider a substance for which the only important reversible work mode is compression or expansion ($P\ dv$ work). The rule says that there will be two independently variable intensive thermodynamic properties for such a substance; specification of the values for any two independent intensive thermodynamic properties will fix the values of all other intensive thermodynamic properties. For example, specification of the specific volume v and the specific internal energy u fixes the intensive thermodynamic state completely; temperature, pressure, and all other intensive thermodynamic properties for that substance are unique functions of u and v,

$$T = T(u, v) \qquad P = P(u, v)$$

Figures B·1 in Appendix B give examples of these relationships for water in graphical form. If we set $u = 1400$ Btu/lbm and $v = 0.3\ \text{ft}^3/\text{lbm}$, then we read from this graph that P is 3000 psia and the temperature is 1200°F, corresponding to $T = 1660$ R.

The other figures and tables in Appendix B give the thermodynamic properties of a variety of substances in a variety of forms. This information was taken from sources available to the practicing engineer, and unfortunately very little of it presently exists in SI form. This fact emphasizes the importance to the engineer of being able to work accurately and comfortably in a variety of unit systems. The student should begin now to become familiar with the graphical and tabular data included in Appendix B.

Later in our study we will encounter another thermodynamic property, *entropy*. We will denote the total entropy of a system by S, and the entropy per unit of mass by s. So, s will be an intensive thermodynamic property and, for substances of the type discussed in the example above, $s = s(u, v)$. The entropy will measure the microscopic chaos of the substance, which we see will be a function of both energy and volume.

As a second application of the state postulate, suppose we consider a substance for which the relevant reversible work modes are volume change ($P\ dv$ work) and electrostatic polarization [$\mathbf{E} \cdot d(v\mathbf{P})$ work]. Such a substance would have three independent intensive thermodynamic properties; the specific volume v, dipole moment per unit volume $\mathbf{P}$, and specific energy u are an independent set. The values for all other intensive thermodynamic properties are dependent on the values of these three properties. Thus

$$P = P(u, v, \mathbf{P}) \qquad T = T(u, v, \mathbf{P}) \qquad \mathbf{E} = \mathbf{E}(u, v, \mathbf{P}) \qquad s = s(u, v, \mathbf{P})$$

The relationships between the properties are called *equations of state*. Examples of such equations in graphical and tabular form constitute Appendix B. For certain idealized substances the equations of state are algebraic (such as $Pv = RT$). Today a modern engineering center will have the equation of state information stored within the memory of a digital computer as tables or fitted equations for use when needed in engineering analysis. The next chapter is devoted to discussion of the equations of state for a very special but extremely important class of substances.

SELECTED READING

Kestin, J., *A Course in Thermodynamics*, chap. 2, Blaisdell Publishing Co., Inc., Waltham, Mass., 1966.

Lee, J. F., and F. W. Sears, *Thermodynamics*, secs. 1-1 through 1-2, Addison-Wesley Publishing Co., Inc., Reading, Mass., 1963.

Van Wylen, G. J., and R. E. Sonntag, *Fundamentals of Classical Thermodynamics*, 2d ed., secs. 2.1–3.1, John Wiley & Sons, Inc., New York, 1973.

Zemansky, M. W., M. M. Abbott, and H. C. Van Ness, *Basic Engineering Thermodynamics*, 2d ed., secs. 1.5–2.2, McGraw-Hill Book Company, New York, 1975.

QUESTIONS

3·1 What concepts were introduced in this chapter? What basic postulates? What definitions?

3·2 Give an example of a property which is relevant to the thermodynamic state, and one which is irrelevant.

3·3 Why is heat not a property?

3·4 In handbooks we sometimes find something called the "heat of the liquid" tabulated as a function of temperature and pressure. Is it possible to tabulate heat as a function of state?

3·5 Give an example of a system which is not in equilibrium, and an example of one which is in a thermodynamic state.

3·6 Heat and energy bear the same relation to matter as rain and water bear to a reservoir. Explain this analogy.

3·7 How would you explain the concepts of property and state to a layperson?

3·8 Why is the differential quantity $đW$ not the differential of a property?

3·9 Can changes in state occur within a system without energy transfer across the boundaries?

3·10 Can the thermodynamic state of a substance be changed without any energy transfer?

3·11 What is a reversible work mode?

3·12 Can you visualize a substance for which shear work would be reversible?

3·13 Devise means for independently varying the pressure and volume of a gas; can the pressure and volume be treated as independent properties?

3·14 What is the difference between intensive and extensive properties?

3·15 Invent a test which will allow you to discover whether a property is intensive or extensive.

PROBLEMS

3•1 The simplest equations of state are obtained when only one work mode is considered important, that is, where there are only two independent intensive properties. Give a set of independent intensive thermodynamic properties for each of the following:
(*a*) Simple compressible substance, $P\,dV$ work only
(*b*) Simple magnetic substance, magnetization only
(*c*) Simple dielectric substance, polarization only
(*d*) Simple surface, surface extension only
(*e*) Simple elastic substance, one-dimensional pure strain only

3•2 For most substances the internal energy can be varied by varying the temperature and density, but for some special substances the internal energy is a function only of temperature. When work is done in an isothermal process on a substance whose internal energy is dependent only on the temperature, how much energy is transferred as heat, and in what direction?

3•3 Consider the generalized forces and displacements for a liquid-vapor-solid capillary system, as derived in Prob. 2•15. Assuming the liquid is incompressible, list a set of independent extensive properties for the system. What must you add if the liquid is compressible? Is the contact angle an intensive or extensive property?

3•4 A system initially consists of 3 lbm of a substance having specific internal energy u of 20 Btu/lbm, and 6 lbm of the same substance having an internal energy of 30 Btu/lbm. 150 Btu of energy is transferred as heat to this system, and it is allowed to come to equilibrium. What will be the specific internal energy of the substance in the final equilibrium state?

3•5 Consider a simple substance for which the temperature may be expressed functionally in terms of energy and volume or, alternatively, in terms of energy and pressure: $T = T(u, v)$ or $T = T(u, P)$. Differentiate these two expressions using the chain rule of calculus; use a subscript on the partial derivates to indicate which variables are kept constant [for example, $(\partial T/\partial u)_P$ for the partial derivative of T with respect to u, with the pressure held constant].

3•6 In a study of the properties of a liquid a 2-kg sample was heated at constant volume from 800 to 850 K. This required an energy input as heat of 11.2 Wh. Calculate the difference in specific internal energy (J/kg) between the initial and final states.

3•7 In a study of the properties of a gas a 3-lbm sample was heated at constant volume from 1100 to 1140°F. This required an energy input as heat of 35.0 Btu. Calculate the difference in specific internal energy between the initial and final states.

3•8 In a study of the properties of a liquid, a 2-kg sample of liquid was heated at a constant pressure of 1 atm from 500 to 600 K. The density of the liquid is 608 and 590 kg/m^3 at the initial and final states, respectively. An energy input as heat of 42 kJ was required. Determine the difference in the specific internal energies between the initial and final states (kJ/kg).

3•9 In order to determine the properties of a dense gas at high pressures, 10 lbm of the gas was heated at 4000 psia from 700 to 740°F. The gas volumes were 0.287 and 0.328 ft^3 in the initial and final states, respectively, and the required energy transfer as heat to the gas was 537 Btu. Determine the difference in the specific internal energy between the initial and final states.

3•10 In order to determine the properties of a gas, 0.05 kg was heated in a 0.1-m^3 container. The amounts of energy input as heat required to achieve temperatures above the initial temperature $T_1 = 500$ K were as follows:

T_2, K	Q_{12}, kJ
600	1.71
700	3.44
800	5.18
900	6.94
1000	8.72
1100	10.54
1200	12.39

Calculate the internal energy of the gas and plot as a function of T for this v.

3·11 The pressure may be thought of as a function of the temperature and specific volume for the perfect gas ($Pv = RT$). Prepare a three-dimensional sketch of this equation of state; project lines of constant temperature onto a P-v plane and sketch the resulting two-dimensional "map." How many coordinates are necessary for a fix on the state?

3·12 Make a table of all the intensive and extensive properties of matter that you know. List some means of measurement for as many as you can.

3·13 A substance for which the only reversible work mode is $-P\,dV$ is made to undergo first an adiabatic expansion process, then a constant-pressure compression process, and finally a constant-volume pressurization process which returns the substance to its initial state. What is the total change in internal energy for this cycle? Is there a net transfer of energy to or from the substance as work for this cycle? Is there any net energy transfer as heat? If so, is it to or from the substance?

3·14 In some presentations of thermodynamics a basic starting point is the hypothesis that for any control mass

$$\oint đW = \oint đQ$$

The integrals are to be taken around a cycle which returns the control mass to its initial state. If $đW$ is interpreted as positive if work is done by the control mass, what must be the interpretation of $đQ$ in the above equation? Derive this equation invoking the first law and the state postulate.

CHAPTER FOUR

STATES OF SIMPLE SUBSTANCES

4·1 THE SIMPLE SUBSTANCE

The simplest descriptions of matter are obtained for substances idealized as having only one relevant reversible work mode. We call such a substance a *simple substance*. The state postulate tells us that the number of independently variable intensive thermodynamic properties of a simple substance is *two*.

The term *simple compressible substance* is applied to any substance for which the only important reversible work mode is volume change ($P\,dv$ work). The theory of simple compressible substances is quite well developed, and considerable data have been accumulated relating the thermodynamic properties of many such substances. No substance is truly simple, but we find that satisfactory engineering analyses can usually be made by treating the substances involved as simple compressible substances.

The simplest descriptions of magnetic substances are obtained when magnetization is considered to be the only significant reversible work mode. Since most magnetic materials are solids, the work of volume change is usually small, and the idealization that a material is a *simple magnetic substance* is often quite valid. Similarly, we can speak of the *simple dielectric substance* as one for which the only important reversible work mode is electric polarization.

Pairs of independent intensive thermodynamic properties for these three simple substances are as follows:

Simple compressible substance, $P\,dv$ work only: (u, v) or (u, P)
Simple magnetic substance, magnetization only: $(u, \mathbf{M})$ or $(u, \mathbf{H})$
Simple dielectric substance, polarization only: $(u, \mathbf{P})$ or $(u, \mathbf{E})$

Thermodynamics concentrates most heavily on the simple compressible substance, which is often called merely a "simple substance."† We shall examine this and other types of simple substances in this chapter.

4·2 EQUATIONS OF STATE

From the state postulate we know that the temperature and pressure of a simple compressible substance can be expressed functionally as

$$T = T(u, v)$$
$$P = P(u, v)$$

These relations imply that we could completely fix the intensive thermodynamic state of a simple compressible substance by specification of any two independently variable intensive thermodynamic properties. As we shall see in a moment, temperature and pressure are not always independently variable. However, temperature and specific volume are always independent properties for a simple compressible substance, and we can alternatively think of the pressure and specific internal energy as being functions of these properties,

$$P = P(T, v)$$
$$u = u(T, v)$$

In certain special cases these equations can be expressed in explicit algebraic form, but in general it is easier to represent them graphically or in tables.‡ The equations, in algebraic, graphical, or tabular form, which relate the intensive thermodynamic properties of any substance are termed the *equations of state* of the substance.

A simple compressible substance can exist in different forms. In the gaseous form the molecules are far apart and move about freely, continually finding that they have new neighbors. Very little of the energy is associated with intermolecular forces. In the liquid form the molecules are much more densely packed but are still free to move about. A considerable amount of energy must be added to a liquid to break the strong force bonds which keep the liquid dense (compared to a gas). In solid forms the molecules are restrained to definite positions in the crystal

† In some texts the term "simple system" is used instead.

‡ Property tabulations are now available for many substances in forms suitable for use in a digital computer.

lattice and consequently always have the same neighbors. Melting of a solid is accomplished by adding enough energy to free individual molecules from one another. There may be different structures in the crystal lattice, hence different forms of the solid.

Some equilibrium states involve the presence of more than one form. For example, solid and liquid forms of water coexist in thermodynamic equilibrium at approximately 0°C. At a certain pressure and temperature near 0°C three forms of water can coexist in thermodynamic equilibrium. Such coexisting forms are called *phases*, and a group of phases is called a *multiphase mixture.*

The properties of each phase of a mixture can be treated separately, and the properties of the mixture then determined by appropriate combination. For example, the energy of the mixture is the sum of the energies of all phases. However, the temperature and pressure of the mixture are the same as the common temperature and pressure of each phase. In graphical representations it is equally convenient to treat the mixture properties directly; the fractions by mass of each phase then become relevant intensive thermodynamic properties.

In engineering we continually use equation-of-state information provided in textbooks, Bureau of Standards brochures, the International Critical Tables, and various engineering, physics, and chemistry handbooks. There are three primary means by which these charts, tables, and equations are developed. First, laboratory measurements provide the main source of numerical values. Second, equations relating various properties can be developed by application of thermodynamic theory (Chap. 8); these play a very important role in checking laboratory data, in interpolation, in extrapolation, and even in evaluation of other thermodynamic properties (such as entropy). A third way that equations of state are obtained is through application of quantum-statistical thermodynamics (as in Fig. B·12). By postulation of an appropriate microscopic model of the substance, application of certain quantum-mechanical fundamentals allows one to predict the equations of state. These must of course be checked against laboratory data and must be consistent with equations obtained from thermodynamic theory. The quantum-statistical thermodynamic method is perhaps the most fundamental but is also the most sophisticated. In fact, only the simplest of molecules can be treated by routine applications of this theory. Hence the large body of equation-of-state information presently available is based primarily on laboratory data and thermodynamic theory.

Laboratory measurement of temperature, pressure, and specific volume is relatively straightforward. Measurement of the specific internal energy requires that the substance be put through some process where the amounts of energy transfer can be measured accurately. The energy transfer as work can be calculated from its basic definition, for example, $đW = -P\,dV$. The energy transfer as heat can be supplied from a device that takes in energy as measurable work and rejects all this energy to the substance as heat. Such a device might consist of an electric generator with a resistive load. If these energy transfers are known, the change in internal energy for the test specimen can be determined. Execution of a large number of such experiments allows one to map out the specific internal energy as a function of the thermodynamic state (say, as a function of the specific volume and temperature), *relative* to the energy at some arbitrarily chosen *datum state*. The value of the energy at the datum state is normally chosen as zero. Since energy-balance analyses always involve energy changes and not absolute values of energy, the use of an arbitrary datum suffices as long as the substance is not involved in chemical reactions with other substances. When reactions are involved, one must be careful to tie the energies of the various substances together properly in order to have a proper energy accounting for the chemical reaction. We shall discuss how this is done in Chap. 11.

4·3 THE GENERAL NATURE OF A SIMPLE COMPRESSIBLE SUBSTANCE

Let us now imagine performing an experiment on a sample of some simple compressible substance. Suppose we carry out a heating process in a piston-cylinder system in a manner that keeps the pressure constant. At each step in the process we add a small amount of energy as heat, allow the substance to expand (thereby doing work) to keep the pressure constant, and then after waiting for equilibrium to be established we record the values for temperature, pressure, and volume. The energy balance then allows us to calculate the incremental change in internal energy between the first and second thermodynamic states. By this experiment we can map out the thermodynamic properties at states having this certain pressure, and these could become part of a basic tabulation of the type given in Appendix B.

A typical outcome of such an experiment is shown in Fig. 4·1. If we start with a solid at state 1, adding energy would raise the temperature to state 2, where the solid would begin to melt. For a given pressure there is one certain temperature at which solid and liquid phases exist in thermodynamic equilibrium. Further energy addition at this pressure results only in more melting, with the intensive thermodynamic states of the solid and liquid remaining constant. This melting would continue until only liquid remains (mixture state 3); for most substances this would be a state of larger mixture specific volume. A further increase in energy would result in increasing the temperature, until at state 4 the liquid would begin to evaporate. From state 4 to state 5 liquid and gas phases would exist together at a constant temperature and pressure, and the volume might undergo a tremendous increase. At state 5 the liquid would finally be completely evaporated,

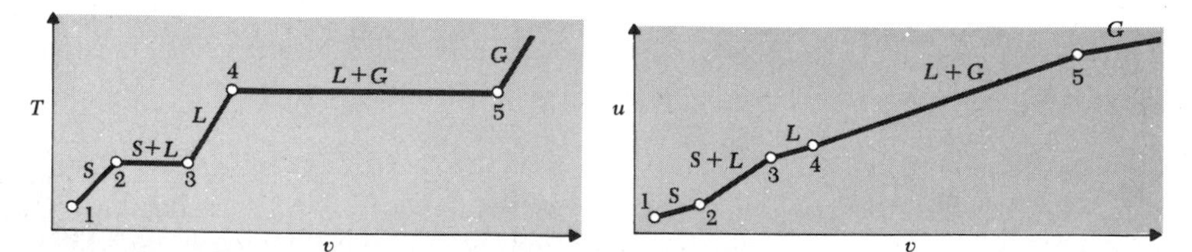

Figure 4·1. History of a constant-pressure heating experiment

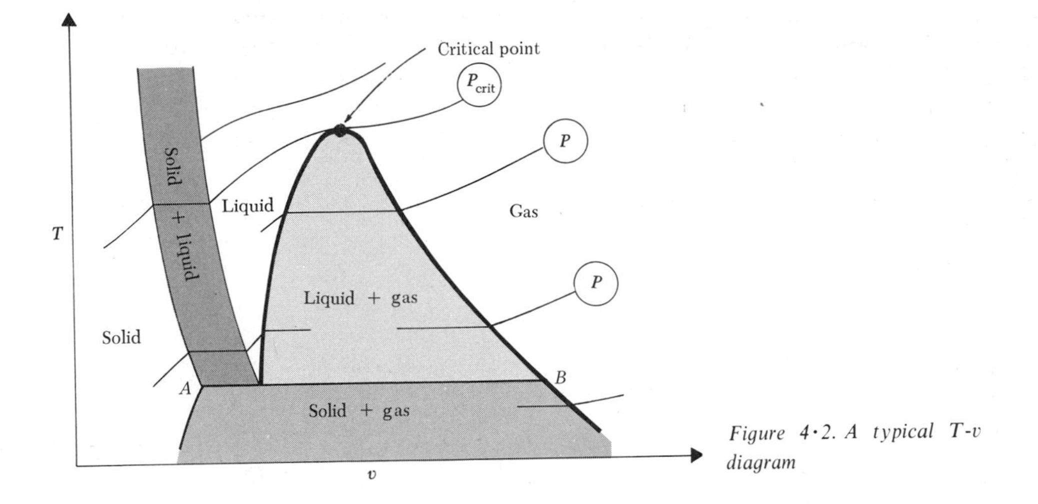

Figure 4·2. A typical T-v diagram

and further additions of energy would cause the temperature of the gas (vapor) to increase. Eventually, at very high temperatures, the molecules would become ionized, and we would call the ion-electron mixture a *plasma*. In plasma states the electrical and magnetic work modes might be important, and idealization that the substance is a simple substance might no longer be realistic.

Repeating the experiment with different starting states, a whole series of constant-pressure lines could be traced out (Fig. 4·2). At pressures below a certain critical pressure we would always encounter a region of coexisting liquid and gas phases; but at greater pressures there would cease to be a well-defined boiling point, and the transition from the liquid form to the gas form would be continuous rather than abrupt.

Figure 4·2 constitutes part of the graphical equation of state for a simple compressible substance. We see clearly that the pressure is a unique function of the temperature and volume. Moreover, the temperature of a multiphase mixture is fixed solely by the pressure. In other words, *in the mixed-phase regions temperature and pressure cannot be specified independently.*

Graphing the functional relation $P = P(T, v)$ would require a three-dimensional map, and the projections of Fig. 4·2 are more convenient for quantitative work. However, it is instructive to examine a qualitative three-dimensional P-v-T surface, and one is shown in Fig. 4·3. The equation of state $P = P(T, v)$ is represented by a surface standing in the P-v-T space. Figure 4·2 represents a view of this surface from above, and the lines of constant pressure are obtained by taking horizontal sections through the surface.

States at which a phase change begins or ends are called *saturation states*. The highest pressure and temperature at which distinct liquid and gas phases can coexist define the *critical point*. The dome-shaped region in Figs. 4·2 and 4·5 bounded by the saturated-liquid line and the saturated-vapor line is referred to as the *vapor dome*. In thermodynamic systems the engineer is primarily concerned

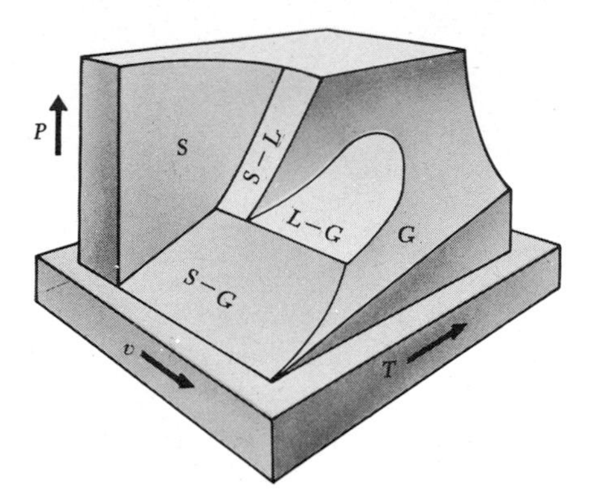

Figure 4·3. P-v-T surface for a substance which expands upon melting

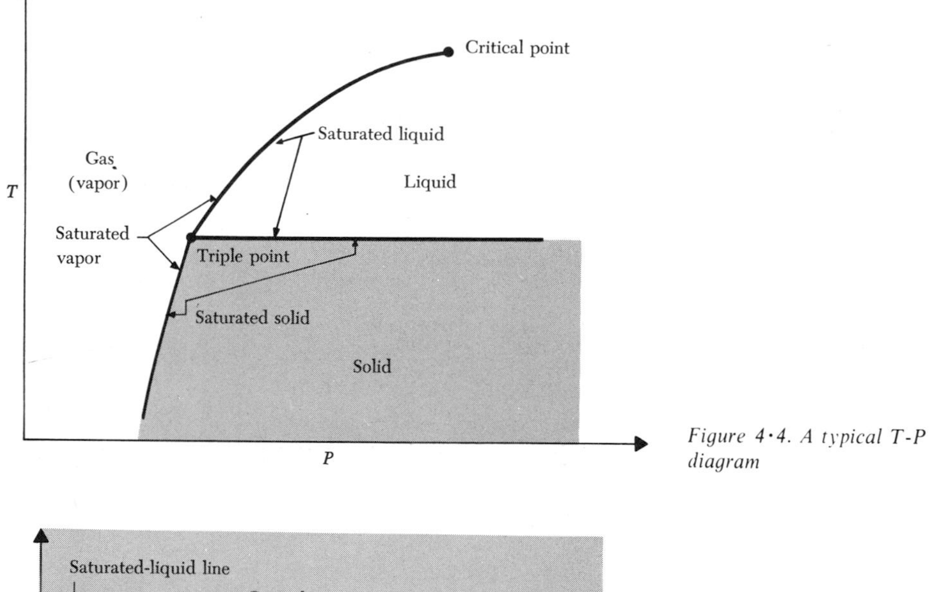

Figure 4·4. A typical T-P diagram

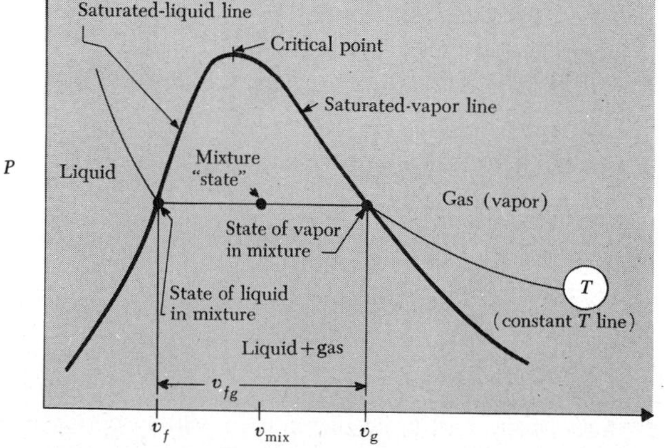

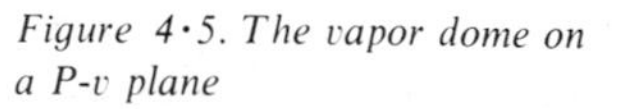

Figure 4·5. The vapor dome on a P-v plane

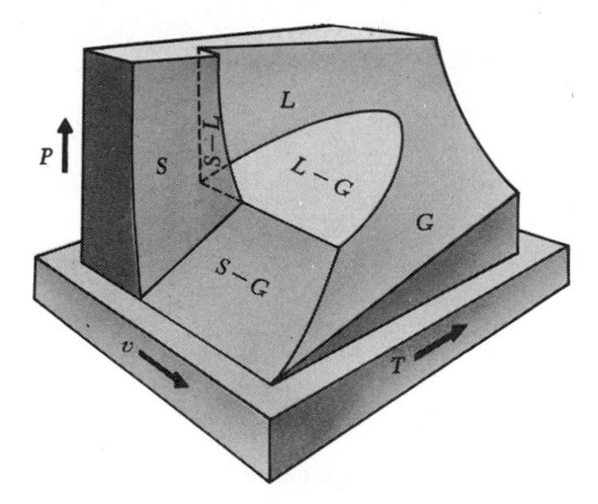

Figure 4·6. P-v-T surface for a substance which contracts upon melting

with liquid, vapor (gas),† and the liquid-vapor mixed-phase region. The area to the right of the vapor dome is commonly called the *superheated-vapor region.* To the left of the saturated-liquid line the substance is said to be in a *subcooled-liquid state.* States at pressures greater than the critical pressure are commonly called *supercritical states.* Note that the three-phase line appears as a point when the surface is viewed in a T-P plane (Fig. 4·4). This point is called the *triple point.*

It is common practice to denote states on the saturated-liquid line by the subscript f, and states on the saturated-vapor line by the subscript g. The difference between the saturated-vapor and saturated-liquid properties is frequently denoted by the subscript fg. For example,

$$v_{fg} \equiv v_g - v_f$$

$$u_{fg} \equiv u_g - u_f$$

Figures 4·1 to 4·3 represent the behavior of a substance that expands upon melting, and most substances are of this nature. However, one of the most important substances, water, contracts upon melting. A three-dimensional P-v-T relationship for such a substance is shown in Fig. 4·6. A typical u-T-P surface would appear as shown in Fig. 4·7. Simple compressible substances differ markedly in their thermodynamic characteristics, as is illustrated by the variations in the critical-point properties shown in Table B·8, Appendix B.

† We shall use the terms "vapor" and "gas" interchangeably.

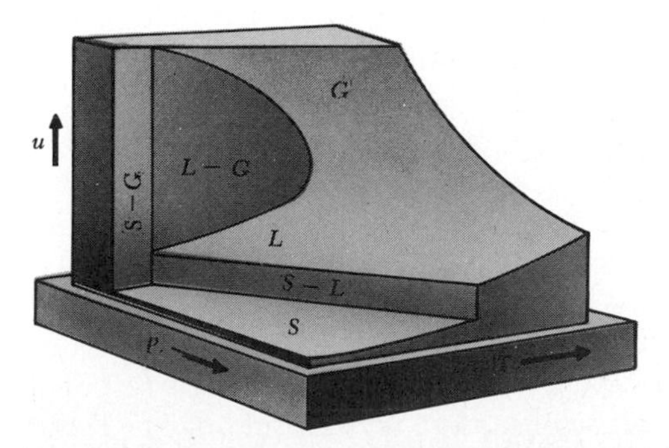

Figure 4·7. A typical u-T-P surface

4·4 USING THE TABULAR AND GRAPHICAL EQUATIONS OF STATE

Measurements of P, v, T, u, and other thermodynamic properties have been obtained for many substances, and graphical or tabular equations of state are now available for use in engineering analysis. Tables B·1, B·3, B·4, and B·5 are typical examples of *saturation tables.* Such tables give the properties of the individual liquid and gas (vapor) phases of a liquid-gas mixture. Since knowledge of either temperature or pressure suffices to fix the state of both the saturated liquid and saturated vapor, saturation tables are of the single-entry type. Note that Tables B·1*a* can be entered on even values of temperature, while Tables B·1*b* are ordered by even values of pressure.

In the United States, saturation tables in old English units are available for most substances, but metric tables are very hard to find. Those metric tables that are available are seldom in the SI. We have included both old English and SI saturation data for water in Table B·1.

As an example of the use of these tables, let's look up the boiling temperature of water at 4×10^5 N/m^2 (0.4 MPa = 4 bars). From the SI version of Table B·1*b*, we read the saturation temperature as 143.6°C. Note that the specific volume v_f of the saturated liquid is 1.084×10^{-3} m^3/kg at this pressure, while the specific volume of the saturated vapor v_g is 0.4625 m^3/kg. The specific internal energies of the saturated liquid and saturated vapor are 604.3 and 2553.6 kJ/kg, respectively.

As a second example, let's find the pressure at which water will boil at a temperature of 500°F. From the old English version of Table B·1*a*, we read a saturation pressure of 680.0 lbf/in^2. Note that each lbm of saturated vapor will take up 0.6761 ft^3.

The last entries in Tables B·1 correspond to the critical point. Note that it occurs at a pressure of 22.088 MPa, equivalent to 3203.6 psia. The temperature at this point is 374.1°C, or 705.4°F.

Tables of the properties of the gaseous form are called *superheat* tables. Since two properties are required to fix each state, these are double-entry tables. Tables B·2 give superheat data for water in both SI and old English units.

For example, suppose we need to know the specific volume of steam at 1 MPa and 300°C. From the SI version of Table B·2, we read $v = 0.2579$ m^3/kg. Similarly, at 100 psia and 1000°F, we read $v = 8.657$ ft^3/lbm from the old English version of Table B·2.

Graphical representations of the equation of state are particularly convenient, because most of the pertinent information can be put on a single page. Such a representation for H_2O is shown in Fig. B·1. Here the internal energy u and volume v are used as the independent coordinates, and the thermodynamic state is completely specified by fixing the values of these two properties. Note that the state can be fixed by specifying values for any two independent properties. For example, water at a pressure of 100 psia and a density of 0.2 lbm/ft^3 (corresponding to $v = 5.0$ ft^3/lbm) will be in the gaseous phase at a temperature of 405°F and will have a specific internal energy of 1139 Btu/lbm. The u-v plot of Fig. B·1 is included here primarily for purposes of illustration. Thermodynamic data are usually presented on other planes, using as coordinates properties that we have yet to define, but we can make use of the charts even before we know the meaning of all the properties. For example, using Fig. B·5, we find the volume of saturated O_2 vapor at 100 K as slightly under 3000 ml/gmole and the saturation pressure as about 2.5 atm. In Fig. B·6, we read the specific volume of CO_2 vapor at 0°F and 100 psia as about 1.2 ft^3/lbm.

Mixture properties. The properties of a liquid-vapor mixture can be read directly from the graphical equations. Alternatively, the mixture properties can be computed from the properties of the individual phases as tabulated in the saturation tables if we know the relative amounts of the two phases present in the mixture. It is convenient to introduce an additional property, the *quality*, defined as the fraction of the total *mass* which is saturated *vapor*. The symbol x is commonly used for this mixture property; $1 - x$ is then the fraction of the mass that is saturated liquid. Then, if M is the total mass, the volume and internal energy are

$$V = (1 - x)Mv_f + xMv_g$$

$$U = (1 - x)Mu_f + xMu_g$$

The specific volume and specific internal energy of a simple compressible substance in the mixed-phase region are then

$$\blacktriangleright \quad v = (1 - x)v_f + xv_g \qquad (4\cdot1)$$

$$\blacktriangleright \quad u = (1 - x)u_f + xu_g \qquad (4\cdot2)$$

The pressure and temperature of the mixture are of course the same as the pressure and temperature of the saturated liquid and vapor.

For example, suppose that we need to calculate the volume and energy of 3 kg of H_2O at 200°C in a two-phase state with a quality of 0.6. The liquid and vapor masses are

$$M_f = (1 - 0.6) \times 3 \text{ kg} = 1.2 \text{ kg}$$

$$M_g = 0.6 \times 3 \text{ kg} = 1.8 \text{ kg}$$

We find values for v_f, v_g, u_f, and u_g in the SI version of Table B·1*a*. The volume occupied by the mixture is then

$$V = Mv = 3.0 \times [(1 - 0.6) \times 0.001156 + 0.6 \times 0.1274] \text{ m}^3/\text{kg} = 0.2307 \text{ m}^3$$

The internal energy of the mixture is

$$U = Mu = 3.0 \times [(1 - 0.6) \times 850.6 + 0.6 \times 2595.3] \text{ kJ/kg} = 5692 \text{ kJ}$$

4·5 METASTABLE STATES IN PHASE TRANSITIONS

It is important to realize that the phase transitions discussed above are not always observed at precisely the points indicated in the tables. Suppose we expand the volume available to a liquid-vapor mixture. The pressure within the liquid might actually be significantly reduced below the saturation pressure before any vapor is formed, for vapor can only be formed at a nucleus, such as a small vapor bubble. Rough walls normally contain many such nuclei, but with a smooth glass container one can virtually eliminate all nuclei, hence prevent the formation of new vapor within the liquid. Water in a clean beaker can be heated several degrees above 100°C without noticeable boiling. The insertion of a roughened glass rod, which provides nucleation sites, gives rise to immediate vapor formation from these sites. The state of the superheated liquid in this demonstration is not given in any of the tables of thermodynamic properties, for the superheated liquid is not in an equilibrium state. A small disturbance (the insertion of the rod) initiates a dramatic change in state (vaporization). A similar effect is observed when water is cooled in a clean beaker at 1 atm. With care, the water can be cooled well below 0°C without solidification. A tap on the beaker then triggers a sudden crystallization throughout the liquid. The "supercooled liquid" is likewise not in an equilibrium state. A related phenomenon occurs during condensation; the vapor requires some small liquid droplets to act as nuclei for the condensation process. This has considerable importance in steam turbine nozzles, where the vapor is usually expanded to somewhat below the saturation pressure before condensation actually occurs.

We can use an analogy to help understand these processes. Consider the equilibrium of a marble on a pedestal (Fig. 4·8). The marble in Fig. 4·8*c* is in a stable state; any departure from its rest position at the bottom of the well will lead to a return to the well. In contrast, the marble in Fig. 4·8*a* is in an unstable state. With the slightest disturbance it will fall off the pedestal. The marble in Fig. 4·8*b* is in a metastable state; it is stable to small disturbances, but unstable to disturbances of sufficient magnitude. Thermodynamic equilibrium states are like the stable state (Fig. 4·8*c*). Supercooled or superheated water, or supercooled steam, are metastable states, and their properties are not included in tables of thermodynamic equilibrium states.

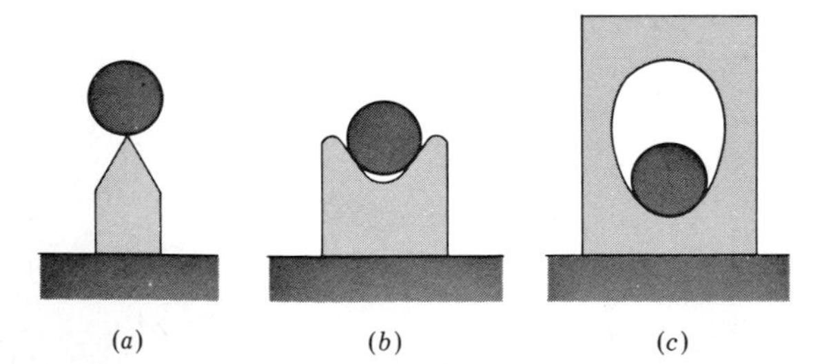

Figure 4·8. Stability as illustrated by a marble. (a) unstable; (b) metastable; (c) stable

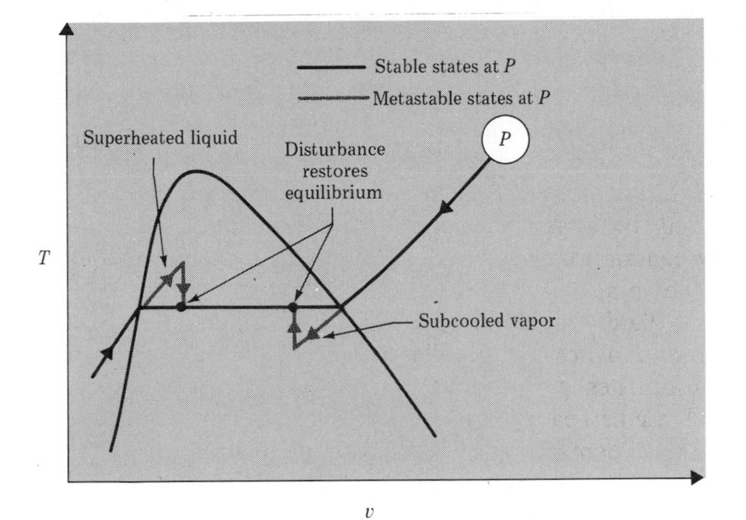

Figure 4·9. Metastable states

Figure 4·9 depicts the process paths for reaching metastable states by heating or cooling at constant pressure. Note that the disturbance will return the system to a thermodynamic equilibrium state. Similar curves can be drawn for changes in pressure at fixed temperature. Indeed, by carefully reducing the pressure on a liquid in an extremely clean container, negative pressures can be reached. Negative pressures correspond to tensions in the fluid, and tensions of thousands of psi have been achieved.

To illustrate what is involved in a quantitative way, suppose we have a tiny bubble in a liquid (Fig. 4·10). The pressure inside the bubble can be related to the pressure of the surrounding liquid and the surface tension σ through a force balance on half of the bubble,

$$P_b \pi r^2 = \sigma 2\pi r + P_l \pi r^2$$

So

$$P_b - P_l = \frac{2\sigma}{r}$$

Now, suppose we heat the liquid at constant pressure, raising its temperature slightly above the saturation value. Suppose that the gas and the liquid remain in thermal equilibrium, so the gas temperature will also rise above the saturation

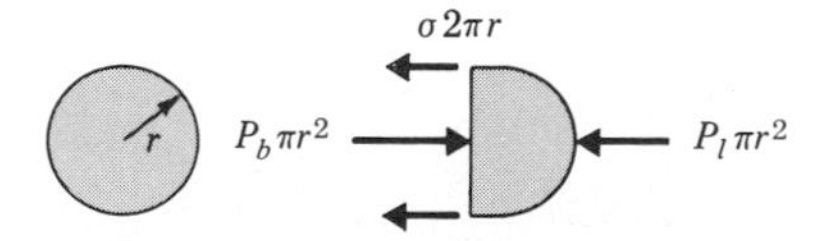

Figure 4·10. Relating the pressure difference across a bubble to the surface tension

temperature *at the liquid pressure*. But if this temperature is below the saturation temperature at the pressure *inside* the bubble, the gas will condense and the bubble will therefore collapse. To get a feel for the numbers, suppose we have water at 0.1 MPa superheated above the saturation temperature of 99.6°C to 100°C. In order for the bubble not to collapse, the gas pressure must be at least as high as the saturation pressure at 100°C, which from the SI version of Table B·1*b* we read as 0.1013 MPa. The surface tension of water is about 50 dyn/cm = 0.05 N/m. So, to sustain a pressure *difference* of 0.0013 MPa, we require a bubble radius of

$$r = \frac{2\sigma}{P_b - P_l}$$

$$2 \times 0.05\ (\text{N/m})/(0.0013 \times 10^6\ \text{N/m}^2) = 7.6 \times 10^{-5}\ \text{m}$$
$$= 0.076\ \text{mm}$$

If the bubble is smaller than this amount, it will collapse. If it is larger, it will grow; the radius will increase, dropping the gas pressure, allowing more liquid to be evaporated, etc., until finally equilibrium is obtained. So, supercooled water under these conditions would be stable in the presence of bubbles smaller than 0.076 mm, but unstable in the presence of larger bubbles.

The actual situation in a beaker or steam turbine is much more complicated. Dissolved air plays an important role in the metastable processes in liquids, and suspended particulates are very important in analysis of the instabilities in supercooled gases. Analysis of these effects is beyond our present scope. Most engineering problems can be adequately treated by assuming that the phase transitions occur exactly at the saturation conditions, and this is the approach we shall take.

4·6 USE OF PROPERTY DATA IN ENGINEERING ANALYSIS

Tabular and graphical equations of state are used in conjunction with the energy balance in analysis of engineering systems. The energy balance provides quantitative information relating the change in internal energy to the energy transfers for the process, and the equations of state relate the energy to other thermodynamic properties whose values may be of interest in the particular system.

An engineering example. Suppose we need to compute the amount of energy which must be transferred as heat to 5 lbm of H_2O initially at 200°F and 2 psia pressure in order to increase the temperature to 1200°F if the heating occurs at constant volume. We begin the solution by defining a control mass consisting of 5 lbm of water. As shown in Fig. 4·11*a*, energy inflow as heat will be considered positive, and there is no work involved. The only energy possessed by

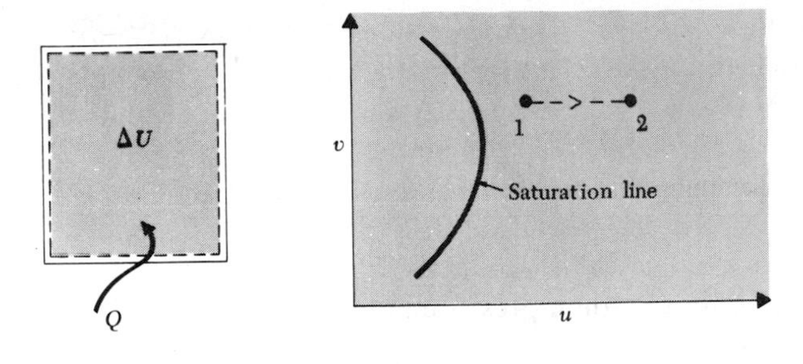

(*a*) *The control mass*

(*b*) *The process representation*

Figure 4·11

the system is its internal energy U, so the energy balance, made over the time for the state change to take place, is

$$\underset{\text{energy input}}{Q} = \underset{\text{increase in energy storage}}{\Delta U}$$

Here
$$\Delta U = M(u_2 - u_1)$$

where u_1 and u_2 represent the internal energy in the initial and final states.

At this point it becomes necessary to make the idealization that the substance is in a state of thermodynamic equilibrium at the start of the process and in another such state at the end of the process. The equation of state for water, treated as a simple compressible substance, can then be employed (see Appendix B). For the initial state, from Table B·2 we find

$$P_1 = 2 \text{ psia} \qquad u_1 = 1077 \text{ Btu/lbm}$$
$$T_1 = 200°\text{F} \qquad v_1 = 196 \text{ ft}^3\text{/lbm}$$

The final state is fixed with the aid of a *process representation*, Fig. 4·11*b*. Such a diagram shows the reader how the problem is solved and helps the analyst in finding a solution. Then, reading from Table B·2, we ıd that the final state is just over 5 psia,

$$v_2 = 196 \text{ ft}^3\text{/lbm} \qquad P_2 \simeq 5 \text{ psia}$$
$$T_2 = 1200°\text{F} \qquad u_2 = 1457 \text{ Btu/lbm}$$

Substituting into the energy balance, we find

$$Q = 5 \text{ lbm} \times (1457 - 1077) \text{ Btu/lbm} = 1900 \text{ Btu}$$

In Chap. 5 we shall see many other examples illustrating the use of equations of state in engineering analysis.

4·7 SOME OTHER THERMODYNAMIC PROPERTIES

Quite frequently other thermodynamic properties will be defined in terms of P, v, T, and u, and we shall now discuss some of these. In single-phase regions, where pressure and temperature are independent, we can think of the volume as being a function of pressure and temperature,

$$v = v(T, P)$$

The difference in specific volume between any two states separated by infinitesimal differences dT and dP is

$$dv = v(T + dT, P + dP) - v(T, P)$$

By Taylor series expansion,

$$v(T + dT, P + dP) = v(T, P) + \left(\frac{\partial v}{\partial T}\right)_P dT + \left(\frac{\partial v}{\partial P}\right)_T dP + \cdots$$

where the partial derivatives are to be evaluated at the point (T, P) and the subscripts denote the variable held constant during the differentiation. Hence, for infinitesimal differences dT and dP,†

$$dv = \left(\frac{\partial v}{\partial T}\right)_P dT + \left(\frac{\partial v}{\partial P}\right)_T dP \tag{4·3}$$

The partial derivative $(\partial v/\partial T)_P$ represents the slope of a line of constant pressure on a v-T plane, and the partial derivative $(\partial v/\partial P)_T$ represents the slope of a line of constant temperature on a v-P plane (Fig. 4·12). These derivatives are themselves intensive thermodynamic properties, since they have definite values at any fixed thermodynamic state. The first represents the sensitivity of the specific volume to changes in temperature at constant pressure, and the second is a measure of the change in specific volume associated with a change in pressure at constant temperature. Two thermodynamic properties related to these derivatives are the *isobaric* (constant-pressure) *compressibility*,

▶
$$\beta \equiv \frac{1}{v}\left(\frac{\partial v}{\partial T}\right)_P \tag{4·4}$$

and the *isothermal* (constant-temperature) *compressibility*,

▶
$$\kappa \equiv -\frac{1}{v}\left(\frac{\partial v}{\partial P}\right)_T \tag{4·5}$$

These compressibility factors are frequently tabulated functions of state. The *coefficient of linear expansion* used in elementary strength-of-materials texts is $\frac{1}{3}\beta$. *Young's modulus of elasticity* is proportional to κ.

In terms of β and κ, Eq. (4·3) becomes

$$dv = \beta v \, dT - \kappa v \, dP \tag{4·6}$$

† Alternatively, this result can be obtained directly from the chain rule of the calculus of a function of two variables.

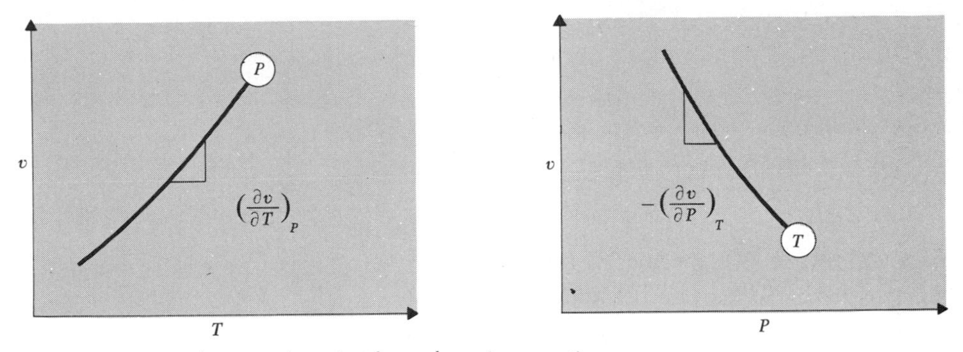

Figure 4·12. The slopes are intensive thermodynamic properties

This represents a differential equation relating specific volume to temperature and pressure on the equation-of-state surface. Its usefulness arises from the fact that β and κ are sometimes slowly varying functions of T and P. Equation (4·6) can be divided by v and integrated to yield

$$\int_{v_0}^{v} \frac{dv}{v} = \int_{T_0}^{T} \beta\, dT - \int_{P_0}^{P} \kappa\, dP$$

If β and κ are constants, we can integrate from some state (T_0, P_0), where the specific volume is v_0, to any other state (T, P), obtaining

$$\ln \frac{v}{v_0} = \beta(T - T_0) - \kappa(P - P_0)$$

This is often an adequate approximation over limited pressure and temperature ranges.

It should be apparent that the calculus of functions of two variables is of considerable utility in the thermodynamics of simple substances. In particular, consider three functions x, y, and z, any two of which may be selected as the independent pair. Then, from calculus, it follows that†

$$\blacktriangleright \qquad \left(\frac{\partial x}{\partial y}\right)_z = \frac{1}{(\partial y/\partial x)_z} \tag{4·7}$$

$$\blacktriangleright \qquad \left(\frac{\partial x}{\partial y}\right)_z \left(\frac{\partial y}{\partial z}\right)_x \left(\frac{\partial z}{\partial x}\right)_y = -1 \tag{4·8}$$

† Consider $x = x(y, z)$ and $y = y(x, z)$. Then, from the chain rule,

$$dx = \left(\frac{\partial x}{\partial y}\right)_z dy + \left(\frac{\partial x}{\partial z}\right)_y dz$$

$$dy = \left(\frac{\partial y}{\partial x}\right)_z dx + \left(\frac{\partial y}{\partial z}\right)_x dz$$

Eliminating dy from these two equations,

$$\left[1 - \left(\frac{\partial x}{\partial y}\right)_z \left(\frac{\partial y}{\partial x}\right)_z\right] dx = \left[\left(\frac{\partial x}{\partial y}\right)_z \left(\frac{\partial y}{\partial z}\right)_x + \left(\frac{\partial x}{\partial z}\right)_y\right] dz$$

But the changes dx and dz are independent. Letting $dz = 0$, Eq. (4·7) follows. Letting $dx = 0$ and using Eq. (4·7), Eq. (4·8) is obtained.

To use these mathematical properties of functions, we note for example that, in regions of P-v-T space where P and T are independent, $v = v(P, T)$, so

$$\left(\frac{\partial P}{\partial T}\right)_v \left(\frac{\partial T}{\partial v}\right)_P \left(\frac{\partial v}{\partial P}\right)_T = -1$$

or

$$\left(\frac{\partial P}{\partial T}\right)_v = -\frac{(\partial v/\partial T)_P}{(\partial v/\partial P)_T} = \frac{\beta}{\kappa} \tag{4·9}$$

Hence knowledge of the isobaric compressibility β and the isothermal compressibility κ would allow us to determine from Eq. (4·9) how the pressure changes with temperature for a constant-volume heating process. We shall make considerable use of the calculus of functions of several variables, and the student is urged to review this material now if necessary.

There are several properties related to the internal energy that are important. Suppose we view the specific internal energy u as being fixed by the specification of T and v,

$$u = u(T, v)$$

The difference in energy between any two states separated by infinitesimal temperature and specific-volume differences dT and dv is then

$$du = \left(\frac{\partial u}{\partial T}\right)_v dT + \left(\frac{\partial u}{\partial v}\right)_T dv \tag{4·10}$$

The derivative $(\partial u/\partial T)_v$ represents the slope of a line of constant v on a u-T thermodynamic plane. The derivative is also a function of state, that is, a thermodynamic property, and is called the *specific heat at constant volume,*†

▶ $$c_v \equiv \left(\frac{\partial u}{\partial T}\right)_v \tag{4·11}$$

Another thermodynamic property which we shall find to be of particular importance is the *enthalpy* h defined by‡

▶ $$h \equiv u + Pv \tag{4·12}$$

† The name given to c_v is somewhat unfortunate in that only for very special conditions is the derivative $(\partial u/\partial T)_v$ related to energy transfer as heat. If a process is carried out slowly at constant volume, no work will be done, and any energy increase will be due solely to energy transfer as heat. For such a process c_v does represent the energy increase per unit of temperature rise (per unit of mass), and consequently historically was called the "specific heat at constant volume." We feel that c_v should be thought of in terms of its definition as a certain partial derivative, and not as being related to energy transfer as heat in the special constant-volume process.

‡ The product of pressure and specific volume has the units of energy per unit of mass, as does u. However, it is customary in the old English system to give u values in Btu/lbm and P and v in mechanical units (lbf and ft). One must be careful to make the two parts of the enthalpy dimensionally equivalent in any numerical computations. This is automatic in the SI. Enthalpy is sometimes called the "heat content," another term with more historical than physical significance.

We leave as an exercise the proof that the enthalpy of a liquid-vapor mixture is

$$h = (1 - x)h_f + xh_g \tag{4·13}$$

The enthalpy of a simple substance is obviously only a function of the thermodynamic state. For states where T and P are independent (single-phase states), we may put

$$h = h(T, P)$$

Then, taking the differential,

$$dh = \left(\frac{\partial h}{\partial T}\right)_P dT + \left(\frac{\partial h}{\partial P}\right)_T dP \tag{4·14}$$

The derivative $(\partial h/\partial T)_P$ is called the *specific heat at constant pressure*,†

$$c_P \equiv \left(\frac{\partial h}{\partial T}\right)_P \tag{4·15}$$

The derivatives c_P and c_v constitute two of the most important thermodynamic derivative functions, and values have been experimentally determined as functions of the thermodynamic state for a tremendous number of simple compressible substances.

As an example, let's estimate the value of c_P for H_2O at 2 MPa and 600°C. From the SI version of Table B·2, we see that this is a gaseous state. Using Table B·2, we can construct a "thinking picture" to guide us in our calculation (Fig. 4·13). We want the value of c_P at point 0, which will be approximately

$$c_P = \frac{h_B - h_A}{T_B - T_A} = \frac{3803.1 - 3578.3}{100} = 2.25 \text{ kJ/(kg·°C)}$$

† c_P is sometimes called the "heat capacity at constant pressure." As with c_v, it is best to think of c_P as a partial derivative. Only in very special processes (constant pressure) is c_P related to energy transfer as heat, yet it is a useful thermodynamic function in many other situations.

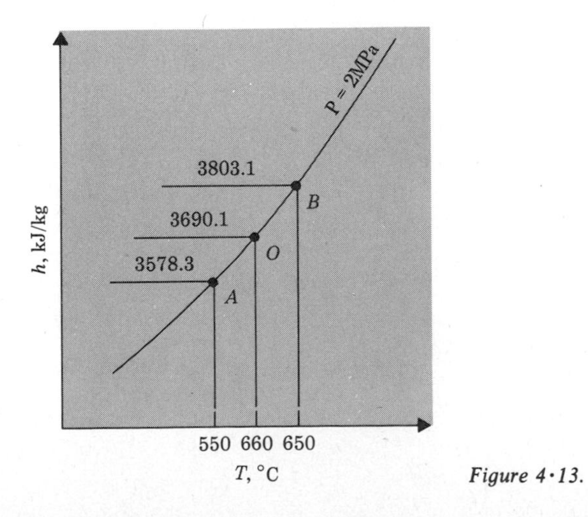

Figure 4·13.

A value 2.25 kJ/(kg·°C) corresponds to 0.537 cal/(g·°C) or 0.537 Btu/(lbm·°F). Since a temperature *difference* appears in the denominator, we could just as well use K in place of °C.

4·8 THE PERFECT GAS

Under appropriate conditions, the equations of state for the vapor of any substance can be approximated by the algebraic equations of state for an *ideal* or *perfect gas*. The defining equation for the perfect gas is

$$Pv = RT \tag{4·16}$$

Here R is a constant for a particular gas, and is related to the *universal gas constant* $\mathscr{R}$ and the molal mass $\hat{M}$ by†

$$R = \frac{\mathscr{R}}{\hat{M}} \tag{4·17}$$

where‡ $\mathscr{R} = 8.314$ kJ/(kgmole·K) = 1545 ft·lbf/(lbmole·R)

Note that T is the *absolute* temperature in K or R.

The value of R can be determined by plotting Pv/T as a function of state. In any region where this value is constant the perfect-gas approximation is adequate. Figure 8·9 shows the value of the *compressibility factor* $Z \equiv Pv/RT$ as a function of the ratio of the temperature and pressure to their values at the critical point; where the ratio Z is near unity the perfect-gas approximation is valid. Note that the perfect-gas model is best at pressures low compared to the critical pressure or at temperatures high compared to the critical temperature. We shall discuss the perfect-gas model in more detail in Chap. 8 and here state only a few of its important features, which will be useful in the next chapter. Of particular importance are the algebraic equations of state, which are very useful in obtaining closed-form analytical solutions in engineering analysis.

One of the important features of a perfect gas is that its internal energy depends only upon its temperature. In Chap. 8 we will show that this must be the case for any gas obeying Eq. (4·16). This feature is also shown nicely in Figs. B·1 and B·4, where the isotherms are seen to coincide with lines of constant u in the vapor range at pressures well below the critical pressure. This type of inspection of the graphical equations of state for different substances provides an easy way to determine if the perfect-gas approximation will be valid in the range of interest.

† The molal mass is the mass of one mole of material. One kgmole is an amount of matter containing the same number of atoms as 12 kg of carbon 12. Thus, for carbon 12 $\hat{M}$ = 12 kg/kgmole = 12 g/gmole = 12 lb/lbmole.

‡ See also Table A·4.

Since the Pv product for a perfect gas also depends only upon temperature, the enthalpy $h = u + Pv$ is also a function only of temperature. Figures B·2, B·7, and B·8 likewise show that the isotherms correspond with lines of constant h for gases at low pressures. The region of perfect-gas behavior can be found by this type of examination. For example, in Fig. B·6 we see that the h and T lines for CO_2 do *not* coincide, indicating that it would not be reasonable to treat CO_2 as a perfect gas in the region covered. However, at higher temperatures CO_2 can be idealized as a perfect gas, provided that the pressure is not too great.

Since u depends only on T for a perfect gas, it follows from Eqs. (4·10) and (4·11) that

$$du = c_v(T)\, dT \tag{4·18}$$

Note that c_v depends only upon T for a perfect gas. Since h depends only on T for a perfect gas, it likewise follows from Eqs. (4·14) and (4·15) that

$$dh = c_P(T)\, dT \tag{4·19}$$

We also have

$$dh = du + d(Pv) = c_v\, dT + R\, dT$$

hence, *for a perfect gas,*

▶ $$c_P = c_v + R \tag{4·20}$$

We emphasize that in general u and h depend on the density as well as temperature; the perfect gas is a very special model.

Figure B·17 shows the values of c_P, c_v, and $k = c_P/c_v$ for several gases. Note the specific heats do not depend very much on pressure; the $P = 0$ limit represents perfect-gas behavior. Note that the low-pressure specific heats are slowly varying functions of temperature, increasing slightly with increasing temperature. At low temperatures the main contribution to the energy of a gas is provided by molecular translation. At higher temperatures there is more and more energy associated with molecular vibration and rotation, and this causes the slight increase in specific heats. Quantum-statistical thermodynamics can be used to evaluate c_p and c_v; cf. Reynolds, *Thermodynamics*, 2d ed., McGraw-Hill.

Since the specific heats are nearly constant over fairly wide ranges in temperature, it is often useful to treat them as constants, which allows us to integrate Eqs. (4·18) and (4·19) in closed form. Replacing c_v and c_P by their averages over the range of temperature in question, we integrate and find

$$u_1 - u_0 = c_v(T_1 - T_0) \tag{4·21}$$

$$h_1 - h_0 = c_P(T_1 - T_0) \tag{4·22}$$

These equations are very useful in calculating internal energy or enthalpy differences. It must be remembered that they hold only for a *perfect gas with constant specific heats;* it is the responsibility of the analyst to establish that this approximation is sufficiently good for the problem at hand. Appreciable errors can result if the gas involved departs significantly from this idealized behavior.

Table B·6 gives the values of R and nominal values for c_v, c_P, and k for several gases. If one needs to consider the variation of the specific heats with temperature, then the integrations must be performed numerically, using the measured functions $c_v(T)$ and $c_P(T)$. This has been done for several gases. Table B·9 gives the values for $h(T)$ and $u(T)$ for air, treated as a perfect gas with variable specific heats. An arbitrary temperature datum point was used for the lower limit of the u integration. Table B·13 gives additional data for the enthalpy *per mole* for several gases. The arbitrary datum for this tabulation is different than that for the other graphs and tables in Appendix B; hence one should use either this data or other data, *but not both,* in any particular analysis. The other entries in these tables will be explained later in the text.

Example. Calculate the density of air at $P = 10$ MPa, $T = 500$ K. From Table B·6, $R = 0.286$ kJ/(kg·K). Then,

$$\rho = \frac{1}{v} = \frac{P}{RT} = \frac{10 \times 10^6 \text{ N/m}^2}{[0.286 \times 10^3 \text{ J/(kg·K)}] \times 500 \text{ K}} = 69.9 \text{ kg/m}^3$$

Engineering problem. To illustrate the use of the perfect-gas equation of state in an engineering problem, suppose we consider the process of slow compression of air in a piston-cylinder system. Since the process is slow, there is adequate time for equalization of the gas and cylinder-wall temperatures through energy transfer as heat. We therefore assume that the process is one of constant temperature and wish to calculate the amount of energy transfer as heat from the gas to the cylinder walls. The system diagram and process representation are shown in Fig. 4·14.

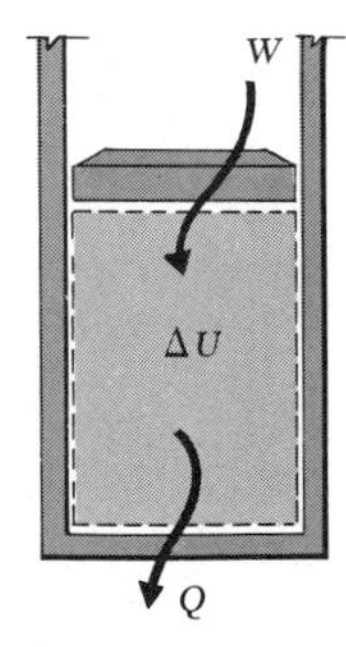

(a) *The control mass*

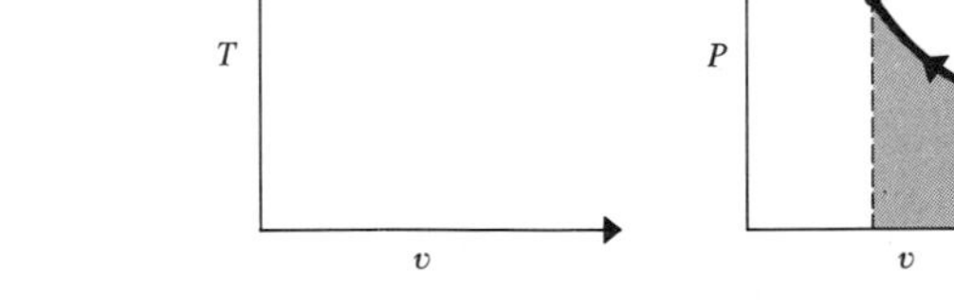

(b) *The process representation*

Figure 4·14.

With some experience in thermodynamic analysis, the student will learn that this type of problem can be handled with reasonable accuracy by treating the air as a perfect gas with constant specific heats. The critical temperature for air is about 132 K, and the critical pressure is about 37 atm. Thus for room temperatures (300 K) and above the ratio $T^* = T/T_{crit}$ exceeds 2.2, and for pressures up to 10 atm the ratio $P^* = P/P_{crit}$ is less than 0.27. Figure 8·9 then indicates that the compressibility factor $Z \equiv Pv/RT$ is very close to unity in this range, and the perfect-gas model is therefore quite adequate for many purposes. Figure B·8 shows that the enthalpy and temperature lines nearly coincide over this range, which again suggests that the perfect-gas model might be adequate. Figure B·17 indicates that the specific heat of air will not vary much over a wide range of temperatures, so let's carry out the analysis treating air as a perfect gas with constant specific heats.

An energy balance on the control mass gives

$$\underset{\text{energy input}}{W} = \underset{\text{energy output}}{Q} + \underset{\text{increase in energy storage}}{\Delta U}$$

where W and Q are the energy transfer as work and heat as defined in Fig. 4·14*a*. Since the initial and final temperatures are identical, the initial and final internal energies of the gas (treated as a perfect gas) are identical; hence $\Delta U = 0$ (*we again remark that this is a feature peculiar to the perfect gas*). We can use the energy balance to calculate Q if W can be evaluated. Assuming that the P-v-T relationship during the process is the same as that for equilibrium states (this will be a good approximation if the process is slow such that at any instant the gas is nearly in thermodynamic equilibrium), the work done on the gas for the isothermal process is

$$W = \int_1^2 -P\,dV = -M\int_1^2 P\,dv = -MRT\int_1^2 \frac{1}{v}\,dv$$

$$= MRT \ln \frac{v_1}{v_2} = MRT \ln \frac{V_1}{V_2}$$

So, if we have 4 lbm of air for which $R = 0.0685$ Btu/(lbm·R) (Table B·6), $T = 480$ R (20°F), and $V_1/V_2 = 2$,

$$W = 4 \times 0.0685 \times 480 \times \ln 2 = 91 \text{ Btu}$$

Then, from the energy balance, since $\Delta u = 0$,

$$Q = 0 + W = 91 \text{ Btu}$$

This example forms a good point for the student to focus sharply on the differences between internal energy, heat, and temperature.

Perfect-gas test. As a second example, let's solve the problem of Sec. 4·6 using the perfect-gas approximation. The energy balance is the same (Fig. 4·11), namely,

$$Q = M(u_2 - u_1)$$

This time we treat the H_2O vapor as a perfect gas with constant specific heats. Since the pressure is low compared to the critical pressure, we might expect the perfect-gas model to be valid. However, the large temperature changes should make us cautious about the constant specific heat assumption. From Table B·6, $c_v = 0.336$ Btu/(lbm·R). Then, using Eq. (4·21),

$$u_2 - u_1 = c_v(T_2 - T_1) = 0.336 \times (1660 - 660) = 336 \text{ Btu/lbm}$$

and

$$Q = 5 \times 336 = 1680 \text{ Btu}$$

Note that this result is in error by approximately 11 percent in comparison to the exact value previously determined as 1900 Btu. Now, the final pressure is calculated from Eq. (4·16), which gives

$$\frac{P_2}{P_1} = \frac{T_2}{T_1}$$

Hence

$$P_2 = \tfrac{1660}{660} \times 2 = 5.03 \text{ psia}$$

The exact analyses gave a pressure just over 5 psia, so the approximate analysis gives a good final pressure.

If we did not have access to the exact property tabulations for water vapor, then we might estimate the specific volume using Eq. (4·16). For H_2O, $R = 85.78$ ft·lbf/(lbm·R) (Table B·6). Then, at state 1,

$$v_1 = \frac{RT_1}{P_1} = 85.78 \times \frac{(200 + 460)}{2 \times 144} = 196 \text{ ft}^3\text{/lbm}$$

which is exactly the tabulated value.

We can check the applicability of the perfect-gas approximation by computing the value of $Z \equiv Pv/RT$ at states 1 and 2. Using the state data collected in Sec. 4·6,

$$Z_1 = \frac{P_1 v_1}{RT_1} = \frac{2 \times 144 \times 196}{85.78 \times (200 + 460)} = 1.00$$

$$Z_2 = \frac{P_2 v_2}{RT_2} = \frac{5 \times 144 \times 196}{85.78 \times (1200 + 460)} = 0.993$$

Note that both values are very close to unity. The main error in the simplified analysis is not the assumption that $P\dot{v} = RT$, but the idealization that c_v is constant over the wide range of temperatures involved, which is responsible for the significant error in Q.

4·9 THE SIMPLE MAGNETIC SUBSTANCE

While magnetic solids are indeed compressible, the work associated with magnetization usually is much more important than the work of compression, and it is often reasonable to idealize that such a substance is a *simple magnetic substance*. The temperature and magnetization (magnetic-dipole moment) may be taken as an independent set of intensive thermodynamic properties, and the internal energy may be expressed as

$$u = u(T, \mathbf{M})$$

A *specific heat at constant magnetization* can be defined as

$$c_{\mathrm{M}} \equiv \left(\frac{\partial u}{\partial T}\right)_{\mathrm{M}} \tag{4·23}$$

Using the manipulations of calculus, we can develop the analog of Eq. (4·9),

$$\left(\frac{\partial \mathbf{H}}{\partial T}\right)_{\mathrm{M}} = -\frac{(\partial \mathbf{M}/\partial T)_{\mathrm{H}}}{(\partial \mathbf{M}/\partial \mathbf{H})_T} \tag{4·24}$$

A graphical equation of state for one common paramagnetic salt is shown in Fig. B·12.

SELECTED READING

Jones, J. B., and G. A. Hawkins, *Engineering Thermodynamics*, chaps. 3 and 5, John Wiley & Sons, Inc., New York, 1960.

Keenan, J. H., F. G. Keys, P. G. Hill, and J. G. Moore, *Steam Tables*, appendix, John Wiley & Sons, Inc., New York, 1969.

Lee, J. F., and F. W. Sears, *Thermodynamics*, chap. 2, Addison-Wesley Publishing Co., Inc., Reading, Mass., 1963.

Van Wylen, G. J., and R. E. Sonntag, *Fundamentals of Classical Thermodynamics*, 2d ed., chap. 3, secs. 5.11 and 5.13, John Wiley & Sons, Inc., New York, 1973.

Zemansky, M. W., M. M. Abbott, and H. C. Van Ness, *Basic Engineering Thermodynamics*, 2d ed., chap. 5, secs. 11.1–11.10, McGraw-Hill Book Company, New York, 1975.

QUESTIONS

4·1 What is state? What is a property? How many independent intensive thermodynamic properties does a simple substance have?
4·2 What is a phase? What is the vapor dome? What is a superheated vapor? What is a supercritical state? What is a saturated liquid?
4·3 What happens when a saturated liquid is heated at constant pressure? What happens when it is cooled at constant pressure?
4·4 What happens when a saturated liquid is heated at constant volume?
4·5 What do you think happens when a saturated vapor is compressed adiabatically? What happens when it is expanded adiabatically?
4·6 What are the definitions of h, c_v, and c_P?
4·7 Lead blocks sink in liquid lead; does lead expand or contract upon melting?
4·8 Why do the liquid and vapor in a mixture of the same substance have the same temperature and pressure? Would two gases in a mixture each exert the same pressure?
4·9 What is your estimate of the pressure of the Freon 12 (liquid and vapor) in the pipes of a refrigerator that has been idle for several days (see Fig. B·7)?
4·10 If ice is thrown into a hot pressure cooker and the mixture is allowed to cool on the stove, will it end up at the triple point?
4·11 Give a molecular explanation of evaporation and one of sublimation.
4·12 What thermodynamic-property data can you find in your chemistry or engineering handbook?
4·13 Why are saturation states simpler to tabulate than superheated-vapor or subcooled-liquid states?
4·14 Why do we arbitrarily select the energy of a substance to be zero at some point? How must this point be described (pressure, temperature, or both)?
4·15 Why is it not true that $\rho = (1 - x)\rho_f + x\rho_g$, where ρ is the mass density ($\rho = 1/v$)?
4·16 How can we tell if a gas behaves like a perfect gas?

PROBLEMS

4·1 What are the enthalpy and internal energy of a mixture of mercury at 0.40 quality at 1000°F? What are the volume fractions of the liquid and vapor in the mixture (see Table B·3)?
4·2 What are the specific volume, enthalpy, and internal energy of a mixture of liquid-vapor H_2O of 0.50 quality at (*a*) 300°C (*b*) 1×10^6 N/m^2 (see Table B·1)?
4·3 Calculate the values of ρ, u, and h for a liquid-vapor H_2O mixture of 0.2 quality at 0.2 MPa (Table B·1*b*).
4·4 Calculate the values of v, u, and h for a liquid-vapor H_2O mixture of 0.2 quality at 180°C (Table B·1*a*).
4·5 Calculate the value of u_{fg} for CO_2 at 100 psia using the data of Fig. B·6.
4·6 Look up the enthalpy of saturated methyl chloride vapor at 100°F in the *Handbook of Chemistry and Physics*.
4·7 Liquid oxygen in a rocket-propellant tank is at a pressure somewhere near 1 atm. From Fig. B·5 estimate the temperature of the lox-vapor mixture.
4·8 Compare the volume changes undergone by 1 lbm of water upon evaporation at 1 atm pressure, at 500°F, and at the critical point.

4·9 In order to illustrate the nature of the critical point, one can place CO_2 in a quartz vial and seal the device. Assuming that $v_{crit} = 0.04$ ft^3/lbm, and that the device is filled at room temperature (60°F), and using Fig. B·6, find: (*a*) the pressure in the vial at room temperature, (*b*) the proper percent liquid mass to ensure passage through the critical point upon heating, (*c*) the percent liquid *volume* associated with part (*b*). (Calculate v_f from x, v_{mix}, and v_g.)

4·10 Using the data of Appendix B, estimate β, κ, h, c_v, and c_P for H_2O at 1000°F and 100 psia.

4·11 Using the data of Table B·7, estimate the density of copper, in lbm/in^3, at 100°F and 5 atm pressure.

4·12 Taking the internal energy of copper to be zero at 50 K and 1 atm pressure, prepare a curve showing the internal energy as a function of temperature at 1 atm, using the data of Table B·7.

4·13 Calculate the approximate value of c_v for the liquid studied in Prob. 3·6.

4·14 Calculate the approximate value for c_v for the gas studied in Prob. 3·7.

4·15 Calculate the value of c_P for the liquid material in Prob. 3·8.

4·16 Calculate the approximate value of c_P for the gas studied in Prob. 3·9.

4·17 Calculate the value of c_v for the gas studied in Prob. 3·10 and plot as a function of temperature at that specific volume.

4·18 Using Fig. B·12 as a guide, sketch a three-dimensional **M**-**H**-T surface for a paramagnetic salt.

4·19 Sketch u-T and v-T diagrams for H_2O, showing lines of constant pressure. Include the liquid, vapor, and mixed-phase regions.

4·20 Verify that Eq. (4·9) holds for a perfect gas by differentiation and substitution.

4·21 Estimate the temperature to which water at the bottom of a 500-ft-deep lake would have to be heated before it would begin to boil.

4·22 Plot the ratio Pv/T versus T for H_2O along a line of 1 atm pressure and show the conditions under which the perfect-gas approximation is reasonable. Taking $\hat{M}$ to be 18.016 kg/kgmole, evaluate the universal gas constant and compare with the accepted value.

4·23 A Curie substance is any magnetic substance obeying the simple equation of state $\mathbf{M} = C\mathbf{H}/T$, where C is the Curie constant. Plot **M** versus **H** for iron-ammonium alum (Fig. B·12) along a line of $T = 1$ K, find the region in which Curie's law is obeyed, and determine the value of the Curie constant.

4·24 What are the values of the specific heat at constant volume and the specific heat at constant pressure of H_2O of 0.30 quality at 100 psia?

4·25 Determine the value of c_{M} for iron-ammonium alum (Fig. B·12) at 1 K in a weak external field. Express the result in Btu/(lbm·R) and compare with the corresponding specific heat of saturated H_2O vapor at 1 atm pressure.

4·26 What are the values of $(\partial v/\partial T)_P$, $(\partial P/\partial v)_T$, $(\partial v/\partial P)_T$, and $(\partial T/\partial P)_v$ for a perfect gas?

4·27 For a Curie substance **M**, **H**, and T are related by $M = C\mathbf{H}/T$, where C is a constant. Evaluate $(\partial \mathbf{H}/\partial T)_{\mathbf{M}}$, $(\partial \mathbf{M}/\partial \mathbf{H})_T$, $(\partial T/\partial \mathbf{H})_{\mathbf{M}}$, and $(\partial T/\partial \mathbf{M})_{\mathbf{H}}$. Express the meaning of each term in words.

4·28 Consider a dielectric substance for which $\mathbf{P} = (\kappa - 1)\varepsilon_0 \mathbf{E}$, where κ is the dielectric constant and ε_0 is a physical constant (Table A·2). κ normally depends upon temperature. Assuming $\kappa = \kappa_0[1 + a(T - T_0)]$, evaluate $(\partial \mathbf{P}/\partial T)_{\mathbf{E}}$, $(\partial \mathbf{P}/\partial \mathbf{E})_T$, $(\partial \mathbf{E}/\partial T)_{\mathbf{P}}$, and $(\partial \mathbf{E}/\partial \mathbf{P})_T$. Express the meaning of each in words.

4·29 Compute the density of gaseous H_2O at each of the following states, using the perfect-gas model, and compare with the actual values given in Table B·2.
1 psia, 200°F   1 atm, 800°F   3000 psia, 800°F

4·30 Compute the density of gaseous CO_2 at each of the following states, using the perfect-gas model, and compare with the actual values given in Fig. B·6.
10 psia, 20°F 150 psia, 60°F 1500 psia, 140°F

4·31 Compute the specific volume of vapor water at each of the following states, using the perfect-gas model, and compare to the actual values from the SI version of Table B·2.
0.01×10^6 N/m², 100°C 0.1×10^6 N/m², 300°C 2×10^6 N/m², 400°C

4·32 Calculate the isothermal compressibility for water vapor at a pressure of 0.1×10^6 N/m² at 200°C, using data in the SI version of Table B·2.

4·33 An experiment on nitrogen results in the tabulated information. Using the perfect-gas model, calculate c_v and c_P and compare to the values given in Table B·6.

v, m³/Mg	u, kJ/kg	T, K
125	230	110
125	272	167

CHAPTER FIVE

ENERGY ANALYSIS

5·1 GENERAL METHODOLOGY

The principle of conservation of energy and the equations of state permit solution of a number of interesting and important technical problems. We shall begin to apply these ideas at this point in order to indicate the manner in which thermodynamics is used in engineering.

Energy analysis is essentially an accounting procedure, in which we take account of energy transfers to and from a system and of changes in energy inside the system. There are two main types of accounting procedures. In *control-mass analysis* we write the conservation-of-energy equation for a specified piece of matter, while in *control-volume analysis* we work instead with specified regions in space.† The latter being somewhat more difficult, we shall begin by illustrating energy analysis for a control mass.

An accounting procedure must be carried out over a set accounting period, and an essential step in control-mass analysis is specification of the *time base*. This might be a given period of time, or the time required for something to happen, or we might specify that our accounting be done on an *instantaneous-rate* basis. Some specification is necessary.

† The terms "system," or "closed system," and "open system" are used by many texts instead of "control mass" and "control volume."

The accounting will be carried out with symbols used to represent energy transfers to and from the control mass, and some sort of sign convention must always be set up for the energy flows. The important thing in energy analysis is to recognize all the energy transfers and changes that take place and to relate these in a proper mathematical manner. The sign convention chosen for W and Q is not particularly important, provided that both the person doing the analysis and the person reading it understand what it is. We shall adopt the policy of selecting whatever convention seems appropriate for the analysis at hand; we will *always* indicate the directions of *positive* energy flows by arrows on the system sketch. This scheme has the additional advantage that the energy flows to be considered are all identified, and where and how the energy crosses the boundary is quite clear. There should always be a one-to-one correspondence between the energy flows shown on the system sketch and those that appear in the energy balance.

In working with energy changes we shall follow the conventions of calculus, in which dx and Δx always represent *increases* in the value of x. ΔE will always represent an *increase* in the energy stored within the control mass.

Having made the energy balance, the next step is to bring in enough other information to permit reduction of the problem to one equation in one unknown. This information might be in the form of equations of state, information about the nature of the process, or other information obtained by applying the principle of conservation of mass, Newton's law, or other fundamental principles to the system.

In any analysis various approximations, or *idealizations*, must be made to reduce the problem to a manageable size. These idealizations must be clearly understood by both the analyst and the reader of the analysis, hence we prefer to list these specifically at the start of the analysis. Sometimes idealizations are implicitly indicated on the system sketch; for example, the absence of an energy transfer term Q would imply that we idealize that energy transfer as heat is negligible during the time period over which the energy balance will be made. Such idealizations should be specifically listed to ensure their communication to the reader.

We cannot emphasize enough the importance of a good system sketch, complete with all the relevant energy-transfer terms. A good sketch can be of great value to the analyst in thinking straight about the process, in being consistent throughout the analysis, and also in seeing what steps must be taken to complete the analysis.

An equally important working diagram is the *process representation*. This is one or more diagrams showing what happens to matter within the system on suitable thermodynamic planes. In most analyses one must find some way to fix the states of the matter at the start or finish of the process, and the process representation is of great value in helping the analyst work through a complex problem. Hence one should always try to draw the process representation *before* carrying out the analysis in order to orient one's thinking properly and efficiently.

In summary, the general methodology for energy-balance analysis is as follows:

1. Define the system carefully and completely, indicating its boundaries on a sketch (control mass or control volume?).
2. List the relevant idealizations.
3. Indicate the flows of energy to be considered in the energy balance and set up their sign convention on the system sketch.
4. Indicate the time basis for the energy balance.
5. Sketch the process representation.
6. Write the energy balance in terms of the symbols shown on the system sketch; there should be a one-to-one correspondence between the terms in the equation and those on the sketch.
7. Bring in equations of state or other information as necessary to allow solution of the problem.

We shall now illustrate this methodology for the control mass by example. The analyses which follow were selected to illustrate different aspects of first-law analysis from the control-mass viewpoint. Different kinds of idealizations are introduced in the various developments, the use of graphical and tabular equations of state are demonstrated, and some important definitions and general consequences are given. Every one of the examples should be studied carefully, even though they may appear to be extremely simple and straightforward (as indeed they are, if a systematic methodology is followed).

5·2 EXAMPLES OF CONTROL-MASS ENERGY ANALYSIS

Evaporation at constant pressure. Three kg of H_2O in a piston-cylinder system is initially in the saturated-liquid state at 0.6 MPa. Energy is added slowly to the water as heat, and the piston moves in such a way that the pressure remains constant. How much work is done by the water, and how much energy must be transferred as heat in order to bring the water to the saturated-vapor state? See Fig. 5·1.

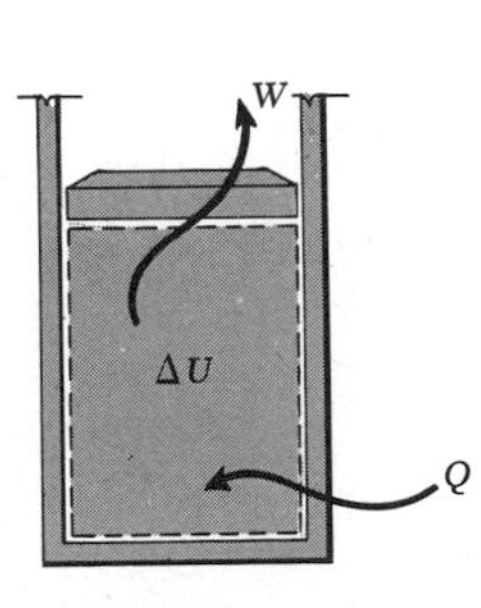

(a) *The control mass*

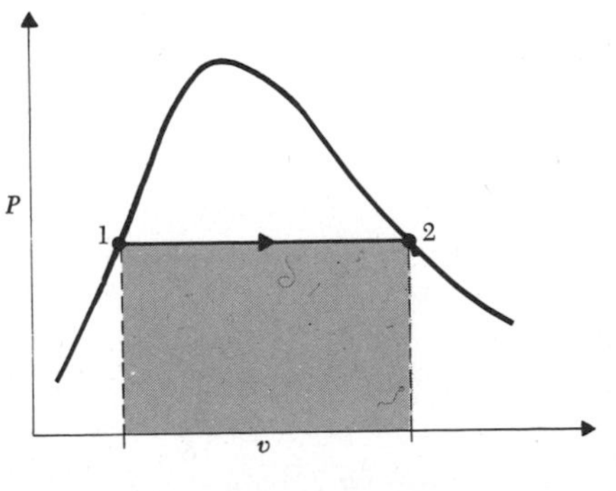

(b) *The process representation*

Figure 5·1.

We could select the water or the water plus the piston as the control mass. Since we are interested in energy transfers to the water, and the piston is merely a means for maintaining the constant pressure, we take only the water as our control mass. The change in state is depicted by the process representation, which is nicely shown on a P-v plane.

In order to solve the problem we idealize that the changes in the gravitational potential energy of the water are negligible in comparison with the changes in the internal energy. We idealize further that the presence of the gravitational field does not significantly alter the behavior of the water molecules, so that the relationships between the thermodynamic properties are the same as if the water were truly a simple compressible substance. In short, we idealize that the water behaves like a simple compressible substance and is in equilibrium states at the start and end of the process.

When we employ these idealizations, the energy balance, made over the period during which the process occurs, is

$$\underset{\substack{\text{energy}\\\text{input}}}{Q} = \underset{\substack{\text{energy}\\\text{output}}}{W} + \underset{\substack{\text{increase in}\\\text{energy storage}}}{\Delta U}$$

where

$$\Delta U = M(u_2 - u_1)$$

Fixing the initial and final states from the SI version of Table B·1*b*, at the initial state (saturated liquid at 0.6 MPa)

$$P_1 = 6 \times 10^5 \text{ N/m}^2 \qquad T_1 = 432 \text{ K } (158.9°\text{C})$$

$$u_1 = 669.9 \text{ kJ/kg} \qquad v_1 = 0.001101 \text{ m}^3/\text{kg}$$

and at the final state (saturated vapor at 0.6 MPa)

$$P_2 = 6 \times 10^5 \text{ N/m}^2 \qquad T_2 = 432 \text{ K } (158.9°\text{C})$$

$$u_2 = 2567.4 \text{ kJ/kg} \qquad v_2 = 0.3157 \text{ m}^3/\text{kg}$$

Note that the volume occupied by the water increases tremendously.

We could now compute the net energy added to the water, but as yet we cannot tell how much is added as heat and how much is taken away as work. The fact that the pressure remains constant during the process allows us to compute the work very simply. The work done by the water is

$$W = \int_1^2 đW = \int_1^2 P\,dV = M \int_1^2 P\,dv$$

Since the pressure is constant,

$$W = MP(v_2 - v_1)$$

The work done is therefore

$$W = 3 \text{ kg} \times 6 \times 10^5 \text{ N/m}^2 \times (0.3157 - 0.0011) \text{ m}^3/\text{kg} = 566{,}300 \text{ J}$$
$$= 566.3 \text{ kJ}$$

We can finally calculate the energy added as heat:

$$Q = 566.3 + 3 \text{ kg} \times (2567.4 - 669.9) \text{ kJ/kg} = 6259 \text{ kJ}$$

The energy transfer as heat could have been calculated directly if we had noticed that it is expressible as

$$Q = M[P(v_2 - v_1) + (u_2 - u_1)] = M(h_2 - h_1)$$

Note that the energy transfer as heat (per unit of mass) to a simple compressible substance during a constant pressure process is equal to the increase in its enthalpy.

Values for h_1 and h_2 could have been obtained from Table B·1*b*. Thus the energy transfer as heat required to evaporate a unit of mass of a simple compressible substance at *constant pressure* is simply $h_g - h_f = h_{fg}$ and is sometimes called the *enthalpy of evaporation*† of that substance. Note that it depends on pressure and vanishes at the critical point.

A Dry-Ice cooler. One lbm of Dry Ice (CO_2) at 1 atm pressure is placed on top of a piece of meat in a cooler. The Dry Ice sublimes at constant pressure as a result of energy transfer as heat from the warmer meat. What is the temperature of the CO_2, and how much energy is transferred as heat from the meat? See Fig. 5·2.

† The term "latent heat of vaporization" is also used.

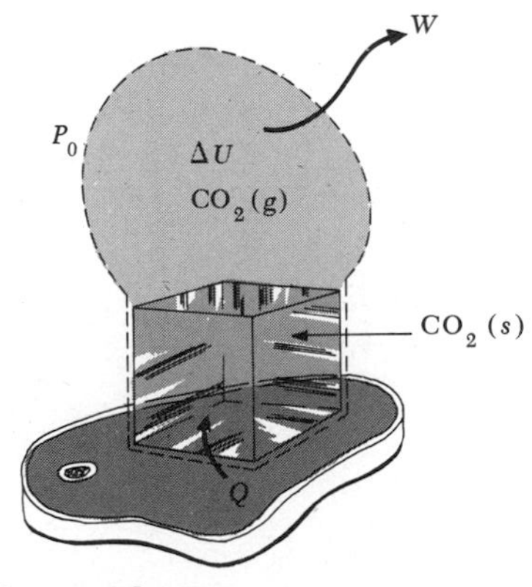

(*a*) *The control mass*

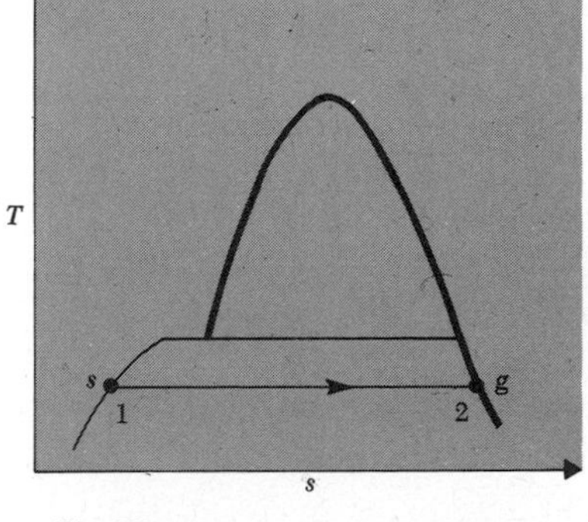

(*b*) *The process representation*

Figure 5·2.

For the control mass we pick the CO_2, including the solid and vapor. The boundary of this control mass moves as the solid sublimes, and so there will be energy transfer as work, in this case from the expanding CO_2 to the environment. We assume no energy transfer as heat from the CO_2 to the surrounding air and neglect mixing of the CO_2 with the air. The energy balance is then

$$\underset{\substack{\text{energy}\\\text{input}}}{Q} = \underset{\substack{\text{energy}\\\text{output}}}{W} + \underset{\substack{\text{increase in}\\\text{energy storage}}}{\Delta U}$$

where

$$\Delta U = M(u_2 - u_1)$$

As in the previous example, the work done by the expanding CO_2 is related to the pressure (constant by assumption) and the volume change,

$$W = MP(v_2 - v_1)$$

Solving for the energy transfer as heat to the CO_2 and expressing it in terms of the change in the enthalpy property,

$$\begin{aligned} Q &= M[(u_2 - u_1) + P(v_2 - v_1)] \\ &= M[(u_2 + P_2 v_2) - (u_1 + P_1 v_1)] = M(h_2 - h_1) \end{aligned}$$

The equation of state for CO_2 is given on a *temperature-entropy plane* in Fig. B·6. The entropy is an important thermodynamic property which we shall discuss in the next chapters. For the time being we can use the equation of state as given in Fig. B·6 without worrying about this property. The initial state is saturated solid at 1 atm, and from Fig. B·6 we read the enthalpy of this state as 31 Btu/lbm. The final state, 2, is saturated vapor at 1 atm, for which we read the enthalpy as 276 Btu/lbm. The positions of these states on the temperature-entropy diagram are indicated on the process representation in Fig. 5·2*b*.

From Fig. B·6 we also read the saturation temperature corresponding to 15 psia as $-108°F$. The energy transferred as heat from the meat to the CO_2 is then

$$Q = 1 \text{ lbm} \times (276 - 31) \text{ Btu/lbm} = 245 \text{ Btu}$$

The difference h_{sg} is called the *enthalpy of sublimation*† and represents the energy which must be added as heat to completely sublime a unit of mass of a substance at constant pressure.

Thermal magnetization. One cm^3 of the paramagnetic substance iron-ammonium alum is magnetized in a constant external field H of 5000 G (gauss) by

† Sublimation is the transformation of a solid directly into a gas at constant pressure. The term "latent heat of sublimation" is sometimes used for $h_{sg} = h_g - h_s$.

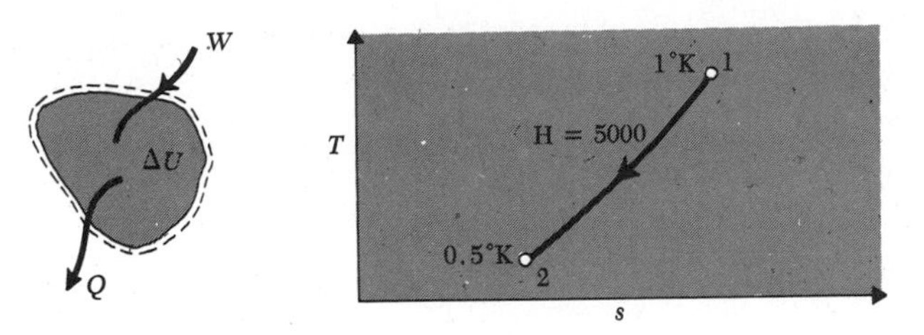

(*a*) *The control mass* (*b*) *The process representation*

Figure 5·3.

cooling. The initial temperature is 1 K and the final temperature is 0.5 K; how much energy must be transferred as heat from the alum? See Fig. 5·3.

This problem is complicated by the fact that the thermodynamic-property data (Fig. B·12) are given in absolute magnetostatic units, while the expressions for work of magnetization given in Chap. 2 are in the rationalized mksc system. The required unit conversions will be good practice.

The energy terms considered to be important are shown on the control-mass diagram; the energy balance, made over the time for the process to occur, is

$$\underset{\text{energy input}}{W} = \underset{\text{energy output}}{Q} + \underset{\text{increase in energy storage}}{\Delta U}$$

where

$$\Delta U = M(u_2 - u_1)$$

From the graphical equation of state we obtain the initial and final internal energies (assuming that the initial and final states are states of thermodynamic equilibrium),

$$u_1 = 1.56 \times 10^4 \text{ ergs/g} \qquad u_2 = 1.39 \times 10^4 \text{ ergs/g}$$

If the alum is idealized as a simple magnetic substance, the work done on the alum by the external field is [see Eq. (2·17)]

$$W = \int_1^2 đW = \int_1^2 V\mu_0 \mathsf{H}\, d\mathsf{M}$$

The work calculation is made particularly simple by the fact that the applied field is held constant. For constant H the work becomes

$$W = \mu_0 V \mathsf{H}(\mathsf{M}_2 - \mathsf{M}_1)$$

From Fig. B·12 we read the initial and final magnetizations as

$$\mathsf{M}_1 = 62 \text{ G} \qquad \mathsf{M}_2 = 84 \text{ G}$$

We choose to convert everything to the rationalized mksc unit system. From Appendix A we find the dimensional equivalents,

$$1 \text{ G of } \mathsf{H} = 79.6 \text{ C/(s·m) of } \mathsf{H}$$

$$1 \text{ G of } \mathsf{M} = 1000 \text{ C/(s·m) of } \mathsf{M}$$

Then, in the rationalized mksc system we have

$$\mathsf{H} = 40 \times 10^4 \text{ C/(s·m)}$$

$$\mathsf{M}_1 = 62 \times 10^3 \text{ C/(s·m)}$$

$$\mathsf{M}_2 = 84 \times 10^3 \text{ C/(s·m)}$$

$$\mu_0 = 1.256 \times 10^{-6} \text{ kg·m/C}^2$$

The energy transfer as work to the alum is therefore

$$\begin{aligned} W &= (0.01 \text{ m})^3 \times 1.256 \times 10^{-6} \text{ kg·m/C}^2 \times 40 \times 10^4 \text{ C/(s·m)} \\ &\quad \times (84 - 62) \times 10^3 \text{ C/(s·m)} \\ &= 0.0111 \text{ kg·m}^2\text{/s}^2 = 0.0111 \text{ J} \end{aligned}$$

The density of the alum is (Fig. B·12) 1.71 g/cm³. The internal-energy increase is therefore

$$\begin{aligned} M(u_2 - u_1) &= 1.71 \text{ g} \times (1.39 \times 10^4 - 1.56 \times 10^4)(\text{g·cm}^2\text{/s})\text{/g} \\ &= -0.3 \times 10^4 \text{ erg} = -0.0003 \text{ J} \end{aligned}$$

So the energy which must be transferred from the alum as heat is

$$Q = 0.0111 - (-0.0003) \text{ J} = 0.0114 \text{ J}$$

This energy is transferred to some environment at a temperature *lower* than 0.5 K.

Thermal mixing at constant pressure. Two lbm of saturated liquid mercury at 1 psia is mixed with 4 lbm of 1400°F mercury vapor at 1 psia. The mixing vessel is such that the pressure remains constant during this process, and no energy transfer as heat occurs between the vessel and the mercury. Determine the equilibrium state reached by this mixture.

In the previous examples the control mass was in a state of equilibrium at the start of the process, so the calculation of the initial energy was quite straightforward. In this example we shall see how to make the corresponding evaluation for a system that is not in equilibrium.

Since the amount of liquid might change during the process, we cannot take only the liquid or only the vapor as the control mass. Instead we take the

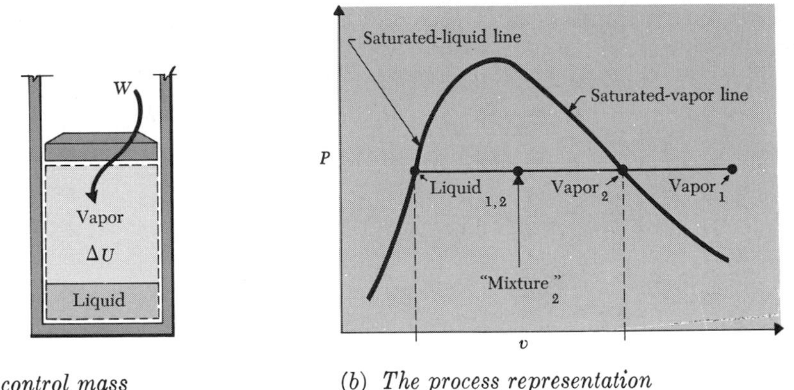

(*a*) *The control mass* (*b*) *The process representation*

Figure 5·4.

entire 6 lbm of mercury. By assumption, no energy transfer as heat occurs, but we do expect the volume to change, resulting in an energy transfer as work. The only energy stored within the control mass is the internal energy of the mercury; the energy balance, made over the time for the process to take place, is therefore (Fig. 5·4)

$$\underset{\text{energy input}}{W} = \underset{\text{increase in energy storage}}{\Delta U}$$

where

$$\Delta U = U_2 - U_1$$

The work calculation is again made easy by the fact that the pressure is constant. When the piston moves in an amount dx, the energy transfer as work from the environment to the control mass is

$$đW = PA\,dx = -P\,dV$$

Integrating,

$$W = \int_1^2 -P\,dV = P(V_1 - V_2)$$

Combining with the energy balance,

$$U_2 + PV_2 = U_1 + PV_1 \qquad (5\cdot1)$$

To evaluate the initial terms we assume that the liquid is in an equilibrium state and the vapor is in an equilibrium state, even though they are not in equilibrium with one another. The graphical and tabular equations of state, Fig. B·10 and Table B·3, may then be employed for each phase. Since the available equation-of-state information is in terms of the enthalpy property, we express the right-hand side of Eq. (5·1) as

$$U_1 + PV_1 = M_{l_1} u_{l_1} + M_{v_1} u_{v_1} + P(M_{l_1} v_{l_1} + M_{v_1} v_{v_1})$$
$$= M_{l_1} h_{l_1} + M_{v_1} h_{v_1}$$

Now, from the tables, the initial liquid enthalpy is (saturated liquid at 1 psia, Table B·3)

$$h_{l_1} = 13.96 \text{ Btu/lbm}$$
$$T_1 = 457.7°\text{F}$$

The initial vapor enthalpy is found from Fig. B·10 as

$$h_{v_1} = 164 \text{ Btu/lbm}$$

Substituting the numbers,

$$U_1 + PV_1 = 2 \times 13.96 + 4 \times 164 = 684 \text{ Btu}$$

The final state is a state of equilibrium, for which

$$U_2 + PV_2 = M(u + Pv)_2 = Mh_2$$

The enthalpy in the final state is therefore

$$h_2 = \frac{684 \text{ Btu}}{6 \text{ lbm}} = 114 \text{ Btu/lbm}$$

The final pressure and enthalpy may be used to fix the final state. Upon inspection of Fig. B·10 we see that the final state is a mixture of saturated liquid and vapor at 1 psia and that the "moisture" $(1 - x)$ is about 21 percent (0.79 quality). Alternatively, we could use the information in Table B·3:

$$114 = (1 - x_2) \times 13.96 + x_2 \times 140.7$$
$$x_2 = 0.79$$

A pneumatic lift. Air is used in a pneumatic lift (Fig. 5·5). The air is initially contained in a 10 ft³ steel tank at 80°F and 100 psia. When lifting is required, the valve is opened and air is bled out into the cylinder. The cylinder is initially filled with air at 1 atm, 60°F, has a cross-sectional area of 1 ft², and a lifting height of 3 ft. Its initial volume is 0.5 ft³. When the pressure in the cylinder reaches 50 psia,

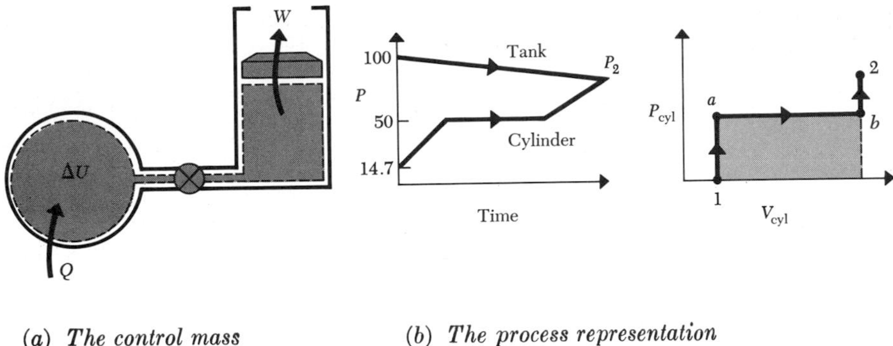

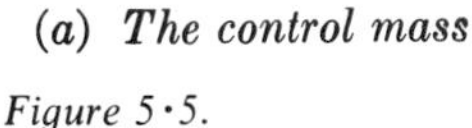

(a) The control mass *(b) The process representation*

Figure 5·5.

the load begins to move upward, maintaining this cylinder pressure until the piston has traveled 3 ft. The load then stops, but air continues to flow into the cylinder. Eventually the pressures in the tank and cylinder are equal, and through energy transfer as heat all the air attains a temperature of 60°F. Calculate the final air pressure and the amount of energy transfer as heat from the tank walls to the air.

In making the energy balance we consider the time period from the start to the end, where the air is in equilibrium throughout the system and the load is at its highest point. We take the air as the control mass and consider the energy flows indicated in Fig. 5·5. The energy balance is

$$\underset{\text{energy input}}{Q} = \underset{\text{energy output}}{W} + \underset{\text{increase in energy storage}}{\Delta U}$$

In order to calculate Q we must first evaluate ΔU and W. The work computation is facilitated by consideration of the process representation. Work will be done by the control mass on the piston only when the piston is moving, and during this period the cylinder pressure is assumed to remain at 50 psia. During this period the force exerted by the gas on the piston is $144 \times 50 = 7200$ lbf, and the piston moves a distance of 3 ft. Hence

$$W = 7200 \times 3 = 21{,}600 \text{ ft}\cdot\text{lbf}/(778 \text{ ft}\cdot\text{lbf/Btu})$$
$$= 27.8 \text{ Btu}$$

Idealizing that the air behaves as a perfect gas, we have enough information to establish the initial and final air states completely. From Table B·6, we obtain the following constants for air:

$$c_v = 0.172 \text{ Btu/(lbm}\cdot\text{R)} \qquad R = 53.3 \text{ ft}\cdot\text{lbf/(lbm}\cdot\text{R)}$$

Initial state

Tank air 80°F(= 540 R), 100 psia

Using Eq. (4·16),

$$\rho = \frac{1}{v} = \frac{P}{RT} = \frac{100 \times 144}{53.3 \times 540} = 0.500 \text{ lbm/ft}^3$$

Hence

$$M_t = 10 \times 0.500 = 5.00 \text{ lbm}$$

Using Eq. (4·21), with $T_0 = 0°\text{F} = 460$ R,

$$u_t = 0.172 \times (540 - 460) = 13.8 \text{ Btu/lbm}$$

Hence

$$U_t = Mu = 5.00 \times 13.8 = 69.0 \text{ Btu}$$

Cylinder air 60°F(= 520 R), 14.7 psia

$$\rho = \frac{P}{RT} = \frac{14.7 \times 144}{53.3 \times 520} = 0.0764 \text{ lbm/ft}^3$$

$$M_c = 0.5 \times 0.0764 = 0.038 \text{ lbm}$$

$$u_c = 0.172 \times (520 - 460) = 10.3 \text{ Btu/lbm}$$

$$U_c = 0.038 \times 10.3 = 0.39 \text{ Btu}$$

So, the total mass and energy in the initial configuration are

$$M = 5.00 + 0.038 = 5.04 \text{ lbm}$$

$$U_1 = 69.0 + 0.39 = 69.4 \text{ Btu}$$

Final state All air at 60°F (520 R)

The total final volume is

$$V = 10.00 + 0.5 + 3.0 = 13.5 \text{ ft}^3$$

Hence the final density is

$$\rho_2 = \frac{M}{V} = \frac{5.04}{13.5} = 0.373 \text{ lbm/ft}^3$$

The final pressure is, from Eq. (4·16),

$$P_2 = \rho_2 R T_2 = 0.373 \times 53.3 \times 520 = 10{,}300 \text{ lbf/ft}^2$$
$$= 71.8 \text{ psia}$$

The final specific internal energy is [Eq. (4·21)]

$$u_2 = 0.172 \times (520 - 460) = 10.3 \text{ Btu/lbm}$$

So
$$U_2 = Mu_2 = 5.04 \times 10.3 = 51.9 \text{ Btu}$$

We now calculate the internal energy increase as

$$\Delta U = U_2 - U_1 = 51.9 - 69.4 = -17.5 \text{ Btu}$$

(The minus sign indicates that the internal energy actually decreases.) Substituting for W and ΔU in the energy balance,

$$Q = 27.8 + (-17.5) = 10.3 \text{ Btu}$$

The student should particularly note the manner in which we used the process representations to orient our thinking and to guide the analysis.

A steady-flow system. The previous examples of control-mass analysis have dealt exclusively with nonflow systems, for which the control-mass method is ideal. Flow systems are of great interest in engineering, and the control-volume view is better suited to analysis of such systems. In a moment we are going to work out a proper energy equation for a control volume, and subsequently we shall consider all flow problems from the control-volume point of view. However, flow problems may be worked by the control-mass method, though this is somewhat awkward. In the interests of motivating the control-volume transformation and clarifying the physics involved, we shall now do a particular flow-system analysis by the control-mass method.

Consider a rather general "black-box" device, where fluid flows in one end and out the other. Energy is transferred as heat to the box, and power is transmitted into the device through a rotating shaft. We define the control mass to be the box and all its contents, and the fluid in the inlet and discharge pipes from section 1 through section 2 at time t. This control mass is indicated in Fig. 5·6.

The analysis will be made over an infinitesimal time interval dt. During this period the control mass will move, and its position at time $t + dt$ is indicated in Fig. 5·6. We shall assume that the inlet and discharge flows are *one-dimensional*, meaning that the velocities and thermodynamic properties are constant across the inlet and exit pipes. We also idealize this as a *steady-flow steady-state* situation,

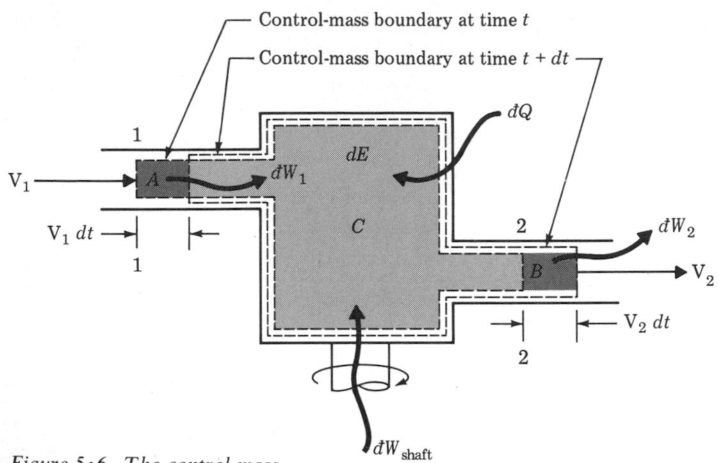

Figure 5·6. The control mass

meaning that the velocities and thermodynamic properties *at each point in space* are unchanging in time.†

Let's illustrate the approach to be used in the energy analysis in a simpler analysis of the mass conservation. Applying the conservation-of-mass principle to the control mass, we obtain

$$dM = 0$$

where M is the mass of the control mass. This mass can be represented in terms of the spatial distribution of matter, that is, in terms of the masses in regions A, B, and C of Fig. 5·6. Using the steady-flow steady-state condition,

$$M(t) = M_A + M_C$$

$$M(t + dt) = M_B + M_C$$

$$dM = M(t + dt) - M(t) = M_B - M_A$$

The volume of A and B are $(A\mathsf{V})_1\, dt$ and $(A\mathsf{V})_2\, dt$, where V_1 and V_2 are the velocities at sections 1 and 2. Hence the *mass balance* becomes

$$dM = (\rho A \mathsf{V})_2\, dt - (\rho A \mathsf{V})_1\, dt = 0$$

The term $(\rho A \mathsf{V})_1\, dt$ represents the amount of mass $đM$ which crosses section 1 in the time interval dt. The *rate of mass flow* $\dot{M}$ is defined as‡

$$\blacktriangleright \qquad \dot{M} \equiv \frac{đM}{dt} = (\rho A \mathsf{V}) \qquad (5 \cdot 2)$$

† The state of a piece of fluid passing through the device does change; and the states of the fluid at various positions within the device may be different.

‡ Note that $\dot{M} \neq dM/dt$. See the Nomenclature at the back of the book.

Note that V is the velocity *normal* to the flow area A. The mass balance can now be interpreted as applying to the *space* C (the *control volume*),

$$\underset{\substack{\text{mass-}\\\text{inflow}\\\text{rate}}}{\dot{M}_1} = \underset{\substack{\text{mass-}\\\text{outflow}\\\text{rate}}}{\dot{M}_2}$$

This merely states that the rate of mass inflow *to the control volume* must equal the rate of mass outflow *from the control volume* under steady-flow steady-state conditions, which is obviously the proper control-volume mass balance. However, the proper energy equation is not as obvious, and we must derive it carefully.

An energy balance on the control mass, made over the time period dt, yields

$$\underbrace{đW_1 + đW_{\text{shaft}} + đQ}_{\text{energy input}} = \underbrace{đW_2}_{\substack{\text{energy}\\\text{output}}} + \underbrace{dE}_{\substack{\text{increase in}\\\text{energy storage}}}$$

The work terms $đW_1$ and $đW_2$ represent energy transfers to and from the control mass due to normal motion of its boundaries in the pipes near sections 1 and 2. To evaluate these terms we assume that the pressure exerted by the fluid on the duct walls is the same as that exerted by the fluid immediately outside the control mass on the fluid just inside the boundary. In other words, the pressure *within the fluid* is presumed to be the same in all directions. Denoting the fluid pressures at sections 1 and 2 by P_1 and P_2 and assuming that these pressures act uniformly over the flow cross sections at those points, we have

$$đW_1 = (PA)_1 \mathsf{V}_1 \, dt$$

$$đW_2 = (PA)_2 \mathsf{V}_2 \, dt$$

The change in the energy of the control mass can be represented in terms of the energies of regions A, B, and C of Fig. 5·6. Using the steady-flow steady-state idealization, we obtain

$$E(t) = E_A + E_C$$

$$E(t + dt) = E_C + E_B$$

$$dE = E(t + dt) - E(t) = E_B - E_A$$

We denote the energy per unit of mass of fluid at 1 and 2 by e_1 and e_2. This represents the internal energy, plus kinetic energy due to motion, plus potential energy due to conservative force fields,

$$e = u + PE + KE$$

Then, the increase in energy within the control mass is

$$dE = e_2\rho_2(A_2\mathsf{V}_2\ dt) - e_1\rho_1(A_1\mathsf{V}_1\ dt)$$

and the energy balance becomes

$$(PA\mathsf{V})_1\,dt - (PA\mathsf{V})_2\,dt + đW_{\text{shaft}} + đQ = (eA\rho\mathsf{V})_2\,dt - (eA\rho\mathsf{V})_1\,dt$$

The *rates* of energy transfer as work and heat are defined as

$$\blacktriangleright \qquad \dot{W}_{\text{shaft}} \equiv \frac{đW_{\text{shaft}}}{dt} \qquad \dot{Q} \equiv \frac{đQ}{dt} \tag{5·3}$$

Dividing by dt and introducing the mass-flow rate $\dot{M} = A\rho\mathsf{V}$, the energy balance may be written as

$$\underbrace{[\dot{M}(e + Pv)]_1 + \dot{W}_{\text{shaft}} + \dot{Q}}_{\text{energy-input rate}} = \underbrace{[\dot{M}(e + Pv)]_2}_{\text{energy-output rate}}$$

The expressions below the equation show how the terms in this equation can be interpreted as applying to the *space C* (the *control volume*).

The energy equation above relates the properties at sections 1 and 2 to the rates of energy transfer as heat and work to the device and consequently is just what we need. Suppose we knew the temperature and pressure at states 1 and 2 and could thereby fix the thermodynamic states of the fluid. If, in addition, we could in some way determine the mass-flow rate and knew the rate of energy transfer as heat, the energy balance could be used to calculate the shaft-power output. This is a typical use for a first-law equation.

As we indicated, the control-mass point of view could be used to handle all flow problems. The analysis would always involve motion of the boundaries in the inlet and exit ducts and a representation of the energy of the control mass in terms of the distribution of energy through space. We might as well do a reasonably general analysis of this type once and then use it whenever it is applicable in flow-system analysis. This brings us now to the control-volume transformation.

5·3 THE CONTROL-VOLUME TRANSFORMATION

A *control volume* is any defined region in space. This region may be moving through space, and its shape and volume may be changing. However, most often we deal with control volumes that are of fixed shape and size and are fixed in the reference frame, so we shall consider this special case first.

Consider a control volume whose boundaries are fixed in space and stationary. Matter flows across the boundaries of this control volume, as indicated in Fig. 5·7. Following our previous analysis, we assume that the flow streams are

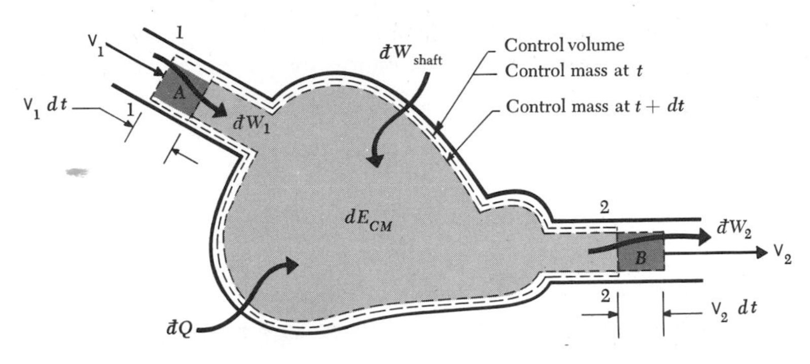

Figure 5·7. A control volume

one-dimensional at points where mass crosses the control-volume boundary. We again assume that the pressure exerted by the fluid on the duct walls is the same as the force per unit of area on an imaginary cut across the duct. This will be a good idealization, except in some cases where viscous shear is important, but there the one-dimensional idealization would be invalid anyway. We allow energy transfer as heat across the boundary of the control volume, but we must be careful in handling such energy transfer at points where mass enters the control volume, and to avoid confusion we shall assume that no energy is transferred as heat at the inlet and exit stations. We also presume that we can choose the control-volume boundaries such that we can neglect work arising from tangential shearing motions of the boundary, except for energy transfer as work through rotating shafts. Thus a number of important restrictions have been made, all of which can be removed by proper extension of the analysis. It is important to realize that our resulting control-volume energy equation will not be fully general, though it will be sufficiently general for most of our purposes.

To develop the proper control-volume energy-conservation expression, we shall consider two infinitesimally differing instants in time, t and $t + dt$, and apply the conservation-of-energy principle to the particular control mass whose boundaries at time t happen to correspond exactly to those of the control volume. Work will be done on the control mass by the rotating shaft and also by the normal motion of the control-mass boundary at the two points where matter flows into and out of the control volume. An energy balance on the control mass, made over the time period dt, gives

$$\underbrace{(PA)_1 \mathsf{V}_1\ dt + đW_{\text{shaft}} + đQ}_{\text{energy input}} = \underbrace{(PA)_2 \mathsf{V}_2\ dt}_{\text{energy output}} + \underbrace{dE_{CM}}_{\text{increase in energy storage}}$$

We want to express all terms in this energy balance in terms of properties of the control *volume* rather than the control mass. The energy contained within the control mass at time t is identical with the energy within the control volume at that instant. At time $t + dt$ the energy of the control mass is equal to the energy of the matter within the control-volume boundaries at $t + dt$, plus the energy of the matter within the shaded portion B, minus the energy of the matter within the shaded portion A. Therefore

$$\begin{aligned} dE_{CM} &= E_{CM}(t + dt) - E_{CM}(t) \\ &= E_{CV}(t + dt) + (A_2 \mathsf{V}_2\, dt)\rho_2 e_2 - (A_1 \mathsf{V}_1\, dt)\rho_1 e_1 - E_{CV}(t) \\ &= dE_{CV} + (A\rho \mathsf{V} e)_2\, dt - (A\rho \mathsf{V} e)_1\, dt \end{aligned}$$

Here we have used e to represent the total energy of the matter per unit of mass. The terms $\text{đ}W_{\text{shaft}}$ and $\text{đ}Q$ represent energy transfers across the common boundaries, that is, across the control-volume boundary, which occur during the time interval dt. Substituting for dE_{CM} and regrouping terms,

$$\text{đ}W_{\text{shaft}} + \text{đ}Q + (A\rho\mathsf{V})_1\left(e + \frac{P}{\rho}\right)_1 dt = (A\rho\mathsf{V})_2\left(e + \frac{P}{\rho}\right)_2 dt + dE_{CV}$$

The term $A\rho\mathsf{V}\, dt$ represents infinitesimal mass transfer across the control-volume boundaries, which we denote by $\text{đ}M$. Then

$$\blacktriangleright\quad \underbrace{\text{đ}W_{\text{shaft}} + \text{đ}Q + [(e + Pv)\, \text{đ}M]_{\text{in}}}_{\text{energy input}} = \underbrace{[(e + Pv)\, \text{đ}M]_{\text{out}}}_{\text{energy output}} + \underbrace{dE_{CV}}_{\text{increase in energy storage}} \qquad (5\cdot4)$$

This can be viewed as a conservation-of-energy equation *for the control volume*, expressed for a definite period of time. Alternatively, we could divide by dt and express the energy balance on a *rate basis* as

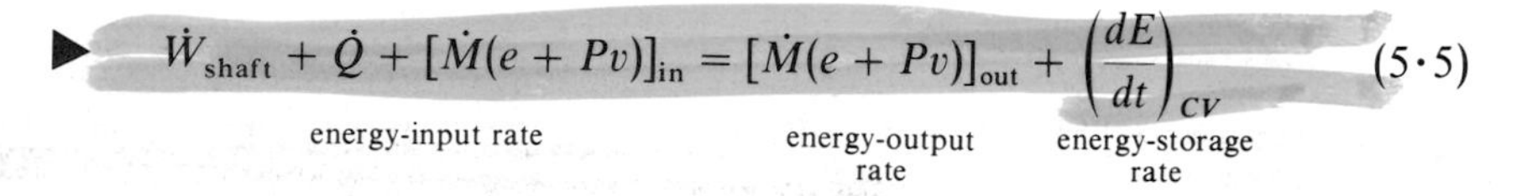

$$\blacktriangleright\quad \underbrace{\dot{W}_{\text{shaft}} + \dot{Q} + [\dot{M}(e + Pv)]_{\text{in}}}_{\text{energy-input rate}} = \underbrace{[\dot{M}(e + Pv)]_{\text{out}}}_{\text{energy-output rate}} + \underbrace{\left(\frac{dE}{dt}\right)_{CV}}_{\text{energy-storage rate}} \qquad (5\cdot5)$$

We see that the conservation-of-energy idea can be retained for the control volume, provided that we adopt a slightly modified picture of energy. When matter flows into a control volume, energy is *convected* in by the matter and, in addition, energy is transferred to the control volume as work. The amount of energy transfer as work associated with a unit of mass is Pv. In a sense, entering matter does work on the matter already within the control volume by "pushing it out of the way." The Pv product, when used in this context, is sometimes called *flow work*. It does not represent any energy contained by the fluid.

Note that the energy of matter is still designated by e. Only when matter crosses the boundaries of a control volume will the additional contribution of the Pv product occur in the energy-balance equation. Furthermore, care must be taken to distinguish between the $P\, dv$ work done by a control mass in expanding and the Pv work done by a unit of mass in flowing into a control volume (which may be of fixed size). Note that e, P, and v appearing in the flow terms are to be evaluated at the point where the matter crosses the boundary.

The energy per unit of mass of substance crossing the boundary is composed of internal energy, bulk potential energy, and bulk kinetic energy. We can therefore write

$$e + Pv = KE + PE + u + Pv$$

The appearance of the combination $h \equiv u + Pv$ suggests further utility for tabulations of the enthalpy property.

The kinetic energy of a unit of mass is simply $\mathsf{V}^2/2g_c$. The potential energy of a unit of mass in a uniform gravitational field may be shown to equal $(g/g_c)z$, where g is the local acceleration of gravity (ft/s^2), g_c is the constant in Newton's law [32.17 ft·lbm/(lbf·s^2)], and z is the height above some arbitrarily selected datum. The electrostatic potential energy of a unit of mass is simply $\mathcal{Q}\mathscr{E}$ where $\mathcal{Q}$ is the charge per unit of mass and $\mathscr{E}$ is the local value of the electrostatic potential above the (arbitrary) ground state. Hence

$$e + Pv = h + \frac{\mathsf{V}^2}{2g_c} + \frac{g}{g_c}z + \mathcal{Q}\mathscr{E} + \cdots$$

For a control volume with matter flowing across the boundaries in several places, and with boundaries which may themselves be moving, the conservation-of-energy statement on the time-interval basis leads to

$$\blacktriangleright \quad \underbrace{\sum_{\text{in}} (e + Pv)\, đM + đW + đQ}_{\text{energy input}} = \underbrace{\sum_{\text{out}} (e + Pv)\, đM}_{\text{energy output}} + \underbrace{đE_{CV}}_{\text{energy storage}} \tag{5·6}$$

and on the rate basis to

$$\blacktriangleright \quad \underbrace{\sum_{\text{in}} (e + Pv)\dot{M} + \dot{W} + \dot{Q}}_{\text{energy-input rate}} = \underbrace{\sum_{\text{out}} (e + Pv)\dot{M}}_{\text{energy-output rate}} + \underbrace{\left(\frac{dE}{dt}\right)_{CV}}_{\text{energy-storage rate}} \tag{5·7}$$

Here $\dot{W}$ is understood to be the sum of shaft-power *input* and power *input* due to normal motion of the control-volume boundaries. $\dot{Q}$ represents the sum of all heat-transfer rates *to* the control volume. If the control-volume boundaries move, $\dot{M}$ must be calculated in terms of the velocity of the fluid *relative and normal to the control surface.*

It is sometimes necessary to replace the sums by integrals, and we then obtain

$$\blacktriangleright \quad \underbrace{-\int_{CS} (e + Pv)\rho \mathbf{V}_{\text{rel}} \cdot d\mathbf{A} + \dot{W} + \dot{Q}}_{\text{net energy-input rate}} = \underbrace{\frac{d}{dt}\int_{CV} \rho e\, dV}_{\text{energy-storage rate}} \tag{5·8}$$

The area integration is to be extended over the control surface CS, and $d\mathbf{A}$ represents an *outward* normal vector. The volume integration is extended over the entire control volume. Note that e appears in the energy-storage term, while $e + Pv$ appears in the energy-convection term. $\mathbf{V}_{rel}$ represents the *outward* velocity *relative to the boundary*, while the *absolute* velocity $\mathbf{V}$ must be used to compute the kinetic energy.

Two important idealizations frequently made in control-volume analysis are (1) that the flow is steady, so that no mass is accumulating within the control volume, and (2) that the state of the matter at each point in space is steady, that is, unchanging in time. This implies that the energy stored within the control volume is unchanging and thereby eliminates the term in the energy equation involving changes in the energy storage. It is therefore not necessary to know the details of what is going on within the control volume. This is very important, because it allows us to analyze intricate systems by examining only the transfers of energy across the control-volume boundary.

The methodology introduced for the control mass may readily be applied in setting up the control-volume energy balance. We shall now illustrate the methodology by examples. In each case it will be necessary to make idealizations in order to render the problem tractable. Such idealizations are usually based on experience, but engineering analysis is both a science and an art and can be learned only by imitation and practice. Thus careful study of the examples will provide the student's first bit of experience and something to imitate. One can often verify the validity of an idealization by making appropriate estimates and should make every attempt to do so whenever a particular assumption is questionable. The examples serve to bring out many important features of systematic control-volume energy analysis and are typical of the simpler types of problems in engineering thermodynamics.

5·4 EXAMPLES OF CONTROL-VOLUME ENERGY ANALYSIS

Flow through a nozzle. Steam enters the nozzle of a steam turbine with a velocity of 10 ft/s at a pressure of 500 psia and a temperature of 1000°F. At the nozzle discharge the pressure and temperature are measured and found to be 300°F and 1 atm. What is the discharge velocity? See Fig. 5·8.

In any such nozzle frictional effects between the fluid and the walls might be important, but by choosing the boundaries as indicated in Fig. 5·8*a* these effects become internal. Then, idealizing the flow as steady and assuming that the state at each point in the control volume is invariant in time (steady state), we do not need to know what is going on within the control volume. We also idealize that the control volume is adiabatic, so that no energy transfer as heat takes place across the control-volume boundaries (this is a reasonable simplifying idealization for most nozzles). This does not mean that no energy transfer as heat takes place within the control volume, but any such transfer would not be involved in a steady-flow steady-state analysis.

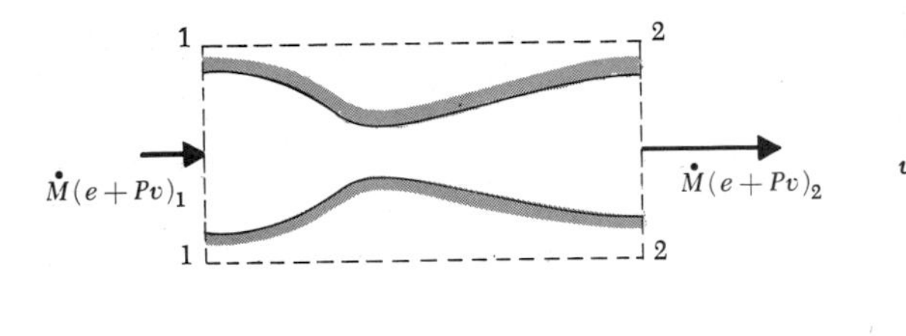

(a) *The control volume*

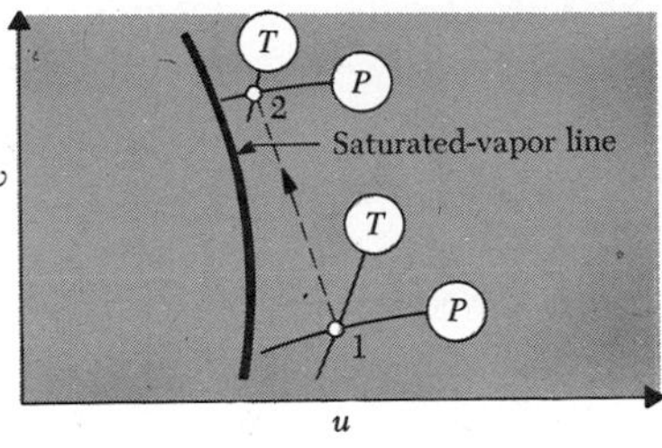

(b) *The process representation*

Figure 5·8.

The conservation-of-mass principle implies that the flow rate is the same at sections 1 and 2. In applying this principle we assume that the flow is one-dimensional at 1 and 2, which means that the properties are uniform across the sections. The energy per unit of mass is taken as

$$e = u + \frac{V^2}{2g_c}$$

In evaluating u as a function of the temperature and pressure we shall assume that the motion in no way alters the thermodynamic equations of state. In microscopic terms, even though the fluid is accelerating, the molecules behave locally as if there were no bulk motion. Satisfactory results are obtained with this idealization, and this experimental support is sufficient justification for its use. We shall also neglect any differences in the potential energy of position of the entering and emergent flows.

The internal process is quite complicated. However, only the states of the fluid at 1 and 2 need to be known, and we can indicate our ignorance of the intermediate states by showing the process as a dotted line (Fig. 5·8*b*).

Several important simplifying idealizations have been made, and it is a good idea to begin any analysis by listing them:

Steady flow steady state
Adiabatic control volume
One-dimensional flow at 1 and 2 (but not necessarily internally)
Equation of state the same as for a simple compressible substance
Changes in the potential energy of position negligible

With these idealizations, application of the conservation-of-energy principle to the control volume gives, on a rate basis,

$$\underbrace{\dot{M}\left(u + Pv + \frac{V^2}{2g_c}\right)_1}_{\text{energy-inflow rate}} - \underbrace{\dot{M}\left(u + Pv + \frac{V^2}{2g_c}\right)_2}_{\text{energy-outflow rate}} = \underbrace{0}_{\text{energy-storage rate}}$$

Solving for the discharge kinetic energy per unit of mass,

$$\frac{V_2^2}{2g_c} = \frac{V_1^2}{2g_c} + [(u + Pv)_1 - (u + Pv)_2]$$

The intensive thermodynamic states are fixed by the temperature and pressure measurements (See Fig. B·1):

$$T_1 = 1000°F \qquad T_2 = 300°F$$

$$P_1 = 500 \text{ psia} \qquad P_2 = 14.7 \text{ psia}$$

$$u_1 = 1360 \text{ Btu/lbm} \qquad u_2 = 1110 \text{ Btu/lbm}$$

$$v_1 = 1.7 \text{ ft}^3\text{/lbm} \qquad v_2 = 31 \text{ ft}^3\text{/lbm}$$

Hence

$$h_1 = (u_1 + P_1 v_1) = 1520 \text{ Btu/lbm}$$

$$h_2 = (u_2 + P_2 v_2) = 1193 \text{ Btu/lbm}$$

Then $$\frac{V_1^2}{2g_c} = \frac{10^2}{2 \times 32.2} (\text{ft}^2/\text{s}^2)/[\text{ft}\cdot\text{lbm}/(\text{lbf}\cdot\text{s}^2)]$$

$$= 1.55 \text{ ft}\cdot\text{lbf/lbm} = 2 \times 10^{-3} \text{ Btu/lbm}$$

$$\frac{V_2^2}{2g_c} = 2 \times 10^{-3} + 1520 - 1193 = 327 \text{ Btu/lbm} = 254{,}000 \text{ ft}\cdot\text{lbf/lbm}$$

$$V_2 = \sqrt{2 \times 32.2 \text{ ft}\cdot\text{lbm}/(\text{lbf}\cdot\text{s}^2) \times 254{,}000 \text{ ft}\cdot\text{lbf/lbm}} = 4050 \text{ ft/s}$$

Note that the enthalpies could have been more conveniently found in Table B·2 or in Fig. B·2. Note also that the inlet kinetic energy is very small and could have been neglected.

A valve. Steam enters a valve as a liquid-vapor mixture at 4 MPa and leaves at 0.01 MPa, 50°C. What is the density of the inlet mixture? See Fig. 5·9.

A valve is a device for producing a pressure drop in a flowing fluid. The valve works by accelerating the flow to high velocity through a narrow restriction; the flow is then decelerated but, because of frictional losses, the discharge pressure is less than the inlet pressure. Now, although the kinetic energy inside the valve is high, the flow kinetic energy at the inlet and discharge will usually be quite small, and the usual assumption in analyzing a valve is to neglect the flow kinetic energy.

The idealizations are as follows:

Steady flow steady state
Adiabatic control volume

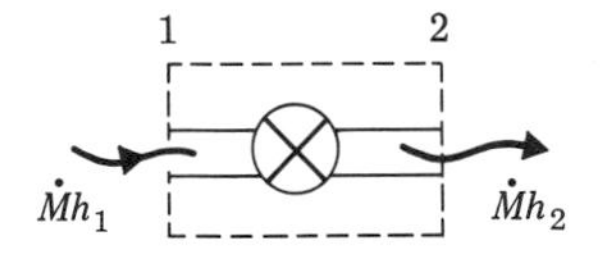

(a) The control volume

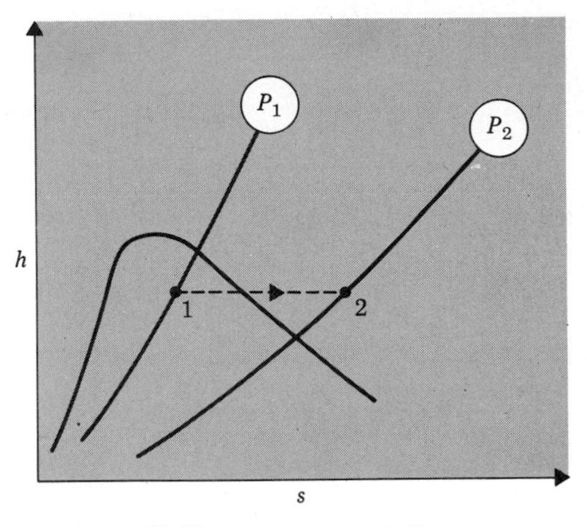

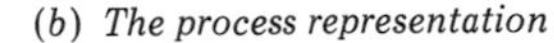

(b) The process representation

Figure 5·9. Valve analysis

One-dimensional flow at 1 and 2
Kinetic and potential energies negligible at 1 and 2
Thermodynamic equilibrium at 1 and 2

The energy balance is then simply

$$\underset{\substack{\text{energy-}\\ \text{input}\\ \text{rate}}}{\dot{M}h_1} = \underset{\substack{\text{energy-}\\ \text{output}\\ \text{rate}}}{\dot{M}h_2}$$

So, the energy balance tells us that the enthalpy of the flow leaving the valve will be the same as that going in, and we have shown this on the process representation of Fig. 5·9*b*.

At the discharge state, we read from the SI version of Table B·2, $h_2 = 2592.6$ kJ/kg. So, h_1 must have this value. As the process representation shows, the intersection of this enthalpy line with the inlet-pressure line fixes the inlet state. Since the h_g at 4 MPa is greater than h_1, the point 1 must lie under the vapor dome. From the SI Table B·1*b*, at the inlet pressure of 4 MPa,

$$h_{f1} = 1087.3 \text{ kJ/kg} \qquad h_{g1} = 2801.4 \text{ kJ/kg}$$

$$v_{f1} = 0.001252 \text{ m}^3\text{/kg} \qquad v_{g1} = 0.04978 \text{ m}^3\text{/kg}$$

Using Eq. (4·13),

$$h_1 = (1 - x_1)h_{f1} + x_1 h_{g1}$$

Solving for x_1

$$x_1 = \frac{h_1 - h_{f1}}{h_{g1} - h_{f1}} = \frac{2592.6 - 1087.3}{2801.4 - 1087.3} = 0.878$$

Using Eq. (4·1),

$$\begin{aligned} v_1 &= (1 - x_1)v_{f1} + x_1 v_{g1} = 0.122 \times 0.00125 + 0.878 \times 0.04978 \\ &= 0.0439 \text{ m}^3/\text{kg} \end{aligned}$$

So, the density is

$$\rho_1 = \frac{1}{v_1} = 1/0.0439 = 22.8 \text{ kg/m}^3$$

Let's investigate the effect of kinetic energy. To obtain a 5 percent error in the density we would need to make a 5 percent error in the quality, which in turn would come from an enthalpy error of 5 percent of h_{fg}, or $0.05 \times (2801.4 - 1087.3) = 86$ kJ/kg. A kinetic energy per unit of mass of $\mathsf{V}^2/2g_c =$ 86 kJ/kg corresponds to a velocity of

$$\begin{aligned} \mathsf{V} &= \sqrt{2 \times 1 \times 86{,}000 \text{ (kg}\cdot\text{m}^2/\text{s}^2)/\text{kg}} \\ &= 415 \text{ m/s} \end{aligned}$$

In typical steam piping the velocity would seldom be this great, and thus our neglect of kinetic energy was therefore reasonable. The engineer can often assess the validity of simplifying assumptions by checks such as this.

A mercury turbine. Mercury enters the turbine of a high-temperature auxiliary power system at 1200°F and 30 psia and emerges as a mixture of liquid and vapor of 0.95 quality at 1 psia. What must the flow rate be if the power output is to be 10 kW? See Fig. 5·10.

A typical turbine consists of (1) a nozzle, which accelerates the flow, converting internal energy and flow work to kinetic energy; (2) a rotor, which slows down the fluid, extracting energy from the fluid as work; and (3) a diffuser, which slows the fluid down again, with a resulting rise in pressure. The internal workings of such a device are quite complicated. However, by judicious selection of the control-volume boundaries and the idealizations of steady flow and steady state, one need worry only about the conditions at the points where the flow crosses the boundaries. This is a great advantage of the control-volume approach to steady-flow steady-state problems.

We take the boundary as shown above and make the following idealizations:

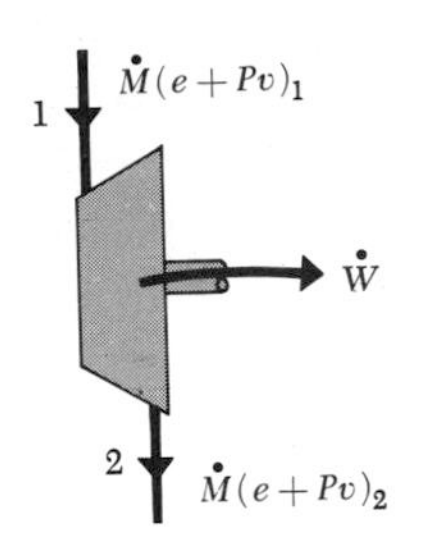

(a) The control volume

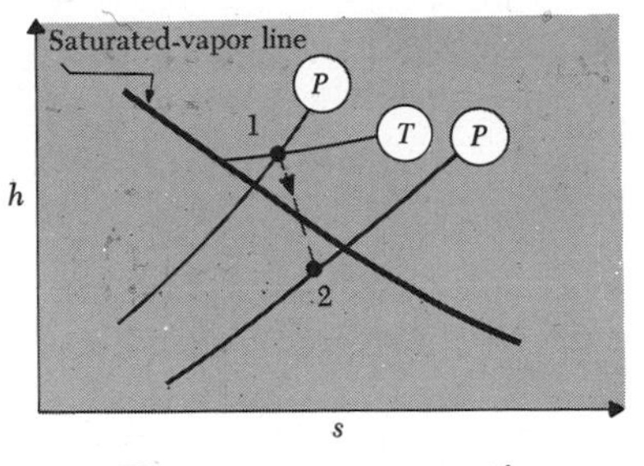

(b) The process representation

Figure 5·10.

Steady flow steady state
Adiabatic control volume
One-dimensional flow at 1 and 2
Kinetic and potential energy negligible at 1 and 2 (but certainly not inside)
Mercury in thermodynamic equilibrium at 1 and 2

Under these idealizations, application of the conservation-of-energy principle to the control volume gives

$$\underbrace{\dot{M}(u + Pv)_1}_{\text{energy-inflow rate}} - \underbrace{[\dot{M}(u + Pv)_2 + \dot{W}_{\text{shaft}}]}_{\text{energy-outflow rate}} = \underbrace{0}_{\text{energy-storage rate}}$$

The shaft-energy output per unit of mass is then

$$W_{\text{shaft}} = \frac{\dot{W}_{\text{shaft}}}{\dot{M}} = h_1 - h_2.$$

The enthalpies may be found directly in Fig. B·10, and we then have

$$W_{\text{shaft}} = (159 - 134)\ \text{Btu/lbm} = 25\ \text{Btu/lbm}$$

The power requirement of 10 kW is equivalent to 34,120 Btu/h, so the required mass-flow rate is

$$\dot{M} = 34{,}120/25 = 1365\ \text{lbm/h} = 0.379\ \text{lbm/s}$$

If the mercury is moving at 100 ft/s (typical for the inlet of a small mercury turbine), the required flow area can be computed from the relation $\dot{M} = A\rho\mathbf{V}$, provided the inlet density is known. The density does not appear on Fig. B·10. However, we can make an estimate of the density by assuming that the vapor behaves like a perfect gas, for which

$$\frac{\rho}{\rho_0} = \frac{P/T}{P_0/T_0}$$

The zero denotes any selected reference state, which we take to be saturated vapor at 20 psia. Then, using the data in Table B·3, we estimate the density at 1 as

$$\rho_1 = \left(\frac{1 \text{ lbm}}{3.09 \text{ ft}^3}\right)\frac{30/(1200 + 460)}{20/(706 + 460)} = 0.34 \text{ lbm/ft}^3$$

At 100 ft/s the inlet-flow area would be

$$A = \frac{0.379}{0.34 \times 100} = 0.0111 \text{ ft}^2 = 1.6 \text{ in}^2$$

We can evaluate the validity of our idealization that the kinetic-energy changes are negligible by some typical numbers. Suppose the discharge velocity is 200 ft/s, with the inlet velocity as 100 ft/s. The difference in the inflow and outflow kinetic energies, per unit of mass, would then be

$$\frac{V_2^2 - V_1^2}{2g_c} = \frac{200^2 - 100^2}{2 \times 32.2} = 465 \text{ ft·lbf/lbm} = 0.6 \text{ Btu/lbm}$$

This represents an error of about 2 percent when compared to the enthalpy difference of 25 Btu/lbm, and our idealization was therefore quite reasonable for first analysis. The analyst is frequently in a position to make such quantitative estimates of the inaccuracies introduced by simplifying assumptions and should do so whenever possible.

It is indeed interesting that the size of the system did not enter the thermodynamic calculation, nor did the particular internal configuration of the device. This is typical of thermodynamic energy analysis, in which important system parameters can be set without the need for detailed knowledge of the design.

Liquefaction of oxygen. Oxygen enters a turboexpander at 200 K and 60 atm. The expander discharges into a separator, from which saturated liquid and saturated vapor at 1 atm emerge as separate streams. The system is heavily insulated. The turboexpander shaft-power output is 1700 W, and the total oxygen-flow rate is 15 g/s. What percentage of liquefaction is achieved, and what is the net liquid-oxygen-production rate? See Fig. 5·11.

The idealizations are as follows:

Steady flow steady state
Adiabatic control volume
One-dimensional flow at 1, 4, and 3
Potential and kinetic energies at 1, 4, and 3 negligible
Thermodynamic equilibrium states at 1, 4, and 3

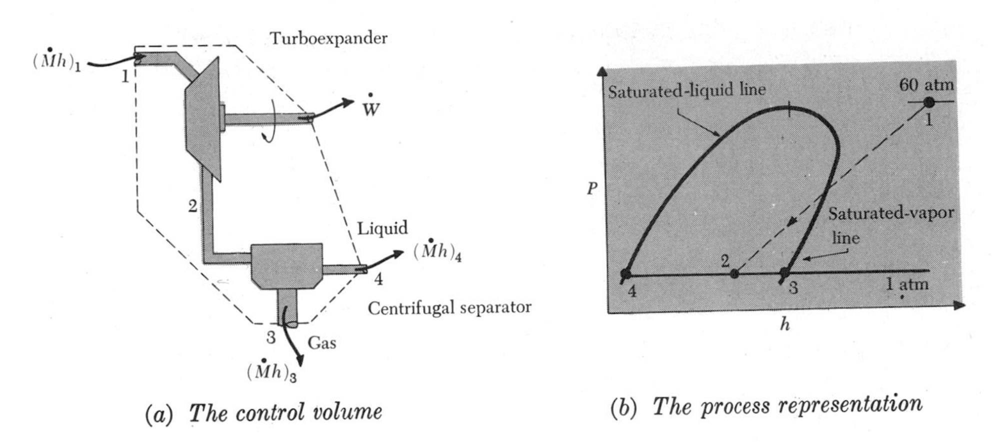

(a) *The control volume* (b) *The process representation*

Figure 5·11.

With these idealizations, an energy balance on the control volume gives, on a rate basis,

$$\underbrace{\dot{M}_1(u + Pv)_1}_{\text{energy-inflow rate}} - \underbrace{[\dot{M}_4(u + Pv)_4 + \dot{M}_3(u + Pv)_3 + \dot{W}]}_{\text{energy-outflow rate}} = \underbrace{0}_{\text{energy-storage rate}}$$

A mass balance will also be required; for the control volume,

$$\underbrace{\dot{M}_1}_{\text{mass-inflow rate}} - \underbrace{(\dot{M}_4 + \dot{M}_3)}_{\text{mass-outflow rate}} = \underbrace{0}_{\text{mass-storage rate}}$$

The problem will be solved if we can establish the relative flow rates. We put $f = \dot{M}_3/\dot{M}_1$, and the energy balance becomes an equation for f,

$$h_1 - [(1 - f)h_4 + fh_3] - \frac{\dot{W}}{\dot{M}_1} = 0$$

The enthalpies are evaluated with the aid of Fig. B·5, and we find (using $\hat{M} = 32.00$ g/gmole)

$$h_1 = \frac{2320}{32.00} = 72.5 \text{ cal/g}$$

$$h_4 = \frac{160}{32.00} = 5.0 \text{ cal/g} \qquad (\textit{saturated liquid})$$

$$h_3 = \frac{1780}{32.00} = 55.6 \text{ cal/g} \qquad (\textit{saturated vapor})$$

The shaft-work output per unit of mass is

$$\frac{\dot{W}}{\dot{M}_1} = \frac{1700 \text{ J/s}}{15 \text{ g/s}} = 1.13 \times 10^2 \text{ J/g} \times \frac{1 \text{ cal}}{4.18 \text{ J}} = 27.2 \text{ cal/g}$$

Solving for f and substituting the numbers,

$$f = \frac{27.1 + 5.0 - 72.5}{5.0 - 55.6} = 0.80$$

The system therefore achieves 20 percent liquefaction and delivers saturated liquid oxygen at 1 atm at the rate of 3.1 g/s.

Do not be led to believe that liquid oxygen can be made "for nothing" in a process from which useful electric power is obtained. Energy is required to separate oxygen from air, to compress it to 60 atm, and to run the refrigerator which precools it to 200 K. Only a portion of this would be recovered from the turboexpander.

CO_2 evaporator. Three hundred lbm/h of CO_2 (liquid) will enter a heat exchanger as a saturated liquid at 100 psia and emerge at 75 psia and 20°F. This evaporation and superheating process is to be accomplished by passing dry air through the other side of the exchanger. The air will enter at 70°F, slightly above atmospheric pressure, and must emerge at 50°F at atmospheric pressure. Specify the air-flow rate required and the heat-transfer rate within the exchanger. See Fig. 5·12.

We make the following idealizations:

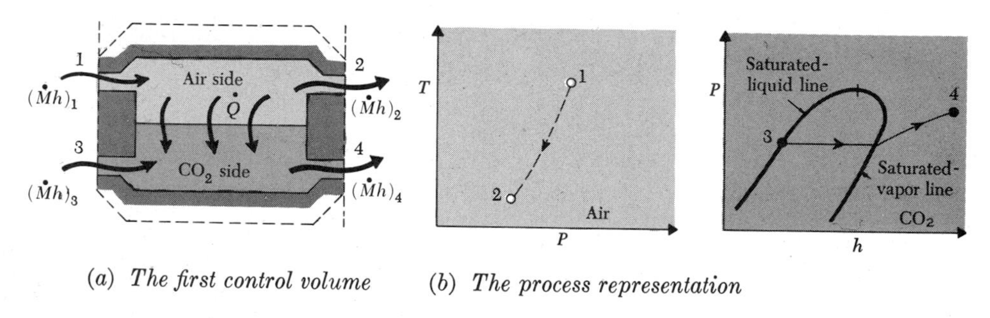

(a) *The first control volume* (b) *The process representation*

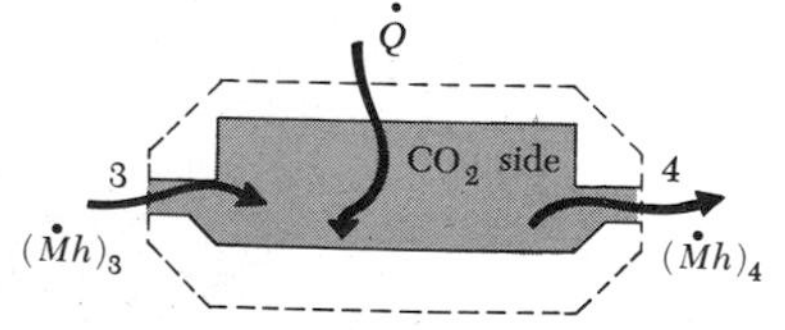

(c) *The second control volume*

Figure 5·12.

Steady flow steady state
Adiabatic control volume (energy transfer as heat occurring inside, and neglecting energy transfer as heat across the boundaries indicated)
One-dimensional flow at 1, 2, 3, and 4
Kinetic and potential energy changes negligible
Equilibrium states at 1, 2, 3, and 4 with properties related by thermodynamic equations of state

With these idealizations, the energy balance, on a rate basis, is

$$\underbrace{\dot{M}_1 h_1 + \dot{M}_3 h_3}_{\text{energy-inflow rate}} - \underbrace{(\dot{M}_2 h_2 + \dot{M}_4 h_4)}_{\text{energy-outflow rate}} = \underbrace{0}_{\text{energy-storage rate}}$$

By conservation of mass, $\dot{M}_1 = \dot{M}_2$ and $\dot{M}_3 = \dot{M}_4$

The enthalpies are found from the given states using Figs. B·6 and B·8:

$$h_3 = 137 \text{ Btu/lbm} \qquad (\textit{saturated liquid at 100 psia, } -55°F)$$

$$h_4 = 299 \text{ Btu/lbm}$$

$$h_1 = 5.5 \text{ cal/g} \qquad (21.5°C, \approx 1 \text{ atm})$$

$$h_2 = 2.5 \text{ cal/g} \qquad (10°C, 1 \text{ atm})$$

The enthalpy differences are then

$$h_4 - h_3 = 299 - 137 = 162 \text{ Btu/lbm}$$

$$h_1 - h_2 = 5.5 - 2.5 = 3.0 \text{ cal/g} = 5.4 \text{ Btu/lbm}$$

The required air-flow rate is therefore

$$\dot{M}_{\text{air}} = \frac{162}{5.4} \times 300 \text{ lbm/h} = 9000 \text{ lbm/h}$$

To determine the heat-transfer rate within the exchanger, a second energy balance is required. Either side of the exchanger may be used. Having the most accurate enthalpy-difference information for the CO_2, we select this side (Fig. 5·12c) and idealize as follows:

Steady flow steady state
One-dimensional flows at 3 and 4
Kinetic and potential energy changes negligible
Thermodynamic equilibrium at 3 and 4

The energy balance, on a rate basis, is

$$\underbrace{\dot{Q} + \dot{M}_{CO_2} h_3}_{\text{energy-input rate}} = \underbrace{\dot{M}_{CO_2} h_4}_{\text{energy-output rate}}$$

Substituting the numbers,

$$\dot{Q} = \dot{M}_{CO_2}(h_4 - h_3) = 300 \times 162 = 48{,}600 \text{ Btu/h}$$

Determination of the physical dimensions of a heat exchanger which could transfer energy from one stream to the other at this rate over the temperature differences involved is a problem for the heat-transfer analyst and requires more than thermodynamics.

A thermoelectric generator. A thermoelectric generator consists of a series of semiconductor elements, heated on one side and cooled on the other. Electric-current flow is produced as a result of energy transfer as heat. A schematic diagram of a typical device is shown in Fig. 5·13. In a particular experiment the current was measured to be 0.5 A and the electrostatic potential at 3 was 0.8 V above that at 4. Energy transfer as heat to the hot side of the generator was taking place at a rate of 5.5 W. Determine the rate of energy transfer as heat from the cold side and the energy-conversion efficiency.

We shall treat this problem by the control-volume method, since mass (electrons) flows across the indicated control surface. The electrons flowing possess electrostatic potential energy and kinetic energy. In addition, the electrons within the "electron gas" passing through the conductors execute randomly oriented motions, giving the "gas" internal energy in exactly the same way that molecules of a gas give it internal energy. Moreover, the electron gas will have a pressure and a specific volume, so it will have an enthalpy. The randomly oriented motions of the electrons within the gas are responsible in part for "noise" in electronic circuits. We shall assume that there is no noise in this system or, in other words, that the internal energy of the electron gas is insignificant compared to the electrostatic potential energy. The kinetic energy is also negligible, since

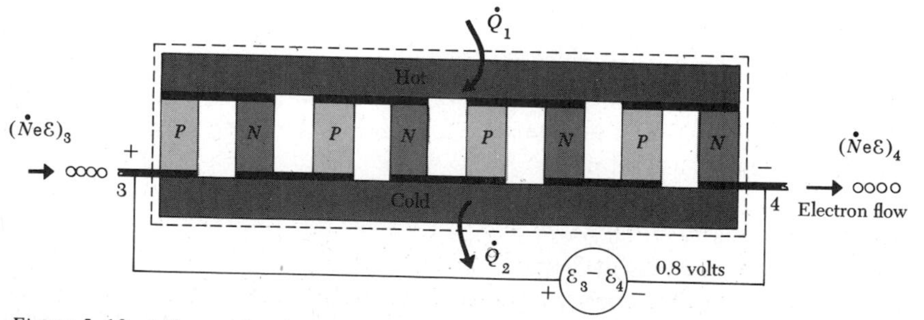

Figure 5·13. A thermoelectric generator

electrons in metallic conductors move at very slow drift velocities. Therefore only the electrostatic potential energy of the electrons will be considered. Idealizations are as follows:

Steady flow steady state
One-dimensional flows
Only electron potential energy important

Denoting the number of electrons crossing the boundary per unit of time by $\dot{N}$ and the charge of an electron by e, an energy balance, made on a rate basis, gives

$$\underbrace{[(\dot{N}\mathrm{e}\mathscr{E})_3 + \dot{Q}_1]}_{\text{energy-inflow rate}} - \underbrace{[(\dot{N}\mathrm{e}\mathscr{E})_4 + \dot{Q}_2]}_{\text{energy-outflow rate}} = \underbrace{0}_{\text{energy-storage rate}}$$

In terms of the current $i = -\dot{N}\mathrm{e}$,

$$i(\mathscr{E}_4 - \mathscr{E}_3) + \dot{Q}_1 - \dot{Q}_2 = 0$$

Solving for $\dot{Q}_2$ and substituting the numbers,

$$\dot{Q}_2 = 5.5 \text{ W} + 0.5 \text{ A} \times (-0.8 \text{ V}) \times 1 \text{ W/VA} = 5.1 \text{ W}$$

The term $i(\mathscr{E}_3 - \mathscr{E}_4)$ represents the useful power output of the device; it is equal to 0.4 W. The *energy-conversion efficiency* is then

$$\eta = \frac{i(\mathscr{E}_3 - \mathscr{E}_4)}{\dot{Q}_1} = \frac{0.4}{5.5} = 0.073$$

This low efficiency is typical of such solid-state direct-energy converters.

An arc heater. Air is to be heated to 8000 K in a steady-flow electric-arc device. The air will enter at 10^{-2} atm at atmospheric temperature and will emerge at the same pressure. Cooling water is provided to maintain low electrode temperatures, and it is estimated that half the electric-energy input will be transferred as heat to the cooling water. The air-flow rate is to be 10,000 lbm/h. Specify the required arc power. See Fig. 5·14.

The air will be ionized but electrically neutral. We assume that thermodynamic equilibrium has been obtained by the time the air emerges. The electric-current flow will again be handled as the flow of mass with electrostatic potential energy. We idealize that the internal energy of the flowing electron "gas" is small compared to its bulk potential energy. Electrons move through solids at very low drift velocities, so the kinetic energy of the electrons at the points at which we draw the control-volume boundaries will be neglected. (It should be noted that the

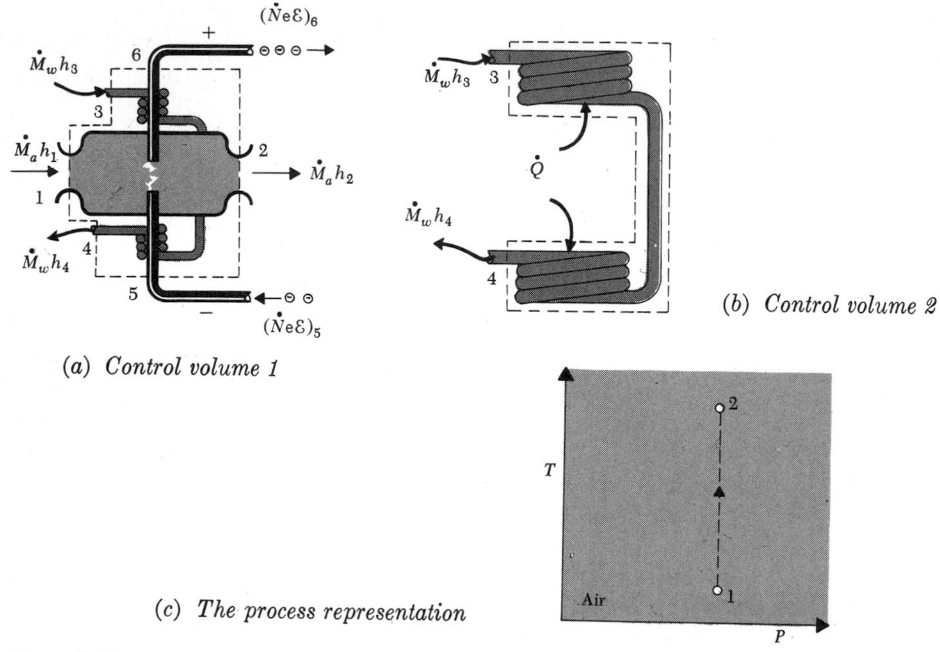

Figure 5·14.

kinetic energy of the electrons striking the anode will be extremely high and could not be neglected were we to make our boundary cut through the air immediately adjacent to the anode.) We make the following idealizations:

Steady flow steady state
One-dimensional flows
Kinetic energy of air and water negligible
Only electrostatic potential energy of electrons important
Control volume 1 adiabatic
Air in thermodynamic equilibrium at 1 and 2
Water in thermodynamic equilibrium at 3 and 4

We again denote the number of electrons flowing per unit of time as $\dot{N}$ and their charge as e. The energy balance, made on a rate basis on control volume 1, gives

$$\underbrace{(\dot{N}e\mathscr{E}_5 + \dot{M}_a h_1 + \dot{M}_w h_3)}_{\text{energy-inflow rate}} - \underbrace{(\dot{N}e\mathscr{E}_6 + \dot{M}_a h_2 + \dot{M}_w h_4)}_{\text{energy-outflow rate}} = \underbrace{0}_{\text{energy-storage rate}}$$

In terms of the current $i = -\dot{N}e$,

$$i(\mathscr{E}_6 - \mathscr{E}_5) = \dot{M}_a(h_2 - h_1) + \dot{M}_w(h_4 - h_3)$$

The first term should be recognized as the electric-power input.

An energy balance on control volume 2 leads to

$$\dot{Q} = \dot{M}_w(h_4 - h_3)$$

Using the design estimate that $\dot{Q} = i(\mathscr{E}_6 - \mathscr{E}_5)/2$, we combine the energy balances and obtain

$$\text{Power} = i(\mathscr{E}_6 - \mathscr{E}_5) = 2\dot{M}_a(h_2 - h_1)$$

The enthalpy at state 2 is obtained from the equation of state, Fig. B·9, as 19,000 Btu/lbm. This value is relative to the enthalpy at absolute zero of a gas obeying the equation of state $h = c_P T$ at low temperatures. For air $c_P \approx 0.24$ Btu/(lbm·R), so an estimate of the enthalpy at state 1 might be $0.24 \times 500 = 120$ Btu/lbm. This is negligible in comparison to the enthalpy of the highly energetic air at state 2. Our power calculation then yields

$$\text{Power} = 2 \times 10{,}000 \text{ lbm/h} \times (19{,}000 - 120) \text{ Btu/lbm}$$
$$= 38 \times 10^7 \text{ Btu/h} = 1.11 \times 10^5 \text{ kW} = 111 \text{ MW}$$

Charging of a high-pressure tank. A gas is pumped into a tank having volume V. The gas in the tank is initially at pressure P_0 and temperature T_0. The inlet temperature is T_1, and the inflow rate $\dot{M}_1$ is constant. Using the perfect-gas equation of state, derive an expression for the temperature in the tank as a function of time, assuming that the gas inside is perfectly mixed and in a state of thermodynamic equilibrium and that no energy transfer as heat takes place from the tank to the gas. See Fig. 5·15.

Our idealizations are as follows:

The gas may be treated as a perfect gas with constant specific heats
Steady inflow (but not steady state!)
Uniform state inside the tank (perfect internal mixing)
Adiabatic control volume
Equilibrium state within tank at every instant, equilibrium state at 1

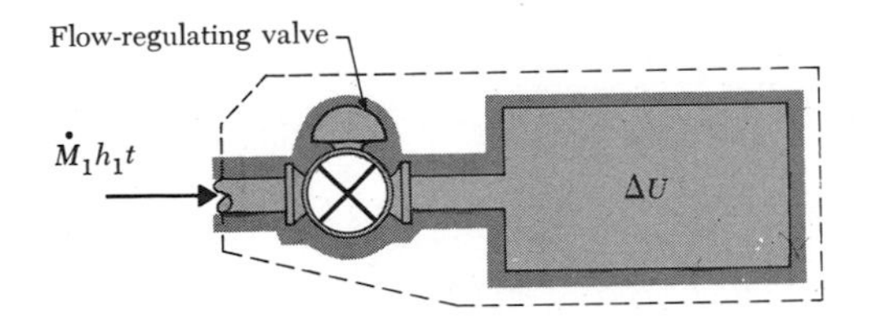

(a) The control volume

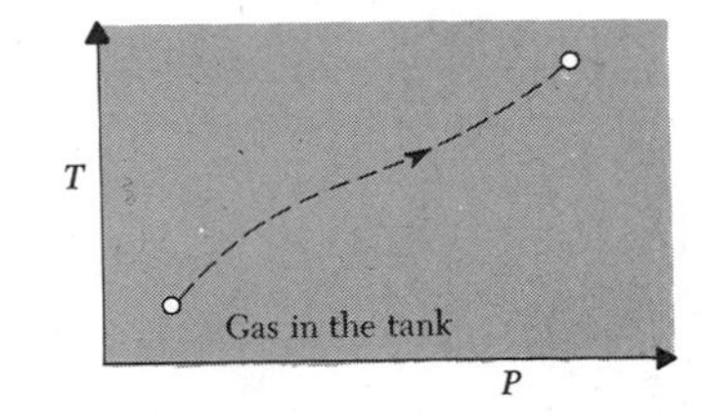

(b) The process representation

Figure 5·15.

Kinetic energy of inflow gas negligible
Inlet state steady (but not the internal state!) and one-dimensional flow at the inlet

The energy balance, made over the time interval from the start of filling (time zero) to some later time t, is

$$\underbrace{\dot{M}_1(u_1 + P_1 v_1)t}_{\text{energy inflow}} - \underbrace{0}_{\substack{\text{energy}\\\text{outflow}}} = \underbrace{\Delta U}_{\substack{\text{increase}\\\text{in energy}\\\text{storage}}}$$

Then,

$$\Delta U = Mu - M_0 u_0$$

A mass balance on the control volume is also needed. Over the same time period,

$$\underbrace{\dot{M}_1 t}_{\text{mass inflow}} - \underbrace{0}_{\substack{\text{mass}\\\text{outflow}}} = \underbrace{M - M_0}_{\substack{\text{increase in}\\\text{mass storage}}}$$

Solving for M and substituting into the energy balance,

$$\dot{M}_1 h_1 t = (M_0 + \dot{M}_1 t)u - M_0 u_0 \tag{5·9}$$

Note that the enthalpy property is involved in the energy-inflow term, while only the internal energy is involved in the energy-storage term.

The Pv product does not represent energy of matter; it is simply an extra term in the energy balance necessitated by our taking the control-volume rather than control-mass point of view.

We next bring in the equation of state. With the datum at 0 R in Eqs. (4·21) and (4·22), we introduce these into Eq. (5·9), obtaining

$$\dot{M}_1 c_P T_1 t = (M_0 + \dot{M}_1 t)c_v T - M_0 c_v T_0$$

The temperature T of the gas in the tank can be computed from this equation as a function of time. It is interesting that for very large t the temperature will approach the value

$$T_{\text{lim}} = \frac{c_P}{c_v} T_1 = kT_1$$

Since the ratio k is greater than unity, the limiting internal temperature will be somewhat greater than the temperature of the incoming gas.

A hydroelectric power plant. Suppose we wish to assess the capabilities of a foreign hydroelectric power plant and have acquired the following field data:

Estimated water flow, 40 m^3/s
River inlet 1 atm, 10°C
Discharged at 1 atm, 10.2°C, 200 m below the intake

We know from experience that the energy transfer as heat to such a power plant is negligible. The enthalpy of water is given approximately by

$$h_2 - h_1 = c_v(T_2 - T_1) + \frac{1}{\rho}(P_2 - P_1)$$

where $c_v = 4.2$ kJ/(kg·K) and $\rho = 1000$ kg/m^3. We also estimate that the inlet and discharge ducts have the same flow area, so that the kinetic energy of the water entering the plant is essentially the same as that leaving. With this information, what is our estimate of the power output? See Fig. 5·16.

We make the following idealizations:

Change in kinetic energy of flow streams negligible
Steady flow steady state
Adiabatic control volume

Note there is no assumption about friction or turbulence or losses within the control volume; these are all internal effects and need not be considered explicitly in a steady-flow steady-state analysis. They are implicitly considered through their effect on state 2. We would have to make some such assumption if we wanted to *predict* state 2. The energy balance, on a rate basis, is

$$\underbrace{\dot{M}\left(h_1 + \frac{g}{g_c}z_1\right)}_{\text{energy-input rate}} = \underbrace{\dot{W} + \dot{M}\left(h_2 + \frac{g}{g_c}z_2\right)}_{\text{energy-output rate}}$$

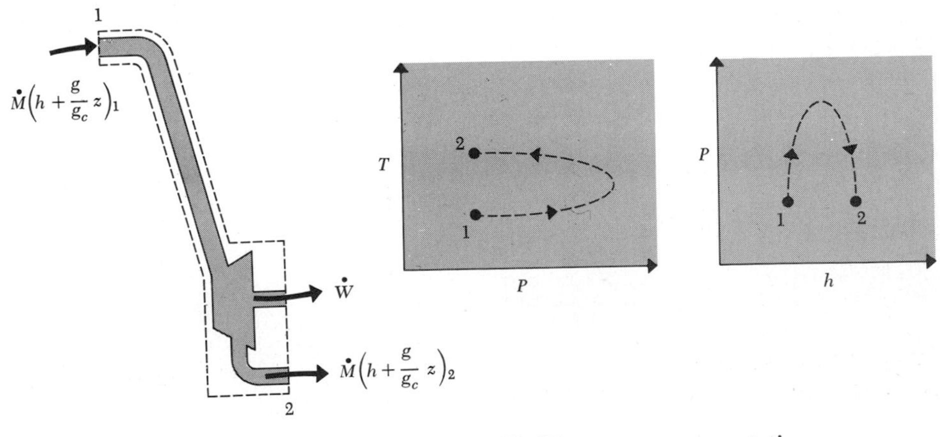

(*a*) *The control volume* (*b*) *The process representation*

Figure 5·16.

If we take the discharge state 2 as the elevation datum, then $z_2 = 0$ and $z_1 = 200$ m. The various terms in the energy balance are then ($g = 9.8$ m/s²)

$$\frac{g}{g_c} z_1 = 9.8 \text{ m/s}^2 \times 200 \text{ m} = 1960 \text{ J/kg} = 1.96 \text{ kJ/kg}$$

$$h_2 - h_1 = c_v(T_2 - T_1) + \frac{1}{\rho}(P_2 - P_1) = 4.2 \times (10.2 - 10.0) = 0.84 \text{ kJ/kg}$$

Hence
$$\frac{\dot{W}}{\dot{M}} = \frac{g}{g_c} z_1 - (h_2 - h_1) = 1.96 - 0.84 = 1.12 \text{ kJ/kg}$$

Then, $\dot{W} = 1.12 \times 1000 \times 40 = 44{,}800 \text{ kJ/s} = 44{,}800 \text{ kW} = 44.8 \text{ MW}$

A heat pump. The heat pump shows promise of becoming a common household heating system; a simple heat pump consists of the components shown in Fig. 5·17. In an experiment with Freon 12 as the working fluid, the following measurements were recorded:

$P_2 = P_3 = 140$ psia
$P_4 = P_1 = 10$ psia
$T_2 = 260°\text{F}$
$T_1 = -20°\text{F}$
Saturated liquid at 3
$\dot{W} = 10$ kW

Determine the condenser and evaporator rates of energy transfer as heat and the flow rate of the Freon 12.

In this experiment no pressure drop through the heat exchangers was measurable. This is frequently true, and a good idealization is that the process undergone by a fluid in passing through a heat exchanger is one of *constant*

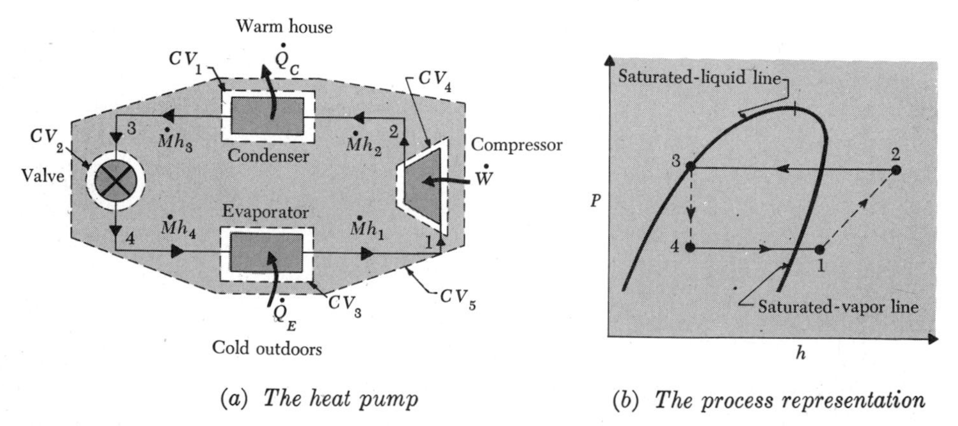

(a) *The heat pump* (b) *The process representation*

Figure 5·17.

pressure. We begin our analysis of this rather complicated problem by a process representation (Fig. 5·17) to orient our thinking. The reason for the positioning of state 4 will be evident momentarily.

We have enough information to establish the condenser heat-transfer rate per unit of mass. For control volume 1 we idealize

Steady flow steady state
Kinetic and potential energies negligible
Equilibrium states at 2 and 3
One-dimensional flows at 2 and 3

The energy balance for $CV1$, made on a rate basis, gives

$$\underset{\text{energy-input rate}}{\dot{M}h_2} = \underset{\text{energy-output rate}}{\dot{M}h_3 + \dot{Q}_C}$$

Then, using Fig. B·7,

$$\frac{\dot{Q}_C}{\dot{M}} = h_2 - h_3 = 115 - 31 = 84 \text{ Btu/lbm}$$

We next analyze the valve (control volume 2), with the idealizations

Steady flow steady state
Kinetic and potential energies negligible
Adiabatic control volume
Equilibrium states at 3 and 4
One-dimensional flows at 3 and 4

The energy balance for $CV2$, on a rate basis, then is simply

$$\underset{\text{energy-input rate}}{\dot{M}h_3} = \underset{\text{energy-output rate}}{\dot{M}h_4}$$

or

$$h_3 = h_4 = 31 \text{ Btu/lbm}$$

Note that the energy balance here tells us something about the end states of a process; with these idealizations the enthalpy of the fluid emerging from a valve will be the same as the enthalpy of the entering fluid (though the enthalpy of the fluid within the valve may be quite different).

An energy balance (rate basis) on control volume 3, with idealizations as for control volume 1, yields

$$\underset{\text{energy-input rate}}{\dot{Q}_E + \dot{M}h_4} = \underset{\text{energy-output rate}}{\dot{M}h_1}$$

So the amount of energy transferred as heat to each lbm of Freon 12 passing through the evaporator is

$$\frac{\dot{Q}_E}{\dot{M}} = h_1 - h_4 = 76 - 31 = 45 \text{ Btu/lbm}$$

Now for control volume 4 (the compressor) we idealize

Steady flow steady state
Kinetic and potential energies negligible
Adiabatic control volume
Equilibrium states at 1 and 2
One-dimensional flows at 1 and 2

The energy balance for $CV4$, on a rate basis, then gives

$$\underbrace{\dot{W} + \dot{M}h_1}_{\text{energy-input rate}} = \underbrace{\dot{M}h_2}_{\text{energy-output rate}}$$

So the compressor work per lbm of fluid which it handles is

$$\frac{\dot{W}}{\dot{M}} = h_2 - h_1 = 115 - 76 = 39 \text{ Btu/lbm}$$

We can check our calculations by an overall energy balance (control volume 5), made on a rate basis, assuming steady flow and steady state:

$$\underbrace{\dot{W} + \dot{Q}_E}_{\text{energy-input rate}} = \underbrace{\dot{Q}_C}_{\text{energy-output rate}}$$

$$\dot{M}(39 + 45 - 84) = 0$$

The coefficient of performance (cop) of a heat pump is defined as

$$\text{cop} \equiv \frac{\text{rate of energy transfer to house}}{\text{compressor shaft power}}$$

For our system

$$\text{cop} = \frac{\dot{Q}_C}{\dot{W}} = \frac{\dot{M}(84 \text{ Btu/lbm})}{\dot{M}(39 \text{ Btu/lbm})} = 2.15$$

Note that the rate of energy transfer into the house is more than twice the electrical power which must be paid for to obtain this heating rate. Note also that the cop could be evaluated without reference to the amount of flow or the total energy-transfer rates. We now compute these from the measured shaft power:

$$\dot{M} = \frac{10\text{ kW} \times 3412\text{ Btu/(kW}\cdot\text{h)}}{39\text{ Btu/lbm}}$$

$$= 875\text{ lbm/h}$$

The rates of energy transfer as heat are therefore

$$\dot{Q}_E = 875 \times 45 = 39{,}400\text{ Btu/h}$$

$$\dot{Q}_C = 875 \times 84 = 73{,}500\text{ Btu/h}$$

The figure 39,400 Btu/h is comparable with that of a small commercial freezer. The energy pumped into the house (73,500 Btu/h) is comparable to that provided by a household furnace.

Heat pumps and refrigerators are sometimes rated in terms of tonnage. One ton of refrigeration is defined as 12,000 Btu/h and is roughly the rate of energy transfer as heat required to freeze 1 ton of ice in a day. Considered as a heat pump, this unit would have a rating of $73{,}500/12{,}000 \approx 6$ tons.

5·5 PRODUCTION BOOKKEEPING

In all our past examples we have done the energy bookkeeping in whatever way seemed most convenient at the time. For steady-state problems we simply wrote

$$\begin{matrix}\text{Energy} \\ \text{input} \\ \text{rate}\end{matrix} = \begin{matrix}\text{energy} \\ \text{output} \\ \text{rate}\end{matrix}$$

while for unsteady problems we usually wrote

$$\begin{matrix}\text{Energy} \\ \text{input}\end{matrix} = \begin{matrix}\text{energy} \\ \text{output}\end{matrix} + \begin{matrix}\text{increase in} \\ \text{energy storage}\end{matrix}$$

These expressions both reflect the idea that *energy can never be produced.* There is one particular way to do bookkeeping that is especially helpful in second-law analysis, where we will deal with entropy bookkeeping. It happens that entropy *can* be produced, but can *never* be destroyed. We will do all our entropy bookkeeping using the concept of *production:*

$$\text{Production} = \text{output} - \text{input} + \text{increase in storage}$$

As logical as this definition is, some students find it confusing at first. An easy way to obtain the signs on the right-hand side correctly is to examine each term with the other two set to zero:

1. If the input and storage are zero, surely if anything comes out it must have been produced inside, so output has a plus sign.
2. If something goes in, but nothing comes out and nothing accumulates inside, then something must disappear (negative production). So, input has a negative sign, which is reasonable since input is negative output.
3. If nothing goes in or comes out, but the amount stored inside gets bigger, some must have been produced. So, storage has a positive sign.

For example, the production bookkeeping for energy is

$$\begin{matrix}\text{Energy} \\ \text{production}\end{matrix} = \begin{matrix}\text{energy} \\ \text{output}\end{matrix} - \begin{matrix}\text{energy} \\ \text{input}\end{matrix} + \begin{matrix}\text{increase in} \\ \text{energy storage}\end{matrix} = 0$$

Throughout the remainder of the text we shall use the symbol $\mathscr{P}$ to denote production; $\mathscr{P}_E$ will denote the amount of production of energy, and $\mathscr{P}_S$ will denote the amount of production of entropy. The first law of thermodynamics says

$$\mathscr{P}_E = 0 \tag{5·10}$$

The second law of thermodynamics says

$$\mathscr{P}_S \geq 0 \tag{5·11}$$

These laws are easy to remember in this basic form. Then to use them all one needs to do is to keep one's energy and entropy books properly.

In dealing with infinitesimal processes we shall use the notation $đ\mathscr{P}_E$ and $đ\mathscr{P}_S$. The $đ$ symbol is appropriate because $đ\mathscr{P}$ represents an *amount* of production, not the *change* in production.

The first and second laws are then

$$đ\mathscr{P}_E = 0 \qquad đ\mathscr{P}_S \geq 0$$

In flow problems we will work with the *rates* of energy and entropy production, which we shall denote by $\dot{\mathscr{P}}_E$ and $\dot{\mathscr{P}}_S$. Then, the first and second laws are

$$\dot{\mathscr{P}}_E = 0 \qquad \dot{\mathscr{P}}_S \geq 0$$

The conservation-of-mass principle says that the mass production, or rate of mass production, is zero, $\mathscr{P}_M = 0$ or $đ\mathscr{P}_M = 0$ or $\dot{\mathscr{P}}_M = 0$. In mechanics, Newton's second law can be framed by stating that the rate of momentum production is equal to the applied force. Similarly, the rate of angular momentum production is equal to the applied moment.

The student should look back over the energy balance examples presented earlier and recast them in production bookkeeping. Many people find that they prefer to write all the physical principles mentioned above in production form.

SELECTED READING

Jones, J. B., and G. A. Hawkins, *Engineering Thermodynamics*, secs. 2.7–2.13, John Wiley & Sons, Inc., New York, 1960.

Mooney, D. A., *Introduction to Thermodynamics and Heat Transfer*, chap. 6, Prentice-Hall, Inc., Englewood Cliffs, N.J., 1955.

Van Wylen, G. J., and R. E. Sonntag, *Fundamentals of Classical Thermodynamics*, 2d ed., secs. 5.4–5.10, John Wiley & Sons, Inc., New York, 1973.

QUESTIONS

5·1 What is the difference between a control mass and a control volume?

5·2 How would you distinguish between positive and negative energy transfers?

5·3 Why, in calculus, does dx represent an increase in x? What is the meaning of Δx?

5·4 In the first control-mass example, why was it necessary to idealize that states 1 and 2 were states of thermodynamic equilibrium, and where was this idealization actually used? Why was it not necessary to make a similar idealization about intermediate states?

5·5 What is a process representation?

5·6 In the first control-mass example, why was it necessary to idealize that the alteration in molecular behavior introduced by the gravitational field is negligible, and where was this idealization actually used? What change in the center of mass of the H_2O would produce a change in the potential energy of position equal in magnitude to the change in the internal energy for the process? What is the difference between the idealization that the bulk potential-energy change for the process is negligible and the idealization that the gravitational field does not alter the molecular behavior?

5·7 Why is the enthalpy change equal to the energy transfer as heat to a unit of mass undergoing a constant-pressure process, but not for other processes?

5·8 What is the latent heat of vaporization as related to thermodynamic properties?

5·9 Why do the conversion factors between G and C/(s·m) differ for **H** and **M**?

5·10 Explain the last statement of the thermal-magnetization example.

5·11 What was assumed about the state of the paramagnetic material *during* the thermal-magnetization process?

5·12 Why does the term Pv appear in the control-volume energy balance and not in that for the control mass?

5·13 Explain why Pv is not energy stored in matter.

5·14 What is the difference between Pv and $P\,dv$, and when do terms of these types appear in energy-conservation equations?

5·15 Why is enthalpy a useful property to have tabulated?

5·16 Derive the continuity equation $M = A\rho V$.
5·17 Why is the control-volume transformation useful?
5·18 What is steady state? What is steady flow? Why do these idealizations greatly simplify control-volume analysis?
5·19 What is a one-dimensional flow?
5·20 In the first control-volume example, why was it necessary to idealize that the bulk motion does not alter the molecular behavior as seen by an observer riding with the fluid, and where was this idealization actually used?
5·21 What is a rate basis for an energy balance?
5·22 In the nozzle example, the energy stored in the fluid changes as it passes through the control volume. Why, then, is the term marked "energy-storage rate" equal to zero?
5·23 In the turbine example, why is it not necessary to know the state of the fluid inside the turbine?
5·24 Do the idealizations for the oxygen-liquefaction example rule out the possibility of any energy transfer as heat taking place to the fluid inside the control volume?
5·25 Why is the control volume of Fig. 5·12*a* idealized as adiabatic, while that of Fig. 5·12*c* is not?
5·26 At what speed do electrons move in a copper wire (see your physics book)?
5·27 Why do we handle electron-flow problems by the control-volume method?
5·28 In the tank-charging example, why is the idealization made that the gas in the tank is in equilibrium at every instant, and where is it used?
5·29 Explain the statement immediately following Eq. (5·9).
5·30 What can you say about the house and outdoor temperatures in the heat-pump example?
5·31 Outline a good energy-analysis methodology.
5·32 Why does it pay to be judicious in selecting a control volume or control mass? Is there sometimes more than one choice?

PROBLEMS

5·1 Two lbm of copper at 300 K are cooled to 250 K at 1 atm. How much energy is transferred as heat, and how much work is done on the copper (Table B·7)?
5·2 How much energy transfer as heat is required to evaporate completely 1 lbm of nitrogen for a constant-temperature process, with $T = 161.09$ R (Fig. B·4)?
5·3 Nitrogen in a 1-ft^3 container is heated from the saturated-vapor state at 50 psia to 300 R. How much energy transfer as heat is involved (Fig. B·4)?
5·4 Three lbm of nitrogen is heated at constant pressure from the saturated-liquid state at 100 psia to a temperature of 400 R. How much energy transfer as heat is required? How much work is done by the nitrogen, if any, and on what (Fig. B·4)?
5·5 One lbm of Freon 12 is heated in a constant-volume container from the critical point to 700°F. How much energy is transferred as heat to the Freon, and from where (Fig. B·7)?
5·6 Iron-ammonium alum is magnetized at 1 K by increasing the external field **H** slowly from zero to 10,000 G. How much energy is transferred as heat for this process, and in which direction? How much work is done on the alum, and by what (Fig. B·12)?
5·7 One lbm of copper and 0.5 lbm of saturated H_2O vapor at 1 atm pressure are initially in equilibrium. The water is then heated at constant pressure and held at 400°F until

equilibrium between the copper and the water is again obtained. How much energy transfer as heat takes place between the water and its vessel, and how much between the water and the copper (Figs. B·1 to B·3, Table B·7)?

5·8 One lbm of CO_2 is expanded adiabatically in a piston-cylinder system from 400 psia and 100°F to the saturated-vapor state at 100 psia. How much work is done by the CO_2 and on what (Fig. B·6)?

5·9 One lbm of O_2 is compressed adiabatically in a piston-cylinder system from the saturated-vapor state at 2.5 atm to a pressure of 17.5 atm and a temperature of 175 K. How much work is done on the O_2 and by what (Fig. B·5)?

5·10 One-half lbm of CO_2 is heated in a 0.5-ft^3 vessel from 0 to 200°F. Determine the initial and final states, and the amount of energy transfer as heat to the CO_2.

5·11 One-half lbm of CO_2 is heated at constant pressure from the saturated-vapor state at 100 psia to 200°F. Determine the amounts of energy transfer as heat and work involved in this process.

5·12 An adiabatic piston-cylinder system contains a 1000-W immersion heater and 4 kg of H_2O initially at 1 atm and 96 percent quality. The heater is operated for 7 min, during which the pressure is held constant. Find the final volume of the system.

5·13 A simple engine uses a perfect gas as the working fluid in a piston-cylinder system. The gas is first heated at constant pressure from state 1 to state 2, then cooled at constant volume to state 3 where $T_3 = T_1$, and then cooled at constant temperature, thereby returning to state 1. Derive expressions for the amounts of energy transfer as work and heat (per lbm of gas) for each process in terms of the temperatures and pressures at each state and the constants of the gas. Suppose $T_1 = 300$ K, $P_1 = 0.2$ MPa, $T_2 = 800$ K, and $k \equiv c_P/c_v = 1.4$. Calculate the cycle efficiency (*net* work output/energy *input* as heat).

5·14 A simple engine uses a perfect gas as the working fluid in a piston-cylinder system. The gas is first heated at constant volume from state 1 to state 2, then heated at constant temperature to state 3 where $P_3 = P_1$, and then cooled at constant pressure, returning to state 1. Derive expressions for the amounts of energy transfer as heat and work (per lbm of gas) for each process in terms of the temperatures and pressures at each state and the constants of the gas. Assuming $T_1 = 300$ K, $P_1 = 0.2$ MPa, $T_2 = 800$ K, and $k \equiv c_P/c_v = 1.4$, calculate the cycle efficiency (*net* work output/energy *input* as heat).

5·15 When the engine of Prob. 5·13 is reversed, it becomes a refrigeration device. Calculate the amount of energy transfer as heat from the cold space for this cycle (per kg of gas), and the cycle cop (energy transfer as heat from cold space/net work input).

5·16 Do Prob. 5·15 for the engine cycle of Prob. 5·14.

5·17 A low-temperature refrigeration cycle uses iron-ammonium alum (Fig. B·12) as the working substance. The alum is first heated at constant $\mathbf{H} = 3000$ G from state 1 where $\mathbf{M}_1 = 90$ G to state 2, where $\mathbf{M}_2 = 70$ G. It is then heated at constant $\mathbf{M}$ until $\mathbf{H} = 10{,}000$ G at state 3, and then cooled at constant $\mathbf{H}$ to state 4 where $\mathbf{M}_4 = \mathbf{M}_1$. A final cooling process at constant $\mathbf{M}$ returns the alum to state 1. Calculate the amounts of energy transfer as heat for the four processes (J/kg), and the cycle cop (energy transfer as heat from cold space/net energy input as work).

5·18 An engine operating on the reverse cycle to that described in Prob. 5·17 is proposed for use in a deep space probe which can have no part at greater than 2 K. Calculate the net work output per g of alum, the cycle efficiency (net work output/energy input as heat), and the required amount of alum for 1 W of power if the cycle can be repeated every 22 s.

5·19 Consider the piston-cylinder-spring system shown below.

Piston area 1 in^2
Vacuum in spring chamber
Spring force $F = ka$, $k = 300$ lbf/in

The cylinder initially contains CO_2 at 80°F, and a is initially 2 in. The cylinder walls are then slowly cooled, and the piston moves to the left until $a = \frac{1}{2}$ in. Calculate the energy transfer as heat from the gas for this process.

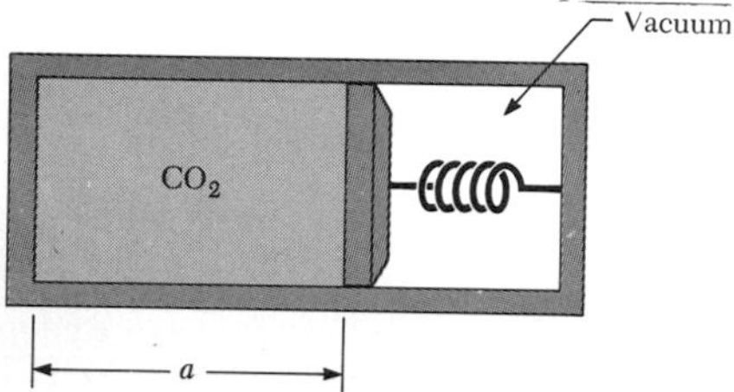

Prob. 5·19.

5·20 Twenty lbm/min of mercury vapor enters a condenser at 800°F, 0.5 psia. The mercury emerges as a saturated liquid at 0.4 psia. Calculate the rate of energy transfer as heat from the mercury side of the condenser (Fig. B·10, Table B·3).

5·21 Oxygen passes through an adiabatic steady-flow compressor at the rate of 1000 kg/h, entering as a saturated vapor at 2.5 atm and emerging at 17.5 atm and 175 K. Determine the shaft work per unit of mass of O_2 (compare Prob. 5·9) and the required motor hp.

5·22 Five hundred lbm/min of CO_2 passes through an adiabatic steady-flow turbine, entering at 400 psia and 100°F and emerging as a saturated vapor at 100 psia. What is the shaft-work output per lbm of CO_2 (compare Prob. 5·8), and what is the power (kW) delivered by the turbine?

5·23 Ten lbm/min of mercury enters a small turbine at 30 psia and 1400°F and emerges at 2 psia and 800°F. What is the shaft-power output if the heat losses (energy transfer as heat from the turbine casing) are negligible? What is the shaft-power output if the heat losses amount to 20 percent of the power output for the adiabatic device (Fig. B·10)?

5·24 Steam flows through a small steam turbine at the rate of 10,000 kg/h, entering at 600°C and 2.0 MPa and emerging at 0.01 MPa with 4 percent moisture. The flow enters at 50 m/s at a point 2 m above the discharge and leaves at 80 m/s. Compute the shaft-power output, assuming that the device is adiabatic but considering kinetic and potential energies. How much error would be made were these secondary terms neglected? What are the diameters of the inlet and discharge pipes (Fig. B·2, Tables B·1 and B·2)?

5·25 If the state at the discharge of the nozzle in the turbine of Prob. 5·24 is 6 percent moisture and 0.01 MPa, what is the velocity at this point?

5·26 Air enters an adiabatic nozzle at 3 atm pressure and 100°F and emerges at 1 atm and 30°F. The inlet velocity is negligible, and the nozzle is adiabatic. What is the discharge velocity (Fig. B·8)?

5·27 The collecting panels of a solar boiler receive energy as heat at the rate of approximately 300 Btu/h per ft^2 of collecting surface during the day. How many square feet of collector would be required for a small desert power plant producing 10 kW of electric power if the electric power output is 8 percent of the collected solar energy? (This is a typical energy-conversion efficiency for such a system.)

5·28 Oxygen flows at the rate of 100 kg/s through a line to a large booster rocket. The lox enters the line as a saturated liquid at 2 atm, and the pressure drop is negligible. Specify the maximum permitted heat-transfer rate ("heat leak in") to the lox if a maximum of 0.5 percent vapor can be tolerated at the booster end of the line (Fig. B·5).

5·29 The throttling calorimeter is a device for measuring the state of a liquid-vapor mixture. The procedure is to bleed off a little of the mixture, throttle it through a valve, and make the measurements shown. Explain why P_1 and T_1 do not fix the state

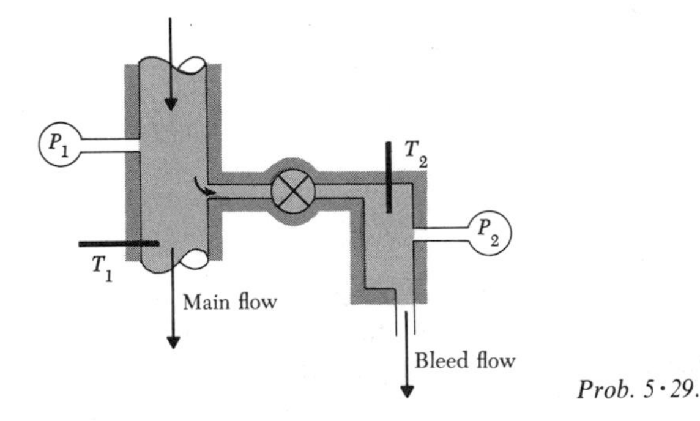

Prob. 5·29.

of the "wet" mixture, and how P_2 and T_2 measurements allow one to determine state 1. How much throttling is necessary for this scheme to work? Compute the quality at state 1 for the measurements below, assuming the fluid is water (Tables B·1 and B·2).

$$P_1 = 100 \text{ psia} \qquad P_2 = 5 \text{ psia}$$

$$T_1 = 327.81°\text{F} \qquad T_2 = 200°\text{F}$$

5·30 A valve in an insulated liquid-oxygen fuel line causes a pressure drop of 0.2 MPa. The inlet state is saturated liquid at 2 atm. What is the discharge quality, and how much is the temperature reduced (Fig. B·5)?

5·31 Mercury emerges from the nuclear boiler of a space power system at 500 psia and 0.95 quality and is "flashed" by passing through a flow restriction (valve) into the superheat region to a pressure of 50 psia. What is the temperature of the mercury in this superheated state (Fig. B·10, Table B·3)?

5·32 Determine the air-flow rate that could be used in the air-CO_2-exchanger example if the CO_2 states and the inlet air state are as given but the air emerges as cold as possible.

5·33 A mixture of CO_2 containing 20 percent solid by mass, 20 percent liquid, and 60 percent vapor enters a pipe and emerges all vapor at the inlet pressure and 0°F. The flow rate is 4 lbm/s. What is the rate at which energy is transferred as heat to the pipe?

5·34 Work the nozzle example using the control-mass approach.

5·35 Work the mercury-turbine example using the control-mass approach.

5·36 Work the tank-charging example using the control-mass approach.

5·37 Ten lbm/min of saturated liquid mercury at 415°F and 2 lbm/min of mercury vapor at 700°F and 1 psia enter an adiabatic mixing device; the mercury emerges mixed in a single stream at 0.3 psia. What is the temperature of the emergent stream, and (if it is a mixture) what is its quality (Fig. B·10)?

5·38 Oxygen is used in a low-temperature refrigerator. The hardware is similar to that of the heat-pump example. The condenser pressure is 20 atm and the O_2 is evaporated at 2 atm. Liquid emerges from the condenser in the saturated state and vapor from the evaporator at a temperature of 100 K. The compressor outlet temperature is 225 K. Determine the cop of this refrigerator, where

$$\text{cop} = (\text{rate of energy removal from cold space})/(\text{compressor power input})$$

What is the required oxygen-flow rate for a total cooling rate $\dot{Q}_E$ of 3000 W?

5·39 A small solar engine for desert water pumping uses steam as the working fluid. The hardware is shown in the figure. Water enters the pump as a saturated liquid at 50°C

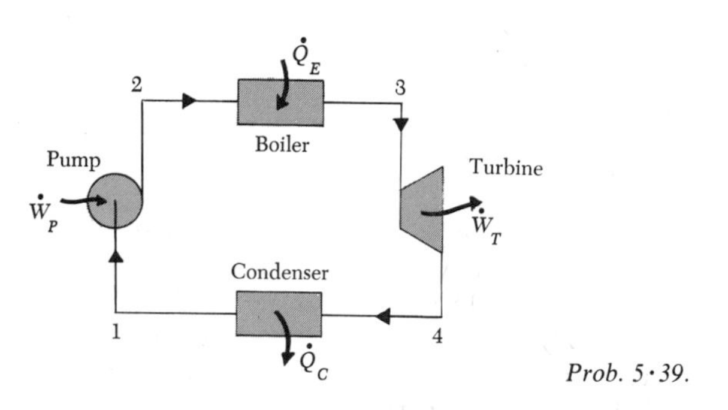

Prob. 5·39.

and is pumped up to 0.2 MPa by a small centrifugal pump. The boiler evaporates the water at 0.2 MPa, and saturated vapor at this pressure enters the small turbine. The steam leaves the turbine with 6 percent moisture at 50°C and is subsequently condensed. The flow rate is 140 kg/h and the pump is driven by a $\frac{1}{2}$-hp motor operating at full load. Making suitable idealizations, determine the power output of this plant (net hp) and the energy-conversion efficiency (net shaft-work output/energy transfer to fluid in boiler), and estimate the area of solar collectors that would be required, assuming that the collectors can pick up 800 W/m^2.

5·40 A proposed nuclear power system for space use employs mercury as the working fluid, using hardware similar to that in Prob. 5·39 above, except that a nuclear reactor replaces the sun as the energy source. The mercury enters the pump as a saturated liquid at 800°F, is compressed, and is then evaporated at 400 psia and superheated to 1400°F in the reactor-boiler. The turbine discharge is at 800°F with 3 percent moisture; pressure drops through the reactor and condenser may be neglected for this thermodynamic analysis. The power required to operate the pump will be 2 percent of the turbine-shaft power. Determine the energy-conversion efficiency for this system (net shaft power/reactor-power input), and the mercury-flow rate and reactor power required for 10 kW of net electric power, assuming a 95-percent-efficient electric generator will be employed (Fig. B·10, Table B·3).

5·41 Consider an electric resistor through which flows a steady current i. The resistor is cooled in order to maintain a steady state. From thermodynamic considerations, show that the rate at which energy must be transferred from the resistor as heat is $\dot{Q} = i\,\Delta\mathscr{E}$, where $\Delta\mathscr{E}$ is the voltage drop across the resistor. What are the restrictions on this result imposed by the idealizations?

5·42 In a linear accelerator electrons are accelerated until their kinetic energy is 10 MeV per electron. If the beam is accidentally diverted, it will strike the 3-mm-thick copper walls of the cavities and can cause serious damage. Assuming that a beam carrying

10^{18} (electrons/s)/cm^2 moving at right angles to the wall strikes the wall and all electrons are stopped, what cooling rate [W/m^2] must be provided to prevent a rise in the temperature of the wall? The electrostatic potential energy of the electrons at the point of impact is negligible compared to their kinetic energy.

5·43 Freon 12 will be used as the propellant for a small portable jet rocket for maneuvering parts during the assembly of a space station. The device will consist of a small spherical tank containing saturated Freon 12 at 80°F, a valve, and a nozzle. The user

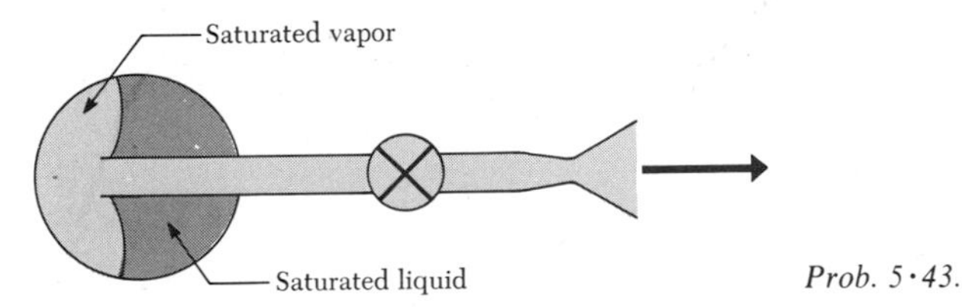

Prob. 5·43.

will attach the jet to the package, open the valve momentarily, and let some gas escape through the nozzle. Energy transfer as heat from the walls to the Freon 12 will keep the pressure constant in the cylinder during a short burst, and the flow through the nozzle may be considered steady. The valve will throttle the Freon 12 to 20 psia, and the nozzle discharge state will be 2 psia and −80°F. The thrusting force is approximately given by $F = \dot{M}V/g_c$, where $\dot{M}$ is the mass-flow rate and V the exit velocity. Use requirements call for a thrust of 10 lbf. Determine the specific impulse of this device, $SI = F/\dot{M}$, in lbf/(lbm·s). By comparison, a good chemical rocket has an SI of the order of 250 lbf/(lbm·s). Specify the size of tank required to hold enough Freon 12 for 1000 s of operation, assuming that the heat-transfer rate is sufficient to keep the pressure constant for short-duration bursts. Compute the total amount of energy which must be transferred from the walls to the Freon 12 during this 1000 s of operation in order to maintain the pressure (Fig. B·7, Table B·5). This requires a careful energy analysis.

5·44 A small power system proposed for use in the Arctic has the hardware of Prob. 5·39 and uses CO_2 as the working fluid. However, a supercritical cycle is employed; condensation occurs underground at −60°F, and saturated CO_2 liquid leaves the condenser at this temperature. The pump raises the pressure to 1600 psia, and the liquid emerges from the pump at −40°F. A solar collector is used to heat the CO_2 to 160°F, and the pressure drop through the "supercritical boiler" is 200 psia. The CO_2, at 1400 psia and 160°F, then enters the turbine and emerges at −40°F and 0.90 quality. Find the energy-conversion efficiency (net shaft-energy output/solar-energy input) for this system. Specify the pump power, turbine power, and solar-energy-collection rate required for a 10-kW plant, assuming that the electric generator has an efficiency of 92 percent. Make appropriate idealizations to allow solutions (Fig. B·6).

5·45 The hardware for a proposed SNAP system for generation of electric power for space vehicles is shown on the following page.

The boiler is a nuclear reactor in which the working fluid is heated by passage over the nuclear fuel elements. $\dot{Q}_B$ represents the rate of energy transfer as heat from the rods to the fluid and is equal to the reactor power. The condenser is a heat exchanger in which energy transferred as heat from the fluid is radiated away to space at the rate $\dot{Q}_C$. Pressure drop through the boiler, condenser, and ducting are negligible, and the pump, valve, turbine, and ducting are adiabatic devices.

Mercury is proposed as the working fluid. The mercury will leave the condenser as a saturated liquid at 30 psia. The pump will raise the pressure to 400 psia, where evaporation will occur. The mercury will leave the reactor at 95 percent quality (at 400 psia). The valve provides a pressure drop for load control.

The manufacturer of the electromagnetic pump can provide a pump that will do the job for a work input of about 0.5 Btu/lbm of mercury pumped. The turbine manufacturer can build a turbine with an isentropic efficiency of 60 percent for the range of flows probably involved. The isentropic efficiency is

$$\eta_s = \frac{h_4 - h_5}{h_4 - h_{5s}}$$

where h_4 and h_5 are the enthalpies entering and leaving the actual turbine, and h_{5s} is the enthalpy at a hypothetical reference state fixed by the inlet entropy (s) and outlet pressure.

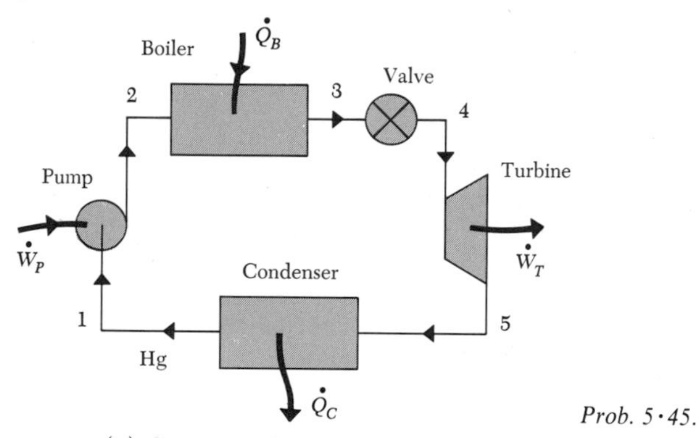

Prob. 5·45.

(a) Suppose the component manufacturers can indeed deliver as they promise. Determine the component energy flows per lbm of mercury, assuming that the valve is wide open so that $P_3 = P_4$. Determine the *net* work output per unit of mass and the overall system efficiency (efficiency of conversion of the reactor power to mechanical shaft power). Then, specify the required mass-flow rate if the net shaft power is to be 10 kW. Calculate the pump power, reactor power, turbine power, and condenser cooling rate (kW) for this flow; discuss briefly the distribution of the input reactor energy.

(b) Suppose the velocity of the stream at section 5 is limited to a maximum of 100 ft/s. Calculate the required duct diameter at this point.

(c) Suppose the valve is partially closed such that the turbine inlet pressure is reduced to 100 psia, while the same evaporating and condensing pressures and flow are maintained. Determine the net power output in this configuration, the condenser cooling rate, and the overall system efficiency.

5·46 A geothermal power plant in California uses steam produced underground by natural sources. Steam enters the adiabatic turbine at 180°C and 0.6 MPa and emerges at 0.01 MPa with 8 percent moisture (92 percent quality). It is condensed at 0.01 MPa and then pumped back up to atmospheric pressure. The plant produces 12.5 MW of electric power. Assuming an electric generator efficiency of 0.95, and neglecting the pump power, determine the steam flow rate and the condenser heat-transfer rate. Why do you think the system has a condenser? Why does it need the pump?

5·47 It is desired to reduce the pressure of cesium vapor from 50 psia at 2400 R to 1.9 psia. Suggest a means for doing this in a steady-flow device without any energy transfer as heat or shaft work to or from the system. For your device, what is the exit temperature at 1.9 psia?

5·48 You are required to determine the state of a mixture of liquid and vapor mercury used in a power plant for a space application. Only P and T measurements can be made. Design a system to fix the mixture state. The system must be reliable, inexpensive, and simple.

5·49 Freon-12 vapor flows at 10 ft/s at 300 psia and 360°F. It is desired to accelerate the flow to 500 ft/s at a pressure of 200 psia in a steady-flow device. What hardware is required and what is the exit temperature?

5·50 Consider the steady, one-dimensional, adiabatic, frictionless flow of an ideal gas in a passage of variable area. Suppose that the gas emerges from a chamber at temperature T_0 with negligible kinetic energy and is accelerated to velocity V in a passage of appropriate design. At this point it now has temperature T, sound speed a, and Mach number $M \equiv \mathsf{V}/\mathsf{a}$. For the ideal gas, $\mathsf{a}^2 = g_c kRT$. Show that

$$\frac{T_0}{T} = 1 + \frac{k-1}{2} M^2$$

5·51 The figure shows a gas-cooled fast breeder reactor. Helium enters the steam generator at 640°C with a flow rate of 1×10^6 kg/h. It leaves at 330°C. Water enters the steam generator at 40°C, 21×10^6 N/m^2 ($h = 185$ kJ/kg). The steam leaves at

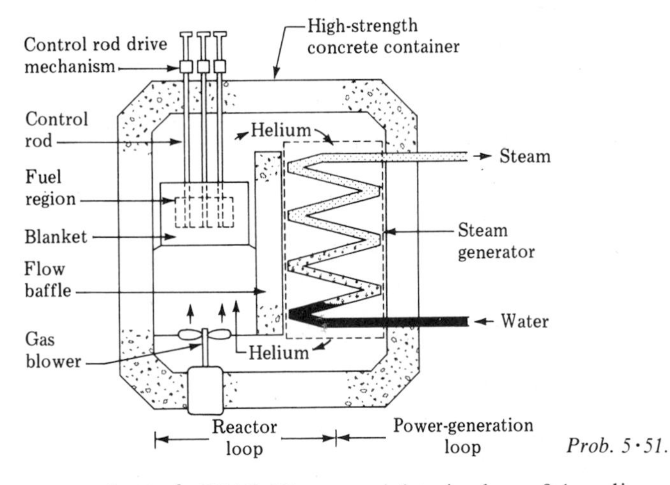

Prob. 5·51.

20×10^6 N/m^2, 550°C. The water inlet pipe has a 0.1 m diameter, and the steam exit pipe is 0.18 m in diameter and 6 m above the inlet. Neglecting heat losses, determine the steam-flow rate and the flow velocities at the inlet and exit. Are the kinetic and potential energy changes negligible? Use the perfect-gas equation of state for the helium, and tabulated properties for steam.

CHAPTER SIX

ENTROPY AND THE SECOND LAW

6·1 THE ESSENCE OF ENTROPY

The central idea in science is that nature behaves in a manner which is predictable. We have seen how energy-balance analysis is used to predict the change in state of a system due to transfers of energy as heat and work or to spontaneous internal changes. But we also know from experience that, while certain spontaneous changes in state can occur in isolated systems, the reverse changes are never observed. Oxygen and hydrogen readily react to form water; but who has ever seen water spontaneously separate into its two basic elements? By itself, first-law analysis cannot reveal the possibility or impossibility of a process; it cannot point the direction of time. A falling object will warm up when it is brought to rest by striking the ground; but have you ever seen an object cool down and leap up? The ability to rule out impossible processes is clearly essential to any complete predictive theory of nature. The *second law of thermodynamics* provides the necessary structure for this second type of analysis.

To explore this idea further, let's consider the system of Fig. 6·1. It consists of a flywheel surrounded by a gas, in a rigid adiabatic container; it is an isolated system. Suppose we start out with the system in state A, with the flywheel spinning and the gas and wheel fairly cool. As time passes we expect that collisions between the gas molecules and the wheel will eventually cause a transfer of energy from the flywheel to the gas molecules, so the wheel will slow down and the gas will start to

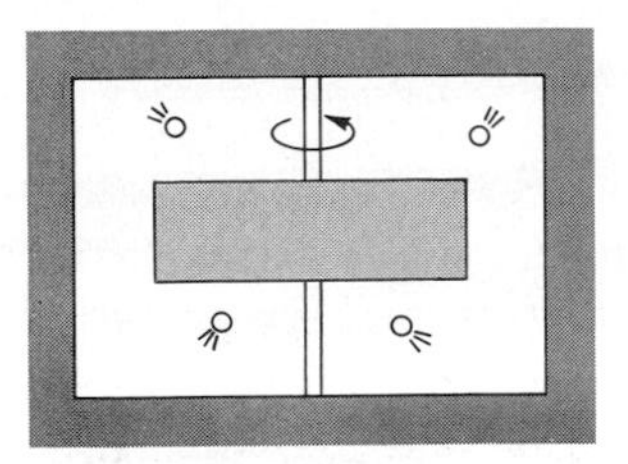

State A: flywheel spinning, system cool

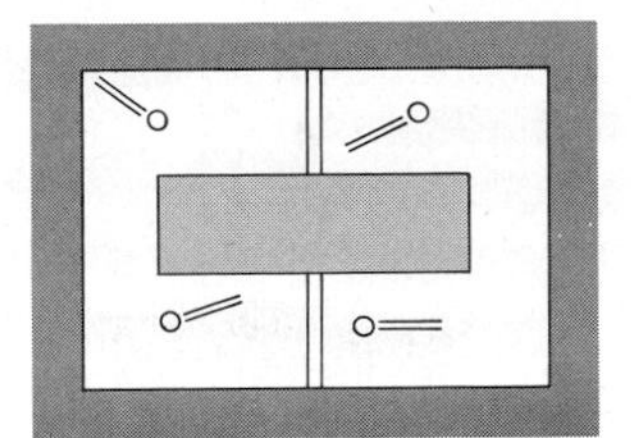

State B: flywheel stationary, system warm

Figure 6·1. Which state came first?

spin around in the box. The random motions of the gas molecules will tend to randomize the energy of the organized swirl of the gas, and eventually both the wheel and the gas will come to rest (macroscopically), and all the energy will reside in the random motions of the gas and wheel molecules, i.e., in the form of internal energy, so the gas and wheel will be warmer. We denote this by state B. The energy balance for this process is

$$\underset{\text{initial energy}}{(U + KE)_A} = \underset{\text{final energy}}{U_B} \tag{6·1}$$

where U is the total internal energy of the system, and KE is the rotational kinetic energy of the flywheel. All this is very rational.

Now, let's propose a process with less credence. Suppose that we *start out* with the system in state B, i.e., with a warm, motionless system, and allow the system to change to state A, where everything is cooler and the wheel is spinning. The energy balance for this process is

$$\underset{\text{initial energy}}{U_B} = \underset{\text{final energy}}{(U + KE)_A} \tag{6·2}$$

Equations (6·1) and (6·2) are identical, so if Eq. (6·1) is satisfied, so is Eq. (6·2). Certainly this second process will never happen, but the conservation-of-energy principle, i.e., the first law of thermodynamics, does not tell us that the process cannot happen. *The first law is insensitive to the direction of the process.*

Let's consider this example further. In the initial state A most of the energy is in a highly organized form. All the molecules of the flywheel are rotating around the axis together, and this organization makes it possible to extract that energy quite easily as useful work. We can simply attach a generator to the flywheel, generate electricity, and use this energy to raise weights, run trains, etc. But once the system has reached state B, where all the energy is microscopically disorganized, it is much more difficult to extract the energy as useful work; we certainly cannot do it just with a generator. Something has been lost by the process that resulted in a randomization of organized energy; we have lost some ability to do useful work. Something has also been produced: a higher state of molecular chaos.

This loss and gain always go hand in hand; *whenever molecular chaos is produced, the ability to do useful work is reduced.*

Entropy is a property of matter that measures the degree of randomization or disorder at the microscopic level. The natural state of affairs is for *entropy to be produced* by all processes. Associated with entropy production is a loss of ability to do useful work. Energy is degraded to a less useful form, and it is sometimes said that there is a decrease in the *availability* of energy. The notion that *entropy can be produced, but never destroyed,* is the *second law of thermodynamics.*

It is also useful to think about entropy as a measure of our uncertainty about the microscopic state. As the molecules move about, collide, and change directions, the microscopic state is continually changing. At any instant a system could be in any one of billions of billions of microscopic states, and we never can be very sure exactly which state exists at a particular time. The magnitude of the entropy reflects uncertainty about the microscopic state.

If there is no uncertainty about the microscopic state, we should be able to capture all the molecular energy. For example, if we knew precisely the position and velocity of each molecule in a gas at each instant, we would know exactly where to hook in little molecule catchers to catch their kinetic energy. So, the fact that we are uncertain about microscopic details makes it impossible for us to convert all the molecular energy to useful work; with increased uncertainty, in other words, with increased entropy, our ability to do useful work with a given amount of energy is reduced.

Another important aspect of the concept of entropy is that entropy is extensive. In other words, the entropy of a complex system is the sum of the entropies of its parts. This cannot be proven; it is as basic to the concept of entropy as the commutative law is to arithmetic. Expressed in microscopic terms, the disorganization of a large system is the sum of the disorganization of its components. Or, our uncertainty about the microscopic state of a large system is the sum of our uncertainties about the microscopic states of the parts.

A measure of the microscopic disorder should be zero when the molecules are perfectly organized, that is, when they are always in a unique microscopic state. In a perfect crystal of a single pure substance at a temperature of absolute zero, the molecules are completely motionless and stacked precisely in accord with the crystal structure and remain that way. This is a fully organized condition and is the only microscopic state that crystal ever exists in at this temperature. There is no uncertainty at all about the microscopic state, and so the entropy of the crystal is zero. This aspect of entropy is called the *third law of thermodynamics.*

Let's apply the second law to the isolated system of Fig. 6·1. Entropy is usually denoted by the symbol S, and we will denote the production of entropy by $\mathscr{P}_S$. In the next chapter we will learn how to obtain values for S. Since this is an isolated system, no entropy can flow in or out, hence any entropy change inside must be due to entropy production inside. So, the entropy bookkeeping is

$$\underbrace{\mathscr{P}_S}_{\text{entropy production}} = \underbrace{S_{\text{final}} - S_{\text{initial}}}_{\text{increase in entropy storage}} \qquad (6 \cdot 3)$$

The second law requires that the entropy production be zero or greater, so

$$S_{\text{final}} - S_{\text{initial}} \geq 0 \tag{6·4}$$

Once we learn how to evaluate the entropy of matter as a function of its state, we can calculate the entropies of states A and B, and we will find that $S_B > S_A$. Therefore it is possible for the system to go from state A to state B, but it is impossible for the system to go from state B to state A; the second law of thermodynamics just will not allow it.

When we know a little more about the second law, we will be able to do even more. For example, we will be able to calculate the maximum useful work that could be obtained from the system of Fig. 6·1. You can already estimate this available energy for state A as simply the kinetic energy of the flywheel; but even in state B there may be some available energy, *if* we can allow the system to interact with the environment. Just how much is available is typical of the important questions answered by the second law of thermodynamics.

As with the system of Fig. 6·1, for any *isolated* system the amount of entropy produced is simply equal to the change in entropy storage in the system. So, for an *isolated* system we can rephrase the second law as

$$\Delta S \geq 0 \qquad \text{isolated system} \tag{6·5}$$

In general, the inequality will hold, and the processes inside the isolated system will produce entropy. The limiting special case where the equality holds corresponds to an idealized process in which entropy is conserved. Equation (6·5) is a very common mathematical form of the second law of thermodynamics.

In the system of Fig. 6·1 we already know what can and cannot occur, so we really do not need the second law. However, the second law becomes tremendously useful when our intuition and experience fail. In particular, the second law can be used in analysis of chemical reactions to find out which reactions can or cannot occur. In engineering, the second law is vital in determining the best that can be achieved in any given situation. For example, we will soon be able to set a maximum on the amount of electric power that can be produced from a given supply of geothermal steam, or the least amount of power input required to run a heat pump, or the maximum amount of electric energy that can be obtained from a fuel cell, all through careful application of the first *and* second laws of thermodynamics. To do these things well the student must develop a good understanding of entropy and the second law, which is what this chapter is all about.

6·2 REVERSIBLE AND IRREVERSIBLE PROCESSES

Processes that do not violate the second law can be classed as reversible or irreversible. The concept of a reversible process is very important in thermodynamics, and the ability to recognize, evaluate, and reduce irreversibilities in a process is essential to a competent engineering thermodynamicist.

Suppose the system of interest is an isolated system. The second law says that any process that would reduce the entropy of the isolated system is impossible. Suppose a process takes place within the isolated system in what we shall call the forward direction. If the change in state of the system is such that the entropy increases for the forward process, then for the backward process (that is, for the reverse change in state) the entropy would decrease. The backward process is therefore impossible, hence we say that the forward process is *irreversible*. If the entropy is unchanged by the forward process, then it will be unchanged by the reverse process, and the process could go in either direction without violating the second law; such a process is called *reversible*. The key idea of a reversible process is that it does not produce any entropy.

Since a reversible process does not produce any entropy, the total molecular disorganization within the isolated system remains constant. It is impossible to tell which state in a reversible process came first; the reversible process leaves no "footprints in the sands of time." The reversible process is an idealization, like frictionless pulleys and resistanceless wires. It can be approached to a very high degree; for example, there is some evidence that at very low temperatures electric current flow can become exactly reversible. Currents started in loops of superconducting materials have been observed to persist for very long times without measurable decay. These experiments are perhaps the closest we have come to truly realizing a reversible process and may even be fully reversible. Some other processes that the student has probably idealized as being reversible in other courses are shown in Fig. 6·2. Note that each of the isolated systems of Fig. 6·2 happens to be an oscillator in which the oscillations persist indefinitely. At the end of each cycle the system is returned to its initial state, thereby "undoing" the state change of the first half cycle. Hence if the entropy increased during a portion of the

(*a*) *Resistanceless current flow*

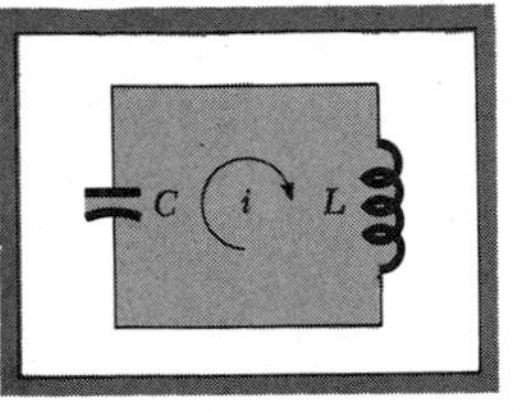

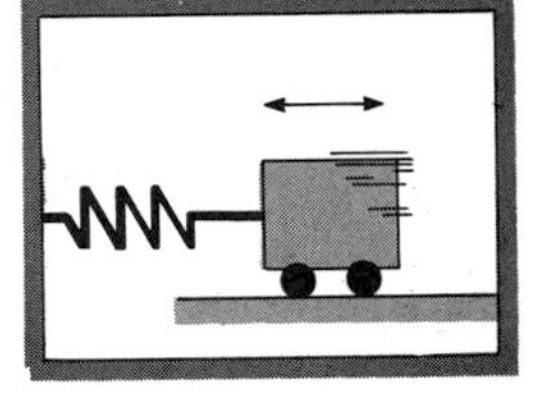

(*b*) *Frictionless motion*

(*c*) *Pneumatic spring*

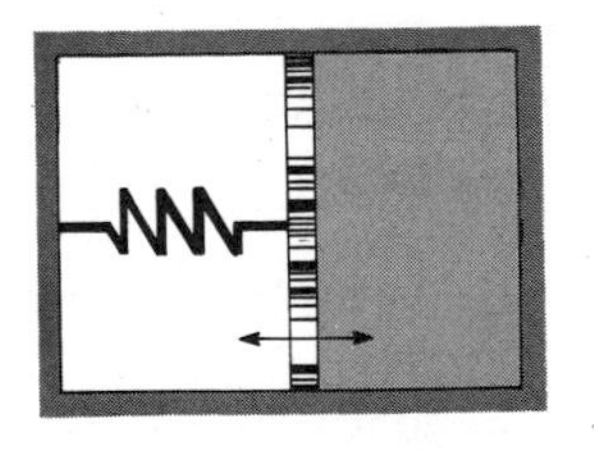

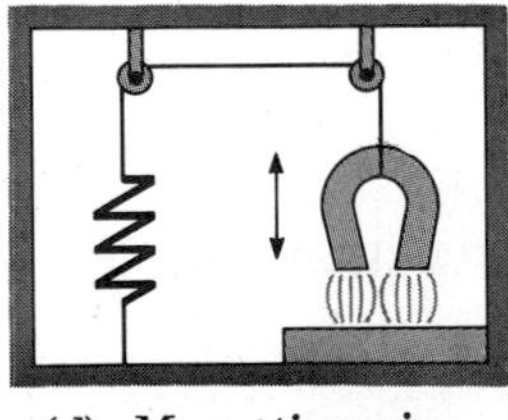

(*d*) *Magnetic spring*

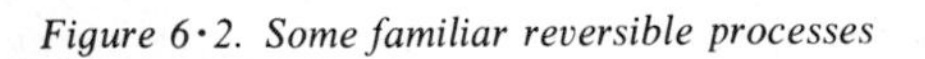

Figure 6·2. Some familiar reversible processes

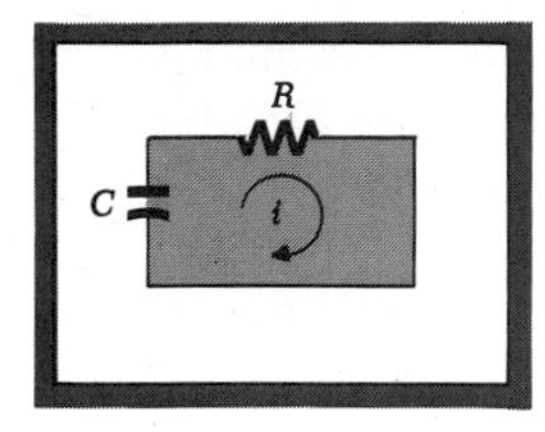

(*a*) *Current flow through a resistance*

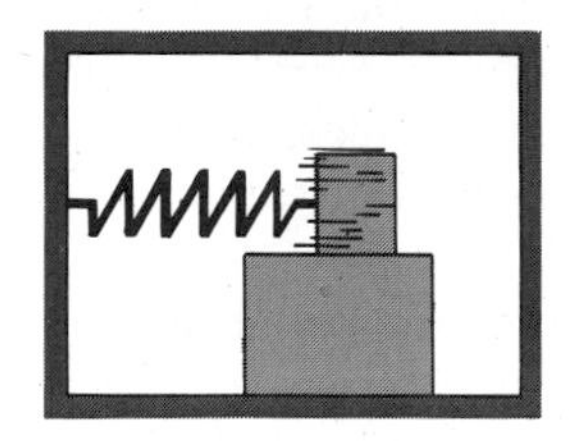

(*b*) *Motion with friction*

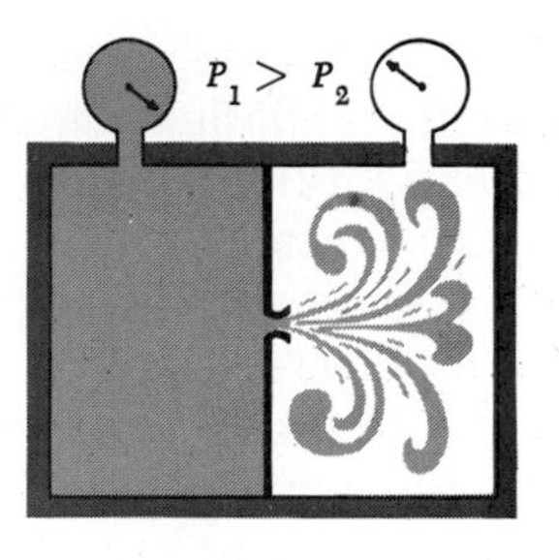

(*c*) *Unrestrained expansion*

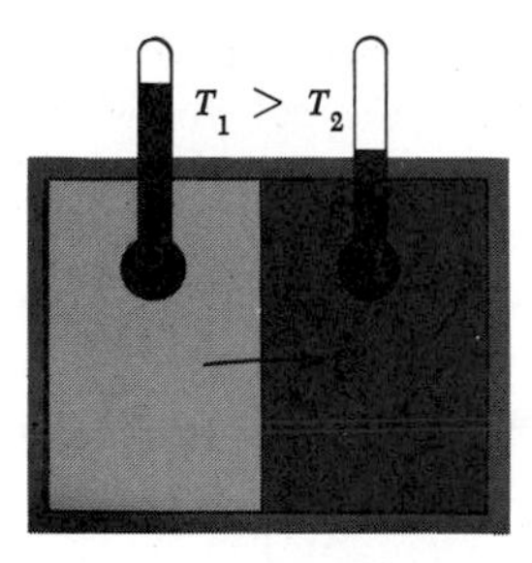

(*d*) *Energy transfer as heat*

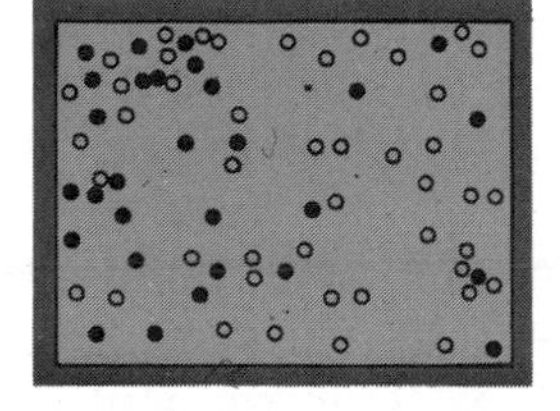

(*e*) *Diffusion*

(*f*) *Spontaneous chemical reaction*

Figure 6·3. Some irreversible processes

cycle, it would have to decrease during another portion in order to return to its initial value. But this would violate the second law, and thus we conclude that the entropy remains constant during each process. The processes are therefore (ideally) reversible.

An irreversible process is one that is not reversible, that is, one that *produces* entropy. All real processes (with the possible exception of superconducting current flows) are in some measure irreversible, though many processes can be analyzed quite adequately by assuming that they are reversible. Some processes that are clearly irreversible include mixing of two gases, spontaneous combustion, friction, and the transfer of energy as heat from a body at high temperature to a body at low temperature. These and some other obviously irreversible processes are shown in Fig. 6·3. Each process of Fig. 6·3 results in some increase in the entropy of the isolated system, that is, in some entropy production. Qualitative microscopic interpretations of this entropy production will now be given.

In Fig. 6·3*a* an electric current flows through a resistance following the closing of a switch. We know from experience that the capacitor eventually becomes completely discharged, the wires become warmer, and finally the current flow ceases entirely. The electrons, which initially were concentrated on the plates of the capacitor, are now distributed throughout the system, and the dielectric in the capacitor has lost the molecular organization which gave it a net dipole moment. This randomization accounts for the entropy increase.

The spring-mass system in Fig. 6·3*b* oscillates at smaller and smaller amplitudes, until finally friction brings the mass to rest. Our experience tells us that the temperature of the system increased during the decay of the oscillation. The highly directed kinetic energy of the mass has been converted into the randomly oriented motion of the molecules, and this extra randomness accounts for the entropy increase.

In Fig. 6·3*c* we see gas leaving a high-pressure container and flowing into one of lower pressure. The entropy increase required by the second law is reflected by the increased uncertainty as to which half of the box a molecule is in.

Energy transfer as heat from a hot system to a cooler one is also irreversible (Fig. 6·3*d*). If this were not the case, we could expect to watch an isothermal system and eventually observe an energy transfer that would produce a nonuniformity of temperature. This would indeed be handy, for we could get hot water from ice for nothing. The energy transfer as heat tends to spread the energy over the system, and this increases the uncertainty about the microscopic state.

The system in Fig. 6·3*e* initially has oxygen on one side and nitrogen on the other. The molecules diffuse through one another, and eventually we have a homogeneous mixture. The mixing increases the randomness and uncertainty about the microscopic state, and this is reflected in the increase in entropy.

The spontaneous chemical reaction within the system of Fig. 6·3*f* is irreversible. Oxygen and hydrogen might explode in an isolated container and form water, but the reverse process never takes place. The spontaneous combustion process converts electron and molecular binding energy into randomly oriented translational energy, and this randomization is evidenced by the increase in entropy.

Recognition of the irreversibilities in a real process is especially important in engineering. Irreversibility, or departure from the ideal condition of reversibility, reflects an increase in the amount of disorganized energy at the expense of organized energy. The organized energy (such as that of a raised weight) is easily put to practical use; disorganized energy (such as the random motions of the molecules in a gas) requires "straightening out" before it can be used effectively. And since we are always somewhat uncertain about the microscopic state, this straightening can never be perfect. Consequently the engineer is constantly striving to reduce irreversibilities in systems in order to obtain better performance.

To summarize:

Processes that are usually idealized as *reversible* include

- Frictionless movement
- Restrained compression or expansion
- Energy transfer as heat due to infinitesimal temperature nonuniformity
- Magnetization, polarization
- Electric current flow through a zero resistance
- Restrained chemical reaction
- Mixing of two samples of the same substance at the same state

Processes that are *irreversible* include

Movement with friction
Unrestrained expansion
Energy transfer as heat due to large temperature nonuniformities
Magnetization or polarization with hysteresis
Electric current flow through a nonzero resistance
Spontaneous chemical reaction
Mixing of matter of different composition or state

Irreversible processes always produce entropy and result in a degradation of energy.

6·3 A QUANTITATIVE EXAMPLE

Although we have not yet developed fully the concept of entropy, it will be helpful now to work through a quantitative illustration.

As we will see in the subsequent chapter, the dimensions of entropy are energy/temperature, or J/K, Btu/R, etc. We have noted that entropy is an extensive property; it is often convenient to deal with the specific entropy, or entropy per unit of mass,

$$s \equiv \frac{S}{M}$$

Thus s will have units such as kJ/(kg·K) or Btu/(lbm·R). Values of s for many substances are included in the graphs and tables of Appendix B. We will find out where these values come from in the next chapter. We will use the tabulated values for water in our calculation.

The system we will consider is shown in Fig. 6·4. It consists of a rigid, insulated container (an isolated system) initially filled with 4.22 kg of saturated water vapor at 10 MPa and 5.78 kg of saturated liquid water at 1 MPa. This system is clearly not in equilibrium initially, for there are severe pressure and temperature variations inside. The interactions that will occur spontaneously within the isolated system will smooth out these temperature and pressure differences; some of the vapor will condense, and this will lower the pressure and vapor temperature. The liquid will warm up, and finally the system will come to rest with

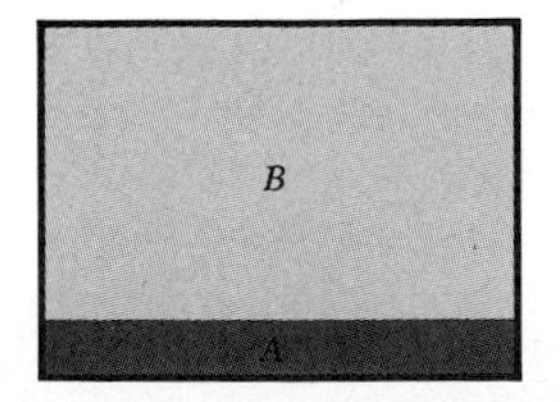

Figure 6·4. The system is isolated; the irreversible process produces entropy

a uniform temperature and pressure inside. Calculation of the final state requires only the ideas of equilibrium and a first-law analysis. The second law can be applied to see if the entropy increases within the isolated system; if it does, the process is irreversible, and the reverse process would be impossible. Intuition tells us that this is the case; indeed, so does the second law, which is really just a generalization of this kind of intuition. We shall carry out the first- and second-law analyses in some detail to illustrate the computation of entropy from the tabular and graphical equations of state and to confirm the expectation that the process does produce entropy.

Let's first evaluate the initial state. We assume that the liquid A and the vapor B are each in equilibrium states, hence we can use the tables for the thermodynamic properties of water (Table B·1*b*) to evaluate their properties.

Liquid A: 5.78 kg of saturated liquid at 1 MPa (180°C)

$$u = 761.7 \text{ kJ/kg}$$
$$v = 0.001127 \text{ m}^3\text{/kg}$$
$$s = 2.1391 \text{ kJ/(kg·K)}$$

Vapor B: 4.22 kg of saturated vapor at 10 MPa (311°C)

$$u = 2544.4 \text{ kJ/kg}$$
$$v = 0.01803 \text{ m}^3\text{/kg}$$
$$s = 5.6149 \text{ kJ/(kg·K)}$$

The total energy, volume, and entropy of the system in its initial state are therefore

$$U_1 = 5.78 \times 761.7 + 4.22 \times 2544.4 = 15{,}140 \text{ kJ}$$
$$V_1 = 5.78 \times 0.001127 + 4.22 \times 0.01803 = 0.0826 \text{ m}^3$$
$$S_1 = 5.78 \times 2.1391 + 4.22 \times 5.6149 = 36.07 \text{ kJ/K}$$

The next job is to determine the final state (2) using a first-law analysis. Assuming that the system is isolated, the energy balance gives

$$U_2 = U_1$$

Also, the volume is fixed, hence

$$V_2 = V_1$$

We assume that the entire system is in an equilibrium state at the end of the process. The internal energy and volume will therefore fix state 2. The specific internal energy and specific volume in the final state will be

$$u_2 = \frac{U_2}{M} = \frac{15{,}140}{10} = 1514 \text{ kJ/kg}$$

$$v_2 = \frac{V_2}{M} = \frac{0.0826}{10} = 0.00826 \text{ m}^3/\text{kg}$$

Now, u_2 and v_2 fix the final state. If we had a metric version of Fig. B·1, we could read off the final pressure and temperature directly. The value of u_2 is such that we can be sure the final state is a liquid-vapor mixture, in equilibrium at some pressure, probably between 10 and 1 MPa. By trial with a number of pressures, one finds that the final pressure is 6 MPa, as follows:

At 6 MPa,

$$v_f = 0.001319 \text{ m}^3/\text{kg} \qquad v_g = 0.03244 \text{ m}^3/\text{kg}$$

$$u_f = 1205.4 \text{ kJ/kg} \qquad u_g = 2589.7 \text{ kJ/kg}$$

$$s_f = 3.0273 \text{ kJ/(kg·K)} \qquad s_g = 5.8900 \text{ kJ/(kg·K)}$$

So, from v_2 we calculate the final quality as

$$(1 - x)(0.001319) + x(0.03244) = 0.00826 \qquad x = 0.223$$

At this quality the internal energy is

$$u = 0.777 \times 1205.4 + 0.223 \times 2589.7 = 1514 \text{ kJ/kg}$$

which agrees with u_2. Hence the final state is

$$x = 0.223 \qquad P = 6 \text{ MPa}$$

The entropy of the system in this state is

$$S_2 = 10 \times (0.777 \times 3.0273 + 0.223 \times 5.8900) = 36.66 \text{ kJ/K}$$

Since the system is isolated, there can be no transfers of entropy, and the entropy production is therefore just the increase in entropy storage; so, the entropy bookkeeping is

$$\underset{\text{entropy production}}{\mathscr{P}_S} = \underset{\text{increase in entropy storage}}{S_2 - S_1}$$

The second law requires that the entropy production be positive, or at the very least zero,

$$\mathscr{P}_S = S_2 - S_1 \geq 0$$

We would certainly be surprised if this were not the case for the problem at hand; let's check:

$$\mathscr{P}_S = 36.66 - 36.07 = 0.59 \text{ kJ/K}$$

The fact that $\mathscr{P}_S$ is small compared to either S_1 or S_2 has no particular significance, since the entropy datum of Fig. B·1 was arbitrarily chosen. Since the process does produce entropy, it is irreversible, as expected.

6·4 ENTROPY TRANSFER AND CHANGE

In the previous section we concentrated on isolated systems, in which entropy can only increase. But what about nonisolated systems? Consider the freezing of a liquid (Fig. 6·5*a*). In the initial state the liquid molecules move about rather freely in a somewhat disorganized way. By transferring energy out of the liquid as heat, we can cause freezing to occur, and in the solid state there is clearly much less molecular disorganization. It seems as though the entropy of a substance should decrease when it is frozen; how can this be consistent with the idea that entropy must always be produced?

To find the answer, it is helpful to consider both the system that is frozen and the environment to which the energy taken from the liquid is transferred. Imagine that the environment is simply another solid substance at its melting point (Fig. 6·5*b*). Then, as energy is transferred out of the first substance into the second, the first freezes, but some melting occurs in the second. So, as the first substance becomes more organized microscopically and its entropy decreases, the molecules of the second substance become less organized and the entropy of this substance increases. The second law requires only that the *total* entropy of the

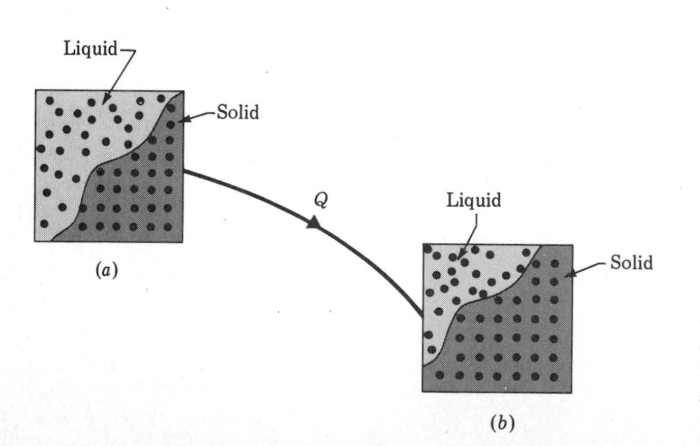

Figure 6·5. When a substance is frozen, its entropy is reduced. But the combined entropy of the substance and its environment is not reduced

isolated system increase; in this case, the entropy of one part goes down, the entropy of another part goes up, and the total entropy increases.

The same sort of behavior is observed in evaporation and condensation. Evaporation (boiling) requires energy addition as heat and increases the entropy. For example, for water at 100°C, from the metric version of Table B·1*a* the specific entropies of the saturated liquid and vapor are

$$s_f = 1.3071 \text{ kJ/(kg·K)} \qquad s_g = 7.3557 \text{ kJ/(kg·K)}$$

Indeed, s_g is greater than s_f. But we can also condense the steam back to water by cooling, reducing the entropy from s_g to s_f. The energy we put elsewhere to do this will result in an increase in entropy elsewhere, and the total increase in entropy will be positive.

Now let's return to the question of entropy change in a system that is not isolated. We have just seen an example where cooling results in a reduction in entropy of the system cooled at the expense of an increase in entropy of its environment. We can think of this as resulting from a *transfer of entropy* from the system to its environment. Transfers of entropy are always associated with energy transfers as heat. After all, heat is a disorganized energy-transfer process (see Fig. 2·11), and so one certainly would expect some disorganization to "flow" with the energy. In engineering analysis we will find it particularly convenient to use this concept of *entropy transfer with heat.* In the next chapter we will develop an expression for the magnitude of this entropy transfer.

What about entropy transfer associated with energy transfer as work? Well, work is a microscopically organized energy transfer, and therefore you might correctly guess that there is no entropy transfer associated with work.† Figure 6·6 presents an argument in support of this. We imagine that the system delivers work to a flywheel, where the energy is stored in a fully recoverable form. The flywheel molecules are simply put into rotation around the axis in a perfectly organized manner, and there is no increase in the entropy of the flywheel. We will develop the idea that "work carries no entropy" in the next chapter.

To summarize, in a nonisolated system the entropy of the system may increase or decrease, and entropy can be transferred in or out, so long as the net entropy production is not less than zero. We will explore these matters more in the next chapter.

† The view that work is "entropy-free" is really the only way one can intelligently distinguish between heat and work. Work is energy transfer *without* associated entropy transfer, and heat is energy transfer *with* associated entropy transfer. These concepts of heat and work are rather sophisticated for the present stage of the student's development, but reflection upon them at a later time is recommended.

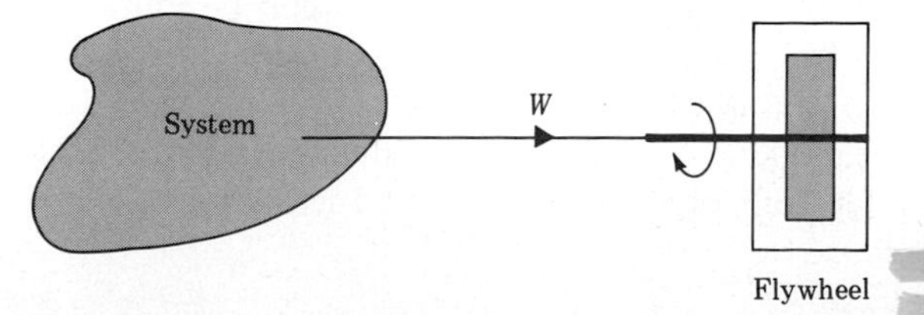

Figure 6·6. There is no transfer of entropy associated with transfer of entropy as work. Work can be "stored" without changing the entropy of the device that receives the energy

6·5 SUMMARY OF THE ENTROPY CONCEPT

We have developed four basic ideas that are all essential to the concept of entropy:

1. Every system has entropy; entropy measures the degree of microscopic disorganization, our uncertainty about the microscopic state.
2. Entropy is extensive; the entropy of a system is the sum of the entropies of its parts.
3. Entropy can be produced, but never destroyed. Consequently, the entropy of an isolated system can never decrease.
4. The entropy of a system that always exists in a unique microscopic state is zero.

Statement 3 is a form of the second law of thermodynamics, which we shall explore in greater detail in the next chapter. Statement 4 is equivalent to the third law of thermodynamics, which is discussed in Chap. 11.

We have also introduced two important definitions:

1. *Reversible process*—one that produces no entropy
2. *Irreversible process*—one that produces entropy

Reversible processes are ideals that form useful limits to what can really be accomplished by human technology. The notion that energy is degraded to less useful forms by irreversible processes was discussed and will be examined in depth in the next chapter.

The amount of entropy production is denoted by $\mathscr{P}_S$. For an infinitesimal process, the infinitesimal amount of entropy production will be denoted by $đ\mathscr{P}_S$. Since the entropy production is a function of the process and not a function of the system states, the notation $đ$ is appropriate. Note that entropy production represents the strength of the inequality in the second law. The second law says that entropy production cannot be negative; entropy production is zero for a reversible process:

$$đ\mathscr{P}_S > 0 \qquad \textit{irreversible process}$$

$$đ\mathscr{P}_S = 0 \qquad \textit{reversible process}$$

$$đ\mathscr{P}_S < 0 \qquad \textit{impossible process}$$

In order to use the second law in an analysis we must first learn how to evaluate entropy quantitatively as a function of state. In statistical thermodynamics entropy is defined in terms of probabilities of the microscopic states. Entropy can then be computed from this definition if the probabilities can be found. In macroscopic thermodynamics the basic concepts of what entropy is and does are employed in order to learn how to evaluate entropy. While the microscopic ideas greatly assist understanding, most practical analysis does not require consideration of the microscopic view. In fact, entropy can be evaluated from macroscopic laboratory data without any explicit consideration of microscopic states. We shall see how this is done in Chap. 7.

In Chap. 7 we will start with the four ideas listed above and use the microscopic ideas only to provide additional insight. The student who does not wish to delve more deeply into the statistical underpinnings of the entropy concept can jump to Chap. 7 from this point without difficulty.

The remainder of this chapter is devoted to further consideration of the statistical view of entropy. The objective of this material is to provide the student with a better understanding of entropy, which we feel strengthens one's ability to deal with the second law in practical analysis. We emphasize that the ability to work quantitatively with the statistical ideas is *not* required now.

6·6 THE STATISTICAL DEFINITION OF ENTROPY

As with energy, the understanding of entropy can be greatly enhanced by consideration of the microscopic nature of matter. An analogy will be helpful in grasping the microscopic concepts underlying the second law. Suppose we have a large tray containing a large number of "educated jumping beans," half red and half white. Imagine that we have taught the beans to jump in pairs, thereby trading places, and we set up the tray initially with the white beans on one side and red on the other. Observers viewing the tray from a great distance could not see individual beans and would say, "That object is white on one side and red on the other." Now we let the beans start to jump; soon our observers might comment, "The red stuff is diffusing through the white; and the system is becoming pink." After a while the tray would appear to our observers a uniform pinkish hue, and it would seem that all changes had stopped. From our closer proximity we would see continual change, with the red and white beans relatively evenly distributed. Occasionally we would see momentary concentrations of one color in various spots, but these would disappear quickly and would not be noticed by the sluggish eye of our distant observers. We could repeat this experiment many times; each time the observers would note that the red diffused through the white, and that after a while an equilibrium condition was reached. If the observers noticed this reproducibility, they might attempt to construct a mathematical theory that would explain the diffusion and perhaps even predict its rate. Even in their ignorance of the beans, they might make a rather good theory. But certainly consideration of the beans would allow a better and more understandable theory.

We might permit our observers to experiment on the tray. They could enlarge it, distort it, shake it, but we would allow the observers to have no permanent control over the behavior of *individual* beans. The beans would remain free to act as they pleased, jumping randomly from spot to spot. Learning of the existence of the beans, our observers might try to make a statistical theory by postulating that, left to themselves, the bean arrangement tends to become more and more random. As a consequence the observers would be uncertain about the instantaneous arrangement of the beans, and as time went on any uncertainty the observers had would surely increase. To make these ideas quantitative would require some *numerical measure* of the randomness or of the uncertainty about the detailed bean arrangement when the tray is viewed from afar. When the tray looked half white and half red, the beans were relatively well organized and our observer could be relatively certain as to the bean arrangement. As time passed and the tray became pink, the bean arrangements would become more disorganized, and our distant observers would be much less certain about the instantaneous detailed state. The bean randomness or disorder, and the uncertainty, would increase, and the observers' theory could be based on the postulate that the value of the randomness-uncertainty measure could never decrease.

This analogy has many points of contact with the notions of entropy and the second law of thermodynamics. We can replace the beans by atoms of argon and helium, the tray by an isolating wall, and the observer by ourselves. The same sort of diffusion process would be observed. The quantitative measure of the microscopic randomness, of our uncertainty as to the exact microscopic state when we know only the macroscopic state, is the entropy; we find we can explain the directions of all processes observed in nature with theory developed from the postulate that the microscopic randomness, that is, the *entropy*, can be produced but is never destroyed.

Quantum states. What are the possible microscopic states of a system? A key idea that we take from quantum mechanics is that the states which atoms, molecules, and entire systems have are *discretely quantized.* For example, a photon associated with radiation of a particular frequency ν can have but one energy. This is given by the Einstein-Planck equation,

$$\varepsilon = \mathsf{h}\nu \tag{6·6}$$

The constant h is *Planck's constant*, 6.626×10^{-34} J·s. The energy traveling with radiation is therefore said to be *quantized.*

Other evidence for quantization is provided by measurements of the energy of electrons orbiting atoms; it is found that the electrons can possess only very particular energies. For example, the electronic quantum states of hydrogen are shown in Fig. 6·7.

The quantum theory of matter and energy seemingly meets every test

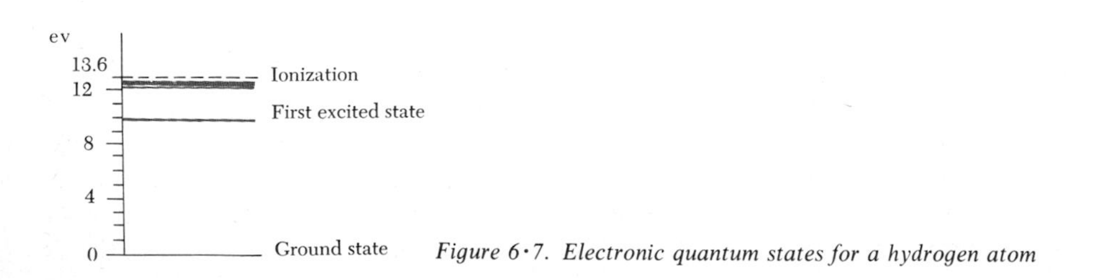

Figure 6·7. Electronic quantum states for a hydrogen atom

humans have been able to devise. By postulating that any particle or system of particles constrained in some manner (such as being required to orbit a nucleus, being in a magnetic field, or being in a box of specified size) can exist only in certain *allowed quantum states*, we obtain amazingly complete and precise descriptions of all natural phenomena. The allowed states are determined by the nature of the particles and the circumstances in which they find themselves; an equation, known as the *Schrödinger equation*, has been postulated as the basis for calculation of the allowed quantum states. For the present we need only the idea that the allowed states of a system are quantized, and it is not necessary to be able to evaluate these states.

It is important to appreciate that quantization refers to very detailed *microscopic* descriptions of state. At any instant a piece of matter must be in one of its allowed quantum states. Macroscopic instruments, which average over time, will reflect averages over the sequence of quantum states; hence macroscopic descriptions of state are very much less detailed. For example, a microscopic description of state might require values for the energy and position of *each particle*, while macroscopic descriptions might involve only the total energy, volume, the temperature, pressure, and of course the entropy.

Quantum-state probabilities. Systems of many interacting molecules undergo continual change in their quantum state. Macroscopic instruments are unable to follow these rapid changes, hence tend to average over a long sequence of quantum states. Since we have essentially no control over individual molecules, it is really quite impossible for us to force the system to be in a particular quantum state at the start of an experiment; hence if we repeat the experiment a number of times a different initial quantum state will be involved in each repetition. Consequently a different sequence of quantum states will be traced out in each experiment. A large set of such repetitions (real or hypothetical) is called an *ensemble* of experiments.

Traces of the quantum-state numbers (no one has ever measured these in a real experiment) might look something like those in Fig. 6·8. The sequences of Fig. 6·8*a* are nearly identical; at each time a narrow band of quantum states is found. In contrast, the sequences of Fig. 6·8*b* are much more random; at each instant there is a wide variation in quantum states between experiments. Consequently, a guess about the instantaneous quantum state in a particular experiment might be quite close for sequence (*a*), but may be way off for sequence (*b*). A *quantitative* measure of the *likelihood* of a particular quantum state existing at some time would be useful. The *quantum-state probability* $p_i(t)$ provides this measure. $p_i(t)$ is defined as the fraction of the ensemble of experiments in which quantum state i would be realized at time t. If the realized quantum states (those which actually occur) are few, the p_i will be high. If the realized quantum states are many, the p_i will be low. If in every experiment the same quantum state exists at time t, the probability of that quantum state will be unity and the probability of all other quantum states will be zero. The set of all the p_i is called the *distribution;* it reflects

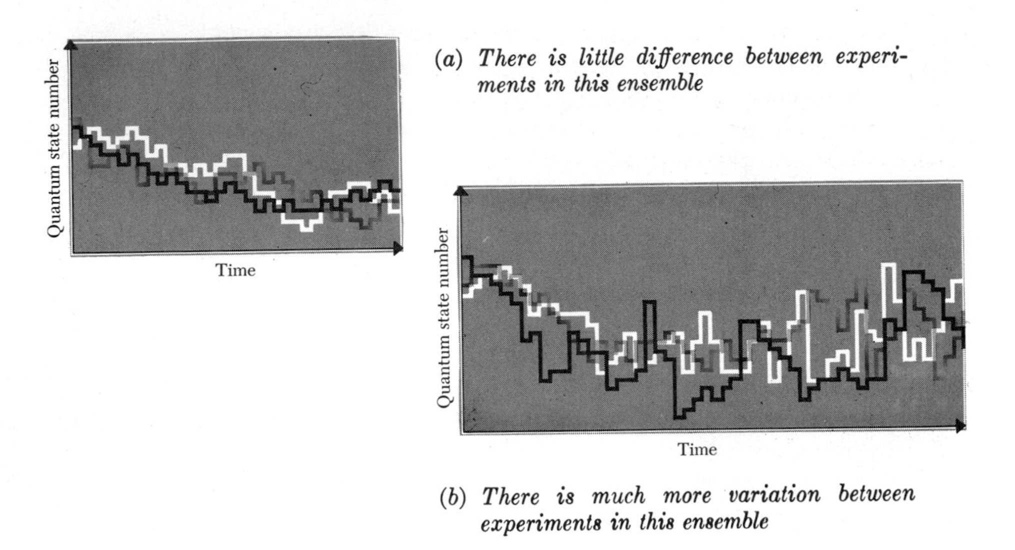

Figure 6·8. Quantum-state sequences; each trace represents a different experiment

the relative likelihood of particular quantum states. Since the sum of the probabilities of all the allowed quantum states must be unity,

$$\sum_i p_i = 1 \tag{6·7}$$

Suppose we allow a system to reach equilibrium with its environment. Thereafter there are no macroscopic changes, though the microscopic state undergoes continual change. In equilibrium there is nothing special about any point in time, and consequently the p_i must be independent of time. The resulting distribution is called the *equilibrium distribution.* In equilibrium p_i can be taken as the fraction of *time* a single system spends in the ith allowed quantum state. The time-fraction definition of probability is valid only in equilibrium; for nonequilibrium the ensemble-fraction definition must be used.

The system properties used in macroscopic analysis are averages over the continuous sequence of microscopic states. Since in equilibrium the p_i represent the fraction of time spent in each state, if we know any property G_i of each quantum state, we can calculate the average G as

$$\langle G \rangle \equiv \sum_i p_i G_i \tag{6·8}$$

Thus, for example, if ε_i is the energy of the ith quantum state, the average energy is

$$\langle E \rangle = \sum_i p_i \varepsilon_i \tag{6·9}$$

This type of average is also used for systems that are not in equilibrium, that is, where the p_i depend upon time. The average can then be interpreted as an average over an ensemble (real or hypothetical) of experiments; the ensemble average reflects the expected value of a property in any particular experiment, hence ensemble averages are sometimes called "expectation values."

The probabilities indicate the randomness of the equilibrium quantum states. For example, suppose we have a system that can exist in one of three quantum states. If the equilibrium distribution is

$$p_1 = 1 \qquad p_2 = 0 \qquad p_3 = 0 \tag{a}$$

we know that the system is always in quantum state 1; the randomness is zero. If asked to guess the instantaneous quantum state, we could say "one" with absolute certainty. Suppose that the equilibrium distribution is

$$p_1 = 0.8 \qquad p_2 = 0.2 \qquad p_3 = 0 \tag{b}$$

This says that 80 percent of the time the system is in quantum state 1, and 20 percent of the time it is in quantum state 2. Asked to guess the instantaneous state we would again say "one," and in the long run we would be correct 80 percent of the time. This distribution shows more randomness than case (*a*) above. Our uncertainty as macroscopic observers about the instantaneous microscopic state is certainly greater in (*b*) than in (*a*). If the equilibrium distribution is

$$p_1 = 0.8 \qquad p_2 = 0.1 \qquad p_3 = 0.1 \tag{c}$$

or

$$p_1 = 0.1 \qquad p_2 = 0.8 \qquad p_3 = 0.1 \tag{d}$$

the quantum-state sequence will show even more variation or randomness, and we shall be even more uncertain about the instantaneous quantum state. However, the same uncertainty will apply to both (*c*) and (*d*). The most randomness is indicated by the distribution

$$p_1 = p_2 = p_3 = \tfrac{1}{3} \tag{e}$$

Here each allowed quantum state is equally likely, and we can only guess the instantaneous state with 33 percent chance of being correct. Note that the broader the distribution is, the more uncertain we must be about the instantaneous quantum state.

Entropy definition. We have seen that the distribution conveys a qualitative impression of the randomness of the sequence of quantum states, that is, of our uncertainty as to the instantaneous microscopic state. A list of all the p_i would convey a very good picture of the randomness and uncertainty. This list would be far too long to be usable; it would be much more convenient if we could use a *single number* to measure the amount of randomness and uncertainty reflected by the entire list of p_i.

Since the entire set of p_i reflects the quantum-state randomness, it would seem appropriate to look for some function of all the p_i that would serve as a single measure of randomness. We shall identify this randomness measure with the *entropy* and denote it by S. Since entropy by concept is extensive, the entropy of system C, composed of parts A and B, must be given by

$$S_C = S_A + S_B \tag{6·10}$$

The randomness measure must be constructed so as to give greater values of S when the system is more random; that is, it should agree with the qualitative ideas about randomness. We are now going to use these ideas to discover an appropriate definition of S in terms of the probability distribution.

Now, we ask what function of the p_i might be useful as a single numerical measure of the randomness? The product of the probabilities $p_1 \times p_2 \cdots \times p_n$ will not do, because it will be zero when *any* one of the p_i is zero. The sum of the p_i will not do either, because it always has the value unity. The *average* probability is sometimes used in statistics as a measure of randomness, so let's see if it will work. The average probability is [see Eq. (6·8)]

$$\langle p \rangle = \sum_i p_i p_i = \sum_i p_i^2$$

When $\langle p \rangle$ is high, only a few quantum states are realized, and there is little randomness. When it is low, many quantum states are realized, and there is much randomness and uncertainty. Hence the average probability does seem to fit the qualitative requirements of a randomness measure. Is it extensive? To investigate this we first make use of Eq. (6·10). Suppose there are n allowed states for system A and m allowed states for system B. If the allowed states in each part are independent of those in the other, the quantum state for the combined system C can be described by specifying the quantum states of A and B; we denote these by i and j, respectively, and suppose that they are numbered such that $i = 1, 2, \ldots, n$, $j = 1, 2, \ldots, m$. The allowed states for C can then be denoted by the number pair i, j. For example, C-state 2,6 is the state where part A is in its state $i = 2$ and part B is in its state $j = 6$. There are $n \cdot m$ allowed states for C, corresponding to the $n \cdot m$ points in Fig. 6·9.

The probability of each C state can be computed from the individual probabilities for A and B states. If p_i is the probability of A-state i, and p_j the probability of B-state j, then

$$p_{ij} = p_i \cdot p_j \tag{6·11}$$

is the probability of the C-state i, j. This follows from a basic part of the concept of probability, namely, that probabilities are such that the joint probability of two

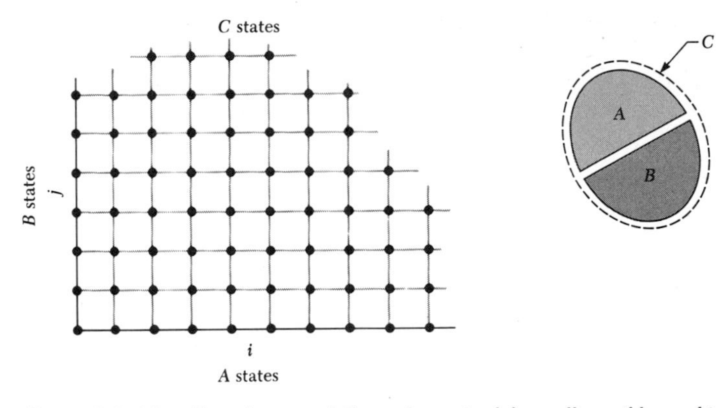

Figure 6·9. The allowed states of C are determined from all possible combinations of allowed states of A and B. Each point represents a C state

independent events (the realization of *A*-state *i* and the realization of *B*-state *j*) is the *product* of their individual probabilities.†

Now, the average probabilities of *A* and *B* are

$$\langle p \rangle_A = \sum_{i=1}^{n} p_i^2 \qquad \langle p \rangle_B = \sum_{j=1}^{m} p_j^2$$

and the average probability of *C* is

$$\langle p \rangle_C = \sum_{i=1}^{n} \sum_{j=1}^{m} p_{ij}^2 = \sum_{i=1}^{n} \sum_{j=1}^{m} (p_i p_j)^2$$

Rearranging the sums,

$$\langle p \rangle_C = \sum_{i=1}^{n} \left[p_i^2 \left(\sum_{j=1}^{m} p_j^2 \right) \right] = \sum_{i=1}^{n} p_i^2 \langle p_B \rangle = \langle p_A \rangle \langle p_B \rangle$$

Noting that

$$\langle p \rangle_C \neq \langle p \rangle_A + \langle p \rangle_B$$

we conclude that the average probability is *not* extensive; as yet we have not found an appropriate definition for entropy.

Let's be more systematic and ask: What function of the probability *does* have an average which is extensive? This leads us to look for a definition of entropy *S* in the form

$$S = \langle f \rangle = \sum_i p_i \, f(p_i) \tag{6·12}$$

† For example, if there is only an 0.2 chance that *A*-state 2 will exist, and only an 0.3 chance that *B*-state 6 will exist, then there is only an 0.06 chance of them both existing at the same time, that is, *C*-state 2,6 has a probability of 0.06.

The function $f(p_i)$ remains to be found. Then

$$S_A = \sum_{i=1}^{n} p_i\, f(p_i) \qquad S_B = \sum_{j=1}^{m} p_j\, f(p_j)$$

$$S_C = \sum_{i=1}^{n}\sum_{j=1}^{m} p_{ij}\, f(p_{ij}) = \sum_{i=1}^{n}\sum_{j=1}^{m} p_i p_j f(p_i p_j)$$

Applying Eq. (6·10),

$$\sum_{i=1}^{n}\sum_{j=1}^{m} p_i p_j f(p_i p_j) = \sum_{i=1}^{n} p_i\, f(p_i) + \sum_{j=1}^{m} p_j\, f(p_j) \tag{6·13}$$

f must be such that this is true regardless of the values of the p_i and p_j. A function that will work is $f(\ ) = \ln(\ )$. With this choice, the left-hand side of Eq. (6·13) becomes [recall that $\ln(a \cdot b) = \ln a + \ln b$]

$$\sum_{i=1}^{n}\sum_{j=1}^{m} p_i p_j \ln p_i + \sum_{i=1}^{n}\sum_{j=1}^{m} p_i p_j \ln p_j$$

Rearranging the sums, this becomes

$$\sum_{i=1}^{n}\left[p_i \ln p_i\left(\sum_{j=1}^{m} p_j\right)\right] + \sum_{j=1}^{m}\left[p_j \ln p_j\left(\sum_{i=1}^{n} p_i\right)\right]$$

since

$$\sum_{i=1}^{n} p_i = 1 \qquad \sum_{j=1}^{m} p_j = 1$$

the left-hand side of Eq. (6·13) may be expressed as

$$\sum_{i=1}^{n} p_i \ln p_i + \sum_{j=1}^{m} p_j \ln p_j$$

Hence Eq. (6·13) is precisely satisfied for *any* p_i, p_j, n, and m with this choice for f. It may be shown that† the most general $f(p_i)$ whose average is an extensive property is $f = C \ln p_i$, where C is an arbitrary constant; since p_i is less than unity, we choose the constant to be negative to make the entropy positive. Thus we have discovered the statistical definition of entropy,

► $$S \equiv -\mathrm{k} \sum_{i} p_i \ln p_i \tag{6·14}$$

† To show this, consider the special case where $p_i = 1/n$ and $p_j = 1/m$. Then Eq. (6·13) becomes

$$\sum_{i=1}^{n}\sum_{j=1}^{m} \frac{1}{n}\frac{1}{m} f\left(\frac{1}{n}\frac{1}{m}\right) = \sum_{i=1}^{n} \frac{1}{n} f\left(\frac{1}{n}\right) + \sum_{j=1}^{m} \frac{1}{m} f\left(\frac{1}{m}\right)$$

(*Continued*)

The constant k is chosen as the *Boltzmann* constant,

$$k = 1.380 \times 10^{-23} \text{ J/K}$$

This choice makes the thermodynamic temperature (defined in Chap. 7) the same as the empirical temperature used in previous chapters.

Characteristics of entropy. We must now become convinced that the entropy has the proper qualitative features of a randomness or uncertainty measure. Arguments in support of this contention will now be given.

Since the uniform probability distribution reflects the greatest randomness, no system with n allowed states should ever have a greater entropy than when each state is equally likely ($p_i = 1/n$). Indeed, the maximum value of S occurs for this case. We can see this easily for $n = 2$. Then

$$S = -k(p_1 \ln p_1 + p_2 \ln p_2)$$

Since $$p_1 + p_2 = 1$$

Carrying out the sums

$$f\left(\frac{1}{n} \cdot \frac{1}{m}\right) = f\left(\frac{1}{n}\right) + f\left(\frac{1}{m}\right)$$

The function $f(\ )$ must therefore satisfy the *functional equation*

$$f(xy) = f(x) + f(y)$$

Functional equations are usually solved by obtaining a differential equation which the function must satisfy. To do this, we differentiate the functional equation with respect to x, obtaining

$$y \frac{df(xy)}{d(xy)} = \frac{df(x)}{dx}$$

Differentiating this with respect to y,

$$\frac{df(xy)}{d(xy)} + (xy) \frac{d^2f(xy)}{d(xy)^2} = 0$$

Letting $z = xy$, this is

$$z \frac{d^2f}{dz^2} + \frac{df}{dz} = 0$$

The general solution of this linear second-order differential equation is

$$f(z) = C_1 + C_2 \ln (z)$$

The functional equation will not be satisfied unless $C_1 = 0$. Hence the most general solution to the functional equation is indeed $f(z) = C \ln (z)$.

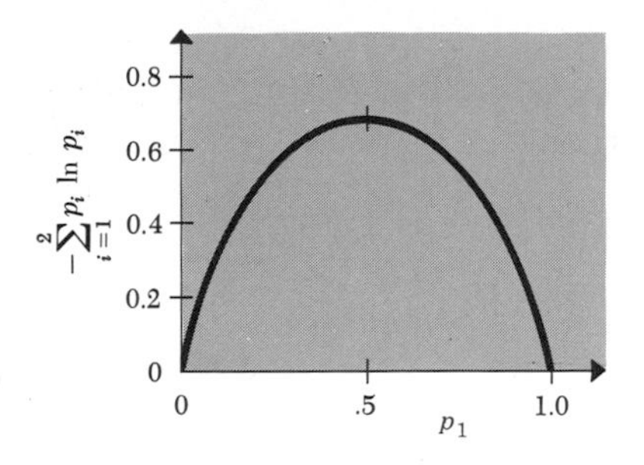

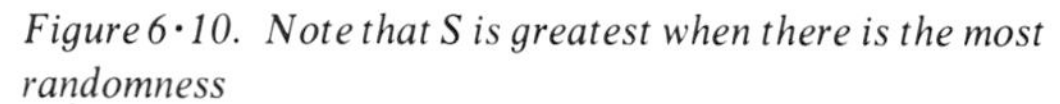
Figure 6·10. Note that S is greatest when there is the most randomness

p_1 and p_2 cannot both be varied independently. Figure 6·10 shows the variations of S with p_1. Note that S is indeed maximum where $p_1 = p_2 = \frac{1}{2}$. One may indeed show that for any n the uniform distribution has the greatest entropy.

Next, consider that class of distributions where each quantum state is equally likely. Suppose there are Ω equally likely quantum states, so $p_i = 1/\Omega$,

$$S = -\mathsf{k} \sum_{i=1}^{\Omega} \frac{1}{\Omega} \ln \frac{1}{\Omega} = -\mathsf{k}\Omega\left(\frac{1}{\Omega} \ln \frac{1}{\Omega}\right) = -\mathsf{k} \ln \frac{1}{\Omega}$$

$$= \mathsf{k} \ln \Omega \qquad (6 \cdot 15)$$

Hence the larger the number of possible states, the greater is the entropy. This seems to be a proper qualitative feature of a randomness-uncertainty measure.

Consider the probability distributions (a), (b), (c), (d), and (e) used as examples earlier. These distributions, and their entropies, are

$$p_1 = 1 \qquad p_2 = 0 \qquad p_3 = 0 \qquad (a)$$

$$S = -\mathsf{k}(1 \ln 1 + 0 \ln 0 + 0 \ln 0) = 0$$

$$p_1 = 0.8 \qquad p_2 = 0.2 \qquad p_3 = 0 \qquad (b)$$

$$S = -\mathsf{k}(0.8 \ln 0.8 + 0.2 \ln 0.2 + 0 \ln 0) = 0.5004\mathsf{k}$$

$$p_1 = 0.8 \qquad p_2 = 0.1 \qquad p_3 = 0.1 \qquad (c)$$

$$p_1 = 0.1 \qquad p_2 = 0.8 \qquad p_3 = 0.1 \qquad (d)$$

$$S = -\mathsf{k}(0.8 \ln 0.8 + 0.1 \ln 0.1 + 0.1 \ln 0.1) = 0.6390\mathsf{k}$$

$$p_1 = p_2 = p_3 = \tfrac{1}{3} \qquad (e)$$

$$S = -\mathsf{k}(\tfrac{1}{3} \ln \tfrac{1}{3} + \tfrac{1}{3} \ln \tfrac{1}{3} + \tfrac{1}{3} \ln \tfrac{1}{3}) = \mathsf{k} \ln 3 = 1.099\mathsf{k}$$

Distribution (a) reflects no randomness or uncertainty. Distribution (b) suggests we know for sure that state 3 is never realized; the higher entropy of distributions (c) and (d) suggests slightly greater randomness and uncertainty about the state than distribution (b). Note that distributions (c) and (d) have the same entropy.

Distribution (*e*), which reflects the greatest randomness and uncertainty, yields the greatest entropy. These all seem to be proper qualitative aspects of a randomness-uncertainty measure; Eq. (6·14) seems to be a plausible definition of entropy.

To summarize, we define the entropy, or microscopic randomness-uncertainty measure, as

$$\blacktriangleright \qquad S \equiv -\mathsf{k} \sum_i p_i \ln p_i \qquad (6\cdot 16)$$

The p_i are the probabilities of individual system quantum states. If all quantum states are equally likely, and there are Ω such equally probable states, we have seen that

$$\blacktriangleright \qquad S = \mathsf{k} \ln \Omega \qquad (6\cdot 17)$$

This expression is sometimes taken as the basic definition of entropy. It should be noted that Eq. (6·17), the *Boltzmann* definition of entropy, is appropriate only if each quantum state is equally likely. This is the case for some statistical models of matter, and Eq. (6·17) is widely used in elementary texts on statistical thermodynamics. We shall also make some use of it in subsequent chapters. Equation (6·16), the *Gibbs* definition of entropy, is more general, hence is preferable as the fundamental definition. The Gibbs definition applies equally well for both equilibrium and nonequilibrium; the Boltzmann definition is only applicable to isolated systems in equilibrium.†

6·7 APPROACH TO EQUILIBRIUM

We have shown that entropy as defined by Eq. (6·16) is extensive and has its greatest value when the system quantum state is most random, that is, when we are most uncertain about the instantaneous quantum state. As time passes, we expect the disorder, randomness, and uncertainty, that is, the entropy, of an *isolated* system to increase and to approach a constant as the isolated system approaches equilibrium. A direct test of the increase in entropy as defined by Eq. (6·16) is difficult to perform in the laboratory but easy to perform on a computer. We shall now describe such an experiment on an interesting, but nonphysical, system.

Consider a system of 10 particles, each of which can exist in one of 5 states. We denote the particles by A, B, C, ..., J, and the particle states by their energies 0,

† For an excellent discussion of the virtues of Eq. (6·16), see R. C. Tolman, *The Principles of Statistical Mechanics*, p. 562, Oxford University Press, London, 1938. See also L. D. Landau and E. M. Lifshitz, *Statistical Physics*, p. 25, Pergamon Press, New York, 1958; J. H. Keenan and G. N. Hatsopoulos, *Principles of General Thermodynamics*, p. 606, John Wiley & Sons, Inc., New York, 1965; F. Reif, *Fundamentals of Statistical and Thermal Physics*, p. 219, McGraw-Hill Book Company, New York, 1965; M. Tribus, *Thermostatics and Thermodynamics*, p. 84, D. Van Nostrand Company, Inc., Princeton, N.J., 1961.

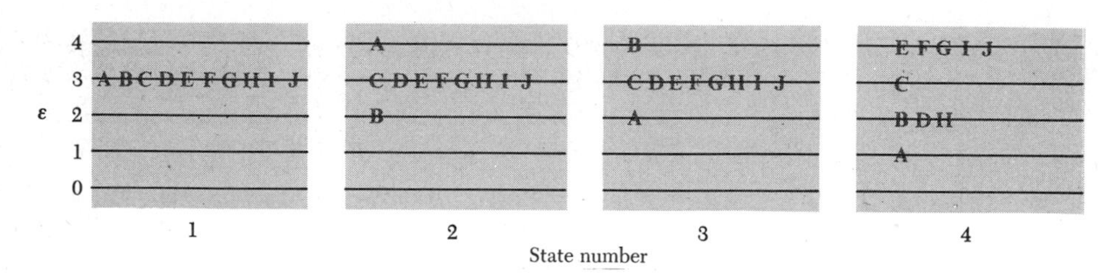

Figure 6·11. Some of the allowed states; note each state has a total energy of 30

1, 2, 3, and 4. Suppose the system is isolated and has total energy 30. We shall first examine the allowed states of the system. There are many possible system states, some of which are shown in Fig. 6·11. Any state in which the total energy is 30 is possible. All possible combinations were considered with the help of a high-speed computer, and it was found that there are 72,403 possible quantum states. We could not hope to list these here; however, they can be collected into groups of states where the states in each group have the same number of particles in each

Table 6·1 The allowed state groups†

	Number of particles with each energy					Number of quantum states	Equilibrium group probability
Group k	$\varepsilon = 0$	1	2	3	4		p_k
1	0	1	2	3	4	12,600	0.1740
2	0	2	1	2	5	7,560	0.1044
3	1	0	2	2	5	7,560	0.1044
4	1	0	1	4	4	6,300	0.0870
5	0	1	1	5	3	5,040	0.0696
6	0	1	3	1	5	5,040	0.0696
7	1	1	0	3	5	5,040	0.0696
8	1	1	1	1	6	5,040	0.0696
9	0	0	3	4	3	4,200	0.0580
10	0	0	4	2	4	3,150	0.0435
11	0	2	0	4	4	3,150	0.0435
12	0	0	2	6	2	1,260	0.0174
13	0	2	2	0	6	1,260	0.0174
14	2	0	0	2	6	1,260	0.0174
15	1	0	0	6	3	840	0.0116
16	1	0	3	0	6	840	0.0116
17	0	3	0	1	6	840	0.0116
18	0	1	0	7	2	360	0.0050
19	1	2	0	0	7	360	0.0050
20	2	0	1	0	7	360	0.0050
21	0	0	5	0	5	252	0.0035
22	0	0	1	8	1	90	0.0012
23	0	0	0	10	0	1	$0.0_4 14$

† Total number of quantum states = 72,403.

particle state. For example, quantum states 2 and 3 of Fig. 6·11 would fall in the same group (22). There are only 23 different groups and these are given in Table 6·1.

We can assign probabilities p_i to the 72,403 quantum states. The distribution that maximizes the entropy has each $p_i = 1/72{,}403$, and this is the equilibrium distribution that should be approached in any experiment. Corresponding to any set of $72{,}403 p_i$ is a set of 23 *group* probabilities p_k, each of which is simply the sum of the p_i in that group. Thus, for example, for the equilibrium distribution the probability that the system is in *one* of the 90 states of group 22 is 90/72,403, while the probability that it is in one of the 12,600 states of group 1 is much higher, namely, 12,600/72,403. The group probabilities corresponding to the equilibrium distribution are shown in Table 6·1.

Now, suppose the system is in some state in a particular group. We consider changes in state resulting from two-particle interactions. For example, suppose the system is in quantum state 1 (state-group 23) of Fig. 6·11. Two particles come together, interact, and may or may not change the system state. There are 45 possible particle pairs for this interaction, and we assume that each pair is equally likely to interact. For each pair there are two interactions that take the system to state 22, and one that leaves the system in state 23. We assume that, for a given particle pair, the possible transitions are equally likely. Hence the probability that a given pair interaction will take the system to state 22 is $\frac{2}{3}$, while the probability that a pair interaction will leave the system in state 23 is $\frac{1}{3}$. The transitions from the other state groups are more complicated, but can be computed according to the transition model (equal probability for pair selection and equal probability for the possible transitions for a given pair). Analysis of all pair interactions permits us to develop a table of *transition probabilities* for passage from one quantum state group to another. The results are shown in Table 6·2.†

In the computer experiments we started the system out in state 23, and then followed a long series of interactions chosen randomly in accordance with the transition probabilities of Table 6·2. The experiment was repeated 10,000 times, with a different group history being traced each time. The fraction of the experiments in which each group occurred at time t was used to calculate the group probabilities $p_k(t)$ at each time. The entropy at each time was calculated for the group distribution p_k. The results are shown in Figs. 6·12 to 6·14. Since the energy of the system was fixed, these results correspond to experiments in an isolated system.

† The number of possible particle pairs from a particular pair of energy levels, divided by the total number of particle pairs, gives the *particle selection probabilities* for interaction of two particles from those particular energy levels. Then, all possible movements of two particles from those energy levels must be counted to find the *movement probabilities* for particles chosen from particular levels. The product of the particle selection and movement probabilities gives the probability of transition from one group to another by a particular particle selection-movement combination. Summing over all possible selection-movement combinations that take the system from one quantum-state group to another, the total probability of passing from the first to the second group is obtained. Table 6·2 was calculated in this manner.

Table 6·2 Transition probabilities

Initial group	Final group 1	2	3	4	5	6	7	8	9	10	11
1	0.7052	0.0667	0.0267	0.0222	0.0593	0.0444	0.0089	0	0.0444	0.0133	0.0089
2	0.1111	0.7244	0.0148	0	0	0.0178	0.0222	0.0356	0	0	0.0370
3	0.0444	0.0148	0.7452	0.0741	0	0.0222	0.0089	0.0444	0	0.0222	0
4	0.0444	0	0.0889	0.7319	0.0356	0	0.0444	0	0.0178	0	0.0074
5	0.1482	0	0	0.0444	0.6630	0	0	0	0.0222	0	0.0556
6	0.1111	0.0267	0.0333	0	0	0.7000	0	0.0267	0	0.0556	0
7	0.0222	0.0333	0.0133	0.0556	0	0	0.7600	0.0444	0	0	0.0444
8	0	0.0533	0.0667	0	0	0.0267	0.0444	0.7548	0	0	0
9	0.1333	0	0	0.0267	0.0267	0	0	0	0.6578	0.0889	0
10	0.0533	0	0.0533	0	0	0.0889	0	0	0.1185	0.6711	0
11	0.0356	0.0889	0	0.0148	0.0889	0	0.0711	0	0	0	0.7007
12	0	0	0	0	0.1333	0	0	0	0.2222	0	0
13	0	0.0889	0	0	0	0.1333	0	0.0444	0	0	0
14	0	0	0.0533	0	0	0	0.1067	0.0444	0	0	0
15	0	0	0	0.2222	0.0667	0	0	0	0	0	0
16	0	0	0.1333	0	0	0.0533	0	0.0267	0	0	0
17	0	0.2000	0	0	0	0	0	0.0444	0	0	0
18	0	0	0	0	0.3111	0	0	0	0	0	0
19	0	0	0	0	0	0	0	0.1556	0	0	0
20	0	0	0	0	0	0	0	0.1244	0	0	0
21	0	0	0	0	0	0.0889	0	0	0	0.1852	0
22	0	0	0	0	0	0	0	0	0	0	0
23	0	0	0	0	0	0	0	0	0	0	0

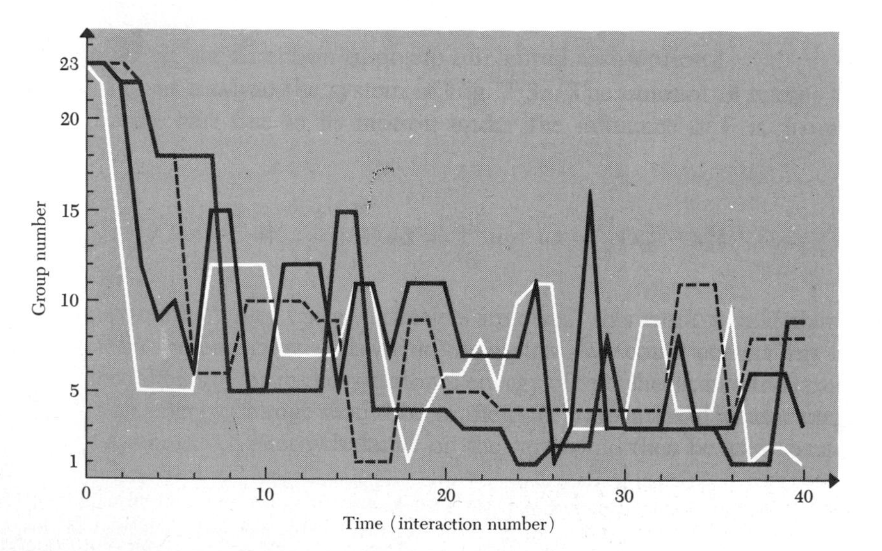

Figure 6·12. Four members of the ensemble of experiments

Initial group	Final group											
	12	13	14	15	16	17	18	19	20	21	22	23
1	0	0	0	0	0	0	0	0	0	0	0	0
2	0	0.0148	0	0	0	0.0222	0	0	0	0	0	0
3	0	0	0.0089	0	0.0148	0	0	0	0	0	0	0
4	0	0	0	0.0296	0	0	0	0	0	0	0	0
5	0.0333	0	0	0.0111	0	0	0.0222	0	0	0	0	0
6	0	0.0333	0	0	0.0089	0	0	0	0	0.0044	0	0
7	0	0	0.0267	0	0	0	0	0	0	0	0	0
8	0	0.0111	0.0111	0	0.0044	0.0074	0	0.0111	0.0089	0	0	0
9	0.0667	0	0	0	0	0	0	0	0	0	0	0
10	0	0	0	0	0	0	0	0	0	0.0148	0	0
11	0	0	0	0	0	0	0	0	0	0	0	0
12	0.5970	0	0	0.0089	0	0	0.0089	0	0	0	0.0296	0
13	0	0.7007	0	0	0.0148	0.0089	0	0.0089	0	0	0	0
14	0	0	0.7807	0	0	0	0	0	0.0148	0	0	0
15	0.0133	0	0	0.6711	0	0	0.0267	0	0	0	0	0
16	0	0.0222	0	0	0.7111	0	0	0	0.0267	0.0267	0	0
17	0	0.0133	0	0	0	0.7156	0	0.0267	0	0	0	0
18	0.0311	0	0	0.0622	0	0	0.5733	0	0	0	0.0222	0
19	0	0.0311	0	0	0	0.0622	0	0.7363	0.0148	0	0	0
20	0	0	0.0519	0	0.0622	0	0	0.0148	0.7467	0	0	0
21	0	0	0	0	0.0889	0	0	0	0	0.6370	0	0
22	0.4148	0	0	0	0	0	0.0889	0	0	0	0.4889	0.0074
23	0	0	0	0	0	0	0	0	0	0	0.6667	0.3333

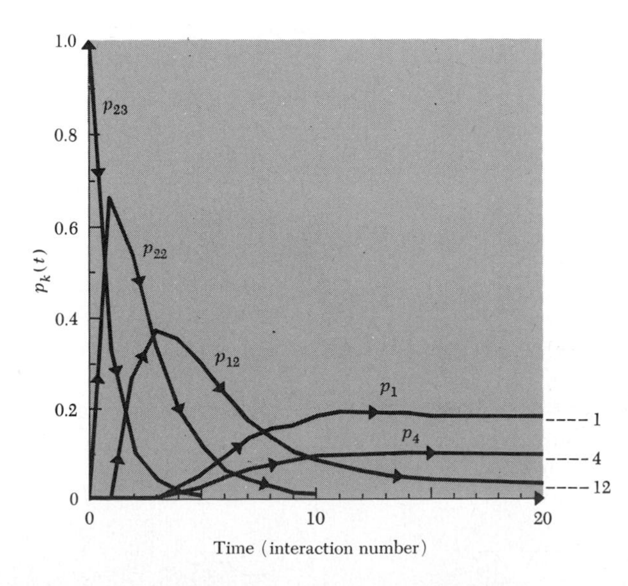

Figure 6·13. Some of the $p_k(t)$ for the ensemble of 10,000 experiments

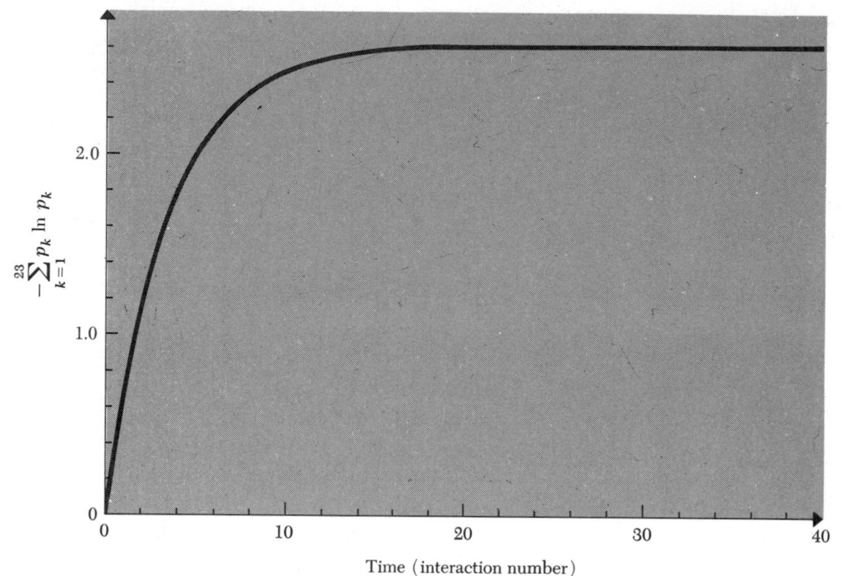

Figure 6·14. Group-distribution entropy for the ensemble of 10,000 experiments

Figure 6·12 shows four members of the ensemble of 10,000 experiments. Each time unit corresponds to one interaction. Note that there is initially no randomness, with all systems in group 23. After the first interaction some remain in group 23 (one-third) and some have moved to group 22 (two-thirds). At this time there is a slightly greater randomness. The randomness rapidly increases, and before long there are some systems in each state at each time. After many interactions the equilibrium condition is reached, and the quantum-group sequence for any particular experiment becomes quite random.

Figure 6·13 shows the evolution of some of the $p_k(t)$. Note that p_{23} drops rapidly, while p_{22} and p_{12} first increase and then relax toward their equilibrium values. Note that after the first interaction p_{22} was up to 0.6666, and p_{23} had dropped to 0.3333. This is indeed in accordance with the transition probabilities.

Figure 6·14 shows the group-distribution entropy as a function of time. If our postulate about the natural tendency for the microscopic randomness of an isolated system to increase is correct, the group-distribution entropy should increase. Indeed, the entropy did increase as time went on, and finally approached the equilibrium value computed from the p_k of Table 6·1.

Another interesting calculation can be made using the table of transition probabilities. Since the probability of a transition sequence is the product of the individual step transition probabilities, the transition

23 22 12 9 1

has the probability

$$0.6667 \times 0.4148 \times 0.2222 \times 0.1333 = 8.191 \times 10^{-3}$$

On the other hand, the reverse transition,

1 9 12 22 23

has the probability

$$0.0444 \times 0.0667 \times 0.0296 \times 0.0074 = 6.5 \times 10^{-7}$$

We see that there is a tremendous probability that the system will move toward and persist in the quantum-state groups having high equilibrium probabilities. Once a system has moved from group 23 to group 1, there is only a very slight chance it will ever return to group 23. This overwhelming likelihood of particular tendencies is exhibited even more dramatically by systems of very many particles. The probability is extremely small that a system of 10^{20} particles will move from a quantum state having a broad distribution of particle energies to a quantum state where all particles have the same energy. Such an event would probably not occur even once in the lifetime of the observer and is far too rare to be useful in an engineering system.

6·8 CONCLUDING REMARKS

The simple calculations presented above are intended to help the student understand the statistical concept of entropy as a well-defined measure of our uncertainty about the microscopic state, i.e., of the randomness or disorder of the microscopic state. It is interesting that this particular idea of entropy, as defined by Eq. (6·16), also arises in other fields, such as information theory, biology, decision theory, and even sociology. In these fields the constant k is taken as unity, and the entropy then becomes a dimensionless measure of the uncertainty represented by a particular message, decision-set probability, or social structure. There is no underlying physical connection with thermodynamic entropy, but the underlying uncertainty concepts are in each case the same. So, the student may find some familiar ideas when studying in these areas.

In the next chapter we return to the development of thermodynamic theory from a macroscopic point of view, but we will occasionally find it quite helpful in understanding these developments to make further reference to the microscopic and statistical ideas presented in this chapter. We emphasize again that a quantitative ability to work with the statistical concepts is not required or expected at this time. However, as the student delves further into thermodynamics, he or she may wish to return to this chapter and study the material in more detail in order to acquire a deeper understanding of the subject.

SELECTED READING

Fast, J. D., *Entropy*, introduction, McGraw-Hill Book Company, New York, 1963.

Goldman, S., *Information Theory*, sec. 1.8, Prentice-Hall, Inc., Englewood Cliffs, N.J., 1953.

Reif, F., *Fundamentals of Statistical and Thermal Physics*, secs. 3.1–3.4, 6.1–6.6, McGraw-Hill Book Company, New York, 1965.

———, *Statistical Physics*, Berkeley Physics Course, vol. 5, secs. 1.1–1.4, 2.1, 2.2, 2.4, McGraw-Hill Book Company, New York, 1967.

Sonntag, R. E., and G. J. Van Wylen, *Fundamentals of Statistical Thermodynamics*, 2d ed., sec. 4.3, John Wiley & Sons, Inc., New York, 1973.

Zemansky, M. W., M. M. Abbott, and H. C. Van Ness, *Basic Engineering Thermodynamics*, 2d ed., sec. 15.6, McGraw-Hill Book Company, New York, 1975.

QUESTIONS

6•1 Why do we need a "second law" of thermodynamics?
6•2 What is the fundamental conceptual property underlying the second law?
6•3 Does the first law rule out the possibility of water spontaneously becoming hydrogen and oxygen in an isolated container?
6•4 Can the entropy of something ever be reduced?
6•5 Is entropy transferred with heat? With mass? With work?
6•6 How do allowed quantum states differ from macroscopic states?
6•7 What is the "probability" of a quantum state?
6•8 What role does the concept of quantization play in the development of entropy?
6•9 Is entropy an intensive or extensive property?
6•10 Why is the logarithm involved in the statistical definition of entropy?
6•11 Can the entropy of a system ever decrease?
6•12 What do concepts of uncertainty have to do with thermodynamics?
6•13 How would you explain entropy to a layperson?
6•14 Look up entropy in a nontechnical dictionary. Can you reconcile this definition with your concept of entropy?
6•15 When did you first learn about energy? When did you first begin to really understand the concept of energy? Do you understand the concept of entropy?

PROBLEMS

6•1 Consider the following sets of quantum-state probabilities for a system with five allowed states.
(*a*) 0, 0.1, 0.2, 0.3, 0.4
(*b*) 0.2, 0.2, 0.2, 0.2, 0.2
(*c*) 0.5, 0.5, 0, 0, 0
(*d*) 0.3, 0.2, 0.3, 0.2, 0
(*e*) 0.3, 0.3, 0.1, 0.1, 0.3
(*f*) 1, 0, 0, 0, 0
(*g*) 0.1, 0.2, 0.4, 0.3, 0

One of these is not proper as a probability set. Which one, and why not? Assuming that the others represent the distributions at six consecutive moments in an ensemble of experiments on an isolated system, which was the initial distribution and which was the final? Rank the others properly between the initial and final distributions.

6·2 Consider a rectangular box containing N indistinguishable molecules. Suppose that each molecule can be on either side of the box with equal probability. What is the probability that all N molecules are on one side? Compute this number for $N = 1, 2,$ 20, and 10^{20}.

6·3 Set up a system of 20 molecules as described in Prob. 6·2, using any suitable markers for the molecules. Place them all initially on the left side of the box. Flip a coin and move one particle to right or left (if possible), depending on whether the coin shows heads or tails, respectively. Record the system state during a sequence of such flips. Discuss the outcome of this experiment. What would be the meaning of entropy here, and how would it be calculated?

6·4 Write a digital computer program to perform 10,000 experiments of the type described in Prob. 6·3. Use a random-number generator instead of the coin. Record the number of molecules on each side after each toss and compute the entropy evolution over the first 100 tosses.

6·5 Repeat the experiment of Prob. 6·3, this time starting the system with 10 molecules on each side. Record the number on the left after each flip. Continue until all molecules are on one side; if you become worn out before then, compute the fraction of the time that such a state would be observed over a very long run of experiments. Does the second law say that such a state could never happen?

6·6 Consider a system of four atoms (A, B, C, and D) that can exist with nuclear spin "up" or "down" (u or d). List all the possible spin states of this system. Group them according to the number of atoms with spin up. Assuming that each spin state is equally likely, determine the probability of each spin-state group. Calculate the entropy of the group probability distribution (take $\mathsf{k} = 1$ for simplicity) and compare to $\ln \Omega$, where Ω is the total number of system spin states, and to $\ln \Omega_g$, where Ω_g is the number of spin-state groups. Why do these three entropies differ? (In which description do you know most about the system?)

6·7 Consider a system of three indistinguishable molecules, each of which can have energy 0, 1, 2, 3, or 4. Find all system states having a total energy of 6 (one is shown in the figure). Assuming that each system state is equally likely, calculate the probability p_k that energy level k is occupied by at least one particle. Why do these probabilities not add up to unity? Suppose we define a property, g, as the sum of the squares of the energies of the three molecules. What is the ensemble average g?

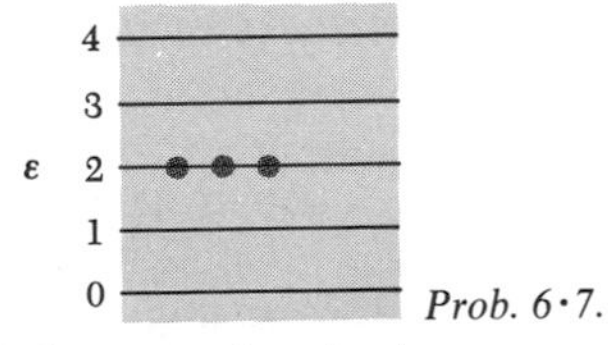

Prob. 6·7.

6·8 Suppose that in the system of Prob. 6·7 two-particle interactions take the system to other states (or leave the system in the same state) with equal probability for all possible transitions from a given initial state. Make up a table of transition probabilities similar to Table 6·2. If the system is started in the state shown in Prob. 6·7, what is the probability of being in each state after 1, 2, 3, and 4 interactions? Compute the entropies (with $\mathsf{k} = 1$) of the probabilities after 0, 1, 2, 3, and 4 interactions for an ensemble of such experiments, and compare with the equilibrium entropy.

6·9 Write a digital computer program to repeat the experiment of Prob. 6·8 1000 times, carrying each experiment through 10 interactions. Use a random-number generator and the transition probability table to find the state sequence in each experiment. Plot the experimentally determined evolution of the probability distribution and the entropy history; compare with the analytic predictions of Prob. 6·8.

6·10 Suppose you have a coin on a table, heads up. Imagine an experiment in which you roll a die and turn the coin over if the die shows a 1 but leave the coin as it stands if 2 to 6 is rolled. What is the probability of turning the coin over on the first roll? On the second roll? What is the probability that the coin will be tails up after the second roll of the die? Compute the probability distribution for the two coin states (H or T) after each of the first 10 rolls of the die. Calculate the entropy of the distribution that would be obtained after each of the first 10 tosses (take $k = 1$ for simplicity). What distribution is obtained after a large number of tosses, and what is its entropy?

6·11 Perform 100 of the experiments as described in Prob. 6·10 (stop after three rolls) and compare the experimental evolution of the distribution to that predicted theoretically.

6·12 Write a digital computer program to carry out an ensemble of 10,000 experiments as described in Prob. 6·10. Use a random-number routine instead of the die. Compare the experimentally determined evolution of the distribution and entropy to that predicted theoretically.

6·13 Calculate the entropy of 1 gmole of gas assuming that there are as many system quantum states as there are particles, and that each state is equally likely.

6·14 Repeat Prob. 6·13 assuming there are 10^{20} times as many system quantum states as particles.

6·15 The entropy of 1 gmole of O_2 at 600°R and 1 atm is 49.7 cal/K. Assuming each quantum state is equally likely, calculate the number of system quantum states for 1 gmole and the ratio of this number to the number of molecules.

6·16 Repeat Prob. 6·15 at 5000 R and 1 atm where the entropy of 1 gmole of O_2 is 67.2 cal/K. What happens to the number of allowed states as the temperature of O_2 increases?

6·17 The entropy of 1 kg of H_2O changes by 6.05 kJ/K when it is evaporated at 1 atm. Assuming that in either the saturated liquid or saturated vapor states the allowed quantum state is equally likely, calculate the number of system quantum states for 1 gmole and the ratio of this number to the number of molecules.

6·18 A two-compartment water tank contains saturated water vapor at 4 MPa on one side and water vapor at the same temperature but at 0.1 MPa in the other compartment. Each compartment has a volume of 0.2 m³. A valve connecting the two compartments is opened, and the water is allowed to come to equilibrium adiabatically. Calculate the entropy change for this isolated system, using entropy data in Table B·1.

6·19 A two-compartment tank contains saturated water vapor at 4 MPa on one side and water vapor at 20 MPa and 800°C on the other side. Each compartment has a volume of 0.5 m³. Energy is transferred as heat through the fixed dividing wall, and the system is allowed to reach equilibrium. Calculate the entropy change for this isolated system, using entropy data in Table B·2.

6·20 Calculate the entropy change for Prob. 6·19, treating the wall as a movable piston with negligible energy storage.

CHAPTER SEVEN

SOME CONSEQUENCES OF THE SECOND LAW

7·1 INTRODUCTION

In the previous chapter we developed the concept of entropy and the ideas behind the second law of thermodynamics. It is indeed very helpful to view entropy as a measure of the randomness in, or uncertainty about, the microscopic state of a system. But it is equally clear that the calculation of entropy from its statistical definition [Eq. (6·16)] is rather involved and requires physical input about the microscopic nature of the system. Such calculations fall in the domain of statistical thermodynamics. Before becoming involved in such calculations, we should see what we can find out about entropy from purely *macroscopic* considerations. This is the central theme of the present chapter.

We shall see that it is possible to relate differences in entropy between two thermodynamic states to other macroscopically measurable properties. This allows us to determine the entropy of a substance as a function of state (relative to an arbitrary datum). The entropy per unit of mass can then be tabulated, graphed, or stored in a computer with the other thermodynamic properties for use in engineering calculations. Our first objective in this chapter is the development of the framework required for the macroscopic evaluation of entropy.

In the initial discussion we shall develop the thermodynamic definition of temperature mentioned in earlier chapters, and a thermodynamic definition of pressure. We shall see that these are equivalent to the empirical temperature and mechanical pressure used earlier. The thermodynamic temperature scale is independent of any arbitrary choice of thermometric substance, hence the thermodynamic definition of temperature places the temperature concept on really firm ground.

Once we have established the macroscopic basis for entropy evaluation we shall turn to discussion of the role played by entropy and the second law in engineering analysis.

Let's recapitulate the four basic aspects of entropy developed in the last chapter:

1. All systems have entropy; entropy measures the degree of microscopic disorganization, or our uncertainty about the microscopic state.
2. Entropy is extensive; the entropy of a complex system is the sum of the entropies of its parts.
3. Entropy can be produced, but never destroyed; consequently, the entropy of an isolated system can never decrease.
4. The entropy is zero for any system that always exists in only one microscopic state.

Together these form a solid postulational basis for a macroscopic approach to the second law. We will also use the notion of an irreversible process as one that produces entropy, and the idea that a reversible process is a limiting case in which no entropy is produced. These are the ideas upon which we will build the developments in this chapter.

7·2 ENTROPY AS A FUNCTION OF STATE

In a thermodynamic equilibrium state, the entropy is a thermodynamic property of the substance, and from the state postulate we know that it is a function of only a few macroscopic properties. For example, the equilibrium entropy of a given amount of a simple compressible substance is some function of the energy and volume,

$$S = S(U, V) \tag{7·1}$$

If in addition the substance can be electrically polarized, the equilibrium entropy will also depend on the total dipole moment,

$$S = S(U, V, \mathbf{P}V)$$

Our goal here is to find some *macroscopic* means to determine these functions.

Just as tabulation of internal energy is facilitated by considering the internal energy per unit of mass, it is convenient to "intensify" the entropy. For example, for a simple compressible substance we can use the specific entropy, or entropy per unit of mass; the symbol s is usually used for the intensified entropy, while S is used for the total (extensive) entropy;

$$s \equiv \frac{S}{M}$$

Then, for a simple compressible substance,

$$\blacktriangleright \qquad s = s(u, v) \qquad (7 \cdot 2)$$

As we shall see, the dimensions of entropy are energy/temperature, that is, J/K, cal/K, or Btu/R. Hence the dimensions of s are energy/(temperature·mass), for example, J/(kg·K), cal/(g·K), or Btu/(lbm·R).

The specific entropy of a liquid-vapor mixture can be expressed in terms of the quality and the specific entropies of the liquid and vapor,

$$\blacktriangleright \qquad s = (1 - x)s_f + xs_g \qquad (7 \cdot 3)$$

The student may derive this equation following the steps used in obtaining Eq. (4·1).

In the next few sections we shall examine the nature of the functional relationships among entropy, energy, and volume for a simple compressible substance. Treatment of more general substances follows exactly these lines of argument, and the details will be omitted here.

7·3 THE THERMODYNAMIC DEFINITION OF TEMPERATURE

Let us first review the concept of temperature. If we bring two masses into contact, and there is some energy transfer as heat between them, we say that their temperatures were different. We also view the energy as passing from the "hotter" mass to the "colder" one. We say that the masses are in thermal equilibrium if they are at the same temperature, and no energy transfer as heat will occur when they are brought together. These notions show that temperature is basically thought of as an indicator of the direction of energy transfer as heat. Differences in temperature reflect lack of equilibrium, and this suggests that we might be able to form a basic definition of temperature through considerations of thermal equilibrium.

Our approach will be to consider two masses, each in a thermodynamic equilibrium state. If we bring them together, and any energy transfer as heat takes place between them, then they were not initially at the same temperature. We then shall look for conditions under which *no* energy transfer as heat takes place when they are brought together and try to find a *property* that has the same value for both masses under this condition of thermal equilibrium. This will lead us to the basic definition of temperature.

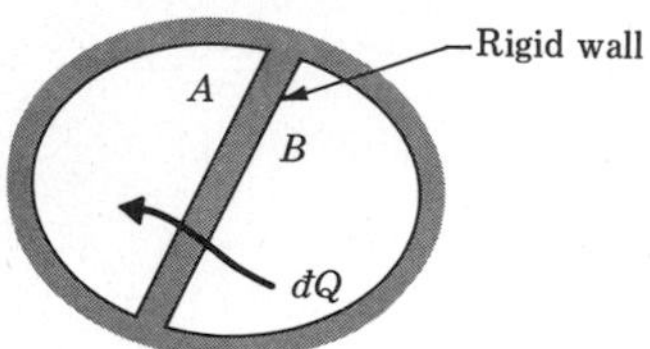

Figure 7·1. The combined system C is isolated

Consider two control masses, A and B, each initially in a state of thermodynamic equilibrium (Fig. 7·1), though A and B are not necessarily in equilibrium with each other. For simplicity assume that each mass is a simple compressible substance. Let C denote the combined system A and B. The entropy of C is

$$S_C = S_A + S_B \tag{7·4}$$

Now, since A and B are each in some equilibrium state,

$$S_C = S_A(U_A, V_A) + S_B(U_B, V_B) \tag{7·5}$$

This expression holds as long as A and B are each in some equilibrium state, even when A and B are not in equilibrium with each other. Note that the individual masses of A and B are also specified and fixed.

We now imagine letting A and B interact through a rigid wall, which prevents changes in V_A and V_B. This rules out the possibility of any energy transfer as work between A and B, and any energy transfer must take place as heat. We also imagine isolating the combined system C, so that the only interactions are between A and B, internal to C. Since system C is isolated, its entropy must not decrease; any interaction between A and B will serve to increase the entropy of C. S_C will attain its largest value when A and B are finally in thermal equilibrium, that is, when they are at the same temperature. Hence the state of C that maximizes S_C is the equilibrium state.

Since C is isolated, the total internal energy $U = U_A + U_B$ must be "shared" by A and B. If we define r to be the fraction of the total internal energy contained by A, then

$$U_A = rU \qquad U_B = (1 - r)U \tag{7·6}$$

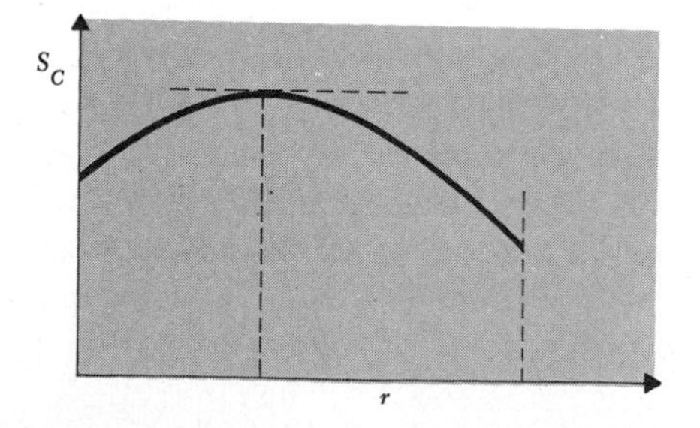

Figure 7·2. The entropy of the combined isolated system is a maximum when A and B are in equilibrium

Since for a given isolation U, V_A, and V_B are fixed, the only free parameter left to vary S_C is r [see Eq. (7·5)]. For given U, V_A, and V_B, the value of r which maximizes S_C determines the distribution of internal energy between A and B in the equilibrium state. Figure 7·2 shows the variation of S_C with r schematically.

The point of maximum S is found by setting $dS_C/dr = 0$. Using the chain rule of differentiation, we differentiate Eq. (7·5),

$$\frac{dS_C}{dr} = \left(\frac{\partial S_A}{\partial U_A}\right)_{V_A} \frac{dU_A}{dr} + \left(\frac{\partial S_B}{\partial U_B}\right)_{V_B} \frac{dU_B}{dr}$$

or

$$\frac{dS_C}{dr} = \left(\frac{\partial S_A}{\partial U_A}\right)_{V_A} U + \left(\frac{\partial S_B}{\partial U_B}\right)_{V_B} (-U) \tag{7·7}$$

This vanishes only when

$$\left(\frac{\partial S_A}{\partial U_A}\right)_{V_A} = \left(\frac{\partial S_B}{\partial U_B}\right)_{V_B} \tag{7·8}$$

Equation (7·8) fixes the state of maximum S_C, that is, the condition under which A and B will be in thermal equilibrium. Note that the derivatives in Eq. (7·8) are *properties* of A and B, respectively; the derivative on the left depends only on the state (U, V) of A, and that on the right depends only on the state of B.

When the *property* $(\partial S/\partial U)_V$ of A is equal to that for B, A and B are in thermal equilibrium. This suggests that we should define temperature in terms of this property. Any function of this property would be such that two systems in equilibrium each have the same value of the function. What function should we choose? A choice which retains the other conceptual aspect of temperature, namely, that energy transfer as heat should take place from the warmer body to the cooler one, is

$$\blacktriangleright \qquad T \equiv \frac{1}{(\partial S/\partial U)_V} \tag{7·9}$$

As we shall show, this choice also makes the thermodynamic temperature scale coincide with the perfect-gas temperature scale, hence makes the thermodynamic temperature equivalent to what we have been using for temperature all along. Equation (7·9) is the *thermodynamic definition of temperature* for a simple compressible substance. Since both S and U are extensive, their ratio, hence T, is *intensive*. Note also that T is defined in terms of the equilibrium equation of state $S(U, V)$, thus T does not have any meaning for a system not in a thermodynamic equilibrium state.

From Eq. (7·9) we can see that entropy has the dimensions of energy/temperature, as stated in Sec. 7·2; this is consistent with Eq. (6·16).

Let us next establish that energy transfer as heat takes place from the hotter to the cooler body. Equation (7·7) can be written as

$$dS_C = U \cdot dr\left(\frac{1}{T_A} - \frac{1}{T_B}\right) \tag{7·10}$$

Now, if in the initial state $T_B \neq T_A$, then S_C must increase as a result of the interaction, and $dS_C > 0$. If $T_B > T_A$, this requires $dr > 0$, which corresponds to an increase in U_A and a decrease in U_B. Hence the energy indeed flows from the warmer body (B) to the cooler one (A).

We can learn still more about the thermodynamic temperature from the analysis of the condition for equilibrium between A and B (Fig. 7·1). In order for the point at which $dS_C/dr = 0$ to be a *maximum*, and not a minimum or inflection point in the $S_C(r)$ curve, the *second* derivative d^2S_C/dr^2 must be negative. Differentiating Eq. (7·7) again with respect to r, we have

$$\frac{d^2S_C}{dr^2} = \left[\left(\frac{\partial^2 S}{\partial U_A^2}\right)_{V_A} U^2 + \left(\frac{\partial^2 S_B}{\partial U_B^2}\right)_{V_B} U^2\right] < 0 \tag{7·11}$$

Suppose A and B are identical systems. Then, the necessary condition for a maximum yields

$$\left(\frac{\partial^2 S}{\partial U^2}\right)_V < 0 \qquad \textit{for A or B} \tag{7·12}$$

This derivative is merely the first derivative of $1/T$, hence $T(U, V)$ must be such that

$$\left(\frac{\partial(1/T)}{\partial U}\right)_V < 0$$

or

$$\left(\frac{\partial T}{\partial U}\right)_V > 0 \tag{7·13}$$

Equation (7·13) says that the temperature must be a monotonically increasing function of the energy (Fig. 7·3); increasing the energy at fixed volume must increase the temperature. This is in accordance with experience cited previously.

A physical interpretation of temperature in terms of the microscopic concepts of entropy discussed in the last chapter is quite helpful. The reciprocal of temperature describes the sensitivity of the entropy to internal energy changes at fixed volume. If a small increase in internal energy greatly changes the microscopic randomness and the uncertainty about the microscopic state, the quantity $1/T$ is large, and T is small. Absolute zero T corresponds to a state in which any infinitesimal increase in internal energy will make a finite increase in the microscopic randomness. At higher and higher temperatures $1/T$ becomes smaller and

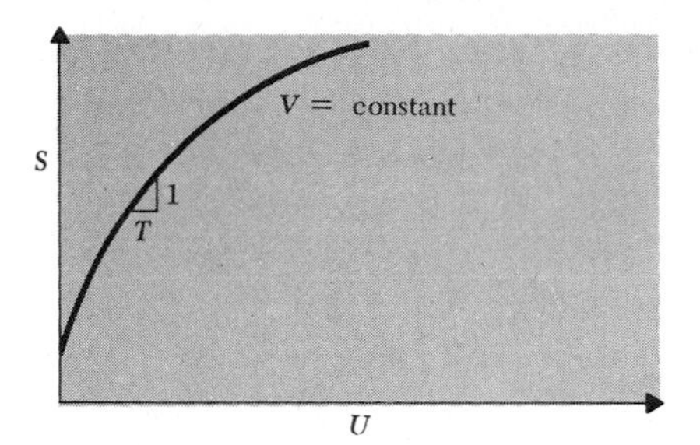

Figure 7·3. Temperature is a monotonically increasing function of internal energy

smaller, hence the amount of additional microscopic randomness or uncertainty produced by adding some energy at constant volume becomes less and less at high temperatures. For simple compressible substances only positive absolute temperatures are of interest; if $T > 0$, any increase in internal energy (at constant volume) will result in an increase in entropy, that is, in an increase in the microscopic randomness and uncertainty. This seems quite reasonable, indeed.

We started with the idea that the entropy of a given amount of a simple compressible substance can be expressed as some function of its internal energy and volume, $S = S(U, V)$. If we take the differential of this function we obtain

$$dS = \left(\frac{\partial S}{\partial U}\right)_V dU + \left(\frac{\partial S}{\partial V}\right)_U dV \tag{7·14}$$

We now know that the derivative coefficient of dU in Eq. (7·14) is $1/T$, which we know as a function of state (presuming that the thermodynamic and empirical temperatures are indeed equivalent as we have stated and will shortly prove). If we can determine the derivative coefficient of dV in Eq. (7·14) as a function of measurable properties, Eq. (7·14) would become a differential equation for the entropy as a function of state. Then, by integration of this equation by appropriate analytical, numerical, or graphical means, we could determine the entropy as a function of the state of the substance, relative to the entropy in some arbitrary datum state. Our next task is therefore to examine $(\partial S/\partial V)_U$, which brings us to the thermodynamic definition of pressure.

7·4 THE THERMODYNAMIC DEFINITION OF PRESSURE

What is the "pressure" of a piece of matter? We think of pressure in mechanical terms as the force per unit area exerted by the matter on its boundaries. A microscopic interpretation of gas pressure in terms of molecular collisions with a wall is very vivid, even though we cannot really see the molecules. Thermodynamics provides an entirely different way to define pressure, and in many ways this definition is more fundamental. Fortunately, the thermodynamic and mechanical pressures are equivalent under certain circumstances, so we can use force-measuring instruments to determine the thermodynamic pressure.

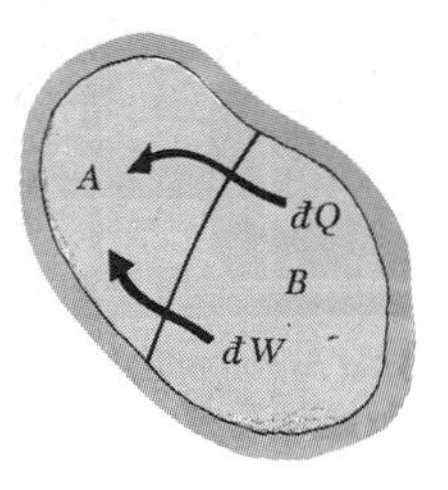

Figure 7·4. The combined system C is isolated

We discovered the thermodynamic definition of temperature by examining the conditions of thermal equilibrium. Following this line of thought, the thermodynamic definition of pressure is obtained from considerations of *mechanical* equilibrium. Pressure is conceived to be a property that two systems have in common when they are in mechanical equilibrium.

Consider two control masses A and B, which together form an isolated system C. We presume A and B are free to expand or contract, thereby exchanging energy as work, and that they can also exchange energy as heat. The sign conventions for our analysis of these interactions are defined in Fig. 7·4. We again treat only the case of simple compressible substances.

The conditions of isolation require that the energy and volume of the combined system C remain fixed,

$$U_C = U_A + U_B = \text{constant} \tag{7·15a}$$

$$V_C = V_A + V_B = \text{constant} \tag{7·15b}$$

Assuming that A and B are each in some thermodynamic state, that the masses of A and B are fixed, and that their allowed quantum states are independent, the entropy of the combined system is again given by

$$S_C = S_A(U_A, V_A) + S_B(U_B, V_B) \tag{7·16}$$

Upon isolation of C, systems A and B will interact, exchanging both energy and volume. This interaction will lead to a state of equilibrium within C; for mechanical equilibrium the pressures of A and B must be equal, and for thermal equilibrium the temperatures must be equal. The conditions of equilibrium are obtained by seeking the state that maximizes the entropy of the isolated system C.

Since the total energy and volume are fixed, we cannot independently vary U_A, U_B, V_A, and V_B in seeking the maximum of S_C. It is therefore convenient to define the fraction of the energy and volume contributed by part A as r_U and r_V, respectively. Then,

$$U_A = r_U U \qquad U_B = (1 - r_U)U \tag{7·17a,b}$$

$$V_A = r_V V \qquad V_B = (1 - r_V)V \tag{7·17c,d}$$

The fractions r_U and r_V can be considered the two free variables to be adjusted in the maximization of S_C. $S_C(r_U, r_V)$ can be thought of as a *surface* above the $r_U - r_V$ plane; the form of this surface is shown in Fig. 7·5.

The state of maximum S_C is determined by the two conditions

$$\left(\frac{\partial S_C}{\partial r_U}\right)_{r_V} = 0 \qquad \left(\frac{\partial S_C}{\partial r_V}\right)_{r_U} = 0 \tag{7·18a,b}$$

Using Eqs. (7·17), these two conditions give

$$\left(\frac{\partial S_A}{\partial U_A}\right)_{V_A} U - \left(\frac{\partial S_B}{\partial U_B}\right)_{V_B} U = 0 \tag{7·19a}$$

$$\left(\frac{\partial S_A}{\partial V_A}\right)_{U_A} V - \left(\frac{\partial S_B}{\partial V_B}\right)_{U_B} V = 0 \tag{7·19b}$$

Noting that the derivatives in Eq. (7·19*a*) are the reciprocal temperatures of A and B, we see that Eq. (7·19*a*) is simply the requirement of thermal equilibrium, that is, that the temperatures of A and B must be equal. Equation (7·19*b*) is the condition of mechanical equilibrium, and these derivatives must somehow be related to the pressure. Noting that the dimensions of entropy are energy/temperature, the derivative $\partial S/\partial V$ has the dimensions of energy/(volume · temperature) = force · length/(length^3 · temperature) = pressure/temperature. Hence this derivative must be either plus or minus the pressure divided by T. The proper choice for sign can be obtained by noting that, if A and B are not in equilibrium the system with the larger pressure should expand, and the entropy of C should thereby increase. The infinitesimal change in S_C associated with an infinitesimal dr_V is, when $T_A = T_B$,

$$dS_C = \left[\left(\frac{\partial S_A}{\partial V_A}\right)_{U_A} - \left(\frac{\partial S_B}{\partial V_B}\right)_{U_B}\right] V\, dr_V \geq 0$$

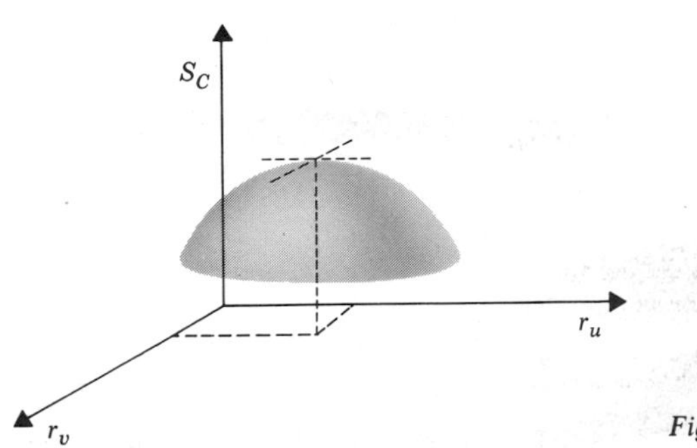

Figure 7·5. The S_C surface

Hence, if $dr_V > 0$, which corresponds to expansion of A,

$$\left(\frac{\partial S_A}{\partial V_A}\right)_{U_A} > \left(\frac{\partial S_B}{\partial V_B}\right)_{U_B}$$

Since for $dV_A > 0$, $P_A > P_B$, the plus sign must be chosen. The thermodynamic definition of pressure is therefore

$$\blacktriangleright \qquad \frac{P}{T} \equiv \left(\frac{\partial S}{\partial V}\right)_U \qquad (7 \cdot 20)$$

Since both S and V are extensive, P/T, hence P, is *intensive*. Note that the thermodynamic pressure is defined in terms of the *equilibrium* equation of state $S = S(U, V)$, hence P has no meaning for a substance not in a thermodynamic equilibrium state.

Following the thoughts of our microscopic interpretation of temperature, we can interpret the pressure as a measure of the sensitivity of the entropy to volume changes at fixed internal energy. If a system of gas molecules expands at fixed internal energy, we become more uncertain about the locations of individual molecules, for they can roam over a larger space. Hence $(\partial S/\partial V)_U$ and P should always be positive, as is indeed the case for a gas. In a solid, an expansion at fixed energy may be accompanied by an ordering of the molecular structure, which reduces the randomness and uncertainty about the microscopic state. Hence in solids the pressure can be negative (which corresponds to tension).

The thermodynamic pressure P has the qualitative aspects of the mechanical pressure, but it is not obvious that they are identical. We shall show that they are identical shortly.

7·5 INTENSIVE REPRESENTATIONS AND SOME EXTENSIONS

We developed the definitions of T and P in terms of the extensive properties S, U, and V. Since the mass M is fixed for any piece of matter, we can divide numerators and denominators in these defining ratios by M, and thereby express T and P in terms of the intensive properties s, u, and v. The definitions of T and P in terms of $s(u, v)$ become

$$\blacktriangleright \qquad \frac{1}{T} \equiv \left(\frac{\partial s}{\partial u}\right)_v \qquad (7 \cdot 21)$$

$$\blacktriangleright \qquad \frac{P}{T} \equiv \left(\frac{\partial s}{\partial v}\right)_u \qquad (7 \cdot 22)$$

The differential of the specific entropy $s(u, v)$ is

$$ds = \left(\frac{\partial s}{\partial u}\right)_v du + \left(\frac{\partial s}{\partial v}\right)_u dv$$

or

$$ds = \frac{1}{T}\,du + \frac{P}{T}\,dv \tag{7·23}$$

This is a very important equation; it provides the means for evaluation of the entropy of substances from macroscopic laboratory data. It is called the *Gibbs equation* for a simple compressible substance.

The definitions above apply for simple compressible substances only, but extension to other classes of substances is quite straightforward. For example, the thermodynamic definition of temperature for a simple magnetic substance is

$$\frac{1}{T} \equiv \left(\frac{\partial s}{\partial u}\right)_{\mathbf{M}} \tag{7·24}$$

while for a simple dielectric substance it is

$$\frac{1}{T} \equiv \left(\frac{\partial s}{\partial u}\right)_{\mathbf{P}} \tag{7·25}$$

Following the development of the thermodynamic pressure, thermodynamic definitions for the electric and magnetic fields can be developed. The Gibbs equations for these substances can then be used to calculate entropy as a function of state. For example, the Gibbs equation for a simple magnetic substance is, in intensive form,

$$ds = \frac{1}{T}\,du - \mu_0 \frac{v\mathbf{H}}{T} \cdot d\mathbf{M} \tag{7·26}$$

The development of this and other Gibbs equations are left as problems for the student.

The temperature T as used here is the absolute temperature, which would be measured on the Rankine (R) or Kelvin (K) scale. The relative temperatures (°F, °C) can be used when temperature *differences* are involved but must not be used when temperature *levels* are needed [such as in Eqs. (7·21) to (7·26)].

Let's consider for a moment the possibility that the absolute temperature T might be negative for some system. Since $1/T$ measures the sensitivity of the entropy to energy changes at fixed constraints (fixed volume, magnetic moment, and so on), a *negative* temperature state would be one for which an *increase* in energy would *reduce* the entropy; by adding energy we could make the system become more ordered, that is, we could become less uncertain about the microscopic state. Most systems have more allowed quantum states at high energy than at low energy, hence do not exhibit the peculiarity of negative absolute temperature. However, imagine a system for which there is an upper limit on the energy of allowed quantum states for individual particles. As the system energy is increased, more and more particles must occupy the state of greatest energy until finally at the system state of maximum energy all the particles are in their maximum energy state. This is a highly ordered condition; if we know that the total (macroscopic) energy is this highest possible amount, we know the microscopic state with no uncertainty. The entropy must therefore be zero in this state. Hence, as we add energy in approaching this state, the entropy would decrease, which corresponds to a situation of negative absolute temperature. Note that negative temperatures would occur at *higher* energies than positive temperatures. Nuclear spin systems in crystals exhibit this negative temperature behavior. The spin state adjusts rapidly to changes in comparison to the lattice vibration, hence the spin states can be treated as a separate thermodynamic system. The operation of solid-state masers depends upon the existence of such spin-system states of negative T.†

7·6 MACROSCOPIC EVALUATION OF ENTROPY

Later in this chapter we shall show that thermodynamic and empirical temperatures are identical, and that thermodynamic and mechanical pressures are identical for a substance in an equilibrium state. Let's accept these equivalences for the moment and begin to do something quantitatively useful with the second law and entropy ideas. This will help the student to get started and will make it easier to see where we are going and why.

Accepting these equivalences, we see that the coefficients of du and dv in the Gibbs equation [Eq. (7·23)] are directly measurable in the laboratory. The Gibbs equation therefore provides a differential equation from which the entropy of a substance can be quantitatively evaluated as a function of thermodynamic state, relative to an arbitrarily chosen reference state. Integration of this equation between any two states 1 and 2 yields

$$s_2 - s_1 = \int_1^2 \frac{du}{T} + \int_1^2 \frac{P}{T}\, dv \tag{7·27}$$

The integrations can be carried out along any path connecting the two states. For example, suppose we take the path shown in Fig. 7·6*a*. We imagine plotting $1/T$ versus u and P/T versus v along this path as shown in Figs. 7·6*b* and 7·6*c*. The integrals under these curves can be computed numerically or graphically, and the

† A. E. Siegman, *Microwave Solid-State Masers*, McGraw-Hill Book Company, New York, 1964.

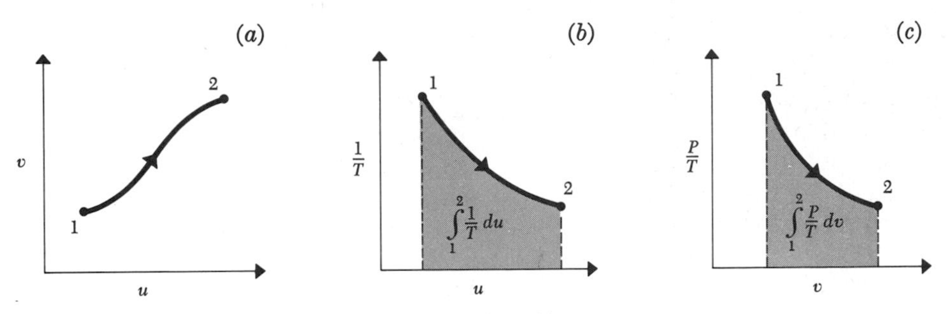

Figure 7·6. Computing an entropy difference from the Gibbs equation

difference $s_2 - s_1$ thereby evaluated from Eq. (7·27). If state 1 is the datum state, we set $s_1 = 0$, and s_2 then is the entropy relative to the datum state. It is *not* the *absolute entropy* as defined by Eq. (6·16). It is possible to evaluate the absolute entropy using the idea that the entropy is zero for a system that exists in only one microscopic state. This is the case for pure substances at $T = 0$ (Chap. 11). So, if we could obtain data down to 0 K, we could use this point as state 1 and then s_2 would be the absolute entropy at state 2. However, only entropy *differences* and *changes* are required for most second-law analysis; hence as long as we work with a single substance we can use its entropy as evaluated relative to the datum state.

As an example of this type of entropy evaluation, let's compute the entropy difference between saturated vapor and saturated liquid water at 0.1 MPa. The necessary u-v-P-T data are taken from Table B·1*b* (note we use the absolute temperature).

State 1: saturated liquid at 0.1 MPa

$$T = 372.8 \text{ K } (99.6^\circ\text{C})$$

$$v = 0.001043 \text{ m}^3/\text{kg}$$

$$u = 417.3 \text{ kJ/kg}$$

$$P = 10^5 \text{ N/m}^2$$

State 2: saturated vapor at 0.1 MPa

$$T = 372.8 \text{ K}$$

$$v = 1.694 \text{ m}^3/\text{kg}$$

$$u = 2506.1 \text{ kJ/kg}$$

$$P = 10^5 \text{ N/m}^2$$

We could take any path of integration connecting two states. The path of constant

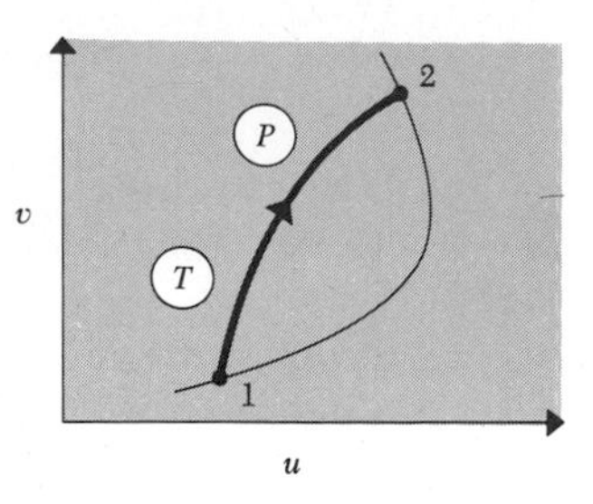

Figure 7·7. The path of integration (see also Fig. B·1)

pressure is easiest, for along this path both the temperature and pressure are constant (Fig. 7·7). The two integrals are then

$$\int_1^2 \frac{du}{T} = \frac{1}{T}(u_2 - u_1) = \frac{2506.1 - 417.3}{372.8} = 5.603 \text{ kJ/(kg·K)}$$

$$\int_1^2 \frac{P}{T}\,dv = \frac{P}{T}(v_2 - v_1) = \frac{10^5 \times (1.694 - 0.001043)}{372.8}$$

$$= 454.1 \text{ kJ/(kg·K)} = 0.454 \text{ kJ/(kg·K)}$$

Hence $$s_2 - s_1 = 5.603 + 0.454 = 6.057 \text{ kJ/(kg·K)}$$

Making allowance for round-off errors, this agrees with s_{fg} at 0.1 MPa as read from Table B·1*b*.

Taking the entropy of state 1, relative to the triple-point datum, from Table B·1*b* as $s_1 = 1.303$ kJ/(kg·K), we see that the entropy at state 2, relative to the triple point, is

$$1.303 + 6.057 = 7.360 \text{ kJ/(kg·K)}$$

Evaluation of the *absolute* entropy requires material from Chap. 11. But to complete the picture here, we state that the absolute entropy of state 1 is about 4.86 kJ/(kg·K), hence the absolute entropy of state 2 is

$$4.86 + 6.06 = 10.92 \text{ kJ/(kg·K)}$$

The graphs and tables in Appendix B give the specific entropy, relative to some datum state, for a variety of substances. Most of these values were obtained by the type of integration outlined here. For some of the more exotic substances, such as cesium vapor (Fig. B·11) and the paramagnetic salt iron-ammonium alum (Fig. B·12), the entropy is based on a combination of statistical thermodynamics and macroscopic thermodynamic equations of the sort discussed in the next chapter. We will use these entropy data in second-law analysis. An illustrative analysis for an isolated system was given in Sec. 6·3.

7·7 TWO IDEALIZED SYSTEMS

One way to make a second-law analysis of nonisolated systems is to replace their actual environments with idealized environments in which there is no entropy production. Then, the system plus its hypothetical environment form an isolated system, and any entropy production in this isolated system must be due to processes within the system of interest. This procedure requires certain hypothetical devices from which the hypothetical non-entropy-producing environment can be constructed. Since the interactions between the system and its environment include heat and work, we need to conceive of two "reservoirs" to or from which energy can be transferred reversibly as work or heat.

We shall conceive of a *thermal energy reservoir* (TER) as some system of fixed mass and volume that can exchange energy with its environment only as heat. Any energy transferred into the TER will appear as an increase in its internal energy. The TER is further idealized as having a uniform internal temperature; a TER is always in an equilibrium state. We usually conceive of the TER as being very large, so that the temperature remains constant for the interactions we consider. Internal energy represents disorganized molecular energy, and energy transfer as heat can be viewed as disorganized microscopic work. The TER can therefore be thought of as a source or sink for disorganized energy.

In contrast, the *mechanical energy reservoir* (MER) is some system that possesses energy only in some fully organized mechanical form, such as in a raised weight. The only energy-transfer mode for a MER is reversible *work;* whatever force acts on the MER is independent of the direction or rate of change of the energy. The force on the MER is conceived to be adjustable by design to fit the analysis at hand. All motions within a MER are assumed to be frictionless, so that any energy put into the MER as work can be completely recovered as work. The MER can be thought of as a source or sink for "fully organized energy." A MER can have but one state for a given energy; given its energy, we know the microscopic state of the MER *exactly*, without uncertainty.

These two conceptual systems would be difficult to build exactly, but can be closely approximated, hence are reasonable concepts. A block of copper can be a good approximation to a TER; it has a large capacity to store internal energy and can be idealized as being incompressible. A dead weight on the end of a frictionless pulley and a dead mass for the weight to gravitate toward form a reasonable conceptualization for a MER (Fig. 7·8*b*).

The MER and TER are useful concepts because they are two nonisolated systems for which we can easily compute the entropy change. The TER is a chunk of matter held at fixed volume. The infinitesimal increase in its entropy associated with an infinitesimal increase in its internal energy is, from Eq. (7·23),

$$dS = \frac{1}{T}\,dU \qquad \textit{for a TER} \tag{7·28}$$

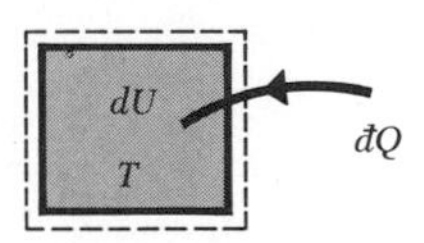

(a) *The TER is a reservoir for "disorganized" energy*

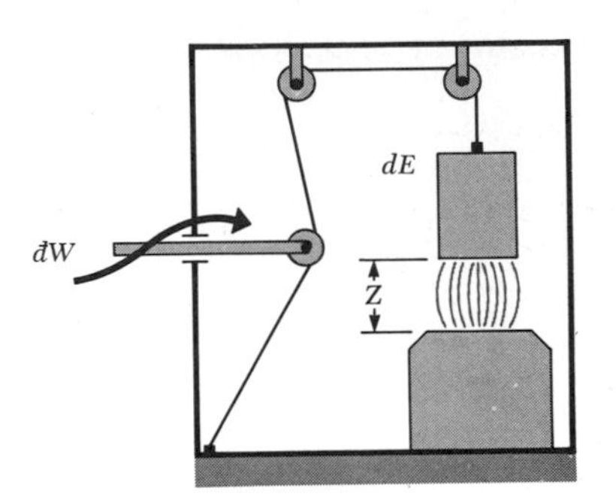

(b) *The MER is a reservoir for "organized" energy*

Figure 7·8. Two conceptual systems

An energy balance (Fig. 7·8*a*) reveals that the internal energy change is due solely to the energy transfer as heat *to* the TER

$$dU = đQ$$

hence the entropy change of a TER can be calculated from

$$dS = \frac{đQ}{T} \quad \textit{for a TER, đQ is energy input} \tag{7·29}$$

Thus the disorganized transfer of energy to a TER will result in an increase in the randomness inside, that is, in an increase in our uncertainty about its microscopic state. The entropy of a TER can be decreased by removal of energy as heat ($đQ < 0$). This will reduce the randomness inside, hence the entropy will decrease ($dS < 0$).

In contrast, a MER is conceptually a *perfectly organized* system. It has *one* microscopic state for each energy; given the energy, the microscopic state of a MER is *precisely* known without any uncertainty, and therefore its entropy is exactly zero. Hence

$$dS = 0 \quad \textit{for a MER} \tag{7·30}$$

An energy balance on a MER (Fig. 7·8*b*) reveals that the energy transfer to the MER as work appears as an increase in the fully organized mechanical energy stored inside. This organization makes the energy fully recoverable as work.

Imagine an isolated system consisting of two interacting MERs (Fig. 7·9). The entropy production in this isolated system will be

$$\underset{\text{entropy production}}{đ\mathscr{P}_S} = \underset{\text{increase in entropy storage}}{dS_{\text{MER1}} + dS_{\text{MER2}}}$$

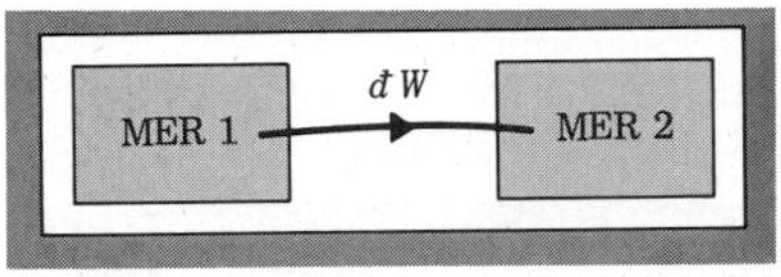

Figure 7·9. Energy exchange between two MERs produces no entropy

which is zero. So, MERs do not produce any entropy; they are reversible devices.

Consider now an isolated system composed of two interacting TERs (Fig. 7·10). We know that energy transfer as heat will not take place "up the temperature hill," hence heat transfer "down the temperature hill" is irreversible. However, in the limit where the temperature difference vanishes, the energy transfer process can be treated as reversible. To show this, let the temperature of TER B be T, and that of TER A be $T(1-\varepsilon)$, where ε is a small number. The entropy production in the combined system C as a result of the transfer of an amount of energy as heat $\text{đ}Q$ from B to A is

$$\underset{\text{entropy production}}{\text{đ}\mathscr{P}_S} = \underset{\text{increase in entropy storage}}{d(S_A + S_B)} \tag{7·31}$$

Considering the signs for positive energy transfer in Figs. 7·8 and 7·9, we note that the entropy changes of A and B are

$$dS_A = +\frac{\text{đ}Q}{T(1-\varepsilon)} = +\frac{\text{đ}Q}{T}(1 + \varepsilon + \varepsilon^2 + \cdots) \tag{7·32a}$$

$$dS_B = -\frac{\text{đ}Q}{T} \tag{7·32b}$$

Adding, we have, to first order,

$$\text{đ}\mathscr{P}_S = \varepsilon\frac{\text{đ}Q}{T} + \cdots = \varepsilon\, dS_A + \cdots \tag{7·33}$$

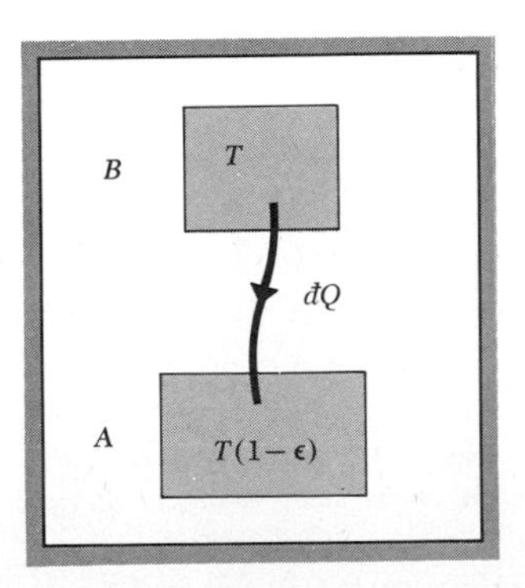

Figure 7·10. As $\varepsilon \to 0$ the process approaches reversibility

Now, when the temperatures of A and B are very nearly equal, ε will be very small, and we see that the entropy production, that is, $đ\mathscr{P}_S$, becomes very small in magnitude in comparison to dS_A or dS_B and vanishes in the limit $\varepsilon \rightarrow 0$. Hence, while the transfer of energy as heat across a finite temperature difference is irreversible and produces entropy, in the limit of zero temperature difference the process produces no entropy, hence becomes reversible. By reducing the temperature difference, the amount of entropy production can be made as small as one desires. Thus we can consider the process of energy transfer as heat between a system at some temperature T and another (say a TER) at some *infinitesimally* different temperature to be a reversible process.

We shall next use the TER and MER to form ideal reversible environments in order to study the entropy production due to processes within a control mass.

7·8 ENTROPY CHANGE AND PRODUCTION FOR A CONTROL MASS

We can now derive an expression for the entropy change of a control mass that undergoes an irreversible process. Suppose the system A receives energy as heat and work during the process; we imagine that these energies come from a TER and a MER, respectively. In order to ensure the reversibility of all processes within the hypothetical environment, the temperature difference between the TER and the point on the control mass where the energy is received as heat must be infinitesimal. The control mass and its hypothetical environment form an isolated system, shown in Fig. 7·11.

Denoting the control-mass entropy by S, we note that the entropy bookkeeping for the isolated system C is

$$\underbrace{đ\mathscr{P}_S}_{\text{entropy production}} = \underbrace{d(S_{\text{TER}} + S_{\text{MER}} + S)}_{\text{increase in entropy storage}}$$

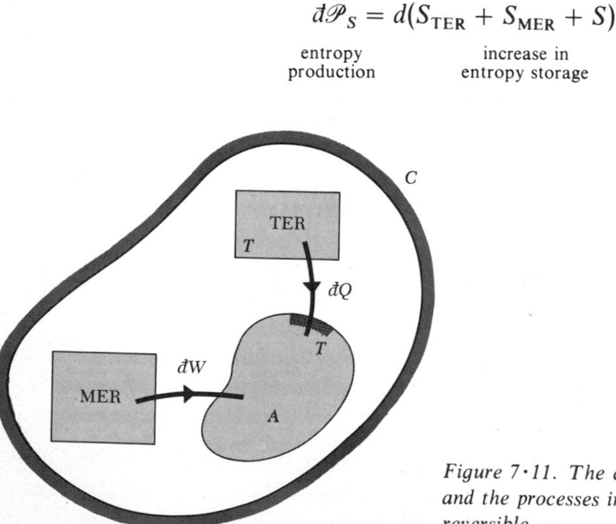

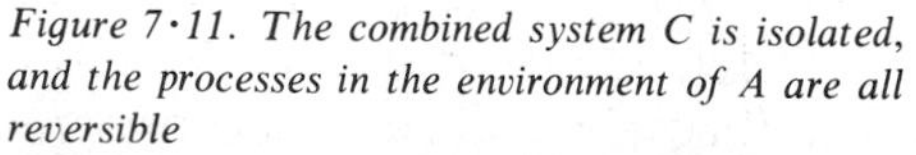
Figure 7·11. The combined system C is isolated, and the processes in the environment of A are all reversible

Now, the entropy of the MER does not change. Noting the difference in the direction of positive energy transfer as heat in Figs. 7·8 and 7·11, the entropy change for the TER is

$$dS_{\text{TER}} = \frac{-\text{đ}Q}{T}$$

Therefore the entropy production is

$$\text{đ}\mathscr{P}_S = dS - \frac{\text{đ}Q}{T} \qquad (7\cdot 34)$$

Solving for dS,

$$dS = \frac{\text{đ}Q}{T} + \text{đ}\mathscr{P}_S$$

The second law requires that $\text{đ}\mathscr{P}_S \geq 0$. Hence the entropy change of the control mass must satisfy

$$\blacktriangleright \quad dS \geq \frac{\text{đ}Q}{T} \quad \begin{array}{c}\textit{control mass, } \text{đ}Q\\ \textit{is energy input}\end{array} \qquad (7\cdot 35)$$

The equality holds for the reversible process, and the inequality for the irreversible process. Since the process within an ideal environment is reversible, any irreversibility must be due to the process within control mass A.

In the special case where the process within A is reversible, the equality in Eq. (7·35) gives

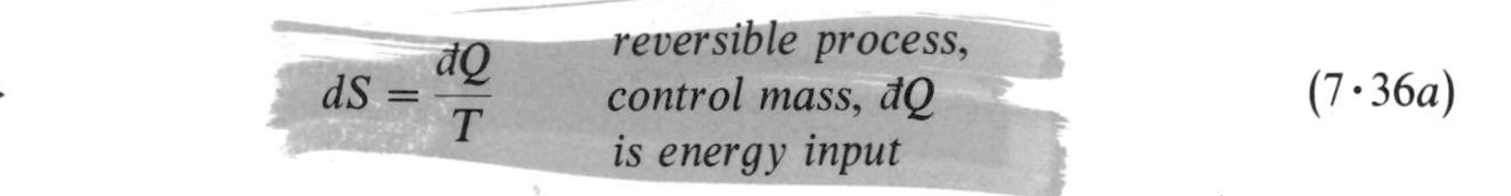

$$\blacktriangleright \quad dS = \frac{\text{đ}Q}{T} \quad \begin{array}{l}\textit{reversible process,}\\ \textit{control mass, } \text{đ}Q\\ \textit{is energy input}\end{array} \qquad (7\cdot 36a)$$

Integrating between an initial state 1 and a final state 2,

$$\blacktriangleright \quad \Delta S = \int_1^2 \frac{\text{đ}Q}{T} \quad \begin{array}{l}\textit{reversible process,}\\ \textit{control mass, } \text{đ}Q\\ \textit{is energy input}\end{array} \qquad (7\cdot 36b)$$

The restrictions are listed because they are important and often forgotten. This equation is very important in thermodynamics, for it provides a means for evaluating the entropy change from laboratory measurements of temperature and energy transfer, provided that the experimental process can be made sufficiently close to reversible that $\text{đ}\mathscr{P}_S$ is small compared to dS. Equation (7·36a) is taken as the definition of entropy in many classical macroscopic thermodynamic treatments.

A reversible process of special importance is the *reversible adiabatic process* ($đQ = 0$). Applying Eqs. (7·36) we see that the entropy of a control mass undergoing a reversible adiabatic process will not change; such a process of constant entropy is called an *isentropic process*.

$$dS = 0 \quad \text{for any control mass, reversible adiabatic process} \tag{7·37a}$$

or

$$\Delta S = 0 \quad \text{for any control mass, reversible adiabatic process} \tag{7·37b}$$

In a reversible adiabatic process the only energy-transfer mechanism is work. Reversible work constitutes perfectly ordered energy transfer and will not increase the molecular randomness. For example, the process of gas compression in a piston-cylinder system can be made very nearly reversible and adiabatic, and this process is often treated as an isentropic process. As the gas is compressed, its energy is increased. If the volume were not also reduced, this increased energy would increase the randomness and our uncertainty about the microscopic state. However, the reduction in volume means that we are more certain about the location of the molecules, and the constancy of the entropy can be regarded as a balance between the increased uncertainty due to the greater energy and the reduced uncertainty due to the more concentrated volume.† Another example of a (nearly) reversible adiabatic process is the rapid demagnetization of a paramagnetic salt at low temperatures. (This process is often used to produce very low temperatures.) If this process is idealized as reversible and adiabatic, the reduction in uncertainty due to the removal of energy in the demagnetization process can be viewed as being balanced by the increase in uncertainty due to the disalignment of the magnetic dipoles. The second law says that the net effect is to maintain constant entropy during the reversible adiabatic demagnetization.

Since the processes outside our control mass are all reversible and produce no entropy, all the entropy production can be charged directly to irreversibilities within the control mass. This means that we can interpret Eq. (7·34) as an entropy accounting *for the control mass*,

$$\underset{\substack{\text{entropy}\\\text{production}\\\text{in control}\\\text{mass}}}{đ\mathscr{P}_S} = \underset{\substack{\text{increase in}\\\text{entropy}\\\text{storage in}\\\text{control mass}}}{dS} - \underset{\substack{\text{entropy}\\\text{inflow to}\\\text{control}\\\text{mass}}}{\frac{đQ}{T}} \tag{7·38}$$

† These arguments are not quite correct, but are made in the proper spirit. Actually, quantum theory says that compression of the gas reduces the number of allowed states, and it is this which really compensates for the uncertainty increase due to energy addition.

We now regard the term $đQ/T$ as the amount of *entropy transfer with heat* into the control mass. This allows us to focus only on the control mass in making future second-law analyses; whenever energy enters a control mass as heat, we consider there to be an entropy *inflow* with heat in the amount $đQ/T$, where T is the temperature of the control mass at the point where the energy $đQ$ is received as heat. If energy is transferred *out* of the control mass, then we think of an entropy *outflow* with heat in the amount $đQ/T$, where T is the temperature of the control mass at the point where the energy $đQ$ is removed as heat. If there is a net inflow of entropy with heat, the entropy of the control mass must increase by at least this amount. The excess of the actual control-mass entropy increase over the net entropy inflow to the control mass represents the entropy that must have been produced within the control mass by irreversible action.

The microscopic view is very helpful in grasping the idea of entropy transfer with heat. We view heat as energy transfer that takes place as work on the microscopic scale, but in a random, disorganized way. This disorganized transfer carries some chaos with it, which makes for an entropy flow.

We also note that there is no term in Eq. (7·38) involving the work W. So there is no entropy transfer associated with energy transfer as work; *work is entropy-free*. Microscopically this is because work is an organized energy-transfer process.

7·9 EXAMPLES OF CONTROL-MASS SECOND-LAW ANALYSIS

The following examples will illustrate the application of the second law in control-mass analysis. The first two show how the ideas are used to help fix end states of processes, and the last two show the use of the second law in determining the best that can be achieved in a system *without knowing exactly what the system is*. The approach in each case is to first calculate the entropy production using control-mass entropy bookkeeping, combine this with the energy balance on the control mass, and then invoke the second law by requiring the entropy production to be nonnegative. Important ideas are introduced with these examples, so each should be studied carefully.

Irreversible compression. Consider the control mass of Fig. 7·12*a*. The cylinder initially contains 0.2 lbm of Freon 12 vapor at $-20°F$ (440 R) with an initial volume of 1.0 ft^3. The piston is suddenly pushed in very rapidly, causing shock waves to propagate through the Freon; after a short period of adjustment, the Freon comes to equilibrium at the new volume (0.2 ft^3). The process is sufficiently rapid that we can neglect any energy transfer as heat to the Freon. Find the possible terminal states for this process.

There is no entropy flow with heat for this adiabatic process, hence any increase in Freon entropy must be due to irreversible effects within the Freon.

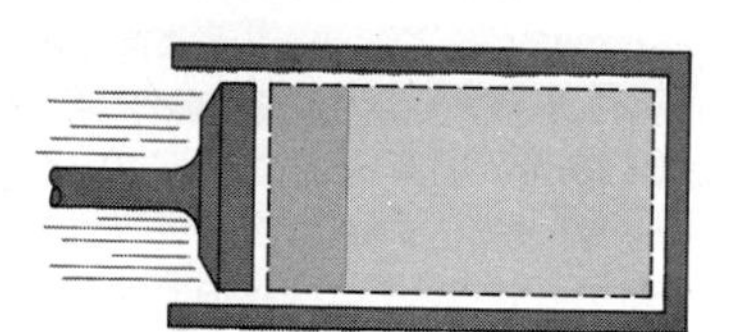

(*a*) *The rapid compression process forms a shock wave, which is an irreversible phenomenon*

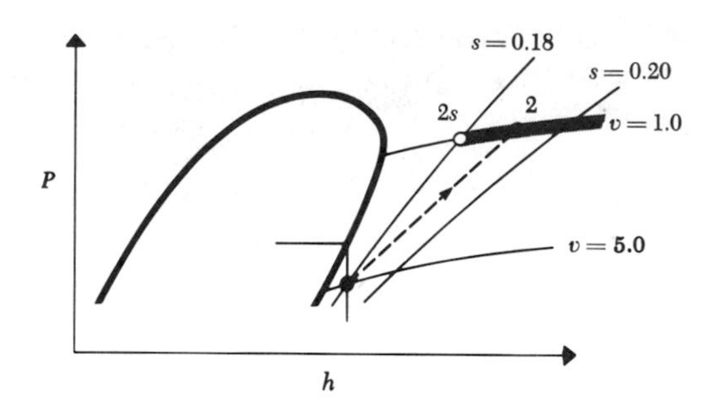

(*b*) *The shaded band shows the possible final states*

Figure 7·12.

Denoting the initial and final states by 1 and 2, respectively, we see that the entropy bookkeeping for the control mass is

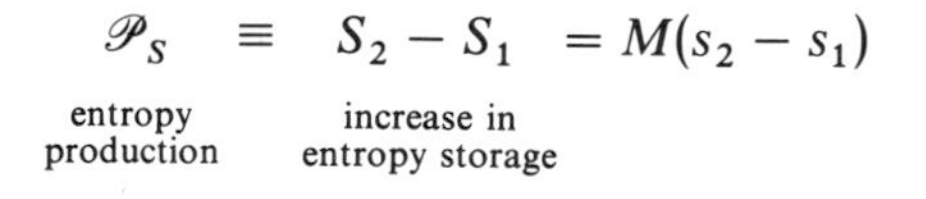

$$\underset{\text{entropy production}}{\mathscr{P}_S} \equiv \underset{\text{increase in entropy storage}}{S_2 - S_1} = M(s_2 - s_1)$$

The second law requires $\mathscr{P}_S \geq 0$. Thus any state with specific entropy greater than the initial state will be a possible final state. From Fig. B·7,

$$T_1 = 440 \text{ R}(-20°\text{F}) \qquad v_1 = 1.0/0.2 = 5 \text{ ft}^3/\text{lbm}$$

$$P_1 = 7.6 \text{ psia} \qquad s_1 = 0.184 \text{ Btu/(lbm·R)}$$

Following up a line of constant entropy to the final specific volume of $0.2/0.2 = 1.0 \text{ ft}^3/\text{lbm}$, we find the state 2 corresponding to no entropy production (reversible compression, isentropic process) as

$$T_{2s} = 93°\text{F} \qquad v_2 = 1.0 \text{ ft}^3/\text{lbm}$$

$$P_{2s} = 47 \text{ psia} \qquad s_{2s} = 0.184 \text{ Btu/(lbm·R)}$$

The other possible final states lie along the line of final specific volume, as shown in Fig. 7·12*b*. Note that we can conclude that the pressure obtained by this process will be at least 47 psia, and possibly greater if the process is irreversible. A first-law analysis could be used to calculate the work required for the compression if state 2 is known. One would find that the *least* amount of work is required when state 2 is state 2*s*, that is, when the compression process is reversible. This conclusion would be of obvious importance in the design of a Freon compressor for a large refrigeration system.

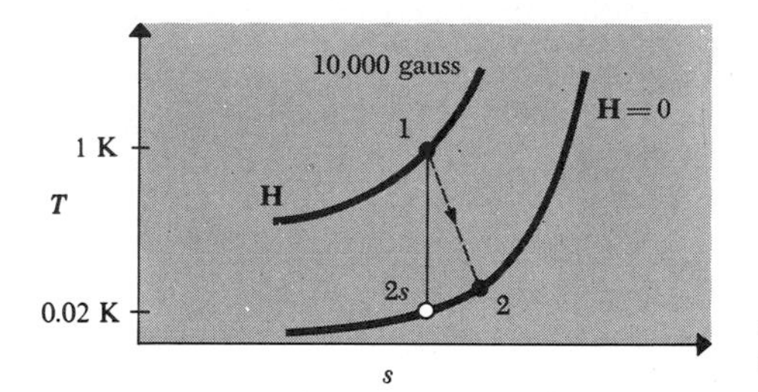

Figure 7·13. Adiabatic demagnetization is an effective way to obtain very low temperatures

Adiabatic demagnetization. Iron-ammonium alum is placed in a magnetic field **H** of 10,000 G at 1 K. The magnetic field is suddenly removed. What is the lowest temperature that can be attained?

We assume the process is adiabatic. Therefore the entropy bookkeeping for the control mass is

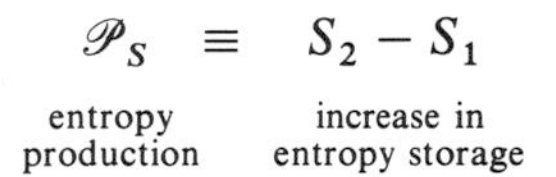

$$\underset{\text{entropy production}}{\mathscr{P}_S} \equiv \underset{\text{increase in entropy storage}}{S_2 - S_1}$$

The second law requires $\mathscr{P}_S \geq 0$. Looking at Fig. B·12, we see that the final state (2) must lie on the line $\mathbf{H} = 0$ to the right of the point where $s = 2.5 \times 10^5$ ergs/(g·K). The lowest temperature is therefore obtained with *reversible* adiabatic demagnetization and is 0.02 K. The actual value of T_2 would be greater (Fig. 7·13).

Heat pump. A house requires 500,000 kJ/day to keep it at 21°C when the outdoor temperature is 10°C. If a heat pump is used to supply this energy, what is the least amount of work that must be supplied to operate the heat pump for 1 day?

Figure 7·14 shows the control mass that we shall analyze. The first and second laws are both required. We assume that there is no change in the state of the heat pump over the cycle of 1 day, so there are no changes in the energy or entropy stored within the heat pump. Then, an energy balance gives

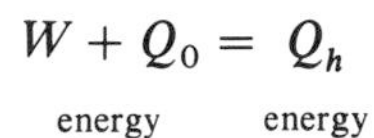

$$\underset{\text{energy input}}{W + Q_0} = \underset{\text{energy output}}{Q_h}$$

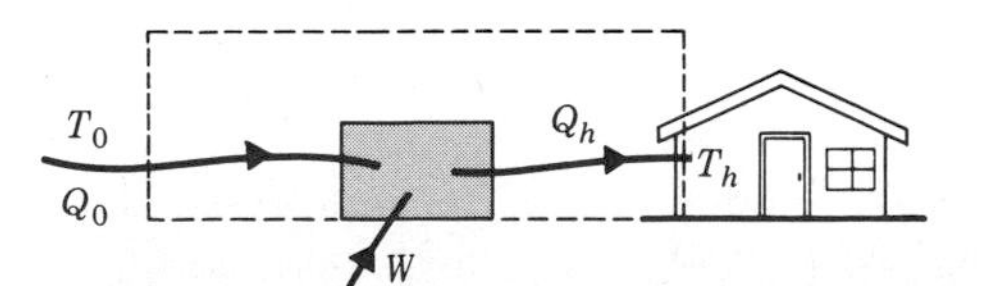

Figure 7·14. Heat pump system

The entropy bookkeeping for the control mass is

$$\underset{\text{entropy production}}{\mathscr{P}_S} = \underset{\text{entropy output}}{\frac{Q_h}{T_h}} - \underset{\text{entropy input}}{\frac{Q_0}{T_0}}$$

So

$$Q_0 = \frac{Q_h \cdot T_0}{T_h} - T_0 \mathscr{P}_S$$

Combining with the energy balance,

$$W = Q_h - Q_0 = Q_h\left(1 - \frac{T_0}{T_h}\right) + T_0 \mathscr{P}_S$$

Since $T_0 \geq 0$, and the second law requires that $\mathscr{P}_S \geq 0$,

$$W \geq Q_h\left(1 - \frac{T_0}{T_h}\right) \tag{7·39}$$

We must remember to use *absolute* temperatures; $T_0 = 283$ K and $T_h = 294$ K.

$$W \geq 5 \times 10^5 \times (1 - 283/294) = 1.87 \times 10^4 \text{ kJ}$$

This minimum work requirement is only about 4 percent of the energy the house requires. The challenge to the engineer is now to design a device that can perform as close to this optimum as possible, at reasonable expense.

The equality holds in the above expressions if the processes within the heat pump are all reversible, so that it produces no entropy. Note that the actual work will differ from the minimum work by $T_0 \mathscr{P}_S$. Therefore this term represents the "extra" work required because of the irreversibilities in the actual system.

Available energy. Suppose we have a control mass with an initial energy E and entropy S. What is the maximum amount of useful work that can be obtained from this control mass if it can interact in any way possible with the environment?

Again this requires a combination of the first and second laws, which we will apply to the control mass of Fig. 7·15. Since we have not ruled out anything,

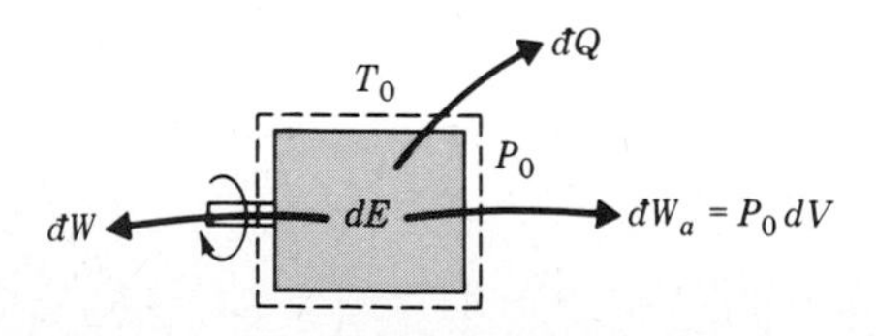

Figure 7·15. The control mass expands against the atmosphere

we must allow energy exchange as heat between the control mass and its environment, and we must allow the control mass to expand or contract, and in addition to put out useful work, say through a shaft. The sign convention for the energy terms are shown in the figure. Note the placement of the boundaries; the pressure and temperature at these boundaries are the atmospheric values, P_0 and T_0.

For an infinitesimal step of the process, the energy balance gives

$$\underset{\text{energy input}}{0} = \underset{\text{increase in energy storage}}{dE} + \underset{\text{energy output}}{đW + P_0\,dV + đQ}$$

The entropy bookkeeping for the control mass is

$$\underset{\text{entropy production}}{đ\mathscr{P}_S} = \underset{\text{increase in entropy storage}}{dS} + \underset{\text{entropy outflow}}{\frac{đQ}{T_0}}$$

Combining, and solving for the increment of useful work,

$$đW = -dE - P_0\,dV - T_0(đ\mathscr{P}_S - dS)$$

Since P_0 and T_0 are constants, we can integrate, obtaining

$$W_{12} = \int_1^2 đW = (E + P_0 V - T_0 S)_1 - (E + P_0 V - T_0 S)_2 - T_0 \mathscr{P}_S$$

where 1 and 2 denote the initial and final states of the control mass, respectively, and $\mathscr{P}_S$ is the total entropy production for the process. Finally, we invoke the second law, which requires that the entropy production $\mathscr{P}_S$ be positive, or at the very least zero; so,

$$W_{12} \le (\mathsf{A}_1 - \mathsf{A}_2) = -\Delta\mathsf{A}$$

where $\mathsf{A} = E + P_0 V - T_0 S$. A is called the *control-mass availability function.* Note that it depends both upon the state of the control mass and its environment. We have shown that the maximum useful work that can be obtained for any specified change in the state of the control mass is the *decrease* in the availability function. We emphasize that the control-mass availability function is different from the steady-flow availability function which we shall discuss shortly.

The end state 2 that has the lowest availability function gives the largest maximum-work output. If the control mass is just a simple compressible substance, E is just the internal energy U. If we assume that it is in an equilibrium state at 2, then $S = S(U, V)$ at state 2. So, in this case

$$\mathsf{A} = U + P_0 V - T_0 S(U, V) = \mathsf{A}(U, V, T_0, P_0)$$

To find the minimum of A at fixed T_0, P_0 we set

$$\left.\frac{\partial A}{\partial U}\right)_V = 0 \qquad \left.\frac{\partial A}{\partial V}\right)_U = 0$$

which gives

$$1 - T_0 \left.\frac{\partial S}{\partial U}\right)_V = 0 \qquad P_0 - T_0 \left.\frac{\partial S}{\partial V}\right)_U = 0$$

Recalling the thermodynamic definitions of temperature and pressure, Eqs. (7·9) and (7·20), we see that these conditions are satisfied when the control mass has the same temperature and pressure as the environment at state 2. The largest maximum work is possible when the control mass ends up in equilibrium with the environment. This makes sense; once the system has equilibrated with the environment, it is impossible to extract further energy as useful work.

Note that, as in the previous example, $T_0 \mathscr{P}_S$ again represents "lost" work associated with the irreversibility of the process.

7·10 EQUIVALENCE OF THE MECHANICAL AND THERMODYNAMIC PRESSURES

We can now show the conditions under which the mechanical and thermodynamic pressures are equal. Consider the control mass of Fig. 7·16. The entropy change of the control mass can be related to the energy and volume changes using the Gibbs equation (7·23), which in extensive form is

$$dS = \frac{1}{T}\,dU + \frac{P}{T}\,dV \tag{7·40}$$

An energy balance on the control mass of Fig. 7·16 gives

$$dU = đQ + đW \tag{7·41}$$

In Chap. 2 we derived an expression for the energy transfer as work for fluid compression; there we used the mechanical pressure, which we denote for the moment as P_m. The work done *on* the substance is

$$đW = -P_m\,dV \tag{7·42}$$

Figure 7·16. The control mass is a simple compressible substance

Finally, the entropy bookkeeping is, from Eq. (7·38),

$$\text{đ}\mathscr{P}_S = dS - \frac{\text{đ}Q}{T} \qquad (7\cdot43)$$

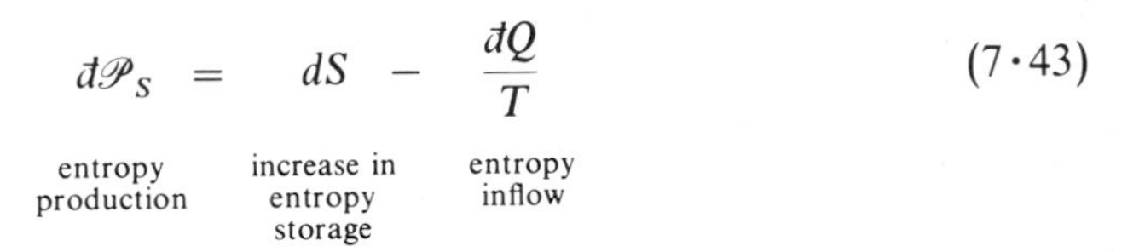

Using Eq. (7·40) to substitute for dS in Eq. (7·43), and Eq. (7·41) to substitute for $\text{đ}Q$, and Eq. (7·42) for $\text{đ}W$,

$$\text{đ}\mathscr{P}_S = \frac{1}{T}\,dU + \frac{P}{T}\,dV - \frac{dU}{T} + \frac{\text{đ}W}{T} = \frac{P\,dV + \text{đ}W}{T} = \left(\frac{P - P_m}{T}\right) dV$$

Since the second law requires $\text{đ}\mathscr{P}_S \geq 0$,

$$\left(\frac{P - P_m}{T}\right) dV \geq 0 \qquad (7\cdot44)$$

Now, if the process is reversible, the amount of entropy production $\text{đ}\mathscr{P}_S$ will be zero, hence we conclude that $P = P_m$ *for a reversible process.* During any such reversible process the control mass must always be in an equilibrium state (otherwise there would be some irreversibility associated with the relaxation toward equilibrium). We conclude that the thermodynamic and mechanical pressures are identical *in any equilibrium state.*

If the process is irreversible, then $\text{đ}\mathscr{P}_S > 0$, hence

$$(P - P_m)\,dV > 0$$

If the substance is expanding, doing work, $dV > 0$, hence $P > P_m$. The pressure measured during an irreversible expansion would therefore be less than the pressure the system would have in equilibrium at the same energy and volume. Hence less work is done by the system than might be. In contrast, if the substance is being compressed, $dV < 0$, and $P < P_m$. The pressure exerted during compression is therefore greater than the equilibrium pressure. Hence more work must be done on the system than for the reversible process.

A microscopic interpretation will be helpful. Consider the compression of a gas in a piston-cylinder system. If the compression process is very rapid, the molecules will bunch up near the piston, exerting a larger pressure (P_m) than if they were distributed uniformly over the cylinder (P). Hence more work will be required for rapid than for slow compression. If the piston moves rapidly away from the gas, there will be fewer molecules near the piston than for a slower expansion ($P_m < P$), and less work will be done by the gas. Figure 7·17 shows a hydraulic analogy that is helpful in grasping these ideas.

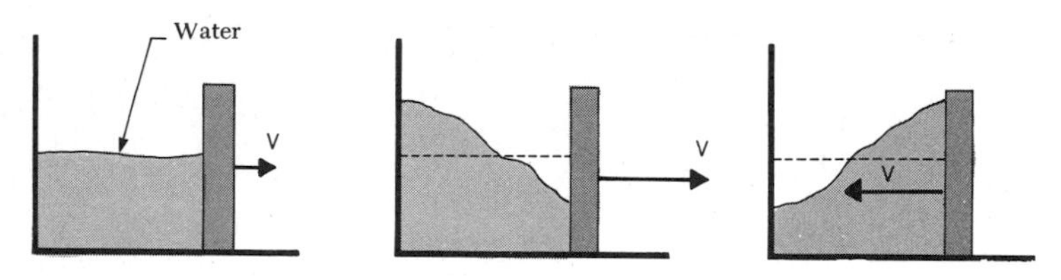

Figure 7·17. Work in reversible and irreversible processes

7·11 EQUIVALENCE OF THE THERMODYNAMIC AND EMPIRICAL TEMPERATURES

Now that we know $P = P_m$ in any equilibrium state, we can use the fact that the entropy is a function of state, together with certain experimental observations, to establish the correspondence between the thermodynamic and empirical temperature scales. The experimental evidence required is that the gases used in empirical thermometers apparently have the property that at very low pressures their internal energies u and their pressure-volume products depend solely on temperature (they behave like perfect gases). This behavior is also predicted by quantum-statistical thermodynamic theory. The fact that the internal energy becomes a function only of temperature at very low pressures is nicely shown in Figs. B·1 and B·4. Conversely, at low pressures, specification of the internal energy is sufficient to determine the temperature, or the temperature is a function only of the internal energy. Consequently, for gases at sufficiently low pressures, we put

$$T = T(u)$$

Here T denotes the thermodynamic temperature. For the Pv product,

$$Pv = f(T) \tag{7·45}$$

where $f(T)$ represents some unknown function of the thermodynamic temperature.

In an equilibrium state the specific entropy is a function only of the specific internal energy and the specific volume,

$$s = s(u, v)$$

Then, from the calculus

$$\left\{\frac{\partial[(\partial s/\partial u)_v]}{\partial v}\right\}_u = \left\{\frac{\partial[(\partial s/\partial v)_u]}{\partial u}\right\}_v$$

Expressed in terms of the thermodynamic temperature and pressure,

$$\left[\frac{\partial(1/T)}{\partial v}\right]_u = \left[\frac{\partial(P/T)}{\partial u}\right]_v$$

Since the temperature of the low-pressure gas is independent of volume along a line of constant internal energy, the term on the left is zero, and consequently

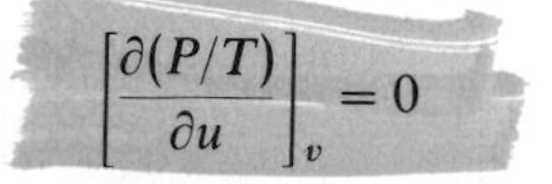

$$\left[\frac{\partial(P/T)}{\partial u}\right]_v = 0$$

We see that P/T is independent of u along a line of constant v. However, since u depends only on T, the ratio P/T must be independent of T along this line. Therefore we conclude that low-density gases have the property that

$$\left[\frac{\partial(P/T)}{\partial T}\right]_v = 0 \tag{7·46}$$

Using Eq. (7·45) in Eq. (7·46), we obtain

$$\frac{1}{v}\left[\frac{\partial(f/T)}{\partial T}\right]_v = \frac{1}{v}\frac{d(f/T)}{dT} = 0$$

The ratio $f(T)/T$ must therefore be *a constant*, which means that $f(T)$ must be of the form

$$Pv = f(T) = \text{constant} \times T$$

This is exactly the form that was arbitrarily selected for the empirical temperature; the empirical and thermodynamic temperature scales therefore must coincide, apart from a multiplicative constant. By judicious selection of the constant in the defining equation for the entropy, the two scales can be made identical. Henceforth we shall consider that the thermodynamic and empirical *absolute* temperatures† are exactly the same.‡

7·12 APPLICATIONS OF THE SECOND LAW TO ENERGY-CONVERSION SYSTEMS

We now have in hand the tools necessary to study the limitations imposed by the second law of thermodynamics on any process. One of the important areas

† On the Kelvin or Rankine scale.

‡ A more accurate experimental test of this could be made using the *Clapeyron equation*. This experiment will be discussed in the next chapter.

of application of thermodynamics is in the study of energy-conversion systems. Nature has been generous in providing large sources of energy, but she has managed to keep most of it tied up in randomly oriented microscopic forms and left humans with the task of devising means for converting it to usable macroscopic forms. However, the second law places certain limitations on the performance of energy-conversion systems; for example, we find that it is not possible to convert all the energy obtained from a nuclear reaction into useful mechanical work. The second law can be used to derive an expression for the maximum possible energy-conversion efficiency for any continuously operating converter, and the same limit holds for both steam power plants and thermoelectric converters. Such sweeping generality is indeed impressive, and very characteristic of thermodynamic theory.

We shall define a *heat engine* as any control mass to and from which energy is transferred as heat and from which energy is transferred as work. We further require that the processes undergone by the matter within the engine be cyclic, or continuous, such that after some period all the matter within the engine is returned to its initial state.

A special type of heat engine is useful. Any heat engine to which energy is transferred as heat at one temperature and from which energy is transferred at a lower temperature is called a *2T engine* (Fig. 7·18).

There are many ways by which one can devise a $2T$ engine, and it would be well to illustrate one at this point. Suppose we have a piston-cylinder system filled with a gas. By causing this system to execute the four-process cycle, shown schematically in Fig. 7·19, we can obtain work as a result of transfers of energy as heat to and from the engine. The process representation shows the temperature of the gas as a function of the piston position. Observe the cyclic nature of this process. By proper control of the piston displacement, the processes can in principle be carried out in such a way that the gas temperature is constant while energy is being transferred as heat. The processes inside the engine can in principle be made reversible. However, if we tried to reverse the engine (this would then be called a heat pump), $T_{A'}$ would have to rise above T_A for there to be a transfer of heat to environment A. But if the process is carried out slowly enough, $T_{A'}$ need be only infinitesimally different than T_A, and $T_{B'}$ only infinitesimally different than T_B, so that the heat-pump process can, in the limit, be made the exact reverse of the heat-engine process (Fig. 7·20); this we call a *reversible 2T engine* (R2T). This particular engine operates on a Carnot cycle and is thus called a *Carnot engine*. The reversible $2T$ engine is of course only an idealization, but it is fully as useful in thermodynamic theory as the frictionless pulley is in mechanics.

The *energy-conversion efficiency* of a heat engine is defined as the ratio of the useful work output to the energy input as heat,

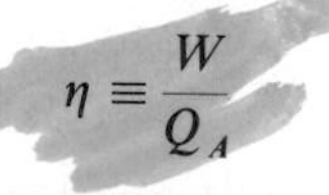

$$\blacktriangleright \qquad \eta \equiv \frac{W}{Q_A} \qquad (7\cdot47)$$

This ratio will not remain constant when the engine is reversed, except for a

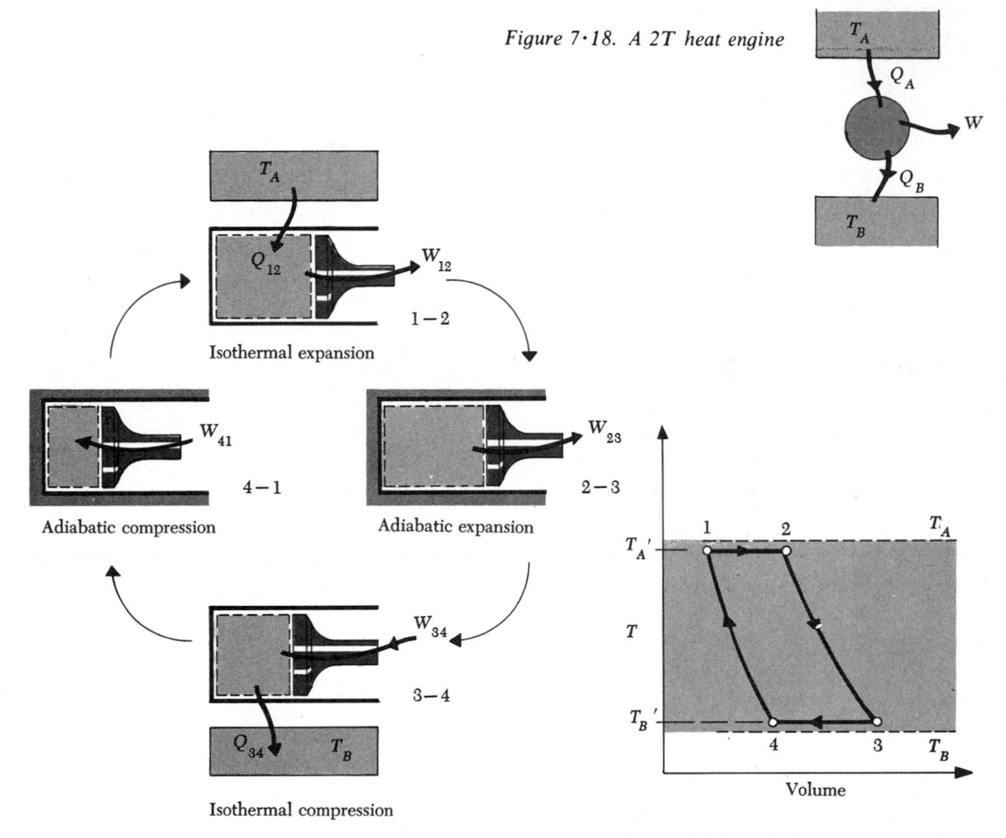

Figure 7·18. A 2T heat engine

Figure 7·19. A 2T-heat-engine cycle

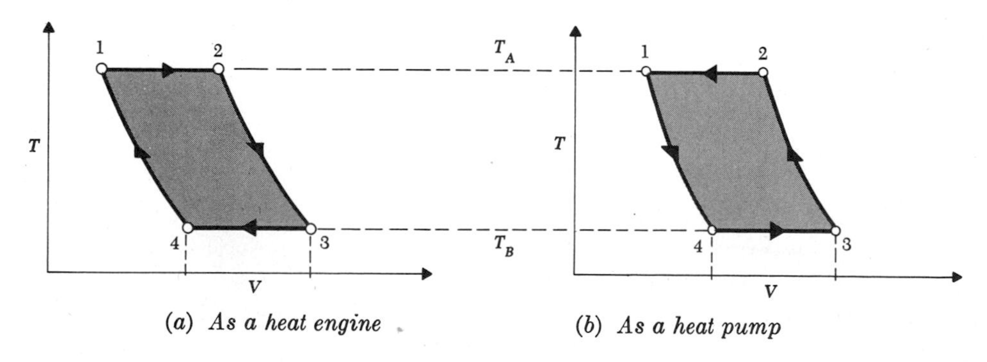

(a) As a heat engine *(b) As a heat pump*

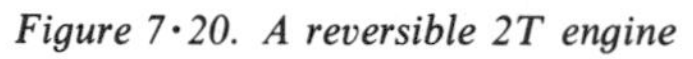

Figure 7·20. A reversible 2T engine

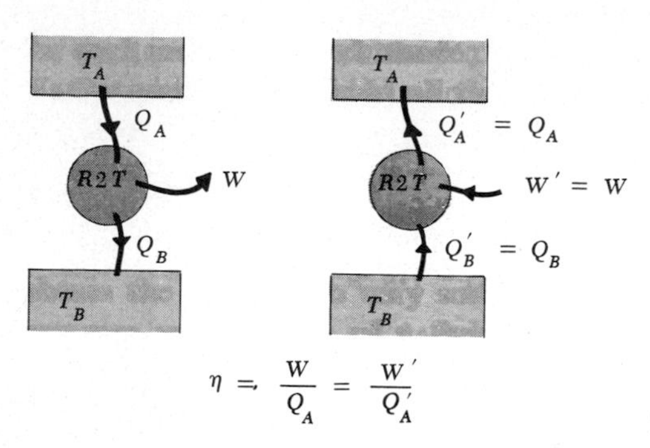

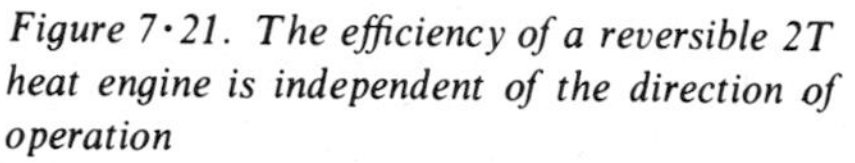
Figure 7·21. The efficiency of a reversible 2T heat engine is independent of the direction of operation

reversible $2T$ engine (see Fig. 7·21). Note that no energy-storage terms have been shown in any of these figures. The symbols Q and W are to be interpreted as being "for a cycle" and, since the engines operate cyclically, there is no change in the energy within an engine over a cycle.

A simple expression for the limiting efficiency of a $2T$ engine can be derived by considering the control mass of Fig. 7·22. The engine receives energy as heat from a source at temperature T_s, rejects energy as heat to the environment at temperature T_0, and puts out work. The engine operates cyclically, or in a steady state, so that over the period of one cycle there is no change in the entropy stored in the engine. The entropy bookkeeping for the control mass is

$$\underset{\text{entropy production}}{\mathscr{P}_S} = \underset{\text{entropy outflow}}{\frac{Q_0}{T_0}} - \underset{\text{entropy inflow}}{\frac{Q_s}{T_s}} \tag{7·48}$$

If the engine is a reversible device, $\mathscr{P}_S = 0$; otherwise $\mathscr{P}_S > 0$. For a reversible engine,

$$\blacktriangleright \quad \frac{Q_0}{Q_s} = \frac{T_0}{T_s} \quad \textit{for a reversible 2T engine} \tag{7·49}$$

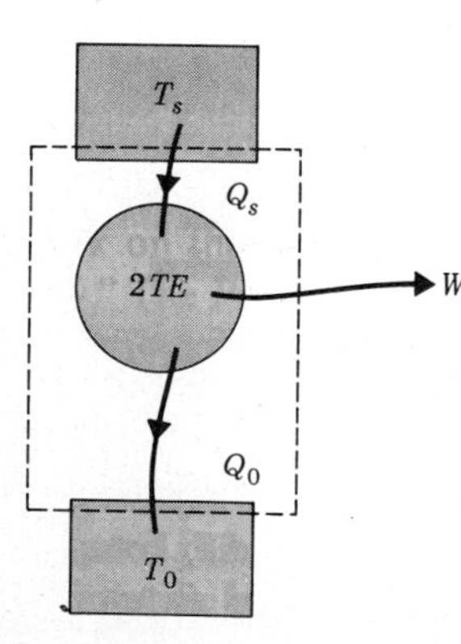

Figure 7·22. Determining the efficiency of a reversible 2T engine

Now, an energy balance on the control mass gives

$$W = Q_s - Q_0 \tag{7·50}$$

Solving Eq. (7·48) for Q_0 and substituting in Eq. (7·50),

$$W = Q_s\left(1 - \frac{T_0}{T_s}\right) - T_0\mathscr{P}_S$$

The second law requires that $\mathscr{P}_S \geq 0$, so

$$W \leq Q_s\left(1 - \frac{T_0}{T_s}\right)$$

Again we see that $T_0\mathscr{P}_S$ represents the reduction in work output caused by irreversibilities. Finally, the efficiency is†

▶ $$\eta_{2T} \equiv \frac{W}{Q_S} \leq 1 - \frac{T_0}{T_s} \tag{7·51}$$

The maximum efficiency, obtained for a reversible engine, is

▶ $$\eta_{R2T} = 1 - \frac{T_0}{T_s} \tag{7·52}$$

It is common practice to call this the *Carnot efficiency*, since the Carnot engine is one type of reversible $2T$ engine. However, we have said nothing at all about the nature of the engine, so the expression actually holds for *any* reversible $2T$ engine.

The Carnot efficiency is an upper limit for the performance of any real heat engine. Highest efficiencies will be obtained when the ratio T_0/T_s is as small as possible; one would like to add the energy as heat at as high a temperature as possible and reject energy as heat at the lowest possible temperature. However, nature places physical limitations on man's capabilities. The energy which is rejected as heat must flow to an environment which is cooler than T_0; this means that we are limited to T_0 of the order of 285 K. The energy transfer as heat must come into the engine from a region at a temperature greater than T_s. Temperatures of the order of 2000 K can be obtained by combustion reactions, but metallurgical considerations normally require that the device be kept much cooler. Modern steam power plants operate at about 900 K (not on the Carnot cycle). Advanced nuclear power systems are being designed and tested in the range 1300 to 1700 K. These high-temperature systems employ exotic metals and are not

† Note that $Q_0/Q_s \geq T_0/T_s$. Also, $\eta = (Q_s - Q_0)/Q_s \leq 1 - T_0/T_s$.

intended for long life or for production of low-cost electric power. At 1000 K the Carnot-cycle efficiency is

$$1 - 285/1000 = 0.72$$

Even under such extreme conditions this most ideal cycle could convert only 72 percent of the energy transferred in as heat to useful work. Realistic devices of present technology operate in the range 15 to 40 percent, and a heat engine having a thermal energy-conversion efficiency of even 50 percent has yet to be built.

It is evident that it would be impossible to devise a 100 percent efficient heat engine even if we really had reversible processes at our disposal. This can be seen by setting $Q_0 = 0$ in Eq. (7·48), for we then find that Q_s must also be zero or negative for any cyclically operating device. A 1*T* engine is therefore impossible, though it is not at all difficult to dream up a continuously operating device *to* which energy is transferred as work and *from* which energy is transferred as heat. It is also possible to make a system which will receive energy as heat, converting all this energy to work, but which will *not* operate *indefinitely*. For example, we can heat the gas in a piston-cylinder system and allow it to expand, removing the same amount of energy we put in, but eventually we would run out of cylinder, the piston would fall out, and the device would cease to operate. A noncyclic converter with 100 percent efficiency is possible, but a continuously operating one is not. The statement that a continuously operating 1*T* engine is impossible is known as the *Kelvin-Planck statement* and is taken as the starting point in many classical developments of the second law. The restrictions placed on heat engines by the second law are illustrated in Fig. 7·23.

The second law also places limitations on systems that continuously transfer energy as heat from a region of low temperature to one of higher temperature (a refrigerator, or a heat pump). In particular, it may easily be shown that it is impossible to devise any system that will do this cyclically or continuously without any energy input as work. Such a device would constitute a refrigerator which

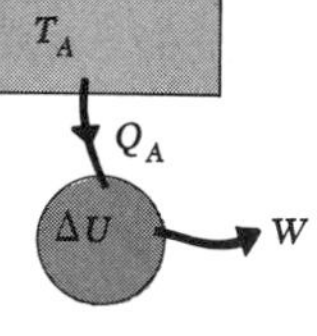

(*b*) *Possible for a short time*

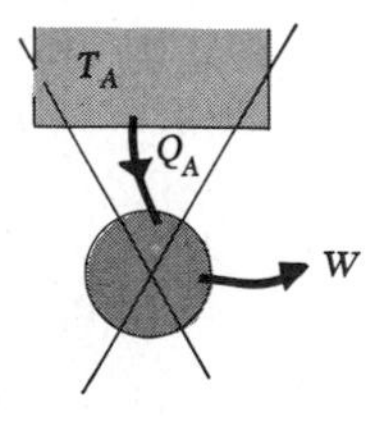

(*a*) *Impossible for continuous or cyclic operation*

(*c*) *Possible for continuous or cyclic operation*

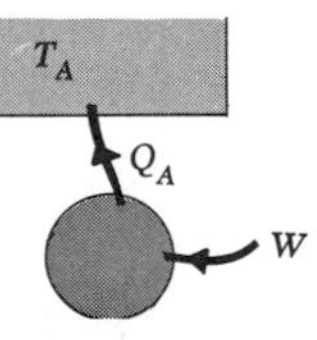

Figure 7·23. A 1T heat engine is impossible

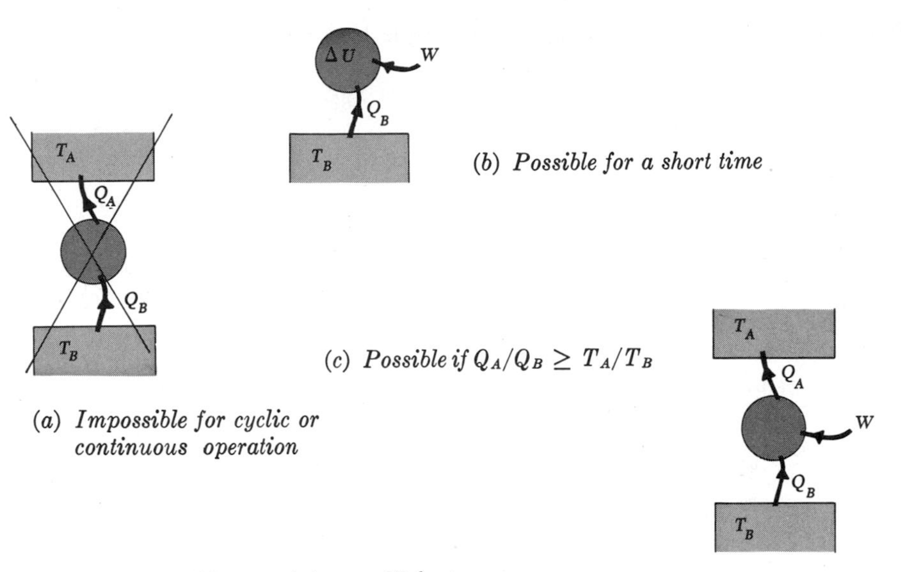

Figure 7·24. Second-law restrictions on 2T heat pumps

requires no input power, and it certainly seems reasonable that this is an impossibility. However, it is possible to devise a system that will do this for a short time, or to have energy flow through a device from high temperature to low temperature without power input. The restrictions imposed by the second law on heat pumps are illustrated in Fig. 7·24. The statement that a zero-work heat pump is impossible is called the *Clausius statement*, and it too is often taken as a starting point for the development of the second law in classical thermodynamics. Its proof from entropy-production considerations is left as an exercise.

7·13 MEASUREMENT OF THE THERMODYNAMIC TEMPERATURE

The fact that all reversible $2T$ engines operating between the same two temperatures have the same efficiency can be used in principle to obtain a direct measurement of the thermodynamic temperature. We imagine using any reversible $2T$ engine as a thermometer and letting it operate between the unknown temperature T_s and the H_2O triple-point temperature $T_0 \equiv 273.16$ K. We measure the energy transfers Q_s and Q_0 and then use Eq. (7·49) to calculate the thermodynamic temperature of the test environment. It is important to appreciate that the thermodynamic temperature scale is independent of the nature of any thermometric substance; we can now look upon the gas thermometer as a means for measuring the thermodynamic temperature rather than as a means for defining an empirical temperature scale.

7·14 THE SECOND LAW FOR A CONTROL VOLUME

So far we have limited our analysis to control masses. We shall now extend the idea of entropy production to a control volume, for it is in this form that the second law is most useful for engineering analysis.

Consider the control volume of Fig. 7·25. We permit energy transfer as heat to take place at various points on the boundary, except where mass crosses the control surface. Energy transfer as work may also occur. For simplicity we assume that there are only the two mass flows shown and that they are one-dimensional at sections 1 and 2. To make the control-volume transformation we consider a control mass which occupies the control volume at time t and examine this control mass over an infinitesimal time interval dt. The change in its entropy is

$$dS_{CM} = S_{CM}(t + dt) - S_{CM}(t)$$

The entropy within the control mass at time t is exactly the entropy within the control volume at that time, since the two then coincide. The value of $S_{CM}(t + dt)$ can be expressed in terms of the entropy within the control volume at $t + dt$ and the entropies of the shaded portions A and B as

$$S_{CM}(t + dt) = S_{CV}(t + dt) + S_B - S_A$$

The entropies within A and B may be expressed in terms of the specific entropies of the substances flowing across the control surface and the masses

$$S_A = đM_A s_1 \qquad S_B = đM_B s_2$$

Thus

$$dS_{CM} = dS + S_B - S_A$$

or

$$dS_{CM} = dS + (s\,đM)_B - (s\,đM)_A$$

where $dS = S_{CV}(t + dt) - S_{CV}(t)$. The energy transfers as heat $đQ_i$ may now be

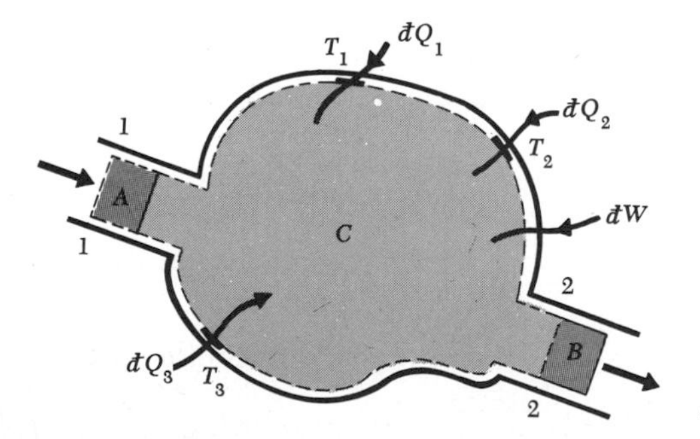

Figure 7·25. The control volume

considered as going to either the control mass or the control volume. The entropy bookkeeping *for the control mass* is then

$$\underset{\text{entropy production}}{\text{đ}\mathscr{P}_S} \equiv \underset{\text{increase in entropy within the control mass}}{[dS + (s\,\text{đ}M)_B - (s\,\text{đ}M)_A]} - \underset{\text{entropy inflow}}{\sum_{\text{in}} \frac{\text{đ}Q_i}{T_i}}$$

Note that this equation is written in terms of the properties of the control volume, hence can be viewed as entropy bookkeeping *for the control volume.*

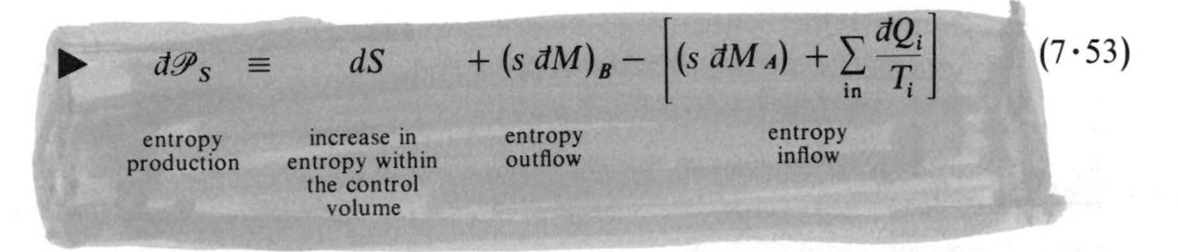

$$\blacktriangleright \quad \underset{\text{entropy production}}{\text{đ}\mathscr{P}_S} \equiv \underset{\text{increase in entropy within the control volume}}{dS} + \underset{\text{entropy outflow}}{(s\,\text{đ}M)_B} - \underset{\text{entropy inflow}}{\left[(s\,\text{đ}M_A) + \sum_{\text{in}} \frac{\text{đ}Q_i}{T_i}\right]} \qquad (7\cdot53)$$

Note that T_i should be interpreted as the temperature of the control surface at the point where $\text{đ}Q_i$ enters the control volume. The terms $s\,\text{đ}M$ represent *convective entropy flow* associated with mass transfer across the boundaries of the control volume. The second law requires that the entropy production be positive, or at the very least zero,

$$\blacktriangleright \qquad \text{đ}\mathscr{P}_S \geq 0 \qquad (7\cdot54)$$

For a control volume with several inflows and outflows, the definition of entropy production can readily be extended and expressed on a *rate* basis as

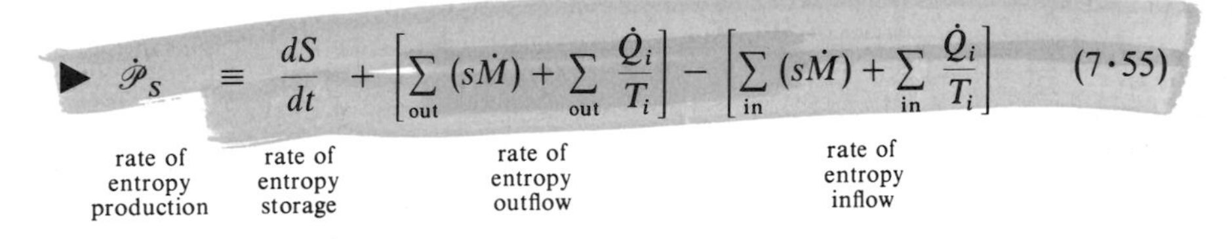

$$\blacktriangleright \quad \underset{\text{rate of entropy production}}{\dot{\mathscr{P}}_S} \equiv \underset{\text{rate of entropy storage}}{\frac{dS}{dt}} + \underset{\text{rate of entropy outflow}}{\left[\sum_{\text{out}} (s\dot{M}) + \sum_{\text{out}} \frac{\dot{Q}_i}{T_i}\right]} - \underset{\text{rate of entropy inflow}}{\left[\sum_{\text{in}} (s\dot{M}) + \sum_{\text{in}} \frac{\dot{Q}_i}{T_i}\right]} \qquad (7\cdot55)$$

Here $\dot{\mathscr{P}}_S$ is the *rate of entropy production* within the control volume. The terms $s\dot{M}$ represent *rates of convective entropy transfer*, and the terms $\dot{Q}/T$ represent *rates of entropy transfer with heat*. The second law requires that the rate of entropy production be positive,

$$\blacktriangleright \qquad \dot{\mathscr{P}}_S \geq 0 \qquad (7\cdot56)$$

The equality is again associated with reversible processes within the control volume, and the inequality with irreversible processes.

7·15 EXAMPLES OF CONTROL-VOLUME SECOND-LAW ANALYSIS

We shall now illustrate the methodology of second-law analysis for control volumes with several examples. The general approach is always the same and follows the pattern we developed for energy analysis:

1. Define the system, list idealizations, define energy symbols on a sketch, and write the energy balance as per energy-balance methodology.
2. Make an entropy accounting for the control volume and develop an expression for the entropy production (or production rate) within the control volume in terms of the properties at state points, mass flows, heat flows, and temperatures indicated on the sketch.
3. Combine the energy balance and entropy accounting to express the desired quantity. For nonadiabatic systems it is usually most convenient to combine these two equations to eliminate the energy transfer as heat to the environment and then to express the quantity of interest in terms of known quantities and the unknown entropy production rate.
4. Invoke the second law by demanding that the entropy production, or entropy production rate, be nonnegative. This allows one to determine the best possible performance, which always corresponds to an ideal system in which all processes are reversible.

In the examples that follow a number of important ideas are developed, hence each example should be studied carefully.

A compressor. Freon 12 enters an adiabatic compressor at 30 psia and 40°F and is compressed to 140 psia. What is the shaft-work input per lbm of Freon 12 for the best adiabatic compressor which might be devised? See Fig. 7·26.

We make the following idealizations:

Steady flow steady state
Kinetic and potential energy changes negligible

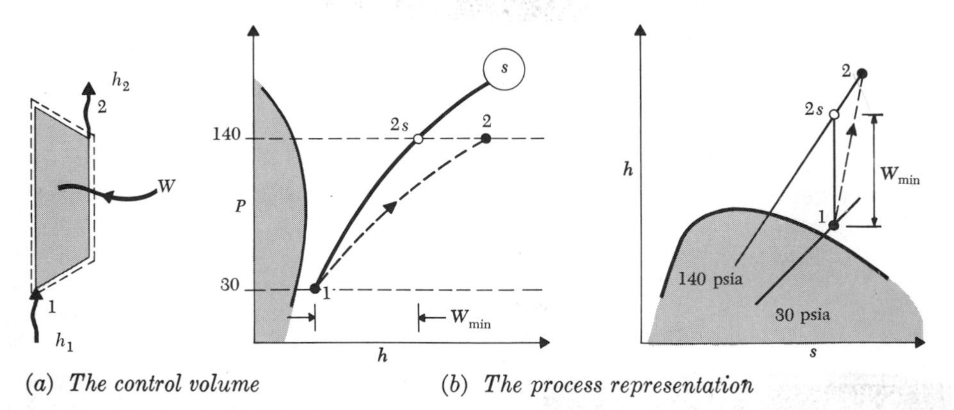

(a) *The control volume* (b) *The process representation*

Figure 7·26. Compressor analysis

An energy balance allows us to relate the work input W, *per unit of mass throughflow*,† to the enthalpy rise across the compressor,

$$W = h_2 - h_1$$

The second law requires that the entropy production be positive. Since we have assumed that steady-flow, steady-state conditions prevail, the entropy within the control volume is not changing and, because of our adiabatic idealization, no entropy flows with heat across the boundary. Only the convected entropy flows are involved, so the entropy bookkeeping for the control volume is

$$\underbrace{\dot{\mathscr{P}}_S}_{\text{rate of entropy production}} = \underbrace{\dot{M}(s_2 - s_1)}_{\text{net rate of entropy outflow}}$$

The second law requires $\dot{\mathscr{P}}_S \geq 0$, so

$$s_2 - s_1 \geq 0$$

The entropy of the fluid must increase as it passes through the device. This means that the outlet state must lie to the right of state $2s$ on the P-h diagram of Fig. 7·26*b* (see Fig. B·7). It is evident that the least enthalpy change, that is, the least work input, is required by a device in which there is no entropy production. This work is $h_{2s} - h_1$, and its magnitude can be fixed with the aid of Fig. B·7. Denoting the work required by this ideal isentropic compressor by W_s, we find

$$W_s = h_{2s} - h_1 = 96 - 83 = 13 \text{ Btu/lbm}$$

The work required by a real irreversible adiabatic compressor would be greater than W_s. It is customary to define the *isentropic efficiency* of a compressor as

$$\blacktriangleright \quad \eta_s = \frac{\text{work required by ideal isentropic compressor}}{\text{work required by actual compressor}} = \frac{W_s}{W_{\text{act}}}$$

Note that the isentropic efficiency compares an actual process to an ideal process and is not an energy-conversion efficiency. Both the actual and ideal devices are assumed to have the same inlet state and the same discharge pressure. A typical value for η_s might be 0.90. Then

$$W_{\text{act}} = \frac{W_s}{\eta_s} = \frac{13}{0.90} = 14 \text{ Btu/lbm}$$

† $W \equiv \dot{W}/\dot{M}$

The actual discharge enthalpy would be

$$h_2 = h_1 + W_{\text{act}} = 83 + 14 = 97 \text{ Btu/lbm}$$

This value, together with the discharge pressure, suffices to determine the discharge state.

An inventor's claim. An inventor claims to have devised a steady-flow compressor which requires no shaft-power input. It is claimed that CO_2 at 200 psia and 120°F can be compressed to 300 psia, where it will emerge at 20°F, simply by a transfer of energy as heat from this device. The patent application states that the device will handle 2 lbm CO_2/s and is driven by a "cold source" at −140°F. Energy is transferred as heat from this device to the cold space at the rate of 60 Btu/s. He further states that the CO_2 enters and leaves the device at very low velocity, and that no significant elevation changes are involved. Can these claims be valid?

The device will be impossible if it violates either the first or second laws of thermodynamics. According to his claims, we may neglect kinetic and potential energies in the fluid streams and assume that steady-flow steady-state conditions prevail. We further assume that the flows are one-dimensional at the inlet and exit and that the CO_2 is in equilibrium states at these points. From Fig. B·6 we read the following:

$$\begin{aligned} T_1 &= 120°\text{F} & T_2 &= 20°\text{F} \\ P_1 &= 200 \text{ psia} & P_2 &= 300 \text{ psia} \\ h_1 &= 318 \text{ Btu/lbm} & h_2 &= 288 \text{ Btu/lbm} \\ s_1 &= 1.315 \text{ Btu/(lbm·R)} & s_2 &= 1.240 \text{ Btu/(lbm·R)} \end{aligned}$$

Figure 7·27 shows the control volume we shall analyze. We assume steady

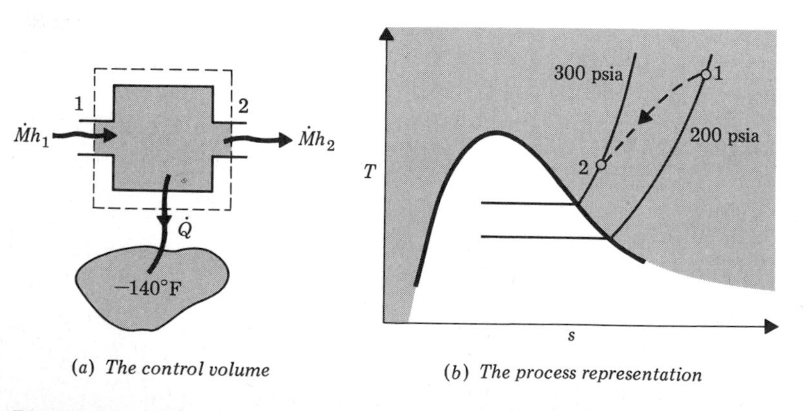

(a) *The control volume*

(b) *The process representation*

Figure 7·27. Analysis of the inventor's claim

flow and steady state and neglect the kinetic and potential energies of the flow. Then the energy balance gives

$$\dot{M}(h_2 - h_1) + \dot{Q} = 2 \times (288 - 318) + 60 = 0 \text{ Btu/s}$$

His device is not an energy producer, so it does not violate the first law.

The entropy bookkeeping for the control volume is

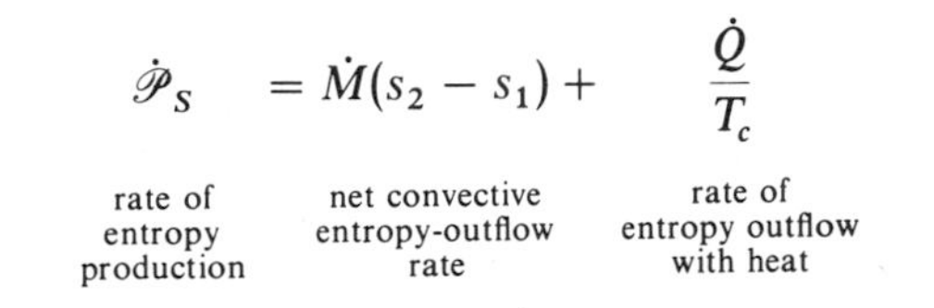

where T_c is the temperature of the cold space ($-140°F = 320$ R). Then

$$\dot{\mathscr{P}}_S = 2 \times (1.240 - 1.315) + \frac{60}{320} = -0.150 + 0.188 = +0.038 \text{ Btu/(R·s)}$$

The rate of entropy production is indeed positive, so the device does not violate the second law. We conclude that it is theoretically possible.

The availability of energy in a steady-flow device. A fluid enters a device at state 1 and emerges at state 2. The device can communicate with a large environment at temperature T_0, from which it can obtain energy as heat. What is the maximum available power output from this device?

This type of problem falls in the general area of *availability analysis*, a technique of great interest in advanced applied thermodynamics courses. The general approach here is to apply the first and second laws, thereby obtaining an upper limit on the amount of power which could be obtained from a device, given the inlet and discharge states.

The system we consider is shown in Fig. 7·28. We allow it to extract energy as heat from the environment at the rate $\dot{Q}_0$, and presume that it puts out useful work at the rate $\dot{W}$. Assuming steady flow and steady state and neglecting kinetic or potential energy of the flow streams, first-law analysis yields

$$\underbrace{\dot{M}h_1 + \dot{Q}_0}_{\text{energy-inflow rate}} = \underbrace{\dot{M}h_2 + \dot{W}}_{\text{energy-outflow rate}}$$

The entropy bookkeeping for the control volume is

$$\underbrace{\dot{\mathscr{P}}_S}_{\substack{\text{rate of}\\\text{entropy}\\\text{production}}} = \underbrace{\dot{M}s_2}_{\substack{\text{entropy-}\\\text{outflow}\\\text{rate}}} - \underbrace{\left(\dot{M}s_1 + \frac{\dot{Q}_0}{T_0}\right)}_{\text{entropy-inflow rate}}$$

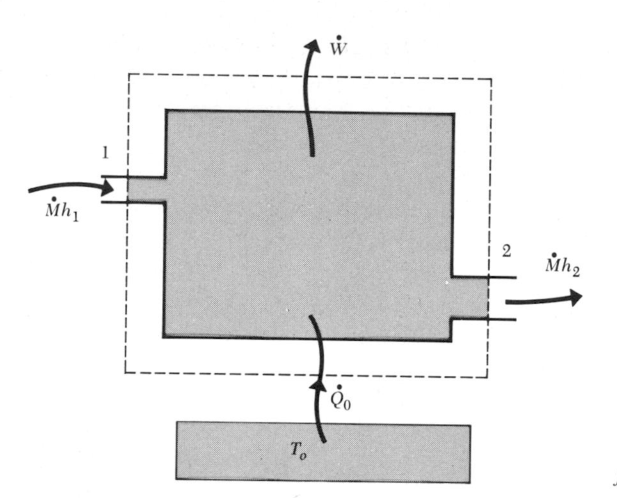

Figure 7·28. The control volume for the availability analysis

Combining these two equations, and solving for $\dot{W}$,

$$\dot{W} = \dot{M}[(h_1 - T_0 s_1) - (h_2 - T_0 s_2)] - T_0 \dot{\mathscr{P}}_S$$

The second law requires $\dot{\mathscr{P}}_S \geq 0$. So,

$$\dot{W} \leq \dot{M}(b_1 - b_2)$$

where

$$b = h - T_0 s$$

is the *steady-flow availability function* (per unit mass); note that it is a function of both the fluid and the environment. The availability analysis tells us that the work output per lbm of fluid, $\dot{W}/\dot{M}$, cannot exceed the decrease in the availability function. If the equality holds, the processes within the control volume must all be reversible, including the energy-transfer process between the device and the environment at T_0. The maximum work output is therefore obtained for a *reversible* process.

The difference between the maximum possible useful work output and the actual work output is sometimes called the *irreversibility* of the process. In the analysis of complex engineering systems one can locate the primary sources of irreversibility, which all cost money, by calculating the amount of irreversibility associated with each component in the system or each step in the process. Efforts to improve performance can then be concentrated in areas where the greatest gains stand to be made. The rate of loss of availability energy is seen to be $T_0 \dot{\mathscr{P}}_S$, which is called the *irreversibility rate* in some engineering literature. The irreversibility rate represents useful mechanical power that could have been obtained but was not.

The maximum power output calculated above was for a given discharge state 2. One can vary this state, and the maximum power output will vary. You might correctly guess that the greatest maximum is obtained when the discharge is at atmospheric temperature and pressure. If there is a temperature difference, it can be used to run a little heat engine and generate more power. If the discharge pressure is greater than atmospheric, a turbine can be used to extract additional power; and if the discharge pressure is less than atmospheric, the flow will not come out. It is possible to show that maximum maximum power is indeed obtained when the discharge is in equilibrium with the environment. This follows the arguments used in the control-mass available energy example done earlier in this chapter and requires some thermodynamic property relations we will develop in the next chapter. The student should try this proof after the completion of Sec. 8·6.

Improving industrial use of energy. As an application of the previous analysis, suppose we have 2 kg/s of steam at 2 MPa and 600°C, and we want to change the state to 1 MPa, 300°C, before using it in a chemical process. We could simply run the flow through a valve, dropping the pressure, thence through a heat exchanger, dropping the temperature. But in the interests of effective use of energy we decide to explore the possibilities of generating power as a side benefit of this steam-state change. How much could we get? From the metric Table B·2,

State 1:

$$h_1 = 3690.1 \text{ kJ/kg}$$

$$s_1 = 7.7032 \text{ kJ/(kg·K)}$$

State 2:

$$h_2 = 3051.2 \text{ kJ/kg}$$

$$s_2 = 7.1237 \text{ kJ/(kg·K)}$$

Taking T_0 as 295 K,

$$b_1 = 3690.1 - 295 \times 7.7032 = 1417.7 \text{ kJ/kg}$$

$$b_2 = 3051.2 - 295 \times 7.1237 = 949.7 \text{ kJ/kg}$$

So

$$\dot{W}_{max} = 2 \text{ kg/s} \times (1417.7 - 949.7) \text{ kJ/kg}$$

$$= 936 \text{ kW}$$

We see that the valve system would be "wasting" 936 kW, which at 2.5¢/kW·h amounts to over $200,000 annually. A good engineer could easily design a system to generate 500 kW of electric power. The cost of this system would have to be weighed against the utility bill saving in deciding whether or not to implement this change.

This type of analysis, sometimes called "availability analysis," has become increasingly important as a tool for identifying the components in a complex industrial facility that are most wasteful of energy. Availability analysis provides guidance to decide which components should be upgraded first in order to improve the facility, and an engineer who can execute such analyses competently can play a major role in helping the nation develop more energy-efficient technology.

Geothermally driven cooling system. In a remote desert area a small geothermal well produces 50 kg/h of saturated steam vapor at 150°C. The environment temperature is 45°C, and it is thought that a clever engineer might be able to devise a system to use the geothermal steam to produce cooling for homes at 23°C (Fig. 7·29*a*). The steam will emerge from this system as condensate at 1 atm. Figure 7·29*b* shows the process representation; the dotted line indicates that we really are not committed to any particular process connecting the inlet and discharge states. What is the maximum cooling rate, in kJ/h, that could be provided by such a system?

The control volume we analyze is shown in Fig. 7·29*a*. We assume steady-flow, steady-state, one-dimensional flows at the inlet and exit, and neglect kinetic and potential energy changes of the flow. The energy balance gives

$$\underbrace{\dot{Q}_c + \dot{M}h_1}_{\text{energy input rate}} = \underbrace{\dot{Q}_0 + \dot{M}h_2}_{\text{energy output rate}}$$

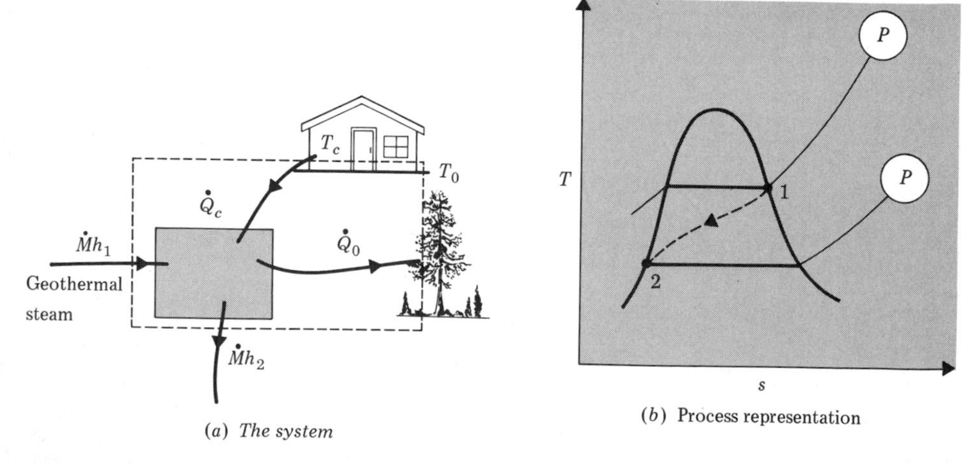

(*a*) *The system*

(*b*) Process representation

Figure 7·29. Geothermal cooling system

The entropy bookkeeping is

$$\underset{\substack{\text{rate of}\\\text{entropy}\\\text{production}}}{\dot{\mathscr{P}}_S} = \underset{\substack{\text{rate of}\\\text{entropy}\\\text{outflow}}}{\left(\frac{\dot{Q}_0}{T_0} + \dot{M}s_2\right)} - \underset{\substack{\text{rate of}\\\text{entropy}\\\text{inflow}}}{\left(\frac{\dot{Q}_c}{T_c} + \dot{M}s_1\right)}$$

Combining, and solving for $\dot{Q}_c$,

$$\dot{Q}_c = \frac{\dot{M}[(h_1 - T_0 s_1) - (h_2 - T_0 s_2)] - T_0 \dot{\mathscr{P}}_S}{(T_0/T_c) - 1}$$

The second law requires that $\dot{\mathscr{P}}_S \geq 0$. Therefore, for a given discharge state 2, the maximum that $\dot{Q}_c$ can be is

$$\dot{Q}_{c\max} = \frac{\dot{M}(b_1 - b_2)}{(T_0/T_c) - 1}$$

where again

$$b = h - T_0 s$$

is the specific steady-flow availability function. Note that any irreversibilities in the device will make $\dot{\mathscr{P}}_S > 0$ and will reduce $\dot{Q}_c$ below the maximum given above.

From the SI version of Table B·1 we read

State 1:

$$T_1 = 423 \text{ K } (150°\text{C}), \text{ saturated vapor}$$
$$h_1 = 2746.4 \text{ kJ/kg}$$
$$s_1 = 6.8387 \text{ kJ/(kg·K)}$$

State 2:

$$T_2 = 373 \text{ K } (100°\text{C}), \text{ saturated liquid}$$
$$h_2 = 419.0 \text{ kJ/kg}$$
$$s_2 = 1.3071 \text{ kJ/(kg·K)}$$

So, since $T_0 = 318$ K (45°C),

$$b_1 = h_1 - T_0 s_1 = 2746.4 - 318 \times 6.8387 = 571.7 \text{ kJ/kg}$$
$$b_2 = h_2 - T_0 s_2 = 419.0 - 318 \times 1.3071 = 3.3 \text{ kJ/kg}$$

Then, with $T_c = 296$ K,

$$\dot{Q}_{c_{\max}} = \frac{50 \text{ kg/h} \times (571.7 - 3.3) \text{ kJ/kg}}{\frac{318}{296} - 1}$$

$$= 3.82 \times 10^5 \text{ kJ/h} = 106 \text{ kW}$$

Any actual device would have a lower cooling capacity, because of irreversibilities. A clever engineer could easily come up with a design that could provide 50 kW of cooling. You might try this after you finish Chap. 9. We again note that the maximum performance was determined without any particular choice of process; the fact that this can be done testifies to the great power of fundamental thermodynamics.

7·16 A SUMMARY OF THE FIRST AND SECOND LAWS

At this point it would be a good idea to summarize the working expressions for the basic laws, lest there be any confusion about what is applicable only to an isolated system or only to a control mass. The formalism of production is convenient and easy to recall; in general, *production* is defined as follows:

> *Production equals the increase in the amount stored within the system plus the excess of amount which flows out over that which flows in.*

Similarly, the *rate of production* of something is defined as follows:

> *Rate of production equals the rate of increase in the amount stored within the system plus the excess of the outflow rate over the inflow rate.*

We applied the production concept in formulating the mathematics of the second law; we stated that entropy can be produced but never destroyed. The formalism is equally useful in stating the ideas of the first law, for energy can be neither produced nor destroyed. We can therefore state the first and second laws of thermodynamics in a very compact and physically appealing way; for an infinitesimal process,

▶ $$đ\mathscr{P}_E = 0 \qquad (7·57)$$

▶ $$đ\mathscr{P}_S \geq 0 \qquad (7·58)$$

Alternatively, on a rate basis,

▶ $$\dot{\mathscr{P}}_E = 0 \qquad (7·59)$$

▶ $$\dot{\mathscr{P}}_S \geq 0 \qquad (7·60)$$

Table 7·1 Summary of the working forms of the first and second laws

	First law	Second law
Basic principle	$d\!\!{}^{-}\mathscr{P}_E = 0$ $\dot{\mathscr{P}}_E = 0$	$d\!\!{}^{-}\mathscr{P}_S \geq 0$ $\dot{\mathscr{P}}_S \geq 0$
Isolated system 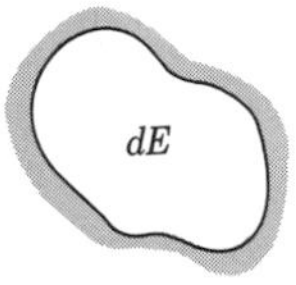 (dE)	$d\!\!{}^{-}\mathscr{P}_E = dE$	$d\!\!{}^{-}\mathscr{P}_S = dS$
Control mass 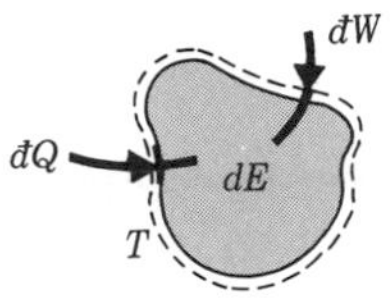 ($d\!\!{}^{-}Q$, $d\!\!{}^{-}W$, dE, T)	$d\!\!{}^{-}\mathscr{P}_E = dE - d\!\!{}^{-}Q - d\!\!{}^{-}W$	$d\!\!{}^{-}\mathscr{P}_S = dS - d\!\!{}^{-}\mathscr{K}$
Control volume ($\dot{W}$, $\frac{dE}{dt}$, $\dot{Q}$, T) 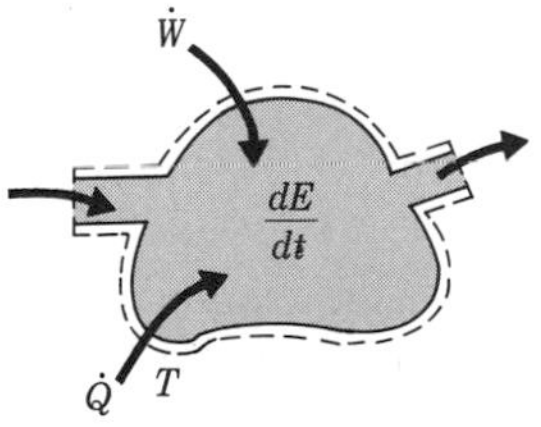	$\dot{\mathscr{P}}_E = \frac{dE}{dt} - \dot{W} - \dot{Q} + \sum_{\text{out}} (e + Pv)\dot{M} - \sum_{\text{in}} (e + Pv)\dot{M}$	$\dot{\mathscr{P}}_S = \frac{dS}{dt} - \sum_{\text{in}} \dot{\mathscr{K}} + \sum_{\text{out}} s\dot{M} - \sum_{\text{in}} s\dot{M}$

Note: $e \equiv u + \frac{V^2}{2g_c} + \frac{g}{g_c} z + \mathcal{Q}\mathscr{E} + \cdots$ $\quad d\!\!{}^{-}\mathscr{K} \equiv \frac{d\!\!{}^{-}Q}{T}$ $\quad \dot{\mathscr{K}} \equiv \frac{\dot{Q}}{T}$

Table 7·1, a summary of these ideas for an isolated system, a control mass, and a control volume, uses the notation $\mathscr{K}$ for entropy transfer with heat,

$$d\!\!{}^{-}\mathscr{K} \equiv \frac{d\!\!{}^{-}Q}{T} \tag{7·61}$$

Here $d\!\!{}^{-}Q$ is an energy transfer as heat to the system under study and is received by the system at a point on the boundary where its temperature is T. The notation $d\!\!{}^{-}$ is used, since $d\!\!{}^{-}\mathscr{K}$ represents an infinitesimal amount of entropy transfer with heat and does not in general represent the change in a property.

SELECTED READING

Callen, H. B., *Thermodynamics*, secs. 1.1, 1.9, 2.4, 2.5, 2.7, John Wiley & Sons, Inc., New York, 1960.

Reif, F., *Fundamentals of Statistical and Thermal Physics*, secs. 3.5, 3.6, 3.10–3.12, McGraw-Hill Book Company, New York, 1965.

Tribus, M., *Thermostatics and Thermodynamics*, chap. 5, D. Van Nostrand Company, Inc., Princeton, N.J., 1961.

Van Wylen, G. J., and R. E. Sonntag, *Fundamentals of Classical Thermodynamics*, 2d ed., secs. 6.4, 6.7, 7.8, 7.9, 7.12, 7.13, chap. 8, John Wiley & Sons, Inc., New York, 1973.

QUESTIONS

7·1 What is the conceptual basis and definition of the thermodynamic temperature?

7·2 Is $-10°F$ a negative thermodynamic temperature?

7·3 Which have meaning in nonequilibrium states, which do not, and why: temperature, energy, entropy?

7·4 Suppose a substance could exist in equilibrium at negative temperatures; what would this mean with regard to the energy and entropy (on a microscopic scale)?

7·5 What is the thermodynamic definition of pressure?

7·6 What is a reversible process?

7·7 How is heat different from work?

7·8 What is an irreversible process?

7·9 One reversible $2T$ engine uses mercury as the working fluid, and another uses steam. When operating between the same two temperatures, how will their efficiencies compare?

7·10 Think of a device which receives energy as heat, rejects energy as work (these are the only energy transfers), yet does not violate the Kelvin-Planck statement.

7·11 Does a $1T$ engine violate the first law, the second law, or both?

7·12 What is entropy production? What is entropy flow with heat?

7·13 Under what conditions will the entropy change of matter undergoing an adiabatic process be zero?

7·14 Suppose the thermodynamic temperature had been defined as $(\partial s/\partial u)_v$. Which way would heat flow, and would positive or negative temperatures be of most interest?

7·15 Give a qualitative microscopic explanation for the entropy change of a gas which undergoes reversible adiabatic expansion.

7·16 Give a qualitative microscopic explanation for the entropy change of a paramagnetic salt which undergoes reversible adiabatic magnetization.

7·17 What will happen to the entropy, energy, and temperature of a gas which is polarized reversibly and adiabatically at constant volume?

7·18 Under what circumstances would measurements of energy transfer as heat and temperature allow you to deduce the change in entropy of a control mass?

7·19 What happens to the entropy produced in a steady-flow, steady-state system?

7·20 What is an availability analysis?

PROBLEMS

7·1 Calculate the thermodynamic temperature of saturated water vapor at 100 psia from the *u-s-v* data in Appendix B and compare the result with the empirical temperature at this state.

7·2 Using the data on Fig. B·12, calculate the thermodynamic temperature for the paramagnetic substance iron-ammonium alum when M = 70 G and H = 1000 G and compare your result with that given by Fig. B·12.

7·3 Calculate the thermodynamic pressure for saturated water vapor at 100°C and compare your result with the tabulated saturation pressure at this state.

7·4 Derive the analog of Eq. (7·13) for a simple magnetic substance.

7·5 Using the method of Lagrange multipliers, show that the conditions for equilibrium for N interacting but independent pieces of simple compressible substances are that their temperatures and pressures all be identical.

7·6 Consider a control mass which executes a cycle, returning to its initial state. Prove that, if $đQ$ inflow is positive,

$$\oint \frac{đQ}{T} \leq 0$$

(This is the *inequality of Clausius*, and is an important intermediate theorem in classical developments of the second law.)

7·7 Consider two interacting systems, one at a negative T and the other at a positive T. Show that any energy transfer as heat will take place from the system at negative T to the one at positive T, hence the negative T state is "hotter."

7·8 Consider two interacting systems at negative T. Show that any energy transfer as heat will take place from the system at the least negative value of T to the one at a more negative value of T. Which state is "hotter"?

7·9 It has been proposed that energy be taken as heat from the atmosphere around Chicago and used to run a power plant. Energy would be rejected as heat to Lake Michigan. Estimate the maximum efficiency of conversion of thermal energy to electric power which could be obtained in such a plant.

7·10 An inventor claims to have perfected an engine which will produce power from energy transferred as heat from a 1300-K flame. Energy will be rejected as heat to the ground at 290 K. A 75 percent energy-conversion efficiency is claimed. As chief patent officer, would you issue a patent for this engine, and if so, would you invest money in this operation?

7·11 An inventor claims to have a device that is able to convert to work all energy transferred to it as heat, but makes no claims as to how long it will work. Could it work at all?

7·12 An inventor claims to have a device which receives 1000 W of energy as heat but puts out only 750 W of electric power. The rest of the energy is put out as mechanical work and dissipated outside the device by friction. Discuss the validity of these claims.

7·13 Thermal power systems for use in space normally reject energy as heat by radiating it into space. Since fluid-filled radiators are generally quite heavy, it has been suggested that this energy be converted to electricity and the current run through lighter resistors to dissipate the energy as heat. What do you think of this idea? Energy transfer as heat from a space vehicle must take place by radiation. The weight of a space radiator is proportional to its area, which is determined by the rate at which the energy must be radiated as heat; the rate of energy radiation is proportional to the product of the area and the fourth power of the radiator temperature. Consider a reversible $2T$ engine giving a fixed amount of power and operating with a fixed source temperature. Show that the least radiator weight is obtained when the radiator temperature is 0.75 times the source temperature.

7·14 The amount of solar energy received at the earth's surface is approximately 1000 W/m^2. Not all this energy can be used in a solar power plant, because of the reradiation of energy by the collector surface. Assume that the reradiation is described by $\dot{Q} = A\sigma\varepsilon T^4$, where A is the collector area, $\varepsilon = 0.5$, and $\sigma = 5.67 \times 10^{-8}$ W/(m^2·K^4) is the Stefan-Boltzmann constant. Find the maximum collector surface temperature (at which all the incident solar energy is reradiated). Suppose the power plant has Carnot efficiency. If T_0 is the environmental temperature to which energy is rejected as heat, and T_c is the collector temperature, at what temperature of T_c should the system operate to achieve the most power output with a collector of fixed size ($T_0 \approx 25°C$)?

7·15 Examine the examples in Sec. 5·2. Indicate which processes might be reasonably idealized as reversible and which are inherently irreversible.

7·16 Determine the amount of entropy production by the control mass of the thermal-mixing example in Sec. 5·2.

7·17 Calculate the entropy-production rate for the control volume of the nozzle example in Sec. 5·4 if $\dot{M} = 10{,}000$ lbm/h.

7·18 Calculate the rate of entropy production for the heat-exchanger example in Sec. 5·4.

7·19 Mercury flows through an adiabatic device. At one end the mercury is a saturated vapor at 400°F, and at the other end it is a mixture of 0.30 quality at 40 psia. The flow rate is 10 lbm/s. Determine the rate at which energy is transferred to the device as work (power input or output) and the direction of flow. Kinetic and potential energies are negligible.

7·20 Calculate the maximum possible output power (kW) from an adiabatic steam turbine handling 10 kg/s, where the inlet state is 2 MPa, 400°C, and the discharge pressure is 1 atm.

7·21 Mercury enters a small adiabatic turbine as a saturated vapor at 1200°F and is discharged at 100 psia. Determine the minimum possible quality of the discharge stream.

7·22 Determine the lowest temperature that could be obtained by adiabatic demagnetization of iron-ammonium alum (Fig. B·12) if the applied field **H** were suddenly dropped from 10,000 G to zero, assuming that the initial temperature is 2 K.

7·23 Devise a Carnot power cycle to operate with iron-ammonium alum, receiving energy at 2 K and rejecting energy at 1 K. State the processes that make up this cycle and give a symbolic energy analysis of each process. What is the cycle efficiency?

7·24 Steam enters an adiabatic diffuser at 400 m/s as a saturated vapor at 0.4 MPa. What is the maximum possible discharge pressure?

7·25 Freon 12 enters an adiabatic nozzle at 100 psia, 120°F, and emerges at 10 psia. What is the maximum possible discharge velocity?

7·26 Compute the maximum amount of work that could be obtained from each lbm of Freon 12 if the valve of Fig. 5·17 were replaced by an adiabatic turbine. Why is the valve usually used?

7·27 Compute the maximum percent liquefaction (by mass) of O_2 that can be achieved by expanding it adiabatically in a piston-cylinder system from the saturated vapor state at 10 atm to twice the volume.

7·28 Compute the minimum amount of power required by an adiabatic compressor that handles 10 lbm/min of Freon 12, compressing it from -20°F, 10 psia, to 60 psia.

7·29 Calculate the amount of useful power output which could be obtained from each component of the heat pump system of Fig. 5·17, assuming that the components have been replaced by devices which keep the inlet and discharge states the same, and that the replacement components can interact with an environment at 60°F. Which components are the most costly in terms of the available energy they waste? Where do you think that efforts should be spent on design improvements in order to realize the most appreciable gain?

7·30 In a certain processing plant, 10,000 kg/h steam is changed from 1 MPa, 400°C, to 0.2 MPa, 200°C, in a steady-flow device that can exchange energy as heat with the environment at 20°C. This is accomplished by first cooling the steam in a heat exchanger and then dropping the pressure by passing the steam through a valve. An energy-conscious plant manager thinks he might replace this hardware by something that could generate electric power while producing the same change in steam state. Calculate the maximum amount of electric power (kW) that could be derived.

7·31 A solar-powered heat pump receives energy as heat from a solar collector at T_H, rejects heat to the atmosphere at T_A, and pumps heat from a cold space at T_C. The three heat transfer rates are $\dot{Q}_H$, $\dot{Q}_A$, and $\dot{Q}_C$, respectively. Derive an expression for the *minimum* ratio $\dot{Q}_H/\dot{Q}_c$, in terms of the three temperatures. If $T_H = 350$ K, $T_A = 290$ K, $T_C = 200$ K, and $\dot{Q}_C = 10$ kW, what is the minimum $\dot{Q}_H$? If each m^2 of collector captures 0.2 kW, what minimum collector area (in m^2) is required?

CHAPTER EIGHT

THE THERMODYNAMICS OF STATE

8·1 INTRODUCTION

We now have the fundamental tools for thermodynamic analysis of engineering systems: the first law and second law. However, little quantitative work can be done without specific knowledge of state of the working substances. Therefore, before we analyze more complex systems, we must spend some time on properties and the relations between properties.

The examples in the previous chapters required thermodynamic data for a variety of substances. The perfect-gas model introduced in Chap. 4 provided a limited theoretical framework for this data. This chapter concentrates on the more general theoretical basis for thermodynamic equations of state.

Over the years a considerable body of equation-of-state information has been obtained by careful laboratory measurements, and these data are extremely important to the analyst. However, experiments are time-consuming and costly, and frequently we cannot wait for the availability of extensive data on a substance of interest. The first and second laws themselves can be extremely helpful in constructing a complete equation of state from a limited amount of data. Furthermore, thermodynamic theory is needed to provide relationships between the properties, so that those which are not directly measurable can be deduced. In this chapter we shall develop some of the more important relations between the intensive thermodynamic properties of some simple substances and indicate how these are used in building equations of state from basic laboratory data.

Equations of state can also be obtained from microscopic theories such as those of quantum-statistical mechanics, if a correct model is employed.† Relationships between the properties, derived by applications of the first and second laws, are essential in order to connect the results of these theories with measurable quantities.

The goals of this chapter are to provide a background in using the first and second laws to provide general equation-of-state information, and to introduce some special model equations of state.

8·2 THE GIBBS EQUATION

The equations of state may be expressed functionally in terms of the properties considered to be independent. In particular, for the simple compressible substance‡ we can put

$$s = s(u, v)$$

Taking the differential,

$$ds = \left(\frac{\partial s}{\partial u}\right)_v du + \left(\frac{\partial s}{\partial v}\right)_u dv$$

Then, using the thermodynamic definitions of temperature and pressure, we find

$$\blacktriangleright \qquad ds = \frac{1}{T}\,du + \frac{P}{T}\,dv \tag{8·1}$$

This is the *Gibbs equation*, which we derived in the same manner in the previous chapter. It is a differential equation of state which is extremely important in the thermodynamic theory of simple compressible substances. Since it relates the difference in entropy between any two infinitesimally separated states to the infinitesimal differences in internal energy and volume between those states, the difference in entropy between two states can be found by integration,

$$s_2 - s_1 = \int_1^2 \frac{du}{T} + \int_1^2 \frac{P}{T}\,dv \tag{8·2}$$

† Figure B·12 was obtained in this manner.

‡ Remember that a simple compressible substance is one in which the only relevant reversible work mode is compression. The state postulate tells us that the thermodynamic state of a simple compressible substance is completely determined by specification of two independent thermodynamic properties.

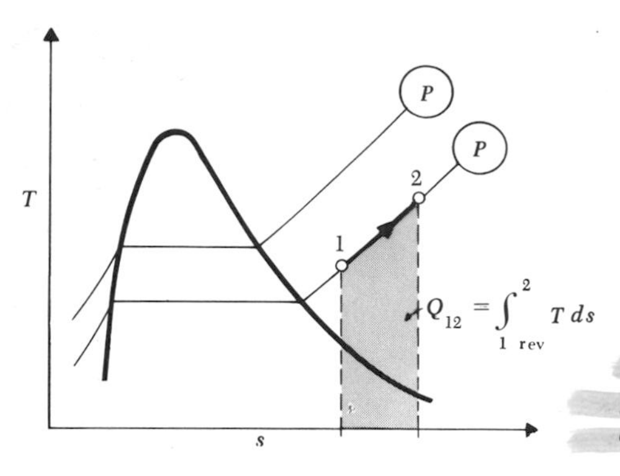

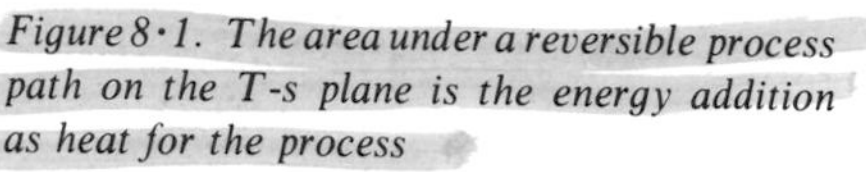

Figure 8·1. The area under a reversible process path on the T-s plane is the energy addition as heat for the process

In analyses not involving chemical reaction only the differences in entropy are involved. Consequently, we can select some state arbitrarily and refer the entropy for that substance to its value in this datum state. The charts and tables in Appendix B employ a variety of datum states. Having selected the datum state, we may perform the integrations of Eq. (8·2) from that point to any other state along any convenient path in the u-v plane. An illustration showing one simple integration is included in Sec. 7·6. Graphical or numerical integrations are usually employed, and we must of course know T and P as functions of u and v in order to perform the integration. In this manner we can determine the entropy of any substance, relative to the datum state.

Knowing the specific entropy of a substance as a function of the thermodynamic state, we can put lines of constant entropy on the graphical equations of state and enter values for the entropy in the tables. We can also use the entropy as a coordinate in a graphical equation of state; such graphs are particularly convenient in engineering analysis, especially where reversible adiabatic (isentropic, or constant-entropy) processes are involved. The temperature-entropy diagram is of special importance, for the area under a curve tracing out a reversible process on a T-s plane represents the energy transfer as heat to a unit of mass of the substance for the process (Fig. 8·1). Extensive use is made of T-s planes in engineering for both quantitative analysis and qualitative portrayal. We shall illustrate some of these ideas in numerical examples after we have developed some further relations between properties.

8·3 EQUATION OF STATE FOR THE PERFECT GAS

A *perfect gas*† is defined as any gas whose P-v-T relationship is of the form

$$Pv = RT \tag{8·3}$$

† The student may wish to review the brief discussion of the perfect gas in Sec. 4·8.

The gas constant R is related to the *universal gas constant* $\mathscr{R}$ and the molal mass $\hat{M}$ by†

$$R = \frac{\mathscr{R}}{\hat{M}} \tag{8·4}$$

$\mathscr{R}$ has the experimentally determined value

$$\mathscr{R} = 8.3143 \text{ kJ/(kgmole·K)} = 1545.33 \text{ ft·lbf/(lbmole·R)}$$

The Boltzmann constant k is defined in terms of $\mathscr{R}$ and Avogadro's number N_0 by

$$\mathsf{k} = \frac{\mathscr{R}}{N_0} \tag{8·5}$$

Sometimes k is called the gas constant per molecule. Values of $\hat{M}$ and R for several gases are given in Table B·6.

It is very important to realize that the idea of a perfect gas is an idealization to *real-gas* behavior. Thus the behavior of a real gas, such as air, will only be approximated by the perfect-gas equation.

A commonly employed microscopic model of the perfect, or ideal, gas is one in which the gas molecules consist of infinitesimally small, hard, round spheres that take up negligible volume and do not exert forces on one another except upon collision. This model ignores both the finite volume the molecules do in fact occupy and also the actual intermolecular forces the molecules exert on each other at long range.

Under what conditions would we expect the above assumptions to be valid? If the molecules are far apart on the average, then there should be little long-range interaction. In addition, the molecules will not occupy much of the total volume. The molecules will be far apart when the gas has a low density, which corresponds to conditions of *low pressure* and *high temperature* (Fig. 8·2). But low pressure compared to what pressure, and high temperature compared to what temperature?

Just what pressure and temperature give ideal-gas behavior depends on the gas and the amount of deviation from Eq. (8·3) that one will accept. Figure 8·3 shows the deviation from perfect-gas behavior for several gases. Note that for the same pressure a higher temperature reduces the deviation from perfect-gas behavior. In the next section we shall discuss how to analyze real-gas behavior. In

† The molal mass is the mass per mole of material. For carbon 12 the molal mass is 12 kg/kgmole = 12 g/gmole = 12 lbm/lbmole. The mole is defined such that one kgmole of a substance contains the same number of molecules as 12 kg of carbon 12; likewise, one lbmole contains the same number of molecules as 12 lbm of carbon 12, and one gmole contains the same number of molecules as 12 g of carbon 12; Avogadro's number N_0 is the number of molecules in a gmole.

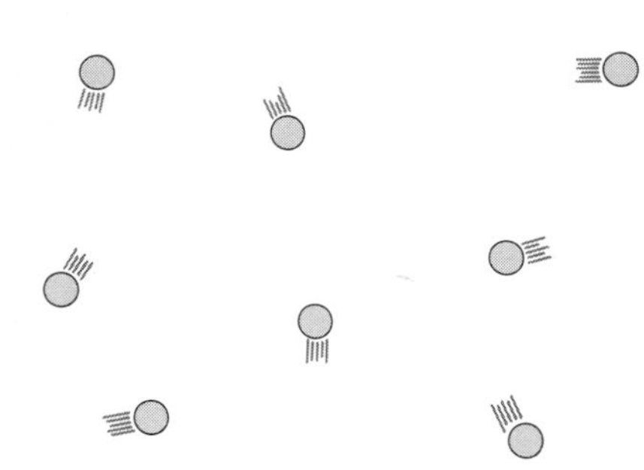

Figure 8·2. In the ideal gas model the molecules are far apart and interact only upon collision

general one can say that, if the temperature is well above the critical temperature and the pressure well below the critical pressure, the perfect-gas model is accurate.

The defining equation $Pv = RT$ can be put in other useful forms. If M is the mass of a sample of gas occupying volume V, multiplication by the mass yields

$$PV = MRT \tag{8·6}$$

Denoting the number of moles of the gas by $\mathcal{N}$ and observing that the mass M is related to the number of moles and the molal mass $\hat{M}$ by

$$M = \mathcal{N}\hat{M}$$

we can express Eq. (8·6) as

$$PV = \mathcal{N}\hat{M}RT = \mathcal{N}\mathcal{R}T \tag{8·7}$$

Other forms are given in Table 8·1.

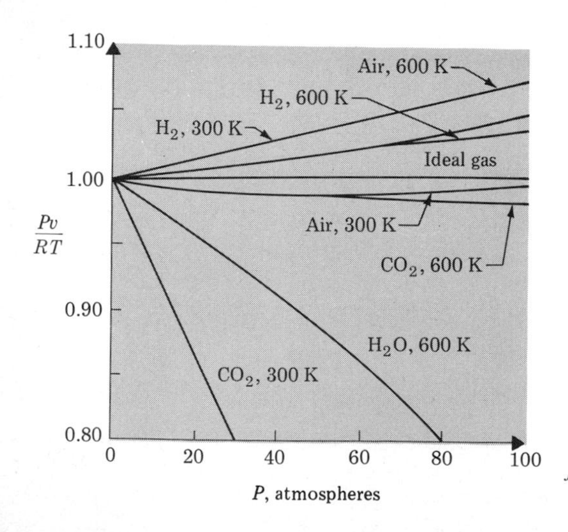

Figure 8·3. Test of the perfect-gas approximation for several gases (by permission from P. M. Morse, Thermal Physics, *rev. ed., W. A. Benjamin, Inc., New York, 1964)*

Table 8·1 Forms of the perfect-gas equation

$Pv = RT$	R = particular gas constant
$Pv = \dfrac{\mathscr{R}}{\hat{M}} T$	$\mathscr{R}$ = universal gas constant
	$\hat{M}$ = molal mass, mass/mole
$P\hat{v} = \mathscr{R}T$	$\hat{v} = \hat{M}v$, volume/mole
$PV = \mathscr{N}\mathscr{R}T$	V = volume of $\mathscr{N}$ moles
$PV = MRT$	V = volume of mass M

In Chap. 7 we showed that thermodynamic and empirical (perfect-gas) temperatures are identical, using the additional information that the internal energy of gases used in empirical thermometers depends only on temperature as P approaches zero. It may easily be shown that the internal energy u of any substance obeying Eq. (8·3) is a function only of temperature. One proof can be made starting directly from the Gibbs equation,

$$ds = \frac{1}{T}\,du + \frac{P}{T}\,dv$$

Then, since ds is exact,

$$\left(\frac{\partial(1/T)}{\partial v}\right)_u = \left(\frac{\partial(P/T)}{\partial u}\right)_v = \left(\frac{\partial(R/v)}{\partial u}\right)_v = 0$$

T is therefore independent of v along any line of constant u, and consequently $T = T(u)$, or $u = u(T)$.

The differential of u is also expressible solely in terms of temperature changes. Recalling the definition of the specific heat at constant volume, we have

$$du = c_v\,dT \tag{8·8}$$

Since u depends only on T, c_v must be a function only of temperature,

$$c_v = c_v(T)$$

The enthalpy is also a function only of temperature, since

▶ $$h = u + Pv = u(T) + RT = h(T)$$

Following the previous line of argument, we can show that

$$\begin{aligned} dh &= c_P\,dT \\ c_P &= c_P(T) \end{aligned} \tag{8·9}$$

The differential equation for the entropy is the Gibbs equation,

$$ds = \frac{du}{T} + \frac{P}{T}\,dv = \frac{c_v}{T}\,dT + R\frac{dv}{v} \tag{8·10}$$

Alternatively, using $dh = du + P\,dv + v\,dP$,

$$ds = \frac{dh}{T} - \frac{v}{T}\,dP = \frac{c_P}{T}\,dT - R\frac{dP}{P} \tag{8·11}$$

We may now integrate these differential equations of state. Integrating Eq. (8·8),

▶ $$u_2 - u_1 = \int_{T_1}^{T_2} c_v(T)\,dT \tag{8·12}$$

Integrating Eq. (8·9),

▶ $$h_2 - h_1 = \int_{T_1}^{T_2} c_P(T)\,dT \tag{8·13}$$

Equations (8·10) and (8·11) integrate to give

▶ $$s_2 - s_1 = \int_{T_1}^{T_2} \frac{c_v(T)}{T}\,dT + R\ln\frac{v_2}{v_1} \tag{8·14}$$

▶ $$s_2 - s_1 = \int_{T_1}^{T_2} \frac{c_P(T)}{T}\,dT - R\ln\frac{P_2}{P_1} \tag{8·15}$$

In order to carry the algebraic equations of state further it would be necessary to know how the specific heats varied with temperature. An important relation between the specific heats is

▶ $$c_P - c_v = R \tag{8·16}$$

which may easily be shown from Eqs. (8·3), (8·8), and (8·9). It is therefore necessary to determine only one of c_v or c_P for any ideal gas and, since c_P is more easily measured, it is customary to work with c_P. Figure B·17 shows c_P and c_v for several gases. The curves labeled $P = 0$ correspond to the perfect-gas approximation and can be used in the equations above.

An approximate equation that fits the experimental data over any fairly wide temperature range is

$$c_P = a + bT + cT^2$$

a, b, and c are experimentally determined coefficients. Table B·16 gives values for several gases. These equations fit the data with an average deviation of less than 1 percent and a maximum deviation of about 2 percent over the temperature range indicated.

It is convenient to define a property ϕ by

$$\phi(T) = \int_{T_0}^{T} \frac{c_P(T)}{T} dT \tag{8·17}$$

where T_0 is some reference temperature and ϕ is a function of temperature only. Using Eq. (8·15), one may express the entropy as:

$$s_2 - s_1 = \phi_2 - \phi_1 - R \ln \frac{P_2}{P_1} \tag{8·18}$$

Values of the enthalpy and the property ϕ as functions of temperature have been obtained from measurements on many gases,† and a tabulation of these properties for air‡ is given in Table B·9. $T_0 = 0$ R forms the datum for these compilations. Note that the simple quadratic variation of c_P is invalid as $T \to 0$, so the lower portion of the integration for $\phi(T)$ must be performed using other, more accurate, expressions.

The function $\phi(T)$ allows one to calculate the dependence of the entropy on T. In the air tables (Table B·9), the *reduced pressure* p_r and *reduced volume* v_r are defined as

$$\ln p_r \equiv \frac{\phi(T)}{R} \tag{8·19}$$

$$\ln v_r \equiv -\frac{1}{R} \int_{T_0}^{T} \frac{c_v}{T} dT \tag{8·20}$$

Note that both are dimensionless quantities and are functions of T only.

The reduced pressure is particularly useful in the analysis of isentropic processes. Setting the entropy difference to zero in Eq. (8·18), we obtain

$$\ln \frac{P_2}{P_1} = \frac{\phi_2 - \phi_1}{R} \qquad \textit{for an isentropic process}$$

† See, for example, J. Keenan and J. Kaye, *Gas Tables*, John Wiley & Sons, Inc., New York, 1948.

‡ Air at low pressures may be treated as a mixture of perfect gases. In Chap. 10 we shall show that any such mixture behaves like a perfect gas.

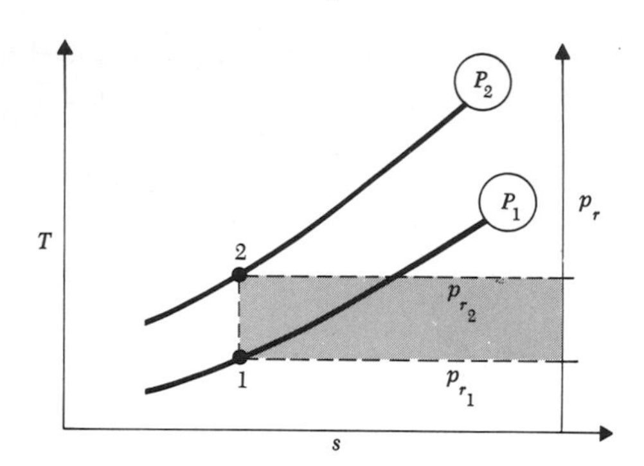

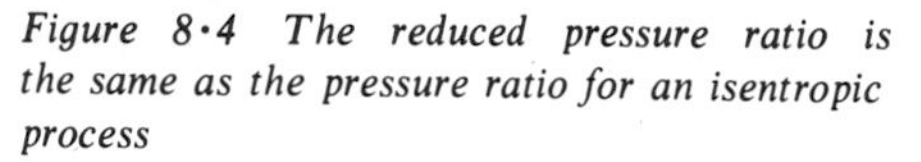

Figure 8·4 The reduced pressure ratio is the same as the pressure ratio for an isentropic process

In terms of the reduced pressures, the right-hand side becomes

$$\ln p_{r_2} - \ln p_{r_1} = \ln \frac{p_{r2}}{p_{r1}}$$

and consequently we see that

$$\blacktriangleright \quad \frac{P_2}{P_1} = \frac{p_{r2}}{p_{r1}} \quad \textit{for an isentropic process} \tag{8·21}$$

Similarly it may be shown that

$$\blacktriangleright \quad \frac{v_2}{v_1} = \frac{v_{r2}}{v_{r1}} \quad \textit{for an isentropic process} \tag{8·22}$$

To illustrate the use of reduced pressure, suppose we want to compress air at 60°F isentropically from 1 atm to 5 atm and need to know the final temperature (see Fig. 8·4). From Table B·9, at 60°F $p_r = 1.215$. Then

$$\frac{p_{r2}}{p_{r1}} = \frac{P_2}{P_1} = \frac{5}{1}$$

$$p_{r_2} = 6.08$$

Reading in Table B·9 for $p_r = 6.08$, we see that the final temperature is slightly greater than 360°F.

A simplified form of the perfect-gas equation of state is obtained if c_P is constant. For this special case it follows from Eq. (8·16) that c_v is also constant, and the equations of state are as follows for a perfect gas with constant specific heats:†

† Equation (8·28) follows from a combination of Eqs. (8·27) and (8·23).

$$Pv = RT \tag{8·23}$$

$$u_2 - u_1 = c_v(T_2 - T_1) \tag{8·24}$$

$$h_2 - h_1 = c_P(T_2 - T_1) \tag{8·25}$$

$$s_2 - s_1 = c_v \ln \frac{T_2}{T_1} + R \ln \frac{v_2}{v_1} \tag{8·26}$$

$$s_2 - s_1 = c_P \ln \frac{T_2}{T_1} - R \ln \frac{P_2}{P_1} \tag{8·27}$$

$$s_2 - s_1 = c_P \ln \frac{v_2}{v_1} + c_v \ln \frac{P_2}{P_1} \tag{8·28}$$

$$c_P - c_v = R \tag{8·29}$$

It is important to remember that Eqs. (8·24) to (8·28) pertain only to *a perfect gas with constant specific heats*. They are especially useful in gas dynamics, where neat closed-form algebraic expressions for the properties of a one-dimensional gas-flow field can be derived.

To illustrate the use of these simplified equations, let's recompute the temperature rise for the isentropic compression of air from 1 atm, 60°F, to 5 atm. Using Eq. (8·27), we set $s_1 = s_2$ and have

$$\frac{T_2}{T_1} = \left(\frac{P_2}{P_1}\right)^{R/c_P} = \left(\frac{P_2}{P_1}\right)^{(k-1)/k}$$

where $k = c_P/c_v$ is the ratio of the specific heats. For air $k = 1.4$, hence

$$\frac{T_2}{T_1} = \left(\frac{5}{1}\right)^{0.286} = 1.585$$

Then, with $T_1 = 520$ R,

$$T_2 = 1.585 \times 520 = 824 \text{ R}$$

Recall that the gas table calculation of the previous example predicted a final temperature of 361°F, or 821 R. The slight difference represents the error associated with the assumption of constant specific heats.

We could compute the work required to carry out the compression process. If it takes place in an adiabatic steady-flow compressor (Fig. 8·5), then an energy balance gives

$$\dot{W} = \dot{M}(h_2 - h_1)$$

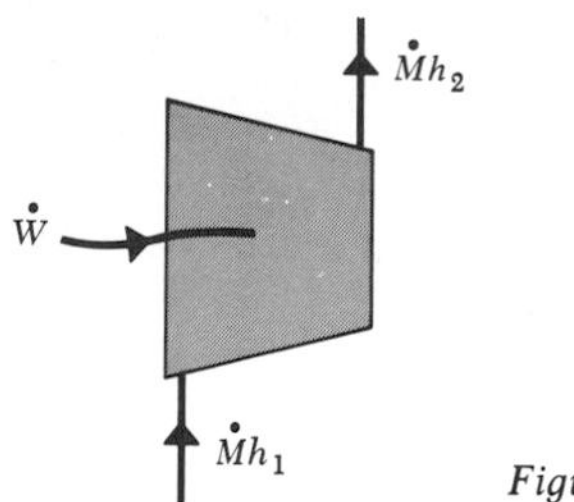

Figure 8·5. Adiabatic compressor analysis

Then, using Eq. (8·25),

$$\frac{\dot{W}}{\dot{M}} = h_2 - h_1 = c_P(T_2 - T_1) = 0.24 \times (824 - 520) = 73 \text{ Btu/lbm}$$

Note that the 3 R error associated with the constant specific heat assumption gives an error in the compressor work of 1 percent. Whether this error is tolerable for an engineering analysis depends on the objectives of the analysis.

A useful relation between the pressures and volumes at two states can be derived from Eq. (8·28). Dividing by c_v, we have

$$\frac{s_2 - s_1}{c_v} = \ln\left[\left(\frac{v_2}{v_1}\right)^k \frac{P_2}{P_1}\right]$$

Taking the antilog, we obtain

$$P_2 v_2^k = P_1 v_1^k \exp\left(\frac{s_2 - s_1}{c_v}\right) \tag{8·30}$$

From the equation above we see that two states of the same entropy will have the same values for their property Pv^k, provided the idealizations we have made are satisfied. Only a *reversible* adiabatic process is isentropic, and this is an important restriction to bear in mind. Further, the gas must obey $Pv = RT$ and, in addition, have constant specific heats. If all these restrictions are met, it is indeed true that, *for a perfect gas with constant specific heats undergoing a reversible adiabatic process,*

▶ $$Pv^k = \text{constant} \tag{8·31}$$

We have now seen two examples of equations $Pv^n = \text{constant}$. For the perfect gas with constant c_P the equation $Pv^k = \text{constant}$ represents the isentropic process, and the equation $Pv = \text{constant}$ the isothermal process. The general process represented by Pv^n = constant is called the *polytropic process.* The polytropic process is a useful generalization;

Isobaric process	$n = 0$
Isothermal process	$n = 1$
Isentropic process	$n = k$
Constant-volume process	$n \to \infty$

Other processes can often be approximated with an appropriate value for the *polytropic exponent n.*

If Pv^n = constant during some process, it follows that

$$P_1 v_1^n = P_2 v_2^n$$

or

$$\frac{P_2}{P_1} = \left(\frac{v_1}{v_2}\right)^n \tag{8·32}$$

Using the perfect-gas equation we can also show that

$$\frac{T_2}{T_1} = \left(\frac{P_2}{P_1}\right)^{(n-1)/n} = \left(\frac{v_1}{v_2}\right)^{n-1} \tag{8·33}$$

This expression is particularly important for the isentropic case when $n = k$.

The equations of state (8·23) to (8·29) allow calculation of the differences in internal energy, enthalpy, and entropy between any two states. Sometimes it is convenient to set the internal energy and entropy to zero at some arbitrarily chosen datum state. In thermodynamic analyses not involving chemical reaction, we work exclusively with differences in internal energy, enthalpy, and entropy, so the use of a datum state is entirely permissible. However, entropy then differs from the true entropy by the value of the absolute entropy at the datum state. Since internal energy is relative anyway, no artificiality is introduced by the use of an internal energy datum. When chemical reactions are involved, it becomes necessary to tie the internal energies and entropies of all elements and compounds together properly in a common way; we shall do this in Chap. 11.

Sometimes it is more convenient to select a datum state for enthalpy rather than for internal energy. Since the values of h and u differ by Pv at the datum temperature, it is not permissible to select data for both enthalpy and internal energy; only one of these may be set as zero at the datum state.

The equations of state for a perfect gas may be represented graphically on several thermodynamic planes, some of which are shown qualitatively in Fig. 8·6. Note that lines of constant T, u, and h coincide, so an h-s diagram would differ from the T-s diagram only by a scale stretching. Familiarity with the general nature of these graphical representations is especially helpful in the analysis of engineering systems involving idealized gases. The student should examine the

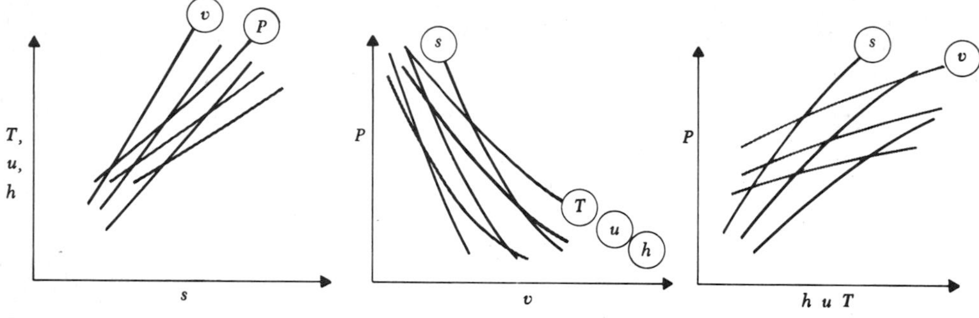

Figure 8·6. Graphical form of the perfect-gas equations

figures in the Appendix, in particular B·1, B·2, and B·6, to see what regions on the plots exhibit perfect-gas behavior. These are regions in which h and T lines, or u and T lines, are the same. It is seen that these occur at high temperatures and low pressures, as previously discussed.

8·4 OTHER P-v-T EQUATIONS FOR GASES

A number of equations of state have been proposed in order to allow for real-gas behavior. The perfect-gas equation has the advantage of great simplicity, but we have seen from the figures in the Appendix that even at moderate pressures it does not accurately predict the true behavior of gases. In fact, our model of the perfect gas has left out two important effects, namely, the intermolecular forces and the volume the gas molecules themselves occupy.

Van der Waals equation. On the basis of molecular arguments, Van der Waals suggested a modification of the perfect-gas equation that helps to account for weakly attractive intermolecular forces. A molecule about to strike the wall will experience a net attraction by molecules within the gas, and this will lessen the impulse it delivers to the wall (Fig. 8·7). The attractive force is proportional to the number of molecules per unit of volume as is the number of particles that strike the wall in a unit of time. Thus the effect of intermolecular attraction should be to lessen the pressure by an amount roughly proportional to the square of the gas density. This suggests that we should replace P in the perfect-gas expression by $P + (a/v^2)$, where a is a constant. In addition, the molecules of a dense gas occupy some volume, which suggests that v should be replaced by $v - b$, where b is a constant roughly indicative of the volume occupied by a unit of mass in a dense (liquid) state. With these qualitative considerations, Van der Waals was led to propose the approximate equation which bears his name,

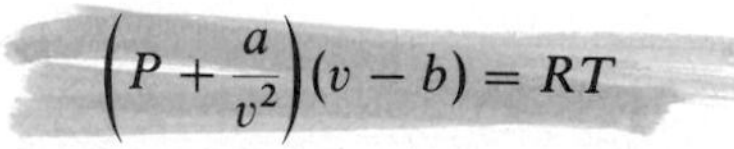

$$\left(P + \frac{a}{v^2}\right)(v - b) = RT \tag{8·34}$$

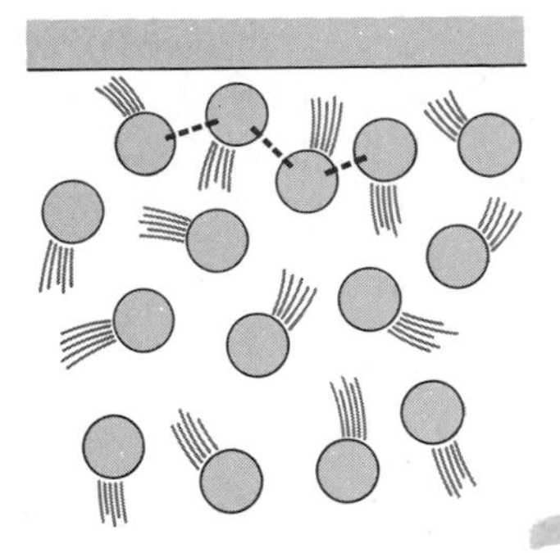

Figure 8·7. Forces between gas molecules will tend to reduce the pressure

The constants a and b have been determined for many gases as those which provide the best fit with experimental data. The fit is not very good over a wide range, and more accurate algebraic representations are usually employed where precise information is required.

The Van der Waals equation is a *cubic* in v and consequently, to any given P and T there will correspond either one or three real values of v. When one plots the equation on a P-v plane, the isotherms are found to appear as shown in Fig. 8·8. We know that for perfect-gas behavior $Pv = RT$, hence the isotherms must have a negative slope on the P-v plane. We therefore expect that the Van der Waals equation is not realistic in the region where the isotherms have a positive slope. However, there is one particular isotherm having an inflection point where $(\partial P/\partial v)_T = 0$; we might interpret this point as the critical point, and it can be determined very simply. We first rewrite Eq. (8·34) as

$$P = \frac{RT}{v - b} - \frac{a}{v^2}$$

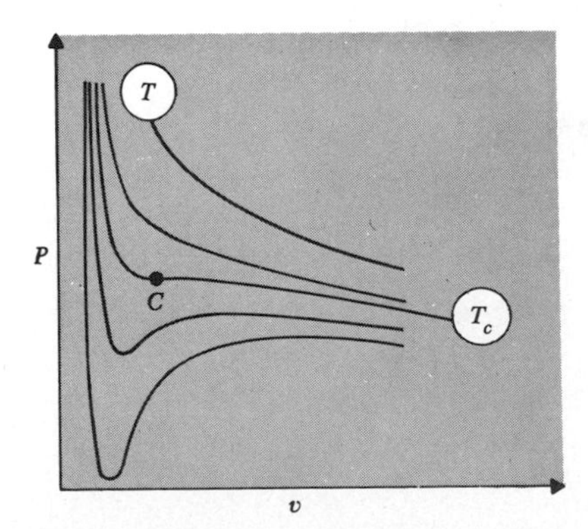

Figure 8·8. Van der Waals isotherms

Both the slope and curvature of the critical isotherm must vanish at the inflection point. Thus

$$\left(\frac{\partial P}{\partial v}\right)_T = 0 = \frac{-RT_c}{(v_c - b)^2} + \frac{2a}{v_c^3}$$

$$\left(\frac{\partial^2 P}{\partial v^2}\right)_T = 0 = \frac{2RT_c}{(v_c - b)^3} - \frac{6a}{v_c^4}$$

Then solving for a, b, and Z_c, using also Eq. (8·34) at the critical point,

$$a = \frac{27}{64}\frac{R^2T_c^2}{P_c} \tag{8·35a}$$

$$b = \frac{RT_c}{8P_c} \tag{8·35b}$$

$$Z_c = \frac{P_c v_c}{RT_c} = \frac{3}{8} = 0.375 \tag{8·35c}$$

We see that the Van der Waals equation predicts that the constants a and b are determined solely by the critical state, and that $P_c v_c / RT_c$ has the same value for all substances. While this is not quite verified experimentally, it is true that Z_c has a value of the order of 0.25 to 0.3, as can be seen in Table B·8. Values of a and b for several gases are given in Table B·17.

The Van der Waals equation of state may be rewritten in terms of the *compressibility factor* $Z = Pv/RT$, a dimensionless thermodynamic property which can be graphed or tabulated as a function of state. Equation (8·34) may be written as

$$\left(\frac{Pv}{RT} + \frac{27}{64}\frac{R^2T_c^2}{P_c RTv}\right)\left(1 - \frac{RT_c}{8P_c v}\right) = 1$$

Now, we define the *reduced pressure*† and *reduced temperature* by

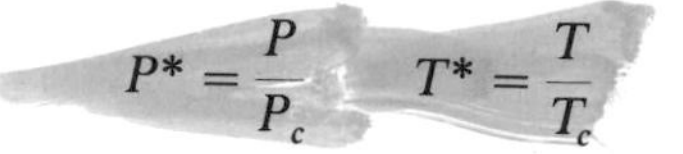

$$P^* = \frac{P}{P_c} \qquad T^* = \frac{T}{T_c}$$

The Van der Waals equation then becomes

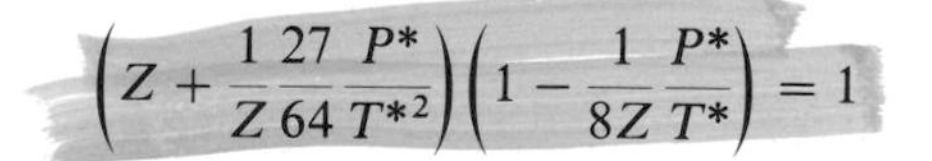

$$\left(Z + \frac{1}{Z}\frac{27}{64}\frac{P^*}{T^{*2}}\right)\left(1 - \frac{1}{8Z}\frac{P^*}{T^*}\right) = 1$$

† Not to be confused with the reduced pressure p_r used in the perfect-gas tables.

which suggests that a plot of the form

$$Z = Z(P^*, T^*)$$

might correlate the P-v-T equation of state for real gases. This brings us to the principle of corresponding states.

Principle of corresponding states. It is found experimentally that the compressibility factor Z is very nearly the same function of the reduced pressure and temperature for many gases. This fact is known as the *principle of corresponding states*; it is not a basic principle in the sense of the first and second laws, but merely a convenient approximation. Figure 8·9 shows the correlation for a number of gases. The Z-P^* plane is called a *generalized compressibility chart*; it is very useful in predicting the properties of substances for which more precise equation-of-state data have not yet been obtained and also provides a convenient way to represent the properties of many substances within the memory of a digital computer. A more accurate generalized compressibility chart is included as Fig. B·14.

In terms of the generalized compressibility charts, we see that at very low pressures Z approaches unity, the value appropriate for the perfect-gas approximation. Also, we see that, when the temperature is much larger than the critical temperature ($T^* \gg 1.0$), perfect-gas behavior can be expected even up to large pressures ($P^* \gg 1.0$).

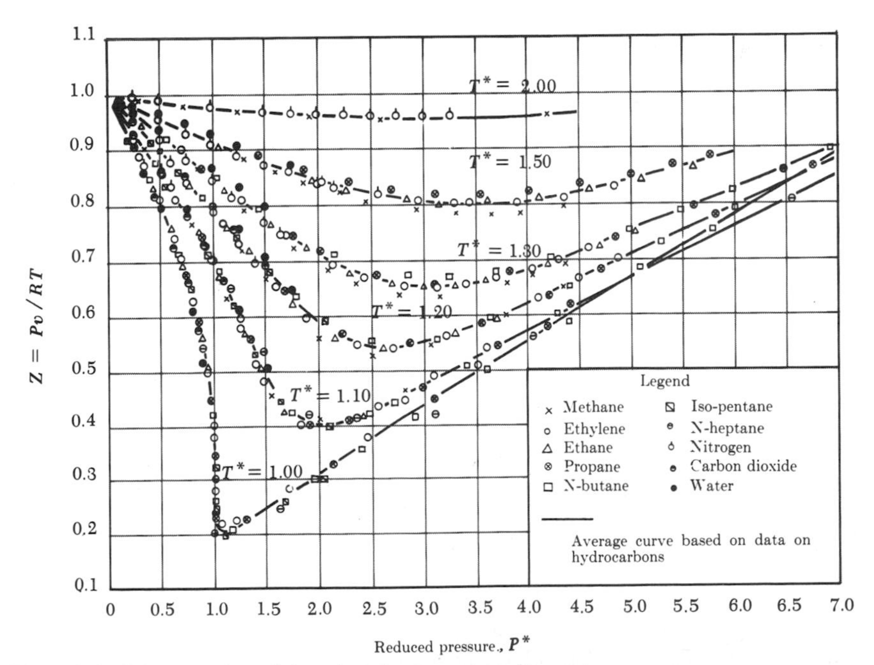

Figure 8·9. Demonstration of the principle of corresponding states

Use of Z. To illustrate use of the generalized compressibility chart, let's find the specific volume of carbon dioxide at a pressure of 800 psia and a temperature of 100°F and compare with the value obtained from Fig. B·6 and from the perfect-gas equation.

For CO_2 (from Table B·8),

$$T_c = 304.20 \text{ K } (= 547 \text{ R})$$

$$P_c = 72.90 \text{ atm } (= 1070 \text{ psia})$$

so

$$T^* = \frac{560}{547} = 1.02 \qquad P^* = \frac{800}{1070} = 0.748$$

From Fig. B·14, $Z = 0.70$. Then

$$v = \frac{ZRT}{P} = \frac{0.70 \times 35.1 \times 560}{800 \times 144} = 0.119 \text{ ft}^3/\text{lbm}$$

From Fig. B·6 we interpolate to read

$$v = 0.120 \text{ ft}^3/\text{lbm}$$

From the perfect-gas equation we find a significant error:

$$v = \frac{RT}{P} = \left(\frac{ZRT}{P}\right)\frac{1}{Z} = \frac{0.119}{0.7} = 0.170 \text{ ft}^3/\text{lbm}$$

We conclude that the generalized compressibility chart gives reasonably accurate information and is particularly useful when the perfect-gas equation is not applicable and data for the substance of interest are not available.

Determination of specific volume. A further example will compare the results calculated from the perfect-gas equation, the Van der Waals equation, and the generalized compressibility model.

Suppose 5000 lbm of nitrogen is to be stored at 300 R and 1000 psia and we need to determine the size of the tank required. We can solve this problem by finding the specific volume for nitrogen under these conditions. From the perfect-gas equation we have

$$v = \frac{RT}{P} = \frac{55.15 \times 300}{1000 \times 144} = 0.115 \text{ ft}^3/\text{lbm}$$

For nitrogen, from Table B·6,

$$T_c = 126.2 \text{ K } (227 \text{ R})$$

$$P_c = 33.5 \text{ atm } (492 \text{ psia})$$

For the generalized compressibility equation,

$$T^* = \frac{T}{T_c} = \frac{300}{227} = 1.32 \qquad P^* = \frac{P}{P_c} = \frac{1000}{492} = 2.03$$

Using Fig. B·14*a*, we find for these conditions $Z = 0.72$, so the generalized compressibility chart predicts

$$v = \frac{ZRT}{P} = 0.72 \times 0.115 = 0.0828 \text{ ft}^3/\text{lbm}$$

The Van der Waals equation of state requires the constants a and b, which are found in Table B·17. For nitrogen, $\hat{M} = 28.02$ lbm/lbmole and

$$a = 346.0 \text{ atm·ft}^6/\text{lbmole}^2 \qquad b = 0.618 \text{ ft}^3/\text{lbmole}$$

In the peculiar units that we are using,

$$\mathscr{R} = 0.7302 \text{ atm·ft}^3/(\text{lbmole·R})$$

Noting that 1000 psi is equal to 68 atm, we see that Eq. (8·34) is

$$\left(68 + \frac{346}{v^2}\right)(v - 0.618) = \mathscr{R}T = 0.73 \times 300 = 219 \text{ atm·ft}^3/\text{lbmole}$$

This is a cubic equation for v, which can be solved by trial and error. As a first choice we can try the result from the generalized compressibility equation (0.0828 ft^3/lbm, which corresponds to 2.32 ft^3/lbmole). Substituting into the equation we find

$$\left(68 + \frac{346}{(2.3)^2}\right)(2.30 - 0.618) = 224 \neq 219$$

Further trials show that the solution is given by $v = 2.19$ ft^3/lbmole, or 0.0782 ft^3/lbm as the van der Waals prediction.

Figure B·4 can also be used to determine the specific volume for nitrogen under these conditions. Reading carefully we find $v = 0.081$ ft^3/lbm. (The tabular reference for Fig. B·4, NBS TN-129A, gives a value of 0.0828 ft^3/lbm.) We see that for these conditions the perfect-gas equation would give a value substantially in error. The Van der Waals equation represents a considerable improvement, and the generalized compressibility result is quite accurate.

Determination of pressure. Let us suppose that 100 kg of oxygen is to be stored in a 1-m^3 tank at 160 K and we want to find the required pressure. If oxygen were a perfect gas under these conditions we would have

$$P = \frac{RT}{v} = \frac{0.260 \text{ kJ/(kg·K)} \times 160 \text{ K}}{1 \text{ m}^3/100 \text{ kg}} = 4.16 \times 10^6 \text{ N/m}^2 = 4.16 \text{ MPa} = 41.0 \text{ atm}$$

For the given conditions we have (see Table B·8)

$$T^* = \frac{160}{154.8} = 1.034 \qquad P^* = \frac{P}{50.14 \text{ atm}} = \frac{P}{5.08 \text{ MPa}}$$

Using the generalized compressibility equation we have

$$Pv = ZRT \qquad \text{or} \qquad P^* = \frac{ZRT}{vP_c}$$

Substituting the known information, we find (noting that RT/v has already been evaluated)

$$P^* = Z \times 4.16/5.08$$

or

$$Z = 1.22P^*$$

We now need to find a value of P^* that will satisfy this relationship subject to the condition that $T^* = 1.034$. One method to do this is to plot the Z-P^* equation on the generalized compressibility chart and note the point where the line crosses the required T^* value (Fig. 8·10). Doing this, we find that $P^* = 0.63$ and $Z = 0.76$. Thus the pressure is

$$P = 0.63 \times 5.08 = 3.20 \text{ MPa} = 31.5 \text{ atm}$$

Using the perfect-gas relation would have produced a substantial error under these conditions.

Beattie-Bridgeman equation. A number of other algebraic P-v-T equations have been proposed, and perhaps the best known and most widely used is the Beattie-Bridgeman equation

$$\blacktriangleright \qquad P = \frac{\mathscr{R}T(1-\varepsilon)}{\hat{v}^2}(\hat{v} + B) - \frac{A}{\hat{v}^2} \tag{8·36}$$

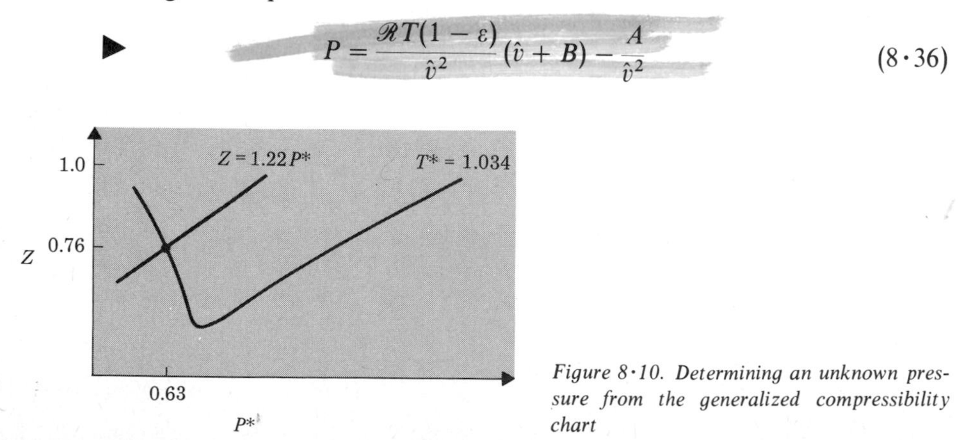

Figure 8·10. Determining an unknown pressure from the generalized compressibility chart

A, B, and ε are functions of state,

$$A = A_0\left(1 - \frac{a}{\hat{v}}\right) \qquad B = B_0\left(1 - \frac{b}{\hat{v}}\right) \qquad \varepsilon = \frac{c}{\hat{v}T^3}$$

where A_0, B_0, a, b, and c are constants which must be determined experimentally for each gas. Table B·17 gives these constants for a few gases. Values for many common gases have been determined and are available in the literature. However, much laboratory work remains to be done. For example, accurate thermodynamic properties of many high-temperature alkali-metal vapors, which are of interest for advanced thermal power systems, are not now generally available.

Generalized charts for the internal energy, enthalpy, and entropy of dense gases can be determined from ideal-gas specific heat data and real-gas P-v-T data with the help of thermodynamic relations. We shall see how this is done later in this chapter.

8·5 ALGEBRAIC EQUATION OF STATE FOR AN INCOMPRESSIBLE LIQUID

An algebraic equation of state can be developed for a *liquid* under the assumption that it is incompressible; that is, $v =$ constant. This idealized equation of state is of particular utility in analysis of liquid pumps, nozzles, heaters, and so on, operating well below the critical pressure or over limited pressure ranges. It is also applicable to a solid that is idealized as incompressible.

If the substance is essentially incompressible, its pressure can be increased a finite amount by an infinitesimal decrease in volume; such an increase can occur without a significant amount of energy transfer as work. The only means for reversibly changing the internal energy of such an idealized liquid is by transfer of energy as heat.† The incompressible liquid is therefore a degenerate case in which there are no reversible work modes, and consequently only one independent thermodynamic property. While the pressure is involved in energy transfers to the bulk fluid (flow work), which show up in increased bulk kinetic and potential energy, it is not involved in energy transfer to the "hidden" microscopic modes, and consequently really is not a relevant thermodynamic property for this idealized liquid. The properties that are purely thermodynamic in nature include internal energy, entropy, and temperature; from the state postulate we see that specification of any one of these suffices to fix the thermodynamic state.

We may express the internal energy as

$$u = u(T)$$

† The internal energy can also be changed by viscous friction, but this is not a reversible work mode. The fluid can also have kinetic energy due to its bulk motion and potential energy due to its position in a gravitational field.

Differentiating and defining a specific heat for the liquid,

$$du = c\,dT$$

where

$$c = \frac{du}{dT} = c(T) \tag{8·37}$$

Integrating between two states,

▶ $$u_2 - u_1 = \int_1^2 c(T)\,dT \tag{8·38}$$

To determine the entropy, imagine heating a unit of mass of the incompressible liquid reversibly. Application of the first and second laws then yields

$$đQ = du \qquad đQ = T\,ds$$

so that the Gibbs equation is

$$ds = \frac{du}{T}$$

as expected, since $dv = 0$.

Substituting for du and integrating,

▶ $$s_2 - s_1 = \int_1^2 \frac{c(T)}{T}\,dT \tag{8·39}$$

If it is reasonable to assume that c is constant, Eqs. (8·38) and (8·39) further reduce to

▶ $$u_2 - u_1 = c \cdot (T_2 - T_1) \tag{8·40}$$

▶ $$s_2 - s_1 = c \ln \frac{T_2}{T_1} \tag{8·41}$$

We shall illustrate the use of these equations in a moment.

In control-volume analysis of engineering systems involving fluids, the enthalpy invariably arises. In the case of the idealized incompressible liquid enthalpy is a mixture of thermodynamic and mechanical properties. For an incompressible liquid with constant specific heat, we find

▶ $$h_2 - h_1 = c \cdot (T_2 - T_1) + (P_2 - P_1)v \tag{8·42}$$

The enthalpy of an incompressible liquid, unlike the enthalpy of a perfect gas, is seen to be a function of both temperature and pressure. Note that for an incompressible substance $c_P = c_v = c$. This may be seen as follows; from the definition of enthalpy, $h = u + Pv$, so

$$dh = du + P\,dv + v\,dP$$

For the incompressible substance $dv = 0$, hence

$$dh = c\,dT + v\,dP$$

so

$$c_P \equiv \left(\frac{\partial h}{\partial T}\right)_P = c$$

Equations (8·40) to (8·42) are useful for estimating the properties of subcooled liquids when tabulations of the real thermodynamic properties are not available. Since internal energy is a function only of temperature, the value of u for a subcooled liquid at T will be the same (idealizing the liquid as incompressible) as for the saturated liquid at the same temperature. A similar statement can be made for the entropy. The enthalpy can be computed in a number of ways, two of which we now describe.

Suppose we wish to estimate the values of u, h, and s for H_2O at 80°F and 1 atm pressure. This is a subcooled liquid state shown as 1 on Fig. 8·11. From Table B·1*a* we can read the values along the saturated liquid line as

$$h_3 = 48.1 \text{ Btu/lbm} \qquad v_3 = 0.01607 \text{ ft}^3\text{/lbm}$$

$$s_3 = 0.0933 \text{ Btu/(lbm·R)} \qquad P_3 = 0.507 \text{ psia}$$

From Eq. (8·41) we see that s is a function of T only for an incompressible liquid. Therefore our estimate of s_1 is made by moving *along a line of constant temperature* to state 3 so that $s_1 = s_3 = 0.0933$ Btu/(lbm·R).

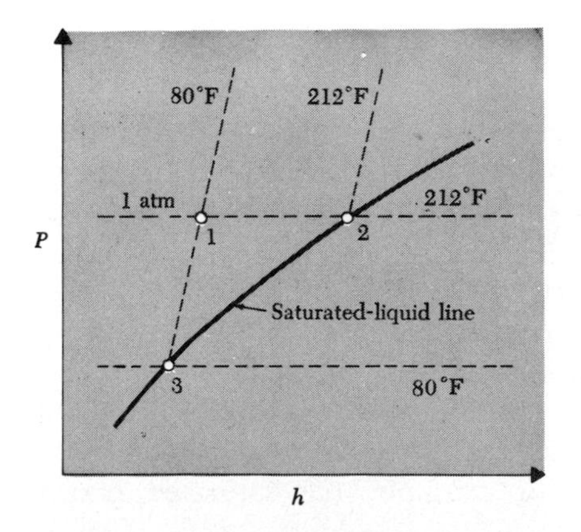

Figure 8·11. Estimation of liquid properties using the incompressible liquid equation of state

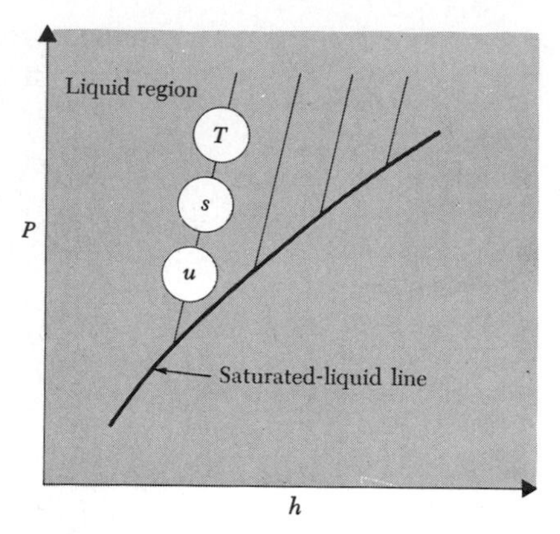

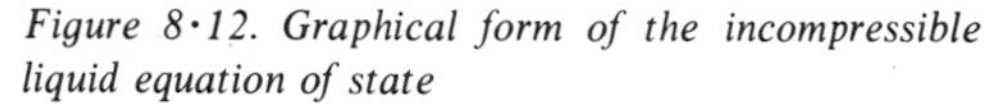
Figure 8·12. Graphical form of the incompressible liquid equation of state

For the internal energy, which is also only a function of temperature:

$$u_1 = u_3 = 48.1 \text{ Btu/lbm}$$

Note that Pv is so small that $h = u$ to at least one decimal.

The enthalpy at 1 is estimated by

$$\begin{aligned} h_1 - h_3 &= u_1 - u_3 + v(P_1 - P_3) = v(P_1 - P_3) \\ &= 0.01607 \times (14.7 - 0.5)\tfrac{144}{778} = 0.042 \text{ Btu/lbm} \\ h_1 &= 48.1 + 0.042 = 48.14 \text{ Btu/lbm} \end{aligned}$$

Alternatively, we could compute h_1 by working from state 2. In this case $h_2 = 180.1$ Btu/lbm and, since we are going along a line of constant pressure,

$$\begin{aligned} h_1 - h_2 &= c(T_1 - T_2) = 1 \times (80 - 212) = -132 \text{ Btu/lbm} \\ h_1 &= 180.1 - 132 = 48.1 \text{ Btu/lbm} \end{aligned}$$

Since our first calculation involves the least extrapolation, we would probably consider it the better estimate if the disagreement had been significant.

The equations of state for an incompressible liquid must be portrayed graphically on a mixed plane, since there is only one independent thermodynamic property. The P-h plane is usually used, and a typical example is shown in Fig. 8·12.

The incompressible liquid is an approximation to the behavior of real liquids, just as the perfect-gas equation is an approximation to real-gas behavior. The approximations become poorer in both cases as we extend the range of application. Tables B·1 and B·3 provide information that can be used to estimate the compressibility of liquid water and mercury. Mercury is seen to be well approximated as an incompressible liquid. For example, the saturated liquid

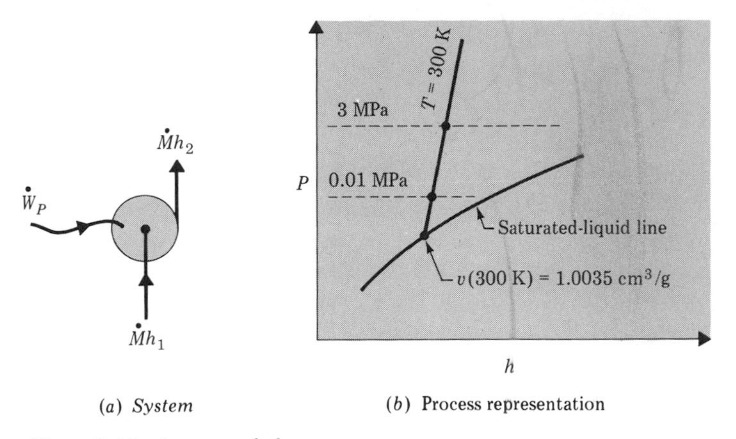

(*a*) *System* (*b*) Process representation

Figure 8·13. An energy balance on a pump

specific volume only changes from 1.25×10^{-3} ft^3/lbm at 10 psia to 1.35×10^{-3} at 1000 psia. On the other hand, for water the saturated liquid specific volume changes by several percent between 14.7 psia and 1000 psia. Values for the thermodynamic properties of compressed liquid water are available in the literature.†

Application to a pump. We make use of the equation of state for an incompressible liquid in estimating the work required to pressurize water. Suppose the first-stage boiler feedwater pump on a power plant takes liquid water at 300 K at 0.01 MPa and raises the pressure to 3 MPa. Let's determine the least work required to do this if the process is idealized as adiabatic.

The energy balance, Fig. 8·13*a*, yields

$$\dot{W} = \dot{M}(h_2 - h_1)$$

The entropy bookkeeping for the adiabatic system gives

$$\dot{\mathscr{P}}_S = \dot{M}(s_2 - s_1)$$

The second law says $\dot{\mathscr{P}}_S \geq 0$, so

$$\dot{M}(s_2 - s_1) \geq 0$$

From Eq. (8·41), for an incompressible liquid,

$$s_2 - s_1 = c \ln\left(\frac{T_2}{T_1}\right)$$

† J. H. Keenan et al., *Steam Tables*, John Wiley & Sons, Inc., New York, 1969.

Equation (8·42), for the enthalpy, allows us to write the first-law energy balance as

$$\dot{W}_s = \dot{M}[c(T_2 - T_1) + v(P_2 - P_1)]$$

Since we know the quantity $P_2 - P_1$, we see that the least work for the adiabatic pump results when $T_2 = T_1$, that is, when the process is isothermal. But from Eq. (8·41) this also means that $s_2 = s_1$. Thus, as we might have anticipated, the least work required to pressurize the water occurs when the process is isentropic. If the entropy does not change, neither does the temperature. So, the isentropic power $\dot{W}_s$ is

$$\frac{\dot{W}_s}{\dot{M}} = v(P_2 - P_1)$$

When using the incompressible-liquid model one should use the specific volume at the saturated-liquid line in the temperature range of interest. From the SI Table B·1, $v \simeq 0.001$ m^3/kg. Hence

$$W_s = \frac{\dot{W}_s}{\dot{M}} = v(P_2 - P_1) = 0.001 \text{ m}^3/\text{kg} \times (3 - 0.01) \times 10^6 \text{ N/m}^2$$

$$= 3.0 \text{ kJ/kg} = 1.29 \text{ Btu/lbm}$$

Note that relatively little work is required to pressurize a liquid, compared to that required to pressurize a gas over the same pressure change.

8·6 DIFFERENTIAL EQUATIONS OF STATE

The most easily measured properties are P, v, and T. However, we also need properties like h, s, u, and quantities like h_{fg}, s_{fg}. We shall now develop some relationships between properties that will permit them to be calculated from the most easily measured variables.

A list of the important properties for a simple compressible substance is given in Table 8·2. In this table we introduce the Helmholtz function a and the Gibbs function g, two thermodynamic properties whose importance will be made clear in Chaps. 11 and 12. The fact that a and g are properties, that is, functions of state, is itself very important, as we shall show next. The isentropic compressibility, a derivative property relating volume changes to pressure changes for an isentropic process, is also defined in Table 8·2. The isentropic compressibility is related to the speed at which pressure waves (sound) travel through a substance and is consequently of considerable importance.

The Gibbs equation may be written in another useful form,

► $$du = T\,ds - P\,dv \tag{8·43}$$

Other differential equations of state can be obtained by combining the Gibbs equation with the differentials of the enthalpy, Helmholtz, and Gibbs functions.

Table 8·2 Intensive thermodynamic properties of a simple compressible substance

Property	Definition
Conceptual	
Internal energy u	energy associated with molecular and atomic motions and forces
Entropy s	a measure of the average disorder on the microscopic scale
Defined	
Temperature T	$T \equiv 1/(\partial s/\partial u)_v$, equal to the empirical temperature
Pressure P	$P \equiv T(\partial s/\partial v)_u$, equal to the mechanical pressure
Specific volume v	volume per unit of mass
Density ρ	mass per unit of volume
Enthalpy h	$h \equiv u + Pv$
Helmholtz function a	$a \equiv u - Ts$
Gibbs function g	$g \equiv h - Ts$
Isobaric compressibility β	$\beta \equiv \frac{1}{v}\left(\frac{\partial v}{\partial T}\right)_P$
Isothermal compressibility κ	$\kappa \equiv -\frac{1}{v}\left(\frac{\partial v}{\partial P}\right)_T$
Isentropic compressibility α	$\alpha \equiv -\frac{1}{v}\left(\frac{\partial v}{\partial P}\right)_s$
Specific heat at constant volume c_v	$c_v \equiv \left(\frac{\partial u}{\partial T}\right)_v$
Specific heat at constant pressure c_P	$c_P \equiv \left(\frac{\partial h}{\partial T}\right)_P$

The enthalpy $h = u + Pv$ can be differentiated, yielding

$$dh = du + P\,dv + v\,dP$$

Combining with Eq. (8·43) results in a second differential equation of state,

$$dh = T\,ds + v\,dP \tag{8·44}$$

This expresses enthalpy differences between two infinitesimally separated states in terms of the infinitesimal differences in the entropy and pressure. Moreover, we see that the coefficients T and v are merely partial derivatives of $h(s, P)$,

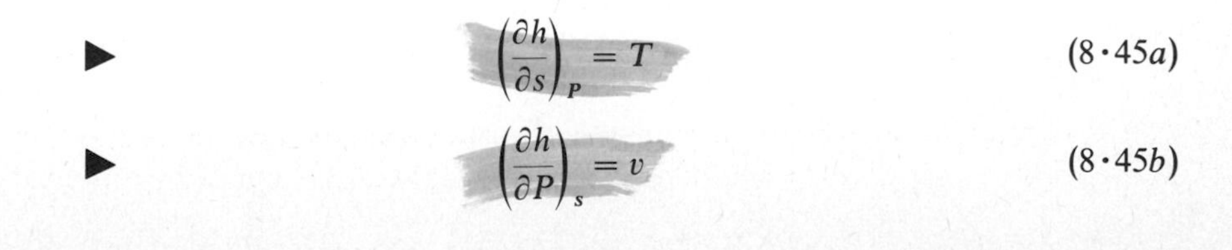

$$\left(\frac{\partial h}{\partial s}\right)_P = T \tag{8·45a}$$

$$\left(\frac{\partial h}{\partial P}\right)_s = v \tag{8·45b}$$

Since $v > 0$, an isentropic increase in pressure will always result in an increase in the enthalpy. In other words, adiabatic compressors require work.

In a similar manner the Gibbs equation can be combined with the differential of the Helmholtz function $a = u - Ts$ to give

$$da = -P\,dv - s\,dT \tag{8·46}$$

The coefficients $-P$ and $-s$ are the partial derivatives of $a(v, T)$, so

$$\left(\frac{\partial a}{\partial v}\right)_T = -P \tag{8·47a}$$

$$\left(\frac{\partial a}{\partial T}\right)_v = -s \tag{8·47b}$$

Similarly, using the Gibbs function $g = h - Ts$, we may show that

$$dg = v\,dP - s\,dT \tag{8·48}$$

Consequently,

$$\left(\frac{\partial g}{\partial P}\right)_T = v \tag{8·49a}$$

$$\left(\frac{\partial g}{\partial T}\right)_P = -s \tag{8·49b}$$

The importance of the Gibbs equation cannot be overemphasized. As an aid in being able to rederive it from first principles, we shall now make an alternative derivation. Consider a control mass consisting of a unit of mass of a simple compressible substance undergoing a reversible process, as shown in Fig. 8·14. Applying the first law over an infinitesimal part of the process, we have

$$du = đW + đQ$$

Since the process is reversible, the second law gives

$$đQ = T\,ds$$

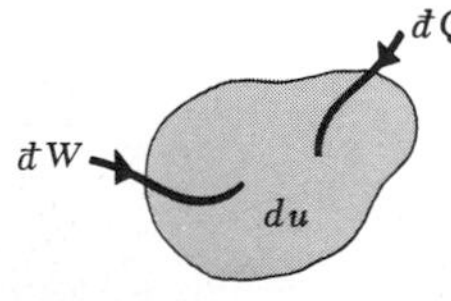

Figure 8·14. System for derivation of the Gibbs equation

Moreover, the work is given by

$$đW = -P\,dv$$

Combining, we again obtain the Gibbs equation,

$$du = T\,ds - P\,dv$$

It should be easy to recall that $đQ = T\,ds$ for a reversible process, from which the Gibbs equation can be rederived if needed, or developed for other classes of substances.†

8·7 SOME IMPORTANT PROPERTY RELATIONS

The four differential equations of state developed in Sec. 8·6 provide the basis for numerous important relationships among the thermodynamic properties of a simple compressible substance. Note that each is of the form

$$dz(x, y) = M\,dx + N\,dy$$

where

$$M = \left(\frac{\partial z}{\partial x}\right)_y \qquad N = \left(\frac{\partial z}{\partial y}\right)_x$$

Mathematically we would say that dz is an exact differential, which simply means that z is a continuous function of the two independent variables x and y. Since the order in which a second partial derivative is taken is unimportant, it follows that

$$\left(\frac{\partial M}{\partial y}\right)_x = \left(\frac{\partial N}{\partial x}\right)_y$$

This result may be applied to each of the four differential equations of state, giving

▶ $$\left(\frac{\partial T}{\partial v}\right)_s = -\left(\frac{\partial P}{\partial s}\right)_v \qquad \textit{from Eq. (8·43)} \qquad (8·50a)$$

▶ $$\left(\frac{\partial T}{\partial P}\right)_s = \left(\frac{\partial v}{\partial s}\right)_P \qquad \textit{from Eq. (8·44)} \qquad (8·50b)$$

▶ $$\left(\frac{\partial P}{\partial T}\right)_v = \left(\frac{\partial s}{\partial v}\right)_T \qquad \textit{from Eq. (8·46)} \qquad (8·50c)$$

▶ $$\left(\frac{\partial v}{\partial T}\right)_P = -\left(\frac{\partial s}{\partial P}\right)_T \qquad \textit{from Eq. (8·48)} \qquad (8·50d)$$

† While this derivation has used reversible processes, the Gibbs equation holds for any change in state regardless of the processes' reversibility.

Equations (8·50) are known as the *Maxwell relations*. They are relations between the derivatives of the thermodynamic properties which must hold for any simple compressible substance; similar relations may be derived for other types of substances.

The internal energy of a simple compressible substance is generally expressible as

$$u = u(T, v)$$

Differentiating and bringing in the definition of the specific heat at constant volume, we find

$$du = c_v \, dT + \left(\frac{\partial u}{\partial v}\right)_T dv \tag{8·51}$$

Equating Eqs. (8·43) and (8·51) and solving for ds,

$$ds = \frac{c_v}{T} \, dT + \frac{1}{T}\left[\left(\frac{\partial u}{\partial v}\right)_T + P\right] dv \tag{8·52}$$

Thinking of the entropy as a function of temperature and volume, we take its differential,

$$ds = \left(\frac{\partial s}{\partial T}\right)_v dT + \left(\frac{\partial s}{\partial v}\right)_T dv$$

and comparing this with Eq. (8·52), we see that

$$\left(\frac{\partial s}{\partial T}\right)_v = \frac{c_v}{T} \tag{8·53}$$

Using the third Maxwell relation, Eq. (8·50*c*),

$$\left(\frac{\partial s}{\partial v}\right)_T = \left(\frac{\partial P}{\partial T}\right)_v = \frac{1}{T}\left[\left(\frac{\partial u}{\partial v}\right)_T + P\right]$$

From this we obtain

▶ $$\left(\frac{\partial u}{\partial v}\right)_T = T\left(\frac{\partial P}{\partial T}\right)_v - P \tag{8·54}$$

This important equation expresses the dependence of the internal energy on the volume at fixed temperature solely in terms of the measurables T, P, and v. Since these properties are easily measured, Eq. (8·54) is very useful in the construction of equation-of-state information from laboratory data.

As an example let us apply Eq. (8·54) to the perfect gas. We have

$$Pv = RT$$

$$\left(\frac{\partial P}{\partial T}\right)_v = \frac{R}{v}$$

So, Eq. (8·54) gives

$$\left(\frac{\partial u}{\partial v}\right)_T = T\frac{R}{v} - P = P - P = 0$$

This implies that, for a perfect gas, the internal energy is independent of the density and depends only upon temperature.

From the previous discussion it follows that

$$ds = \frac{c_v}{T}\,dT + \left(\frac{\partial P}{\partial T}\right)_v dv \tag{8·55}$$

By considering the entropy as a function of the temperature and pressure and bringing in the differential of h in the form of Eq. (8·44), it may similarly be shown that†

$$ds = \frac{c_P}{T}\,dT - \left(\frac{\partial v}{\partial T}\right)_P dP \tag{8·56}$$

Subtracting Eq. (8·55) from Eq. (8·56) and solving for dP,

$$dP = \frac{c_P - c_v}{T(\partial v/\partial T)_P}\,dT - \frac{(\partial P/\partial T)_v}{(\partial v/\partial T)_P}\,dv$$

Then, considering P as a function of T and v, we see that

$$\frac{c_P - c_v}{T(\partial v/\partial T)_P} = \left(\frac{\partial P}{\partial T}\right)_v$$

It may be shown from the calculus [see Eq. (4·9)] that

$$\left(\frac{\partial P}{\partial T}\right)_v = \frac{(\partial v/\partial T)_P}{-(\partial v/\partial P)_T}$$

Combining and solving for the difference in the specific heats,

▶ $$c_P - c_v = -\frac{T[(\partial v/\partial T)_P]^2}{(\partial v/\partial P)_T} \tag{8·57}$$

† This requires use of a Maxwell relation. The development is left for an exercise.

This is one of the more important equations of thermodynamics, and it tells us a great deal. For example, the derivative $(\partial v/\partial P)_T$ is negative for all stable substances, and consequently c_P can never be less than c_v. Furthermore, whenever the derivative $(\partial v/\partial T)_P = 0$, the two specific heats will be equal (for example, water at 4°C). Finally, experiments indicate that as T approaches absolute zero $(\partial v/\partial P)_T$ does not vanish, and so the two specific heats must approach one another at very low temperatures.

Equation (8·57) can be expressed in terms of the isothermal and isobaric compressibilities,

$$c_P - c_v = \frac{Tv\beta^2}{\kappa} \tag{8·58}$$

It is difficult to measure c_v with any precision for a solid or liquid, and the above equation is quite useful for obtaining c_v from more easily measured quantities.

Let's verify Eq. (8·57) for a perfect gas. If

$$Pv = RT$$

then

$$\left(\frac{\partial v}{\partial P}\right)_T = -\frac{RT}{P^2} \qquad \left(\frac{\partial v}{\partial T}\right)_P = \frac{R}{P}$$

Substituting, we correctly find [Eq. (8·16)]

$$c_P - c_v = \frac{TR^2P^2}{P^2RT} = R$$

Equation (8·57) can also be used to compare c_P to c_v for copper (Table B·7),

$$c_P - c_v = \frac{Tv\beta^2}{\kappa}$$

at 200 K for copper

$$\beta = 45.6/\text{K}$$
$$v = 7.029\ \text{cm}^3/\text{gmole}$$
$$\kappa = 0.748\ \text{cm}^2/\text{dyn}$$

Substituting,

$$c_P - c_v = \frac{200 \times 7.029 \times (45.6)^2}{0.748} = 3.9 \times 10^6\ \text{ergs/(gmole·K)}$$
$$= 0.093\ \text{cal/(gmole·K)}$$

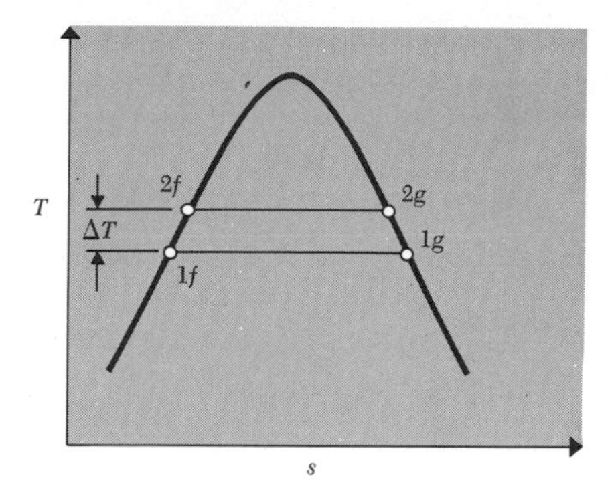

Figure 8·15. The Clapeyron equation relates saturation conditions to the slope of the saturation pressure–saturation temperature curve

From Table B·7 at 200 K,

$$c_P - c_v = 5.41 - 5.32 = 0.09 \text{ cal/(gmole·K)}$$

Since copper is practically incompressible, we indeed expect $c_P \approx c_v$.

By manipulating with the previous equations, we may show that the ratio of specific heats is related to the isentropic and isothermal compressibilities by

$$\blacktriangleright \qquad k \equiv \frac{c_P}{c_v} = \frac{\kappa}{\alpha} \tag{8·59}$$

The isentropic compressibility is related to the speed at which sound waves travel in the substance, and such speed measurements have been used as a means for determining the ratio of the specific heats.

A useful expression describing the variation of saturation pressure with saturation temperature can be derived with the help of Eq. (8·48). Since the liquid and vapor phases of a mixture in equilibrium have the same temperature and pressure, integration of Eq. (8·48) along an isotherm (a line of constant temperature) from the saturated-liquid line to the saturated-vapor line indicates that the two phases also have the same Gibbs function,† that is, $g_f = g_g$. Since g_f and g_g are functions of the saturation temperature only, it then follows that (Fig. 8·15)

$$\frac{dg_f}{dT_{\text{sat}}} = \frac{dg_g}{dT_{\text{sat}}}$$

Again using Eq. (8·48), we then find

$$v_f\left(\frac{dP}{dT}\right)_{\text{sat}} - s_f = v_g\left(\frac{dP}{dT}\right)_{\text{sat}} - s_g$$

or

$$\left(\frac{dP}{dT}\right)_{\text{sat}} = \frac{s_g - s_f}{v_g - v_f} = \frac{s_{fg}}{v_{fg}} \tag{8·60}$$

† This is a condition for equilibrium between phases of a mixture, as we shall show in Chap. 12.

This expresses the change in the saturation pressure with respect to changes in temperature in terms of the properties of the saturated liquid and vapor. Now, since $g_f = g_g$,

$$h_f - Ts_f = h_g - Ts_g$$

or

$$s_g - s_f = \frac{h_g - h_f}{T}$$

Combining with Eq. (8·60), we find

$$\blacktriangleright \qquad \left(\frac{dP}{dT}\right)_{\text{sat}} = \frac{h_{fg}}{Tv_{fg}} \tag{8·61}$$

This is known as the *Clapeyron equation;* it relates the variation of pressure with temperature along the saturated-vapor (or liquid) line to the enthalpy and volume of vaporization. The Clapeyron equation is extremely useful in constructing a graphical or tabular equation of state from a minimum of experimental measurements. It also tells us that $(dP/dT)_{\text{sat}}$ is positive, since enthalpies of vaporization are always positive, as are the volume changes undergone during the vaporization phase change. Consequently, the vapor pressure of any substance increases with temperature.

As an illustration, let's calculate h_{fg} for water at 160°C from data for P, T, and v only. We have, from Table B·1*a*,

$$v_{fg} = 0.3060 \text{ m}^3/\text{kg}$$

To calculate $(dP/dT)_{\text{sat}}$, let's use the P and T data at 150°C and 170°C. We have

$$\Delta P = (0.7916 - 0.4758)\text{MPa} = 0.32 \text{ MPa}$$

so

$$\left(\frac{dP}{dT}\right)_{\text{sat}} \approx \frac{0.32 \times 10^6 \text{ N/m}^2}{20 \text{ K}} = 0.16 \times 10^5 \text{ N/(m}^2\cdot\text{K)}$$

so, using the Clapeyron equation,

$$h_{fg} = Tv_{fg}(dP/dT)_{\text{sat}} = (160 + 273) \times 0.3060 \times 0.16 \times 10^5 = 2120 \text{ kJ/kg}$$

This differs slightly from the value in the table of 2082.6 kJ/kg, because of approximation made in the differentiation.

A Clapeyron equation for fusion may be derived in an analogous manner. Denoting the saturated-solid state by a subscript *s*, the result would be

$$\blacktriangleright \qquad \left(\frac{dP}{dT}\right)_{\text{sat sol}} = \frac{h_{sf}}{Tv_{sf}} \tag{8·62}$$

The enthalpy of melting $h_{sf} = h_f - h_s$ is positive for all known substances. The volume change $v_{sf} = v_f - v_s$ is positive for most substances, indicating that the freezing temperature increases with increasing pressure. However, water has the peculiar property that v_{sf} is negative and therefore has a fusion curve with a negative slope. Hence water will freeze at a lower temperature if the pressure is increased, or melt at a lower temperature if the pressure is increased.

The Clapeyron equations provide a means for direct measurement of the thermodynamic temperature T; this has been proposed, but never tried. Equation (8·61) can be integrated along the saturation line to give

$$\ln \frac{T_2}{T_1} = \int_1^2 \frac{v_{fg}}{h_{fg}}\, dP$$

The terms in the integral can be measured by mechanical and electrical means, without any reference to temperature. Hence, in principle we could compute the ratio of $T_{\text{sat}}(P)$ to the triple point for water by performing the integration. This would require very accurate v_{fg} and h_{fg} data. Thus, through the Clapeyron equation it is possible to measure the thermodynamic temperature without resort to perfect-gas thermometers.

8·8 GENERALIZED THERMODYNAMIC CHARTS

In the absence of better data, the enthalpy of a real gas can be estimated using a combination of thermodynamic equations, perfect-gas theory, and the generalized compressibility correlation. Considering $h(T, P)$, we have

$$dh = \left(\frac{\partial h}{\partial T}\right)_P dT + \left(\frac{\partial h}{\partial P}\right)_T dP$$

The first partial derivative is simply c_P, and the second may be calculated solely from P-v-T data using an equation analogous to Eq. (8·54),

$$\blacktriangleright \qquad \left(\frac{\partial h}{\partial P}\right)_T = v - T\left(\frac{\partial v}{\partial T}\right)_P \tag{8·63}$$

The derivation of this important relation follows that of Eq. (8·54) and is left as an exercise. The enthalpy can therefore be determined from

$$dh = c_P(T, P)\, dT + \left[v - T\left(\frac{\partial v}{\partial T}\right)_P\right] dP$$

We emphasize that in general c_P depends upon both T and P. However, at low

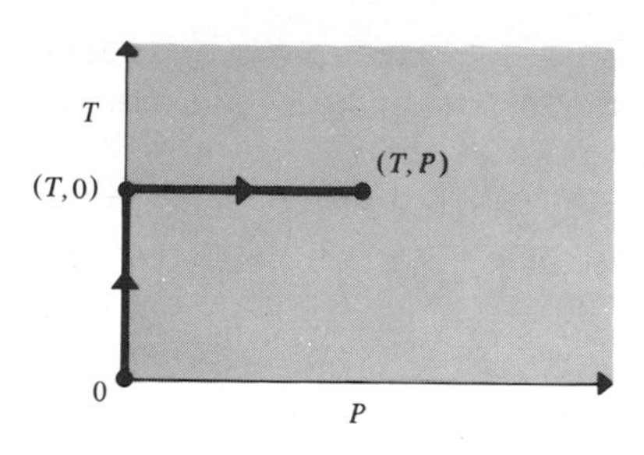

Figure 8·16. The path of integration

pressures the perfect-gas limit is approached. The enthalpy of a perfect gas can therefore be viewed as

$$h_{pg}(T) = \int_0^T c_P(T, 0)\, dT \tag{8·64}$$

Now, we imagine carrying out the integrations from the reference state $T = 0$, $P = 0$. The path of integration is shown in Fig. 8·16. On the first part $dP = 0$. The contribution to h from this path is precisely the perfect-gas term, Eq. (8·64). On the second part $dT = 0$, and the contribution to h can be evaluated solely from P-v-T data. Then,

$$\blacktriangleright \qquad h = h_{pg}(T) + \int_0^P \left[v - T\left(\frac{\partial v}{\partial T}\right)_P \right] dP \tag{8·65}$$

The temperature variation of c_P should of course be considered when evaluating h_{pg}. Note that Eq. (8·65) permits determination of $h(T, P)$ from P-v-T data and low-pressure c_P data.

The entropy can be found in a similar manner. We write Eq. (8·56) as

$$ds = \frac{c_P(T, P)}{T}\, dT - \left(\frac{\partial v}{\partial T}\right)_P dP$$

In the perfect-gas limit for small P,

$$ds_{pg} = \frac{c_P(T, 0)}{T}\, dT - \frac{R}{P}\, dP$$

We now subtract these two equations and again carry out the integrations along the path of Fig. 8·16. Since there is no contribution from the dT terms on the second path, and they precisely cancel on the first path, we have

$$\blacktriangleright \qquad s - s_{pg}(T, P) = \int_0^P \left[\frac{R}{P} - \left(\frac{\partial v}{\partial T}\right)_P \right] dP \tag{8·66}$$

Since the integrand approaches zero as $P \to 0$, the integral presents no problems near the lower limit. Equation (8·66) permits calculation of the entropy of real gases from P-v-T data and low-pressure c_P data.

If we now had some convenient source of P-v-T property information which was of general validity, we could set up charts for h and s which would also be of general validity. The $Pv = ZRT$ information allows us to do this. Using information like the more complete compressibility chart included in Appendix B as Figs. B·14*a* and B·14*b*, the corresponding enthalpies and entropies have been developed from Eqs. (8·65) and (8·66). Values for the generalized enthalpy and entropy can be read from the charts (Figs. B·15 and B·16). It should be emphasized that these charts are based on the principle of corresponding states, which is only an approximation. Whenever possible, one should use equation-of-state data for the particular substance of interest; the generalized properties charts provide at least one way to work beyond the range of available data, which is often a necessity in modern engineering analysis.

Example As an example of the use of Figs. B·15 and B·16, consider a gas with the following known properties:

$$\hat{M} = 31 \text{ lbm/lbmole} \qquad P_c = 900 \text{ psia} \qquad T_c = 540 \text{ R}$$

$$c_P = 0.50 \text{ Btu/(lbm·R)} \textit{ at low pressure}$$

It is desired to run the gas through a heat exchanger at an inlet pressure of 700 psia and an inlet temperature of 540 R. The exit conditions are 650 psia at 1000 R. How much energy must be added as heat to the gas and what will be the entropy change?

Let's first check the inlet state 1 to see if the perfect-gas model is appropriate. We have

$$\frac{P}{P_c} = \frac{700}{900} = 0.779$$

$$\frac{T}{T_c} = \frac{540}{540} = 1.000$$

Checking on Fig. B·14, we find $Z \simeq 0.6$, so that the conditions are not those of the perfect gas ($Z = 1.0$) and we must use Fig. B·15 to calculate the enthalpy change. For the exit state 2 we have

$$\frac{P}{P_c} = \frac{650}{900} = 0.723$$

$$\frac{T}{T_c} = \frac{1000}{540} = 1.85$$

An energy balance gives the heat input per unit mass as

$$Q = h_2 - h_1$$

From Fig. B·15 at state 1,

$$\frac{\hat{h}_{pg} - \hat{h}_1}{T_c} = 2.7 \text{ cal/(gmole·K)} = 2.7 \text{ Btu/(lbmole·R)}$$

Then $$h_1 = \frac{\hat{h}_{pg} - 2.7T_c}{\hat{M}} = h_{pg1} - \frac{2.7 \times 540}{31} = (h_{pg1} - 47) \text{ Btu/lbm}$$

Similarly,

$$h_2 = h_{pg2} - \frac{0.55 \times 540}{31} = (h_{pg2} - 10) \text{ Btu/lbm}$$

So $Q = h_2 - h_1 = (h_{pg2} - h_{pg1}) - 10 + 47$

$$Q = c_P(T_2 - T_1) + 37 = 0.5 \times (1000 - 540) + 37 = 267 \text{ Btu/lbm}$$

Note that h_{pg} is evaluated using the *low-pressure* (ideal-gas) value of c_P, which we have assumed here not to vary over our temperature range of interest. We see that, had we used the perfect-gas approximation, our error would have been about 15 percent. The entropy change can be calculated in a similar manner from Fig. B·16. We find

$$s_2 - s_1 = (s_2 - s_1)_{pg} + \frac{1.9 - 0.27}{31}$$
$$= 0.36 \text{ Btu/(lbm·R)}$$

The student should carry out this calculation using Eq. (8·27) for $(s_2 - s_1)_{pg}$.

8·9 THERMODYNAMICS OF A SIMPLE MAGNETIC SUBSTANCE

Thermodynamics is not restricted to simple compressible substances. In fact, some of the most interesting and important uses of thermodynamics involve other kinds of substances; we shall examine the simple magnetic substance to illustrate again the manner in which thermodynamics is used to obtain information about the properties of matter.

For a simple magnetic substance the state postulate tells us that there are only two independent intensive thermodynamic properties; we may take these as

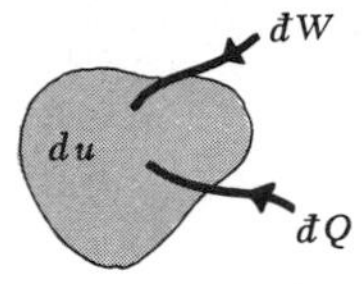

Figure 8·17. Deriving the Gibbs equation for a simple magnetic substance

the internal energy and dipole moment per unit of mass, $v\mathbf{M}$. The magnetic Gibbs equation may be obtained by considering a unit of mass of the material undergoing a reversible process between two infinitesimally separated thermodynamic states (Fig. 8·17). An energy balance gives

$$du = đQ + đW$$

The energy transfer as work per unit of mass [see Eq. (2·17)] is†

$$đW = \mu_0 v\mathbf{H} \cdot d\mathbf{M}$$

Since the process is reversible, the second law requires that

$$ds = \frac{đQ}{T}$$

Combining, the Gibbs equation is found to be

▶ $$du = T\,ds + \mu_0 v\mathbf{H} \cdot d\mathbf{M} \tag{8·67}$$

Inspecting the coefficients, it follows that

▶ $$\mu_0 v\mathbf{H} = \left(\frac{\partial u}{\partial \mathbf{M}}\right)_s \tag{8·68}$$

▶ $$T = \left(\frac{\partial u}{\partial s}\right)_{\mathbf{M}} = \frac{1}{(\partial s/\partial u)_{\mathbf{M}}} \tag{8·69}$$

The latter is merely the thermodynamic definition of temperature for magnetic systems.

It is convenient to define a magnetic enthalpy and a magnetic Gibbs function as

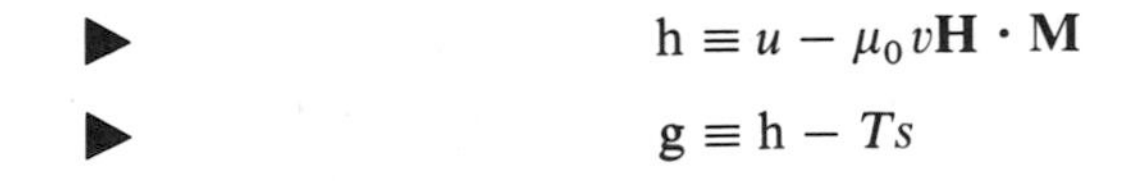

▶ $$\mathrm{h} \equiv u - \mu_0 v\mathbf{H} \cdot \mathbf{M}$$

▶ $$\mathrm{g} \equiv \mathrm{h} - Ts$$

† The rationalized mksc system is employed throughout this section (see Appendix A). However, the equation-of-state information of Fig. B·12 is in the absolute magnetostatic system.

Specific heats at constant **M** and **H** are the analogs of c_v and c_P; they are defined as

$$c_{\mathbf{M}} \equiv \left(\frac{\partial u}{\partial T}\right)_{\mathbf{M}}$$

$$c_{\mathbf{H}} \equiv \left(\frac{\partial \mathrm{h}}{\partial T}\right)_{\mathbf{H}}$$

A set of Maxwell relations for magnetic substances can be obtained. For example, the analog of Eq. (8·50*c*) is

$$\mu_0 v \left(\frac{\partial \mathbf{H}}{\partial T}\right)_{\mathbf{M}} = -\left(\frac{\partial s}{\partial \mathbf{M}}\right)_T \qquad (8\cdot70)$$

An expression relating the difference in specific heats may be found by the methods used in obtaining Eq. (8·57),

$$c_{\mathbf{H}} - c_{\mathbf{M}} = \frac{T\mu_0 v[(\partial\mathbf{M}/\partial T)_{\mathbf{H}}]^2}{(\partial\mathbf{M}/\partial\mathbf{H})_T} \qquad (8\cdot71)$$

If $(\partial\mathbf{M}/\partial\mathbf{H})_T$ is positive for any substance, as seems always to be the case, $c_{\mathbf{H}}$ must be larger than $c_{\mathbf{M}}$.

Superconductors are known to undergo a transition from superconducting to normally conducting states at well-defined values of the applied field. This "threshold field" is a function of temperature, and experimental data suggest that the equation of state might appear as shown in Fig. 8·18.

It may be shown from energy considerations that the energy which must be transferred as heat to effect a constant-T, constant-**H** transition is equal to the difference in the magnetic enthalpies,

$$\mathrm{h}_{sn} \equiv \mathrm{h}_n - \mathrm{h}_s$$

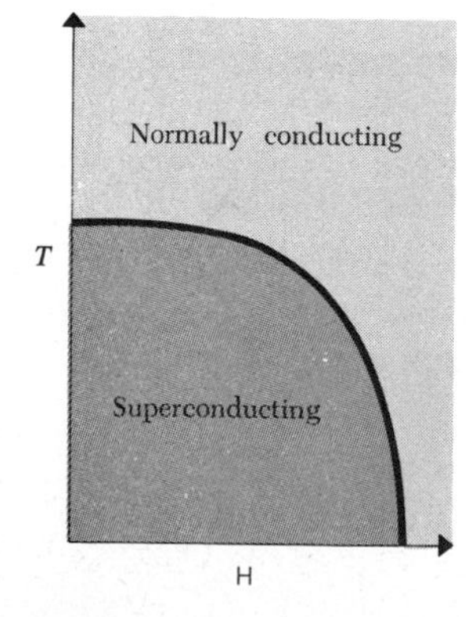

Figure 8·18. Phase transitions in superconductors at low temperatures

Here we use the subscripts n and s to denote the normal and superconducting states, respectively. Following the development of the Clapeyron equation, it may be shown that the threshold field is related to the temperature and to the enthalpy of transformation h_{sn} by

$$\left(\frac{d\mathbf{H}}{dT}\right)_{\text{threshold}} = -\frac{\mathrm{h}_{sn}}{T\mu_0 v(\mathbf{M}_n - \mathbf{M}_s)}$$

In normally conducting states the magnetic moment **M** is zero. Superconduction occurs when the magnetization exactly cancels the applied field **H**, so that electrons passing through the material experience no magnetic field. Setting $\mathbf{M}_n = 0$ and $\mathbf{M}_s = -\mathbf{H}$, the above equation reduces to

$$\left(\frac{d\mathbf{H}}{dT}\right)_{\text{threshold}} = \frac{-\mathrm{h}_{sn}}{T\mu_0 v\mathbf{H}} \tag{8·72}$$

We see that the threshold field decreases with increasing temperature, which is indeed found to be the case.

8·10 ALGEBRAIC EQUATION OF STATE FOR A CURIE SUBSTANCE

A Curie substance is any simple magnetic substance obeying the equation of state

$$\mathbf{M} = C\frac{\mathbf{H}}{T} \tag{8·73}$$

Paramagnetic salts at temperatures which are not too low and in fields which are not too strong behave in this general manner.

It may be shown that the internal energy is a function only of temperature for any simple magnetic substance for which $\mathbf{H}/T = f(\mathbf{M})$. Consider the magnetic Gibbs equation (8·67), written in the form

$$ds = \frac{1}{T}\,du - \mu_0 v\left(\frac{\mathbf{H}}{T}\right) d\mathbf{M}$$

If $\mathbf{H}/T = f(\mathbf{M})$, then since ds is exact,

$$\left[\frac{\partial(1/T)}{\partial \mathbf{M}}\right]_u = -\mu_0 v\left[\frac{\partial(\mathbf{H}/T)}{\partial u}\right]_{\mathbf{M}} = 0$$

T is therefore independent of **M** along a line of constant u, and $T = T(u)$. Inverting this relationship, we have $u = u(T)$. It must be emphasized that this holds only for the special circumstances where $\mathbf{H}/T = f(\mathbf{M})$ and is not true for paramagnetics in general.

The energy of a Curie substance is therefore given by

$$du = c_{\mathbf{M}}(T)\,dT$$

The entropy may be found from the magnetic Gibbs equation, which for a Curie substance is

$$ds = \frac{c_{\mathbf{M}}}{T}\,dT - \frac{\mu_0 v}{C}\,\mathbf{M}\cdot d\mathbf{M}$$

For a Curie substance with constant $c_{\mathbf{M}}$ the above equations integrate to give

$$u_2 - u_1 = c_{\mathbf{M}}(T_2 - T_1) \tag{8·74}$$

$$s_2 - s_1 = c_{\mathbf{M}} \ln \frac{T_2}{T_1} - \frac{\mu_0 v}{2C}(\mathbf{M}_2^2 - \mathbf{M}_1^2) \tag{8·75}$$

SELECTED READING

Lee, J., and F. Sears, *Thermodynamics*, 2d ed., chap. 2, secs. 7.1–7.7, Addison-Wesley Publishing Co., Inc., Reading, Mass., 1962.

Van Wylen, G., and R. Sonntag, *Fundamentals of Classical Thermodynamics*, 2d ed., chaps. 3 and 10, John Wiley & Sons, Inc., New York, 1973.

Wark, K., *Thermodynamics*, 2d ed., chap. 13, McGraw-Hill Book Company, New York, 1971.

Zemansky, M. W., *Heat and Thermodynamics*, 5th ed., chaps. 11, 13, and 14, McGraw-Hill Book Company, New York, 1968.

QUESTIONS

8·1 What role do equations of state play in the analysis of engineering systems?

8·2 Of what value are the differential equations of state?

8·3 Explain how entropy can be found (relative to some datum state) without enumerating quantum states.

8·4 What is the Gibbs equation?

8·5 Starting from the fundamental definition of a partial derivative, can you derive the chain rule of calculus?

8·6 Why is it essential to put a subscript on a partial derivative to indicate the property held constant?

8·7 Of what utility is the Clapeyron equation?

8·8 Explain how c_v can be determined without ever being measured directly.

8·9 On what diagram is c_P the slope of a line, and what line?

8·10 Considering the Clapeyron equation, what peculiar characteristic of H_2O makes ice skating possible?

8·11 Is the specific heat of a substance obeying $Pv = RT$ necessarily constant? Is it necessarily a function only of temperature?

8·12 If c_v is constant for a perfect gas, must c_P also be constant?

8·13 What are the meaning and utility of p_r and v_r in the air tables (Table B·9)?

8·14 Under what conditions does Pv^k = constant?

8·15 What is the difference between c_v and $\hat{c}_v$?

8·16 When can $Pv = RT$ be expected to apply?

8·17 What is the approximate value of Pv/RT near the critical point?

8·18 Why is thermodynamics useful in the study of magnetic substances?

8·19 What is the principle of corresponding states?

1 K° = 1.8 R°

PROBLEMS

8·1 Starting with the Gibbs equation, show that for a perfect gas undergoing an isentropic process $(T_2/T_1) = (P_2/P_1)^{(k-1)/k}$, where $k = c_P/c_v$.

8·2 A well-designed nozzle is idealized as isentropic. If the inlet conditions are 16 psia, 600 R, and the exit pressure is 14.7 psia, find the exit velocity assuming the air obeys the perfect-gas equations of state and neglecting the inlet kinetic energy.

8·3 In designing the air-conditioning system for an airplane, it is desired to determine the work required to compress air in an isentropic steady-flow process from 0.5 atm to 5 atm. What is the exit temperature if the inlet temperature is 0°F?

8·4 Consider steam flow through an adiabatic valve at (*a*) an input state of 1000 psia, 850°F, and an exit pressure of 300 psia, and (*b*) an input state of 60 psia, 800°F, with an exit at 1 atm. Examine and discuss the validity of using the perfect-gas model in analyzing these two flows.

8·5 A test of an air compressor reveals that the compression stroke can be modeled by a polytropic process (Pv^n = const) with $n = 1.35$. The initial temperature is 70°F, the initial pressure 13.5 psia. If the final pressure is 81.0 psia, find the final temperature, the work done on the gas, and the heat transferred per lbm of air.

8·6 A fixed mass of air is contained in a piston-cylinder system. Initially the gas occupies 1 ft^3 at 70°F and 15 psia. After a compression process the pressure is 100 psia. Find (*a*) the mass of gas in the cylinder, (*b*) the final temperature, (*c*) the work required, (*d*) the change in u, (*e*) the heat transferred, if any, if the process is isothermal. Rework the above for a process of $Pv^{1.25}$ = constant. Treat air as a perfect gas.

8·7 Compute the complete algebraic equation of state for a perfect gas for which $c_v = a + bT$, where a and b are constants.

8·8 The general process represented by Pv^n = constant is called the polytropic process. Consider a fixed mass of a perfect gas in a piston cylinder undergoing a process for which Pv^n = constant. Show that the work done per lbm in such a process is given by $W = (P_2 v_2 - P_1 v_1)/(n - 1)$. If the process is isentropic, show that this reduces to $W = c_v(T_2 - T_1)$.

8·9 Derive expressions for the slopes of constant-pressure and constant-volume lines on a T-s plane for a perfect gas with constant specific heats and show that the volume line is steeper.

8·10 Derive the complete algebraic equation of state for a perfect gas for which $c_P = a + bT$, where a and b are constants.

8·11 Show that c_v is a function only of temperature for a Van der Waals gas. Derive an expression for $c_P - c_v$ for the gas. Is c_P also a function only of temperature? Derive algebraic expressions for the internal energy and entropy of a Van der Waals gas, with constant c_v, as a function of temperature and specific volume. Compare these with the perfect-gas equations.

8·12 Calculate the pressure as a function of molal volume for CO_2 at 400 K using the Beattie-Bridgeman equation of state. Cover the range 0 to 100 atm. Compare with values which would be obtained from the generalized compressibility chart.

8·13 Derive the complete algebraic equation of state for an incompressible liquid for which $c = a + bT$, where a and b are constants.

8·14 Estimate the enthalpy and entropy of mercury at room conditions relative to the datum used in Table B·3.

8·15 Using data from Appendix B, find the entropy, enthalpy, and internal energy for (*a*) water at 70°F, 100 psia, (*b*) mercury at 875°F, 800 psia, and (*c*) Freon at 114 psia, −10°F.

8·16 Liquid natural gas (LNG) ships have a boil-off rate of about 0.3 percent per day. For a ship of 50,000-m^3 capacity, calculate the natural gas released by boil-off, treating the gas as methane, if the temperature is 15°C at 1 atm pressure.

8·17 Find the enthalpy of liquid water for (*a*) saturated liquid at 300°F, (*b*) subcooled liquid at 300°F and 267 psia, and (*c*) subcooled liquid at 300°F and 3400 psia.

8·18 Find the enthalpy of liquid water for (*a*) saturated liquid at 150°C, (*b*) subcooled liquid at 150°C and 1.7 MPa, and (*c*) subcooled liquid at 150°C and 24 MPa.

8·19 A small pump takes saturated liquid water at 30°C and pressurizes it to a pressure of 2 MPa. Calculate the least work required per kg of flow if the pump is idealized as adiabatic. If the flow rate is 400 kg/min find the minimum power required.

8·20 The last-stage pump on a 750-MW supercritical steam cycle raises the water pressure from 4 MPa at 200°C to 24 MPa. Calculate the least work required per kg of flow if the pump is idealized as adiabatic. If the flow rate is 400 kg/s, find the minimum power required.

8·21 A large power plant requires two pumps in the circulating water system. Each pump is rated at 135,500 gal/min when producing a pressure rise equal to a head of 66 ft of water. Treating the pumps as adiabatic, calculate the minimum hp required to drive each pump.

8·22 Saturated liquid water at an absolute pressure of 3.5 in of mercury is pumped to 390 psia at a rate of 5100 gal/min. Calculate the minimum power required if the pump is adiabatic.

8·23 A boiler feedwater pump takes subcooled water at 415°F and 600 psia and raises the pressure to 3850 psia. The flow rate is 6900 gal/min. Determine the least amount of work per lbm of water and the least hp required from the steam turbine which drives the adiabatic pump.

8·24 Argon is compressed to 20 atm from an inlet condition of −40°F at 2 atm. Calculate the exit temperature and the work required in kJ/kg for the adiabatic process which requires the least work.

8·25 Calculate the work required per kg to compress helium isentropically from 10^6 N/m^2 to 3.5×10^6 N/m^2 if the inlet temperature is 200°C.

8·26 Argon is compressed *isothermally* from 100°C at 1 atm to 4 atm. Calculate the final temperature and the work required per kg of flow. Compare this to an isentropic compression from the same inlet conditions to the same final pressure.

8·27 Compare the specific volume of steam at 40 MPa, 800 K, as given by the perfect-gas equation, the compressibility equation, and the steam tables.

8·28 Compare the specific volume of steam at 15 MPa, 700 K, as given by the perfect-gas equation, the compressibility equation, and the steam tables.

8·29 A new gas-cooled nuclear reactor utilizes helium at 1600 psia at a turbine inlet temperature of 1340°F. Calculate the gas density at these conditions.

8·30 Sulfur dioxide, SO_2, enters a pollution clean-up device at a temperature of 200°C and a pressure of 10 MPa. At that point the flow rate is 4 kg/s through a 4-cm-diameter pipe. Design criteria call for the flow velocity to be less than 15 m/s. Will these conditions meet the design?

8·31 Use Van der Waals equation to verify the accuracy of the pressure calculation for O_2 (160 K, 1 m^3, 100 kg) in the example in this chapter.

8·32 One-hundred lbm of CO_2 is stored in a 10-ft^3 vessel at 100°F. Determine the pressure in the container using the perfect-gas, Van der Waals, Beattie-Bridgeman, and $Pv = ZRT$ equations.

8·33 Forty kg of CO_2 is stored in a 0.2-m^3 container at 100°C. Compare the required pressure using the perfect-gas and generalized compressibility equations.

8·34 Propane is to be stored at 4 MPa, 100°C. Estimate the specific volume at these conditions. How large a tank is required to store 1000 kg?

8·35 Propane is to be stored at 40 atm, 200°F. Estimate the specific volume at these conditions. How large a tank is required to store 2500 lbm at these conditions?

8·36 Methane has been compressed to a pressure of 15×10^6 N/m^2 and a specific volume of 6 m^3/Mg. Find the temperature.

8·37 Natural gas is pumped through a pipeline at 1800 psia, 100°F. Find the specific volume. Treat the gas as methane.

8·38 The pump in a nuclear power system pressurizes mercury from saturated liquid at 20 psia to 600 psia. Find the work required, treating the pump as adiabatic and the process as isentropic.

8·39 Using the generalized compressibility chart, determine the pressure required to fix a specific volume of 0.2 ft^3/lbm at a temperature of 120°F for CO_2. Compare with the CO_2 chart.

8·40 Nitrogen is stored in a thick-walled pressure tank at about 2200 psia. Find the specific volume at 2200 psia, 60°F, using the generalized compressibility chart. Compare to the value of Fig. B·4. If the tank has a volume of 2.2 ft^3, what is the mass of nitrogen stored?

8·41 It is desired to store 10 lbm of O_2 in a 1-ft^3 tank at 300 R. To design the tank we need to determine the pressure under these conditions. Estimate the pressure in atmospheres from (*a*) the perfect-gas equation, (*b*) the Van der Waals equation, and (*c*) the Beattie-Bridgeman equation of state.

8·42 Determine the specific volume of steam at a pressure of 5000 psia and a temperature of 800°F from the generalized compressibility chart and compare to that value given in the steam tables. (Some power plants now operate at these pressures.)

8·43 Test the Clapeyron equation for *fusion* for CO_2 at −100°F using the data of Fig. B·6.

8·44 Vapor pressure is often approximated as an exponential function of temperature. Show how this approximation is suggested by the Clapeyron equation.

8·45 The Joule-Kelvin coefficient μ is defined by $\mu = (\partial T/\partial P)_h$. Prove that

$$\mu = \frac{1}{c_P}\left[T\left(\frac{\partial v}{\partial T}\right)_P - v\right]$$

Discuss how this equation might be used.

8·46 Show that the Joule-Kelvin coefficient for a perfect gas is zero.

8·47 Show that

$$\left(\frac{\partial c_v}{\partial v}\right)_T = T\left(\frac{\partial^2 P}{\partial T^2}\right)_v \qquad \left(\frac{\partial c_P}{\partial P}\right)_T = -T\left(\frac{\partial^2 v}{\partial T^2}\right)_P$$

Discuss the use of these equations.

8·48 Prove that

$$\left(\frac{\partial u}{\partial P}\right)_T = -T\left(\frac{\partial v}{\partial T}\right)_P - P\left(\frac{\partial v}{\partial P}\right)_T$$

How might this be used?

8·49 Prove that

$$\left(\frac{\partial P}{\partial T}\right)_s = \frac{c_P}{Tv\beta}$$

Describe an experiment that would use this equation to determine c_P without any energy measurements.

8·50 Describe an experiment by which the latent heat of vaporization of a substance could be determined without any energy measurements.

8·51 Estimate the vapor pressure of mercury at 1600°F using the Clapeyron equation.

8·52 Using appropriate differential equations of state and taking the datum for entropy to be saturated liquid at 14.7 psia, determine the entropy of nitrogen at the following states, using those data of Fig. B·4 [graphical integrations are required for (*b*) and (*c*)]; use the most convenient differential equation of state: (*a*) saturated vapor, 14.7 psia; (*b*) 500 R, 14.7 psia; (*c*) the critical point.

8·53 Show that the isothermal compressibility is always greater than or equal to the isentropic compressibility.

8·54 The properties of ice at 1 atm may be found in the *Handbook of Chemistry and Physics*. Estimate the melting temperature of ice at 1000 psia.

8·55 Derive an expression analogous to Eq. (8·65) for $u - u_{pg}$. Derive an equation for $u - u_{pg}$ using this expression and the Beattie-Bridgeman equation.

8·56 Derive an expression analogous to Eq. (8·66) for $s - s_{pg}$ using Eq. (8·55). Derive an equation for $s - s_{pg}$ using this expression and the Beattie-Bridgeman equation.

8·57 Nitrogen is throttled from 2700 lbf/in^2 at -160°F to 270 lbf/in^2 in an adiabatic steady-flow process. Determine the temperature at the exit state.

8·58 Steam is throttled through an adiabatic valve from an inlet state of 1000 psia, 850°F, to an exit pressure of 300 psia. Find the exit state using the generalized enthalpy chart and compare with results from the Mollier chart.

8·59 Methane is heated in a constant-pressure heat exchanger from 50°F, 700 psia, to 250°F. Calculate the energy required per mole of gas assuming that the specific heat is constant over this temperature range.

8·60 Methane is cooled in a constant-pressure process from 120°C at 4.7 MPa to 10°C. Calculate the energy removal per mole of methane assuming that the specific heat is constant over this temperature range.

8·61 Derive the analog of the Maxwell relations for a simple magnetic substance.

8•62 Derive the complete algebraic equation of state for a Curie substance for which $c_M = aT^2$, where a is a constant.

8•63 Consider a simple dielectric substance for which the polarization is related to the electric field and temperature by $\mathbf{E} = AT\mathbf{P}$, where A is a constant. Show that the energy of this dielectric is a function only of temperature. Derive a complete algebraic equation of state for the case where the specific heat at constant polarization $c_{\mathbf{P}} \equiv (\partial u/\partial T)_{\mathbf{P}}$ is constant.

8•64 Consider a simple surface for which the only possibly reversible work mode is surface extension. Derive the Gibbs equation $T\,dS = dU - \sigma\,dA$, where A is the surface area and σ is the surface tension. Derive the relation

$$\left(\frac{\partial U}{\partial A}\right)_T = \sigma - T\left(\frac{\partial \sigma}{\partial T}\right)_A$$

Show that, if the surface tension is a function only of temperature, the energy per unit of area also depends only on temperature, and consequently that

$$\frac{U}{A} = \sigma - T\frac{d\sigma}{dT}$$

The surface tension is easily measured, but it is difficult to measure the surface energy directly. How else might it be determined?

8•65 Consider a simple elastic solid for which the only significant possibly reversible work mode is pure strain. Develop the Gibbs equation $T\,ds = du - \sigma\,d\varepsilon$, where s and u are the entropy and energy per unit of volume, σ is the stress (positive in tension), and ε is the strain. Following the development of the Maxwell relations, derive an expression for $(\partial\varepsilon/\partial s)$ in terms of $(\partial T/\partial\sigma)_s$. Take a rubber band, stretch it rapidly, and hold it to your lip. Which way does the temperature change? Check your result by rapidly releasing the strain. On the basis of your theory and experiment, will the elongation of a rubber rod under fixed stress increase or decrease when the rod is heated? (You might try to verify this prediction experimentally.) Unstressed metal rods normally expand when heated. What will happen to the temperature of a metal rod when it is rapidly stretched?

8•66 If the strain in the thin wire of Prob. **8•65** is small, it is convenient to put $l = l_0(1 + \varepsilon)$, where l_0 is the unstretched length of the wire and ε is the strain (extension per unit length). The isothermal Young's modulus is $Y \equiv (\partial\sigma/\partial\varepsilon)_T$. Show that, if Y is constant, so is $\alpha \equiv (\partial\sigma/\partial T)_\varepsilon$. The volumetric specific heat at constant strain is $c_\varepsilon \equiv (\partial u/\partial T)_\varepsilon$, where u is the internal energy per unit volume. Derive algebraic equations for the energy and entropy per unit of volume (u and s) for a slightly stretched wire with constant Y and c_ε.

8•67 Consider an idealized liquid for which $P = A(v_0 - v)$, where A and v_0 are constants. Derive algebraic equations for energy, enthalpy, and entropy, assuming c_P is constant. Compare these with the equations for the incompressible liquid.

CHAPTER NINE

CHARACTERISTICS OF SOME THERMODYNAMIC SYSTEMS

9·1 ANALYSIS OF THERMODYNAMIC SYSTEMS

Thermodynamics is invaluable in the analysis of any system involving energy transfers; the most common and practical uses of thermodynamics in engineering are in analysis of systems containing some sort of working substance, usually in a liquid or gaseous phase, which is flowing or circulating through the device. In this chapter we shall examine the characteristics of a number of thermodynamic systems, mostly of this variety, and, in addition, we shall look at the behavior of some more unusual devices which show promise of practical utility in the future. These characteristics may be predicted from a combination of thermodynamic analysis and experience with operating hardware. For the most part we shall consider systems sufficiently idealized that analysis of their performance is within the range of our studies thus far.

The systems of chief interest here are those which effect some sort of *energy conversion*. In power-generating systems we are interested in converting the internal energy of hydrocarbon fuel molecules, or the atomic energy of uranium or plutonium, into electric or mechanical energy. In refrigeration systems we are interested in keeping some area cool by continual removal of energy as heat from that area. Most such systems involve a *working fluid*, such as water or air, which is circulated through the system in a *cycle*. In steam power plants (Fig. 9·1) the cycle

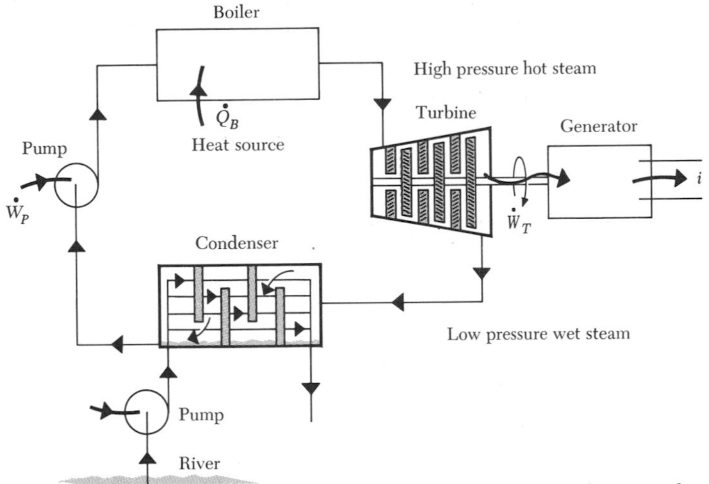

Figure 9·1. Schematic of a simple steam power plant

is usually *closed*, while in gas power systems, such as the turboprop engine (Fig. 9·2), the cycle is often *open* (closed by the atmosphere).

The general methodology for energy analysis of a thermodynamic system was presented and illustrated in Chap. 5, and this chapter will afford additional practice with these important tools. In addition, we now have the many consequences of the second law at our disposal and are in a position to employ them

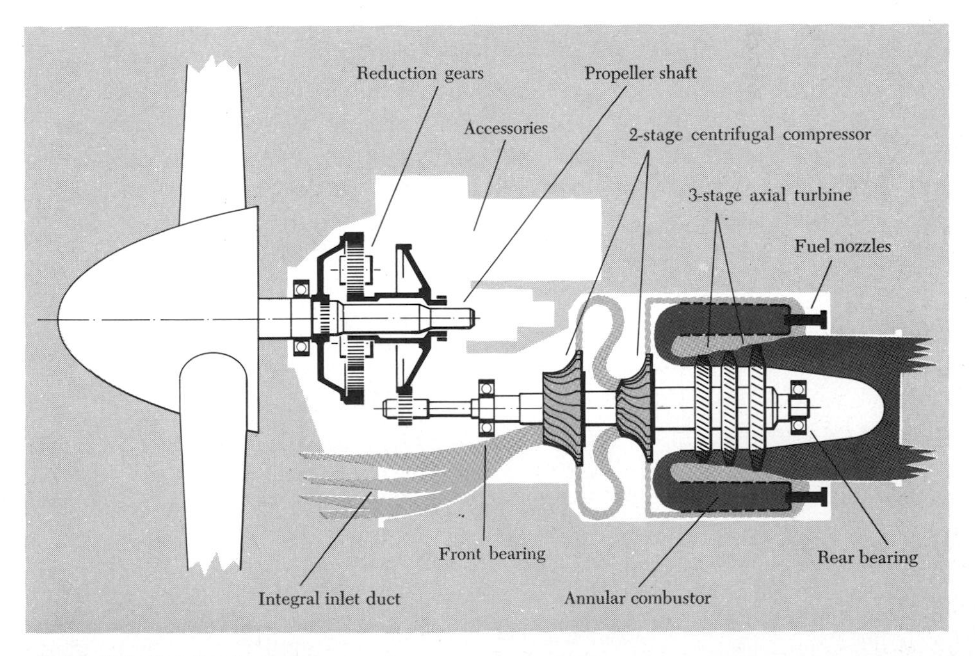

Figure 9·2. Details of a turboprop engine

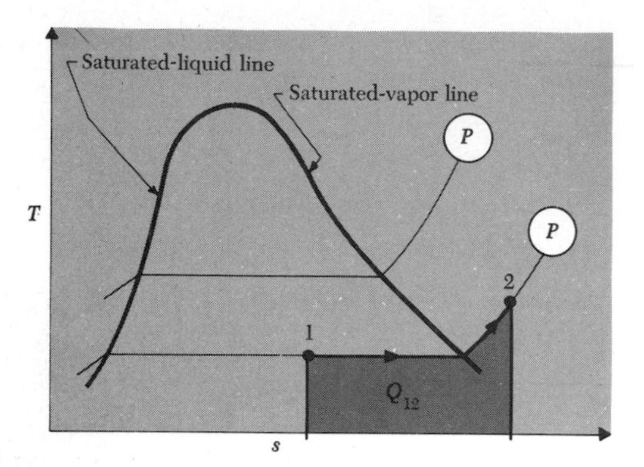

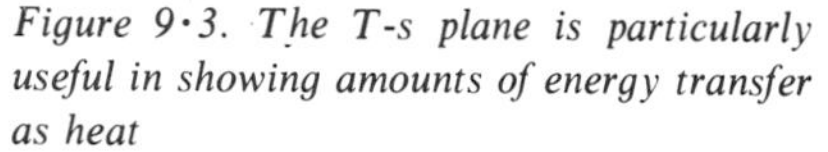

Figure 9·3. The T-s plane is particularly useful in showing amounts of energy transfer as heat

in discussing the performance of thermodynamic systems. The second law places strong limitations on the performance of thermal energy-conversion and thermal transfer systems,† and we shall examine these in the following discussions.

The value of the process representation in the analysis of a thermodynamic system was emphasized in Chap. 5. Especially important is the temperature-entropy plane; when matter undergoes a reversible process, the sequence of states through which it passes traces out a line on the T-s plane (Fig. 9·3). Since the process is reversible, the energy transferred as heat *to* a unit of mass of the substance is represented by the area under the curve on the T-s plane. If the substance undergoes a cyclic process, there will be no net change in its internal energy over a cycle, and consequently the net energy transferred to a unit mass of the substance as heat during the cycle must equal the net energy transfer as work from the substance (work done), and both equal the area enclosed by the reversible path on the T-s plane (Fig. 9·4). The T-s process representation can therefore be a very graphic aid in comparing and evaluating thermodynamic systems, and we shall make extensive use of it in this chapter. Other thermodynamic planes are also quite descriptive, as will be seen.

† Refrigerators, heat pumps, and so on.

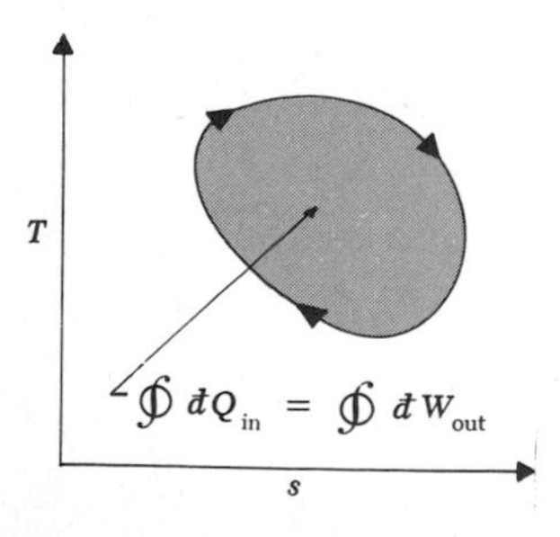

Figure 9·4. The cyclic integral of $T\,ds$ represents the net energy transfer as heat to the substance

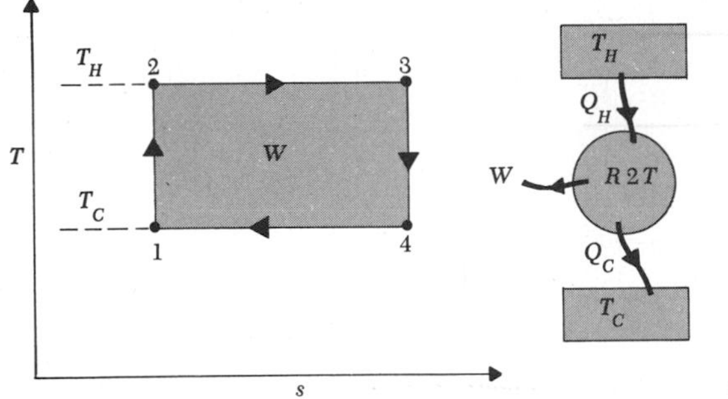

Figure 9·5. The Carnot engine

9·2 THE CARNOT CYCLE

The *Carnot cycle* is the reversible cycle defined by two isothermal processes and two isentropic processes (Fig. 9·5). Since a reversible isentropic process is adiabatic, the only energy transfer as heat to a piece of substance undergoing a Carnot cycle occurs during the isothermal processes. The Carnot cycle constitutes a reversible $2T$ engine, and consequently the ratios of the energy transfers as heat defined in Fig. 9·5 are given by†

$$\blacktriangleright \qquad \frac{Q_H}{Q_C} = \frac{T_H}{T_C}$$

Its energy-conversion efficiency is therefore

$$\blacktriangleright \qquad \eta = \frac{W}{Q_H} = \frac{Q_H - Q_C}{Q_H} = 1 - \frac{T_C}{T_H} \tag{9·1}$$

Highest efficiencies will be obtained when the ratio T_C/T_H is as small as possible. One would like to *add the energy as heat at as high a temperature as possible and reject energy as heat at the lowest possible temperature;* the practical limitations on T_H and T_C were discussed in Chap. 7.

The Carnot cycle operates as a refrigerator when reversed (Fig. 9·6). The area enclosed by its T-s process path would represent the work required per cycle of operation, and we should like this to be as small as possible. This suggests that

† This relation can easily be derived at will from $đQ_{\text{rev}} = T\,dS$, since

$$\frac{Q_H}{Q_C} = \frac{T_H(S_3 - S_2)}{T_C(S_4 - S_1)} = \frac{T_H}{T_C}$$

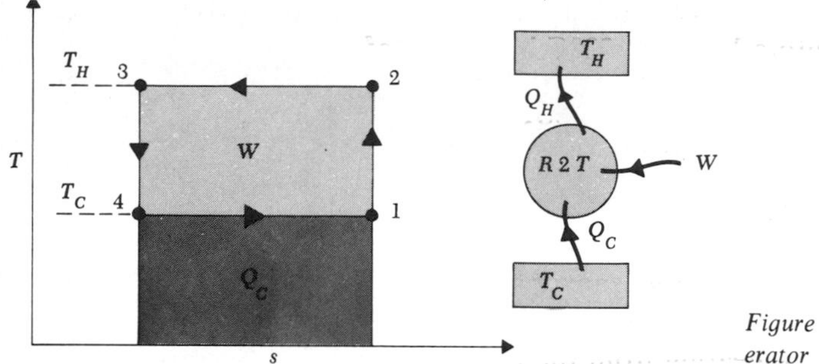

Figure 9·6. The Carnot refrigerator

having T_H as close as possible to T_C is most desirable. A refrigeration cycle is rated in terms of its cop:†

$$\text{cop}_{\text{refrig}} = \frac{Q_C}{W} \tag{9·2}$$

For the Carnot refrigerator,

$$\text{cop} = \frac{Q_C}{Q_H - Q_C} = \frac{T_C}{T_H - T_C} \tag{9·3}$$

Unlike the efficiency, the cop can range from zero to infinity. For a Carnot refrigerator extracting energy as heat from a cold space at 0°F and transferring energy as heat to an environment at 60°F,

$$\text{cop} = \frac{460}{520 - 460} = 7.7$$

Real refrigeration systems operating between the same two temperatures have cop values of the order of 2 to 3.

An interesting use of the refrigerator is as a *heat pump*. Here the objective is not to keep a region cool but instead to keep a region (such as a house) warm. The energy transfer to the hot space is then of prime interest, and it is customary to define the cop as

$$\text{cop}_{\text{heat pump}} = \frac{Q_H}{W} \tag{9·4}$$

Then, for a Carnot heat pump,

$$\text{cop} = \frac{Q_H}{Q_H - Q_C} = \frac{T_H}{T_H - T_C} \tag{9·5}$$

† Coefficient of performance.

For example, a Carnot heat pump taking energy from the outdoors at $-10°C$ and transferring energy into a house at 20°C has a cop of

$$\text{cop} = \frac{293}{293 - 263} = 9.7$$

This means that the homeowner would be getting energy into the house equal to almost 10 times the electric energy showing up on his utility bill. A practical heat pump may have a cop of 3 to 4, but even this makes a heat pump far superior to direct electric-resistance heating from the point of view of utility costs.†

The Carnot cycle is very useful in estimating limits of efficiency for given operating temperatures; no real system limited by the same temperatures could exceed the performance of a Carnot cycle, since the efficiency of any irreversible $2T$ engine is less than Carnot-cycle efficiency. Unfortunately, it is very difficult to build a real device to operate on the Carnot cycle,‡ and its chief value is as a standard of comparison for real energy-conversion and refrigeration systems.

9·3 PROCESS MODELS

In thermodynamic analysis one often knows the initial state of a substance and can compute the work done or the energy transferred as heat only if something is known about the process undergone by the substance. The analyst frequently has to make appropriate idealizations in order to render the analysis tractable; the idealized processes employed are frequently referred to as *model processes.* For example, if a gas is being compressed very slowly in a massive container, there will be plenty of time for energy to be transferred as heat, and a good idealization might be that the gas is always at the container temperature, that is, that the process is isothermal. In this case the isothermal process represents a simplified model of what actually might occur.

When a substance undergoes a process very rapidly, there is little time for energy transfer as heat to occur, and we can make the idealization that the process is adiabatic. If, in addition, the irreversibilities discussed in Chap. 7 are not too important, it may also be appropriate to idealize that the process is reversible. The second law tells us that the entropy of matter undergoing a process which is both reversible and adiabatic will not change. Such an isentropic process is the model process normally employed for

1. Restrained adiabatic compression or expansion of a gas in a piston-cylinder system

† The capital cost of the heat pump would, however, be significantly greater.

‡ The inherent difficulty in building an operating Carnot cycle arises from the difficulty in obtaining isothermal compression and expansion by proper control of the energy transfers as heat and work.

2. Compression or expansion of a fluid in a steady-flow adiabatic compressor, turbine, or pump
3. Frictionless adiabatic flow of a substance through a duct
4. Adiabatic magnetization or demagnetization
5. Adiabatic polarization or depolarization

The actual process may not be very close to the model process, and in such cases the analyst often introduces some sort of *performance parameter* to account at least partially for the departures from the idealized behavior. For example, in analysis of *turbines* we introduce the *isentropic efficiency*, defined as

$$\eta_s \equiv \frac{W}{W_s} \tag{9·6}$$

Here W represents the work output which would be measured from an actual adiabatic turbine, and W_s is the theoretical work output of an isentropic-process turbine *operating with the same inlet state and discharge pressure*. The isentropic efficiency is not an energy-conversion efficiency, but rather a parameter which compares an actual device to an ideal (model) device. Second-law analysis indicates that $\eta_s \leq 1$. The value of η_s depends upon the design of the turbine blades, nozzle, and diffuser, and prediction of efficiencies requires some fairly sophisticated fluid-mechanical analysis. Small turbines have isentropic efficiencies of the order of 60 to 80 percent; large steam and gas turbines with isentropic efficiencies of the order of 96 percent have been built in recent years.

The designer of a power system uses an estimate of the turbine isentropic efficiency in a system analysis. This estimate is based on experience with typical turbines of comparable power rating. The turbine manufacturer has to meet this minimum efficiency specification as part of the contract, so the turbine engineers will make rather complex fluid-mechanical calculations in an attempt to predict the performance of trial designs. After the best design has been determined, a prototype is fabricated and tested to determine its operating characteristics, including its isentropic efficiency. If the efficiency is better than the original specified minimum, the power system designer will certainly set higher minimum specifications in the next power plant design. So goes engineering.

To illustrate the evaluation of the isentropic efficiency, let's suppose that test data on a steam turbine yield

inlet state	*outlet state*
$T_1 = 700°F$	$T_2 = 500°F$
$P_1 = 600$ psia	$P_2 = 200$ psia

From the equations-of-state information for H_2O (Appendix B), we have

$$h_1 = 1350.6 \text{ Btu/lbm} \qquad h_2 = 1268.8 \text{ Btu/lbm}$$

$$s_1 = 1.5874 \text{ Btu/(lbm·R)} \qquad s_2 = 1.6240 \text{ Btu/(lbm·R)}$$

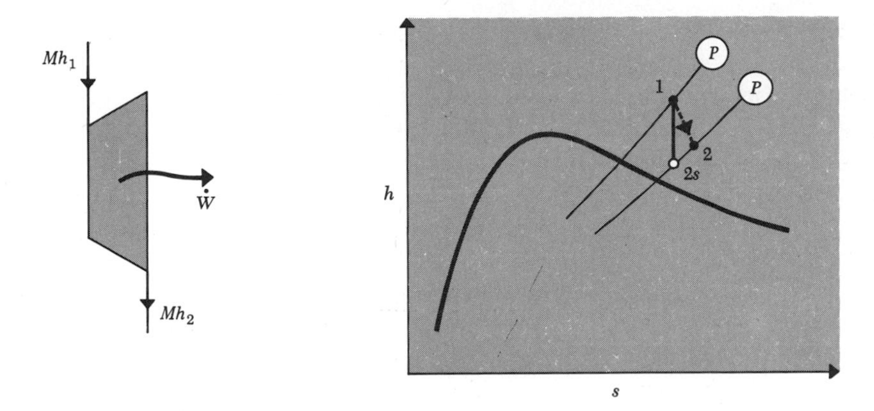

Figure 9·7. The isentropic efficiency can be determined from laboratory data

An energy balance on the turbine, assuming it is adiabatic and neglecting kinetic and potential energy changes, gives (Fig. 9·7),

$$\dot{W} = \dot{M}(h_1 - h_2)$$

so the work output per unit mass is

$$W \equiv \frac{\dot{W}}{\dot{M}} = h_1 - h_2$$

Hence, if the turbine process were isentropic, we would have

$$W_s = h_1 - h_{2s}$$

where state 2*s* is fixed by the *inlet* entropy and the *exit* pressure (see Fig. 9·7). With the use of Fig. B·2, the reference state 2*s* is

$$s_{2s} = s_1 = 1.5874 \text{ Btu/(lbm}\cdot\text{R)}$$

$$P_{2s} = P_2 = 200 \text{ psia}$$

$$h_{2s} = 1230 \text{ Btu/lbm}$$

$$T_{2s} = 430°\text{F}$$

Hence

$$W = 1351 - 1269 = 82 \text{ Btu/lbm}$$

$$W_s = 1351 - 1230 = 121 \text{ Btu/lbm}$$

$$\eta_s = \frac{W}{W_s} = \frac{82}{121} = 0.68$$

Examples illustrating the use of the isentropic efficiency in design calculations are included with the system analyses in this chapter.

Compressors and pumps are handled in a similar way. The isentropic efficiency of an *adiabatic* compressor, or pump, is defined as

$$\eta_s \equiv \frac{W_s}{W} \tag{9·7}$$

where again W denotes work per unit of mass flow, and W_s the work per unit of mass flow for an isentropic process. Since the work required to produce a given pressure rise in an actual adiabatic compressor is greater than that for an isentropic compressor, η_s is always† less than 1, as we showed from the second law in Sec. 7·15. The ideal and actual devices are considered to have the same inlet state and the same discharge pressure. Small centrifugal hydraulic pumps have efficiencies of the order of 40 to 60 percent, and larger pumps do somewhat better; the axial and centrifugal compressors employed in gas-turbine power plants and in jet engines have isentropic efficiencies in the range 75 to 85 percent. The fact that small gas-turbine power systems have become practical realities in recent years is largely a result of advances in technology which have produced these levels of compressor efficiency.

Isentropic efficiencies of *nozzles* are defined in terms of the discharge kinetic energy of the actual device compared to that for an ideal isentropic nozzle,

$$\eta_s \equiv \frac{(\mathsf{V}^2/2g_c)}{(\mathsf{V}^2/2g_c)_s} \tag{9·8}$$

The ideal and actual nozzles are considered to have the same inlet state and the same discharge pressure. Nozzle efficiencies are typically quite high (of the order of 90 to 95 percent); the main irreversibility is due to friction on the walls, and this is usually of minor importance, especially in very large nozzles.

Fluid flowing through a heat exchanger, nuclear reactor, or combustion chamber is retarded by friction. However, in well-designed systems the resulting pressure drop can be made quite small, and consequently the pressure is very nearly equal at every point in the flow.‡ The model process for fluid being heated in steady flow is therefore one of *constant pressure*. The analyst will often assume some pressure drop, based on experience or other calculations, in order to make a slightly better calculation. However, the constant-pressure model process is usually quite satisfactory for preliminary system studies.

The use of model processes and component efficiencies allows us to make thermodynamic calculations of system performance without knowledge of the

† If energy is transferred as heat from the actual compressor, η_s can exceed unity. Reversible adiabatic compression requires less work than irreversible adiabatic compression, but reversible isothermal compression requires even less work for a given pressure increase.

‡ For a constant-area duct.

details of the hardware construction. We can learn a great deal about a proposed system from the process representation for fluid which circulates through the device. Temperature-entropy diagrams are particularly useful for reasons already mentioned. The *Mollier diagram* (h-s plane) is also very useful; vertical distances on this diagram are proportional to the energy transfer as work per unit mass for an adiabatic compressor or turbine and to the energy transfer as heat for a steady-flow heater. Furthermore, the isentropic process conveniently appears as a vertical line, allowing the end state for an idealized adiabatic process to be found easily. In approaching a new system it is generally a good idea first to make a simple schematic flow diagram and then to sketch the process representation on an appropriate thermodynamic plane. Working out the process representation provides an understanding of how the system works and of any limitations imposed by the second law, and orients thinking as to how to proceed with first-law analysis. We shall illustrate these ideas in the sections to follow.

9·4 A PARTICULAR VAPOR POWER SYSTEM

Thermodynamics developed as a result of nineteenth-century work with reciprocating steam engines. Present-day applications of thermodynamics go well beyond this single area, but vapor power systems remain a major source of electric power and continue to grow in importance with the perfection of nuclear-reactor boilers.

Modern vapor power systems employ rotating rather than reciprocating machinery, for numerous practical reasons. The flow diagram and process representation for a simple *Rankine cycle* are shown in Fig. 9·8. We treat a somewhat idealized system for simplicity. Liquid is compressed by the pump and fed to the boiler. The boiler evaporates the fluid and delivers high-pressure vapor to the power-producing turbine. In the system shown, the turbine discharges to the atmosphere, and the only fluid we can possibly afford to throw away in this manner is water.

To get some feeling for vapor power systems and to illustrate the general method of system analysis, we shall now analyze this system. It will be convenient to work with energy transfer as heat and work *per unit of mass flow*, because the analysis can then be made without regard to the system power or the mass-flow rate. We assume the following:

Working fluid, water
State 1, liquid at 30°C and 1 atm
State 2, 0.80 MPa
State 3, saturated vapor at 0.8 MPa
State 4, 1 atm pressure
$\eta_s = 0.60$ for pump, $\eta_s = 0.80$ for turbine
Kinetic and potential energies negligible at states 1, 2, 3, and 4; pump and turbine adiabatic
Steady flow, steady state; water in thermodynamic equilibrium at 1, 2, 3, and 4

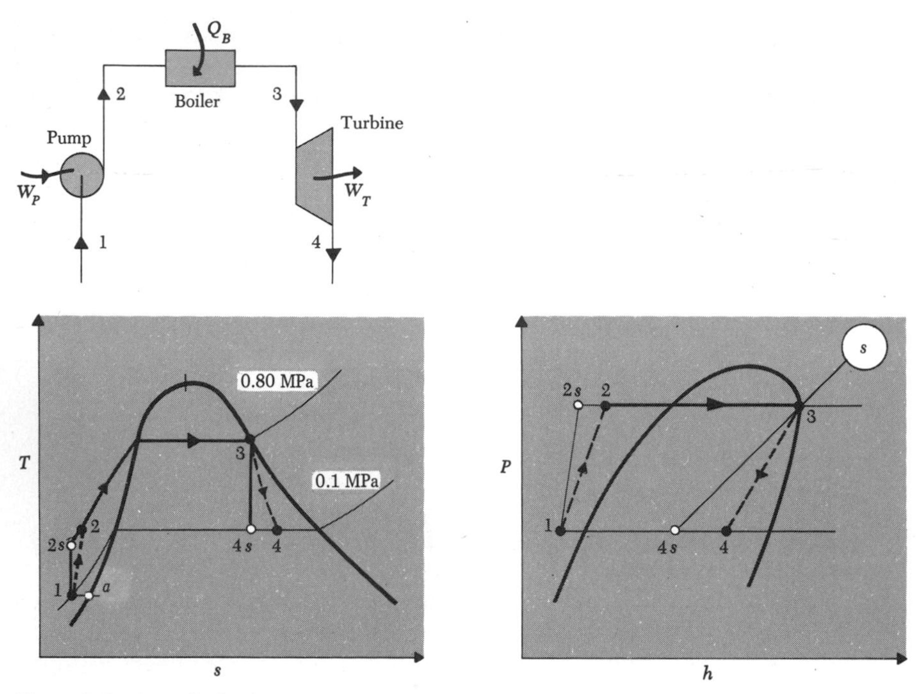

Figure 9·8. A simple Rankine vapor power system

An energy balance on the pump yields

$$W_P \equiv \dot{W}_P/\dot{M} = h_2 - h_1$$

where W_P is the shaft-work input for each unit of mass handled by the pump. Having no tabular or graphical equation-of-state information for subcooled liquid water, we shall treat it as an incompressible liquid.† Using Eq. (8·42), the W_P for an ideal adiabatic pump is (isentropic process)

$$W_{Ps} = v(P_2 - P_1)$$

From Table B·1, $v = 0.00100$ m^3/kg at 30°C.

$$W_{Ps} = 0.00100 \times (0.8 - 0.101) \times 10^6 = 699 \text{ J/kg}$$

The actual device then requires

$$W_P = \frac{W_s}{\eta_s} = \frac{699}{0.6} = 1170 \text{ J/kg} = 1.17 \text{ kJ/kg}$$

† For an incompressible liquid undergoing an isentropic process, $s_2 = s_1$, and therefore $T_{2s} = T_1$. Thus point 2s in Fig. 9·8 lies directly on top of point 1 in the incompressible-liquid model.

The value of h_1 may be found by employing the incompressible-liquid equation of state in conjunction with the tabulated thermodynamic properties using the reference state as shown in Fig. 9·8:

$$h_1 = h_a + v(P_1 - P_a) = 125.8 + 0.001 \times (0.101 - 0.004) \times 10^3 = 125.9 \text{ kJ/kg}$$

Therefore

$$h_2 = h_1 + W_P = 125.9 + 1.2 = 127.1 \text{ kJ/kg}$$

T_2 may be found from Eq. (8·42); for water, $c = 4.186$ kJ/(kg·K), so

$$T_2 = T_1 + \frac{(h_2 - h_1) - v(P_2 - P_1)}{c} = 30 + \frac{1.17 - 0.70}{4.186} = 30.1°\text{C}$$

Note the very small work requirement of the pump and the very slight temperature rise.

From Table B·1*b* we find $h_3 = 2769.1$ kJ/kg. An energy balance on the boiler then gives

$$Q_B \equiv \frac{\dot{Q}_B}{\dot{M}} = h_3 - h_2 = 2769.1 - 127.1 = 2642.0 \text{ kJ/kg}$$

where Q_B is the energy transfer as heat to the boiler per kg of water passing through.

An energy balance on the turbine yields

$$W_T \equiv \frac{\dot{W}_T}{\dot{M}} = h_3 - h_4$$

where W_T is the turbine work output per kg of fluid handled. To determine state 4 we must first fix state 4*s*. From Table B·1*b* we read

$$s_3 = 6.6636 \text{ kJ/(kg·K)}$$

and, from Table B·1*a*,

$$s_{f4} = 1.3071 \text{ kJ/(kg·K)} \qquad s_{fg4} = 6.0486 \text{ kJ/(kg·K)} \qquad \textit{at } 100°\textit{C or 1 atm}$$

Since $s_{4s} = s_3$ and $s_{4s} = s_{f4} + x_{4s}s_{fg4}$,

$$x_{4s} = \frac{6.6636 - 1.3071}{6.0486} = 0.886$$

Again in Table B·1*b* we find (at 1 atm or 100°C)

$$h_{f4} = 419.0 \text{ kJ/kg}$$

$$h_{fg4} = 2257.0 \text{ kJ/kg}$$

Thus $\quad h_{s4} = h_{f4} + x_{4s} h_{fg4} = 419.0 + 0.886 \times 2257.0 = 2418.7 \text{ kJ/kg}$

Consequently,

$$W_{Ts} = h_3 - h_{4s} = 2769.1 - 2418.7 = 350.4 \text{ kJ/kg}$$

Then $\quad W_T = W_{Ts}\eta_s = 350.4 \times 0.8 = 280.3 \text{ kJ/kg}$

$$h_4 = h_3 - W_T = 2769.1 - 280.3 = 2488.8 \text{ kJ/kg}$$

The discharge quality is then

$$x_4 = \frac{h_4 - h_{f4}}{h_{fg4}} = \frac{2488.8 - 419.0}{2257.0} = 0.917$$

Note that the turbine discharge temperature is 100°C, so that the steam thrown away is quite energetic.

To summarize, we have found

$$W_T = 280.3 \text{ kJ/kg}$$

$$W_P = 1.2 \text{ kJ/kg}$$

$$Q_B = 2642.0 \text{ kJ/kg}$$

Part of the turbine work is required to drive the pump. One of the attractive features of vapor power plants is that this is a small fraction of the turbine work output. In this particular case the bwr† is

$$\text{bwr} = 1.2/280 \simeq 0.004$$

The energy-conversion efficiency of this plant is

$$\eta = \frac{W_{\text{net}}}{Q_B} = \frac{280.3 - 1.2}{2642.0} = 0.106$$

Note that only 10 percent of the energy input in the boiler is converted to useful work. The balance appears as an increase in the energy of the working fluid. Denoting this increase by Q_C,

$$Q_C = h_4 - h_1 = 2488.8 - 125.9 = 2362.9 \text{ kJ/kg}$$

† Back work ratio.

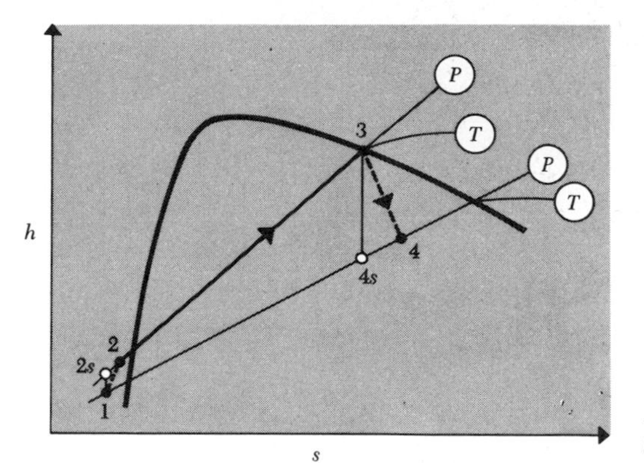

Figure 9·9. Process representation on the h-s plane

Since $P_1 = P_4$ in this example, we could close the cycle with the addition of a condenser. If states 1 and 2 were fixed, this additional device would not influence the efficiency of the system. The energy that would be transferred as heat from the condenser per unit mass of flow handled would be Q_C, or 2362.9 kJ/kg.

As a check on the analysis, an overall energy balance gives

$$Q_C = Q_B - W_T + W_P = 2642.0 - 280.3 + 1.2 = 2362.9$$

which checks.

The series of processes is also shown on an h-s plot in Fig. 9·9. This plot graphically shows the small bwr as the respective enthalpy changes for the pump and turbine processes.

It is instructive to compare the energy-conversion efficiency of this system to that of a Carnot cycle operating between the same temperatures ($T_3 = 170.4°\text{C}$) and $T_1 = 30°\text{C}$. For the Carnot cycle,

$$\eta = 1 - \frac{273 + 30}{273 + 170.4} = 0.317$$

Some of the difference can be attributed to irreversibilities in the turbine, and some is due to the fact that the temperatures at which energy is transferred as heat to and from the working fluid are more widely separated for the Carnot cycle (see Fig. 9·10). Note that the area enclosed by the vapor-cycle process path is not equal to the net work output per kg of fluid since the processes were not all reversible. The performance of this plant is dismally poor by modern standards. In the next sections we shall examine ways of obtaining better performance, making considerable use of process representations to show why improvements are obtained.

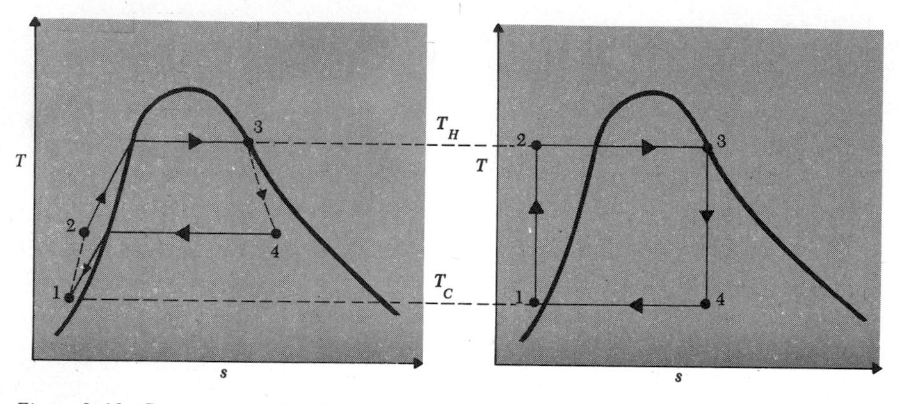

Figure 9·10. Comparison with a Carnot cycle

9·5 MODERN UTILIZATION OF THE RANKINE CYCLE

Two simple changes can provide considerable improvement in the energy-conversion efficiency of the system discussed in the previous section. Noting that the turbine work output would be enhanced if the turbine discharge pressure were lower, we could try to create a low-pressure region into which the turbine can discharge. This is accomplished with the addition of a condenser (Fig. 9·11), in which the steam can be condensed at a temperature closer to the environment temperature. For example, if river water is available at 15°C, it would not be difficult to condense the steam at 35°C by transferring energy as heat to the cooler river water. The steam side of the condenser would operate at a pressure of 5628 N/m^2 (Table B·1*a*), considerably less than atmospheric, and by inspection of the *h-s* diagram of Fig. 9·11 we see that the turbine work output per kg of steam could be nearly doubled. The condensed H_2O (the *condensate*) would have to be

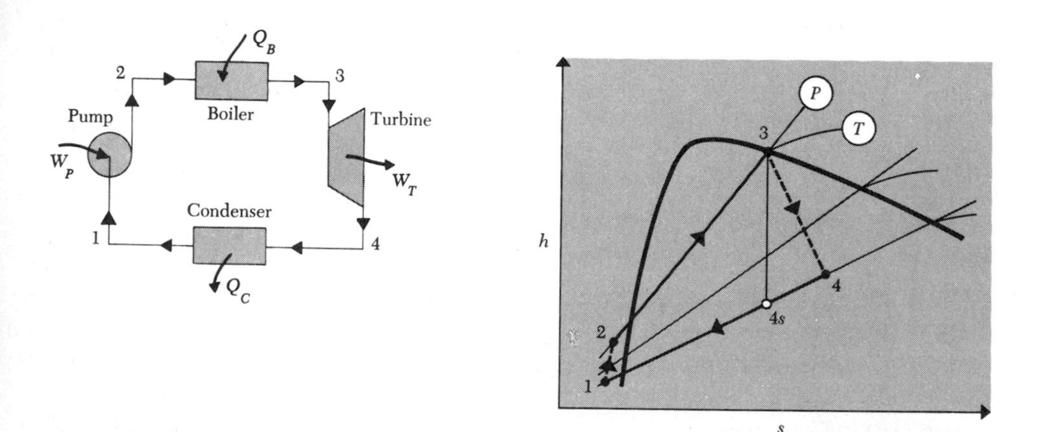

Figure 9·11. By condensing at a low pressure, the turbine work output can be markedly increased

pumped from this low pressure to the boiler inlet pressure, but as the work required to pressurize a liquid is very small compared to the work obtained in the expansion of a gas (the steam), a considerable net gain can be obtained.

Using a condenser permits continual circulation of the same working fluid, which means that purified water (which is less corrosive than tap water) can be used as the working fluid. This is a practical advantage of the *closed Rankine cycle*. A practical problem with condensing water at temperatures less than 100°C (212°F) is that the pressure inside the condenser is less than 1 atm. This means that any leaks in the condenser shell will permit air to contaminate the working fluid, making it difficult to maintain the low condensing pressure. Consequently, condensing systems are always equipped with vacuum pumps that initiate the low pressure when the system is started and remove small amounts of air which inevitably leak into the condenser. Steam-driven *jet pumps*, or jet ejectors, are often used. In these devices a small amount of high-pressure steam is bled from the main line and fed through a venturi-like device in which the steam accelerates until its pressure is reduced to somewhat below the condensing pressure. The air can then be sucked up by the "venturi action," and the steam then slowed down with a resulting pressure rise and then discharged to the atmosphere. Figure 9·12 shows a schematic of a typical steam condenser and the jet pumps used to maintain the low pressure against leakage from the atmosphere.

When the turbine operates over a large pressure range, the density variation within the turbine can be enormous. Such turbines are usually made in multiple stages, with each stage operating over a more modest pressure range (say, 2 : 1). Sometimes the stages are separate machines on the same shaft, and sometimes the stages are all incorporated into a single large machine. Because of the wide density variation, the high-pressure stages are much smaller than the low-pressure stages, though the power outputs from the various stages are typically about the same. Modern power plants have last-stage turbine blades (buckets) as long as 4 ft and produce over 1000 MW from a single unit.

If the quality of the steam in the turbine drops too low, the liquid droplets can impact upon the blades, causing serious damage. Hence it is usually the practice to maintain at least 90 percent quality ($x = 0.9$) in the low-pressure stage, and this places a lower limit on the condenser operating pressure. Cycle

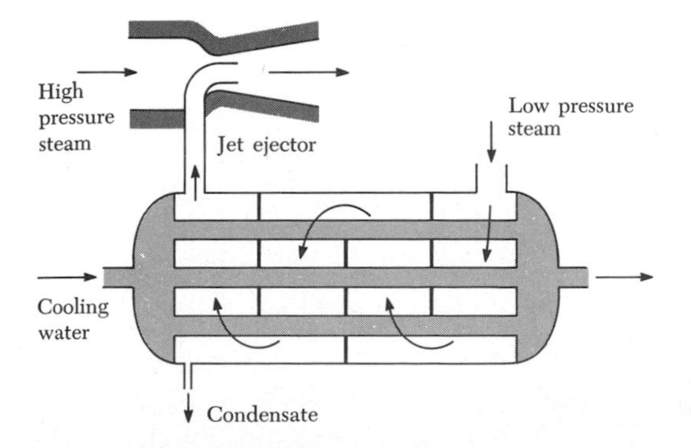

Figure 9·12. Schematic of a steam condenser with jet ejector

modifications, which we are about to discuss (superheat and reheat), help solve this practical problem and permit turbine operation over wider pressure ranges.

We can think about the improvement offered by condensing in a somewhat different way. The open Rankine cycle (Fig. 9·8) is essentially closed by the atmosphere, and energy is rejected from the H_2O as heat at the condensing temperature of 100°C. By adding the condenser we permit the energy to be rejected from the closed Rankine system, which is a heat engine, at lower temperatures. Our discussions of the implications of the second law, and in particular of the Carnot cycle, indicate that improved efficiencies are obtained by rejecting energy as heat at as low a temperature as possible, and by adding energy as heat at as high a temperature as possible. The condenser in effect permits heat rejection at a lower temperature.

The energy addition as heat can also be put in at a higher (average) temperature by operating the boiler at a higher pressure. This will bring some practical problems of the sort discussed above, particularly in regard to the quality of the turbine steam. But we are not limited to saturated vapor at the boiler discharge. We can add further energy as heat to the steam, either in the boiler or in a separate heat exchanger (a *superheater*), and raise the temperature of the steam still further (at constant pressure in the model process). Figure 9·13 shows the *h-s* process representations for three turbines operating at the same discharge pressure. Turbine *A* operates over a small pressure range with saturated vapor at the inlet. *B* operates over a larger pressure range, also with saturated vapor at the inlet, and we see that the discharge quality is considerably lower in *B* than in *A*. However, the work output per lbm of fluid $(h_3 - h_4)$ is greater in *B* than in *A*, so that while *B* is practically "worse" it is in principle "better" than *A*. Turbine *C* operates with superheated steam at the inlet and is better than either *B* or *A*, for not only is the work output per lbm better but the turbine discharge steam can be maintained at a much higher quality. Hence, in modern power plants, turbine operating conditions are almost always like *C*.

Substantial gains in performance can be obtained by the simple modifications discussed above. To illustrate, let's analyze the system of Fig. 9·14.

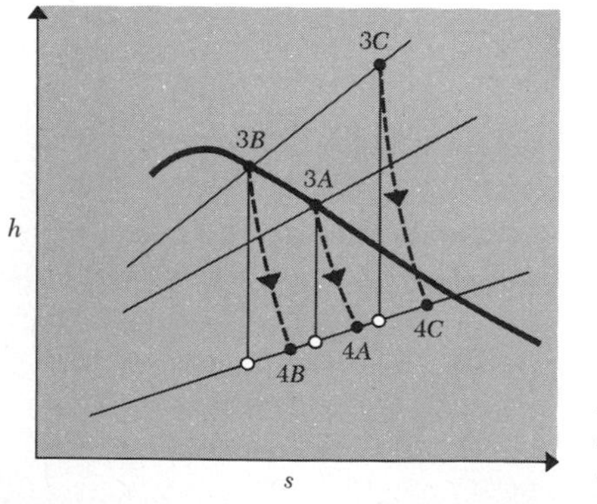

Figure 9·13. Note that superheating increases the turbine work output and reduces problems caused by low quality

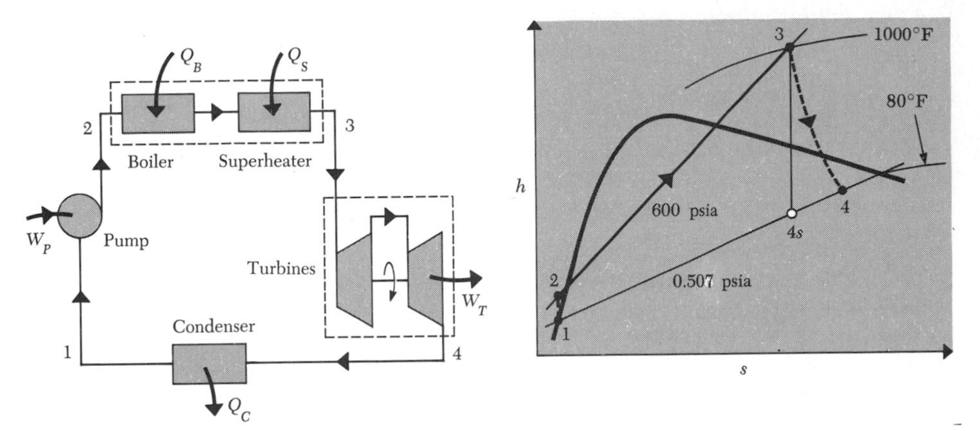

Figure 9·14. The model power system

Note that we have set the turbine inlet state in the superheat region at 600 psia and 1000°F, and the condensing temperature at 80°F; these temperature values are typical of operating power plants. Newer plants now go to 3800 psia and higher.

The idealizations we shall make in the analysis are that the pump and turbines are adiabatic devices, that the heating and cooling processes take place at constant pressure, and that the flow stream kinetic and potential energy changes are negligible. We shall also assume that the isentropic efficiencies of the pump and turbine are 0.8 and 0.9, respectively, typical values for reasonably large present-day machines.

By these assumptions, the pump inlet is saturated liquid at a pressure of 0.507 psia (Table B·1*a*), and its discharge pressure is 600 psia. An energy balance on the pump yields for the pump work per lbm of liquid flow

$$W_P = h_2 - h_1$$

State 1 is fixed, and we shall establish state 2 and W_P, treating the liquid as incompressible. From Table B·1*a*,

$$h_1 = 48.1 \text{ Btu/lbm}$$

Using the incompressible-liquid equation of state,

$$h_{2s} = h_1 + v(P_2 - P_1) = 48.1 + 0.01607 \times (600 - 0.5)\tfrac{144}{778}$$
$$= 49.9 \text{ Btu/lbm}$$

Hence, were the pump isentropic,

$$W_{Ps} = 49.9 - 48.1 = 1.8 \text{ Btu/lbm}$$

For our actual pump,

$$W_P = \frac{W_{Ps}}{\eta_s} = \frac{1.8}{0.8} = 2.2 \text{ Btu/lbm}$$

This establishes the boiler inlet enthalpy as

$$h_2 = h_1 + W_P = 48.1 + 2.2 = 50.3 \text{ Btu/lbm}$$

The superheater discharge enthalpy is fixed by the specification of the turbine inlet state. From Table B·2

$$h_3 = 1517.8 \text{ Btu/lbm} \qquad s_3 = 1.7157 \text{ Btu/(lbm·R)}$$

An energy balance on the boiler-superheater combination gives

$$Q_{\text{in}} \equiv Q_B + Q_s = h_3 - h_2$$

where Q_B and Q_S are the boiler and superheater heat inputs per lbm of flow. Hence

$$Q_{\text{in}} = 1517.8 - 50.3 = 1467.5 \text{ Btu/lbm}$$

The student may make appropriate assumptions about the boiler discharge state and calculate the amounts of Q_B and Q_S separately.

Now, an energy balance on all the turbine stages gives

$$W_T = h_3 - h_4$$

where W_T is the total turbine work output per lbm of steam. To establish state 4 we first must establish the reference state 4*s*. This state is fixed by its entropy and pressure. Looking at the *h*-*s* diagram, Fig. B·2, we see that state 4*s* lies in the vapor dome. Table B·1*a* confirms this, for the saturated-vapor entropy at 80°F is 2.0358, which is indeed greater than s_3. We can use the value of $s_{4s} = s_3$ to calculate the quality at state 4*s*, and then use this quality to calculate h_{4s}. We have

$$s_{4s} = [(1 - x)s_f + xs_g]_{4s}$$

so

$$x_{4s} = \frac{s_{4s} - s_{f4s}}{s_{fg4s}} = \frac{1.7157 - 0.0933}{1.9425} = 0.835$$

Then

$$h_{4s} = [(1 - x)h_f + xh_g]_{4s}$$

$$= 0.165 \times 48.1 + 0.835 \times 1096.4 = 923 \text{ Btu/lbm}$$

Note that we could have read this value from Fig. B·2, though not to the same accuracy. The work output from an isentropic turbine is therefore

$$W_{Ts} = 1518 - 923 = 595 \text{ Btu/lbm}$$

The work output from our actual device is therefore

$$W_T = W_{Ts}\eta_s = 595 \times 0.9 = 535 \text{ Btu/lbm}$$

The actual discharge enthalpy is therefore

$$h_4 = h_3 - W_T = 1518 - 535 = 983 \text{ Btu/lbm}$$

This enthalpy, together with the known pressure P_4, fixes state 4. From Fig. B·2 we see that the turbine discharge quality is about 0.90, which is probably acceptable.

Finally, the condenser heat transfer can be calculated from an energy balance on the condenser. This is left to the student as an exercise.

We now can calculate the plant performance. The net work output per lbm of fluid circulated is

$$W_{\text{net}} = W_T - W_P = 535 - 2.2 = 533 \text{ Btu/lbm}$$

Note that again a rather small amount of the turbine power is required by the pump. The plant energy-conversion efficiency is then

$$\eta = \frac{W_{\text{net}}}{Q_{\text{in}}} = \frac{533}{1467} = 0.363$$

This efficiency of 36.3 percent is a considerable improvement over the 10 percent of the previous example, and would reduce the fuel consumption and cut the cost of power by about a factor of 4. This efficiency is typical of a relatively small central power station, and efficiencies of the order of 40 percent can be obtained with further cycle modification to be discussed shortly.

The energy supplied as heat per unit of net work output is termed the *heat rate* by some power plant engineers. Since the net work is usually expressed in kW·h and the heat input in Btu,

$$\text{Heat rate} = [3413 \text{ Btu/(kW}\cdot\text{h)}]/\eta$$

Thus a plant operating at 36 percent efficiency has a heat rate of 9480 Btu/(kW·h). When the world thinks only in SI, the heat rate will be simply $1/\eta$ or, here, 2.78 J/J. However, if you tell a present-day power plant operator that your design gives a heat rate of 2.78, he (or she) will probably not understand what you mean.

The designer of a power plant carries out the calculations described above, which we note were on a per lbm basis. After these are completed, the mass-flow rate can be determined by the design power requirements for this system. For example, if 1000 MW is the electric power output required,

$$\dot{W}_{\text{elect}} = 10^9 \times 3.41 = 3.41 \times 10^9 \text{ Btu/h}$$

If the efficiency of the electric generator is, say, 0.98, then the plant shaft power will have to be

$$\dot{W} = 3.41 \times \frac{10^9}{0.98} = 3.5 \times 10^9 \text{ Btu/h}$$

Since the net turbine shaft work is 533 Btu/lbm, the flow requirement is

$$\dot{M} = \frac{\dot{W}}{W_{\text{net}}} = 3.5 \times \frac{10^9}{533} = 6.56 \times 10^6 \text{ lbm/h}$$

Hence the energy transfer rates for the several components are

$$\dot{W}_P = \dot{M} W_P = 6.56 \times 10^6 \times 2.2 = 14.4 \times 10^6 \text{ Btu/h} = 5.66 \times 10^3 \text{ hp}$$

$$\dot{W}_T = \dot{M} W_T = 6.56 \times 10^6 \times 535 = 3.50 \times 10^9 \text{ Btu/h} = 1.37 \times 10^6 \text{ hp}$$

$$\dot{Q}_{\text{in}} = \dot{M} Q_{\text{in}} = 6.56 \times 10^6 \times 1467 = 9.6 \times 10^9 \text{ Btu/h}$$

The power plant designer then sends out for bids on the components. The specifications might be as follows:

Pump
Inlet, 0.5 psia, saturated liquid
Discharge to 600 psia
Minimum isentropic efficiency, 0.8
Design flow rate, 6.56×10^6 lbm/h

Turbine
Inlet, 1000°F, 600 psia
Discharge to 0.5 psia
Minimum isentropic efficiency, 0.9
Design flow rate, 6.56×10^6 lbm/h at 1800 rpm

Boiler-superheater
Operating pressure, 600 psia
Inlet at 80°F, discharge superheated steam at 1000°F
Design flow, 6.56×10^6 lbm/h (nominal duty 10^{10} Btu/h)

It is then up to the manufacturers of the components to design and deliver components meeting these specifications. The general plant designer coordinates these designs to ensure that they fit into the space available in a reasonable and economic way. Specifications would probably include piping sizes, component maximum sizes, weights, and so on, estimated on the basis of previous experience with similar systems with the help of appropriate engineering calculations.

9·6 OTHER RANKINE CYCLE MODIFICATIONS

A Rankine cycle modification normally employed in large central power stations is *reheat* (Fig. 9·15). The steam discharging from the first-stage turbine is reheated before being fed to the second-stage turbine and, as a study of the *h-s* diagram reveals, this provides the second-stage turbine with a larger possible enthalpy change. While the work output per lbm will be enhanced, the efficiency may be reduced or increased, depending on the reheat temperature range. To illustrate, suppose we reheat between the turbine stages in the example of the previous section, with a reheater pressure of 60 psia. We will reheat to 1000°F and assume that the isentropic efficiencies of both turbine stages are 0.90. By appropriate thermodynamic analysis as illustrated previously we can establish the intermediate states as follows (see Fig. 9·15):

$$h_3 = 1518 \text{ Btu/lbm} \qquad h_4 = 1265 \text{ Btu/lbm}$$

$$h_5 = 1533 \text{ Btu/lbm} \qquad h_6 = 1110 \text{ Btu/lbm}$$

The work outputs per unit mass for the two turbine stages are then

$$W_1 = h_3 - h_4 = 1518 - 1265 = 253 \text{ Btu/lbm}$$

$$W_2 = h_5 - h_6 = 1533 - 1110 = 423 \text{ Btu/lbm}$$

and the reheater energy input per unit mass is

$$Q_R = h_5 - h_4 = 1533 - 1265 = 268 \text{ Btu/lbm}$$

The boiler heat input and pump work are as before. The net work output and heat input per unit mass are therefore

$$W_{\text{net}} = 253 + 423 - 2.2 = 674 \text{ Btu/lbm}$$

$$Q_{\text{in}} = 1467 + 268 = 1735 \text{ Btu/lbm}$$

Note that the work output has increased from 535 Btu/lbm to 674 Btu/lbm. The cycle efficiency is now

$$\eta = \frac{W_{\text{net}}}{Q_{\text{in}}} = \frac{674}{1735} = 0.388$$

The increase of 2.5 percentage points in efficiency amounts to nearly a 10 percent increase in efficiency, hence nearly a 10 percent reduction in fuel costs for a given power output. In a small power station this gain might not offset the additional capital investment in system hardware, maintenance, and other items, but in a large power station the gain would most certainly be worth the investment. Large

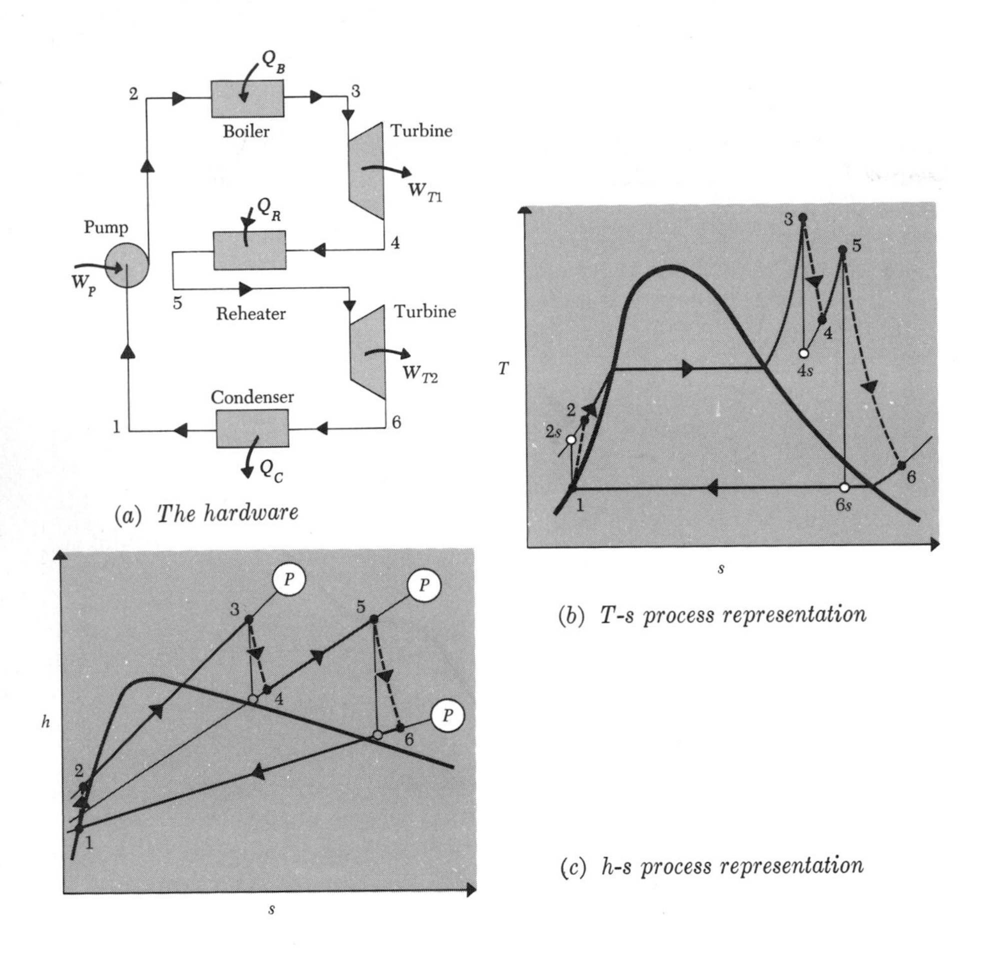

(a) *The hardware*

(b) *T-s process representation*

(c) *h-s process representation*

Figure 9·15. Reheat in the Rankine cycle

modern power plants normally employ one or two stages of reheat, each of which inches the efficiency upward by a small but not insignificant amount.

Another modification normally employed in large power stations is *extraction and regeneration*, or *feedwater heating* (Fig. 9·16). Part of the flow emerging from the first-stage turbine is bled off and condensed. The energy transferred from this fluid is added as heat to the low-temperature liquid emerging from the main pump. The only energy that must be added as heat from *outside* the system is that required to take the total stream from state 5 to state 6, and a considerable improvement in efficiency can be obtained.

Two basic types of regenerative heaters are used. Figure 9·16 shows a *closed* feedwater heater (the regenerator). Figure 9·17 shows an *open* feedwater heater, in which the extracted steam is mixed directly with liquid from the low-

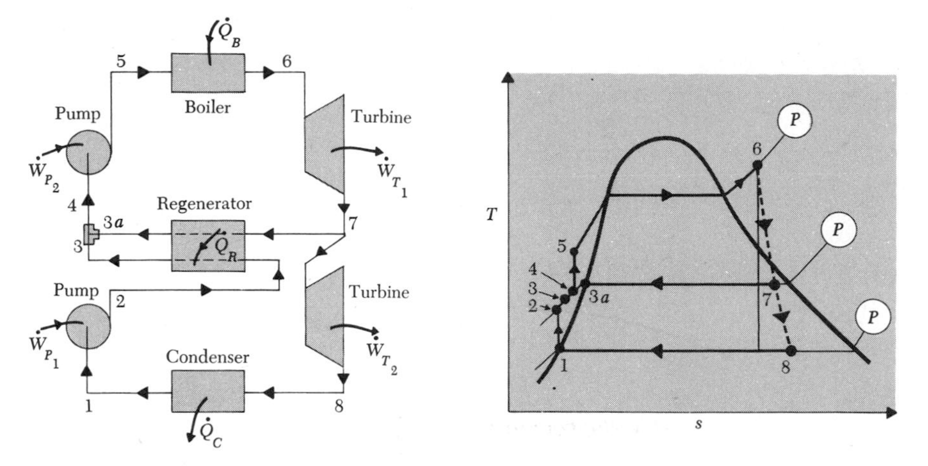

Figure 9·16. Regenerative Rankine cycle with a closed feedwater heater

pressure condenser. To illustrate the effects of regeneration, let's add an open feedwater heater to the Rankine cycle (with reheat) analyzed above. The pressure between turbine stages will again be taken as 60 psia, and we shall extract just enough steam at the reheat pressure to heat the water emerging from the first pump to the saturation line. The pressure drop in the feedwater heater will be neglected. Both pumps have an isentropic efficiency of 0.8. With these assumptions, we have enough information to fix the states indicated in Fig. 9·17, by appropriate thermodynamic analysis, and these are tabulated below. You should verify that these are correct.

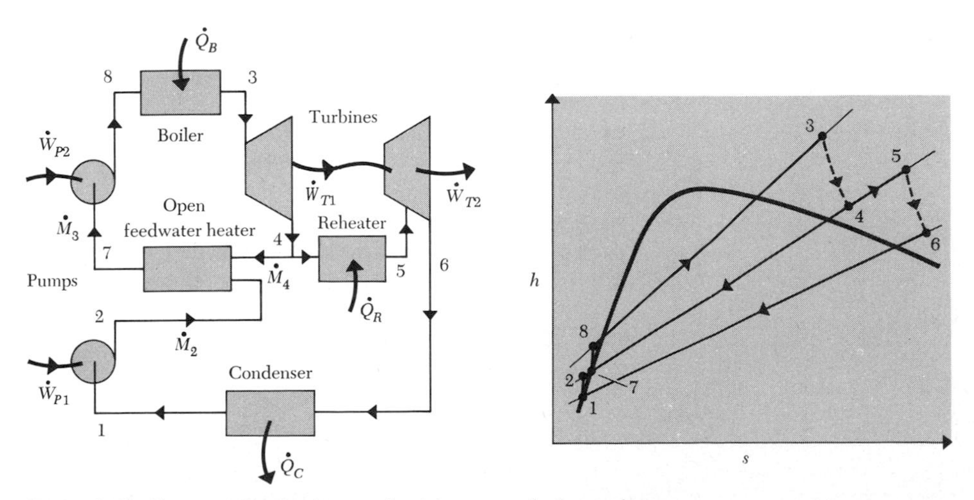

Figure 9·17. Regenerative Rankine cycle with an open feedwater heater

State	P, psia	T, °F	h, Btu/lbm
1	0.5	80	48.1
2	60	80	48.3
3	600	1000	1518
4	60	475	1265
5	60	1000	1533
6	0.5	115	1110
7	60	293	262
8	600	292.5	264

We next calculate the *extraction fraction* $\dot{M}_4/\dot{M}_3$ from mass and energy balances on the open regenerator (what idealizations do the equations below imply?)

$$\dot{M}_2 h_2 + \dot{M}_4 h_4 = \dot{M}_3 h_7$$

$$\dot{M}_2 + \dot{M}_4 = \dot{M}_3$$

Combining,

$$\frac{\dot{M}_4}{\dot{M}_3} = \frac{h_7 - h_2}{h_4 - h_2} = \frac{214}{1217} = 0.176$$

Then

$$\frac{\dot{M}_2}{\dot{M}_3} = 1 - 0.176 = 0.824$$

Energy balances on the components then allow us to express their energy transfers per lbm of *total* flow,

$$\frac{\dot{W}_{T1}}{\dot{M}_3} = h_3 - h_4 = 253 \text{ Btu/lbm}$$

$$\frac{\dot{W}_{T2}}{\dot{M}_3} = \left(\frac{\dot{M}_2}{\dot{M}_3}\right)(h_5 - h_6) = 0.824 \times 423 = 349 \text{ Btu/lbm}$$

$$\frac{\dot{W}_{P1}}{\dot{M}_3} = \left(\frac{\dot{M}_2}{\dot{M}_3}\right)(h_2 - h_1) = 0.824 \times 0.2 = 0.16 \text{ Btu/lbm}$$

$$\frac{\dot{W}_{P2}}{\dot{M}_3} = h_8 - h_7 = 2 \text{ Btu/lbm}$$

$$\frac{\dot{Q}_B}{\dot{M}_3} = h_3 - h_8 = 1254 \text{ Btu/lbm}$$

$$\frac{\dot{Q}_R}{\dot{M}_3} = \left(\frac{M_2}{M_3}\right)(h_5 - h_4) = 0.824 \times 268 = 221 \text{ Btu/lbm}$$

The net work output and heat input, per lbm of total flow, are then

$$\frac{\dot{W}_{net}}{\dot{M}_3} = 253 + 349 - 0.16 - 2 = 600 \text{ Btu/lbm}$$

$$\frac{\dot{Q}_{in}}{\dot{M}_3} = 1254 + 221 = 1475 \text{ Btu/lbm}$$

Note that these have both been reduced by the extraction process. The system energy-conversion efficiency is

$$\eta = \frac{\dot{W}_{net}}{\dot{Q}_{in}} = \frac{600}{1475} = 0.408$$

So, from the total performance point of view, the extraction and regeneration have improved the system. The student should particularly note the need for careful consideration of the mass-flow rate variations in analyzing systems in which the flow rate is not the same everywhere in the system.

In theory, if an infinite number of extraction and regeneration stages could be employed, the external addition of energy as heat would take place at one temperature and the external removal of energy as heat would take place isothermally in the low-pressure condenser. The performance of this limiting system is then that of a reversible $2T$ engine, that is, of a Carnot cycle. This is very interesting, for the process representation does not look much like that of the Carnot cycle (see Fig. 9·18). In large central power stations employing reheat, regeneration, and extraction, efficiencies of the order of 35 to 40 percent have been obtained by using several (four to seven) stages of regeneration. Extensive optimization calculations are normally part of the design of any such system.

Another way of obtaining high efficiency is with the *supercritical vapor cycle* (Fig. 9·19). Here the fluid is heated at a pressure in excess of the critical pressure, hence is transformed continuously from a liquid to a gas. With water the critical pressure is about 3200 psia, so a supercritical steam power plant is indeed a very high-pressure system. The two newest and largest units at the Moss Landing power station (Fig. 1·1) are supercritical systems.

In addition to the ideas discussed above, there are energy management techniques outside the steam cycle, which can improve the effectiveness of energy use. The boiler itself is not 100 percent efficient in transferring the energy released in the combustion process to the steam side of the boiler; between 5 and 10 percent of the energy released goes up the stack. Designers are able to use some of this, however, in the *economizer* and *air preheaters*. In the economizer section, feedwater to the boiler is passed through a heat exchanger in which energy from the hot stack gases is used to preheat the boiler feedwater before boiling is initiated. (Why does this improve the cycle efficiency?) The stack gases then pass through a large heat exchanger, often of the rotary type, where energy is trans-

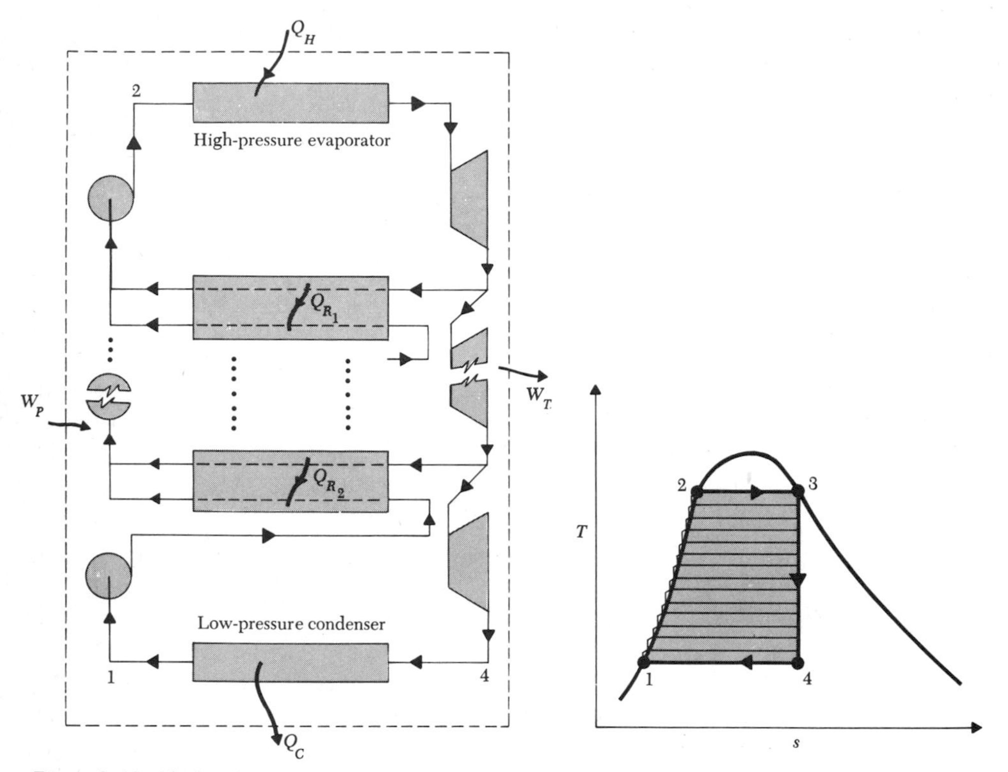

Figure 9·18. Idealized vapor cycle with an infinite number of reheat and regeneration stages

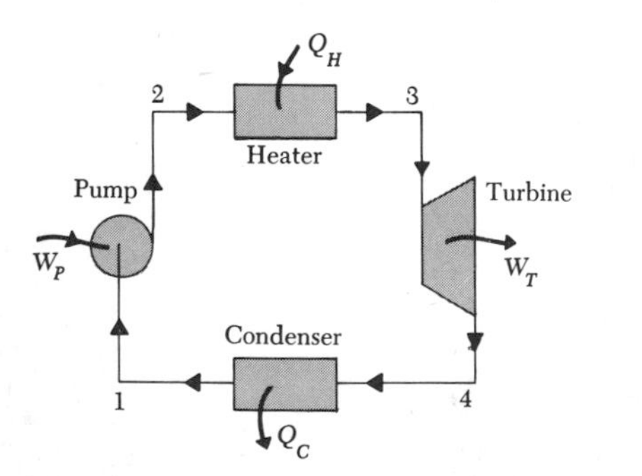

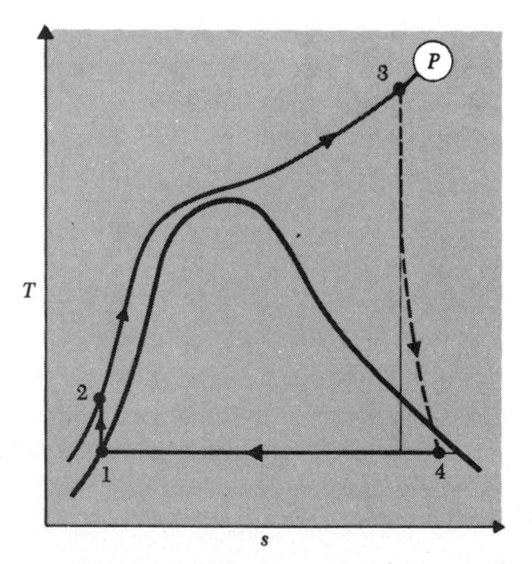

Figure 9·19. The supercritical vapor cycle

ferred to the incoming air entering the boiler. In this manner the stack gas is cooled to about 150°C (300°F) before being released. Because the combustion products contain corrosive gases, notably sulfur dioxide which in the presence of water is converted to sulfuric acid, the temperature is maintained at a high enough level to prevent the water vapor from condensing.

The rotary heat exchanger consists of a metal matrix which rotates slowly about an axis parallel to the gas flow. Hot gases pass through one side to heat the metal portion which then rotates over to the other side where the entering cold air is heated by the warm metal.

In coal-fired power plants hot air is required for drying the coal. The air can be heated via a heat exchanger using the hot stack gases as the energy source. By use of such techniques, utilities attempt to make use of as much of the available energy in the power plant as possible.

We shall close the section on the Rankine cycle with a brief description of the operating conditions for one of the units of an actual modern fossil fuel-fired power plant, the Navaho plant, located near the Grand Canyon at Page, Arizona. This plant is coal-fired with three turbine units each rated at 750 MW. Extraction for six stages of feedwater heating is provided. At the outlet of the superheater the steam is at a pressure of 3590 psig (supercritical) and a temperature of 1005°F. A single stage of reheat to 1002°F at 646 psig is used with a takeoff from the first turbine at 583°F and 676 psig. The pressure drop through the reheater is 30 psi. The steam-flow rate in the superheater is 5.4×10^6 lbm/h, and in the reheater section is 4.85×10^6 lbm/h.

The feedwater system is used to supply the steam generator with heated, chemically treated, deaerated water. Three booster pumps, each driven by a 1750-hp motor, draw feedwater to the actual boiler feed pumps; these are steam turbine-driven centrifugal pumps delivering 415°F feedwater at 3850 psia.

Plant power requirements include power for the soot-blower air compressor rated at 4500 hp, two 2000-hp motors to drive the primary air fans, two 3000-hp motors to drive the forced-draft fans, and four 5500-hp motors to drive the induced-draft fans which push the exhaust up the 755-ft stack. The circulating water pumps on the main steam condenser are driven by two 3000-hp motors. There are numerous other requirements for power in the coal supply system, water supply system, etc. About 7 percent of the total power generated must be used at the site to run the plant.

The preliminary engineering for this plant began in mid-1968 and construction began in April 1970. The first 750-MW unit went on line in June 1974. Other units followed at 1-year intervals.

9·7 APPLICATIONS TO NUCLEAR POWER SYSTEMS

From a thermodynamic point of view the difference between nuclear power and conventional power is that the fossil fuel used as the energy supply for the boiler is replaced by nuclear fuel in the reactor. In a *pressurized-water reactor* (PWR)

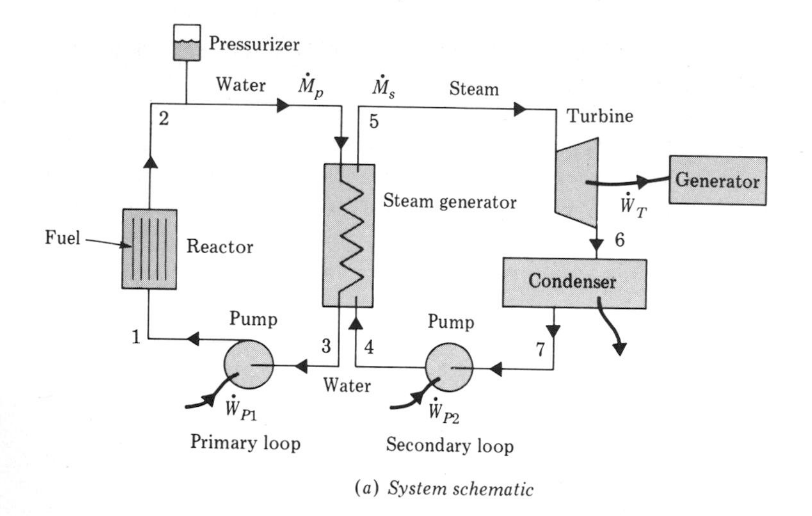

(*a*) *System schematic*

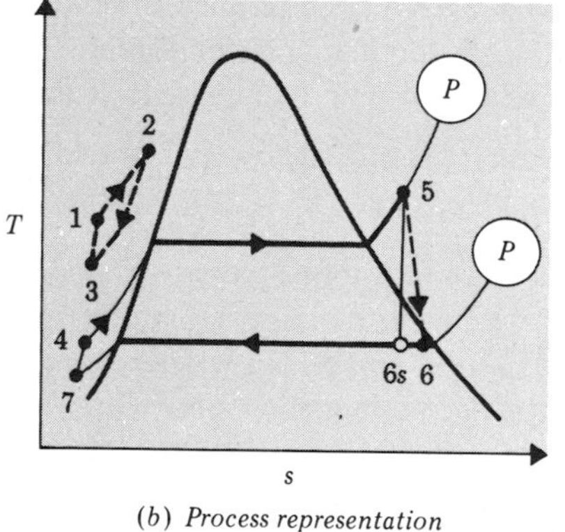

(*b*) *Process representation*

Figure 9·20. Pressurized-water reactor system

boiling does not occur, and the pressure must be above the saturation pressure at the reactor exit temperature. To power the steam turbine, a secondary water-steam loop is required with a steam generator. Figure 9·20 shows a schematic of a PWR system. In addition to the steam generator, which is simply a heat exchanger with boiling occurring on the steam side, there is a surge tank, noted in the figure as the pressurizer. This tank contains steam in the top section, and liquid water in the bottom, and is used to control the primary-loop pressure.

In a *boiling-water reactor* (BWR) boiling is allowed in the reactor core, and a steam separator is used to separate the saturated steam from the liquid water. Such systems have schematics very much like Fig. 9·11. Since boiling is desired, a lower pressure is used in the reactor, and typically pressures are about half those required in a PWR. A high-temperature, gas-cooled reactor (HTGR) uses helium as the reactor coolant in a primary loop. Hot helium (760°C) goes through steam generators to create superheated steam for the turbine.

PWR example. As another example of thermodynamic analysis, and to get a feel for the numbers, let's make some calculations on a representative PWR system. The conditions below are typical; we will use a primary-loop flow rate of 17,500 kg/s.

State	T, °C	P, atm
1	290	150
2	320	143
3		136
4		65
5	300	61.2
6		0.08
7	40	0.08

Component	η_s
primary-loop pump	0.80
secondary-loop pump	0.75
turbine	0.85

We again calculate liquid enthalpies using the incompressible-liquid equation of state and the saturation tables, as discussed in Sec. 8·5:

$$h(T, P) = h_f(T) + v_f(T)[P\text{-}P_{sat}(T)]$$

This produces

$$h_1 = 1300 \text{ kJ/kg} \qquad h_2 = 1466 \text{ kJ/kg} \qquad h_7 = 168 \text{ kJ/kg}$$

For the primary-loop pump,

$$W_{P1} = h_1 - h_3 = \frac{h_{1s} - h_3}{\eta_s} = \frac{v(P_1 - P_3)}{\eta_s}$$
$$= 0.0014 \times (150 - 136) \times 0.1013 \times 10^6/0.8 = 2482 \text{ J/kg}$$

So, the primary-loop pump power requirement is

$$\dot{W}_{P1} = \dot{M}_p W_{P1} = 17{,}500 \times 2482 = 43.4 \text{ MW}$$

This large power requirement is due to the enormous flow rate required and the pressure drop caused by flow friction.

We now know that $h_3 = h_1 - W_p = 1300 - 2.5 = 1297.5$ kJ/kg. The reactor power is

$$\dot{Q}_R = \dot{M}_p(h_2 - h_1) = 17{,}500 \times (1466 - 1300) \times 10^3 = 2905 \text{ MW}$$

This is called the *thermal power rating* of the system.

We can now analyze the conditions in the secondary loop starting with the steam generator. From an energy balance on the steam generator,

$$\dot{M}_p(h_2 - h_3) = \dot{M}_s(h_5 - h_4)$$

Interpolating in Table B·1*b*, we find $h_5 = 2874$ kJ/kg. From an energy balance on the secondary-loop pump,

$$W_{P2} = h_4 - h_7 = \frac{h_{4s} - h_7}{\eta_s}$$

Using the incompressible-liquid equation of state and approximating the liquid density as that at 40°C,

$$W_{P2} = \frac{v(P_4 - P_7)}{\eta_s} = 0.001 \times (65 - 0.08) \times 0.1013 \times 10^6/0.75$$

$$= 8.77 \text{ kJ/kg}$$

Hence $$h_4 = h_7 + W_{P2} = 168 + 9 = 177 \text{ kJ/kg}$$

Now we can calculate $\dot{M}_s$ from the steam-generator energy balance:

$$\dot{M}_s = \frac{\dot{M}_p(h_2 - h_3)}{h_5 - h_4} = 17{,}500 \times \frac{168.5}{2697} = 1093 \text{ kg/s}$$

Note that the primary-loop mass flow rate is *16 times* that in the secondary loop.

Having fixed states 4, 5, and 7, we can now determine the turbine power output. From an energy balance on the turbine,

$$W_T = h_5 - h_6 = (h_5 - h_{6s}) \times \eta_s$$

State 6*s* is fixed by $s_{6s} = s_5$ and $P_{6s} = P_6 = 0.08$ atm. One finds that 6*s* is in the vapor dome with a quality of 0.714, giving $h_{6s} = 1891$ kJ/kg. This produces $W_T = (2874 - 1891) \times 0.85 = 835$ kJ/kg. So, the turbine power output is

$$\dot{W}_T = \dot{M}_s W_T = 1093 \times 835{,}000 = 913 \text{ MW}$$

The secondary-loop pump requires

$$\dot{W}_{P2} = \dot{M}_s(h_4 - h_7) = 1093 \times (177 - 168) = 9.8 \text{ MW}$$

So, the *net* power output from this plant (assuming 100 percent efficient electromechanical coupling of the turbine output and pump drive systems) is

$$\dot{W}_{\text{net}} = 913 - 9.8 - 43.4 = 859.8 \text{ MW}$$

The plant thermal efficiency is

$$\eta = \frac{\dot{W}_{\text{net}}}{Q_R} = \frac{859.8}{2905} = 0.295$$

This design would cause the turbine outlet condition to be far too wet for satisfactory operation. An actual PWR calls for reheat stages between the high-pressure, intermediate, and low-pressure sections of the turbine. In this manner the exit state reaches an acceptable quality. In addition, several stages of boiler feedwater heating with extraction are used to increase the overall thermal efficiency to the 31 to 33 percent range.

9·8 SOME THOUGHTS ON THE WORKING FLUID

We have used water as the working fluid in the vapor power cycle examples in this chapter. Indeed, water is the most common fluid in large central power stations, though by no means is it the only fluid used in vapor power systems. What are desirable properties for the working fluid? In view of our previous discussions, these certainly seem to be important factors:

1. *High critical temperature*—to permit evaporation at a high temperature
2. *Low saturation pressures at the maximum temperatures*—to minimize the pressure vessel and piping costs
3. *Pressure just above 1 atm at condensing temperature*—to eliminate air leakage problems
4. *Rapidly diverging pressure lines on the h-s diagram*—to minimize the bwr and to make reheat modifications most effective
5. *Large enthalpy of evaporation*—to minimize the mass-flow rate for given power output
6. *No degrading aspects*—noncorrosive, nonclogging
7. *No hazardous features*—nontoxic, inflammable
8. *Low cost and ready availability*

It is evident that water is rather good only in regard to items 4, 5, 7, and 8 above. There are many other fluids that are better on the other items, as examination of the data in Appendix B will reveal. But, on an overall basis for ground-based power plants, water remains a predominant choice, hence steam power engineering remains an interesting and important area of applied thermodynamics.

Vapor power cycles for use in space generally use other working fluids, such as mercury and NaK, a sodium-potassium mixture. These fluids have reasonable vapor pressures at high temperatures, which is necessary because space power systems normally operate with maximum temperatures of 2000°F and condensing temperatures of 800 to 1400°F. These high condensing temperatures

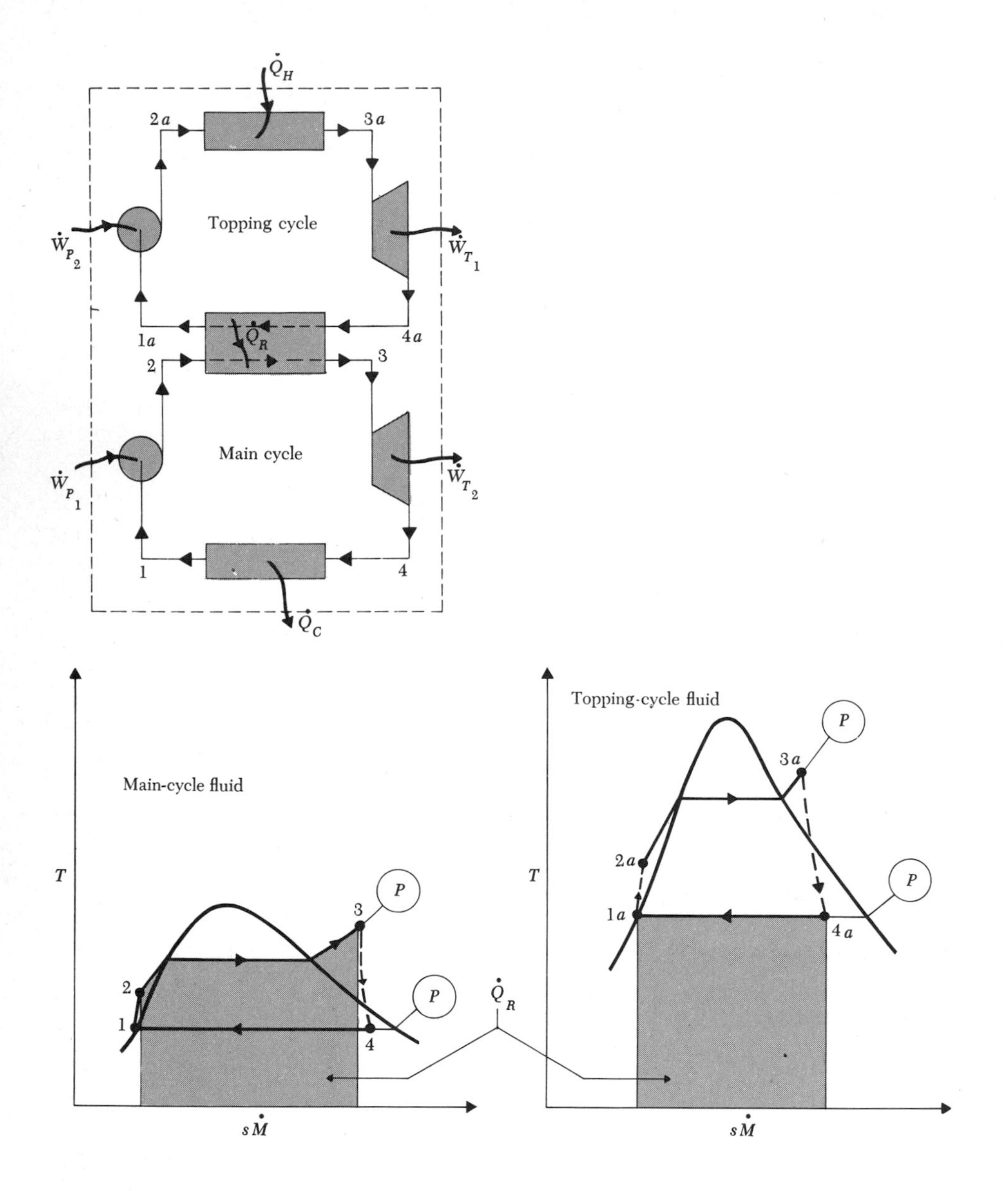

Figure 9·21. The binary vapor cycle

are required to minimize the weight of the condenser, which must be a radiator in a space environment. Vapor power cycles for very cold temperatures may work better with such fluids as CO_2, Freon, or ammonia. The proposed power plant using the temperature difference between ocean surfaces and depths of the sea is one such example. The environment, application, utilization, and budget must all be considered in selecting the working fluid for any power system.

In order to take advantage of the suitability of water in a relatively low temperature Rankine cycle and the preference for other fluids, such as mercury, in high-temperature Rankine cycles, one can use the *binary vapor power system* (Fig. 9·21). Here two separate Rankine systems are combined, with the heat rejection from the high-temperature cycle being used as the energy source for the low-temperature cycle. Many systems of this type have been studied theoretically, but few have been put into operation. One operating system uses mercury in the high-temperature cycle (the *topping cycle*) and steam in the low-temperature cycle, and obtains efficiencies of about 40 percent. As the technology of exotic fluids is developed for space applications, the ability to use fluids such as potassium, rubidium, mercury, and sulfur in topping cycles may well lead to construction of more large central power stations operating on the binary cycle, and theoretical studies show that efficiencies of the order of 50 percent may be realizable in practical power systems of the future.

9·9 VAPOR REFRIGERATION SYSTEMS

To refrigerate an area, we must expose that area to a fluid somewhat colder than the desired cold-space temperature. Energy can then be transferred as heat from the cold space to the colder fluid, and this will maintain the cold-space temperature against energy transfer as heat from the warm environment through the insulated cold-space walls. If we do not wish to discard the fluid, it must be circulated through the system in a manner that permits us to remove the energy taken from the cold space. This is generally done in a second heat exchanger, where energy is transferred as heat from the fluid to the environment. Of course, the fluid at this point must be slightly warmer than the environment in order for this energy transfer as heat to take place.

Recalling that expansion of a low-quality mixture in a throttling valve produces a marked drop in the fluid temperature, we see that the expansion process provides an easy way to achieve a low fluid temperature. After this cold fluid has been evaporated by extracting energy from the region to be cooled, we must pressurize it again to raise its temperature above the environment temperature so that the heat-rejection process can take place. With these notions we have just reinvented the *vapor-compression refrigeration* cycle. The vapor-compression refrigeration cycle is the model cycle for the great majority of refrigeration systems (Fig. 9·22). The cycle is almost the reverse of the Rankine power cycle; the difference is that a valve is used to produce the pressure drop, and no attempt is made to extract useful work from this expansion. The amount of work which could be obtained were the valve replaced by an isentropic turbine would be small in comparison with the compressor work requirement, since the volume changes are quite different. Furthermore, the quality of the mixture at state 4 is normally very low, and it is difficult to make a turbine operate for very long in this range. With

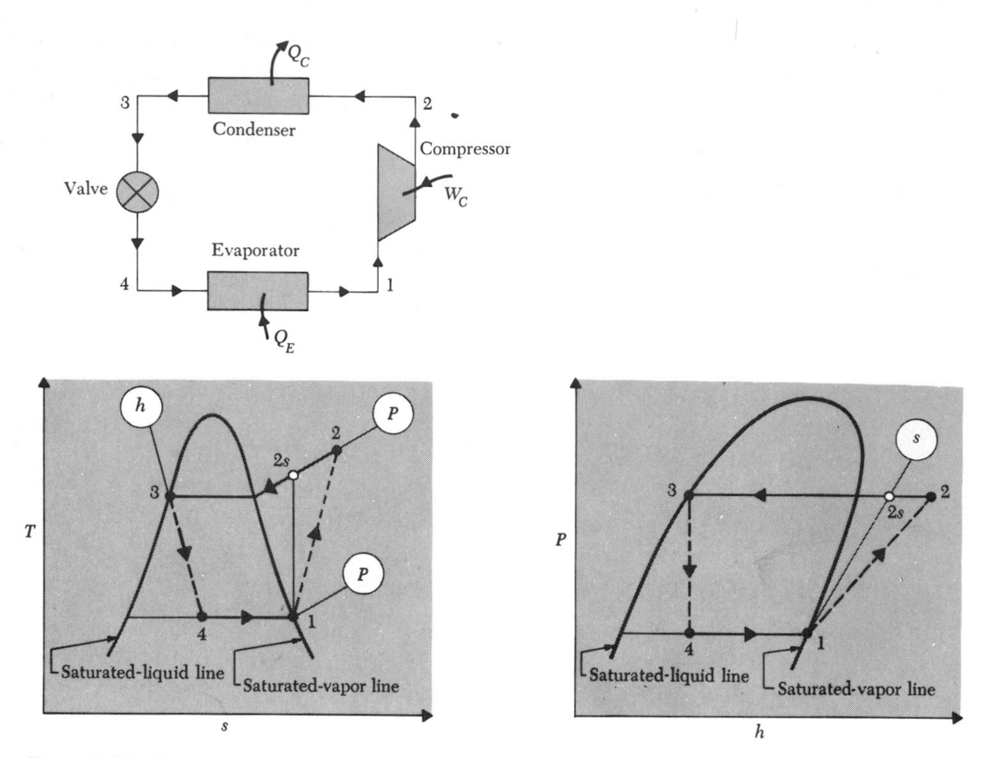

Figure 9·22. Vapor-compression refrigeration system

the valve the cycle can never be reversible, since the throttling process is inherently irreversible.

To illustrate the thermodynamic considerations in the engineering of a refrigeration system, let's design a system for a specific task. Suppose we must maintain a cold space at 15°F in a 70°F environment. Suppose the estimated "heat leak" from the environment to the cold space is 100,000 Btu/h. We must allow some temperature difference for the heat-transfer processes, so let's set the condensing temperature at 85°F and the evaporation temperature at 0°F. Freon 12 will be used as the working fluid and, following normal refrigeration practice, we shall operate the compressor with 10°F superheat at the inlet (meaning 10°F above the saturation temperature at the compressor inlet pressure). The hardware and process representations are as in Fig. 9·22, except that point 1 is slightly in the superheated vapor region. On the basis of experience with reciprocating refrigeration compressors, we assume that a compressor having an isentropic efficiency of 0.85 can be provided.

We first establish the states. An energy balance on the compressor yields (the student should list the implicit idealizations in the margin)

$$W = h_2 - h_1$$

where W is the work input to the compressor per lbm of fluid circulated. By assumption, state 1 is fixed by its pressure and superheat (Fig. B·7, Table B·5):

$$P_1 = 23.85 \text{ psia } \textit{saturated at } 0°F$$

$$T_1 = 10°\text{F}$$

$$h_1 = 79 \text{ Btu/lbm}$$

For the ideal isentropic compressor, state 2*s* is fixed by the entropy (same as state 1) and the pressure,

$$P_{2s} = 106 \text{ psia } \textit{saturated at } 85°F \qquad T_{2s} = 110°\text{F}$$

$$h_{2s} = 90 \text{ Btu/lbm}$$

So, for the hypothetical isentropic compressor,

$$W_s = h_{2s} - h_1 = 90 - 79 = 11 \text{ Btu/lbm}$$

Hence, for our actual machine,

$$W = \frac{W_s}{\eta_s} = \frac{11}{0.85} = 13 \text{ Btu/lbm}$$

So, the discharge enthalpy is

$$h_2 = h_1 + W = 79 + 13 = 92 \text{ Btu/lbm}$$

This value, together with the discharge pressure (106 psia), fixes the compressor discharge state. Note that $T_2 = 120°$F.

State 3 is assumed to correspond to saturated liquid at 85°F, so $h_3 = 27$ Btu/lbm. To fix state 4, we note that an energy balance on the valve gives

$$h_3 = h_4$$

so that state 4 is fixed by the known enthalpy and pressure; from Fig. B·7,

$$P_4 = 23.85 \text{ psia} \qquad T_4 = 0°\text{F}$$

$$h_4 = 27 \text{ Btu/lbm} \qquad x_4 = 0.27$$

Now, energy balances on the evaporator and condenser give, for their energy-transfer rates per lbm of mass flow,

$$Q_C = h_2 - h_3 = 92 - 27 = 65 \text{ Btu/lbm}$$

$$Q_E = h_1 - h_4 = 79 - 27 = 52 \text{ Btu/lbm}$$

and we have

$$W = h_2 - h_1 = 92 - 79 = 13 \text{ Btu/lbm}$$

The cop of this refrigeration system is then

$$\text{cop} = \frac{Q_E}{W} = \frac{52}{13} = 4.0$$

Note that four times as much energy is pumped out of the cold space as is required by the compressor. Note also that all the input energy ($Q_E + W$) is removed in the condenser (the overall energy balance also checks).

Now, our design requirement was for $Q_E = 100{,}000$ Btu/h. Therefore the required mass-flow rate is

$$\dot{M} = \frac{\dot{Q}_E}{Q_E} = \frac{100{,}000}{52} = 1920 \text{ lbm/h}$$

The rating of this unit in *tons of refrigeration* (1 ton = 12,000 Btu/h, approximately the rate required to freeze 1 ton of ice in a day) is

$$\text{Tonnage} = \frac{100{,}000}{12{,}000} = 8.3$$

The compressor power input will be

$$\dot{W} = \dot{M}W = 1920 \times 13 = 25{,}000 \text{ Btu/h} = 9.8 \text{ hp}$$

Note that the hp/ton is 9.8/8.3 = 1.1. A good rule of thumb is that roughly 1 hp of power is required for each ton of refrigeration (over operating temperatures of this range).

We note that this system can be used as a heat pump, in which the 70°F environment into which energy is "pumped" as heat is a house. On a cold day (0°F outside) our heat pump would deliver to the house

$$\dot{Q}_C = \dot{M}Q_C = 1920 \times 65 = 125{,}000 \text{ Btu/h}$$

which is indeed comparable with a household gas-fired furnace. However, the homeowner would only have to pay for the electric power required to run the compressor, which in this case amounts to only 25,000 Btu/h. The dollar savings over a few cold winters could well offset the high capital outlay required to purchase and maintain the heat pump equipment.†

Consideration of the Carnot cycle cop indicates that with refrigeration

† But then, if the neighbors are too close they may complain to the heat pump owner about the refrigeration of an already too cold out-of-doors.

systems the best performance is obtained when the temperature difference over which the system operates is as small as possible. Hence systems designed for a relatively small "lift" (the temperature to which the energy is "lifted" from the cold space) will have high cop values, while those intended for very cold refrigeration or for use in very hot environments will tend to have poorer performance. The cycle lift must always be somewhat larger than the difference between the cold space and environment temperature, for reasons noted above, but the performance can be improved markedly by good heat-exchanger design in the condenser and evaporator. The closer the condensing temperature can be brought to the environment, the better the performance. The 15°F temperature differences used in this example are typical of modern condensers and evaporators.

Another vapor refrigeration scheme is the *vacuum refrigeration system* that can be employed when refrigeration at temperatures slightly above 32°F is desired. The 32°F limit is set by the freezing point of water, the fluid used in such a system. In this system a steam ejector replaces the compressor, and the only moving parts are the pump impellers. The ejector maintains the flash chamber at a low pressure, where the saturation temperature is at the desired low level. A low-temperature supply of water is thus maintained by the flash-evaporation process, and this water can then be run through an air conditioning system, shown as the chiller in the figure. The system schematic and process representation (idealized) are shown in Fig. 9·23. The refrigeration occurs from the heat input to the chiller. The heat-rejection process occurs in the condenser and is noted as Q_C. Energy input as work is required by the pump.

An interesting application of vapor-compression refrigeration is in the production of liquid oxygen. The simplified system diagram and process representation for one type of lox plant are shown in Fig. 9·24. Note that for this system a turboexpander is used to replace the valve, so that useful work is extracted during the expansion process. However, in this case the expansion is from a superheated vapor state to a high-quality mixture state, whereas in the simple refrigerator the expansion occurs from a saturated liquid state to a low-quality mixture state.

The *absorption refrigeration cycle* (sometimes called *gas refrigeration*) is shown in Fig. 9·25. It uses two fluids, a refrigerant and an absorbent. In this system the compressor of a conventional vapor refrigeration system is replaced by an absorber-generator-pump assembly noted within the dotted lines in the figure. The system working fluid, shown here as ammonia, NH_3, is absorbed into water at a low pressure. The NH_3–H_2O solution has a lower enthalpy than the NH_3 and H_2O taken separately, and so energy must be removed as heat ($\dot{Q}_A$) to effect the absorption process. The liquid solution is then pumped to a high pressure, where in the generator the ammonia is driven out of solution by adding energy as heat ($\dot{Q}_G$). The ammonia then continues through the cycle as a conventional refrigerant, and the water is returned to the absorber. This arrangement permits the pressurization process to be applied to a *liquid* rather than to a vapor, hence the pump work requirement is much less than for a conventional vapor-compression refrigerator. While the pumping of the liquid is an advantage over compressing a vapor, this cycle requires a much larger heat-rejection system.

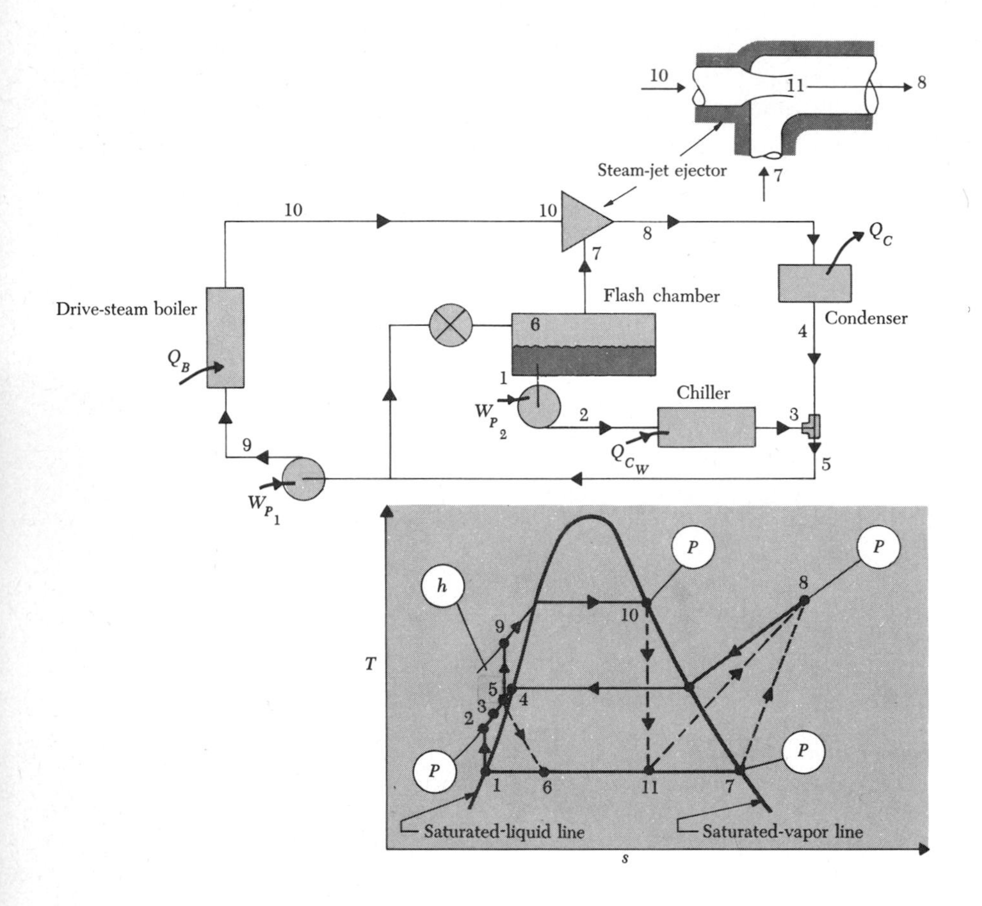

Figure 9·23. A vacuum refrigeration system

In industrial processes where waste heat, i.e., steam, is available as a by-product, the absorption cycle may be an economically attractive method for providing refrigeration. Industrial cycles employ several improvements over the basic cycle, including a heat exchanger where the water-ammonia high-pressure solution is heated by warm water returning to the absorber from the generator. The exchanger thus reduces both $\dot{Q}_G$ and $\dot{Q}_A$. A second refrigeration system of much interest uses lithium bromide as the absorbent and *water as the refrigerant*. The basic cycle is the same as noted on Fig. 9·25 but operates at lower pressures than for the ammonia system. Since water is the refrigerant, the evaporator temperature must be above 32°F.

The lithium bromide system has been suggested as feasible for use with both geothermal energy and solar energy as the source for the generator power Q_G. In one design using solar power, pressurized hot water is heated in solar collectors to 240°F. The hot water then acts as the energy source in the generator for a lithium bromide (absorber)–water (refrigerant) system.

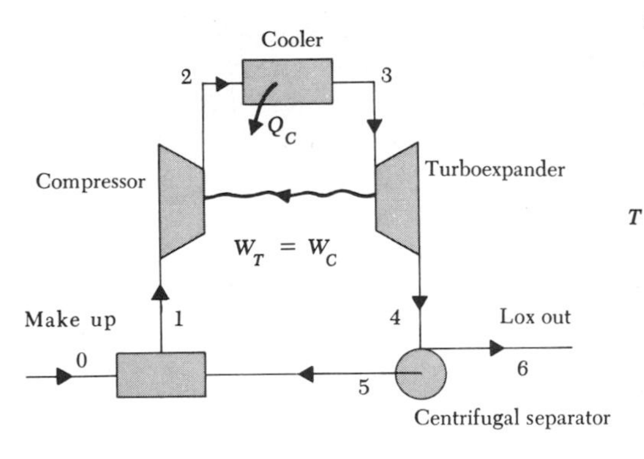

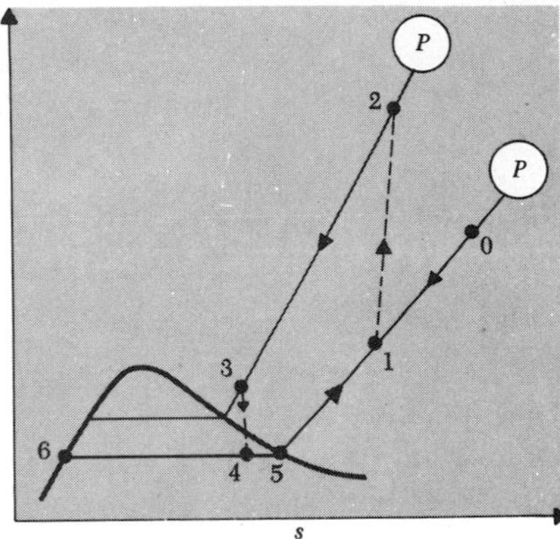

Figure 9·24. Oxygen liquefaction

9·10 SOME PROBLEMS AND ADVANTAGES OF ALL-GAS CYCLES

The cycles studied thus far in this chapter use a working fluid in both the liquid and gaseous phases. The chief advantage of this scheme is that the work required to pressurize the liquid is small in comparison to the work that can be obtained by expanding the gas (vapor), hence the bwr is quite low. Therefore poor efficiency in the compression and expansion processes can be tolerated. If instead we choose to keep the working fluid in the gaseous phase throughout the cycle, we shall be faced with the problem of a high bwr; that is, the compression process will drain off a large fraction of the work obtained in the expansion process. The low-bwr advantage of the Rankine cycle was the primary reason for the early development

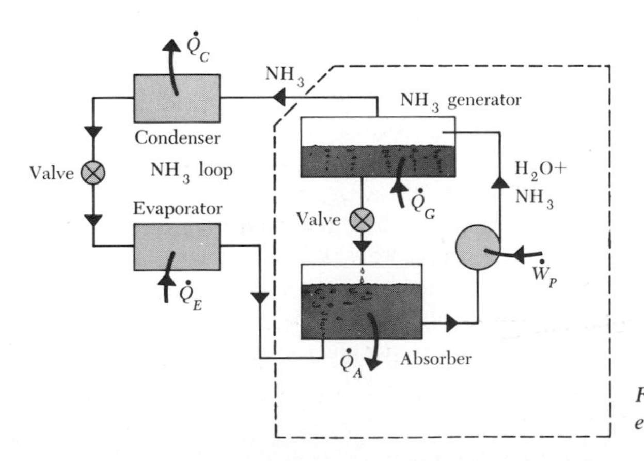

Figure 9·25. The absorption refrigeration cycle

of vapor power system technology, which had reached a rather remarkable state of advancement by the end of the last century.

The first gas power systems to be developed were simple internal combustion engines. In these systems the operating temperature ranges of the cycles are so wide that the low-temperature gas that must be compressed is much more dense than the expanding high-temperature gas, hence only a moderate bwr is required. Because of the reciprocating nature of such engines, the metal parts are exposed to these high temperatures (4000°F) only momentarily, and with sufficient cooling the systems are practical. Gas power cycles operating on a steady-state basis (gas-turbine power systems) are restricted to lower maximum temperatures, and consequently the densities of the fluid which must be compressed and that which can be expanded are more nearly equal. This tends to produce a high bwr, and consequently high compressor and turbine isentropic efficiencies are required if any net work is to be obtained. Because of fluid-mechanical problems, it is more difficult to design an efficient compressor than an efficient turbine; in fact, it was not until the 1940s that compressor efficiencies sufficiently high to provide positive net work from the cycle were obtained. Since that time there have been major advances in the technology of centrifugal and axial flow compressors, and now gas-turbine power systems with efficiencies competitive with vapor power and reciprocating internal combustion systems can be built. Gas turbine systems have become the standard aircraft prime mover, are used in many military vehicles, and will likely become very important in buses, trucks, trains, and possibly even automobiles in the near future.

While all-gas systems have certain theoretical thermodynamic disadvantages, the practical advantages are numerous. The working fluid is clean and can be totally inert (helium is a popular fluid in nuclear gas power systems). Gas-bearing technology has been developed to the point where the working fluid can also be the lubricant, which simplifies the system operation and maintenance and reduces costs. With the advent of fluidic control devices, even the control system can be operated with the working fluid, hence a much more reliable integrated system can be obtained. The turbines are not beset by the erosive problems of wet vapor, and cavitation is not a problem in the compressor. The pressures can be relatively low, hence heavy piping is not required. Hence gas turbine systems tend to be very compact and to have a higher power/weight ratio than vapor power plants, and consequently cost less and can be erected more quickly. Many utility companies now use gas turbine power for load peaking. The practical advantages will certainly make gas-turbine power systems grow in importance in the next decade, particularly for prime-mover applications.

Atmospheric pollution problems are making a definite impact on the nature of prime-mover power systems, particularly for automobiles and trucks. In conventional spark-ignition engines the combination of momentary high temperatures with sudden quenching by the exhaust expansion "freezes" harmful nitrogen oxides and carbon monoxide in the exhaust stream. These, in combination with unburned hydrocarbons resulting from rich (excess fuel) operation, are believed responsible for much of the smog in the great valleys of California. Cycles that run lean (excess air) and operate at lower peak temperatures, such as the gas turbine system, can substantially reduce the smog problem in such areas.

9·11 A PARTICULAR GAS-TURBINE POWER SYSTEM

In order to gain some appreciation for the problems of gas-turbine power production, let us analyze a very simple open-cycle system, shown in Fig. 9·26. We make the following idealizations:

Working fluid, air treated as a perfect gas
State 1, 20°C and 1 atm
State 2, 4 atm
State 3, 525°C and 4 atm
State 4, 1 atm
$\eta_s = 0.65$ for compressor, $\eta_s = 0.87$ for turbine
Kinetic and potential energies negligible at states 1, 2, 3, 4; compressor and turbine adiabatic; steady flow, steady state; equilibrium states at 1, 2, 3, and 4

The analysis consists of determining the various states and computing the energy transfers as heat and work per unit of mass of air. Energy balances on the three components yield

$$W_C = h_2 - h_1 \qquad W_T = h_3 - h_4 \qquad Q_H = h_3 - h_2$$

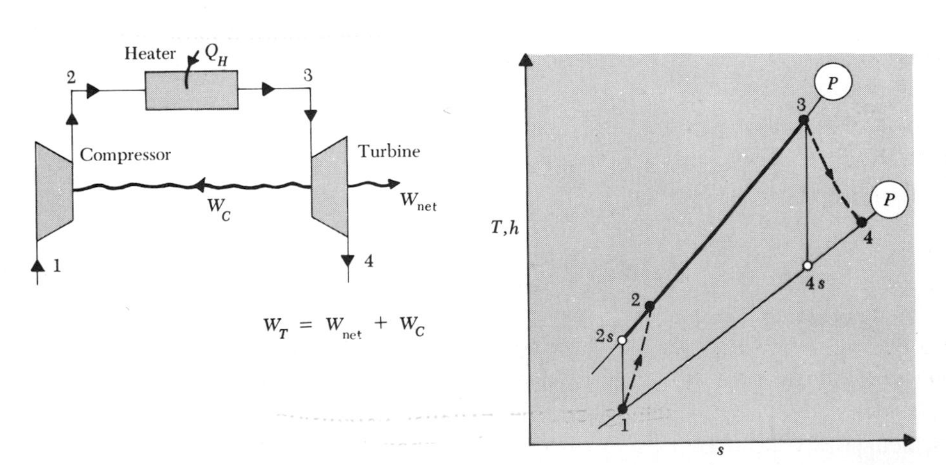

Figure 9·26. A simple gas-turbine power system

Noting that

$$\frac{R}{c_P} = \frac{c_P - c_v}{c_P} = \frac{k-1}{k}$$

we find from Eq. (8·27) that

$$\frac{T_{2s}}{T_1} = \left(\frac{P_{2s}}{P_1}\right)^{(k-1)/k} \qquad \frac{T_3}{T_{4s}} = \left(\frac{P_3}{P_{4s}}\right)^{(k-1)/k}$$

Thus, since $(k - 1)/k = 0.286$,

$$T_{2s} = T_1 \times 4^{0.286} = 293 \times 1.486 = 435 \text{ K}$$

Using Eq. (8·25) to determine enthalpy changes,

$$W_{Cs} = h_{2s} - h_1 = c_P(T_{2s} - T_1) = 1.004 \times (435 - 293) = 142.5 \text{ kJ/kg}$$

$$W_C = \frac{W_{Cs}}{\eta_s} = \frac{142.5}{0.65} = 219.3 \text{ kJ/kg}$$

$$W_C = h_2 - h_1 = c_P(T_2 - T_1)$$

$$T_2 = T_1 + \frac{W_C}{c_P} = 293 + \frac{219.3}{1.004} = 511.4 \text{ K}$$

$$Q_H = h_3 - h_2 = c_P(T_3 - T_2) = 1.004 \times (798 - 511.4) = 287.7 \text{ kJ/kg}$$

$$T_{4s} = T_3 \times 4^{-0.286} = \frac{798}{1.486} = 537 \text{ K}$$

$$W_{Ts} = h_3 - h_{4s} = c_P(T_3 - T_{4s}) = 1.004(798 - 537) = 262 \text{ kJ/kg}$$

$$W_T = W_T \eta_s = 0.87 \times 262 = 228 \text{ kJ/kg}$$

$$W_T = h_3 - h_4 = c_P(T_3 - T_4)$$

$$T_4 = T_3 - \frac{W_T}{c_P} = 798 - \frac{228}{1.004} = 571 \text{ K}$$

The bwr is then

$$\text{bwr} = \frac{W_C}{W_T} = \frac{219.3}{228} = 0.961$$

The net work output is

$$W_{\text{net}} = W_T - W_C = 228 - 219.3 = 8.7 \text{ kJ/kg}$$

and the system energy-conversion efficiency is

$$\eta = \frac{W_{\text{net}}}{Q_H} = \frac{8.7}{287.7} = 0.03$$

We see that compression of a gas takes quite a bit more work than compression of a liquid. Consequently, the bwr for gas power cycles will always be much larger than that for a Rankine cycle. Whereas we can tolerate poor isentropic pump efficiencies in a vapor power system, we must achieve high compressor isentropic efficiency in order to make gas turbine cycles practicable. The compressor efficiency selected in this example was purposely low in order to emphasize this point. A rather substantial research-and-development effort on axial-flow compressors, begun in the 1940s, has brought us to the level of about 85 percent for larger machines. With this value in our example,

$$W_C = 142.5/0.85 = 167.6 \text{ kJ/kg}$$

$$T_2 = 293 + 167.6/1.004 = 460 \text{ K}$$

$$Q_H = 1.004 \times (798 - 460) = 339.3 \text{ kJ/kg}$$

$$\text{bwr} = 167.6/228 = 0.735$$

$$W_{\text{net}} = 228 - 167.6 = 60.4 \text{ kJ/kg}$$

$$\eta = 60.4/339.3 = 0.178$$

Note that this 30 percent gain in compressor efficiency would reduce the fuel consumption by a factor of more than 5. The importance of good turbomachinery design should now be apparent.

The pertinent energy transfers for this system are summarized below. For comparison we show the corresponding values for a cycle with isentropic compression and expansion. Note the large effect on system efficiency η due to irreversibilities in the compressor and turbine.

	W_C	W_T	W_{net}	Q_H	η
Ideal	142.5	262	119.5	364	0.328
Actual	167.6	228	60.4	339.3	0.178

The high level of the turbine outlet temperature (571 K), compared to a Rankine cycle, may come as a surprise, and it may seem that there should be some way to use this energy. We have to live within the restrictions of the second law, but it is possible to make some gains. We shall discuss these shortly.

9·12 SOME REMARKS ON THE STEADY-FLOW COMPRESSION PROCESS

The compression process forms a key part of the gas-turbine power cycle. Gas compressors are also used in many other applications, and it will be profitable to digress from our study of power systems to discuss the gas-compression process in more detail.

We have indicated that the isentropic process is the usual model process for gas compression. This model is appropriate for devices in which the amount of energy transfer as heat is small in comparison to the energy transfer as work. But it is not necessarily true that the isentropic compressor is the "best." In order to study this question, let us consider a model process in which we presume that the pressure and specific volume are related by

$$Pv^n = \text{constant} \tag{9·9}$$

This model process is called the *polytropic process*, and n is the *polytropic exponent*. By choosing n to be various values we can study the work requirements for different compression processes. For example, if the working fluid is a perfect gas with constant specific heats, $n = k$ corresponds to an isentropic compression, and $n = 1$ corresponds to isothermal compression.

Consider now a compressor in which the fluid undergoes *reversible* polytropic compression. An energy balance on the compressor (Fig. 9·27) gives

$$W + Q = h_2 - h_1 \tag{9·10}$$

For the reversible process, the energy transfer as heat can be related to the change in entropy of the fluid. Taking a unit mass of the fluid passing through the compressor as the control mass, the second law requires that

$$Q = \int_1^2 T\,ds \tag{9·11}$$

where Q is the amount of energy transfer as heat to a unit mass of the fluid as it passes through the compressor. Now, the entropy, enthalpy, and pressure changes for any infinitesimal step of the process are related by Eq. (8·44),

$$T\,ds = dh - v\,dP$$

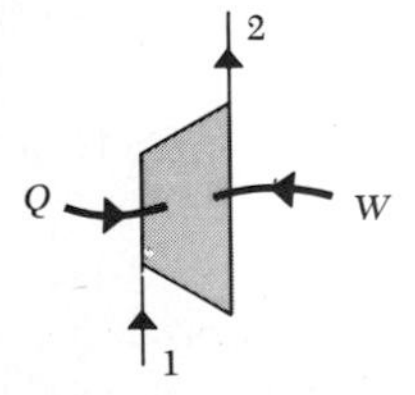

Figure 9·27. Polytropic compression analysis

Substituting in Eq. (9·11), and integrating,

$$Q = h_2 - h_1 - \int_1^2 v\,dP \tag{9·12}$$

Upon combination with the energy balance, we have

$$W = \int_1^2 v\,dP \tag{9·13}$$

The student might compare this with the expression for the work done *by* a unit mass of fluid in a simple expansion process. Note that the smaller the specific volume the less is the work required to achieve a given pressure increase. This observation is behind our earlier remarks on the bwr for Rankine cycles versus gas power cycles.

Using the assumed polytropic relationship between P and v Eq. (9·9), the integration of Eq. (9·13) may be carried out, and the work expressed in terms of the initial state (P_1, v_1), the pressure ratio P_2/P_1, and the polytropic exponent n; one finds

$$W = \frac{n}{n-1} P_1 v_1 \left[\left(\frac{P_2}{P_1} \right)^{(n-1)/n} - 1 \right] \qquad n \neq 1 \tag{9·14a}$$

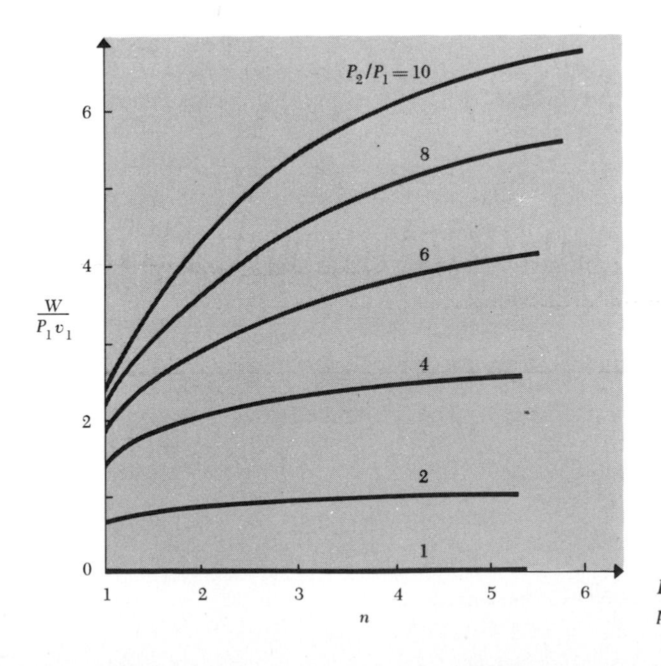

Figure 9·28. Work required for polytropic compression

The case $n = 1$ must be handled separately; this yields

$$W = P_1 v_1 \ln\left(\frac{P_2}{P_1}\right) \qquad n = 1 \tag{9·14b}$$

Figure 9·28 shows the work required to achieve a given compression as a function of the polytropic exponent. Note that as n decreases less work is required, and that the least amount of work for $n \geq 1$ is obtained for $n = 1$, that is, for isothermal compression. Hence, for the compression process, the *isothermal* compressor is really a better ideal, and the *isothermal compression efficiency* η_T is often used to compare the work input for an actual compressor with that for an ideal, reversible isothermal compression process. Note that isothermal compression requires that $Q < 0$ (Fig. 9·27).

9·13 THE BRAYTON CYCLE AND ITS MODIFICATIONS

The pattern cycle for the gas-turbine power system studied in Sec. 9·11 is known as the *Brayton cycle*. The hardware schematic and process representation for a closed Brayton-cycle system is shown in Fig. 9·29. In the idealized Brayton cycle the compressor and turbine processes are isentropic, and pressure drop across the heat exchangers is neglected. If the working fluid is treated as a perfect gas with constant specific heats, a simple expression for the energy-conversion efficiency may be obtained† in terms of the pressure ratio,

$$\eta = 1 - \left(\frac{1}{P^*}\right)^{(k-1)/k} \tag{9·15}$$

where

$$P^* = \frac{P_2}{P_1} = \frac{P_3}{P_4}$$

This relationship for $k = 1.4$ and 1.67 is shown in Fig. 9·30. In particular, for the example of the last section, $P^* = 4$ and $\eta_{\text{ideal}} = 32.8$ percent, about twice what we obtained with the improved compressor. Note that high pressure ratios make the idealized Brayton cycle more efficient. However, compressor efficiency tends to drop off with increased pressure ratio, and this has a compensating effect in a real Brayton cycle. Pressure ratios in the range 4 to 6 are typical of simple Brayton systems.

In gas turbine systems intended for prime-mover use, it is usually important to keep engine weight small, which normally means maximizing the work

† The development is as follows.
Energy balances: $W_T = c_P(T_3 - T_4)$, $W_C = c_P(T_2 - T_1)$, $Q_H = c_P(T_3 - T_2)$
Isentropic processes: $T_2/T_1 = T_3/T_4 = (P^*)^{(k-1)/k} = A$
Conversion efficiency:
$\eta = [(T_3 - T_4) - (T_2 - T_1)]/(T_3 - T_2) = (T_4 A - T_4 - T_1 A + T_1)/(T_4 A - T_1 A)$
$\eta = (T_4 - T_1)(A - 1)/[(T_4 - T_1)A] = 1 - 1/A$

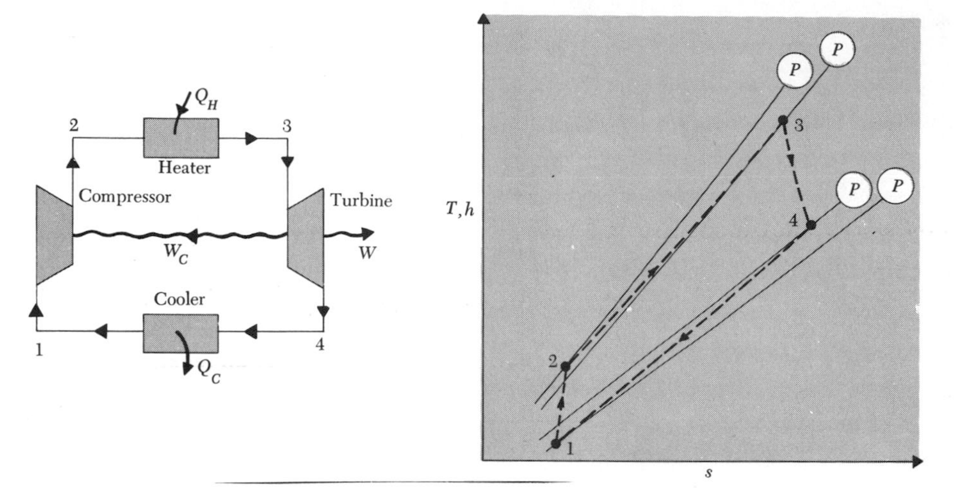

Figure 9·29. The Brayton cycle

output per lbm of fluid. Metallurgical considerations limit the turbine inlet temperature, and current systems operate at about 970 to 1080°C (1800 to 2000°F) although some military gas turbines with blade cooling now have inlet temperatures as high as 1350°C (2500°F). Within this temperature limitation there will be some pressure ratio that yields a maximum efficiency and another that yields the most work output per unit of mass of fluid, and these two ratios will not generally be the same. Figure 9·31 illustrates this in terms of the T-s process representation. Recall that the work output per lbm of fluid is equal to the area enclosed by the T-s process path for a reversible process. The primed cycle in the figure has a greater work output per unit of mass than does the high-pressure cycle, but a lower energy-conversion efficiency. This is seen from the fact that in the primed cycle there is a greater input of energy at the lower temperatures, while in the

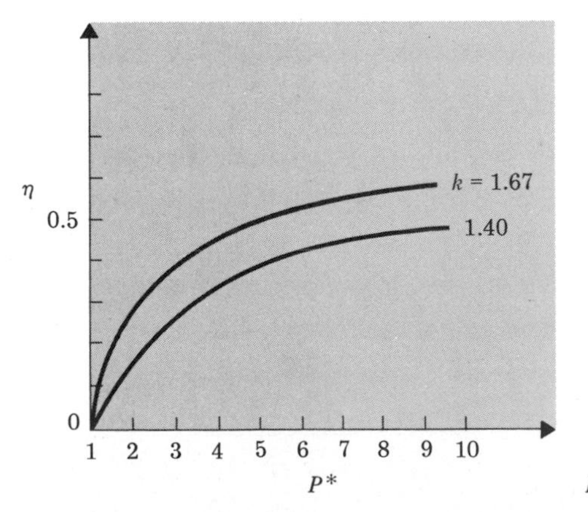

Figure 9·30. Idealized Brayton-cycle efficiency

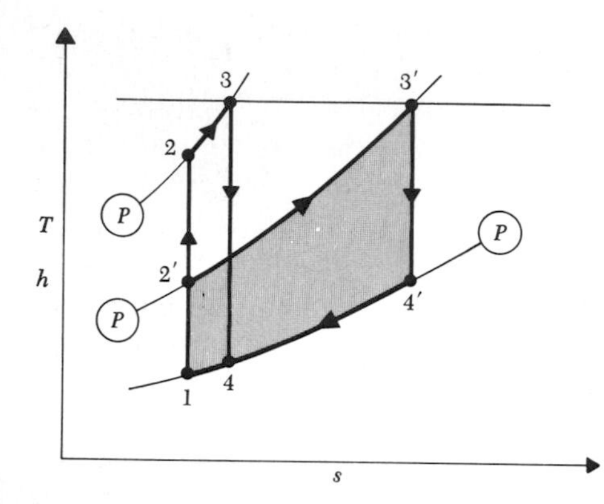

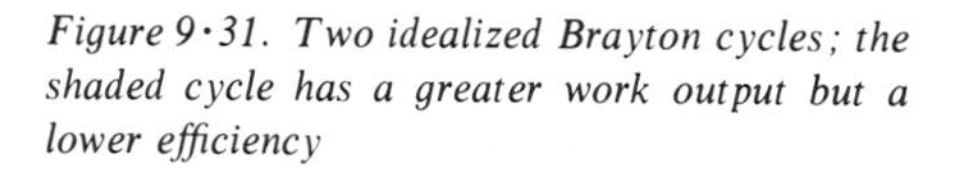

Figure 9·31. Two idealized Brayton cycles; the shaded cycle has a greater work output but a lower efficiency

unprimed cycle all the energy input as heat occurs at a high temperature, which leads to higher efficiency.

It may be shown that the pressure ratio yielding the most work per unit of mass for the idealized Brayton cycle of Fig. 9·31 is

$$P^*_{\text{opt}} = \left(\frac{T_3}{T_1}\right)^{k/[2(k-1)]} \tag{9·16}$$

Where a low power/weight ratio is desirable, it will be necessary to operate near to P^*_{opt}, hence away from the maximum efficiency point.

We have seen that the turbine exhaust temperature of a simple Brayton cycle is quite high; if it exceeds the compressor outlet temperature *regeneration* is possible, as shown in Fig. 9·32 (the ideal cycle is shown there). In the regeneration process the high-temperature exhaust gas is used to preheat the gas coming from the compressor, thus reducing the required energy input from the external heat source. With a counterflow regenerator it is theoretically possible to heat the gas from state 2 to the temperature of state 5 with energy transferred as heat from the

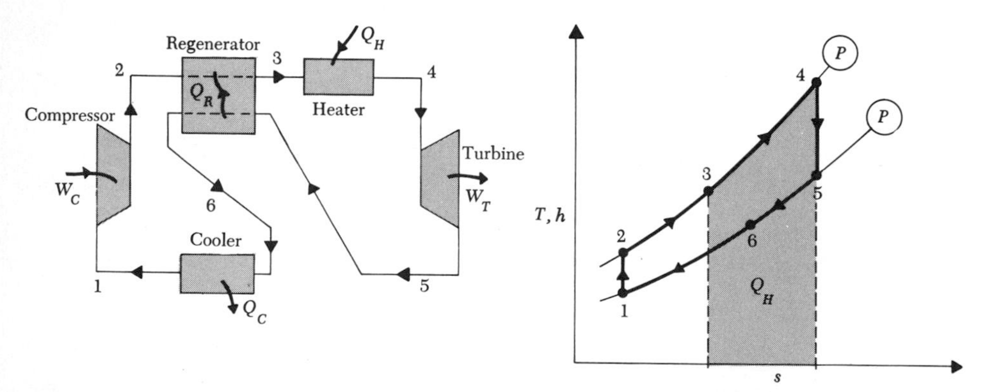

Figure 9·32. Idealized regenerative Brayton cycle

turbine exhaust gases. This results in cooling of the exhaust gases to the compressor outlet temperature (state 2). Systems have been built in which more than 95 percent of the possible regenerative effect has been obtained. The amount of possible regeneration is the largest at low pressure ratios and, in contrast to the simple Brayton cycle, the regenerative cycle yields the highest efficiencies at low pressure ratios. It is much easier to build high-performance compressors for low pressure differences, and this fact makes the regenerative system even more attractive. Nuclear-powered gas turbine systems employing regeneration have been built in the power range 100 to 300 MW, and automotive and shipboard gas turbines that produce 300 to 3000 hp are now in use. Efficiencies in the range 20 to 30 percent are typical with regeneration.

To illustrate the effect of regeneration, let us put a regenerator into the improved Brayton-cycle example of the previous section. We could conceivably warm the fluid leaving the compressor at 460 K to the turbine outlet temperature (571 K), but this would require an immense heat exchanger. However, 85 percent of this maximum possible temperature rise could be obtained with a heat exchanger of reasonable size. We therefore assume that the high-pressure gas leaves the regenerator at $460 + 0.85 \times (571 - 460) = 554$ K. The heater power would then be reduced to $Q_H = 1.004 \times (798 - 554) = 245$ kJ/kg, which would give an energy-conversion efficiency of $\eta = 60.4/245 = 0.246$, a 50 percent improvement over the nonregenerative system. Thus the regenerator would cut fuel costs by about 50 percent.

The development of compact and efficient regenerators may allow the gas-turbine engine to move into the competitive automotive market. Of particular

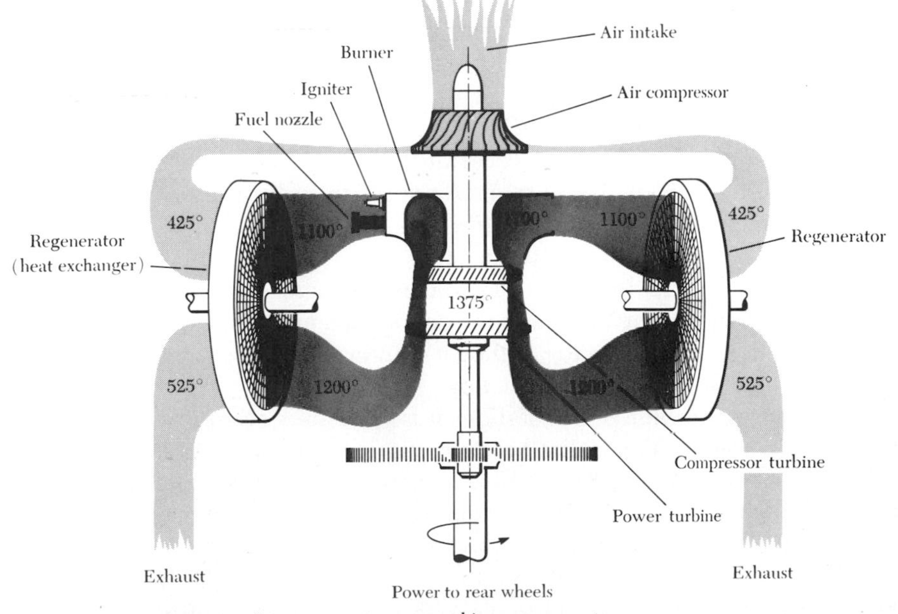

Figure 9·33. Schematic of an automotive gas-turbine power system

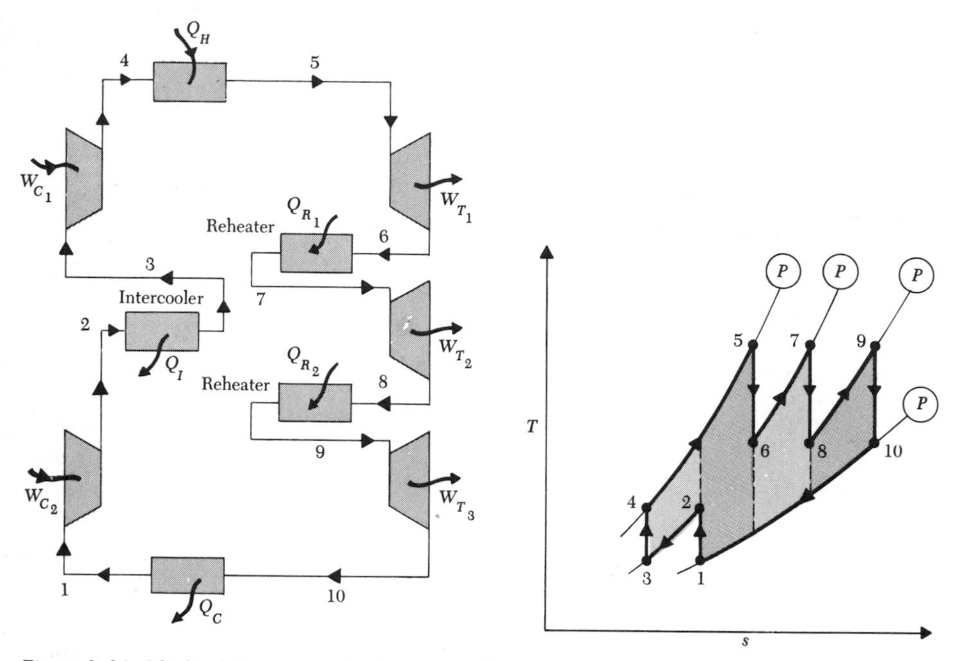

Figure 9·34. Idealized Brayton cycle with intercooling and reheat

importance is the *rotary regenerator*, an unusual type of heat exchanger. A disk of porous metal is rotated through two semicircular ducts. Hot gases in one duct heat up the metal, which is then moved by the disk rotation to the other duct, where the energy is transferred as heat from the metal to the cooler gas. Such exchangers can be made very compact and highly efficient. There are of course numerous interesting engineering problems with such regenerators, such as sealing between the two ducts, thermal expansions that cause distortions, carryover of the hot gases directly into the cold gas duct, and so on. A remarkable development effort by the automotive companies in the past decade has now resolved most of these problems, and most operating automotive gas turbines are now equipped with rotary regenerators. Figure 9·33 shows a schematic of one manufacturer's system. Note particularly the dual turbine disks, and the operating temperatures, in °F, revealed in their diagram. A competitor (or a student) can make a rather thorough assessment of their system from this information.

In larger gas-turbine power stations it is sometimes economical to employ *intercooling and reheat*. An idealized system with one intercooling stage and two stages of reheat is shown in Fig. 9·34. As shown in Fig. 9·35, intercooling between compressor stages reduces the compressor work requirement. The least work is required for isothermal compression, approached as the number of intercooling stages becomes large. Similarly, reheat between the turbine stages increases the work output per unit of mass for a given pressure ratio and turbine inlet temperature. The net work output per unit of mass therefore is increased by

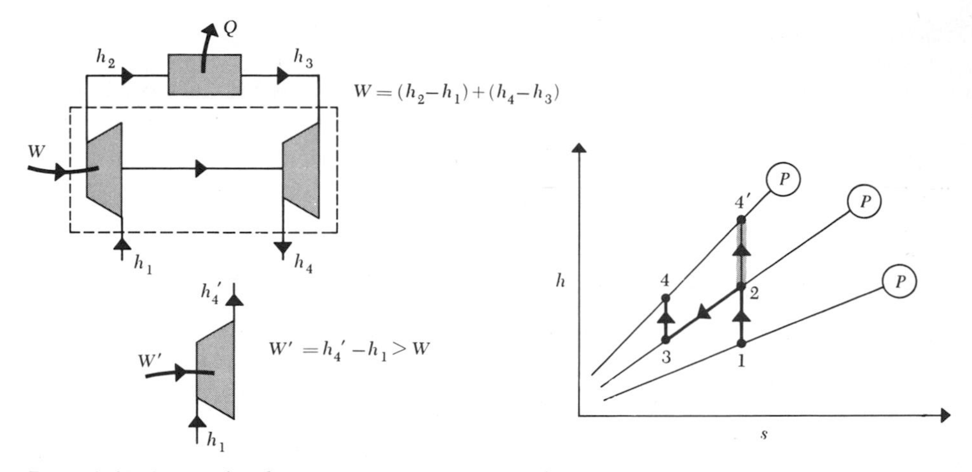

Figure 9·35. Intercooling between compression stages reduces the work requirement

intercooling and/or reheat. However, the efficiency of a Brayton cycle with intercooling and/or reheat is *lower* than the simple Brayton-cycle efficiency, for more energy must be added as heat.† Intercooling and reheat serve to *increase the potential for regeneration* by increasing the final turbine outlet temperature and at the same time reducing the compressor outlet temperature. Intercooling and reheat are therefore usually used in conjunction with regeneration. In theory, if an infinite number of reheat and intercooling stages is employed, and regeneration is also used, all the energy added externally occurs in the reheat exchangers when the fluid is at its maximum temperature, and all the energy rejection as heat takes place in the intercoolers, where the gas is at its lowest temperature. Thus the limiting case is a $2T$ engine (ideally reversible) having an efficiency given by the Carnot-cycle efficiency (see Fig. 9·36). This is very interesting, for the process

† Thinking graphically, the portions of the cycle that are "patched on" to the simple Brayton cycle are lower temperature ratio cycles and therefore are less efficient.

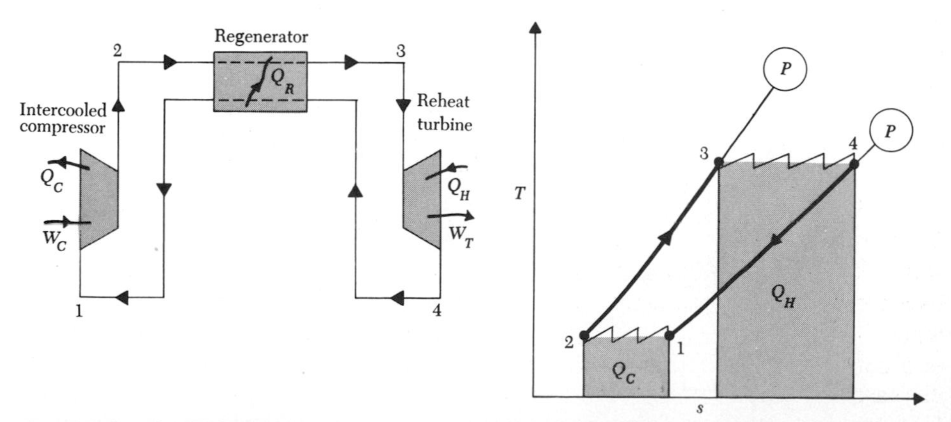

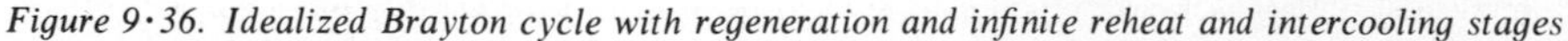
Figure 9·36. Idealized Brayton cycle with regeneration and infinite reheat and intercooling stages

representation is quite unlike that of a Carnot cycle. This limiting cycle is sometimes called the *Ericsson cycle*.

Gas-turbine-generator sets are increasingly being used to meet peak power demands. The gas turbine offers the advantage of a very quick start-up to full load; the vapor power system (Rankine cycle) requires up to 24 h to reach full power from a cold start, largely because of the thermal time constant of the boiler. Many utilities use vapor power systems to meet base-load requirements, and then bring gas turbines in the 25- to 50-MW size range on line as demand builds up during the day.

9·14 APPLICATIONS TO NUCLEAR POWER

We have previously described a nuclear powered Rankine cycle plant. Gas-cooled nuclear reactors may also be used to provide the energy input to vapor power systems. At present no commercial reactors are used to drive a gas turbine directly. However, such systems have been proposed for plants in the near future. As a last example of the use of Brayton cycles, consider the system noted in Fig. 9·37. The operating conditions are typical of proposed designs and include an unusual heat-rejection system. A helium-to-air heat exchanger is proposed so as to minimize water requirements for the system. Such an exchanger would be large and expensive to build but would allow the plant to be placed in an arid area or in an area where the maximum allowable thermal discharge to the local river has already been reached.

Operating conditions are:

Working fluid, helium
State 1, $T = 50°C$, $P = 3.03$ MPa
State 2, $P = 6.06$ MPa
State 3, $T = 1000°C$, $P = 6$ MPa
State 4, $P = 3.10$ MPa
$\eta_s = 0.85$ for the compressor, $\eta_s = 0.91$ for the turbine

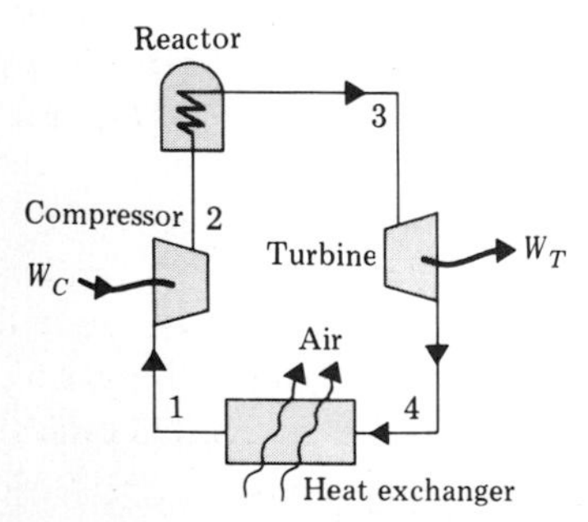

Figure 9·37. Proposed nuclear powered Brayton cycle

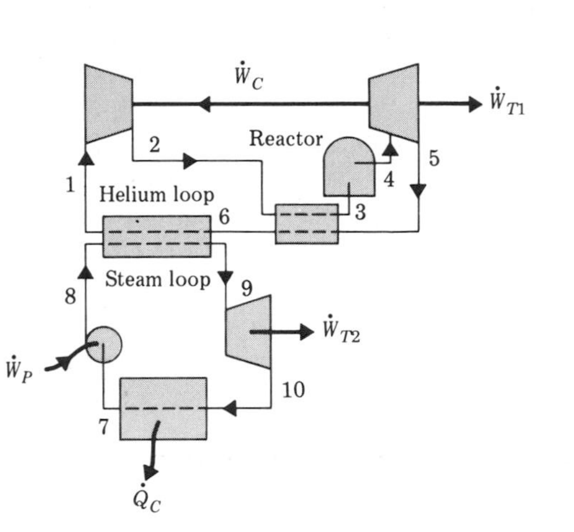

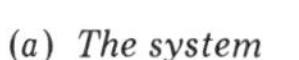

(a) The system

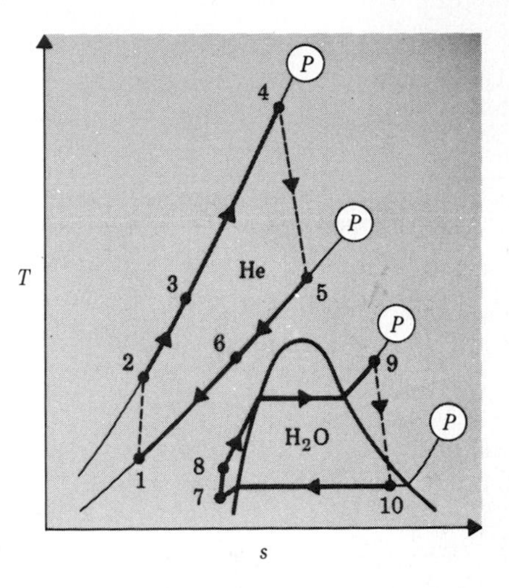

(b) The process representation

Figure 9·38. A combined Brayton-Rankine cycle

You should carry out the calculations to verify the tabulated results given below:

$$W_C = 630 \text{ kJ/kg}$$

$$Q_R = 4311 \text{ kJ/kg}$$

$$W_T = 1396 \text{ kJ/kg}$$

$$\eta = 0.18$$

$$Q_{\text{out}} = 3545 \text{ kJ/kg}$$

Instead of using the helium-to-air heat exchanger to remove the reject energy, could we take advantage of this 3545 kJ/kg? We could put into the cycle a regenerator which would reduce the external heat input and markedly increase η. A second idea could also be considered. Unlike in a Rankine cycle, the rejected energy is available here at a high temperature, $T_4 = 1004$ K (731°C). Even allowing for a 230°C temperature driving force for heat transfer, one could heat water to a temperature of 500°C, which is close to the present-day peak operating temperatures for steam plants. We might therefore use the gas turbine (Brayton) cycle as a topping cycle for a Rankine steam power plant. Figure 9·38 shows such a cycle.

Such combined gas-steam cycles (COGAS) are very popular, since they offer a higher overall efficiency than either simple cycle. Predesigned units can be purchased in the 100- to 400-MW range. Coupling the two cycles enables one to increase the average temperature at which energy is received as heat by the system and, as the Carnot cycle suggests, this improves efficiency. Future combined cycles will very probably have thermal efficiencies of the order of 50 percent.

9·15 OTHER GAS POWER CYCLES

The *Otto cycle* is the model cycle for reciprocating *spark-ignition engines.* The pressure of the gas within the cylinder of an idealized spark-ignition engine is shown as a function of the piston position in Fig. 9·39. With the piston at top dead center (tdc), the intake valve opens and a fresh charge of fuel-air mixture is sucked in. At bottom dead center (bdc) the intake valve closes, and the return stroke causes the gas to be compressed. In the idealized system ignition occurs instantaneously at tdc, causing a rapid rise in temperature and pressure. The gas is then expanded on the outstroke, until at bdc the exhaust valve opens, and the gas "blows down" through the exhaust port. With a fourth stroke the gases are purged out. In the idealized Otto cycle the compression and expansion processes are considered reversible and adiabatic, that is, isentropic, and it is assumed that the pressure within the cylinder during the intake and exhaust strokes is equal to the atmospheric pressure. The work done by the piston on the gas inside the cylinder during the exhaust stroke is exactly equal to the work done on the piston by the gas during the intake stroke, so that useful work output results only from the excess of the work done by the gas during the expansion stroke over that done on the gas during the compression stroke.

The process representation for the fluid during the compression, ignition, and expansion parts of the cycle is shown in Fig. 9·40. The combustion process is idealized in terms of a simple energy addition (as heat), and the changes in the chemical composition of the mixture are neglected. If it is further idealized that the gas is a perfect gas with constant specific heats, appropriate thermodynamic analysis leads to a simple algebraic expression for the efficiency of the Otto cycle in terms of the compression ratio,

$$\eta = \frac{W_{\text{net}}}{Q_{34}} = 1 - \frac{1}{r^{k-1}} \tag{9·17}$$

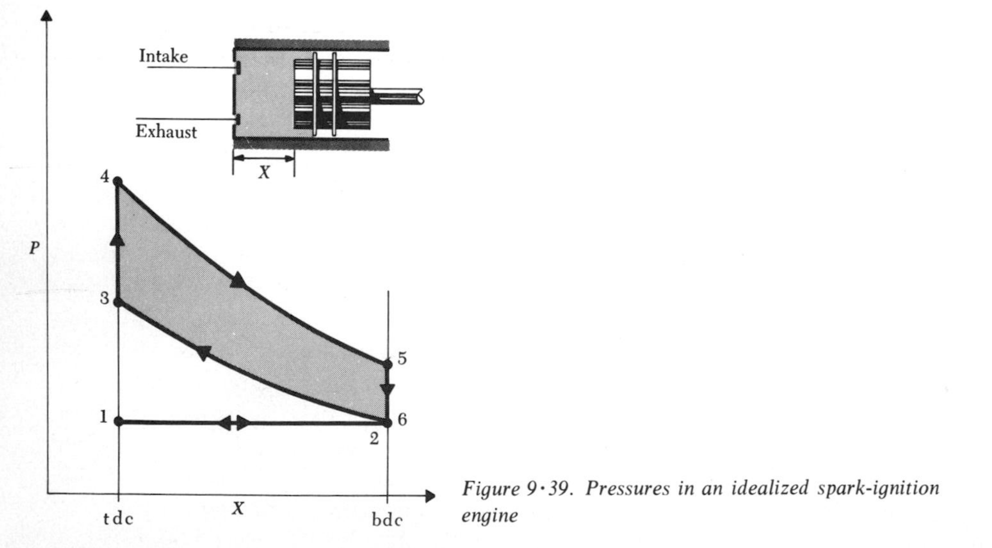

Figure 9·39. Pressures in an idealized spark-ignition engine

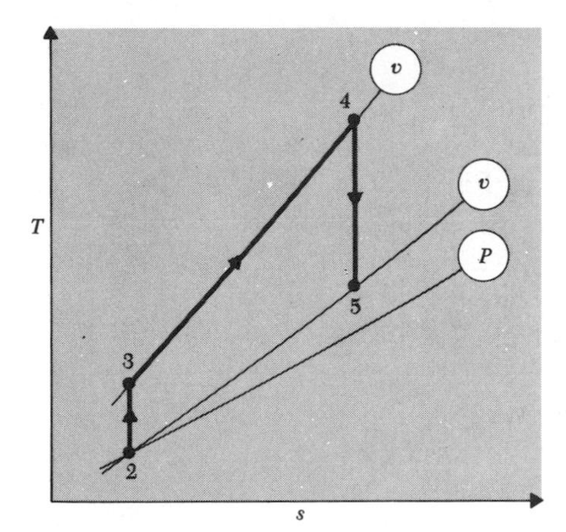

Figure 9·40. Idealized Otto-cycle process representation

where the compression ratio r is

$$r = \frac{v_2}{v_3} = \frac{v_5}{v_4}$$

This relationship for $k = 1.4$ is shown in Fig. 9·41.

A real spark-ignition engine will not meet the performance of the highly idealized Otto cycle. Combustion takes time, and for this reason it is initiated before tdc by "advancing the spark." Furthermore, there will be a pressure drop across the valve during intake and exhaust; the piston must do work on the air to get it out, and this is more than the work done on the piston by the cylinder gases during the intake stroke. Heat transfer is involved, so the compression and expansion processes are not isentropic. The pressure-displacement diagram of a realistic spark-ignition engine is shown in Fig. 9·42.

The peak temperatures in such an engine will be of the order of 4500 to 5000 R (2500 to 2800 K), and peak pressures will be 30 to 40 atm. In the ideal Otto cycle the peak operating conditions are higher. In the real cycle the peak is

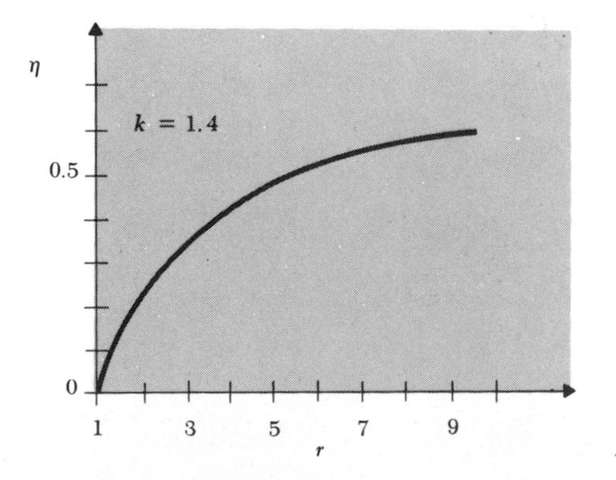

Figure 9·41. Otto-cycle efficiency

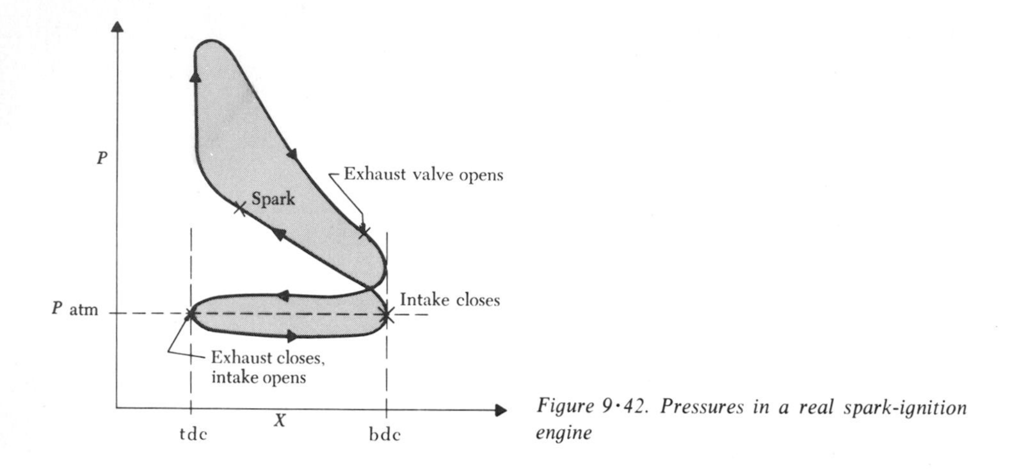

Figure 9·42. Pressures in a real spark-ignition engine

cut off, since the combustion process is not a heat addition at fixed volume, that is, at tdc, but is a chemical reaction occurring during a period of time while the compression and power strokes of the engine take place. The pressure at the engine exhaust is slightly above atmospheric, and the temperature is about 1500 R (not at the tailpipe).

From the curve of efficiency versus compression ratio it is clear that one would like to operate at high compression ratio v_2/v_3. However, in the spark-ignition engine this is often difficult to do, for the fuel-air mixture may undergo *detonation* in which the fuel begins to burn beyond the area ignited by the spark. This causes high-pressure waves in the combustion chamber ("knocking") which may cause severe damage if allowed to continue over a period of time. Fuels with higher octane ratings have higher detonation temperatures, and the higher compression ratios of the automobiles of the late 1960s were due primarily to improved fuel rather than to improved engine technology.

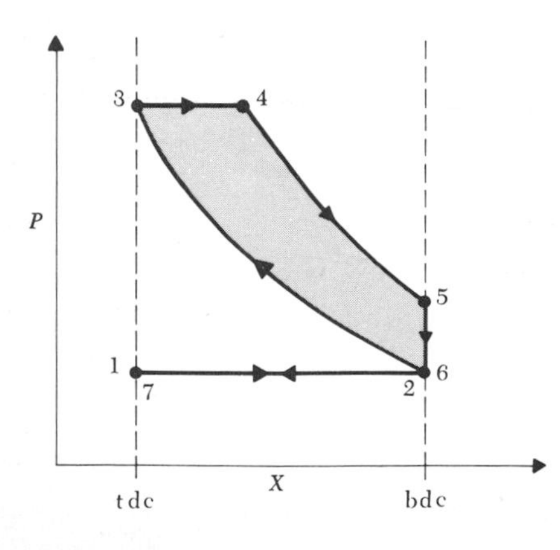

Figure 9·43. Pressures in an idealized diesel engine

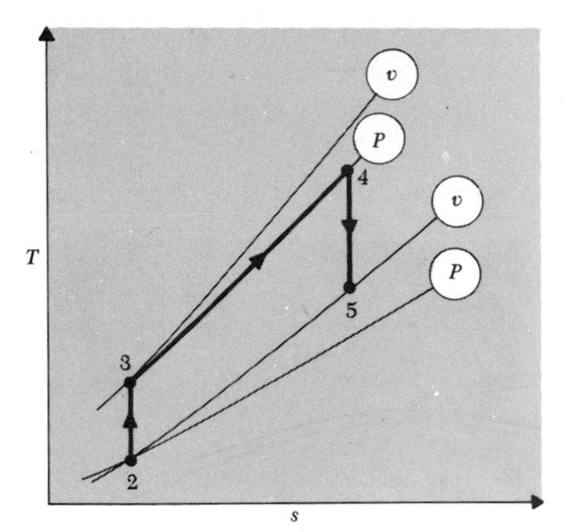

Figure 9·44. Idealized diesel-cycle process representation

One method of avoiding detonation during compression, with its limitation on the compression ratio, is to inject the fuel *after* the compression process. Since there is no fuel in the cylinder during the compression process, higher compression ratios (and therefore higher air temperatures) can be reached without detonation. Then upon injection the fuel ignites spontaneously owing to the high air temperature. This ignition tends to occur fairly uniformly over the cylinder, without harmful blast waves. Such an engine is called a *compression-* (rather than spark-) *ignition* engine.

The *diesel cycle* is the model cycle for reciprocating compression-ignition engines. In the idealized system air is compressed to tdc, at which time fuel is injected, and it is idealized that the combustion process takes place at constant pressure for part of the expansion stroke. The remainder of the expansion stroke and the compression stroke are idealized as isentropic. The pressures within the cylinder of an idealized diesel engine are shown in Fig. 9·43. The process representation is shown in Fig. 9·44.

The efficiency of a diesel cycle may be worked out by thermodynamic analysis. If it is assumed that the gas is a perfect gas with constant specific heats, an algebraic expression for the efficiency is obtained in terms of the compression and cutoff ratios,

$$\eta = 1 - \frac{1}{r^{k-1}} \frac{r_c^k - 1}{k(r_c - 1)} \tag{9·18}$$

$$r = \frac{v_2}{v_3} \qquad \textit{compression ratio}$$

$$r_c = \frac{v_4}{v_3} \qquad \textit{cutoff ratio}$$

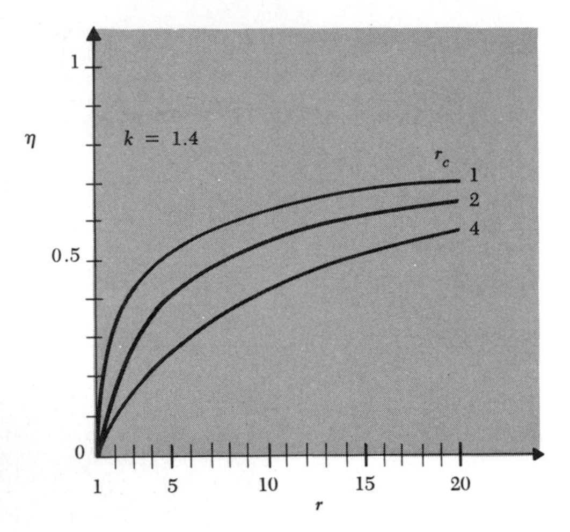

Figure 9·45. Idealized diesel-cycle efficiency

The idealized diesel-cycle efficiency is shown as a function of the compression and cutoff ratios in Fig. 9·45. Note that the efficiency for $r_c = 1$ is the same as for the Otto cycle, and that in general the diesel cycle has a lower efficiency than the Otto cycle *operating at the same compression ratio.*

Compression ratios of practical engines are limited by the pressures and temperatures which can be tolerated within the cylinders. In the spark-ignition engine combustion occurs as the gas is being compressed, while in the injection engine (compression-ignition) combustion begins as the gas is being expanded. As a result, diesel engines can operate at higher compression ratios (of the order of 15 : 1) than can spark-ignition engines (of the order of 8 : 1), and consequently can obtain comparable and even superior efficiencies.

A common modification used with spark-ignition engines for aircraft operation is the *supercharger*, a steady-flow compressor used to compress the air before it enters the reciprocating engine. Supercharging an Otto engine will not improve its efficiency, which is only a function of its compression ratio. In fact, work must be provided to run the supercharger. However, the increase in density of the working fluid produced by the supercharger can result in greater power, since more fluid mass is circulated through the engine. The power/weight ratio is thereby increased.

A second modification, employed on the Wright turbosupercharged engine, is the *exhaust turbine*. The pressure in the cylinder of an Otto-cycle engine at bdc is greater than atmospheric pressure, and consequently work can be obtained by passing the gases through a steady-flow turbine. The result is (ideally) an increase in efficiency, since the work obtained from the turbine is greater than that required to drive the compressor.† A flow diagram and a process representation

† This may be seen easily by examining the *h-s* process representation for the turbosupercharged cycle (Fig. 9·46), since the enthalpy changes are related to work for steady-flow adiabatic devices.

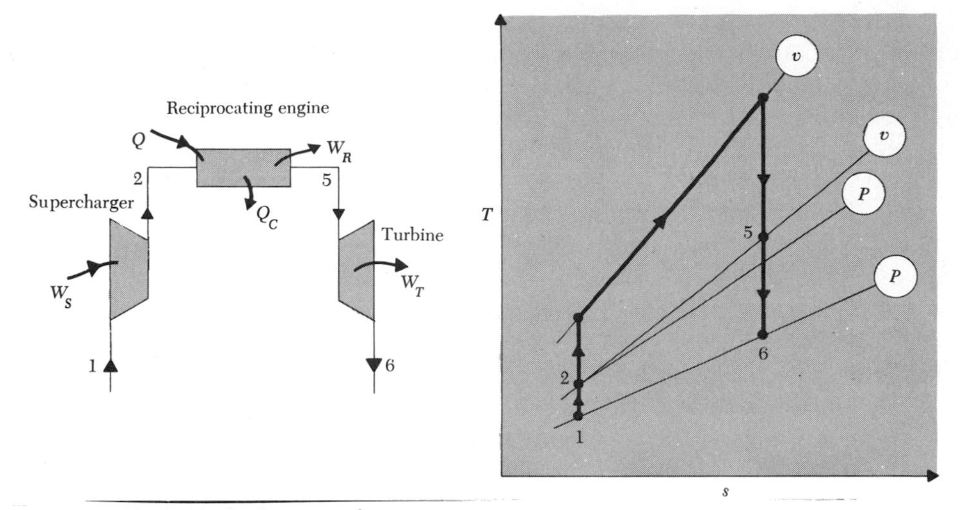

Figure 9·46. An idealized compound engine system

for the idealized *turbocompound engine* is shown in Fig. 9·46. Recently there has been considerable interest in turbosupercharging of diesel and spark-ignition engines. Here a free-running turbine is used to drive the supercharger directly, and useful power is extracted solely from the reciprocating engine.

An interesting hybrid employs the *free-piston gas generator*. This device consists of a pair of free-floating pistons synchronized to move in opposition. Air is compressed between them as they move together, diesel fuel is injected and ignites spontaneously, and the hot high-pressure gases are exhausted through a power-producing turbine. The system schematic and process representation (idealized) are shown in Fig. 9·47. High isentropic efficiencies are more easily obtained with reciprocating (rather than steady-flow) compressors, and this is a practical advantage of the free-piston system.

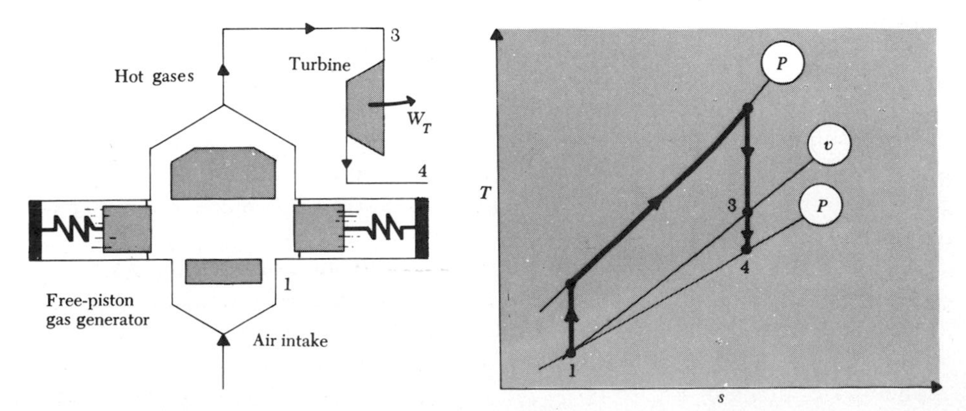

Figure 9·47. Idealized free-piston compound engine system

Another cycle of interest is the *Stirling cycle*. This is of interest because it offers the possibility of reaching (in the ideal cycle) the Carnot efficiency. It consists of an isothermal compression 1-2, a constant-volume heating 2-3, an isothermal expansion 3-4, and a constant-volume cooling 4-1. Note that heat is rejected in process 1-2 and added to the system in process 3-4. Ideally, a regenerator could be placed in the system, and the heat rejected in process 4-1 could be used as the heat input in process 2-3. Under these conditions the ideal system would have an efficiency equal to the Carnot efficiency for the two temperature extremes. Stirling cycles using reciprocating hardware are currently being developed for automotive and small-power systems and have even been considered for artificial heart power. These systems employ air, helium, or some other gas in a *closed* cycle, using an external burner, nuclear reactor, radioisotope pile or other suitable heat source, and rejecting energy as heat to a colder region via an appropriate heat exchanger. A scheme for artificial heart power uses a radioisotope pile at 1200°F, implanted in superinsulation in the abdominal cavity, with the bloodstream acting as the cycle coolant.

The student should note that all these model cycles are highly idealized. They have great merit as an aid to studying the actual systems, but one should not lose sight of the idealizations involved. Real spark-ignition engines do not have energy input exactly at constant volume. Real diesel systems do not have a constant-pressure combustion process, which is merely a simple approximation to the process of heating due to fuel burning during the expansion process. We also remark that in these simplified model cycles the energy delivered to the air from the fuel is considered as if it were a heat input, and no chemical changes in the working fluids are considered. The chemical processes in real engines are of course important and must be considered in accurate engineering calculations for such systems. We shall learn how to do this in Chaps. 11 and 12.

9·16 AIR-CYCLE REFRIGERATOR

The reverse of the Brayton cycle can be used as a refrigeration cycle, and such systems are in common use aboard jet aircraft. Air is bled from the discharge of the main-engine compressor and expanded through a small turbine. The shaft power from the turbine is usually used to drive the fans circulating air through the cabin. The flow schematic and process representation for this *air-cycle refrigeration system* are shown in Fig. 9·48.

To illustrate air-cycle cooling as applied to aircraft, let's design a very simple cabin cooling system for a light jet airplane designed to cruise at 600 mph at an altitude of 10,000 ft. At this altitude the ambient temperature is approximately 35°F and the pressure is about 10 psia. We shall use the hardware and model cycle shown in Fig. 9·49.

It might be thought that we could easily cool the cabin by simply scooping

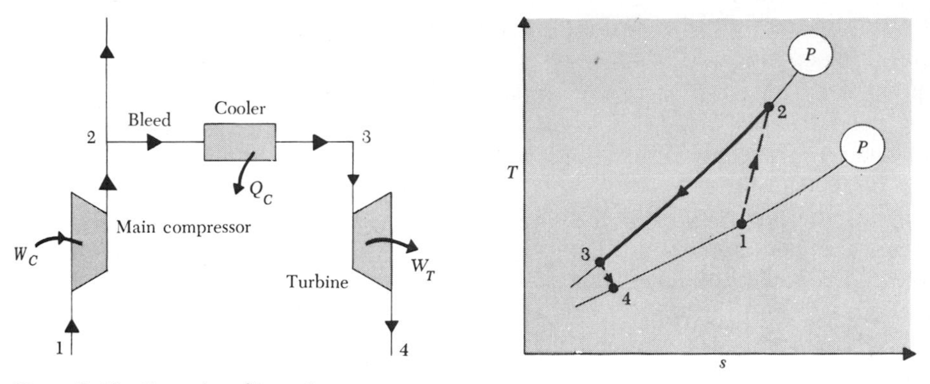

Figure 9·48. Air-cycle refrigeration

in the cold ambient air. But let's see what happens when we try. An energy analysis of the scoop, in which flow relative to the airplane is diffused from 600 mph to very low speeds, can be used to calculate the temperature of the air after it has been brought nearly to rest. Assuming that the scoop is adiabatic, and making other simplifying idealizations (the student may list these in the margin), we have

$$h_1 + \frac{V_1^2}{2g_c} = h_2$$

Treating the air as a perfect gas, the enthalpy difference may be expressed in terms of the temperature difference, and we have

$$T_2 = T_1 + \frac{V_1^2}{2g_c c_P}$$

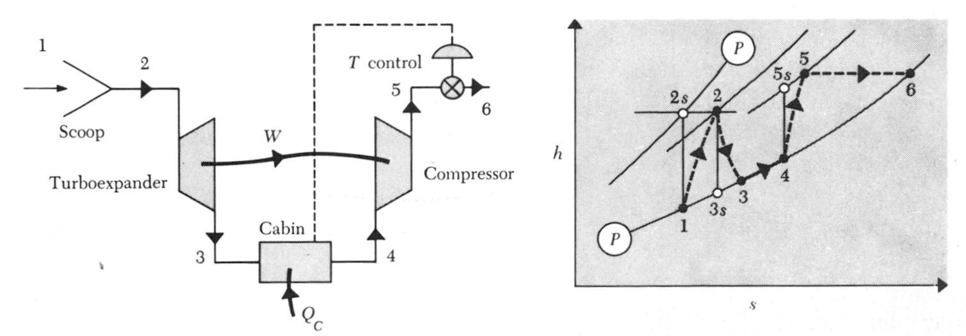

Figure 9·49. Our aircraft cooling system

Putting in the numbers (600 mph = 880 ft/s),

$$T_2 = 495 + \frac{880^2}{2 \times 32.2 \times 0.24 \times 778} = 559 \text{ R}$$

We see that in the process of slowing down the flow we convert the flow stream bulk kinetic energy into molecular kinetic energy and consequently greatly increase the gas temperature. This temperature (99°F) is too hot for use as cooling air, and somehow the temperature must be reduced before the air is admitted to the cabin.

Suppose we drop the temperature of the air by expanding it through a turbine. Energy removed from the air will appear as turbine shaft work, which must somehow be absorbed; we will connect the turbine to a small compressor which will take the air blown through the cabin and pump it to a higher pressure for ejection from the aircraft. The small turboexpander will have an isentropic efficiency of the order of 0.85, and a very crude compressor can be used, for it must only serve to deliver the turbine work to the air just before it leaves the aircraft. System control may be effected by discharging through a control valve, which could regulate the flow so as to maintain a preset cabin temperature.

We must determine the turboexpander inlet pressure. The scoop-diffuser may have a *pressure recovery factor* C_P of about 0.7 where (see Fig. 9·49)

$$C_P \equiv \frac{P_2 - P_1}{P_{2s} - P_1}$$

From the perfect-gas equation of state,

$$\frac{P_{2s}}{P_1} = \left(\frac{T_2}{T_1}\right)^{k/(k-1)}$$

So,

$$P_{2s} = 10 \times \left(\tfrac{559}{495}\right)^{3.5} = 15.2 \text{ psia}$$

Then, calculating P_2 from the assumed value of C_P,

$$P_2 = 10 + 0.7 \times (15.2 - 10) = 13.6 \text{ psia}$$

Let's assume that the cabin is designed to operate at 10 psia, which we take as the turbine discharge pressure. Analyzing the turbine in the usual way we find

$$T_3 = 512 \text{ R } (52°\text{F})$$

$$W_T = 11.3 \text{ Btu/lbm}$$

It would not take much air at 52°F to maintain the cabin of a small airplane at a comfortable 75°F. Suppose, for example, that the cabin heat load Q_C is 10,000 Btu/h (about 3000 W). We analyze the cabin by assuming that the air leaves the cabin at the cabin temperature ($T_4 = 535$ R). Then, the required air-flow rate is

$$\dot{M} = \frac{\dot{Q}_C}{h_4 - h_3} = \frac{\dot{Q}_C}{c_P(T_4 - T_3)}$$

$$= \frac{10{,}000}{0.24 \times (535 - 512)} = 1810 \text{ lbm/h}$$

At the scoop inlet density of 0.0546 lbm/ft^3, and the scoop inlet velocity of 880 ft/s, the scoop frontal area would have to be only

$$A = \frac{\dot{M}}{(\rho_1 \mathsf{V}_1)} = \frac{\frac{1810}{3600}}{0.0546 \times 880} = 0.010 \text{ ft}^2$$

The turboexpander power output is

$$\dot{W}_T = M\dot{W}_T = 1810 \times 11.3 = 20{,}500 \text{ Btu/h}$$

which is about 8 hp.

The student might complete the analysis of the other components in the system, and in particular examine the possibility of using this turbine power in a more direct way. The designer of such a system would carry out an extensive optimization analysis, experimenting with different system state points in an attempt to provide the necessary cooling for the least weight and cost.

9·17 A SIMPLE THRUSTING SYSTEM

Thermodynamics plays an important role in the analysis of propulsion systems. Let us consider one of the simpler types of thrust producers, shown in Fig. 9·50. Gas at high pressure is bled from a storage bottle to a plenum chamber, from which it is exhausted through a nozzle. The pressure forces acting on the chamber walls are shown in Fig. 9·50*a*. Note that there is a net force which would tend to accelerate the walls opposite the direction of flow. This type of *blowdown thruster* is used in low-thrust situations where simplicity and reliability are key factors and where propellant weight is not vital. The attitude-control systems of our space vehicles have used nitrogen thrusters in the range 2 to 100 lbf.

The thrust of a propulsion system is the force the fluids exert on the container as a result of nonuniformities in pressure distribution. Examination of Fig. 9·50*b* indicates that this force is roughly $P_0 A_t$, where P_0 is the chamber, or *stagnation*, pressure, and A_t is the throat area of the nozzle. An exact calculation of thrust requires application of the *momentum principle* to a control volume surrounding the gas and cutting across the exit plane of the nozzle. The momentum

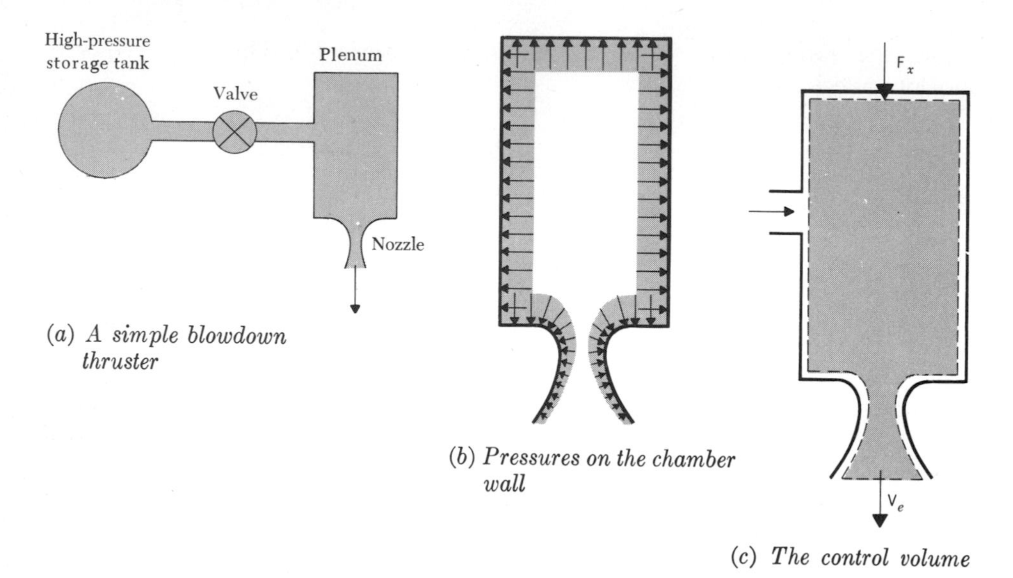

(a) *A simple blowdown thruster*

(b) *Pressures on the chamber wall*

(c) *The control volume*

Figure 9·50.

principle is a generalization of Newton's law arising from a control-volume transformation. We shall develop the momentum principle for a control volume in Chap. 13, and here merely give the result of this transformation in Table 9·1. Here $\dot{\mathscr{P}}_{\mathbf{M}}$ denotes the *rate of production of momentum*, as defined in the table.

Let us now apply the momentum principle to the control volume of Fig. 9·50*c*, assuming that the system is steady-flow, steady-state, one-dimensional at the inlet and exit planes, and fixed in space. We write the momentum equation for the x direction, since this is the component of force we seek. The entering fluid has no momentum in this direction, and the rate of storage of momentum $d(\mathbf{M})/dt$

Table 9·1 The momentum principle

System	Momentum equation†
Isolated system	$\dot{\mathscr{P}}_{\mathbf{M}} = \dfrac{d(\mathbf{M})}{dt} = 0$
Control mass	$\dot{\mathscr{P}}_{\mathbf{M}} = \dfrac{d(\mathbf{M})}{dt} = g_c\mathsf{F}$
Control volume	$\dot{\mathscr{P}}_{\mathbf{M}} = \dfrac{d(\mathbf{M})}{dt} + \sum_{\text{out}} \dot{M}\mathsf{V} - \sum_{\text{in}} \dot{M}\mathsf{V} = g_c\mathsf{F}$

† $\mathbf{M}$ is total momentum of matter within the system. V is the momentum of a unit mass of matter crossing the boundaries. F is the total force exerted on the control volume.

is zero by the steady-flow, steady-state and no-acceleration idealizations. The only terms remaining are

$$\dot{M}\mathsf{V}_e = g_c(\mathsf{F}_x - P_e A_e) \tag{9·19}$$

Here F is the force exerted by the walls on the fluid, and $P_e A_e$ is the force acting across the exit plane. Often the exit pressure force is negligible, and

$$\mathsf{F}_x \approx \frac{\dot{M}\mathsf{V}_e}{g_c} \tag{9·20}$$

Thermodynamics becomes involved when we want to calculate the discharge velocity in order to determine the thrust. Consider the control volume of Fig. 9·51. We assume that the flow is steady, one-dimensional, and adiabatic, and that the kinetic energy of the fluid in the plenum chamber is negligible. An energy balance then gives

$$h_0 = h + \frac{\mathsf{V}^2}{2g_c} \tag{9·21}$$

Let us further assume that the fluid is a perfect gas with constant specific heats. Then, using Eq. (8·25), we can relate the chamber (stagnation) temperature T_0 to the temperature and velocity at any flow section and express the result as

$$\frac{T}{T_0} = 1 - \frac{\mathsf{V}^2}{2g_c c_P T_0} \tag{9·22}$$

If we further idealize that the process undergone by the fluid is isentropic, from Eq. (8·27) we have

$$\frac{P}{P_0} = \left(\frac{T}{T_0}\right)^{k/(k-1)} = \left(1 - \frac{\mathsf{V}^2}{2g_c c_P T_0}\right)^{k/(k-1)} \tag{9·23}$$

The process representation is shown in Fig. 9·51.

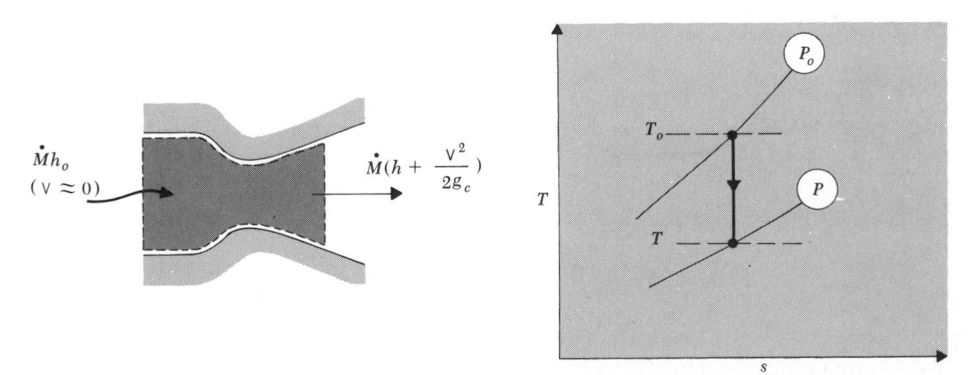

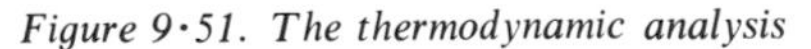
Figure 9·51. The thermodynamic analysis

The dynamics and thermodynamics, plus conservation of mass, allow complete analysis of this thrusting system. Equations (9·21) and (9·22) hold at any section of the flow (provided our idealizations are reasonable), and in particular at the exit plane. They tell us that, the more the gas is expanded in the nozzle, the higher will be the exit velocity, and consequently the greater will be the thrust.

As an example, suppose we consider a system with $P_e/P_0 = 0.1$ and a stagnation temperature of 30°C. For nitrogen, $k = 1.4$ and $c_P = 1.038$ kJ/(kg·K), so the temperature at the exit plane is†

$$T = T_0\left(\frac{P}{P_0}\right)^{(k-1)/k} = 303 \times 0.1^{0.286} = 157 \text{ K}$$

The exit velocity may now be calculated from Eq. (9·22),

$$\mathsf{V} = \sqrt{2 \times 1 \times 1.038 \times 303 \times (1 - 157/303) \times 1000} = 550 \text{ m/s}$$

The thrust calculation would require knowledge of the mass-flow rate. A useful measure of the performance of a propulsion system is the *specific impulse*, or thrust force per unit of mass-flow rate. For our system,

$$SI = \frac{\mathsf{F}_x}{\dot{M}} \approx \frac{\mathsf{V}_e}{g_c}$$

so

$$SI = \frac{550}{1} = 550 \text{ N/(kg/s)}$$

Hence, a 100-N nozzle will require about 0.18 kg of nitrogen per thrusting second.

Inspection of Eq. (9·23) indicates that there is a limit on the velocity which can be obtained by isentropic expansion of a gas. Since the pressure cannot fall below zero,

$$\mathsf{V}_{\max} = \sqrt{2 g_c c_P T_0}$$

which for our case amounts to 793 m/s. Thus the best specific impulse we could hope to obtain with the blowdown thruster is of the order of 793 N/(kg/s). We see that a higher specific impulse with a gaseous flow requires higher stagnation temperatures. In the next section we shall examine the performance of some more powerful propulsion systems, indicating the sorts of specific impulses they attain.

† This extreme cold would tend to freeze any water vapor in the gas, and such little ice crystals are believed to be the cause of the "fireflies" seen by astronauts.

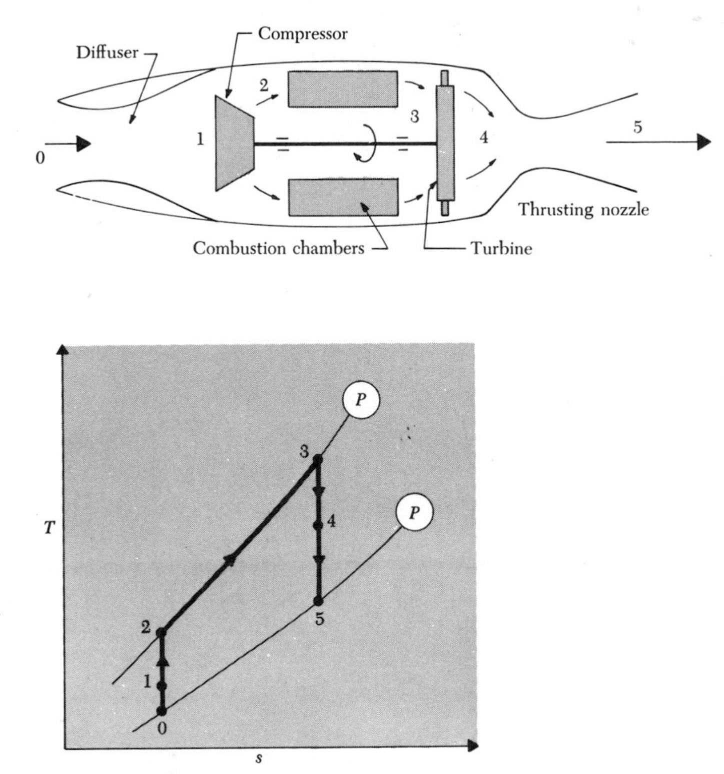

Figure 9·52. An idealized turbojet engine

9·18 OTHER THRUSTING SYSTEMS

The earliest thrusting systems for aircraft propulsion were merely propellers driven by internal combustion engines.† The *turboprop engine*, now used aboard commercial cargo aircraft, is an open-cycle gas turbine engine in which the turbine drives the propeller.

The attainment of higher-speed flight requires higher jet velocities than can be obtained with large propellers, and the *turbojet engine* has become the work-horse of larger commercial and military aircraft. It is essentially a gas turbine plant in which the turbine power output is just sufficient to drive the compressor. The turbine exhaust is fed to a nozzle from which the flow is discharged at high velocity. A system schematic and idealized process representation are shown in Fig. 9·52. Military aircraft employ an afterburner, which is essentially a reheat

† The delay in developing the turboprop and the turbojet engine was in part because of difficulties in developing good compressors. In fact, the early Brayton cycles required more work to their compressors than the turbine would produce. These were, obviously, used for research and further development.

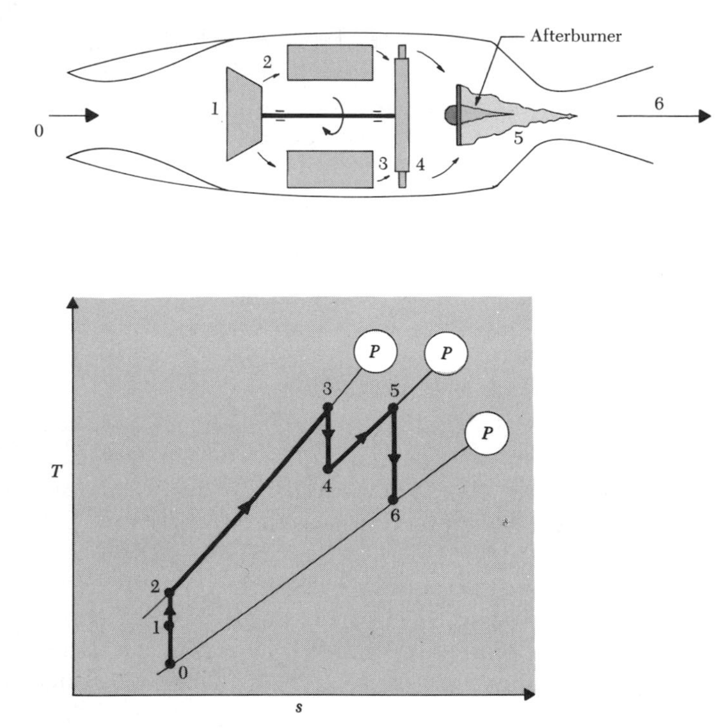

Figure 9·53. Idealized turbojet engine with afterburner

device (Fig. 9·53). The afterburner provides greater nozzle-exhaust velocities and substantially increases the engine thrust. Specific impulses of the order of 600 N/(kg/s) are typical for modern jet engines. The afterburner increases this somewhat but greatly increases fuel consumption, and consequently, afterburners are "cut in" only for short periods of high thrust.

The choice of a propulsion system for aircraft involves many factors. Propeller systems are best suited to low-speed flight (300 mph); these engines derive their thrust by accelerating relatively large amounts of air to a modest velocity. In contrast, high-speed aircraft are best powered by jet engines, which provide thrust by accelerating smaller amounts of air to much higher velocities. In the high subsonic range (600 mph) the hybrid *turbofan engine* has provided the best choice where fuel economy is a prime factor. The turbofan is a jet engine which derives additional thrust from a *ducted propeller*, or *bypass fan*, which is driven off the main turbine shaft. The amount of thrust contributed by the two streams can be adjusted by varying the ratio of the bypass air-flow rate to the engine air-flow rate, and an optimum design thereby obtained for any particular type of mission. Bypass ratios in the range 4 to 8 are typical of current systems. Jumbo jets all use high-bypass engines.

The *ramjet* is a carefully designed flow passage which uses the momentum of the onrushing air for compression, thereby entirely eliminating any moving parts. The aircraft must already be flying at a considerable speed before the ramjet engine can be started. Commercial supersonic transport craft could possibly employ turbojet engines for takeoff, ascent, descent, and landing and switch to ramjets for high-altitude supersonic cruising. A system schematic and process representation are shown in Fig. 9·54. Ideally the diffuser process is isentropic, but boundary-layer separation on the diffuser walls makes it difficult to achieve high isentropic efficiency in real supersonic diffusers.

Chemical-rocket engines are essentially open Rankine cycles in which the turbine is replaced by a rocket nozzle. Since a chemical rocket carries its own oxidizer, it is especially well suited for operation in space. Chemical rockets have specific impulses in the range 1500 to 3000 N/(kg/s). A system schematic and highly idealized process representation are shown in Fig. 9·55.

It may be shown that the highest specific impulse is obtained for the lightest possible particles. Hydrogen has therefore been studied in conjunction with a nuclear-reactor energy source, and it appears that specific impulses of the order 9000 N/(kg/s) can be obtained. The nuclear hydrogen rocket (Fig. 9·56) could become an important engine for future space flights.

Ion engines offer the possibility of high specific impulses and will probably be used on long-term space voyages. An easily ionized element, such as cesium, is evaporated and then brought into contact with a suitable high-temperature surface. Electrons are removed at the surface, and the cesium ions and electrons are accelerated by appropriate electric fields and then mixed to form a neutral beam which is shot from the end of the engine. Specific impulses of the order of 100,000 N/(kg/s) are possible. A schematic of a proposed ion engine is shown in Fig. 9·57. Note that while the ion engine offers high specific impulse, the flow rate would be very small so that the thrust would be small. Consequently such a device would be used for accelerating a vehicle only after the vehicle was already in space and away from the earth's gravitational field. The advantage of such a device is that a very high thrust per kg expended is obtained, so that for long flights extremely high velocities can be achieved, which greatly shorten the flight time.

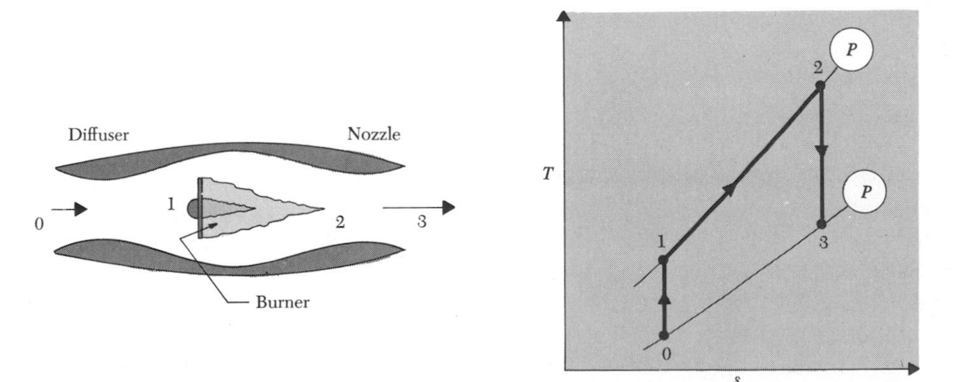

Figure 9·54. An idealized ramjet engine

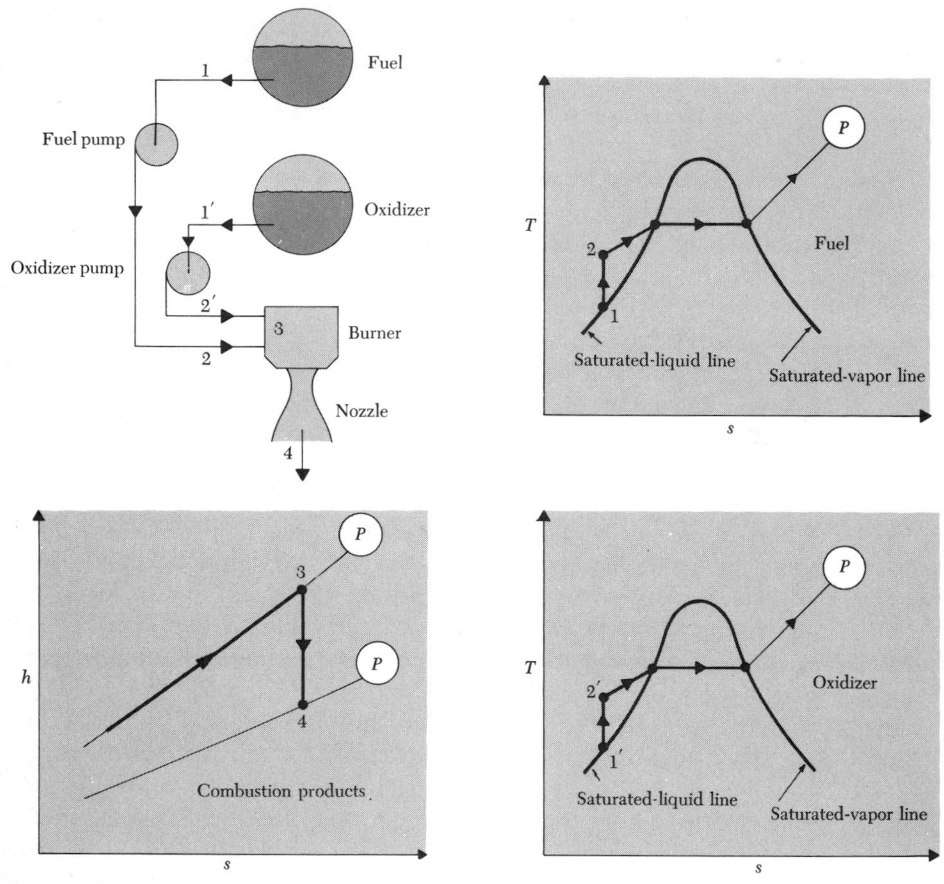

Figure 9·55. A chemical-rocket engine

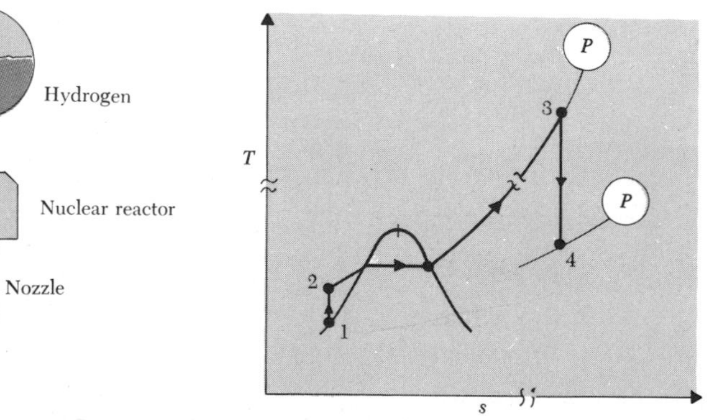

Figure 9·56. The hydrogen-rocket engine

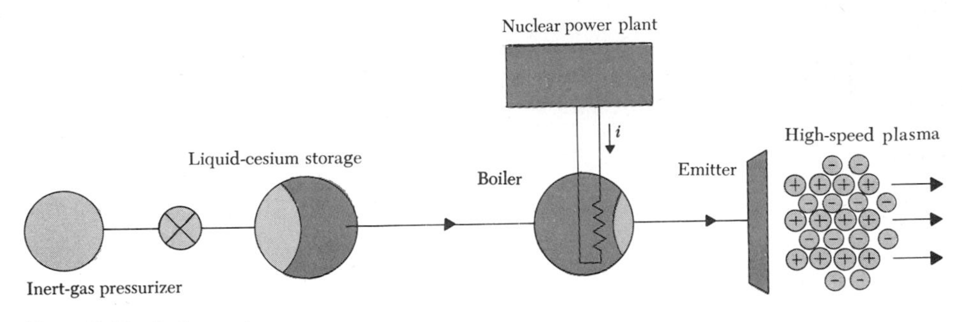

Figure 9·57. An ion engine

9·19 DIRECT ENERGY-CONVERSION SYSTEMS

The conventional devices for electric power generation which we have so far studied are typified by the Rankine vapor power cycle and the Brayton gas power cycle. All involve the circulation and state-manipulation of a working fluid which in a rather indirect way takes the energy from the fuel and helps us convert some of that energy to electric power. This type of power generation requires many heavy bulky components and many moving parts. In addition, we have seen that typical efficiencies are on the order of 30 to 35 percent.

Of growing importance is a class of energy-conversion systems called *direct energy-conversion* (DEC) devices. In these systems the electric energy is obtained more directly from the fuel energy, without the need for circulating and manipulating a working fluid. DEC systems have been developed rapidly as a result of space-age demands, and there are many such systems that show great promise for consumer use. The absence of fluid-handling machinery means that many of the devices work with *no moving parts*, an obvious asset from the reliability point of view.

DEC methods rely on various physical phenomena that permit direct conversion of energy into electricity. Some of the methods use an energy input in the form of heat, while others utilize solar radiation energy or chemical energy. Naturally, the direct conversion devices must still satisfy both the first and second laws of thermodynamics. Analyses of these newer systems require competence in several areas, including thermodynamics, electromagnetics, fluid mechanics, and solid-state physics, and with a few exceptions they are beyond the scope of this text. However, we shall now discuss some of these schemes in a qualitative way, because they have definite and growing importance in engineering.

An interesting type of DEC system already in use in space applications is the *thermoelectric converter*. In 1822, Seebeck found that an electromotive force (emf) is generated between two junctions, of dissimilar metals, at different temperatures. This effect forms the basis of the *thermocouple*, a temperature-measuring device (Fig. 9·58). For metals the emf is quite low, on the order of a few μV per degree of temperature difference. However, the semiconductor materials give

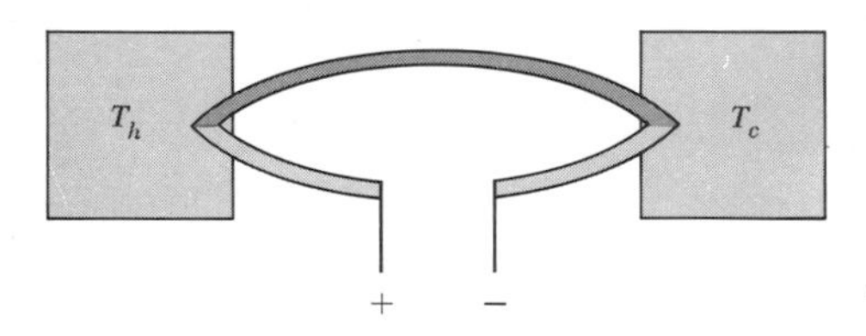

Figure 9·58. A thermocouple circuit

much larger voltage outputs, since *p*- and *n*-type materials can be coupled to take advantage of the difference in behavior between the two materials.†

Since the thermocouple can be represented as a heat engine, the efficiency depends on T_h and T_c as well as on materials parameters of the semiconductors. Figure 9·59 shows a typical thermoelectric converter. Systems producing a few W of power at efficiencies of 5 to 12 percent are reliable realities today.‡ For use at remote locations radioisotope heating has been used successfully as the heat source. Present systems do not come close to Carnot efficiency, but they do represent a reliable source of power with no moving parts.

In 1834, Peltier discovered that a current flow around a loop of two dissimilar metals warms up one end and cools down the other. Again this phenomenon can be enhanced by using semiconductor materials, and in this case a Peltier refrigerator can be made. In fact, one can buy commercial units using this effect. Figure 9·60 shows an example of a thermoelectric refrigerator. A patent§ has even been issued for an electric blanket employing thermoelectric elements which can, in principle, be used either for heating or cooling. Unfortunately, rather high-amperage direct currents are presently required, and it is questionable just when such a blanket will become a consumer item.

A second type of direct energy power device is the *thermionic converter*

† In *n*-type material heating causes electrons to diffuse to the cooler end of the material, so that the hot end becomes positively charged. In *p*-type material the cold end becomes positively charged by diffusion of the "holes."

‡ Thermoelectric converters powered by the waste heat from kerosene lamps are in wide use in rural areas of the Soviet Union, chiefly as a power source for radio receivers.

§ U.S. Pat. No. 3080723.

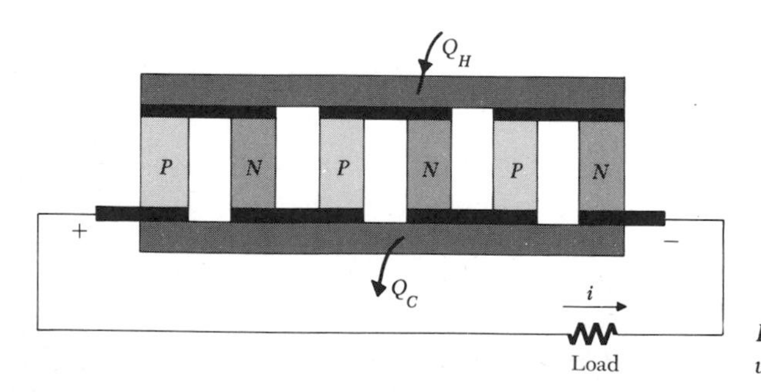

Figure 9·59. A thermoelectric converter

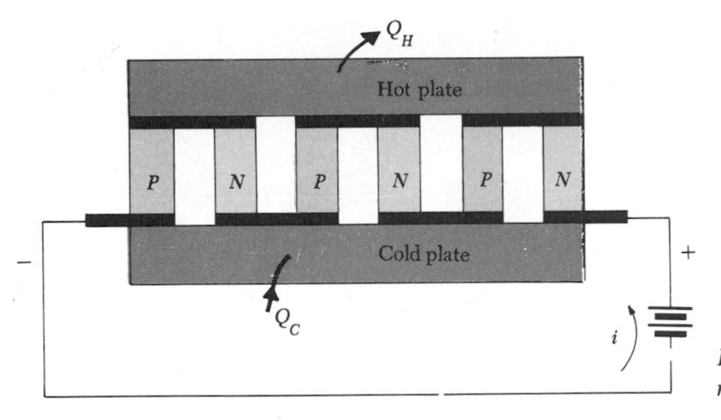

Figure 9·60. A thermoelectric refrigerator

(Fig. 9·61). The converter is essentially a heat engine using electrons as the working fluid, and therefore the Carnot efficiency is applicable as an upper limit. Direct conversion is possible if the "work function" of the anode (cold plate) is less than that of the cathode (hot plate). In order to operate satisfactorily, the vacuum diodes must have very close spacings (of the order of 0.02 mm), for otherwise the retarding potential arising from the distribution of electrons in the gap becomes too great for satisfactory operation. Diodes of this type have been operated with cathode temperatures of the order of 1500 K and with the anode at approximately 900 K, yielding power densities of the order of 2 to 10 W/cm^2. The energy-conversion efficiency of these systems is of the order of 5 to 15 percent compared to a Carnot efficiency of 40 percent with these temperatures. The small spacings are difficult to maintain, especially with large cathode-anode temperature differences, and shortouts are frequent. The gas (plasma) diodes employ easily ionizable elements, such as cesium, whose positively charged ions tend to neutralize the retarding effects of the space charge, allowing larger spacings to be employed (1 mm). Efficiencies of the order of 18 percent have been obtained with this type of plasma diode operated at cathode temperatures of about 2500 K.

The thermionic converter is well suited for the production of electric power in space from a nuclear heat source. Since the energy which must be rejected as heat must be radiated to space, and high temperatures are required for low-weight radiators, space-systems designers can accept low efficiency in the interests of

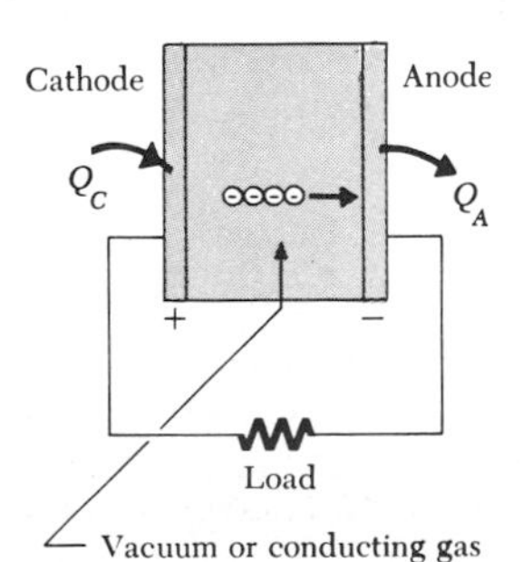

Figure 9·61. A thermionic generator

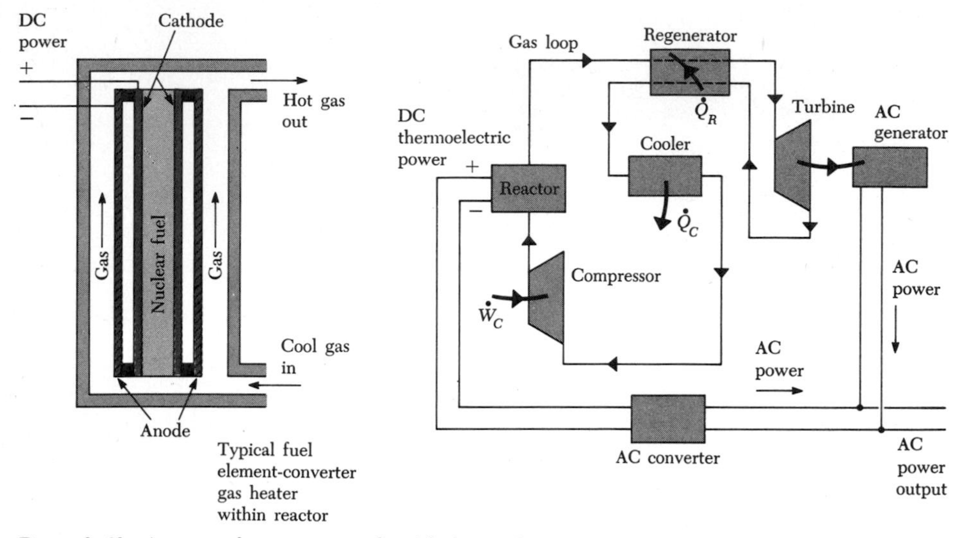

Figure 9·62. A proposed gas power cycle with thermoelectric topping

lightweight nonmechanical power-conversion equipment. The thermionic converter will probably also find application as a topping device for other power systems. Figure 9·62 shows a possible closed-cycle, nuclear-powered gas system using in-pile thermionic converter elements to generate electricity directly.

A somewhat different DEC system is the *photovoltaic device*. It is *not* a heat engine and actually operates efficiently at a low (room) temperature using photons (light) as the primary energy source. In this device light falls on a surface and produces electrons by interaction within the surface. Semiconductor materials are utilized. Energy-conversion efficiencies on the order of 15 percent have been achieved in photovoltaic cells. The most well-known application of these devices has been on space vehicle solar panels. These solar cells are used to recharge batteries which serve as the primary power source.

The *fuel cell* is a device in which a chemical reaction is harnessed directly to produce electric power. Unlike a chemical battery, in which an electrolyte is decomposed into its basic components, a fuel cell utilizes a controlled reaction between components such as hydrogen and oxygen. Since the hydrogen and oxygen can be continuously supplied, the fuel cell does not run down as does a chemical battery. A schematic of a fuel cell is shown in Fig. 9·63. Gaseous hydrogen and oxygen at pressures of the order of 40 atm enter the cell and are brought into contact with porous electrodes. Between the electrodes is a liquid electrolyte, which serves to limit the reaction rate. Hydrogen diffuses through the porous anode, is absorbed on the surface, and reacts with the OH^- ions in the electrolyte, forming water and yielding free electrons. Oxygen diffuses through the cathode, is absorbed by the surface, and reacts with the water to form OH^-

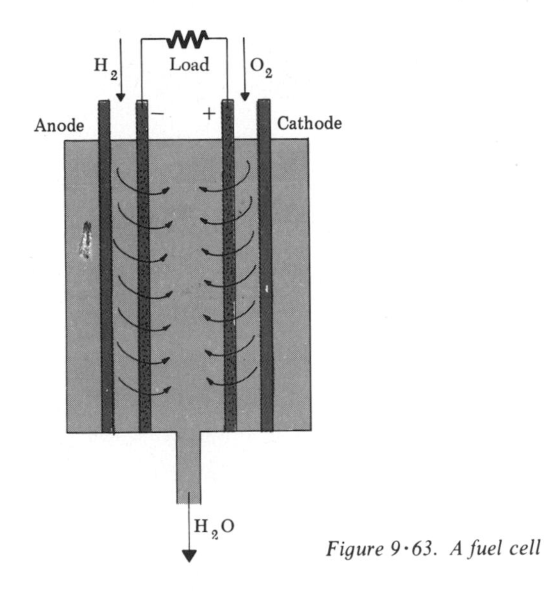

Figure 9·63. A fuel cell

ions. Thus water is continually being formed at the anode and decomposed at the cathode. The reaction rate is controlled by the rate of migration of OH^- ions through the electrolyte. Electrons flow out of the cell through a load and are returned to the cathode. The maximum cell emf is of the order of 1 V and depends to a certain extent on the choice of reactants employed.

Fuel cells have been developed for operation in space vehicles and have many other possible applications. A major difficulty of present cells is the relatively short life span of the electrodes, especially when operated with relatively low-grade fuels. Their lifetime can be markedly increased by use of purified hydrogen, but the costs of the purifying operation are presently prohibitive from the standpoint of consumer use. However, fuel cells may become the power sources for electric cars of the future. Fuel cells operating on natural gas are being developed for use by the electric utility industry.

The fuel cell is not a heat engine (in fact the process is essentially isothermal), and consequently the Carnot efficiency is an irrelevant maximum for fuel-cell performance. However, given a supply of hydrogen and oxygen, more useful power can be obtained with a fuel cell than if the gases were allowed to react spontaneously and a Carnot cycle were run from the high-temperature flame. This is because the voltage of the cell tends to restrain the chemical reaction, much as a piston restrains spontaneous expansion of gas. The maximum theoretical efficiency of the fuel cell, based on the conversion of chemical to electric energy, is 100 percent.

If a fluid that is a good electric conductor flows through a magnetic field, an electric field is induced in the fluid, and power can be obtained. Such a device is

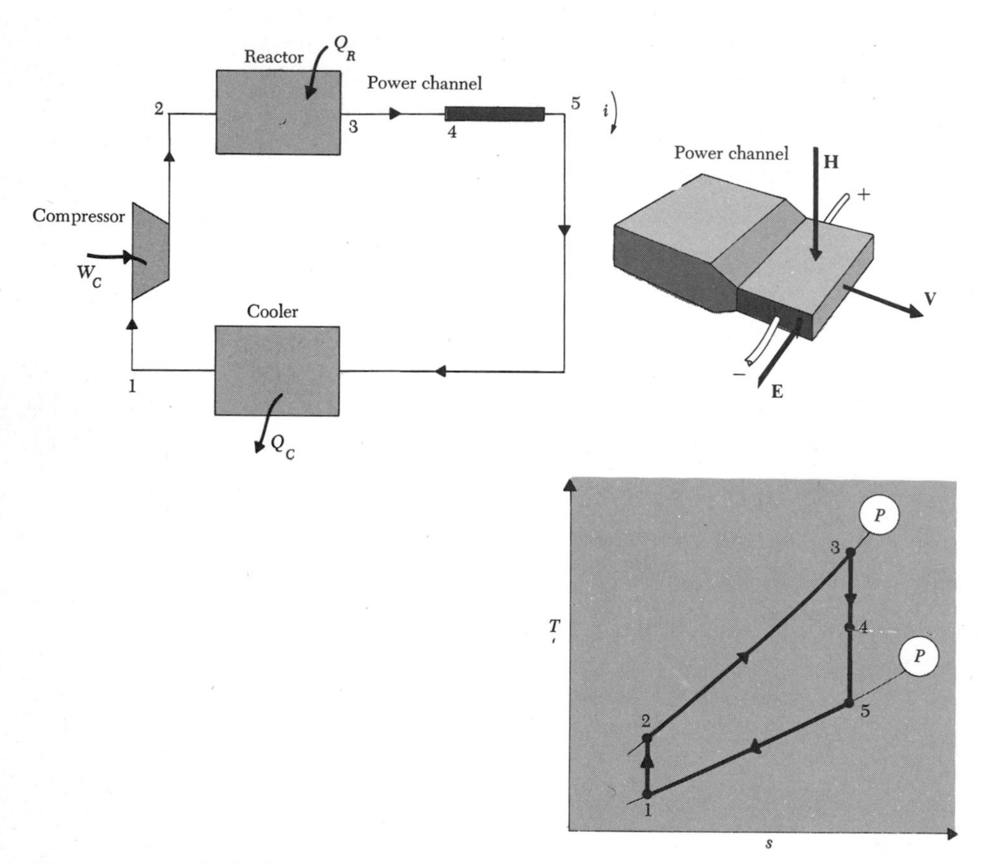

Figure 9·64. Idealized mhd power generator

called a *magnetohydrodynamic* (mhd) *generator*.† While liquid metals such as mercury could be used, ionized gases (plasmas) are much more practical for the job of converting internal energy to electricity. The mhd generators now in existence use the products of combustion of fossil fuels, seeded with easily ionized elements, such as cesium, to form the plasma. Using coal as the energy source in an mhd system looks promising. Fly-ash formation prevents using coal directly in present-day gas turbine systems, but the ash is not a limiting factor in the mhd generator. In the future we can expect to see much more energetic plasmas produced as a result of controlled-fusion reactions. The gas is pressurized, heated, and passed through a nozzle (this looks quite a bit like the Brayton cycle). The high-velocity gases are ducted through a magnetic field, and an electric field perpendicular to the applied magnetic field is developed. A closed-cycle mhd system powered by a nuclear reactor is shown in Fig. 9·64.

† In an mhd device the ionized gas flows at right angles to a magnetic field such that a force is generated at right angles to both the velocity of flow and the magnetic field. The force on the ions and electrons in the device drives them to opposite electrodes, giving rise to a current.

Irreversibilities due to friction, ohmic losses, and heat transfer severely limit the practicability of small mhd generators; the most promising use therefore appears to be in conjunction with large central power stations; plants in the 500-MW range have been studied. The electric conductivity of the gas decreases rapidly with decreasing temperature, which imposes a major limitation on the minimum-operating-temperature range. If the temperature falls too low, the gas will deionize, and the device will open-circuit. The mhd system is therefore best suited to high-temperature operation and looks very attractive as a potential topping system for a binary power plant. The reject energy from the mhd system would thus be used as input energy to a conventional power system. If the plasma were at a temperature of, say, 4500°F, the potential efficiency of a binary cycle with energy rejection at 60°F would be

$$\eta = 1 - \frac{60 + 460}{4500 + 460} = 0.90$$

Design studies suggest efficiencies of about 55 percent may be obtained in large practical systems. The most developed mhd system is a 25-MW experimental system now operating in the Soviet Union. The U.S. effort involves more sophisticated technology, and concentrates on the direct combustion of coal to fire the mhd channel. There are many problems, such as those involving coal slagging, which remain to be solved by research programs such as the substantial mhd program at Stanford University.

An exotic type of refrigerator is the *magnetic refrigerator*, which takes advantage of the magnetic and superconducting properties of materials at low temperatures and has been built for laboratory use.† A sample of material is magnetized slowly while in contact with the high-temperature region, then suddenly (adiabatically) demagnetized, then demagnetized slowly while in contact with the low-temperature region, and finally magnetized adiabatically, returning

† C. V. Heer et al., *Review of Scientific Instruments*, vol. 25, no. 11, p. 1088.

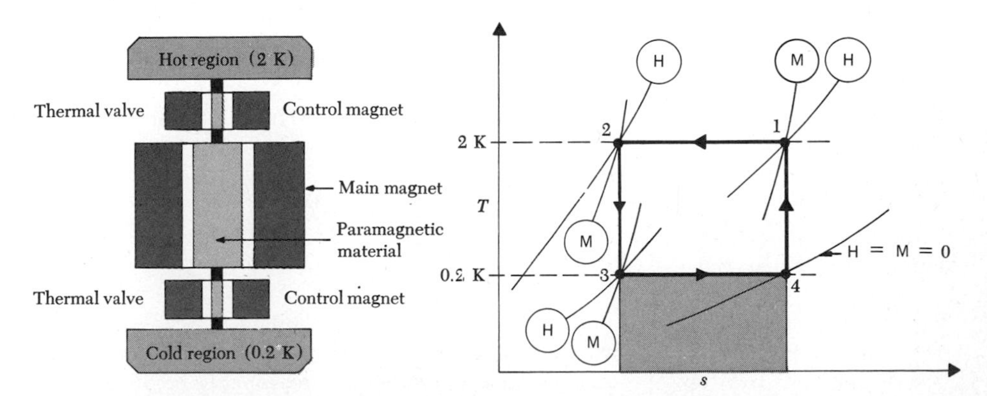

Figure 9·65. Low-temperature magnetic refrigeration

to the initial state. Contact with the low- and high-temperature regions is made through superconductors. Lead is an electric superconductor at low temperatures and a very poor thermal conductor. If a small field (only a few hundred G) is applied to the lead, it undergoes a transition to a normal electric-conduction state, in which it is a good thermal conductor. A "thermal valve" can thereby be made. The schematic diagram and process representation of this type of magnetic refrigerator are shown in Fig. 9·65. The processes undergone by the magnetic substance are in the limit reversible, so that the limiting behavior is that of a Carnot cycle.

9·20 CONCLUDING REMARKS

The student must remember that there are many criteria other than thermodynamics for choosing what systems should be used in a particular application. For space work a low efficiency may be acceptable if a high power/weight ratio can be achieved. Reliability is also of great importance, and the advantage of having no moving parts may offset a high initial cost. In any design a tradeoff will be necessary between the numerous parameters of interest before a final system is chosen. For power generation for large cities, reliability and low fuel cost are important. The problems of site location, air and water pollution, and an aesthetic design of the plant all have to be considered—along with the thermodynamic analysis.

SELECTED READING

Lee, J., and F. Sears, *Thermodynamics*, 2d ed., chaps. 11–13, Addison-Wesley Publishing Co., Inc., Reading, Mass., 1962.

Obert, E., and R. Gaggioli, *Thermodynamics*, 2d ed., chaps. 15, 17, and 18, McGraw-Hill Book Company, New York, 1963.

Van Wylen, G., and R. Sonntag, *Fundamentals of Classical Thermodynamics*, 2d ed., chap. 9, John Wiley & Sons, Inc., New York, 1973.

Wark, K., *Thermodynamics*, 2d ed., chaps. 16 and 17, McGraw-Hill Book Company, New York, 1971.

QUESTIONS

9·1 What is the significance of the area under a line on a T-s plane?

9·2 Derive the Carnot-cycle efficiency from things which you remember.

9·3 What is an isentropic process? What processes might reasonably be idealized as isentropic?

9·4 What is an isentropic efficiency?
9·5 Give an argument for idealizing a steady-flow heating process as isobaric.
9·6 To what matter does the process representation of Fig. 9·5 pertain?
9·7 What is the reason for using reheat in a Rankine cycle?
9·8 How do the bwr values of the Rankine and Brayton cycles compare?
9·9 What is the reason for using a topping cycle?
9·10 What is the reason for supercharging an engine?
9·11 Why do intercooling and reheat lower the efficiency of a nonregenerative gas-turbine power system?
9·12 Explain with the aid of process representations why nonregenerative gas turbine systems operate at higher pressure ratios than regenerative systems.
9·13 Is the thermoelectric converter a thermodynamic system?
9·14 What is the function of the valve in a vapor refrigeration system?
9·15 Can the cop of a heat pump ever exceed unity?
9·16 What is an afterburner?
9·17 Why does a ramjet not have a turbine?
9·18 What is a turboprop engine?
9·19 Why do jet engines have diffusers at the inlet?
9·20 Is a high or a low pressure ratio most desirable for a jet engine?
9·21 Can regeneration be used in jet engines?
9·22 Why does the hydrogen engine have a higher specific impulse than a chemical rocket?
9·23 Why is the ion engine not being considered for launchings, but being planned for long-term inflight propulsion?

PROBLEMS

9·1 Determine the isentropic efficiency of the turbine example of Sec. 5·4.
9·2 Determine the isentropic efficiency of the compressor in the heat-pump example of Sec. 5·4.
9·3 Air is to be compressed from atmospheric pressure at 60°F to 100 psia in a centrifugal compressor (isentropic efficiency about 0.70). The flow rate will be 500 lbm/min. Specify the hp requirement for the driving motor.
9·4 Air is to be supplied steadily at 10 lbm/s, 120 psia, 70°F. Determine the necessary hardware. Specify the compressor efficiency, power input, and the heating (or cooling) requirements of any heat exchangers.
9·5 A steam turbine manufacturer has determined the following data for a small 10-kW turbine: inlet state, 300 psia, 550°F; exit state, 20 psia, 228°F. In the throttling calorimeter (see Prob. 5·29) on the turbine exit $P = 14.7$ psia, $T = 220$°F. Calculate the turbine isentropic efficiency.
9·6 Air enters an adiabatic nozzle at 1000°F and 20 psia and emerges at 1 psia. The isentropic efficiency of the nozzle is known to be 0.96. Determine the discharge velocity.
9·7 Oxygen flows through a flow-metering nozzle located in a 1-in-diameter pipe. The nozzle throat diameter is $\frac{1}{4}$ in. The upstream pressure and temperature are 40 psia and 100°F, and the pressure at the nozzle throat is 10 in of water lower than the upstream pressure. Determine the oxygen-flow rate, assuming $\eta_s = 0.94$.

9·8 What is the value of the efficiency of the adjacent reversible cycle?

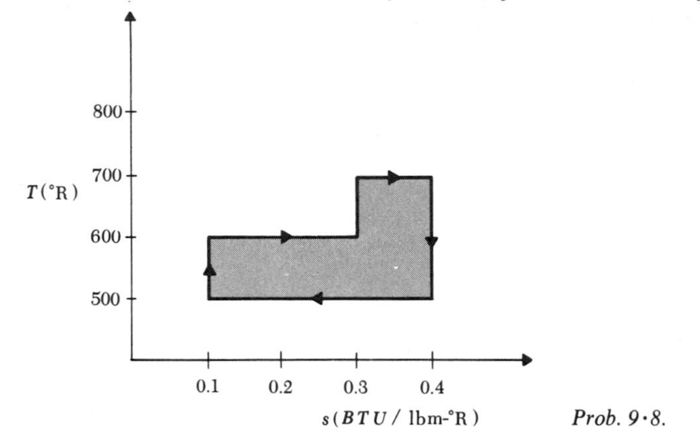

Prob. 9·8.

9·9 A paramagnetic refrigeration system uses iron-ammonium alum in a cyclic process. The refrigeration cycle begins with a slow isothermal magnetization at 2 K, brought about by slowly increasing the external field H to 20,000 G. This is followed by a sudden reduction of the external field to 1000 G, which suddenly drops the temperature of the alum. The alum is then put into thermal contact with the region to be maintained at low temperature, and the external field is slowly reduced to zero. The external field is then suddenly increased, causing the temperature to rise to 2 K and completing the cycle. The sudden processes may be idealized as adiabatic and the other processes are isothermal. Using the equation of state of Fig. B·12, calculate the amount of energy pumped per cycle from the cold space by 15 g of alum, and the cold-space temperature. What net energy input must the external field supply to the alum (as work per cycle)?

9·10 A manufacturer's test results indicate the following data for a small steam turbine:

Inlet state	*Exit state*
$T_1 = 600°\text{C}$	$T_2 = 250°\text{C}$
$P_1 = 4.0 \times 10^6\ \text{N/m}^2$	$P_2 = 0.2 \times 10^6\ \text{N/m}^2$

Determine the isentropic efficiency for the turbine assuming that the heat loss is negligible.

9·11 Three high-pressure turbines have been submitted for testing. Each operates with an inlet state of 1000°F, 3500 psia steam. Turbine X has an exit state of 1000 psia at 660°F and a heat loss of 7 Btu/lbm; turbine Y's exit state is 1000 psia at 635°F with a heat loss of 12 Btu/lbm; and turbine Z's heat loss is 2 Btu/lbm with an exit state of 1000 psia at 680°F. Summarize the test data, calculate the work output and isentropic efficiency of each turbine, and give a purchase recommendation.

9·12 A European company has recommended their steam turbine over that of a U.S. manufacturer. The European turbine operates with an inlet state at $34 \times 10^6\ \text{N/m}^2$ at 540°C and an exit state of $11 \times 10^6\ \text{N/m}^2$ at 370°C. The heat loss is estimated as 25 kJ/kg. The U.S. turbine operates with an inlet state of 1000°F at 5000 psia and exits at 680°F, 1600 psia, with a heat loss of 10 Btu/lbm. Which turbine do you recommend?

9·13 Steam flows through a condenser at 2×10^6 kg/h, entering as saturated vapor at 40°C and leaving at the same pressure as subcooled liquid at 30°C. Cooling water is available at 18°C. Environmental requirements limit the exit temperature to 25°C. Determine the required cooling water–flow rate.

9·14 The heat-rejection system for a proposed helium-cooled, high-temperature nuclear reactor power station is pictured below. Helium enters at 230°C and must leave at 40°C at a flow rate of 7×10^6 kg/h. Water is cooled in the dry cooling tower to 35°C. The water-flow rate is 12×10^6 kg/h. Determine the water exit temperature. If the pressure drop in the water loop is 0.05 MPa, calculate the pumping power required for a pump efficiency of 0.7. The air enters at 25°C and leaves at 75°C. Calculate the required air-flow rate.

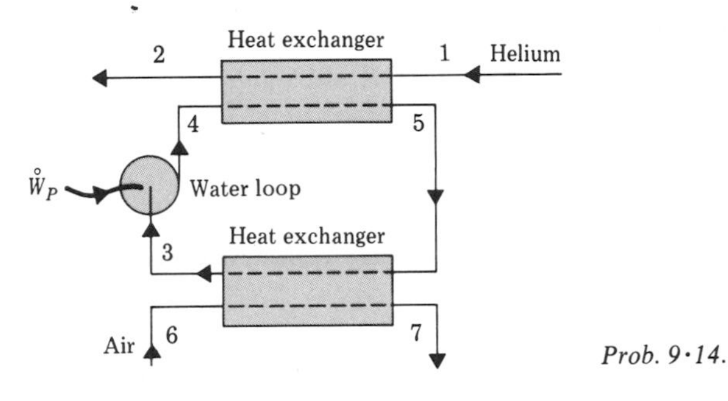

Prob. 9·14.

9·15 The condensate pump on the Navaho plant raises the water pressure from 3.5 in of mercury to 390 psia at a flow rate of 5100 gal/min. If the pump isentropic efficiency is 0.80, calculate the power required in hp.

9·16 The boiler feedwater pumps on the Navaho plant are driven by a steam turbine. The pump inlet is 600 psia at 415°F and the pressure rise is to 3850 psia. If the pump efficiency is 0.85 calculate the work required per lbm of flow. If the flow rate is 6900 gal/min and the isentropic efficiency of the turbine is 0.87, calculate the steam turbine power output.

9·17 The Navaho power plant uses two pumps for circulating cooling water. Each is rated at 135,500 gal/min when producing a pressure rise equal to 66 ft of water. If the isentropic efficiency is 0.75, calculate the hp required.

9·18 A booster pump ($\eta_s = 0.85$) at a large power plant raises the pressure from 3×10^6 N/m^2 at 40°C to 6×10^6 N/m^2. If the flow rate is 20,000 kg/min, calculate the power required in W.

9·19 A secondary air fan at a power plant moves 10,000 ft^3/min and produces a pressure rise equal to a head of 2 in of water. Calculate the power required. The inlet conditions are 70°F at 1 atm.

9·20 The primary air fans in the Navaho power plant move 305,000 ft^3/min at 100°F and produce a pressure rise of 35 in of water. Calculate the hp required if the isentropic efficiency is 80 percent.

9·21 Calculate the thermal efficiency of a simple Rankine cycle for which steam leaves the boiler as saturated vapor at 3×10^6 N/m^2 and is condensed to saturated liquid at 7000 N/m^2. The pump and turbine have isentropic efficiencies of 0.6 and 0.8, respectively. The pump inlet is saturated liquid.

9·22 Calculate the thermal efficiency for a Rankine cycle with turbine inlet conditions of 3×10^6 N/m^2 at 500°C and a condenser pressure of 0.007 MN/m^2. The water leaves the condenser as saturated liquid. The pump and turbine efficiencies are 0.65 and 0.85, respectively.

9·23 A Rankine-cycle power plant has one stage of reheat. The turbine inlet is 3×10^6 N/m^2 at 500°C. After expansion to 0.5×10^6 N/m^2 the steam is reheated to 500°C and expanded in a second turbine to a condenser pressure of 0.007×10^6 N/m^2. The steam leaves the condenser as saturated liquid. Calculate the cycle efficiency, using a pump efficiency of 0.6 and turbine efficiencies of 0.8.

9·24 Redo the previous problem for a pump isentropic efficiency of 0.9 and turbine isentropic efficiencies of 0.88.

9·25 A possible design for a Rankine-cycle steam power plant using a BWR features a valve located between the reactor and the turbine. The water in the reactor is at 6.8 MPa and leaves as slightly superheated vapor at 300°C. The valve drops the pressure by 2 MPa after which the steam enters a turbine of $\eta_s = 0.75$. Condensation occurs at 0.006 MPa. The pump efficiency is 0.85, and the pump inlet temperature is 30°C. Analyze the system for operating conditions, turbine and pump work, cycle efficiency, heat rates, and water-flow rate if the net plant output is 500 MW.

9·26 A Rankine cycle for space uses cesium as the working fluid. The turbine inlet condition is 2000 R at 21.4 psia and is fixed. The pump inlet is saturated liquid, and the pump work is negligible. The turbine efficiency is 0.7. In space the condenser must be a radiator, and weight is of crucial importance. The object of this problem is to determine the operating conditions that produce a minimum-weight radiator for a given turbine power. Calculate and plot the ratio of radiator area to power output (ft^2/kW) versus the condenser temperature. Try a condenser temperature range from 1200 to 1800 R. The condenser heat is radiated according to $Q_c = \sigma T^4 A$, when A is the radiator area and σ is a physical constant, $\sigma = 0.1714 \times 10^{-8}$ Btu/(h·ft^2·R^4).

9·27 Determine the energy-conversion efficiency that could be obtained in the example in Sec. 9·4 if the saturated liquid is fed directly to the pump (no subcooling). Specify the flow rate and boiler-heat input for 1 MW of output power.

9·28 Consider the system of Fig. 9·14. Keeping the turbine inlet and condensing temperatures fixed, and using pump and turbine isentropic efficiencies of 0.8 and 0.9, calculate the performance over the maximum range of boiler pressures and determine the boiler pressure yielding maximum cycle efficiency.

9·29 Determine the energy-conversion efficiency for a supercritical steam power plant where condensation occurs at 100°F and the high-pressure side is at 4000 psia. Assume that the maximum system temperature is 1000°F and that the overall isentropic efficiency of the turbines (several used in series) is 0.90. The pump efficiency is 0.60. Specify the heating rate and steam-flow rate for 100 MW of power output.

9·30 A mercury-vapor power cycle for use in space operates on the Rankine cycle, except that a valve is added between the boiler and the turbine. Condensation occurs at 10 psia, and saturated liquid at this pressure enters the electromagnetic pump, which has an isentropic efficiency of 0.12. The boiler operates at 180 psia and delivers 0.98-quality vapor at this pressure. The valve drops the pressure to 100 psia before the fluid enters the turbine, which has an isentropic efficiency of 0.60. Make a thermodynamic analysis of this system and determine the energy-conversion efficiency, mercury-flow rate, reactor thermal power, and condenser heat-rejection rate for 10 kW of power output. What happens to the energy transferred as heat from the condenser?

9·31 Determine the performance of the system of Prob. 9·30 when the valve is wide open (no pressure drop); i.e., find all the above parameters for this case.

9·32 The valve in Prob. 9·30 is necessary to reduce the moisture content of the vapor in the turbine. Suppose an effective zero-g boiler-superheater could be designed such that the valve could be eliminated and the boiler pressure reduced to 100 psia, keeping the turbine inlet state the same as in Prob. 9·30. Determine the performance of this system.

9·33 A cesium-vapor power plant for use in space operates with a boiler pressure of 50 psia and a condensing temperature of 1000 R. The maximum vapor temperature is 2200 R. The isentropic efficiencies of the pump and turbine are 0.65 and 0.70, respectively, and the system operates on the Rankine cycle. Determine the system performance. Compute the conversion efficiency, reactor heat-transfer rate, flow rate, condenser heat-rejection rate, and turbine hp for 10 MW of power.

9·34 Suppose the cesium-vapor cycle of Prob. 9·33 is used as a topping cycle for the steam power cycle of Prob. 9·29. The flow rates of the two systems will not be the same as in these problems. Determine the efficiency of the combined plant and specify the heat-transfer rates for the cesium boiler, the cesium-steam heat exchanger, and the steam condenser, and the pump and turbine hp for 300 MW of total power, using the same component efficiencies.

9·35 Suppose the mercury-vapor cycle of Prob. 9·30 is used as a topping cycle for a Rankine steam power plant in which the evaporation occurs at 550°F and condensing occurs at 70°F. The steam-turbine inlet state has 75°F of superheat. The isentropic efficiencies of the pump and turbine are 0.65 and 0.90 in the steam system. Determine the system efficiency, the mercury- and steam-flow rates, the heat-transfer rates for the mercury and steam boilers and the condenser, and the pump and turbine hp for 300 MW of total power output.

9·36 A supercritical CO_2 power plant powered by a low-temperature nuclear source has been proposed for operation in the Arctic. The high-pressure side is at 1100 psia, and condensation occurs at −40°F. Assuming pump and turbine efficiencies of 0.50 and 0.85 and a peak cycle temperature of 200°F, determine the system energy-conversion efficiency and the required reactor power and CO_2-flow rate for 2 kW net output.

9·37 The geothermal plant operated by a utility company in northern California utilizes steam produced by natural means underground. Steam wells are drilled to tap this steam supply which is available at 65 psia at 350°F. The steam leaves the turbine at 4-in Hg absolute pressure. The turbine isentropic efficiency is 0.75. Calculate the efficiency of the plant. If the unit produces 12.5 MW what is the steam-flow rate? The 4-in-Hg exhaust pressure is maintained by using a condenser system that requires a cooling tower. How can you justify the capital expense for the condenser system?

9·38 A steam turbine plant has a single stage of regenerative feedwater heating using a closed feedwater heater. Steam enters the turbine at 400 psia and 800°F. The condenser operates at 1 psia. The bled steam is extracted at a temperature halfway between that in the *boiler* and the condenser. Assuming that the turbine stages have $\eta_s = 0.8$, and for the pumps $\eta_s = 0.6$, analyze the cycle in detail.

9·39 A potassium-steam binary vapor power cycle has been proposed as a means of achieving a high-efficiency system.† The energy input to the potassium cycle will be from a nuclear reactor. The steam cycle design is based on the existing supercritical

† W. R. Chambers et al., A Potassium-Steam Binary Vapor Cycle for Nuclear Power Plants, ORNL-3584 (1964).

Eddystone power plant. The object of this series of problems is to analyze the system for efficiency.

The analysis will require location of thermodynamic data in the literature, an experience simulating the problems and frustrations of real engineering practice.

(*a*) Determine the efficiency of the potassium topping cycle under the following conditions: turbine inlet, 1540°F, saturated vapor; turbine exit at 2.4 psia; turbine isentropic efficiency, 0.7; condenser exit conditions, 1100°F at 2.4 psia.

(*b*) Determine the efficiency of the steam cycle, which has two stages of reheat. At the inlet to the first turbine the conditions are 1050°F steam at 4000 psia. The steam exits the turbine at 786°F at 1133 psia. The first reheat stage goes to 1050°F at 1043 psia. The exit from the second turbine is at 705°F at 263 psia. The second stage of reheat goes to 1050°F at 251 psia. The steam leaves the third turbine at 1.5 psia. This turbine has an isentropic efficiency of 0.878. Condensation occurs at 1.5 psia and carries the steam to the saturated-liquid state. Besides determining the overall efficiency, find the isentropic efficiencies of the first two turbines.

(*c*) Now determine the efficiency of the combined cycles with the energy input as heat to the steam cycle supplied by the condensation process of the potassium cycle.

(*d*) Now add a single open feedwater heater to the potassium cycle to operate with bleed conditions of 26 psia at 1510°F. Assume the turbine efficiency remains 0.7. Find the cycle efficiency for the potassium cycle.

(*e*) Write a brief discussion of the advantages, based on your analysis, and disadvantages of this binary system.

9·40 It is proposed that warm water from the Gulf Stream (25°C) be used as an energy source for a power system for an oceanic research station. Cold water from 600 m below the surface will be used as the energy sink (10°C). Design a vapor power system using Freon 12 for this operation, assuming a power requirement of 10 kW. Specify the Freon-flow rate, heat-exchanger energy-transfer rates, and compressor and pump power requirements. Assuming that the cold water leaves the condenser 5°C below the condensing temperature and is discharged at the water surface, estimate the cold-water pumping power required. How might this pumping power be reduced?

9·41 Suppose geothermal steam is available at 75 psia, saturated vapor, at the rate of 10^4 lbm/h. Design a power system to utilize as much of the energy of this steam as you can, assuming that the environmental temperature is 60°F. Specify the required component efficiencies, energy-transfer rates, intermediate states, and so on.

9·42 A major fact contributing to smog is that the high-speed combustion in conventional automobile engines tends to be relatively incomplete. In contrast, the combustion in a steady-flow burner is much more complete, and a significant reduction in automotive smog could be obtained by a national change to engines with continuous burners. Steam engines are now of considerable interest for this reason.

Using reasonable component efficiencies and operating parameters, design an automotive steam turbine power plant (closed cycle) for 100 hp. Specify the burner heat-input rate, radiator heat-output rate, and so on. Discuss the problems of load control and part-load performance.

9·43 Do Prob. 9·42 using a reciprocating steam engine with cam-driven values. Assume adiabatic compression and expansion.

9•44 Large central power plants are often placed on rivers where a good low-temperature sink is available for heat rejection. Estimate and discuss the effects of such plants on the river environments. How much power could one generate on, say, the Mississippi and Sacramento rivers, if the river temperature rise is to be held to 5°C? What is the effect of such thermal pollution on the ecology of our rivers?

9•45 Two identical bodies of constant specific heat are at the same initial temperature, T_1. A refrigerator operates between these bodies until one of them is cooled to a temperature of T_2. Derive an expression for the minimum amount of work required.

9•46 A 2-ton refrigerator system operates between 0 and 100°F. Its cop is 0.8 of the Carnot cop for the same temperatures. What is the required work input in hp?

9•47 Using Freon 12 as the refrigerant, determine the cop for a vapor compression refrigeration cycle with saturated vapor leaving the evaporator at 40°F and saturated liquid leaving the condenser at 80°F. The compressor has an isentropic efficiency of 0.80. Compare to a Carnot refrigerator operating at 80°F and 40°F.

9•48 Using the table of refrigerants provided by the American Society of Refrigerating Engineers, compare the suitability of the following refrigerants on the basis of toxicity and flammability, pressures at operating temperatures of 5 and 86°F, and h_{fg} at 5°F. Comment on effect of a pressure less than atmospheric in the evaporator. Refrigerants: Freon 11, 12, and 27, carrene 7, methyl chloride, ammonia, carbon dioxide.

9•49 It has been suggested that a used household refrigerator might be modified to serve as a heat pump for heating a swimming pool. The purpose of this problem is the evaluation of this suggestion. First estimate the performance of the refrigerator

Working substance, Freon 12
Motor, 1 hp, 1750 rpm
η_M = mechanical efficiency of motor = 0.95
η_s = compressor isentropic efficiency = 0.85
Compressor driven at motor speed
η_{vol} = compressor volumetric efficiency† = 0.70
Condensing temperature, 120°F
Evaporating temperature, 0°F
Compressor inlet, 10°F superheat
Condenser discharge, saturated liquid

Making suitable idealizations and employing proper analysis methodology, determine the cycle state points, flow rate when the motor draws 1 hp of electric power, volume-flow rate (based on compressor inlet state), compressor displacement rate, rate of heat transfer from cold space and to kitchen, heat-pump cop, and refrigerator cop.

Now, if this device is used as a swimming-pool heat pump, the evaporator temperature will be much higher, since the environment temperature will be higher than the refrigerator cold space. Also, the condenser temperature can be lower, owing to improved heat-transfer characteristics when water rather than kitchen air is on the outside of the condenser coils. However, the flow rate will be altered

† η_{vol} = (actual volume-flow rate based on inlet density)/(compressor displacement rate).

because of the different compressor inlet density. Analyze the heat-pump operation, assuming the following conditions:

$\eta_{vol} = 0.70$
Motor speed, 1750 rpm
Condensing temperature, 90°F
Evaporating temperature, 20°F
Compressor inlet, 10°F superheat
Condenser discharge, saturated liquid

Determine the cycle state points, flow rate, required motor power, heat-transfer rate to pool, and heat pump cop. The swimming-pool heating requirement will be of the order of 25,000 to 50,000 Btu/h. What is your evaluation of this suggestion?

9·50 A vacuum refrigeration system is to produce 100 tons of cooling. The water leaves the flash chamber at 45°F and reaches a temperature of 65°F on leaving the chiller. Determine the pressure in the flash chamber and the mass-flow rate through the chiller.

9·51 The objective of this problem is to make a realistic economic analysis of the heat pump concept for application to space heating in a region like your own. In a typical installation the system receives energy as heat from the air outside a house at a temperature somewhat below the prevailing atmospheric temperature and rejects energy as heat to the air inside the house at a temperature somewhat above the inside air temperature. In order to keep the heat-transfer area in the condenser and evaporator from being too large, a reasonable temperature difference for heat transfer must be provided (say 10°F minimum).

(*a*) Choosing a suitable refrigerant and making other assumptions as necessary, carry out the cycle analysis. Take the interior temperature as 70°F and use an outside temperature typical of your area on a cold winter day.

(*b*) Electric energy costs about 4 cents/(kW·h). Gas costs vary widely across the country. Your local gas company can quote you a rough cost per 1000 Btu. Typical household furnaces manage to get about 0.7 of the energy released by combustion into the house. Discuss the relative merits of heat pumps and gas furnaces in your area. If a heat pump is more expensive to operate in your area, in what sort of areas might it be cheaper than gas?

9·52 Compression-distillation units for water purification offer certain advantages over the older type of heat-distillation units. A schematic flow diagram of a simple compression-distillation unit is shown in the sketch. A steady flow of impure water

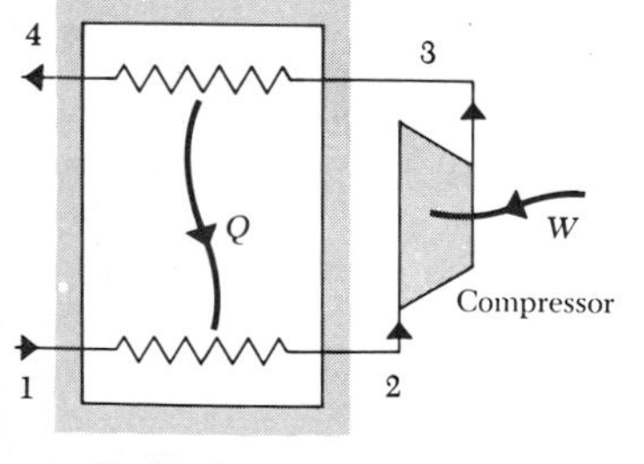

Prob. 9·52.

at 14.7 psia and 60°F (state 1) enters a heat exchanger in which it is evaporated. The purified saturated vapor is compressed and then fed back to the same exchanger for condensation and subcooling. Assuming that the compressor isentropic efficiency is 0.7, and that the purified water emerges at 212°F, what is the pressure level of the condensing process, and what is the work input to the compressor per lbm of distilled water?

9•53 Air is to be compressed from 1 atm to 300 psia. Determine the work required for an isentropic compressor and for a highly cooled compressor in which the air undergoes an isothermal process. What is the isentropic efficiency of the isothermal compressor? What is the *isothermal efficiency* of the isentropic compressor ($\eta_T \equiv W_T/W$)? The inlet temperature is 60°F.

9•54 Determine the compressor efficiency for the example of Sec. 9•11 below which the plant would not be self-sustaining, all other things being equal.

9•55 Using the 0.85-efficient compressor, determine the efficiency of the system of Sec. 9•11 as a function of pressure ratio, keeping the turbine inlet temperature fixed. What pressure ratio must be achieved before the system can be started? Specify the flow rate, heating and cooling rates, compressor and turbine power, and pressure ratio (*a*) at the maximum efficiency point and (*b*) at the minimum flow rate for 300 hp of net shaft output.

9•56 Part of a Brayton cycle using helium as the working fluid includes a two-stage compressor with intercooling. The compressor inlet temperature is 30°C at 25 atm, and the overall pressure ratio is 4. The intercooler operates at 50 atm and reduces the temperature to 40°C. Calculate the work per kg required with and without the intercooler, assuming all isentropic efficiencies are 100 percent.

9•57 In a liquid natural gas plant, one compressor operates at a flow rate of 18,000 kg/min at an inlet temperature of 30°C. The molecular weight of the gas mixture is 32, and the mixture $c_p = 1.68$ J/(g·K). The inlet pressure is 0.55 MN/m^2, and the pressure ratio is 7.0. Calculate the power required from the steam turbine which drives this compressor if $\eta_c = 0.85$ and $\eta_T = 0.90$.

9•58 Many geothermal sources produce hot water instead of steam. The object of this problem is a preliminary design of a Brayton cycle using CO_2 with heat input from geothermal water at 60 psia, 300°F. The peak temperature in the Brayton cycle will be 290°F at a pressure of 1600 psia. The compressor inlet temperature is 90°F with a pressure ratio of 2. For the first trial assume $\eta_c = \eta_T = 0.8$. Calculate the compressor and turbine works, the energy transfers as heat, and the cycle efficiency. What do you think about the idea?

9•59 In the previous problem the net work output from the cycle is 5 Btu/lbm flow. For a 25-MW plant calculate the CO_2-flow rate. Calculate the water-flow rate if a 10°F temperature difference is maintained between the CO_2 and water all along the heat exchanger.

9•60 Geothermal water is available at 250°F. This will be used to supply the energy input to a Freon power system. Assuming cooling water available at 80°F, design an operating plant with a power output of 50 MW using a peak Freon pressure of 500 psia.

9•61 A proposed Brayton cycle will use energy input from a thermonuclear (fusion) reactor employing lithium as an intermediate heat-transfer fluid. The cycle is noted in the figure. Calculate the thermal efficiency of the cycle and the helium flow per MW power required. Treat the working fluid, helium, as a perfect gas with constant specific heats and neglect pressure losses. (See figure on following page.)

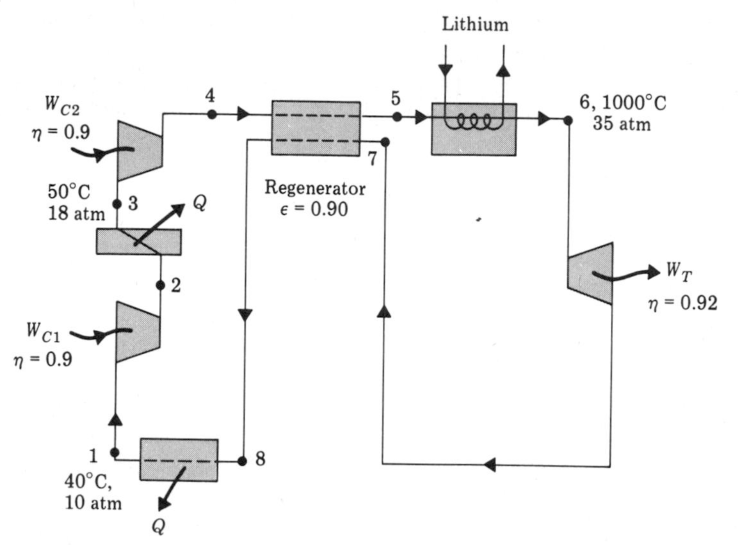

Prob. 9·61.

9·62 A refrigeration system using Freon 12 is to be sized for a southwestern home. The heat load has been calculated as 60,000 Btu/h at an outdoor temperature of 100°F and an indoor temperature of 70°F. Assuming a 20°F temperature difference across the evaporator and a 40°F difference across the condenser, calculate the cop, required hp, and operating cost if electricity is 4 cents/(kW·h). (Assume 10°F superheat at the compressor inlet, an isentropic efficiency of 0.8, and 12 h/day operating time.)

9·63 Consider the example of Sec. 9·14 for the gas-cooled reactor Brayton power system. If a regenerator having an efficiency of 0.90 is added to the cycle, what is the new cycle thermal efficiency?

9·64 Helium, air, sulfur dioxide, and methane are to be compressed from the same inlet pressure and temperature to a pressure four times that of the inlet. Assuming η_s is the same for all four compression processes, calculate the value of $W_c \eta_s / T_1$ for each of these gases. Explain why such a large difference occurs.

9·65 For the simple Brayton cycle, develop an expression relating the ratio of the compressor work input to the turbine work output in terms of P_2/P_1, η_c, η_T, and T_3/T_1.

9·66 Using a P-v plot show graphically that an isothermal compression from P_1 to P_2 requires less work than an isentropic compression between the same two pressures.

9·67 Consider the two-stage compressor shown. Treating the gas as perfect with constant c_p, derive an expression for the work input per lbm in terms of P_1, P_2, P_4, η_1, η_2, T_1, and T_3. Then, for all other parameters fixed, find the value of P_2 which minimizes

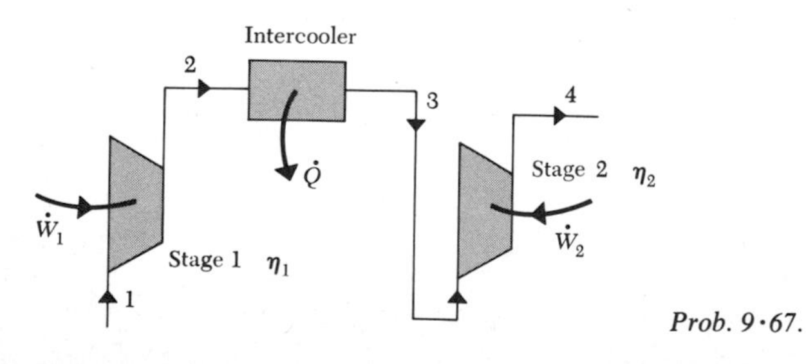

Prob. 9·67.

the work requirement. Show that, when $T_3 = T_1$ and $\eta_1 = \eta_2$, the optimum condition is $P_2/P_1 = \sqrt{P_4/P_1}$.

9·68 A reciprocating air compressor operates as shown in the sketch. The intake valve opens whenever the cylinder pressure drops to 1 atm, and the discharge valve opens when the cylinder pressure rises to 150 psia. The supply air is at 70°F, 15 psia. At tdc the cylinder volume is 1 in^3, and at bdc it is 40 in^3.

(*a*) Sketch the *indicator diagram* (pressure versus cylinder volume).

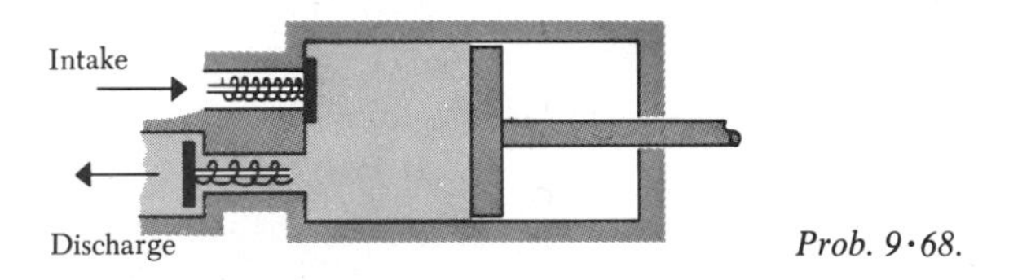

Prob. 9·68.

(*b*) Calculate the indicator state points and the net work input requirement (per lbm of air) assuming isentropic compression and expansion and adiabatic, constant-pressure intake and exhaust.

(*c*) Calculate the overall (quasi-steady flow) isentropic efficiency.

(*d*) Calculate the air-flow rate and power input requirement at 600 rpm.

9·69 Suppose that the cylinder of Prob. 9·68 is water-cooled.

(*a*) What will be the effect of this cooling on the flow rate supplied by the compressor?

(*b*) Approximating the compression process as polytropic, determine the minimum polytropic exponent for successful compression.

(*c*) Suppose that the maximum air temperature attained is 240°F, which is maintained during the discharge stroke, and that when the intake valve opens the air temperature is 120°F. Treating the compression and expansion processes as polytropic, analyze the compression cycle, calculating the net work input and heat output per cycle. Then, calculate the air-flow rate, power input, and cooling rate at 600 rpm.

9·70 Derive an expression for the efficiency of a Brayton cycle analogous to Eq. (9·15) but including compressor and turbine isentropic efficiencies η_c and η_t.

9·71 Use the solution to Prob. 9·70 to derive an expression for the minimum compressor efficiency for which the Brayton cycle will work.

9·72 Develop expressions for the work output per unit mass and the efficiency of an ideal Brayton cycle with regeneration, assuming maximum possible regeneration. For fixed maximum and minimum temperatures, how do the efficiency and work outputs vary with pressure ratio? What is the optimum pressure ratio?

9·73 An ideal Brayton-cycle gas turbine with regeneration but no reheat or intercooling operates at full load with an inlet temperature to the compressor of 60°F and an inlet temperature to the turbine of 1540°F. The compressor pressure ratio is 6.0. The cycle is also to be operated at part load at a pressure ratio of 3.0 with a turbine inlet temperature of 1100°F. Compare the thermal efficiency and net work output per lbm flow for the two operating points if the working fluid (air) is treated as a perfect gas and the maximum possible regeneration is obtained.

9·74 Rework Prob. 9·73 under the same conditions for full load and part load but without the regenerator. What do you conclude about the use of a regenerator?

9·75 Redo Prob. 9·73 using compressor and turbine isentropic efficiencies of 0.8 and 0.85, respectively, and a regenerator effectiveness ($\dot{Q}_{act}/\dot{Q}_{max}$) of 0.85.

9·76 A closed-cycle gas-turbine power plant, such as might be used to provide power for a remote experimental station, is shown below. Typical component performance factors might be compressor isentropic efficiency 0.80, turbine isentropic efficiency 0.82, and *regenerator effectiveness* $(T_3 - T_2)/(T_5 - T_2) = 0.85$.

The objective here is to analyze this system for two cases: (*a*) bypass valve closed, that is, maximum output power, and (*b*) bypass valve opened such that the

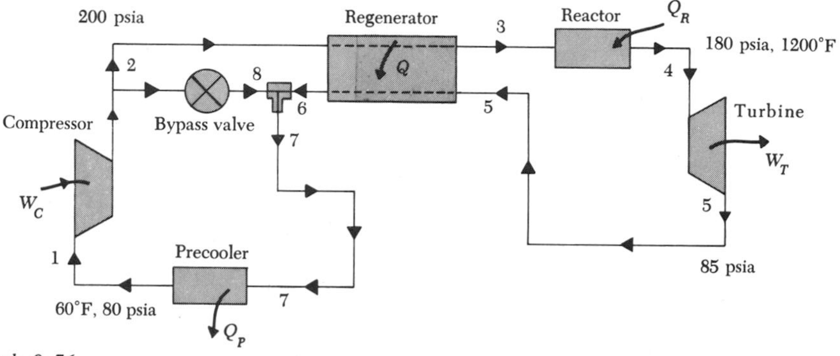

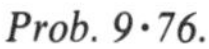
Prob. 9·76.

system is self-sustaining, but no net output power. The assumed compressor pressure ratio of 2.5 : 1 is typical for a regenerative gas-turbine power system. The function of the bypass valve is load control. Making suitable idealizations, determine the state points for the two operating conditions. Note that states 1 and 4 are given, and you can start from these. The working fluid in the system is nitrogen, which may be assumed to be a perfect gas with constant specific heats.

Calculate the overall energy-conversion efficiency, the reactor heat-transfer rate (kW), the compressor and turbine powers (hp), the precooler load (Btu/h), and the flow rates (lbm/h) for the case of (*a*) no bypass flow, net output power of 500 kW, and (*b*) same compressor flow, no net power (zero load, self-sustaining). Keeping states 1 and 4 the same, what will be the efficiency for no bypass flow if the regenerator were eliminated (bigger reactor and precooler)?

9·77 A gas-turbine power plant for vehicular use consists of the components shown. Air enters at state 1 and is compressed, heated, and expanded in a two-stage turbine system. The first turbine is used to drive the compressor, and the second to power the vehicle. A valve for load control is located between the two turbines.

The objective is to make a preliminary study of the feasibility of such a system. You may make the following idealizations:

All turbomachinery adiabatic (isentropic efficiencies as given)
Mass of fuel added in combustion negligible; to be treated simply as a device in which energy is added as heat to the air
Kinetic and potential energies negligible at the numbered state points
Air a perfect gas
Process in combustor isobaric ($P_2 = P_3$)
Maximum permissible air temperature, 1600°F (metallurgical limit)

Determine states 5 and 6 for the two cases of (*a*) full power, valve wide open (state 4 = state 5), and (*b*) no power, no work from turbine 2 (state 5 = state 6).

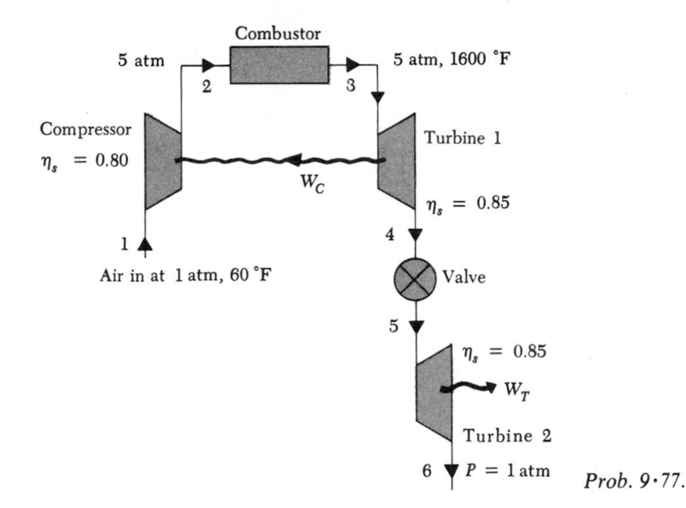

Prob. 9·77.

Show these on your scale *T-s* diagram and add them to your table. Determine the compressor work requirement, the work delivered by turbines 1 and 2 for these two cases, and the heat input required, all per lbm of air. Compute the full-load system efficiency W_{shaft}/Q. For the case where the system is running at full power, delivering 300 hp, what air-flow rate is required? If the energy transfer as heat to the air in the combustor is equivalent to 19,000 Btu/lbm of fuel, what fuel-flow rate is required, what is the air/fuel ratio for this engine, and what is the full-load specific fuel consumption (lbm of fuel per hp·h)?

9·78 A preliminary design is required for a gas-turbine power plant with a single-stage intercooling and a regenerator with an effectiveness of 0.80 (see Prob. 9·76). The turbine inlet temperature is limited to 1400°F, and the compressor inlet conditions are 520 R at 1 atm. Determine the *pressure ratio* for maximum efficiency, assuming $\eta_c = 0.85$ at each stage, that the air emerges from the intercooler at 540 R, that each compressor stage has the same pressure ratio, and that $n_T = 0.9$.

9·79 An idealized air-standard diesel cycle has a compression ratio of 15. The energy input is idealized as a heat transfer of 700 Btu/lbm. The inlet conditions are 70°F at 1 atm. Find the pressure and temperature at the end of each process in the cycle and determine the cycle efficiency.

9·80 The compression ratio on an ideal air-standard Otto cycle is 8.0. The inlet conditions are 1 atm at 70°F. The combustion process is idealized as a heat transfer of 700 Btu/lbm. Find the pressure and temperature at the end of each process and determine the cycle efficiency.

9·81 Discuss the inadequacies in the models of Probs. 9·79 and 9·80. Which idealizations are likely to be the weakest, and what effects will they have on the predicted performance?

9·82 An ideal Otto cycle operates with a compression ratio of 9 : 1 at intake conditions of 14.0 psia and 70°F. The cylinder volume initially is 125 in^3. If 4 Btu of energy is added as heat to the gas during the constant-volume heating process, calculate the cycle efficiency and the pressure and temperature at the end of each process.

9·83 An ideal diesel cycle operates with a compression ratio of 15 : 1 at intake conditions of 14.0 psia and 70°F. The cylinder volume initially is 125 in^3. If 4 Btu of energy is added as heat to the gas during the constant-pressure heating process, calculate the cycle efficiency and the pressure and temperature at the end of each process.

9·84 A small turbojet engine for use in takeoff assist is to be designed to the following specifications:

Static sea-level thrust, 1000 lbf
Fuel, gasoline
Size, maximum diameter not to exceed 18 in

Since the engine is to be used for takeoff assist, it will be operated at sea level, and you may therefore assume that the temperature and pressure at the compressor inlet are 1 atm and 60°F. For a small engine of this type a compressor can probably be built to raise the pressure to 6.5 atm with an isentropic efficiency of 80 percent, so use these values. A small turbine is needed to deliver the work required by the compressor, and one with an 80 percent isentropic efficiency can probably be made. However, the turbine inlet temperature should probably not exceed 2000°F, so take this as T_3 (see Fig. 9·52). If the supersonic nozzle is properly designed, it will exhaust the gas at a pressure of 1 atm, so assume this to be the case. The energy added as heat to the air in the combustor is 20,000 Btu/lbm fuel.

The objective is to make a preliminary analysis of such an engine. Using the assumed values, calculate the compressor work requirement (Btu/lbm), the air/fuel ratio (lbm air/lbm fuel), and the state of the air entering the nozzle (P and T). You may neglect the mass of fuel added to the air. If the nozzle is adiabatic and frictionless and no shock waves occur, the flow in the nozzle may be idealized as being isentropic. Calculate the state of the gas leaving the nozzle and determine the exit velocity.

The Mach number in compressible flow is the ratio of the velocity at any point to the velocity of sound at that point,

$$M = \frac{\mathsf{V}}{\mathsf{c}}$$

For a perfect gas, it may be shown that the speed of sound c is

$$\mathsf{c} = \sqrt{g_c k R T}$$

Calculate the speed of sound and the Mach number at the nozzle exit. At the throat the Mach number is 1. Calculate the velocity, temperature, pressure, and density of the air at the throat.

The specific impulse for this engine may be shown to be

$$SI = \frac{1}{g_c}(\mathsf{V}_5 - \mathsf{V}_1)$$

The velocity at the compressor inlet is of the order of 200 ft/s; assuming this value, calculate the specific impulse. Calculate the air-flow rate required to produce the specified 1000-lbf thrust. Calculate the density at the nozzle discharge and determine the flow area and nozzle diameter at this point. In a similar manner, calculate the throat area and diameter. Assuming that the free-flow area at point 1 is 80 percent of the frontal area and that the outside of the engine has the maximum diameter of 18 in, calculate the velocity and Mach number at the engine inlet. Calculate the fuel required to run this engine at full load for a 1-min takeoff-assist period.

9·85 An aerospace plane uses turbojet engines for takeoff and ramjet engines for upper-stratosphere flight. Assuming that the vehicle is flying at 5000 mph through the air at $-40°F$, determine the temperature of the air entering the combustion chamber of the ramjet engine.

9·86 Estimate the maximum specific impulses for the following thrusting systems:
(*a*) $T_0 = 3000°F$, $k = 1.3$, $c_P = 0.27$ Btu/(lbm·°F)
(*b*) $T_0 = 6000°F$, hydrogen
(*c*) Ion engine, cesium plasma ejected at 1 percent of the speed of light

9·87 Consider a blowdown thrusting system which will use hydrogen as the working fluid. The hydrogen will be stored in a high-pressure tank at 3000 psia, from which it will be bled to a chamber having a regulated pressure of 200 psia. The system temperature will be approximately 80°F. The thrusting nozzles will operate with an exit pressure of 10 psia. Determine the specific impulse. What flow rate is required for a 40-lbf thrust? How large must the storage vessel be if the total thrusting requirement is 1000 lbf·s?

9·88 Estimate the power requirements of a cesium-ion engine which produces 1 lbf of thrust. The ions are ejected at 5 percent of the speed of light. How much power is used for vaporization and how much for acceleration? What is the specific impulse of this engine? How might the required power be generated?

9·89 Consider a nitrogen thrusting system with an exit to stagnation pressure ratio of 0.1 and a stagnation temperature of 30°C. Calculate the exit velocity and the specific impulse in SI units.

9·90 It has been proposed that a small rocket system for orientation of a space satellite could be built schematically as shown below. A two-phase substance would be stored in the tank. The vapor would be bled off to be ejected through the nozzle.

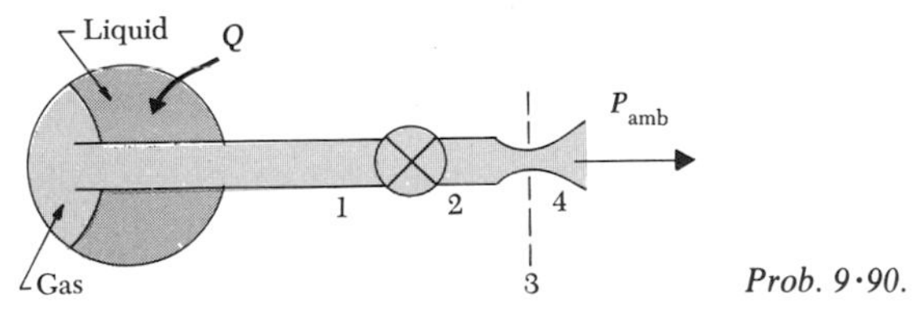

Prob. 9·90.

Energy would be added as heat, so that the temperature of the fluid in the tank remained constant. The valve would provide throttling as necessary prior to the passing of flow through the supersonic nozzle.

The purpose of this problem is the study of this system. Assume the following:

Substance, Freon 12
Design restrictions: minimum temperature in system, $-30°F$; minimum permissible quality in nozzle, 0.98
Flow treated as steady between 1 and 4, state of vapor inside the tank constant (saturated vapor at tank temperature, 100°F)
Nozzle exhausts to free space, so $P_{amb} = 0$ (note that $P_4 > 0$)
Nozzle flow isentropic, V_2 negligible, valve adiabatic

Sketch the process representation in the tank, starting from a completely liquid-filled tank and ending up with only saturated vapor. Sketch the representation of the processes between 1 and 4 on an *h-s* plane for the following three cases:

(*a*) No valve ΔP, state 4 at minimum quality
(*b*) State 4 at minimum temperature, minimum quality
(*c*) Intermediate case between (*a*) and (*b*)
Making suitable idealizations, compute the nozzle exit area A_4 and the flow rate required per lbm of thrust force for each of the three cases. Plot these quantities against the nozzle stagnation pressure P_2. The design requiring the least flow will be the lightest. Perhaps you will have to look at a few additional cases to pinpoint this optimum.

9·91 A stream of CO_2 at 20°F, 50 psia, 1200 ft/s, is needed for a chemical processing plant. The required flow rate is 2 lbm/s. Design a system to obtain this flow. Bottled liquid CO_2 is available at room temperature (60°F).

9·92 A very low-level thruster is required for a deep space probe. It has been suggested that subliming CO_2 be used. The CO_2 will be maintained at −60°F, and the required thrust is 0.01 lbf. Design a system to utilize the CO_2 vapor, which should not go below −140°F. Specify the nozzle exit area, the flow rate, intermediate states, other apparatus, and so on.

9·93 A refrigeration system capable of removing 10,000 W is required for cooling a computer. The cold space will be maintained at 45°F, and the ambient temperature is 75°F. Allowing 10°F temperature difference on either side of the system for heat transfer, design a refrigeration system for this task. Specify component efficiencies, intermediate states, energy-transfer rates, and so on.

9·94 Design a desert power plant that uses solar energy as the energy source. The boiler temperature can be as high as 200°F, and the condenser temperature will be 100°F. The pressure within the system should be greater than atmospheric to avoid contamination by air. The total power requirement is 10 W.

9·95 Design an Arctic power plant that uses solar energy as the power source. The boiler temperature can be as high as 35°C, and the condenser temperature as low as −51°C. The internal pressure must exceed atmospheric to avoid contamination by air. The total power requirement is 5 kW.

9·96 A wind tunnel experiment requires air at 250°C, 0.01 atm, moving at 700 m/s through a 0.02-m^2 test section. Design a system to produce the desired flow from atmospheric air. Specify the hardware required, energy-transfer rates, flow rates, etc.

9·97 Oxygen is available at room temperature and 0.2 atm. Design a system in which the oxygen can be liquefied at a rate of 5 kg/min. The environmental temperature is 15°C, and the liquid oxygen can be stored at 10 atm.

9·98 A mercury vapor jet at 400 ft/s, 1 atm, 1400°F, is required for a commercial process. Design a system to obtain this jet from liquid mercury at room temperature; the required mass-flow rate is 100 lbm/h.

9·99 In a certain power system cesium vapor is available at 50 psia and 2000 R. Vapor is to be bled from this point at 0.2 lbm/min and fed to an auxiliary apparatus at 2200°F, 0.3 psia. Design a system for this job.

9·100 A liquid having $c_P = c_v = 0.4$ Btu/(lbm·°F) is to be cooled in a parallel-flow heat exchanger from 60°F to 30°F for a particular laboratory experiment. The liquid-flow rate will be 100 lbm/h. The cooling is to be accomplished using one of the following bottled gases: (*a*) air at 100 psia, 60°F; (*b*) Freon 12 at 72.4 psia, 60°F; (*c*) CO_2 at 500 psia, 60°F; (*d*) H_2O vapor at 20°F. The gas can be bled through the heat exchanger and will be discharged to the atmosphere. No auxiliary hardware (other than a valve) can be used.

Only one of the gases will work for this purpose. Explain why each of the others is not satisfactory, using appropriate process sketches to make your points. For the gas that will work, specify the system configuration, gas-flow rate, and so on. Use as little of the gas as possible with your system.

9·101 A dentist's drill of new design requires 0.01 hp. It will be driven by a small air turbine in the drill tip, with the discharged air used to blow chips away. The air must emerge from the drill at low velocity, atmospheric pressure, 75°F. Compressed air at 100 psia, 70°F, is available to run the device.

Design a simple system which will work in the manner indicated. Sketch the *h-s* process representation for your system and explain the concept of your design briefly, using this diagram. Then calculate the required air flow, power input to any heaters, pressure drop across any valves, and so on. Use the perfect-gas approximations and the following turbine operating conditions:

$$P_{\text{inlet}}/P_{\text{outlet}} = 3 \qquad \text{isentropic efficiency 60 percent}$$

9·102 It has been proposed that power for an artificial heart be generated by a small reciprocating gas engine implanted in the chest. The body would act as the energy source (99°F), and energy would be rejected as heat to inhaled air (85°F). Investigate the feasibility of this proposal, presuming that 0.02 hp is required and that the cycle rate must be about 80 cycles/min.

9·103 A proposed scheme for powering an artificial heart imagines that breathed air will be expelled from the lungs through a simple mechanical engine. The body itself will then power the heart directly. Making suitable estimates and design calculations, investigate this proposal. The heart requires about 0.02 hp.

9·104 Small radioisotope heat sources are presently under development for implantation in the abdominal cavity. Temperatures of 1200°F seem possible with advanced types of insulation. Design a thermal power system using an inert gas to operate from this source, using the bloodstream as the heat sink. Reciprocating machinery will be most practical, and the shaft-power output should be 0.02 hp (for an artificial heart pump). Estimate the size of your design as best you can.

9·105 In a certain processing plant, 5000 lbm/h of air is available in a steady supply at 400°F, 250 psia. It is to be changed to 200°F, 50 psia, for use in a chemical process.

(*a*) Calculate the maximum amount of useful power (kW) that could be generated in a steady-flow, steady-state device by a system that produces this state change, assuming that it could exchange heat with the environment at 60°F (520 R).

(*b*) Using only *two* components, design a system to produce the desired state change while generating some useful power. Assume that you have available the following components:

Valves
Heat exchangers
Turbines (80 percent isentropic efficiency)
Compressors (70 percent isentropic efficiency)

Specify the power inputs or outputs (hp) to or from the turbomachinery, specify the heat-transfer rates to or from the air in the heat exchangers (Btu/h), and specify the pressure drops across valves (psia) (for whichever of these components you use).

CHAPTER TEN

THERMODYNAMICS OF NONREACTING MIXTURES

10·1 DESCRIPTIONS OF MIXTURES

In the previous chapters we dealt only with a single substance, such as water, oxygen, or mercury. We have treated air as if it were a single substance, which is only reasonable at relatively low temperatures, where its chemical structure is constant, and at temperatures above the point at which one of the gases begins to condense. In this chapter we shall develop the thermodynamic theory for mixtures of substances, and explore various engineering applications of these ideas.

Mixtures have many important applications in thermodynamics. Properties of the air–water-vapor mixture are central to the design of air conditioning systems and cooling towers. A recent Brayton-cycle design proposes to use a helium-xenon gas mixture as the working substance, and of course the properties of our atmosphere are those of a mixture of gases.

A *mixture* is any collection of molecules, ions, electrons, and so on. Each group of particles, distinguishable from the others by virtue of its chemical structure, is called a *constituent* (or *species*) of the mixture. In mixtures where chemical reactions occur the amount of each constituent present cannot necessarily be varied independently. *Components* are those constituents the amounts of which can be independently varied. In this chapter we shall deal exclusively with nonreacting mixtures in which each and every constituent is also a component.

A quantity of matter that is homogeneous in chemical composition and physical structure is called a *phase*. A phase can contain several components; for example, a mixture of gases is a phase, and a mixture of the same substances in liquid form is a different phase. A system having only one phase is called *homogeneous*, and a system with more than one phase is *heterogeneous*. A *pure substance* has the same chemical structure in all states but may have several distinguishable phases.

The composition of a mixture may be described by specification of either the *mass* or the *number of moles* of each constituent. Two ways of specifying the composition independently of the quantity of the mixture are the *mass fraction*,†

$$\Phi_i \equiv \frac{M_i}{\sum_i M_i} = \frac{M_i}{M} \tag{10·1}$$

and the mole fraction,

$$\chi_i \equiv \frac{\mathcal{N}_i}{\sum_i \mathcal{N}_i} = \frac{\mathcal{N}_i}{\mathcal{N}} \tag{10·2}$$

M and $\mathcal{N}$ denote the total mass and total number of moles, respectively. The *molal mass* of the mixture is then

$$\hat{M} = \frac{\sum_i M_i}{\sum_i \mathcal{N}_i} = \frac{\sum_i \hat{M}_i \mathcal{N}_i}{\mathcal{N}}$$

or

$$\hat{M} = \sum_i \chi_i \hat{M}_i \tag{10·3}$$

For example, air is a mixture of approximately 3.76 mole of nitrogen for every mole of oxygen. The mole fractions are therefore

$$\chi_{N_2} = \frac{3.76}{4.76} = 0.79 \qquad \chi_{O_2} = \frac{1.00}{4.76} = 0.21$$

Then, in this approximation,

$$\hat{M}_{air} = 0.79 \times 28.02 + 0.21 \times 32.00 = 28.86 \text{ kg/kgmole}$$

However, real air contains traces of other gases, and the actual $\hat{M}$ is 28.97 kg/kgmole = 28.97 g/gmole = 28.97 lbm/lbmole.

† The subscript i refers to any single unspecified constituent.

In this chapter we shall restrict ourselves to *simple compressible mixtures*, for which the only significant reversible work mode is volume change. According to the state postulate, the extensive thermodynamic state of any simple compressible mixture will be specified solely by its composition, energy, and volume. Its composition is specified by the set of mole numbers $\mathcal{N}_1, \ldots, \mathcal{N}_n$ of the n components. Thus the total entropy could be viewed functionally as

$$S = S(U, V, \mathcal{N}_1, \mathcal{N}_2, \ldots, \mathcal{N}_n)$$

The derivatives of this function define the temperature, pressure, and electrochemical potentials. For temperature T

$$\blacktriangleright \qquad \frac{1}{T} \equiv \left(\frac{\partial S}{\partial U}\right)_{V, \mathcal{N}_1, \ldots, \mathcal{N}_n} \tag{10·4}$$

For pressure P

$$\blacktriangleright \qquad \frac{P}{T} \equiv \left(\frac{\partial S}{\partial V}\right)_{U, \mathcal{N}_1, \ldots, \mathcal{N}_n} \tag{10·5}$$

For the molal *electrochemical potential* $\hat{\mu}_i$ of the ith component,

$$\blacktriangleright \qquad -\frac{\hat{\mu}_i}{T} \equiv \left(\frac{\partial S}{\partial \mathcal{N}_i}\right)_{U, V, \mathcal{N}_1, \ldots, \mathcal{N}_{i-1}, \mathcal{N}_{i+1}, \ldots, \mathcal{N}_n} \tag{10·6}$$

The definitions of temperature and pressure are equivalent to those introduced in Chap. 7. The significance of the electrochemical potential will be explained in Chap. 12.

The Gibbs equation for a mixture is obtained by differentiation of the entropy,

$$\blacktriangleright \qquad dS = \frac{1}{T}\,dU + \frac{P}{T}\,dV - \sum_i \frac{\hat{\mu}_i}{T}\,d\mathcal{N}_i \tag{10·7}$$

We shall use this important equation in Chap. 12. Note that, if the composition does not change, each of the $d\mathcal{N}_i$ is zero, and Eq. (10·7) reduces to Eq. (8·1), the Gibbs equation for a single substance.

10·2 MIXTURES OF INDEPENDENT SUBSTANCES

If we can idealize that the allowed quantum states of a given constituent are not influenced by the presence of the other constituents, the properties of the mixture may be determined from the "private" properties of its constituents. We call this a *mixture of independent substances*. Gas mixtures that are not too dense and some liquid and solid solutions (*ideal solutions*) may be treated in this manner.

In a mixture of independent substances each species has a "private" energy, hence the total energy is

$$U = \sum_i U_i$$

Here U_i is the internal energy of the ith constituent.

Since the allowed quantum states of the constituents are independent, the mixture quantum states are determined from all possible combinations of constituent quantum states. As we saw in Chap. 6, the entropy of such a system is the sum of the entropies of its independent subsystems (the constituents). Hence

The entropy of a mixture of independent substances is the sum of the entropies of the constituents,

$$S = \sum_i S_i$$

This fact is known as the *Gibbs rule*.

Let us consider a mixture of two independent substances A and B, each filling the total volume V. The entropy of the mixture can be expressed functionally as (for given amounts of A and B)

$$S = S_A(U_A, V) + S_B(U_B, V)$$

This is true as long as each constituent is in a thermodynamic equilibrium state, even though the two constituents may not be in equilibrium with one another. We can find the condition of equilibrium between the constituents by isolating the mixture and finding the configuration of maximum entropy. The total internal energy, $U = U_A + U_B$, must remain fixed during this maximization. Differentiating the entropy with respect to U_A and keeping the volume and total internal energy fixed, we find

$$\frac{\partial S}{\partial U_A} = \left(\frac{\partial S_A}{\partial U_A}\right)_V + \left(\frac{\partial S_B}{\partial U_B}\right)_V \frac{dU_B}{dU_A} = \left(\frac{\partial S_A}{\partial U_A}\right)_V - \left(\frac{\partial S_B}{\partial U_B}\right)_V = 0$$

The condition of maximum entropy is then

$$T_A = T_B$$

We see that the equilibrium state will be one in which the temperatures of the two constituents are equal. This demonstration may be extended to mixtures of more than two constituents by the method of undetermined multipliers.

The pressure of the mixture is given by

$$P = T\left(\frac{\partial S}{\partial V}\right)_U = \sum_i \left[T\left(\frac{\partial S}{\partial V}\right)_U\right]_i = \sum_i P_i$$

Here P_i is the pressure the ith constituent would have if it occupied the same volume as the mixture at the mixture temperature; P_i is called the *partial pressure* of the ith constituent. The fact that the total pressure is equal to the sum of the partial pressures for a mixture of independent substances is known as the *Dalton rule*.

A microscopic interpretation of the Dalton rule is helpful. The partial pressures represent the contribution to the average normal force per unit of area acting on the boundaries of the mixture system, resulting from impacts of the various constituents. The fact that the constituents are independent means that the average normal velocity of impacting particles of a constituent is not influenced by the presence of the other species.

To summarize, a mixture of independent substances in equilibrium has the following properties:

▶ $$U = \sum_i U_i \tag{10·8}$$

▶ $$S = \sum_i S_i \tag{10·9}$$

▶ $$T_1 = T_2 = \cdots = T \tag{10·10}$$

▶ $$P = \sum_i P_i \tag{10·11}$$

When the mixture has only one phase, it is convenient to work with the specific, or intensified properties. The specific internal energy of the mixture is

$$u = \frac{U}{M} = \sum_i \frac{M_i}{M} u_i = \sum_i \Phi_i u_i$$

or, on a molal basis,

$$\hat{u} = \sum_i \chi_i \hat{u}_i$$

The molal enthalpy of the mixture is

$$\hat{h} = \hat{u} + P\hat{v} = \sum_i \chi_i \hat{u}_i + \left(\sum_i P_i\right)\hat{v}$$

The molal enthalpy of the ith constituents may be written as

$$\hat{h}_i = \hat{u}_i + P_i\hat{v}_i = \hat{u}_i + P_i \frac{V}{\mathcal{N}_i}$$

Since the molal specific volume of the mixture is

$$\hat{v} = \frac{V}{\sum_i \mathcal{N}_i} = \frac{V}{\mathcal{N}}$$

we can write

$$\hat{h} = \sum_i \chi_i \hat{u}_i + \sum_i \left(\frac{\mathcal{N}_i}{\mathcal{N}} P_i \frac{V}{\mathcal{N}_i} \right) = \sum_i \chi_i \hat{h}_i$$

Similarly, it follows from Eq. (10·9) that

$$\hat{s} = \sum_i \chi_i \hat{s}_i$$

The molal specific heats of a mixture are defined as

$$\hat{c}_v \equiv \left(\frac{\partial \hat{u}}{\partial T} \right)_{\hat{v}, \chi_1, \chi_2, \ldots, \chi_n}$$

$$\hat{c}_P \equiv \left(\frac{\partial \hat{h}}{\partial T} \right)_{P, \chi_1, \chi_2, \ldots, \chi_n}$$

Thus

$$\hat{c}_v = \sum_i \chi_i \left(\frac{\partial \hat{u}_i}{\partial T} \right)_{\hat{v}} = \sum_i \chi_i \hat{c}_{v_i}$$

$$\hat{c}_P = \sum_i \chi_i \left(\frac{\partial \hat{h}_i}{\partial T} \right)_P = \sum_i \chi_i \hat{c}_{P_i}$$

We see that the molal internal energy, enthalpy, entropy, and specific heats for a mixture of independent substances are merely the sum of the contributions for each constituent evaluated at the mixture volume and temperature and weighted by their respective mole fractions. Equivalently, the contributions of the constituents may be evaluated at the mixture temperature and their respective *partial* pressures.

Summarizing,

▶ $$\hat{u} = \sum_i \chi_i \hat{u}_i(T, P_i) \tag{10·12}$$

▶ $$\hat{h} = \sum_i \chi_i \hat{h}_i(T, P_i) \tag{10·13}$$

▶ $$\hat{s} = \sum_i \chi_i \hat{s}_i(T, P_i) \tag{10·14}$$

▶ $$\hat{c}_v = \sum_i \chi_i \hat{c}_{v_i}(T, P_i) \tag{10·15}$$

▶ $$\hat{c}_P = \sum_i \chi_i \hat{c}_{P_i}(T, P_i) \tag{10·16}$$

Since each of the constituents occupies the same volume, the ratio of the specific volumes of any two, based on their partial pressures and the mixture temperature, is inversely proportional to the ratio of their two masses,

$$\frac{v_i(T, P_i)}{v_j(T, P_j)} = \frac{V/M_i}{V/M_j} = \frac{M_j}{M_i} \tag{10·17}$$

As an example, let's find the specific heat at constant pressure for air, treating air as a mixture of nitrogen and oxygen. We know that $\hat{c}_P = \sum_i \chi_i \hat{c}_{P_i}$. Treating air as 1 mole of oxygen for every 3.76 mole of nitrogen,

$$\chi_{O_2} = \frac{1}{4.76} = 0.21 \qquad \chi_{N_2} = \frac{3.76}{4.76} = 0.79$$

From Table B·6 we then find

$$\begin{aligned}\hat{c}_P &= 0.21 \times 29.34 + 0.79 \times 29.08 \\ &= 29.13 \text{ kJ/(kgmole·K)}\end{aligned}$$

Note that the table gives a value of 29.09 kg/(kgmole·K). Air actually contains other gases, and if we account for these contributions, to $\hat{c}_P$, the given value is obtained.

10·3 MIXTURES OF PERFECT GASES

Consider a mixture of perfect gases, each obeying the equation

$$PV = \mathcal{N}\mathcal{R}T \tag{10·18}$$

We idealize that this is a mixture of independent substances, so that the results of the preceding section may be applied directly. In particular, the partial pressure of the ith constituent is

$$P_i = \frac{\mathcal{N}_i}{V}\mathcal{R}T$$

The total pressure is then

$$P = \sum_i P_i = \left(\sum_i \mathcal{N}_i\right)\frac{\mathcal{R}T}{V} = \frac{\mathcal{N}\mathcal{R}T}{V}$$

The ratio of the partial pressure of any constituent to the total pressure, the *pressure fraction*, is therefore equal simply to its mole fraction,

$$\blacktriangleright \qquad \frac{P_i}{P} = \frac{\mathcal{N}_i \mathcal{R} T/V}{\mathcal{N} \mathcal{R} T/V} = \frac{\mathcal{N}_i}{\mathcal{N}} = \chi_i \qquad (10 \cdot 19)$$

The *partial volume* V_i of the ith constituent is defined as the volume that would be occupied by the ith constituent at the mixture temperature and pressure,

$$V_i = \frac{\mathcal{N}_i \mathcal{R} T}{P} = \frac{\mathcal{N}_i}{\mathcal{N}} V = \chi_i V$$

Since $\sum \chi_i = 1$, the total volume is equal to the sum of the partial volumes; this fact is sometimes called *Amagat's rule*.

The *volume fraction* of the ith constituent is defined as

$$\blacktriangleright \qquad \Psi_i \equiv \frac{V_i}{V} = \chi_i \qquad (10 \cdot 20)$$

So, for a perfect-gas mixture, the mole fraction is equal to both the pressure fraction and the volume fraction. This fact is extremely useful in laboratory analysis of gas composition.

Perfect-gas mixtures are sometimes specified by volume fractions and sometimes by mass fractions; either specification determines the other. For example, consider a mixture of O_2, H_2, and CO_2, specified as having volume fractions

$$\Psi_{O_2} = 0.15 \qquad \Psi_{H_2} = 0.60 \qquad \Psi_{CO_2} = 0.25$$

For a perfect-gas mixture the volume and mole fractions are identical [Eq. (10·20)]. The mixture calculation is carried out in Table 10·1. Then, the gas constant for the mixture is

$$R = \frac{\mathcal{R}}{\hat{M}} = \frac{1545}{17.01} = 90.8 \text{ ft} \cdot \text{lbf/(lbm} \cdot \text{R)} = 488.7 \text{ J/(kg} \cdot \text{K)}$$

Table 10·1 Mixture calculation

Component	$\Psi_i = \chi_i$ lbmole/lbmole mixture		$\hat{M}_i$ lbm/lbmole		$M_i/\mathcal{N}$ lbm/lbmole mixture	M_i/M lbm/lbm mixture
O_2	0.15	×	32.0	=	4.80	0.282
H_2	0.60	×	2.016	=	1.21	0.071
CO_2	0.25	×	44.01	=	11.00	0.647
					$\hat{M}$ = 17.01 lbm/lbmole	1.000

Table 10·2 Mixture calculation

Component	M_i/M $\frac{\text{lbm}}{\text{lbm mixture}}$		$\hat{M}_i$ $\frac{\text{lbm}}{\text{lbmole}}$		$\mathcal{N}_i/M$ $\frac{\text{lbmole}}{\text{lbm mixture}}$	χ_i $\frac{\text{lbmole}}{\text{lbmole mixture}}$
O_2	0.282	÷	32.0	=	0.0088	0.15
H_2	0.071	÷	2.016	=	0.0353	0.60
CO_2	0.647	÷	44.01	=	0.0147	0.25
					$1/\hat{M} = 0.0588$ lbmole/lbm	1.00

Note that while H_2 has the largest mole fraction it has the least mass fraction.

If instead the mass fractions were given, the mixture calculation would yield the mole fractions (Table 10·2).

Let's continue this example. Suppose the mixture pressure and temperature are 60 psia, 70°F. The density is then

$$\rho = \frac{P}{RT} = \frac{60 \times 144}{90.8 \times (70 + 460)} = 0.179 \text{ lbm/ft}^3$$

The partial pressures are

$$P_{O_2} = 0.15 \times 60 = 9 \text{ psia}$$

$$P_{H_2} = 0.60 \times 60 = 36 \text{ psia}$$

$$P_{CO_2} = 0.25 \times 60 = 15 \text{ psia}$$

If the mixture is heated, the mole fractions (hence $\hat{M}$ and R) will remain unchanged, and we could treat this mixture as a single simple compressible substance in any engineering thermodynamic analysis.

Now, suppose we put the mixture in a closed tank with a "getter" that absorbs all the oxygen. If it is assumed that the temperature remains 70°F, the pressure will be reduced. Since the constituent partial pressure depends only on the number of moles of the constituent, the volume, and the temperature, the H_2 and CO_2 partial pressures are unchanged, and the new total pressure is

$$P = P_{H_2} + P_{CO_2} = 36 + 15 = 51 \text{ psia}$$

The new *mass* fractions are

$$\frac{M_{H_2}}{M} = \frac{0.071}{1 - 0.282} = 0.099$$

$$\frac{M_{CO_2}}{M} = \frac{0.647}{1 - 0.282} = 0.901$$

The new *mole* fractions can be computed in the manner of the last table. However, here it is easier to compute them from the known partial pressures:

$$\chi_{H_2} = \frac{P_{H_2}}{P} = \frac{36}{51} = 0.706$$

$$\chi_{CO_2} = \frac{P_{CO_2}}{P} = \frac{15}{51} = 0.294$$

Let us now turn to the mixing of perfect gases. Consider a system in an initial state where the constituents are separated, each at the *same temperature and pressure*, with each occupying its partial volume. The gases are allowed to mix within the isolated system. We would like to determine the final system state.

Denoting the final state by 2 and the initial state by 1, and using the fact that the energy of a perfect gas is a function of the temperature only, we find from an energy balance on the isolated system that

$$\hat{u}_1 = \hat{u}_2$$

or, using Eq. (10·12),

$$\sum_i \chi_i \hat{u}_i(T_1) = \sum_i \chi_i \hat{u}_i(T_2) \tag{10·21}$$

Each of the internal energies will increase with an increase in T. Therefore, since the mole fractions are fixed, any temperature change will result in a change in the total internal energy. Since this is fixed, we conclude that the temperature will not change when the ideal gases are allowed to mix in an isolated container. Equivalently, there will be no heat transfer if the gases are allowed to mix in a constant-temperature, constant-volume container.

The pressures for the ith component initially and finally are therefore related by

$$\frac{P_{i2} V_{i2}}{P_{i1} V_{i1}} = \frac{\mathcal{N}_i \mathcal{R} T}{\mathcal{N}_i \mathcal{R} T} \tag{10·22}$$

but $\quad V_{i2} = V_{\text{total}} = V \quad$ and $\quad P_{i1} = P_{\text{initial}} = P_1$

so

$$\frac{P_{i2} V}{P_1 V_{i1}} = 1$$

However, it is also true that

$$V = \sum_i V_{i1}$$

Thus

$$P_2 = \sum_i P_{i2} = \sum_i \frac{V_{i1} P_1}{V} = \frac{P_1}{V} \sum V_{i1} = P_1 \tag{10·23}$$

We see that the final pressure is equal to the initial pressure for this particular mixing process.

This mixing process is of course irreversible and must produce entropy. The entropy change of the ith constituent will be given by [see Eq. (8·18)].

$$s_{i2} - s_{i1} = [\phi_{i2}(T) - \phi_{i1}(T)] - R_i \ln \frac{P_{i2}}{P_{i1}} \tag{10·24}$$

Note that the partial pressure is used to evaluate the entropy, since the gases are assumed to be independent of one another, and therefore the entropy of i is affected only by the partial pressure of i. Then, for the isothermal mixing process under consideration,

$$s_{i2} - s_{i1} = -R_i \ln \frac{P_{i2}}{P_1} = -R_i \ln \chi_i \tag{10·25}$$

On a molal basis this becomes

$$\hat{s}_{i_2} - \hat{s}_{i_1} = -\mathscr{R} \ln \chi_i \tag{10·26}$$

The total change in entropy occurring when *different* gases are allowed to mix in an isolated system at constant temperature and pressure is therefore

$$S_2 - S_1 = -\mathscr{R} \sum_i \mathscr{N}_i \ln \chi_i = \mathscr{R} \sum_i \mathscr{N}_i \ln \frac{1}{\chi_i} \tag{10·27}$$

The mole fractions are all less than unity, and therefore the entropy change within the isolated system is positive. It should be noted that two samples of the same gas cannot be treated as independent substances, and therefore this result applies only to mixtures of different gases. There is *no* entropy change when samples of the same gas at the same pressure and temperatures are mixed.

We remind the student that the mixing process discussed above is started with the gases in thermal and mechanical equilibrium with one another (same temperature and pressure). If other initial conditions are prescribed, the temperature, pressure, or both may change. The final equilibrium state can of course be calculated by appropriate thermodynamic analysis, which includes energy and mass balances, use of the equations of state and assumptions about the confinement of the gases during the mixing process.

Compressing a mixture. Let us consider compression of a mixture of 60 percent methane and 40 percent propane by volume. The mixture is compressed isentropically from 40°C, 0.4×10^6 N/m^2, to 1.2×10^6 N/m^2. Let's find the final temperature of the mixture, the work required per unit mass, and the entropy change for each gas, assuming perfect-gas behavior with constant specific heats.

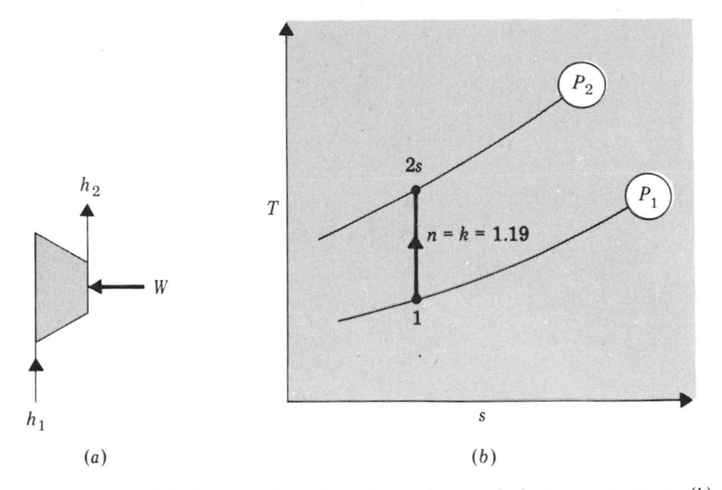

Figure 10·1. (a) Compression of a mixture in an adiabatic compressor; (b) process representation for the mixture

From an an energy balance (Fig. 10·1*a*) we know that

$$W = h_2 - h_1 = c_P(T_2 - T_1)$$

To establish T_2 we require use of the isentropic relations [Eq. (8·33)],

$$\frac{T_2}{T_1} = \left(\frac{P_2}{P_1}\right)^{(k-1)/k}$$

Thus we must first determine k for the mixture. Noting that the mole fraction is the same as the volume fraction, we have, from Table B·6,

$$\hat{c}_P = \sum \chi_i \hat{c}_{Pi} = 0.60 \times 35.72 + 0.4 \times 74.56 = 51.256 \text{ kJ/(kgmole·K)}$$

Similarly, for $\hat{c}_v$,

$$\hat{c}_v = \sum \chi_i c_{vi}$$

However, we can also determine $\hat{c}_v$ from

$$\hat{c}_v = \hat{c}_P - \mathscr{R} = 51.256 - 8.314 = 42.942 \text{ kJ/(kgmole·K)}$$

The mixture molecular weight is given by

$$\hat{M} = \sum \chi_i \hat{M}_i = 0.60 \times 16 + 0.40 \times 44 = 27.2 \text{ kg/kgmole}$$

We then have

$$k = \frac{\hat{c}_P}{\hat{c}_v} = \frac{51.256}{42.942} = 1.193$$

We can determine the final temperature as

$$T_2 = T_1 \left(\frac{P_2}{P_1}\right)^{0.162} = 313 \times (3)^{0.162} = 374.0 \text{ K}$$

From the energy balance

$$\hat{W} = 51.26 \times (374 - 313) = 3127 \text{ kJ/kgmole mixture}$$

On a per kg basis we have

$$W = 114 \text{ kJ/kg}$$

To establish the entropy change† we have the relation

$$\hat{s}_2 - \hat{s}_1 = \hat{c}_P \ln\left(\frac{T_2}{T_1}\right) - \mathscr{R} \ln\left(\frac{P_2}{P_1}\right)$$

For methane,

$$\begin{aligned} \hat{s}_2 - \hat{s}_1 &= 35.72 \ln (374/313) - 8.314 \ln (3) \\ &= -2.7686 \text{ kJ/(kgmole}\cdot\text{K)} \end{aligned}$$

For propane,

$$\begin{aligned} \hat{s}_2 - \hat{s}_1 &= 74.56 \ln (374/313) - 8.314 \ln (3) \\ &= +4.1529 \text{ kJ/(kgmole}\cdot\text{K)} \end{aligned}$$

Then, for the mixture,

$$\begin{aligned} (\hat{s}_2 - \hat{s}_1)_{\text{mix}} &= \sum \chi_i(\hat{s}_2 - \hat{s}_1) = 0.6 \times (-2.7686) + 0.4 \times (4.1519) \\ &= 0.0000 \text{ kJ/(kgmole mixture}\cdot\text{K)} \end{aligned}$$

Since the mixture compression is in fact isentropic, we appear to have made our calculation accurately. However, why has the entropy of the methane

† The entropy calculation is very sensitive to the value of k and to round-off error. The calculation was carried out on a hand calculator using more digits than noted here.

gone down and that for the propane gone up? If we consider the methane and propane as separate control masses, there has been an energy transfer as heat from the methane to the propane. This transfer is reversible if the two gases are always at the same temperature, hence the total entropy change remains zero.

10·4 APPLICATION TO AIR–WATER-VAPOR MIXTURES

Water vapor at pressures below 1 atm may be idealized as a perfect gas, and consequently the theory developed in the previous sections has important application in the study of air–water-vapor mixtures. If the partial pressure of the water vapor corresponds to the saturation pressure of water at the mixture temperature, the mixture is said to be *saturated.*† A closed volume of air in contact with water will, given sufficient time, become fully saturated, and then the partial pressure of the water vapor in the air will be given by the saturation pressure of water at the air temperature, as given by Table B·1. Air that has not been in contact with water for a sufficiently long period of time may not be saturated. The water vapor in such a mixture would then be *superheated.*

If an air–water-vapor mixture that is not saturated is cooled at constant pressure, the mixture will eventually reach the saturation temperature corresponding to the partial pressure of the water vapor. This is called the *dew-point temperature,* because it is associated with the formation of liquid droplets (dew). The constant-pressure cooling process and the dew point are shown in Fig. 10·2.

The composition of an air–water-vapor mixture is often indicated by the *specific humidity,* or *humidity ratio,* defined as the ratio of the mass of water vapor to the mass of air in the mixture,

$$\gamma \equiv \frac{M_w}{M_a} \tag{10·28}$$

† Actually the presence of air alters the saturation pressure by a very small amount. We shall neglect this effect here; its magnitude is calculated in Chap. 12.

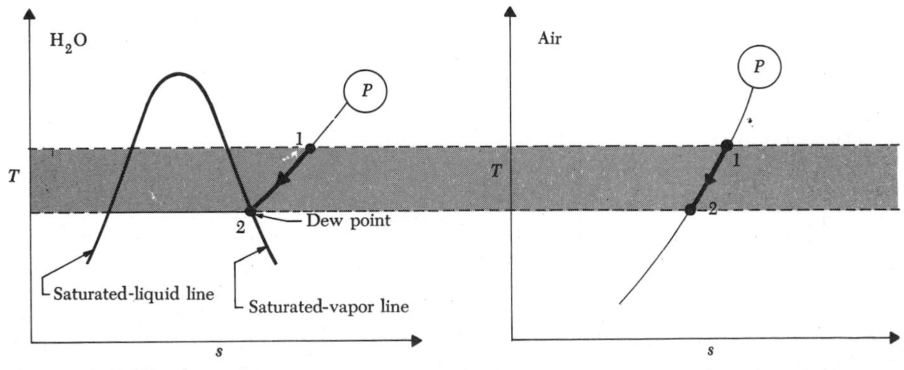

Figure 10·2. The dew point

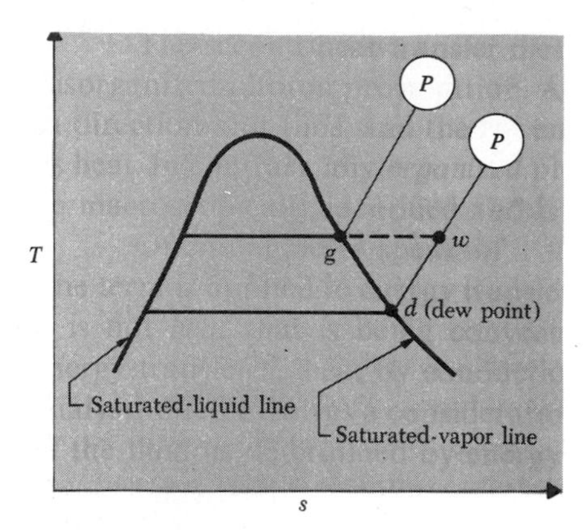

Figure 10·3. Saturation and partial pressure

An alternate specification is the *relative humidity*, which is defined as the ratio of the partial pressure of the water vapor to the saturation pressure at the mixture temperature† (see Fig. 10·3),

$$\phi \equiv \frac{P_w}{P_g} \tag{10·29}$$

Since the mass ratios are inversely proportional to the specific-volume ratios [Eq. (10·17)],

$$\gamma = \frac{M_w}{M_a} = \frac{v_a(T, P_a)}{v_w(T, P_w)} \tag{10·30}$$

We have idealized the vapor as a perfect gas; then, since $T_g = T_w$,

$$\phi = \frac{P_w}{P_g} = \frac{v_g(T, P_g)}{v_w(T, P_w)} \tag{10·31}$$

$$\gamma = \phi \frac{v_a(T, P_a)}{v_g(T, P_g)} \tag{10·32}$$

Using the perfect-gas equation of state, and Eq. (10·30),

$$\gamma = \frac{R_a T/P_a}{R_w T/P_w} = 0.662 \frac{P_w}{P_a} \tag{10·33}$$

† In this section ϕ denotes only the relative humidity and should not be confused with the function $\phi(T)$ in the entropy expression for a perfect gas.

Then, combining with Eq. (10·31),

$$\blacktriangleright \qquad \phi = \frac{\gamma P_a}{0.622 P_g} \tag{10·34}$$

Finally, the total pressure is given by Dalton's rule as

$$\blacktriangleright \qquad P = P_a + P_w \tag{10·35}$$

To illustrate the use of this theory, suppose that the relative humidity of an air–water-vapor mixture is measured as 0.4 at a temperature of 20°C and 1 atm. We wish to determine the specific humidity, the dew point, and the partial pressure of the water vapor and air. We start with Eq. (10·29),

$$\phi = \frac{P_w}{P_g} = 0.40$$

Then, at 20°C, the pressure of saturated water vapor is

$$P_g = 0.002338 \text{ MPa} \qquad \textit{Table B·1a}$$

Hence the actual water partial pressure is

$$P_w = 0.40 \times 0.002338 = 0.000935 \text{ MPa}$$

The air pressure must bring the total pressure up to 1 atm, hence

$$P_a = (1.01325 - 0.00935) \times 10^5 = 1.0039 \times 10^5 \text{ N/m}^2$$

Then, using Eq. (10·33),

$$\gamma = 0.622 \frac{P_w}{P_a} = 0.622 \frac{0.00935}{1.0039} = 0.0058 \text{ kg water/kg air}$$

Finally, the dew point is the temperature of saturated water vapor at the partial water pressure,

$$T_{\text{DP}} = 5.8°\text{C} \ (T_{\text{sat}} \text{ at } 0.000935 \text{ MPa})$$

Hence dew will form if the ambient temperature drops below 5.8°C (42°F). In a humid climate where $\phi = 0.90$ at a temperature of perhaps 85°F the dew point is about 82.5°F. The impact of high humidity on human comfort should be obvious from these numbers.

The humidity can in principle be measured with an *adiabatic saturator*. This is a device into which flows the air–water-vapor mixture of unknown humid-

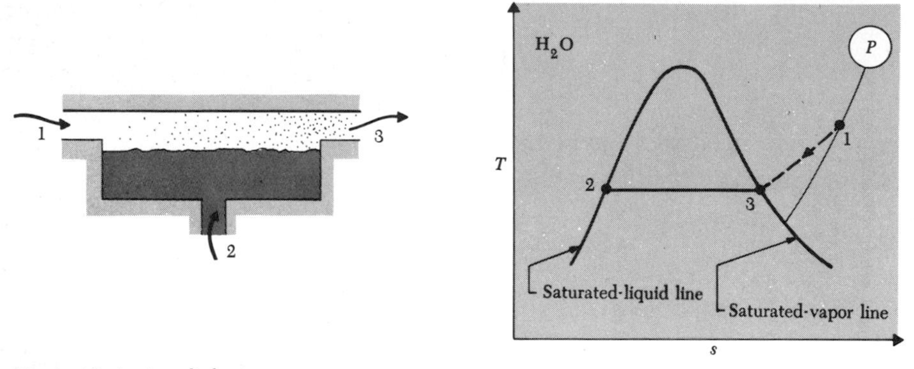

Figure 10·4. An adiabatic saturator

ity, out of which flows a saturated air–water-vapor mixture at some lower adiabatic-saturation temperature, and to which water is added continuously. Schematic and process representations for an adiabatic saturator are shown in Fig. 10·4. The device is sufficiently long that equilibrium between the air–water-vapor mixture is obtained by the time the mixture reaches the exit. Thus the temperature of the mixture at state 3 will be the same as that of water at state 3, and we assume the water temperature is uniform, so that $T_2 = T_3$. Using the properties of perfect-gas mixtures and idealizing the device as adiabatic, energy and mass balances allow us to develop a useful expression:

Conservation of air and water

$$\dot{M}_{a1} = \dot{M}_{a3} \qquad \dot{M}_{w1} + \dot{M}_{w2} = \dot{M}_{w3}$$

Conservation of energy

$$\dot{M}_{a1} h_{a1} + \dot{M}_{w1} h_{w1} + \dot{M}_{w2} h_{w2} = \dot{M}_{a3} h_{a3} + \dot{M}_{w3} h_{w3}$$

By definition,

$$\gamma_1 = \frac{\dot{M}_{w1}}{\dot{M}_{a1}} \qquad \gamma_3 = \frac{\dot{M}_{w3}}{\dot{M}_{a3}}$$

Combining with the water mass balance,

$$\frac{\dot{M}_{w2}}{\dot{M}_{a1}} = \gamma_3 - \gamma_1$$

Substituting into the energy equation, one obtains

$$h_{a1} + \gamma_1 h_{w1} + (\gamma_3 - \gamma_1) h_{w2} = h_{a3} + \gamma_3 h_{w3} \qquad (10 \cdot 36)$$

Measurements of the pressure and temperature allow determination of all the enthalpies and the specific humidity at state 3 from the equations of state of air and H_2O. The one remaining unknown in Eq. (10·36) is the specific humidity at state 1, which may then be calculated from the equation.

We have

$$\gamma_1 = \frac{(h_{a3} - h_{a1}) + \gamma_3(h_{w3} - h_{w2})}{h_{w1} - h_{w2}}$$

or

$$\gamma_1 = \frac{c_P(T_3 - T_1) + \gamma_3(h_{fg})_3}{h_{g1} - h_{f2}}$$

where we have noted that the water enthalpy is evaluated as that of the saturated liquid for state 2 and that of the saturated vapor for state 1.

For example, suppose a mixture of air and water vapor enters an adiabatic-saturation device at 90°F and 1 atm pressure and leaves at 70°F and 1 atm. Let's find the specific and relative humidities at the inlet. Since the air leaving the device is saturated, we know that

$$\phi_3 = 1 \qquad P_{w3} = P_{g3} = 0.363 \text{ psia} \qquad \textit{at } 70°\textit{F}$$

Then, from Eq. (10·33),

$$\gamma_3 = 0.622 \frac{P_w}{P_a} = 0.622 \frac{P_w}{14.7 - P_w}$$

or

$$\gamma_3 = 0.622 \times \frac{0.363}{14.7 - 0.363} = 0.0157 \text{ lbm } H_2O\text{/lbm air}$$

The enthalpies can be read in Appendix B. From Tables B·9 and B·1*a*,

$$h_{a1} = 131.5 \text{ Btu/lbm}$$

$$h_{a3} = 126.7 \text{ Btu/lbm}$$

$$h_{w1} = 1100.7 \text{ Btu/lbm} \qquad \textit{saturated vapor at } 90°\textit{F}$$

$$h_{w3} = 1092.0 \text{ Btu/lbm} \qquad \textit{saturated vapor at } 70°\textit{F}$$

To evaluate h_{w2}, we may use the incompressible-liquid equation of state. This must be properly tied to the other tabulated data for H_2O. If we take the base point at the liquid temperature, we may write

$$h = h_f(T) + \frac{1}{\rho}[P - P_f(T)]$$

where the subscript f again denotes saturated liquid, and T is the liquid temperature (70°F). The pressure term may be neglected as a first approximation (the student should compute the error in the margin), hence

$$h_{w2} = 38 \text{ Btu/lbm} \qquad \textit{saturated liquid at } 70°F$$

Using Eq. (10·36) for γ_1,

$$\gamma_1 = \frac{(h_{a3} - h_{a1}) + \gamma_3(h_{w3} - h_{w2})}{h_{w1} - h_{w2}}$$

$$= \frac{(126.7 - 131.5) + 0.0157 \times (1092.0 - 38)}{1100.7 - 38} = 0.01105$$

Eq. (10·33) may be written as

$$\gamma = 0.622 \frac{P_w}{(P - P_w)}$$

Solving for P_{w1},

$$P_{w1} = \frac{(0.01105/0.622) \times 14.7}{1 + (0.01105/0.622)} = 0.257 \text{ psia}$$

Finally, the relative humidity is, from Eq. (10·31),

$$\phi_1 = \left(\frac{P_w}{P_g}\right)_1 = \frac{0.257}{0.698} = 0.37$$

Now suppose that the makeup water (2) in the adiabatic saturator is not at the pool temperature (T_3). Equation (10·36) still applies, but h_{w2} has a different value. Because of the high h_{fg} for water, h_{w1} is much larger than h_{w2}, hence the γ_1 calculation is very insensitive to h_{w2}. For example, suppose T_2 is instead 60°F, for which $h_{w2} = 28$ Btu/lbm instead of 38 Btu/lbm. The change in γ_1 is less than 1 percent. Hence, for air–water-vapor mixtures at 1 atm pressure the adiabatic-saturation temperature may be idealized as being independent of the makeup water temperature; with this idealization, the adiabatic-saturation temperature depends only on the temperature and humidity of the air–water-vapor mixture.

This idealization provides the basis for simple laboratory measurements of humidity using the *wet-bulb thermometer* (Fig. 10·5). Water brought up the wick by capillary action is evaporated to the air flowing around the bulb, so that the air inside the wick becomes saturated at the temperature of the water in the wick. This temperature will be influenced by the rates of heat and mass transfer from the water to the air, which in turn depend on the configuration of the bulb, air velocity, and other factors. The wet-bulb temperature is therefore not a property

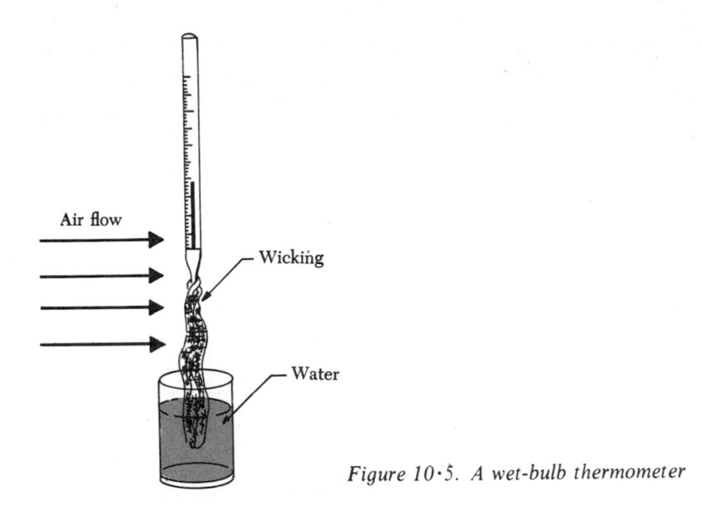

Figure 10·5. A wet-bulb thermometer

of the gas mixtures, since it depends upon the nature of the measuring instrument. However, it has been found experimentally that the wet-bulb temperature is very close to the adiabatic-saturation temperature for *air–water-vapor mixtures*, and thus it provides a satisfactory estimate for this property of the mixture. Note that for other mixtures, e.g., alcohol-air, this is not true.

The solution of Eq. (10·36) can be represented in convenient graphical form on a *psychrometric chart*. Figure 10·6 indicates the form of this chart. Figure B·13 shows complete charts. This chart is based on an adiabatic-saturation process at 1 atm. More complete charts available in handbooks include correction factors for pressure. The value of specific humidity obtained from the charts of Fig. B·13 will be correct to within a few percent at pressures within 1 in Hg of atmospheric pressure. At lower or higher pressures the errors in using these charts become significant, and a more complete psychrometric chart should be consulted.

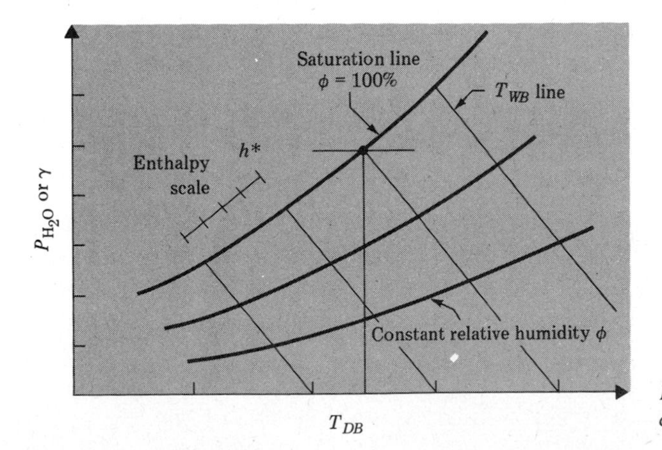

Figure 10·6. The psychrometric chart

Note that the dew point and adiabatic-saturation temperature (wet-bulb temperature) are identical for a saturated mixture. The dry-bulb temperature is simply the mixture temperature. The psychrometric chart has the dry-bulb temperature as the abscissa and the specific humidity as the ordinate. The units on specific humidity are g/g dry air, lbm/lbm dry air, or grains/lbm dry air. One lbm contains 7000 grains.

A *sling psychrometer* has both wet- and dry-bulb thermometers on a common swivel handle. The psychrometer is whirled around and around in the air, providing the necessary air flow over the wet bulb. When equilibrium is reached, the wet- and dry-bulb temperatures are read.

As an example, suppose that wet- and dry-bulb temperatures measured in moist air at 1 atm of pressure are 70 and 90°F, respectively. We shall now find the specific and relative humidities and the thermodynamic properties of the mixture. From Fig. B·13*a* we read, for $T_{DB} = 90°\text{F}$ and $T_{WB} = 70°\text{F}$,

$$\gamma = 78 \text{ grains vapor/lbm dry air}$$

$$P_w = 0.25 \text{ psia}$$

$$\phi = 0.37$$

$$T_{DP} = 60°\text{F}$$

The partial pressure of the air is therefore

$$14.70 - 0.25 = 14.45 \text{ psia}$$

The mole fractions of air and water vapor are then

$$\chi_a = \frac{14.45}{14.70} = 0.983$$

$$\chi_w = \frac{0.25}{14.70} = 0.017$$

The mixture molal mass is then (see Table B·6)

$$\hat{M} = 28.97 \times 0.983 + 18.016 \times 0.017$$

$$\hat{M} = 28.8 \text{ lbm/lbmole} = 28.8 \text{ g/gmole}$$

Note that the presence of water vapor will always decrease the molal mass of the mixture from that for dry air. The mixture density is then

$$\rho = \frac{P}{(\mathscr{R}/\hat{M})T} = \frac{14.70 \times 144}{(1545/28.8) \times 550} = 0.0717 \text{ lbm/ft}^3$$

Since the enthalpy of a perfect gas depends only on temperature, we may read the enthalpy of the water from Table B·1*a*, even though it is not in a saturation state. Thus

$$h_w(90°\text{F}) = h_g(90°\text{F}) = 1100.7 \text{ Btu/lbm}$$

or

$$\hat{h}_w = 1100.7 \times 18.016 = 19{,}830 \text{ Btu/lbmole}$$

The enthalpy of the air may be obtained from the air tables (Table B·9), which assume perfect-gas behavior, as

$$h_a = 131.5 \text{ Btu/lbm}$$

$$\hat{h}_a = 131.5 \times 28.97 = 3810 \text{ Btu/lbmole}$$

The molal enthalpy of the mixture, relative to the datum states used in the steam and air tables, is then

$$\hat{h} = 0.983 \times 3810 + 0.017 \times 19{,}830 = 4080 \text{ Btu/lbmole}$$

so the enthalpy per lbm of *mixture* is

$$h = \frac{4080}{28.8} = 141.6 \text{ Btu/lbm}$$

The molal entropy of the water vapor could be obtained from superheat tables, but they do not extend down to 90°F. Instead we extrapolate from the saturation tables, treating the vapor as a perfect gas. Using Eq. (8·18), we find

$$\hat{s}(T, P) - \hat{s}_g(T) = -\mathcal{R} \ln \frac{P}{P_g(T)}$$

Then, with the help of Table B·1*a*,

$$\hat{s}_w = 2.0085 \times 18.016 - 1.986 \ln \frac{0.25}{0.6989} = 38.23 \text{ Btu/(lbmole·R)}$$

In determining the entropy of the air we must select a datum state. Table B·9 is based on 0 R (extrapolated perfect gas); consequently, we select 1 atm and 0 R for the entropy datum. Then, using Eq. (8·18),

$$s_a(T, P) = \phi(T) - R_a \ln \frac{P}{1 \text{ atm}}$$

From Table B·9,

$$\hat{s}_a = 0.6051 \times 28.97 - 1.986 \ln \frac{14.45}{14.7}$$

$$\hat{s}_a = 17.57 \text{ Btu/(lbmole·R)}$$

Then the mixture entropy, relative to the indicated air and H_2O data, is

$$\hat{s} = 0.983 \times 17.57 + 0.017 \times 38.23 = 17.92 \text{ Btu/(lbmole·R)}$$

$$s = \frac{17.92}{28.8} = 0.622 \text{ Btu/(lbm·R)}$$

Finally, for c_P (see Table B·6),

$$\hat{c}_P = 0.983 \times 6.95 + 0.017 \times 8.04 = 6.97 \text{ Btu/(lbmole·R)}$$

$$c_P = \frac{6.97}{28.8} = 0.242 \text{ Btu/(lbm·R)}$$

The property values calculated above pertain to a unit mass of *mixture*. In practice a different base is often used; since the air-flow rate is usually the same at the inlet and outlet of air conditioning systems, it is convenient to express the *mixture* enthalpy *per lbm of dry air*, which we shall here designate as h^*. In terms of the air and water enthalpies, h^* is readily shown to be

$$h^* = h_a + \gamma h_w \text{ Btu/lbm dry air} \tag{10·37}$$

Values of h^* are shown on most psychrometric charts. The data state is somewhat peculiar, with h_a measured relative to 0°F (not 0 R as in the air tables) and h_w referenced to saturated liquid at 32°F. In terms of h^*, the adiabatic-saturation equation [Eq. (10·36)] becomes

$$h_1^* + (\gamma_3 - \gamma_1)h_{w2} = h_3^*(T_{WB})$$

Note that h_3^* depends only upon the adiabatic-saturation temperature. Since the second term in this equation is typically very small compared to h^* (0.2 Btu/lbm dry air compared with 30 Btu/lbm dry air), a reasonable approximation for many engineering and meteorological applications is that

$$h^* \approx h^*(T_{WB})$$

With this approximation, lines of constant h^* coincide with lines of constant T_{WB}. Simplified psychrometric charts (such as Fig. B·13) portray the h^* information accordingly. More complete psychrometric charts include an enthalpy correction to account for the $\Delta\gamma \cdot h_w$ term.

Let's determine the h^* value at $T_{DB} = 80°F$, $\phi = 60$ percent. From Table B·9 (which has a 0-R datum), we have at 80°F $h_a = 129.1$ Btu/lbm. However, as noted, the psychrometric chart uses an air enthalpy datum of 0°F. From Table B·9 we find $h_a = 109.9$ Btu/lbm at 0°F. Thus, relative to a 0°F datum, $h_a = 129.1 - 109.9 = 19.2$ Btu/lbm. The water enthalpy is taken as the saturation value at 80°F and is 1096.4 Btu/lbm. For a relative humidity of 60 percent at a temperature of 80°F, the γ value is

$$\gamma = 0.622 P_g \phi / P_a = (0.622 \times 0.60 \times 0.507)/(14.7 - 0.6 \times 0.507)$$

$$\gamma = 0.0131 \text{ lbm water/lbm air}$$

or

$$\gamma = 92 \text{ grains water/lbm air}$$

Thus

$$h^* = h_a + \gamma h_w = 19.2 + 0.0131 \times (1096.4)$$

$$h^* = 19.2 + 14.4 = 33.6 \text{ Btu/lbm dry air}$$

Checking on Fig. B·13*a* we find that this value is very close to that indicated on the figure.

Figure B·13*b* is an SI version of the psychrometric chart. In this figure the h^* values use 0°C as the datum for h_a and the usual steam table datum for h_w. The barometric pressure is again 1 atm.

10·5 AIR CONDITIONING

The ideas just developed are applicable to the engineering aspects of air conditioning. We can use the psychrometric chart as the process diagram for these problems. For example, constant-pressure heating will not change the specific humidity but will change the relative humidity. Such a process is shown in Fig. 10·7 (see Fig. B·13). Constant-pressure cooling will not change the specific humidity unless the mixture is cooled to the adiabatic-saturation temperature. Further cooling would then result in condensation and a lowering of the specific humidity. Such a process is shown in Fig. 10·8.

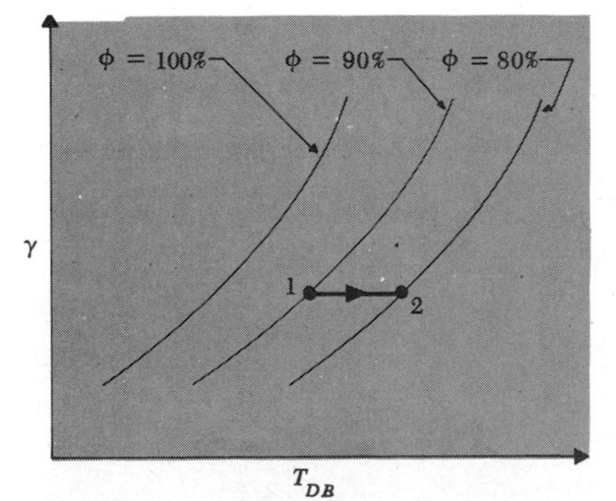

Figure 10·7. Heating reduces relative humidity

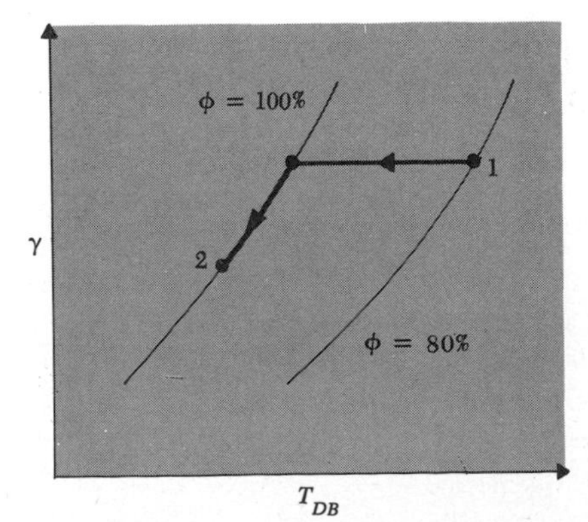

Figure 10·8. A cooling and condensation process

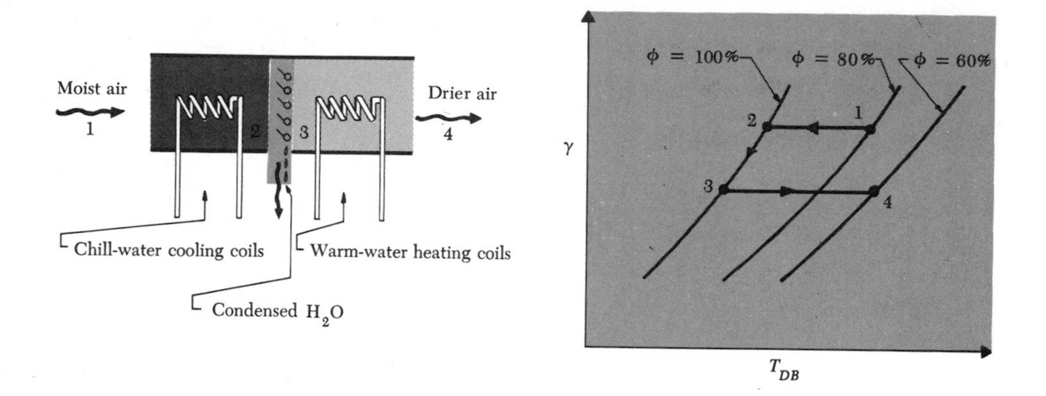

Dehumidification may be accomplished by first cooling, allowing some vapor to condense, and then reheating the mixture. The hardware schematic and process representation (on the psychrometric chart) for a system to accomplish this are shown in Fig. 10·9. Some commercial air conditioners employ spray cooling, in which chilled water is injected into the air, lowering its temperature below the original dew point and allowing a net dehumidification when the mixture is reheated. Such a system is shown in Fig. 10·10.

As an example, let's consider the process required to take air at $\phi_1 = 100$ percent, $T_{DB1} = 85°\text{F}$, to $\phi_3 = 50$ percent, $T_{DB3} = 85°\text{F}$. We can imagine carrying out this process with the technique of Fig. 10·11. The amount of water to be removed can be determined directly from the psychrometric chart,

$$\gamma_1 - \gamma_3 = 184 - 90 = 94 \text{ grains/lbm dry air}$$

Another process of importance in air conditioning is the adiabatic mixing process in which two streams each of wet air are mixed. The final state is to be

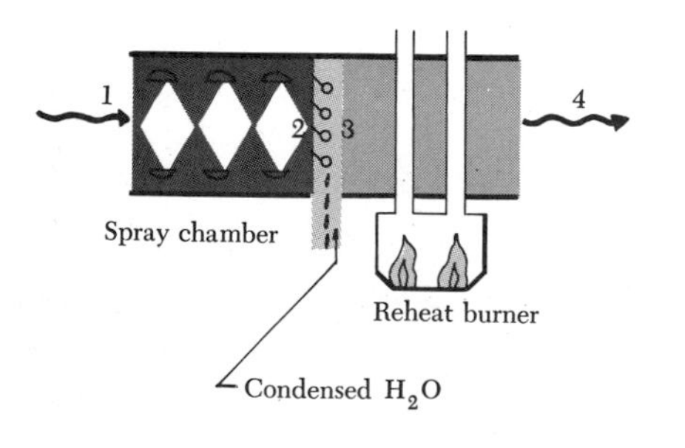

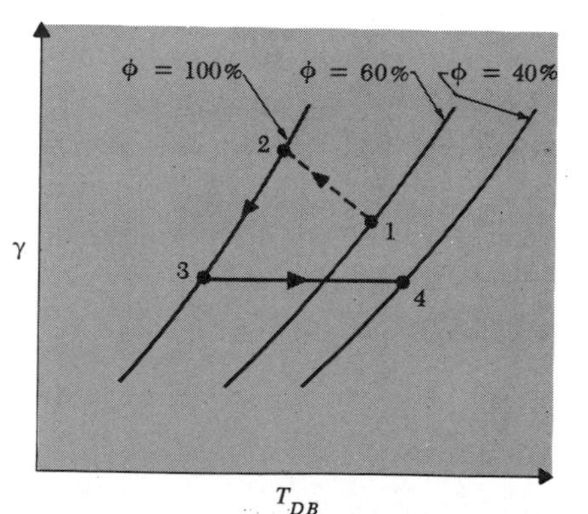

Figure 10·10. Spray dehumidification

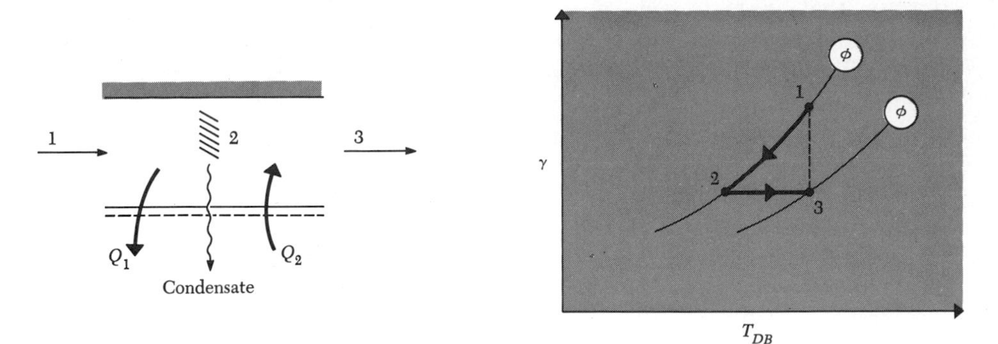

Figure 10·11.

calculated; the control volume is shown in Fig. 10·12. The mass balance for the air gives

$$\dot{M}_{a3} = \dot{M}_{a1} + \dot{M}_{a2} \tag{10·38a}$$

The water mass balance can be written as

$$\dot{M}_{a3}\gamma_3 = \dot{M}_{a1}\gamma_1 + \dot{M}_{a2}\gamma_2 \tag{10·38b}$$

With appropriate idealizations (list these in the margin), the energy balance may be written as

$$\dot{M}_{a1}h_1^* + \dot{M}_{a2}h_2^* = \dot{M}_{a3}h_3^* \tag{10·38c}$$

If the inlet flow rates and states are known, Eqs. (10·38) form three equations with $\dot{M}_{a3}$, γ_3, and h_3^* as three unknowns, and solution is straightforward. For example,

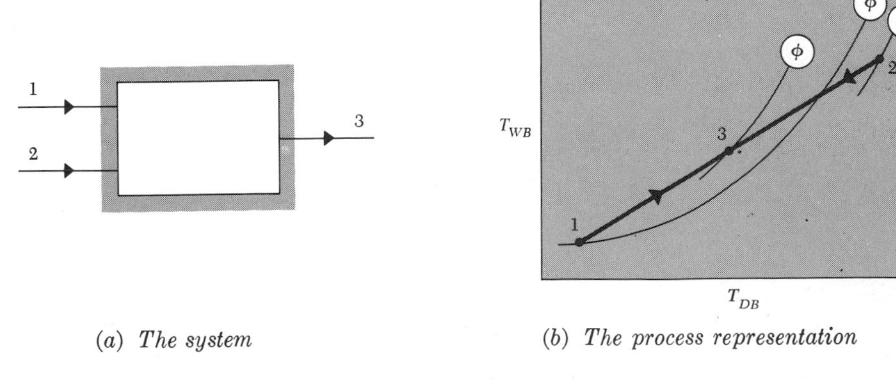

(a) *The system* (b) *The process representation*

Figure 10·12. Adiabatic mixing of two wet air streams

suppose that 2000 ft^3/min of air at 1 atm, 50°F, $\phi = 0.8$, is mixed with 1500 ft^3/min of air at 1 atm, 100°F, $\phi = 0.6$. We must first determine the mass-flow rates of *dry* air in the two streams. To do this we need the air specific volume, which is sometimes available on the psychrometric chart. Our chart does not include this, so we shall calculate it. At 50°F, with $\phi = 0.8$,

$$P_w = 0.80 \times 0.178 = 0.142 \text{ psia}$$

Then
$$v_1 = \frac{RT}{P_a} = \frac{53.3 \times 510}{(14.7 - 0.142) \times 144} = 12.97 \text{ ft}^3\text{/lbm dry air}$$

Similarly, for the 100°F air,

$$v_2 = \frac{53.3 \times 560}{(14.7 - 0.57) \times 144} = 14.66 \text{ ft}^3\text{/lbm dry air}$$

Therefore
$$\dot{M}_{a1} = \frac{2000}{12.97} = 154 \text{ lbm dry air/min}$$

$$\dot{M}_{a2} = \frac{1500}{14.66} = 102 \text{ lbm dry air/min}$$

Now, from the psychrometric chart we find

$$\gamma_1 = 0.0061 \text{ lbm H}_2\text{O/lbm dry air}$$

$$\gamma_2 = 0.0254 \text{ lbm H}_2\text{O/lbm dry air}$$

$$h_1^* = 19 \text{ Btu/lbm dry air}$$

$$h_2^* = 52 \text{ Btu/lbm dry air}$$

Then, from Eq. (10·38*a*),

$$\dot{M}_{a3} = 154 + 102 = 256 \text{ lbm dry air/min}$$

Next, from Eq. (10·38*b*),

$$\gamma_3 = \frac{(0.0061 \times 154) + (0.0254 \times 102)}{256}$$

$$= 0.0138 \text{ lbm H}_2\text{O/lbm dry air} \qquad (= 97 \text{ grains H}_2\text{O/lbm dry air})$$

Finally, from Eq. (10·38*c*),

$$h_3^* = \frac{(19 \times 154) + (52 \times 102)}{256}$$

$$= 32 \text{ Btu/lbm dry air}$$

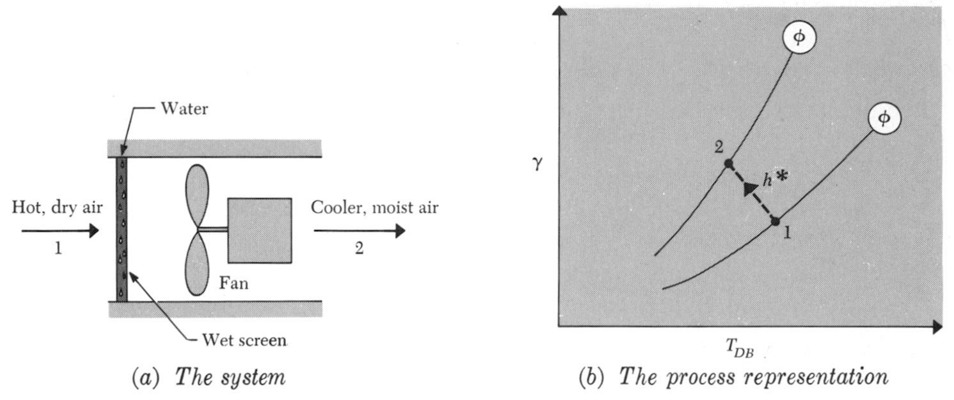

(a) *The system* (b) *The process representation*

Figure 10·13. Swamp coolers provide low-cost air conditioning suitable for dry climates

h_3^* and γ_3 fix state 3 on the psychrometric chart; we find

$$\phi_3 = 0.85 \qquad T_3 = 71°\text{F}$$

Another application of air conditioning to cooling houses in hot and dry climates is *evaporative cooling.* This involves either spraying water into air or blowing air through water-soaked pads. Such a device is shown in Fig. 10·13. This device is essentially an adiabatic saturator, except that complete saturation is not necessarily achieved. We have discussed the validity of the approximation that h^* is constant for this process [see Eq. (10·37) *et seq.*] and that h^* depends only upon the wet-bulb temperature. With these approximations, the process representation is shown in Fig. 10·13*b*. Assuming $T_1 = 105°\text{F}$, $\phi_1 = 0.1$, $T_2 = 80°\text{F}$, the psychrometric chart shows that the discharge air has a relative humidity of about 48 percent. This would be far more comfortable than the very hot dry inlet air. Simple *swamp coolers* of this type are consequently of great importance in desert areas.

The water required for such a device can also be determined from the psychrometric chart. Since $\gamma_1 = 34$ grains/lbm dry air and $\gamma_2 = 74$ grains/lbm dry air, the makeup water is 40 grains or 0.0057 lbm water/lbm air. A typical cooler may have a flow rate of 5000 ft^3/min, and at 105°F the density is 0.070 lbm/ft^3 (neglecting the effect of the water vapor), so the water requirement is

$$\dot{M}_w = (0.0057)(5000)(0.070) = 2.0 \text{ lbm/min}$$

or

$$\dot{M}_w = 120 \text{ lbm/h} = 14.4 \text{ gal/h}$$

For a complete cooling season of 5 months operation at a 60 percent load factor, we find that the water requirements are a surprising 31,000 gal. However, the cooler requires perhaps a 0.5-hp motor instead of the 7-hp motor an equivalent refrigerated air conditioning system would require. In choosing between the two systems, users presumably weight the relative costs and other uses for water and power.

10·6 COOLING TOWERS

Cooling towers are used to provide a quantity of relatively cool water which can then be used to receive energy as heat, for example, from a condenser at a steam power plant. In forced-draft cooling towers air is blown either across falling water or vertically upward against the downward flow of water. A schematic of a counterflow tower is shown in Fig. 10·14. Details of the natural-draft Rancho Seco power plant cooling tower are given in Fig. 10·15. Note that the warm water is cooled by evaporation and the air is humidified, almost to saturation, by picking up the water vapor. Because of the evaporation mechanism it is possible for the water to be cooled *below* the entering-air dry-bulb temperature. In fact, the water could in theory reach the entering-air wet-bulb temperature. Because there is a difference in temperature between the water and the air, energy transfer as heat also occurs, typically from the warm water to the air. However, if the water temperature falls below the air dry-bulb temperature, the energy transfer as heat will be from the air to the water.

Cooling towers are rated in terms of approach and range. The *approach* is the difference in temperature between the cooled-water temperature and the entering-air wet-bulb temperature. The *range* is the temperature difference between the water inlet and exit states. For example, if the warm water enters at 100°F and leaves at 80°F, the range is 20°F. If the air inlet wet-bulb temperature is 70°F, the approach is $80 - 70 = 10$°F. The *makeup* water is the amount of water which must be supplied to "make up" for the water carried away by the air stream.

The analysis of what goes on inside of a cooling tower requires a knowledge of heat transfer, mass transfer, and fluid mechanics. But by looking at the entrance and exit states we can apply the principles of conservation of energy and

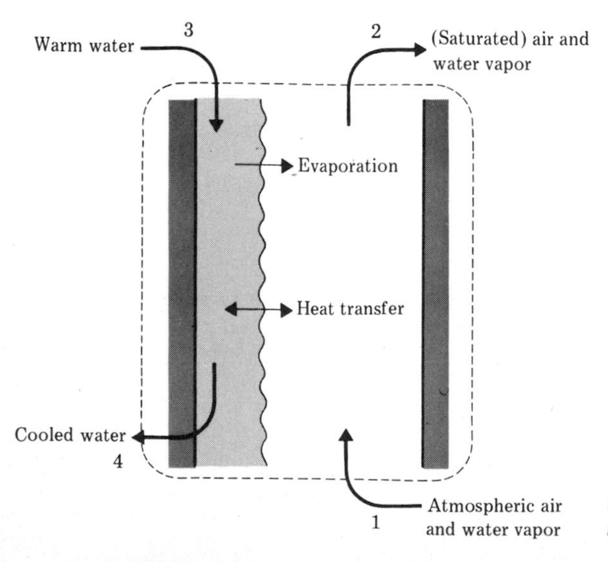

Figure 10·14. A cooling tower schematic

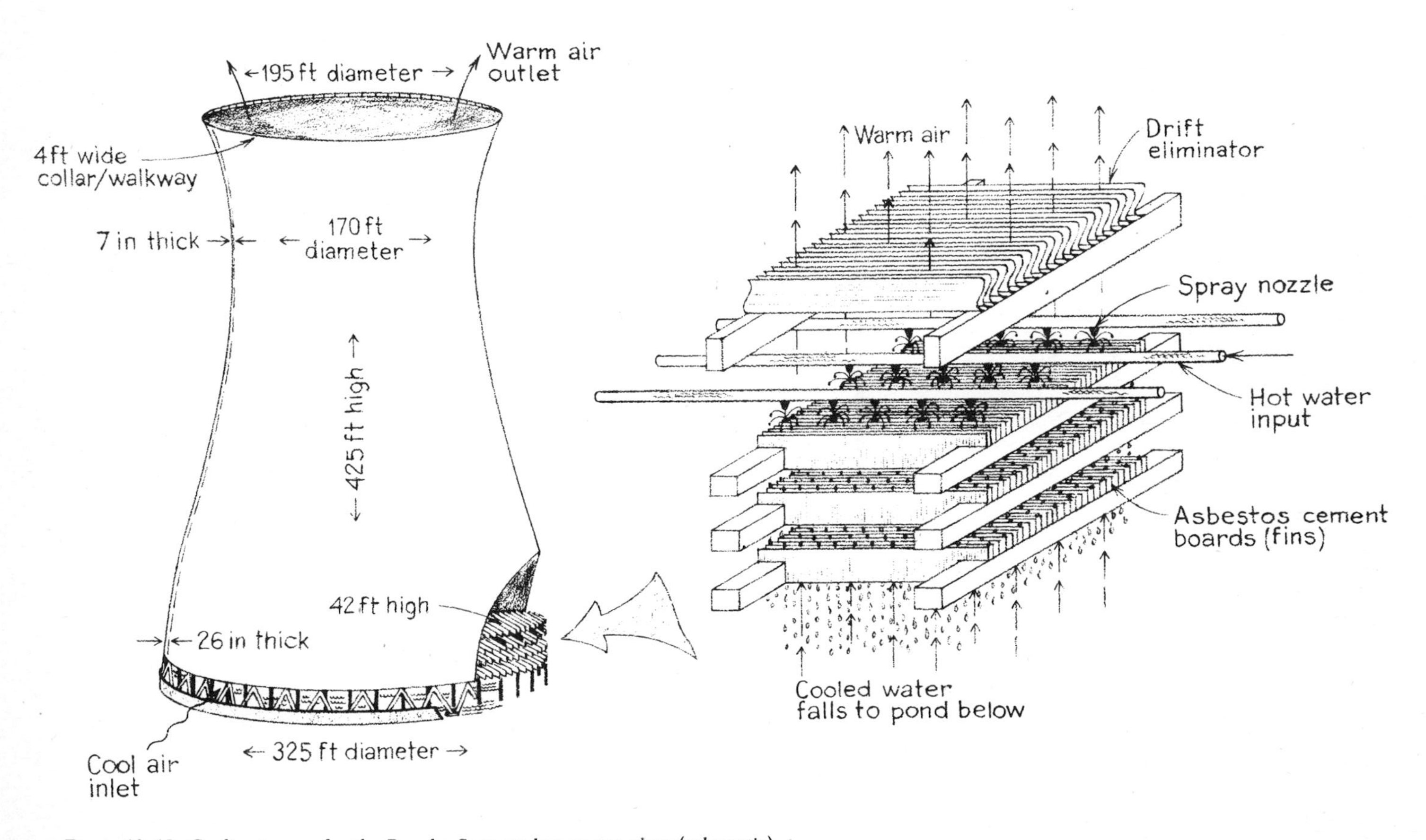

Figure 10·15. Cooling towers for the Rancho Seco nuclear power plant (schematic)

mass to the system as noted in Fig. 10·14 and learn something of its characteristics. We shall illustrate by example.

The water-flow rate from the condenser of an 800-MW power plant is 100×10^6 kg/h. The water is to be cooled in a bank of forced-draft cooling towers from 40°C to 27°C. Atmospheric air enters the tower at 29°C with a relative humidity of 35 percent. The tower is designed to have air leave at 37°C at 98 percent relative humidity. We will determine the makeup water required, the air-flow rate, and the range and approach.

The approach is $T_4 - T_{1WB}$; from the psychrometric chart T_{1WB} is 18.5°C, so the approach is $27 - 18.5 = 8.5$°C. The range is $T_3 - T_4 = 40 - 27 = 13$°C. From an energy balance (what assumptions have been made?) we find

$$\dot{M}_{w3} h_3 + \dot{M}_{a1} h_{a1} + \dot{M}_{w1} h_{w1} = \dot{M}_{w4} h_4 + \dot{M}_{a2} h_{a2} + \dot{M}_{w2} h_{w2} \qquad (10 \cdot 39)$$

Dividing by the air-flow rate, $\dot{M}_a = \dot{M}_{a1} = \dot{M}_{a2}$,

$$\frac{\dot{M}_{w3} h_3 - \dot{M}_{w4} h_4}{\dot{M}_a} = (h_{a2} + \gamma_2 h_{w2}) - (h_{a1} + \gamma_1 h_{w1}) \qquad (10 \cdot 40)$$

From a mass balance on the water

$$\dot{M}_{w3} + \dot{M}_{w1} = \dot{M}_{w4} + \dot{M}_{w2} \qquad (10 \cdot 41)$$

Dividing by $\dot{M}_a$, this becomes

$$(\gamma_2 - \gamma_1) = \frac{\dot{M}_{w3} - \dot{M}_{w4}}{\dot{M}_a} \qquad (10 \cdot 42)$$

Thus the makeup water per unit mass of air is just $\gamma_2 - \gamma_1$. While this is a significant amount, the makeup water rate is much less than the water-flow rate. Therefore in Eq. (10·40) we make the approximation $\dot{M}_{w3} \approx \dot{M}_{w4}$, which gives

$$\frac{\dot{M}_w}{\dot{M}_a}(h_3 - h_4) = h_2^* - h_1^* \qquad (10 \cdot 43)$$

From the given data $h_1^* = 52$ kJ/kg dry air, $h_2^* = 142$ kJ/kg dry air, $\gamma_1 = 0.0085$, and $\gamma_2 = 0.0405$. Thus from the energy balance we find, using saturated-liquid values for h_3 and h_4,

$$\frac{\dot{M}_w}{\dot{M}_a}(167.5 - 113.2) = 142 - 52 = 90$$

So

$$\frac{\dot{M}_w}{\dot{M}_a} = 1.66$$

The air-flow rate is then

$$\dot{M}_a = 100 \times 10^6/1.66 = 60.2 \times 10^6 \text{ kg/h}$$

The makeup water can now be calculated as

$$\dot{M}_{w3} - \dot{M}_{w4} = \dot{M}_a(\gamma_2 - \gamma_1) = 60.2 \times 10^6 \times (0.0405 - 0.0085)$$
$$= 1.93 \times 10^6 \text{ kg/h}$$

The fraction of water evaporated is

$$\frac{\Delta \dot{M}_w}{\dot{M}_{w3}} = 1.93 \times 10^6/100 \times 10^6 = 1.93 \text{ percent}$$

The approximation of Eq. (10·40) by Eq. (10·43) seems justified. If needed, a correction could now be made.

This example had the water exit temperature T_4 equal to 27°C which was below the air inlet temperature of 29°C. As mentioned before, this is possible because the energy transfer via evaporation is greater than that due to direct heat transfer.

What happens to the water evaporated? In some parts of the country this vapor can form a fog which may in turn lead to the icing of roads downwind of the cooling tower under winter weather conditions.

SELECTED READING

Lee, J., and F. Sears, *Thermodynamics*, 2d ed., chap. 10, Addison-Wesley Publishing Co., Inc., Reading, Mass., 1962.

Van Wylen, G., and R. Sonntag, *Fundamentals of Classical Thermodynamics*, 2d ed., chap. 11, John Wiley & Sons, Inc., New York, 1973.

Wark, K., *Thermodynamics*, 2d ed., chaps. 11 and 19, McGraw-Hill Book Company, New York, 1971.

Zemansky, M. W., *Heat and Thermodynamics*, 5th ed., chap. 15, McGraw-Hill Book Company, New York, 1968.

QUESTIONS

10·1 What are mass fractions, volume fractions, and mole fractions, and under what circumstances is one equal to another?

10·2 How many independently variable properties are there for a mixture of two compressible electrically polarized gases?
10·3 What is a phase of a mixture?
10·4 What is a mixture of independent substances?
10·5 Under what circumstances are partial pressures equal to the total pressure times the mole fractions?
10·6 How would you compute the specific heat of a mixture of gases (nonperfect) from the known specific heats of its constituents?
10·7 Why is the datum state for enthalpy of a mixture arbitrary? Why can different datums be used for the different species as long as no chemical changes occur?
10·8 What is the meaning of the term "molal specific heat"?
10·9 Why does the entropy not change when two identical gases are allowed to mix?
10·10 What is a psychrometric chart?
10·11 What is the difference between relative humidity and specific humidity? Which do you think is most related to human comfort?
10·12 How can air be dehumidified?
10·13 A swamp cooler is simply a burlap sack dripping with water through which air is blown. What happens to the humidity and temperature of the air passing through it?
10·14 What are the basic principles employed in the thermodynamic design of air conditioning systems?
10·15 List 10 different engineering systems in which the theory of this chapter is employed.
10·16 What is a typical dew point during the summer months in your locality?
10·17 What is an adiabatic saturator?
10·18 What is a sling psychrometer? How do you use it?
10·19 What are the reference conditions for enthalpy on the usual psychrometric chart?

PROBLEMS

10·1 Air is roughly 21 per cent oxygen and 79 percent nitrogen (by volume). Calculate the specific heat of air from the data in Table B·6 and compare your result with the value given there.
10·2 A gas having a specific-heat ratio k of 1.500 is required for a certain gas-dynamics experiment. Specify a mixture that could be used.
10·3 A gas from an industrial process has the following analysis by volume: $H_2 = 8$ percent, $CO = 22$ percent, $CH_4 = 2$ percent, $CO_2 = 9$ percent, $O_2 = 3$ percent, $N_2 = 56$ percent. Find the analysis on a weight basis.
10·4 A gas mixture has the analysis by weight of 15 percent He, 60 percent N_2, and 25 percent O_2. Find the mole fractions and molecular weight of the mixture.
10·5 A mixture of gases used for a special heat-transfer application consists of 25 percent Ar, 50 percent He, and 25 percent H_2 by weight. For a total pressure of 10 atm determine the partial pressures. Find the c_P for the mixture in kJ/(kg·K) and the mixture molecular weight.
10·6 A nominal model atmosphere for Jupiter is described in the table. Calculate the ratio of specific heats k, the molecular weight of the mixture, and the mole fractions.

	Wt fraction
H_2	0.75348
He	0.23000
CH_4	0.00429
NH_3	0.00109
H_2O	0.00800
Others	0.00314

N. Divine and F. D. Palluconi, *JPL Quarterly Technical Review*, vol. 2, no. 7, 1972.

10·7 The mole fraction model composition for the nominal atmosphere of Saturn is given by the following table.

H_2	0.88572
He	0.11213
CH_4	0.00063
NH_3	0.00015
H_2O	0.00105
Ne	0.00013
Other	0.00019

N. Divine and F. D. Palluconi, *JPL Quarterly Technical Review*, vol. 2, no. 7, 1972.

Using these results calculate $\hat{c}_p$, k, and $\hat{M}$ for the mixture, neglecting the contributions from the smaller constituents.

10·8 Air is made up by volume of 78.09 percent N_2, 20.94 percent O_2, 0.93 percent Ar, and 0.0320 percent CO_2, plus trace constituents such as Ne, He, and Kr. Calculate the specific heats, molecular weight, and k for air.

10·9 A mixture has an analysis by weight of 40 percent argon and 60 percent nitrogen. The mixture is compressed from 1 atm, 20°C, to 4 atm. Find the final temperature, the work required, and the entropy change of each gas.

10·10 A mixture of 40 percent argon and 60 percent hydrogen by volume has been proposed for use as the working fluid in a closed Brayton cycle. The mixture is compressed isentropically from 40°C at 0.4×10^6 N/m^2 to a pressure of 1.2×10^6 N/m^2. Find the final temperature of the mixture and the work required, assuming perfect-gas behavior is applicable.

10·11 Natural gas consists of a mixture of several gases. One example is 84 percent methane, 14.8 percent ethane, 0.70 percent carbon dioxide, and 0.50 percent nitrogen by volume. Calculate the specific heats, the molecular weight, and the k value for this mixture.

10·12 For the mixture in Prob. 10·11, determine the percent by weight for H_2, C, N_2, and O_2.

10·13 One tank contains 2 kg of methane at 2 atm, 10°C. Another tank contains 4 kg of oxygen at 6 atm, −10°C. A valve between the tanks is opened, and the gases mix adiabatically. Find the final mixture temperature and pressure.

10·14 One kg of water vapor and 1 kg of air are contained in a tank with a volume of 1 m^3 at 200°C. Find the mixture pressure.

10·15 A receiver tank from an air compressor contains 40 kg of air and 2 kg of water vapor at a temperature of 200°C. If the tank volume is 3 m^3, compute the pressure of the mixture.

10·16 Oxygen can be separated from air by fractional distillation, which amounts to cooling the air until either nitrogen or oxygen condenses. Which will occur first at 10 atm pressure?

10·17 One ft^3 of helium at 1 atm and 80°F is mixed with 2 ft^3 of oxygen at 1 atm and 100°F. Specify the partial pressures, mole fractions, mass fractions, and specific heats of this mixture at 2 atm and 150°F. Determine the difference in the enthalpy and entropy of the mixture between 50 and 100°F at 1 atm.

10·18 A gas has the following analysis by mass: O_2, 0.2; H_2, 0.08; CO_2, 0.25; and CO, 0.47. Determine the volumetric analysis and the gas constant of the mixture.

10·19 What are the specific heats per mole for a mixture consisting of 1 mole of carbon dioxide, 3 mole of helium, and 1 mole of oxygen?

10·20 How much will the enthalpy of a mixture of 20 percent Freon 12 and 80 percent CO_2 (parts by mass) change when the mixture is heated at constant pressure (100 psia) from 140°F to 200°F?

10·21 At what temperature will fractional distillation occur in the mixture of Prob. 10·20 and which component will be first to condense?

10·22 Eight ft^3 of O_2 at 100°F, 100 psia, are mixed adiabatically with 10 ft^3 of H_2 at 60°F, 80 psia, in a constant-volume container. What are the final pressure and temperature?

10·23 A 10-ft^3 condenser chamber contains saturated air at 90°F, 27 in Hg vacuum. Air leaks into the chamber until the temperature is reduced to 75°F and the vacuum reduced to 21 in Hg. If the barometer reading is 29.92 in Hg, find the amount of air that has leaked in.

10·24 A steady-flow mixing process occurs such that oxygen at 70°F and methane at 150°F are mixed adiabatically to form a mixture which is 50 percent by volume hydrogen. All pressures are 1 atm, and the hydrogen-flow rate is 0.01 lb/s. Determine the exit temperature.

10·25 Nitrogen and hydrogen are mixed in a steady-flow adiabatic process in a ratio of 4 kg of hydrogen per kg of nitrogen. The hydrogen enters at 0.14 MPa, 40°C, and the nitrogen enters at 0.14 MPa, 250°C. The pressure after mixing is 0.12 MPa. Determine the exit temperature and the entropy production per kg of mixture.

10·26 Nitrogen and hydrogen are mixed in a steady-flow adiabatic process in the ratio of 4 kg of nitrogen per kg of hydrogen. The hydrogen enters at 0.14 MPa, 40°C, and the nitrogen at 0.14 MPa at 250°C. The pressure after mixing is 0.12 MPa. Determine the exit temperature and the entropy production per kg of mixture.

10·27 A sling psychrometer reads 22°C dry bulb and 16°C wet bulb. What are ϕ, γ, and h^*?

10·28 An evaporative cooler operates with an inlet of 38°C at a wet-bulb temperature of 20°C. Exit conditions are a dry-bulb temperature of 25°C. How much makeup water is required if the entering flow rate is 150 m^3/min?

10·29 Using the formulas, calculate the relative and specific humidities at a dry-bulb temperature of 15°C and a wet-bulb temperature of 5°C. Compare with the values on the psychrometric chart.

10•30 Verify the position of the 20°C wet-bulb line on the psychrometric chart. Calculate the h^* value at 20°C wet bulb if $h^* = c_P T + \gamma h_w$ and the h_w is taken from the steam tables.

10•31 An air–water-vapor mixture enters a cooling apparatus at 40°C, and $\phi = 80$ percent. Design calls for the mixture to leave at 25°C saturated. How much water must be removed and how much heat transfer must take place if the volumetric flow rate is 600 m^3/min at a pressure of 1 atm?

10•32 A cooling tower cools water from 40°C to 25°C at a flow rate of 100,000 kg/h. Air enters the bottom of the tower at 20°C and $\phi = 40$ percent and leaves at 35°C, saturated. Determine the makeup water required and the air-flow rate.

10•33 Nitrogen and hydrogen are mixed in a steady-flow adiabatic device in a ratio of 4 lbm hydrogen/lbm nitrogen. The hydrogen enters at 20 psia, 100°F, and the nitrogen at 20 psia, 500°F. The pressure after mixing is 18 psia. Determine the exit temperature and the entropy production per lbm of mixture.

10•34 A mixture of air and water vapor containing 0.013 lbm of water vapor and 1 lbm of air occupies a tank at 14.7 psia and 90°F. Determine the dew point, relative humidity, and specific humidity.

10•35 Measurements with a sling psychrometer have fixed $T_{WB} = 60$°F and $T_{DB} = 75$°F at 14.7 psia. Determine the relative humidity and specific humidity without using the psychrometric chart. Compare with the chart.

10•36 Air is supplied to a certain room from the outside, where the temperature is 30°F and the relative humidity is 60 percent. It is desired to supply the air to the room at 70°F and 60 percent relative humidity. How many pounds of water must be supplied to each lbm of air entering the room if these conditions are to be met?

10•37 The temperature of a certain room is 75°F and the relative humidity is 55 percent. The barometric pressure is 29.92 in Hg. Find the partial pressures of the air and water vapor and the specific humidity of the mixture.

10•38 In a steady-flow device, air is heated to 90°F, without the addition of water, from a state where $T_{DB} = 60$°F, $T_{WB} = 55$°F. Find the relative humidity of the original mixture, the dew-point temperature, the specific humidity, the energy added, and the final relative humidity.

10•39 An air–water-vapor mixture at 19°C, 50 percent relative humidity, 1 atm, is heated at constant pressure to 28°C in part of a steady-flow air conditioner. Determine the inlet and discharge specific humidities, the discharge ϕ, and the energy input per kg of dry air.

10•40 We assumed in this chapter that in psychrometric calculations water vapor can be treated as a perfect gas. Calculate a few Z values for typical conditions and verify this assumption.

10•41 Atmospheric air at 90°F and 50 percent relative humidity is to be conditioned to 70°F at 35 percent relative humidity using a cooling coil section, a condensate removal section, and a reheat section. The air-flow rate is 350 lbm/h. At 1 atm find the mass of water removed and the heat transfers.

10•42 An air conditioning system for Great Desert University operates as noted below.

(*a*) Find the compressor power.

(*b*) Find the condenser heat transfer rate.

(*c*) Find the makeup water required, the water-flow rate in the spray tower loop, and the pump power.

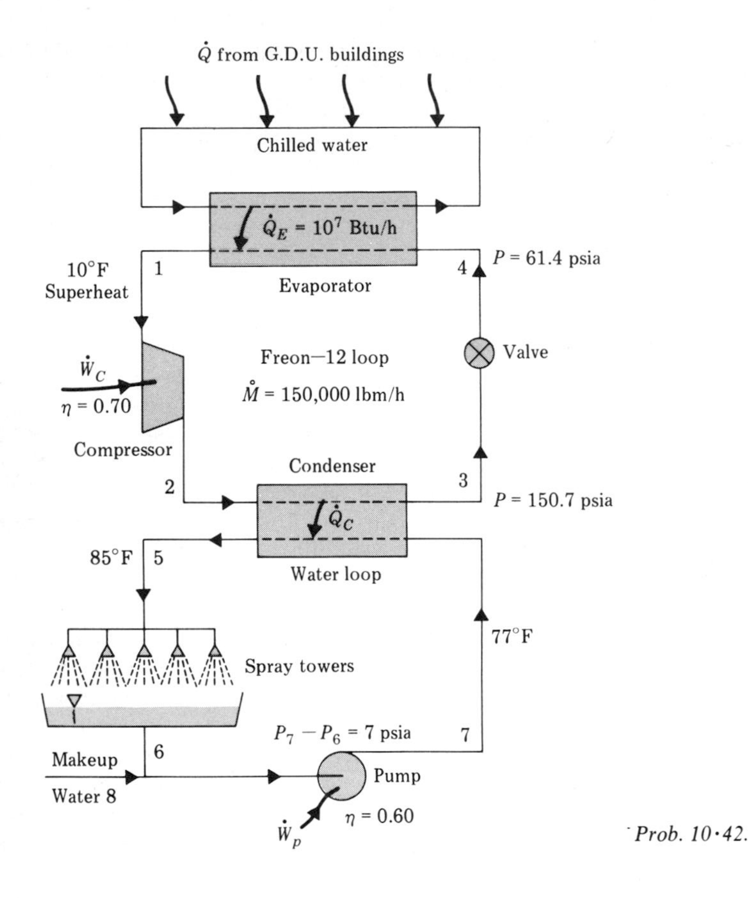

Prob. 10·42.

10·43 An air–water-vapor mixture at 67°F, 50 percent humidity, 1 atm, is heated at constant pressure to 82°F in part of a steady-flow air conditioner. Determine the inlet and discharge specific humidities, the discharge relative humidity, and the energy input as heat per lbm of dry air.

10·44 Two streams of air are mixed in a steady-flow adiabatic device. The first stream is at a temperature of 55°F with $\phi = 0.2$ and a flow rate of 650 ft^3/min. The second stream is at 75°F with $\phi = 0.8$ and a flow rate of 900 ft^3/min. If all pressure is at 1 atm, find the outlet mixture relative humidity, temperature, and specific humidity.

10·45 Atmospheric air at 84°F and $\phi = 0.7$ flows over a set of cooling coils at a rate of 15,000 ft^3/min. The condensed liquid leaves the system at 50°F. The air is then heated to 75°F, $\phi = 0.4$. Determine the rate of energy removal as heat in the cooling section, the condensate-flow rate, and the energy input rate in the heating section.

10·46 For $T_{WB} = 70$°F and $T_{DB} = 90$°F, determine the relative and specific humidities using the adiabatic-saturator model, and compare your results with those obtained from the psychrometric chart.

10·47 Thirty m^3/min of air at 24°C, 1 atm, and 70 percent relative humidity are to be cooled and dehumidified to 20°C and 55 percent relative humidity by cooling and reheating in a steady-flow air conditioner. Find the temperature to which the air must be cooled and the number of tons of refrigeration required.

10•48 Condensation on cold-water pipes often occurs in warm humid rooms. If the water temperature is 50°F and the room temperature is 75°F, what maximum relative humidity can be tolerated if condensation is to be avoided?

10•49 An air compressor takes in air from the atmosphere at 60°F and 75 percent relative humidity and discharges it at 100 psia. The compressor isentropic efficiency is 82 percent. What are the relative and specific humidities of the discharged air?

10•50 Develop a psychrometric chart for a mixture of helium and water vapor at 2 atm. Show the 25, 50, 75, and 100 percent relative-humidity lines over the range 40 to 120°F.

10•51 Develop a psychrometric chart for helium-mercury mixtures at 60 psia for the range 300 to 500°F. Plot lines of 25, 50, 75, and 100 percent relative humidity.

10•52 Air at $T_{DB} = 110°F$, $T_{WB} = 70°F$, is humidified adiabatically with wet steam. The steam contains 16 percent moisture at 20 psia. What is the dry-bulb temperature of the humidified air when sufficient steam is added to humidify the air to 50 percent relative humidity? Assume 1 atm pressure.

10•53 An air–water-vapor mixture enters a cooler-dehumidifier at the rate of 1000 ft^3/min. The mixture enters at 14.7 psia, 100°F, and $\phi = 0.8$. The specific humidity on leaving is decreased to one-third of that on entering. The relative humidity on leaving is 100 percent, and the pressure is 14.7 psia. Determine the required heat-transfer rate.

10•54 Consider an evaporative cooler with atmospheric air (14.7 psia) at 100°F, $\phi = 0.1$. Cooling water at 50°F is sprayed into the air. If the air–water-vapor mixture emerges at 80°F, what will be its relative humidity? What are the disadvantages of this approach to air conditioning?

10•55 A tank contains helium with mercury vapor. Initially the mixture is at 700°F and 2 psia. The mixture is cooled at constant pressure, and at 415°F mercury droplets are observed inside the tank. How much could the pressure of the original mixture have been increased in an isothermal compression before mercury condensation occurred at 700°F? What were the initial relative and specific mercury humidities?

10•56 A mixture of CO_2 and He is cooled at constant volume from 40°F and 500 psia to −20°F, at which point CO_2 condensation is observed. What was the initial mixture composition?

10•57 A mixture of Freon 12 and argon is cooled at constant volume from 200°F and 100 psia to −20°F, at which point Freon condensation is observed. What was the initial mixture composition?

10•58 Specify a mixture of Freon 12 and CO_2 such that both constituents will condense when the mixture temperature is 0°F. At what mixture pressure will this occur? What will happen when a mixture of this proportion is cooled at 100 psia? At 600 psia?

10•59 A mixture consists of 2.7 lbm of helium and 3.0 lbm of nitrogen at 10 psia, 80°F. The mixture is isentropically compressed to 100 psia. Find the change in entropy and enthalpy for the two components during this process.

10•60 A possible device to desalt sea water using solar energy is shown in the figure, where typical operating conditions are also noted. Sea water is heated inside tubes in the condenser where it acts as the energy sink for the energy released in the condensation process. Because air cannot contain as much water at low temperatures, the cooling of the air in the condenser causes a condensation process in which fresh water is released. The now warm salt water is further heated in solar collectors and pumped into a packed-column evaporator. The air from the condenser is forced up

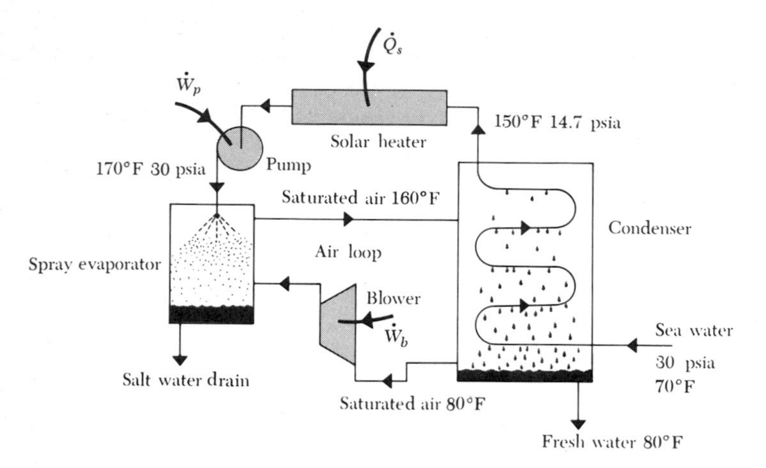

Prob. 10·60.

through the evaporator and becomes saturated with water vapor at a high temperature. In effect the air acts as a transfer medium for water vapor.

The object of this problem is to determine the fresh-water production per lbm of dry air circulated in the system. For a required output of 10,000 gal of fresh water per solar (8-h) day, what air-flow rate must be maintained through the condenser, and what water-flow rate on the water side of the condenser? Under the given conditions what is the energy input from the sun? If the collection rate is an average of 300 Btu/(h·ft²) how many ft² of collector are required?

10·61 A cooling tower is used to cool water from 100°F to 75°F at a water-flow rate of 200,000 lbm/h. The air enters the bottom of the tower at 60°F with a relative humidity of 50 percent and leaves the top at 90°F, saturated. Find the air-flow rate and the makeup water required.

10·62 Find the refrigeration required to cool 20,000 ft³/min of outside air at 90°F dry bulb and 75°F wet bulb to saturated air at 60°F. How much condensate at 60°F must be removed? The pressure is 29.92 in Hg.

10·63 Desert air enters an evaporative cooler at 100°F, 15 percent relative humidity. The air leaves the cooler having reached 67 percent of the possible temperature change. Determine the exit conditions and the heat transfer, if any.

10·64 Calculate the h^* value (*a*) for sea level, $P = 29.92$ in Hg, with a relative humidity of 50 percent and a temperature of 70°F; (*b*) at an altitude of 5000 ft, $P = 24.90$ in Hg, with a relative humidity of 50 percent and a temperature of 70°F.

10·65 Air at 85°F and a relative humidity of 80 percent is to be processed in a steady-flow air conditioner and delivered at 70°F, 50 percent humidity. The air-flow rate is 200 ft³/min. Design a suitable system, specifying any energy-transfer rates, power requirements, temperatures, intermediate states, coolant-flow rates, and so on.

10·66 Cooling air for a large computer must be supplied at the rate of 100 ft³/min. The temperature must be 50°F, and the humidity less than 0.0005 lbm H_2O/lbm dry air. Suppose the supply air is at 80°F and 50 percent relative humidity. Design a suitable air conditioning system, specifying any energy-transfer rates, coolant-flow rates, temperatures, intermediate system states, and so on.

10·67 A simple cooling system using CO_2 sublimation has been proposed for a short-term extravehicular space suit. The CO_2 will be sublimed from a small solid block at

0.67 atm and mixed with the breathing gas at 22°C. The breathing rate is nominally 0.25 m^3/min, and the mixture must leave the suit at 32°C. The total energy removal rate is to be 200 kJ/h. Specify the required CO_2 sublimation rate, the mass requirement for 1 h of operation, and the mass percent of CO_2 in the inlet mixture.

10·68 CO_2 must be removed from the O_2 in an astronaut life support system. Suppose the breathing rate is 6 ft^3/min, and that the exhaled mixture contains CO_2 with 10 percent volume fraction. Design a system to remove 95 percent of the CO_2 from the discharged gas continuously, assuming the mixture is delivered at 90°F to the conditioner and to the space suit at 70°F. Specify the volume flow of makeup oxygen required, the processes, power requirements, intermediate states, and so on, within the conditioner.

10·69 Design a system to extract air from sea water at a depth of 30 m. You may treat the water–dissolved air system as a mixture of independent substances. Finding the necessary solubility data will be good practice for your engineering career.

10·70 Hydrocarbon emissions are considered air pollutants. Estimate the hydrocarbon emitted into the air from the filling of your automobile gas tank. Making some assumption about mileage, determine an equivalent g/mi emission due to filling the tank. EPA standards are intended to reach a limit of 0.41 g/mi. How do your numbers compare?

CHAPTER ELEVEN

THERMODYNAMICS OF REACTING MIXTURES

11·1 SOME CHEMICAL CONCEPTS AND TERMS

In this chapter we shall study mixtures that may be undergoing chemical reaction. The first law of thermodynamics forms the basis for quantitative analysis of chemical reactions. The second law of thermodynamics is used to identify the *direction* of chemical reactions. The *third law of thermodynamics* will be introduced in this chapter. This law is needed to connect properly the entropies of different species involved in the chemical reaction. Thermodynamics also allows us to determine the equilibrium composition of any mixture of chemical substances; this will be discussed in the next chapter. The prediction of *reaction rates* is beyond the scope of thermodynamics; the topic of chemical kinetics, which deals with reaction rates, will not be discussed in this text.

Associated with every chemical reaction is a *chemical equation*, derived by applying the conservation of atoms to each of the atomic species involved in some sort of a "unit reaction." The reaction begins with a collection of certain chemical constituents, called the *reactants*, and the chemical reaction causes a rearrangement of the atoms and electrons to form different constituents, called the *products*. The reaction between the constituent reactants hydrogen and oxygen to form the product water can be expressed as

$$2H_2 + O_2 \rightleftarrows 2H_2O \tag{11·1}$$

This expression indicates that two molecules of hydrogen and one molecule of oxygen can combine to form two molecules of water (the reaction may also go in the opposite direction). The coefficients in the chemical equation are called *stoichiometric coefficients* (2, 1, and 2 in this example). The number of hydrogen atoms is conserved, as is the number of oxygen atoms. The chemical equation can also be interpreted as an equation relating the molal masses involved in a reaction; here 2 moles of hydrogen and 1 mole of oxygen combine to form 2 moles of water. Note that the number of moles of the products may differ from that of the reactants.

A *stoichiometric mixture* of reactants is one in which the molal proportions of the reactants are exactly as given by the stoichiometric coefficients so that no excess of any constituent is present. A *stoichiometric combustion* is one in which all the oxygen atoms in the oxidizer react chemically to appear in the products.

The most common oxidizer is air, which for many purposes may be considered a mixture of 21 percent oxygen and 79 percent nitrogen (mole or volume fractions). The chemical equation for stoichiometric combustion of methane (CH_4) with air is then

$$\underbrace{CH_4 + 2(O_2 + 3.76N_2)}_{\text{reactants}} \rightleftarrows \underbrace{CO_2 + 2H_2O + 7.52N_2}_{\text{products}} \qquad (11\cdot2)$$

If more air is supplied, not all will be involved in the reaction, and the composition of the products will differ from that of a stoichiometric combustion. The additional air supplied is called *excess air*. The term *theoretical air* is also used; 200 percent theoretical air is equivalent to 100 percent excess air. For example, if methane is burned with 125 percent theoretical air, the chemical equation is†

$$\underbrace{CH_4 + 1.25 \times 2(O_2 + 3.76N_2)}_{\text{reactants}} \rightleftarrows \underbrace{CO_2 + 2H_2O + 0.5O_2 + 9.4N_2}_{\text{products}} \qquad (11\cdot3)$$

and the products mixture would not be stoichiometric. In order to *complete* a combustion reaction (to burn all the fuel), excess air is normally supplied.

Power plant boilers normally run about 10 to 20 percent excess air. Natural gas-fired boilers may run as low as 5 percent excess air, and pulverized coal-fired boilers may run as much as 20 percent excess air. Internal combustion engines normally run with a little excess air. Gas turbines run very *lean*, with up to 400 percent theoretical air. Incomplete combustion, typical in rich systems, results in the production of CO instead of CO_2 and the production of unburned or partially burned hydrocarbons. These gases contribute to the urban smog problem.

† Note that this equation is balanced and that the stoichiometric coefficients are not necessarily integer. One can balance the equation (find the product's stoichiometric coefficients) by first considering the C atoms, then the H atoms, and finally the O atoms.

Another important combustion parameter is the *air/fuel ratio* (AFR) in a reaction. For the stoichiometric combustion process of CH_4 we have, on a molal basis,

$$\text{AFR} = \frac{\mathcal{N}_{\text{air}}}{\mathcal{N}_{\text{CH}_4}} = \frac{2 + 7.52}{1} = 9.52 \text{ moles air/mole fuel}$$

Since $\hat{M}_{\text{air}} = 28.95$ lbm/lbmole and $\hat{M}_{\text{CH}_4} = 16$ lbm/lbmole, the AFR may also be expressed on a mass basis as

$$\text{AFR} = 9.52 \times \frac{28.95}{16} = 17.2 \text{ lbm air/lbm fuel}$$

If a reaction occurs in an isolated vessel, the internal energy of the products and reactants will be the same, and the entropy of the products will be greater than that of the reactants. Reactions can be made to occur at constant pressure and temperature by allowing the volume to change and transferring energy as heat from the reacting mixture. The internal energy of the products will then be different from that of the reactants. If the products have less internal energy (at constant P and T), the reaction is said to be *exothermic*, and energy must be transferred as heat from the mixture in order to keep the temperature constant. A reaction at constant P and T for which the opposite is true is called *endothermic*. The exothermic reaction is of particular utility in engineering as a means for supplying energy in the form of heat to thermal power systems.

11·2 FUEL ANALYSIS AND PRODUCT COMPOSITION

In an actual combustion process it may be necessary to determine experimentally both the products of combustion and their respective amounts. Often the fuel itself is unknown. The analysis of gaseous products of combustion is frequently made with an *Orsat apparatus*. A sample of the gas to be analyzed is brought into the Orsat analyzer at a known temperature and atmospheric pressure. The gas is then placed in contact with a liquid that absorbs CO_2, such as KOH. The temperature and the total pressure are maintained constant during the measurement, so that the volume of the gas decreases as CO_2 is absorbed and this decrease is noted. The gas is successively placed in contact with a liquid that absorbs O_2, such as pyrogallic acid, and then with a liquid that absorbs CO, such as cuprous chloride. The further decrease in volume of the gas sample is noted. The change in the volume of the gas sample during each separate absorption test is a measure of the volumetric fraction of that particular gas in the sample. Since the measurements are on a volumetric basis at a constant temperature and pressure and we treat the gases as ideal, the values obtained are also a measure of the mole fraction of each of the measured constituents. If additional gases (other than N_2 and H_2O) are expected to be present in significant amount in the sample, additional absorbers will be required. At each stage of absorption the mixture is presumed to be saturated with water vapor. Since the temperature and pressure remain constant, the partial pressure of the water vapor is unchanged by the absorptions, and consequently an appropriate amount of H_2O condenses out with each absorption. The net effect is that indicated mole fractions are precisely those that would be obtained in a mixture *without* any water vapor (a *dry mixture*); the proof of this is left as a problem for the student. The gas remaining in the analyzer after the overall test is completed determines the nitrogen mole fraction of the *dry* mixture. The actual H_2O content is then determined from a mass balance.

As an example, suppose an Orsat analysis has provided us with the volume measurements given in Table 11·1. We want to find the chemical equation and the AFR from these data. However, the fuel is unknown. Let's assume it is an equivalent hydrocarbon of the form

$$C_nH_m$$

We can use the Orsat data to determine n and m, hence to discover the fuel. We assume the oxidant is air. Then, the chemical equation is assumed to be of the form

$$C_nH_m + a(O_2 + 3.76N_2) \rightleftarrows bCO_2 + cH_2O + dCO + eO_2 + fN_2$$

Since the dry mixture product mole fractions are known, it is convenient to write the chemical equation on the basis of 100 moles of dry mixture. Hence

$$b = 11 \qquad d = 1 \qquad e = 3 \qquad f = 85$$

Then, by conservation of N atoms,

$$a = \frac{85}{3.76} = 22.6$$

Table 11·1 Dry mixture composition from Orsat analysis

	$\Psi_i = \chi_i$
CO_2	0.11
O_2	0.03
CO	0.01
Subtotal	0.15
N_2	0.85
Total	1.00

Conservation of O atoms requires

$$2a = 2b + c + d + 2e$$

Hence $$c = 2 \times 22.6 - 2 \times 11 - 1 - 2 \times 3 = 16.2$$

Conservation of C atoms requires

$$n = b + d = 11 + 1 = 12$$

and conservation of H atoms requires

$$m = 2c = 32.4$$

Therefore the fuel can be viewed as an equivalent hydrocarbon $C_{12}H_{32.4}$. The chemical equation is then

$$C_{12}H_{32.4} + 22.6(O_2 + 3.76N_2) \rightleftarrows 11CO_2 + 16.2H_2O + CO + 3O_2 + 85N_2 \qquad (11{\cdot}4)$$

We could calculate the AFR on a molal basis but, as the fuel molecular configuration is not really known, the result would be meaningless. We can, however, calculate the AFR on a mass basis. The molal mass of our effective $C_{12}H_{32.4}$ fuel is

$$\hat{M} = 12 \times 12.0 + 32.4 \times 1.0 = 176.4 \text{ lbm/lbmole} = 176.4 \text{ kg/kgmole}$$
$$= 176.4 \text{ g/gmole}$$

Then, $$\text{AFR} = \frac{\mathcal{N}_{\text{air}}\hat{M}_{\text{air}}}{\mathcal{N}_{\text{fuel}}\hat{M}_{\text{fuel}}} = \frac{(22.6 \times 4.76) \times 28.97}{1 \times 176.4} = 17.7 \text{ kg air/kg fuel}$$

Let's also calculate the percent excess air. The stoichiometric chemical equation is

$$C_{12}H_{32.4} + 20.1(O_2 + 3.76N_2) \rightleftarrows 12CO_2 + 16.2H_2O + 75.5N_2 \qquad (11{\cdot}5)$$

Then, the excess air fraction is

$$\frac{22.6 - 20.1}{20.1} = 0.124$$

corresponding to 12.4 percent excess air, or 112.4 percent theoretical air.

Because of the high water content in typical combustion products, and the very corrosive effects of high-temperature condensate, knowledge of the dew point of products of combustion is often important. If the product composition is known, the partial pressure of water is readily determined, and this corresponds to the saturation pressure of water at the dew point.† For example, for the products mixture of the previous example,

$$\chi_{H_2O} = \frac{16.2}{11 + 16.2 + 1 + 3 + 85} = 0.139$$

Then, by Eq. (10·19),

$$P_{H_2O} = 0.139 \times 14.7 = 2.04 \text{ psia}$$

Inspecting Table B·1*b*, we see this corresponds to a saturation (dew-point) temperature of about 126°F.

11·3 STANDARDIZED ENERGY AND ENTHALPY

In order to carry out thermodynamic calculations for chemical reactions we need to know the equations of state for all species involved. Equations of state for many substances have been worked up from a combination of laboratory data, thermodynamic relations, and quantum-statistical analyses; we are already familiar with the examples given in Appendix B. In the preparation of these equations of state it is customary to select some arbitrary datum state at which the internal energy (or alternatively, the enthalpy) and the entropy are taken to be zero. These equations of state may be used in any thermodynamic analysis involving one substance only (or mixtures of nonreacting substances), since differences in internal energy and entropy between two states are all that ever enter into consideration. However, with chemically reacting systems it is necessary to use a common basis for evaluation of the thermodynamic properties of all substances involved in any particular analysis. For example, suppose we arbitrarily selected some state T_0 and P_0 as the datum for H_2, O_2, and H_2O. Then, any reaction carried out at this temperature and pressure would appear to result in no change in the internal energy or entropy of the mixture, an obviously incorrect situation.

It might be thought that, since energy is a monotonically increasing function of temperature, the energy of all substances at the absolute zero of temperature is zero. This is certainly not the case, because a considerable amount of energy may be associated with molecular and nuclear binding forces and other energy modes. It does not appear practical to speak of the "absolute energy" of matter. A datum must be selected that will allow us to tie together properly the energy of different substances so that we can make proper energy analyses of chemical reactions.

† To a very good approximation. See Sec. 12·6.

The procedure we follow is to select some arbitrary datum (temperature and pressure) at which we give zero values to the enthalpy of the basic elements.† It would be nice if this could be absolute-zero temperature and, say, 1 atm pressure, but this would require accurate low-temperature data not generally available at the present time. A more practical datum conventionally adopted for chemical thermodynamic tabulations is 25°C (77°F) and 1 atm pressure. This is called the *standard reference state*. By convention, we make the enthalpy of every *elemental* substance zero at the standard reference state.‡ By elemental substance we mean the substance composed of only one kind of atom in the form in which it exists in equilibrium at the standard reference state. For example, the enthalpy of mercury (Hg liquid) is zero at the standard reference state, as is the enthalpy of oxygen (O_2 gas) at this temperature and pressure.§

The *enthalpy of formation* of a compound is defined as the difference between the enthalpy of the compound h°_{comp} and the enthalpy of the elemental substances from which it is formed, all evaluated at the standard reference state. It is conventionally denoted by Δh°_f and, on a molal basis, is

$$\blacktriangleright \qquad \Delta \hat{h}^\circ_f \equiv \hat{h}^\circ_{\text{comp}} - \sum \nu_i \hat{h}^\circ_i \tag{11·6}$$

Here $\hat{h}^\circ_i$ is the molal enthalpy of the ith elemental substance involved in the formation reaction, and ν_i is the number of moles of the ith elemental substance involved in forming a single mole of the compound. The enthalpy of formation can be evaluated by appropriate measurements of energy transfers as heat and work and is a frequently tabulated property. Since by convention the elemental substances have zero enthalpy at the standard reference state, the enthalpy of a compound at the standard reference state is merely its enthalpy of formation. The enthalpy at other states may be established using this datum-point enthalpy. We call this the *standardized enthalpy*, meaning simply that it is properly related to the enthalpy of other elements and compounds. Values for Δh°_f can be determined by laboratory measurement or by advanced methods of statistical thermochemistry (see Table B·12). Knowing Δh°_f for any substance, the standardized enthalpy can be calculated by simple adjustment of the enthalpy data for that substance. Table B·13 was constructed in this manner. Note that the enthalpy of formation

† It is customary to give the enthalpy, rather than the internal energy, a zero value at the datum state; this then fixes the value of the energy at the datum state, since absolute values for P and v are measurable.

‡ Recall that different isotopes of an element have the same chemical properties. If no nuclear reactions are involved, the enthalpy of each isotope can be considered zero at this standard state.

§ These may not agree with the graphical and tabular equations of state often presented. The standard reference state is used primarily in chemical thermodynamic literature, as for example, in some places in the *Handbook of Chemistry and Physics*. The user of equation-of-state information should always check the datum, especially if a chemical reaction is being analyzed.

and standardized enthalpy often have negative values, as a result of the choice of reference state.

The term "heat of formation" is used sometimes to refer to the enthalpy of formation and sometimes to mean the negative of the enthalpy of formation; the user of any tabulation must be sure which (if either) the tabulator had in mind. The term arises because it may be shown that the energy which must be transferred as heat from the mixture to keep the temperature and pressure constant is equal to the enthalpy of formation.

11·4 HEATS OF REACTION, HEATING VALUES

We now have practically all the tools necessary for analysis of the energetics of chemical reactions. Of particular interest is the combustion of fuels, where it is common practice to write the energy balance on a per-mole-of-fuel basis. Consider the combustion of a hydrocarbon fuel C_nH_m with air in a steady-flow burner (Fig. 11·1). It is common to assume that the products contain only H_2O, CO_2, and N_2, with additional O_2 if excess air is supplied. For a stoichiometric reaction the chemical equation is

$$C_nH_m + \left(n + \frac{m}{4}\right)(O_2 + 3.76N_2) \rightleftarrows nCO_2 + \frac{m}{2}H_2O + \left(n + \frac{m}{4}\right)3.76N_2$$

Note that in this form the stoichiometric coefficients give the amounts of species in the products and reactants per mole of fuel. Now, the energy balance on a per-mole-of-fuel basis is

$$\hat{H}_R = \hat{H}_P + \hat{Q}$$

where for brevity we have denoted the *total* enthalpies of the products and reactants, *per mole of fuel*, by $\hat{H}_P$ and $\hat{H}_R$, respectively. $\hat{Q}$ is then the energy transfer as heat from the combustor per mole of fuel burned. It is common practice to evaluate $\hat{H}_R$, $\hat{H}_P$, and $\hat{Q}$ considering the reaction to occur *at the standard reference state* (1 atm, 25°C = 77°F), for this provides a sensible way to compare various fuels.

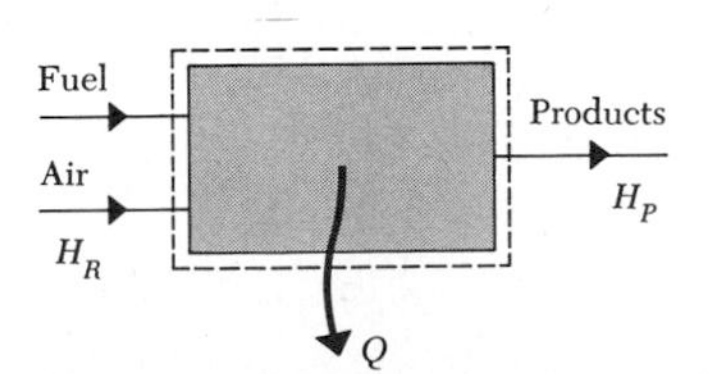

Figure 11·1. The energy terms are per mole of fuel

Under these conditions we have

$$\hat{H}_P^\circ = \left(\sum_i \mathcal{N}_i \hat{h}_i\right)_{\text{products}} \qquad (11\cdot7a)$$

$$\hat{H}_R^\circ = \left(\sum_i \mathcal{N}_i \hat{h}_i\right)_{\text{reactants}} \qquad (11\cdot7b)$$

where the mole numbers $\mathcal{N}_i$ are simply the stoichiometric coefficients in the chemical equation. Then,

$$\hat{Q}^\circ = \hat{H}_R^\circ - \hat{H}_P^\circ \qquad (11\cdot8)$$

$\hat{Q}^\circ$ is termed the *heating value* or *heat of reaction* of the fuel. Note that $\hat{Q}^\circ$ represents the energy that must be transferred as heat from the system, per mole of fuel, in order to maintain the system at constant temperature. The term *enthalpy of combustion* is sometimes used for the *negative* of $\hat{Q}^\circ$, which is then denoted by $\hat{H}_{RP}^\circ$. (The superscript is often omitted, and we include it here to emphasize that the reaction is taken at the standard reference state.)

$$\hat{H}_{RP}^\circ = -\hat{Q}^\circ = \hat{H}_P^\circ - \hat{H}_R^\circ \qquad (11\cdot9)$$

The H_2O in the products may appear in either the liquid or vapor phase. When the H_2O is in its *liquid* phase, $\hat{Q}^\circ$ is termed the *higher heating value* (HHV), while when H_2O vapor is considered, $\hat{Q}^\circ$ is called the *lower heating value* (LHV). Noting that condensation releases h_{fg}, which adds to $\hat{Q}^\circ$, we see that HHV > LHV.

To illustrate, let's evaluate the quantities discussed above for ethane (C_2H_6). The chemical equation is

$$C_2H_6 + 3.5(O_2 + 3.76N_2) \rightleftarrows 2CO_2 + 3H_2O + 13.16N_2$$

Using enthalpy data from Table B·12,

$$\hat{H}_R^\circ = 1 \times (-36{,}401) + 3.5 \times 0 + 3.5 \times 3.76 \times 0$$
$$= -36{,}401 \text{ Btu/lbmole fuel}$$
$$\hat{H}_P^\circ = 2(-169{,}183) + 3(-122{,}976) + 13.16 \times 0$$
$$= -707{,}294 \text{ Btu/lbmole fuel}$$

Note that we consider the H_2O to be *liquid*. Then,

$$\hat{Q}^\circ = (-36{,}401) - (-707{,}294) = 670{,}893 \text{ Btu/lbmole fuel}$$
$$\hat{H}_{RP}^\circ = -\hat{Q}^\circ = \hat{H}_P - \hat{H}_R = -670{,}893 \text{ Btu/lbmole fuel}$$

Since we treated the H_2O as liquid, the HHV is

$$\text{HHV} = 670{,}893 \text{ Btu/lbmole}$$

To find the LHV, we need the molal enthalpy of water vapor at 77°F. Interpolating in Table B·1*a*,

$$h_{fg} = 1050.0 \text{ Btu/lbm at } 77°\text{F}$$

so,

$$\hat{h}_{fg} = 1050.0 \times 18.016 = 18{,}917 \text{ Btu/lbmole}$$

Hence, for water vapor at the standard reference state,

$$\hat{h}° = -122{,}976 + 18{,}917 = -104{,}059 \text{ Btu/lbmole}$$

Alternatively, we could have read from Table B·13

$$\hat{h}° = -103{,}968 \text{ Btu/lbmole}$$

which is less exact because of the perfect-gas approximations employed. Then, with water vapor in the products,

$$\hat{H}_P^\circ = 2(-169{,}183) + 3(-104{,}059) = -650{,}543 \text{ Btu/lbmole}$$

hence the lower heating value is

$$\text{LHV} = (-36{,}401) - (-650{,}543) = 614{,}142 \text{ Btu/lbmole fuel}$$

Often the heating values are quoted on a mass basis. Since the molal mass of ethane is 30.07 lb/lbmole, the lower heating value can be expressed as

$$\text{LHV} = \frac{614{,}142}{30.07} = 20{,}420 \text{ Btu/lbm}$$

In SI units, the enthalpy of formation for liquid water is $-286{,}022$ kJ/kgmole. Since at 25°C, $h_{fg} = 2442.3$ kJ/kg, we have

$$\hat{h}_{fg} = 2442.3 \times 18.016 = 44{,}000 \text{ kJ/kgmole}$$

Thus, for water vapor at 25°C,

$$\hat{h}° = -286{,}022 + 44{,}000 = -242{,}021 \text{ kJ/kgmole}$$

The ethane LHV is then

$$\text{LHV} = \frac{1{,}428{,}494}{30.07} = 47{,}500 \text{ kJ/kg}$$

Table 11·2 presents some typical results for heating values.†

† Our examples here took place at constant pressure. Sometimes such processes are referred to as yielding the constant-pressure heating value. In the case of combustion in a bomb calorimeter the process occurs at constant volume. Here the heat transfer is equal to the change in internal energy. This process gives the constant-volume heating value.

Table 11·2 Typical heating values at 25°C

Compound	H_2O (liquid) in products (HHV)		H_2O (gas) in products (LHV)	
	Btu/lbm fuel	kJ/kg fuel	Btu/lbm fuel	kJ/kg fuel
Methane, CH_4 (gas)	23,861	55,496	21,502	50,010
Ethane, C_2H_6 (gas)	22,304	51,875	20,416	47,484
Hexane, C_6H_{14} (gas)	20,930	48,679	19,391	45,100
Octane, C_8H_{18} (gas)	20,747	48,254	19,256	44,786
Hydrogen, H_2 (gas)	60,958	141,788	51,571	119,954

11·5 SOME ILLUSTRATIVE CALCULATIONS

We shall now go through a sequence of example calculations to illustrate the matters discussed above and to introduce some new ideas.

LHV calculation. Let's first find the LHV and $\hat{H}_{RP}$ for methane (CH_4) burned with air in a stoichiometric reaction at 77°F and 1 atm. The chemical equation is

$$CH_4 + 2(O_2 + 3.76N_2) \rightleftarrows CO_2 + 2H_2O + 7.52N_2$$

An energy balance on the control volume (see Fig. 11·1) on a per-mole-of-fuel basis, gives

$$\hat{H}_R = \hat{H}_P + \hat{Q}$$

Now,

$$\text{LHV} = \hat{Q} = \hat{H}_R - \hat{H}_P$$

Using the chemical equation,

$$\text{LHV} = \hat{h}^\circ_{CH_4} + 2\hat{h}^\circ_{O_2} + 7.52\hat{h}^\circ_{N_2} - \hat{h}^\circ_{CO_2} - 2\hat{h}^\circ_{H_2O} - 7.52\hat{h}^\circ_{N_2}$$

where the H_2O is in vapor form. Using data from Tables B·12 and B·13,

$$\begin{aligned}\text{LHV} &= -32{,}179 + 0 + 0 - (-169{,}183) - 2(-103{,}968) - 0\\ &= 344{,}940 \text{ Btu/lbmole } CH_4\end{aligned}$$

Note the difference with the molal LHV of ethane computed in the previous section. However, the molal mass of methane is only 16.04 lb/lbmole, hence on a mass basis

$$\text{LHV} = \frac{344{,}940}{16.04} = 21{,}500 \text{ Btu/lbm}$$

which is quite similar to that of ethane. Most hydrocarbon fuels have lower heating values in the range 10,000 to 22,000 Btu/lbm.

The enthalpy of combustion of methane is

$$\hat{H}^{\circ}_{RP} = -\text{LHV} = -344{,}940 \text{ Btu/lbmole } CH_4$$

Temperature effects. In order to study the effect of temperature on $\hat{H}_{RP}$, let's now consider the same reaction at an elevated temperature, say, 2000 R. Now,

$$\begin{aligned}\hat{H}_{RP} &= \hat{H}_P - \hat{H}_R \\ &= \hat{h}_{CO_2} + 2\hat{h}_{H_2O} + 7.52\hat{h}_{N_2} - \hat{h}_{CH_4} - 2\hat{h}_{O_2} - 7.52\hat{h}_{N_2}\end{aligned}$$

But now all enthalpies must be evaluated at 2000 R. The pressure is unimportant, for we treat all constituents as perfect gases with $h = h(T)$. The nitrogen contributions cancel in the equation above, since the product and reactant temperatures are identical. From Table B·13 we can find all but $\hat{h}_{CH_4}$ at 2000 R. We can calculate h_{CH_4}, assuming perfect-gas behavior.

$$\hat{h}(2000 \text{ R}) - \hat{h}(537 \text{ R}) = \int_{537 \text{ R}}^{2000 \text{ R}} \hat{c}_P(T)\, dT$$

Over this wide temperature range the variation in $\hat{c}_P$ must be considered. Specific-heat data for methane are given in Table B·16,

$$\hat{c}_P(T) = 3.381 + 0.018044T - 4.3 \times 10^{-6}T^2 \text{ cal/(gmole·K)}$$

where T is in kelvins. Substituting and integrating, taking proper care in the unit conversion, we find

$$\hat{h}(2000 \text{ R}) = \hat{h}(537 \text{ R}) + 20{,}086 \text{ Btu/lbmole}$$

So, for CH_4,

$$\hat{h}(2000 \text{ R}) = -32{,}179 + 20{,}086 = -12{,}093 \text{ Btu/lbmole}$$

Then, combining with the enthalpy data in Table B·13,

$$\begin{aligned}\hat{H}_{RP} &= -152{,}194 + 2(-90{,}797) - (-12{,}093) - 2(11{,}439) \\ \hat{H}_{RP} &= -344{,}574 \text{ Btu/lbmole}\end{aligned}$$

At 77°F we found $\hat{H}^{\circ}_{RP} = -344{,}940$. We see that the enthalpy of combustion is only slightly changed with temperature, which is true for many hydrocarbon fuels. This fact is often useful in making simplified combustion calculations.

Adiabatic flame temperature. Let us further extend the calculations. Suppose the reactants enter at 77°F at 1 atm and the products leave at an unknown temperature at 1 atm. We wish to determine this exit temperature.

Whereas in the previous calculations we considered exothermic reactions at constant temperature, we now assume that the burner is adiabatic. Our procedure will be to make an energy balance and solve for the enthalpy of the products in terms of the enthalpies of the reactants. The enthalpy, compositions, and pressure of the products will then suffice to fix the exhaust temperature. We make the following idealizations:

Products contain only CO_2, H_2O, N_2, and O_2
O_2, N_2, CH_4, and CO_2 are perfect gases
The products of combustion constitute a mixture of independent perfect gases
Steady flow, steady state
Adiabatic control volume
Equilibrium at inlets and exhaust
Kinetic and potential energies negligible

With these idealizations, an energy balance on the control volume (Fig. 11·2) yields

$$(\dot{\mathcal{N}}\hat{h})_{CH_4} + (\dot{\mathcal{N}}\hat{h})_{O_2} + (\dot{\mathcal{N}}\hat{h})_{N_2} = (\dot{\mathcal{N}}\hat{h})_{\text{prod}} \tag{11·10}$$

where $\dot{\mathcal{N}}$ denotes the molal mass-flow rate. The chemical equation is again

$$CH_4 + 2(O_2 + 3.76N_2) \rightleftarrows CO_2 + 2H_2O + 7.52N_2 \tag{11·11}$$

Note that 10.52 moles of products are formed by combustion of 1 mole of fuel.

Hence,

$$\frac{\dot{\mathcal{N}}_{CH_4}}{\dot{\mathcal{N}}_{\text{prod}}} = \frac{1}{10.52} = 0.095$$

$$\frac{\dot{\mathcal{N}}_{O_2}}{\dot{\mathcal{N}}_{\text{prod}}} = \frac{2}{10.52} = 0.190$$

$$\frac{\dot{\mathcal{N}}_{N_2}}{\dot{\mathcal{N}}_{\text{prod}}} = \frac{7.52}{10.52} = 0.715$$

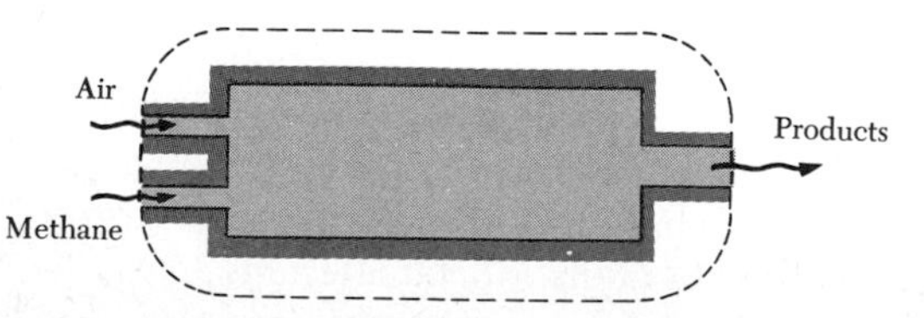

Figure 11·2. The control volume is adiabatic

So, the energy balance can be written as

$$\hat{h}_{prod} = 0.095\hat{h}_{CH_4} + 0.190\hat{h}_{O_2} + 0.715\hat{h}_{N_2}$$

which is equivalent to an energy balance written on a per-mole-of-products basis. Since the state of the reactants is known, the energy balance becomes an equation for $\hat{h}_{prod}$. Using the data of Table B·12, we have

$$\hat{h}_{prod} = 0.095(-32{,}179) + 0 + 0 = -3057 \text{ Btu/lbmole}$$

We now need to determine what temperature will give this value of $\hat{h}_{prod}$. The product enthalpy is related to the individual species enthalpies by [see Eq. (10·13)]

$$\hat{h}_{prod} = \chi_{N_2}\hat{h}_{N_2} + \chi_{CO_2}\hat{h}_{CO_2} + \chi_{H_2O}\hat{h}_{H_2O}$$

Calculating the constituent mole fractions,

$$\chi_{N_2} = \frac{7.52}{10.52} = 0.715$$

$$\chi_{CO_2} = \frac{1}{10.52} = 0.095$$

$$\chi_{H_2O} = \frac{2}{10.52} = 0.190$$

Hence $\quad \hat{h}_{prod}(T) = 0.715\hat{h}_{N_2}(T) + 0.095\hat{h}_{CO_2}(T) + 0.190\hat{h}_{H_2O}(T)$

The component enthalpies are given in Table B·13. A trial-and-error solution will determine the temperature that will give the correct product enthalpy (−3057 Btu/lbmole). The results are tabulated in Table 11·3. Interpolating between 4000 and 4200 R, we find $T = 4187$ R for $\hat{h} = -3057$ Btu/lbmole.

The product temperature is called the *adiabatic flame temperature*, the temperature reached in an adiabatic steady-flow combustion process.

Table 11·3 Product enthalpy calculation

		4000 R		4200 R	
	χ_i	$\hat{h}_i$	$\chi_i\hat{h}_i$	$\hat{h}_i$	$\chi_i\hat{h}_i$
CO_2	0.095	−123,981	−11,780	−121,050	−11,500
H_2O	0.190	−67,747	−12,880	−65,227	−12,400
N_2	0.715	27,599	19,700	29,338	20,980
			−4,960		−2,920

Table 11·4 Some adiabatic flame temperatures†

Fuel	Oxidizer			
	Oxygen		Air	
	K	R	K	R
H_2	3079	5542	2384	4291
CH_4	3054	5497	2227	4009
C_3H_8	3095	5571	2268	4082
C_8H_{18}	3108	5594	2277	4098

† Reactants at 298 K, 1 atm, stoichiometric mixture.

Table 11·4 presents some results for the adiabatic flame temperature for the oxidizers air and O_2. The value for CH_4 differs from the value of 4187 R we have just calculated because the tabular results include as products of combustion the proper amounts of CO, O, H, and OH neglected in our calculation.

Nonstandard reactant states. A typical combustion process would not have reactants exactly at the standard state. To illustrate the effects of the deviations, let's modify the previous example. Suppose the air enters at 90°F and 40 psia, and the CH_4 enters at 60°F and 20 psia. First, we evaluate the standardized enthalpies of the reactants. Since the inlet temperatures are not far from 77°F, the constant-specific-heat assumption is reasonable here. Then [see Eq. (8·25)],

$$\hat{h} = \hat{h}^\circ + \hat{c}_P(T - T_0)$$

So, using data from Table B·6,

$$\hat{h}_{O_2} = 0 + 7.01(90 - 77) = 91.1 \text{ Btu/lbmole}$$

$$\hat{h}_{N_2} = 0 + 6.95(90 - 77) = 90.3 \text{ Btu/lbmole}$$

$$\hat{h}_{CH_4} = -32{,}179 + 8.53(60 - 77) = -32{,}325 \text{ Btu/lbmole}$$

As before, $\hat{h}_{prod}$ is calculated from the energy balance; one finds $\hat{h}_{prod} = -2983$ Btu/lbmole. The temperature can be determined from our previous calculations. Interpolating between 4000 R and 4200 R, we find $T = 4193$ R. Note that the modest change in inlet conditions has only slightly affected the burner discharge temperature.

Excess air. A typical combustion process utilizes excess air to ensure that complete combustion takes place. Excess air will also cause a decrease in the adiabatic flame temperature, since energy will be required to increase the temperature of the nonreacting air. In fact, excess air is often introduced to control the adiabatic flame temperature and maintain it within the limits set by the materials in the system.

Again let us consider the combustion of methane in a steady-flow burner. The oxidizer will be air, and we suppose that twice as much air is supplied as is necessary for the combustion (200 percent theoretical air). The air will enter at 90°F and 40 psia, and the CH_4 at 60°F and 20 psia. The products of combustion emerge at 1 atm. The chemical equation is [compare with Eq. (11·11)]

$$CH_4 + 4(O_2 + 3.76N_2) \rightleftarrows CO_2 + 2H_2O + 2O_2 + 15.04N_2 \qquad (11\cdot 12)$$

The energy balance again produces Eq. (11·10), which we use as the basis for the discharge-temperature calculation. The reactant enthalpies were calculated for these inlet states in the previous example and are summarized below.

$$\hat{h}_{CH_4} = -32{,}325 \text{ Btu/lbmole}$$

$$\hat{h}_{O_2} = 91.1 \text{ Btu/lbmole}$$

$$\hat{h}_{N_2} = 90.3 \text{ Btu/lbmole}$$

Now, the mole fractions of the reactants mixture are indicated by the chemical equation

$$\frac{\dot{\mathcal{N}}_{CH_4}}{\dot{\mathcal{N}}_{prod}} = \frac{1}{20.04} = 0.0499$$

$$\frac{\dot{\mathcal{N}}_{O_2}}{\dot{\mathcal{N}}_{prod}} = \frac{4}{20.04} = 0.1996$$

$$\frac{\dot{\mathcal{N}}_{N_2}}{\dot{\mathcal{N}}_{prod}} = \frac{4 \times 3.76}{20.04} = 0.7505$$

The standardized enthalpy of the product mixture is then

$$\hat{h}_{prod} = 0.0499 \times (-32{,}325) + 0.1996 \times 91.1 + 0.7505 \times 90.3 = -1527 \text{ Btu/lbmole}$$

This value will be used to establish the temperature of the exhaust gases.

The mole fractions of the species in the product mixture are, from the chemical equation,

$$\chi_{CO_2} = \frac{1}{20.04} = 0.0499$$

$$\chi_{H_2O} = \frac{2}{20.04} = 0.0998$$

$$\chi_{O_2} = \frac{2}{20.04} = 0.0998$$

$$\chi_{N_2} = \frac{15.04}{20.04} = 0.7505$$

The enthalpy of the products may again be written as

$$\hat{h}_{prod}(T) = \chi_{CO_2}\hat{h}_{CO_2}(T) + \chi_{H_2O}\hat{h}_{H_2O}(T) + \chi_{O_2}\hat{h}_{O_2}(T) + \chi_{N_2}\hat{h}_{N_2}(T)$$

Data for the standardized enthalpies of the constituents in the product mixture may be obtained from Table B·13. Calculation of the product temperature again requires iteration, which is summarized in Table 11·5. Interpolating, we find $T_{prod} = 2674$ R. Note the dramatic decrease in flame temperature caused by excess air.†

Figure 11·3 indicates the temperature at which the products of combustion are released when methane is burned in air. If no energy is removed from the combustor, we will have the case of the temperature reaching the adiabatic flame value. Suppose we are burning methane as a fuel for gas turbines and the maximum permissible continuous temperature on the turbine blades is 2000 R; we see that a mixture containing about 200 percent excess air would be required to reduce the exit temperature to this value. The other curves on the figure indicate the exhaust temperature for the noted energy extractions from the system.

Treating the internal combustion engine as a steady-flow system of the type shown in Fig. 11·3*a*, we can estimate the exhaust temperature from a car fueled with natural gas (largely methane). In the internal combustion engine the work output is about 27 percent of the fuel energy input. In addition, the energy removed from the engine via the cooling water is about 40 percent of the input.

† The serious student should be aware of the availability of extensive tables for computation of hydrocarbon combustion reactions. In particular, the Keenan and Kaye *Gas Tables* (see Selected Reading list) are very useful, for they include product enthalpy information for 200 percent and 400 percent theoretical air for combustion of a general fuel $(CH_2)_n$.

Table 11·5 Product enthalpy calculation

		2600 R		2700 R	
	χ_i	$\hat{h}_i$	$\chi_i\hat{h}_i$	$\hat{h}_i$	$\chi_i\hat{h}$
CO_2	0.0499	−144,025	−7,187	−142,631	−7,117
H_2O	0.0998	−84,397	−8,423	−83,279	−8,311
O_2	0.0998	+16,586	+1,655	+17,458	+1,742
N_2	0.7505	+15,686	+11,773	+16,517	+12,396
		$\hat{h}_{prod}$ =	−2,182	$\hat{h}_{prod}$ =	−1,290

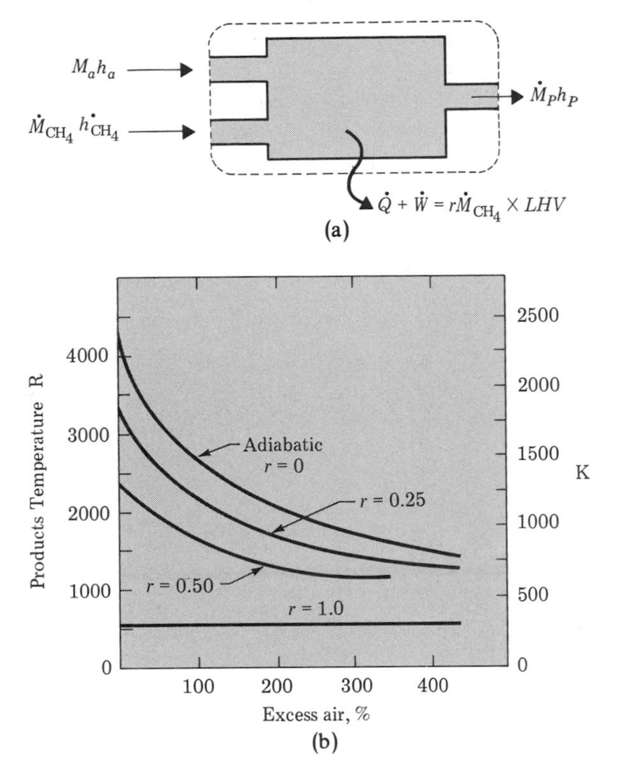

Figure 11·3(a). A combustor control volume; (b) combustion of methane with air at an inlet condition of 25°C

This leaves about one-third of the energy to go out with the exhaust. Interpolating in the figure for $r = 0.67$ and 0 percent excess air, the exhaust temperature should be about 1500 to 1600 R.

11·6 ABSOLUTE ENTROPY AND THE THIRD LAW OF THERMODYNAMICS

Thus far in this chapter we have considered only the first-law aspects of chemical reactions, emphasizing combustion of hydrocarbon fuels by air. We always assumed complete combustion in the examples above and made certain assumptions about the products of reaction. The second law provides the basis for theoretical prediction of the true composition of a product mixture, hence it plays a very important role in chemical thermodynamics. To apply the second law we need to relate properly the entropies of all elements and compounds.

Let us first recall the meaning of zero entropy. As discussed in Chap. 6, a state of zero entropy is one for which a single system quantum state is always observed. We would know the microscopic state of a system at zero entropy precisely, with zero uncertainty. Imagine now that we add one quantum of energy to the system, and that this permits it suddenly to take any one of a very large number of quantum states. Assuming each quantum state is equally likely, the entropy change would be

$$\delta S = \mathsf{k} \ln \Omega - 0$$

where Ω is the number of quantum states available to the system with the single quantum of energy. The internal energy change δU would be simply ε, the very small amount of energy added. Since Ω is likely to be very large, a tiny increase in energy would give a huge increase in entropy. Recalling the thermodynamic definition of temperature, Eq. (7·9), we are led to suspect that T must be very small when the entropy is zero.

We can illustrate this by a numerical example. Consider 1 cm^3 of matter having a mass of the order of 1 g and a molal mass of the order of 20 g/gmole. This system contains roughly 3×10^{22} atoms. The quantum of energy added might be possessed by any one of the atoms, so the entropy in the slightly energized state will be approximately

$$S \approx \mathsf{k} \ln N \approx 1.38 \times 10^{-23} \text{ J/K} \times \ln (3 \times 10^{22})$$
$$\approx 70 \times 10^{-23} \text{ J/K}$$

We imagine adding a quantum of energy by means of a photon having a wavelength equal to 1 cm (this is a reasonable estimate for the least energetic photon that might be captured by 1 cm^3 of matter). Its energy is

$$\varepsilon = \frac{\mathsf{hc}}{\lambda} = \frac{6.62 \times 10^{-34} \text{ J}\cdot\text{s} \times 3 \times 10^8 \text{ m/s}}{0.01 \text{ m}}$$
$$\varepsilon \approx 2 \times 10^{-23} \text{ J}$$

The temperature is therefore estimated to be

$$T \approx \frac{1}{\delta S/\varepsilon} \approx \frac{1}{70 \times 10^{-23}/2 \times 10^{-23}} \approx \frac{1}{35} \text{ K}$$

This should be interpreted as the average temperature over the range $0 < S < 70 \times 10^{-23}$ J/K. Since the temperature is a monotonic function of the entropy, the temperature at the zero entropy state may be expected to be even smaller.

On the basis of these considerations we make the following macroscopic postulate:

The temperature of any pure substance in thermodynamic equilibrium approaches zero as the entropy approaches zero.

Conversely, since the temperature is a monotonic function of the entropy,

The entropy of any pure substance in thermodynamic equilibrium approaches zero as the temperature approaches zero.

Expressed mathematically, for a pure substance in equilibrium,

▶ $$\lim_{T \to 0} S = 0 \qquad (11\cdot13)$$

This particular form of the postulate is called the *third law of thermodynamics*. The third law was developed in the early 1900s, primarily through the efforts of Nernst, and consequently is often called the *Nernst theorem*.

A considerable body of experimental data supporting the third law now exists. We can calculate the absolute entropy of a substance from the measured thermodynamic properties by integrating the differential equations of state from absolute zero; for gases this requires passage from the solid state through the liquid state and vaporization to reach the gaseous phase. It is also possible to calculate the entropy of gases using statistical thermodynamics. The close agreement of the entropies calculated in these different ways provides some of the strongest evidence in support of the third law.

Some exceptions to the agreement between the macroscopically determined and statistically determined absolute entropies have been claimed, and there is some disagreement as to the validity and generality of the third law, though no one denies its usefulness. According to Fast,† all the exceptions seem to be satisfactorily explained. First, the substance must be pure, since the presence of any additional species would immediately give a variety of places where molecules of different types could exist, hence a nonunique microscopic state, and thus a nonzero entropy. Second, it is necessary to extrapolate to absolute zero, and data down to very low temperatures must be provided. Other apparent exceptions arise from the fact that thermodynamic equilibrium is not maintained as the temperature is reduced. None of these contradicts the third law which holds only at absolute zero for pure substances in equilibrium states.

The third law permits determination of the absolute entropy of every substance, including elemental substances (Hg, O_2, ...) and compounds. These entropies are all properly tied to a common base and may therefore be used in analyzing chemical reactions. The absolute entropies of several substances are given in Tables B·12 and B·13.

11·7 A SECOND-LAW APPLICATION

To illustrate one application of the second law, consider again the last example of Sec. 11·5. There we presumed (without justification) that the reaction could in fact occur; let us now test this presumption with the aid of the second law. Applying the second law to the control volume (Fig. 11·2), we find

$$\mathscr{P}_s = (\dot{\mathscr{N}}\hat{s})_{\text{prod}} - (\dot{\mathscr{N}}\hat{s})_{\text{CH}_4} - (\dot{\mathscr{N}}\hat{s})_{\text{O}_2} - (\dot{\mathscr{N}}\hat{s})_{\text{N}_2} \geq 0 \qquad (11\cdot14)$$

where $\hat{s}$ represents the absolute molal entropy. If the entropy production is positive, the reaction is possible; if it is zero, the reaction is reversible, that is, could go in either direction; if a negative value is calculated, the assumed reaction could not occur.

† J. D. Fast, *Entropy*, p. 85, McGraw-Hill Book Company, New York, 1962.

The absolute molal entropies of the reactant gases will be calculated by extrapolation from the entropies at standard reference state; we treat the gases as perfect, with constant specific heats over the small temperature range of the extrapolation. Equation (8·27) is equivalent to

$$\hat{s} = \hat{s}^0 + \hat{c}_p \ln \frac{T}{T_0} - \mathscr{R} \ln \frac{P}{P_0}$$

where $\hat{s}^0$ is the entropy in the standard reference state (T_0, P_0). The partial pressures of the O_2 and N_2 in the air are proportional to their mole fractions.

Thus

$$P_{O_2} = 0.21 \times 40 = 8.4 \text{ psia}$$

$$P_{N_2} = 0.79 \times 40 = 31.6 \text{ psia}$$

The absolute molal entropies of the incoming constituents are then, for CH_4,

$$\hat{s} = 44.47 + 8.53 \times \ln \frac{520}{537} - 1.986 \times \ln \frac{20}{14.7} = 43.58 \text{ Btu/(lbmole·R)}$$

for O_2,

$$\hat{s} = 48.986 + 7.01 \times \ln \frac{550}{537} - 1.986 \times \ln \frac{8.4}{14.7} = 50.27 \text{ Btu/(lbmole·R)}$$

and for N_2,

$$\hat{s} = 45.755 + 6.95 \times \ln \frac{550}{537} - 1.986 \times \ln \frac{31.6}{14.7} = 44.41 \text{ Btu/(lbmole·R)}$$

The molal entropy of the products is computed in Table 11·6 using the data from Table B·13.

Table 11·6 Product entropy calculation

		2600 R		2700 R	
	χ_i	$\hat{s}_i$	$\chi_i \hat{s}_i$	$\hat{s}_i$	$\chi_i \hat{s}_i$
CO_2	0.0499	69.245	3.455	69.771	3.482
H_2O	0.0998	59.414	5.926	59.837	5.972
O_2	0.0998	61.287	6.116	61.616	6.149
N_2	0.7505	57.436	43.106	57.750	43.321
			$\hat{s}_{prod} = 58.603$		$\hat{s}_{prod} = 58.924$

Interpolating,

$$\hat{s}_{\text{prod}} = 58.603 + \frac{2674 - 2600}{2700 - 2600} \times (58.924 - 58.603)$$

$$= 58.839 \text{ Btu/(lbmole}\cdot\text{R)}$$

Substituting into Eq. (11·14),

$$\frac{\dot{\mathscr{P}}_s}{\dot{\mathscr{N}}_{\text{prod}}} = 58.839 - 0.0499 \times 43.58 - 0.1995 \times 50.27 - 0.7505 \times 44.41$$

$$= 13.306 \text{ Btu/(lbmole}\cdot\text{R)} \geq 0$$

Since a positive entropy-production rate is obtained, the process assumed does not violate the second law. Note that the reaction is not reversible (because it produces entropy).

We assumed that the products of combustion contain no CO, NO, or other compounds and that the reaction was complete. The second-law structure also provides a means for determining the equilibrium composition of a reacting mixture, and we shall consider this problem in the next chapter.

SELECTED READING

Keenan, J., and J. Kaye, *Gas Tables*, John Wiley & Sons, Inc., New York, 1948.

Lee, J., and F. Sears, *Thermodynamics*, 2d ed., chap. 14, Addison-Wesley Publishing Co., Inc., Reading, Mass, 1962.

Van Wylen, G., and R. Sonntag, *Fundamentals of Classical Thermodynamics*, 2d ed., chap. 12, John Wiley & Sons, Inc., New York, 1973.

Wark, K., *Thermodynamics*, 2d ed., chap. 14, McGraw-Hill Book Company, New York, 1971.

Zemansky, M. W., *Heat and Thermodynamics*, 5th ed., chaps. 17 and 18, McGraw-Hill Book Company, New York, 1968.

QUESTIONS

11·1 What are reactants, products, and chemical equations?
11·2 What is a stoichiometric mixture?
11·3 Why is it essential that the enthalpies and entropies of all substances be tied to a common base when chemical reactions are considered?
11·4 Is the energy of a substance ever really zero? Is its entropy?
11·5 What basic principles are employed in a calculation of the temperature following combustion in an adiabatic steady-flow device?
11·6 What is the heat of formation?

11·7 What is the enthalpy of formation?
11·8 What is the enthalpy of reaction?
11·9 What do the terms excess air and theoretical air mean?
11·10 What is the difference between exothermic and endothermic?
11·11 What does an Orsat analysis determine?
11·12 What are the values of T and P at the standard reference state?
11·13 To what does the Nernst theorem refer?
11·14 Does the first law hold true for a chemical reaction? Does the second law?
11·15 What is the difference between the higher heating value and the lower heating value?
11·16 Why does the introduction of excess air reduce the adiabatic flame temperature?

PROBLEMS

11·1 Write out the chemical equation for the reaction of octane (C_8H_{18}) with a stoichiometric reaction with air. Determine the theoretical air/fuel ratio for this reaction on both a molal and mass basis.

11·2 Write out the chemical equation for the reaction of octane (C_8H_{18}) with (*a*) 100 percent excess air and (*b*) 250 percent theoretical air.

11·3 Write the chemical equation for the reaction of *n*-butane (C_4H_{10}) with a stoichiometric reaction with air. Determine the theoretical air/fuel ratio for this reaction on both a molal and mass basis.

11·4 An oil fuel contains 84 percent C and 16 percent H_2 by mass. Find the stoichiometric air for complete combustion of 1 lbm of fuel. Determine the air/fuel ratio.

11·5 Compute the air/fuel ratio by mass used in an engine if the exhaust gas dry analysis in percent by volume is CO_2, 0.124; O_2, 0.032; CO, 0.001; H_2, 0.002.

11·6 Determine the dew point of the products when octane (C_8H_{18}) is burned with 400 percent theoretical dry air if the pressure is 14.7 psia.

11·7 Compute the standardized enthalpy and entropy of a mixture of equal parts by mass of CO_2 and H_2O at 2 atm pressure and 100°F.

11·8 Determine from the *Handbook of Chemistry and Physics* the standardized enthalpy of ozone and atomic oxygen at 18°C and 2 atm.

11·9 Determine the dew point of the products when *n*-butane (C_4H_{10}) reacts with stoichiometric dry air if the pressure is 18 psia. What is the effect on the dew point if 100 percent excess air is applied to the reaction?

11·10 Liquid benzene (C_6H_6) is burned in a stoichiometric reaction with dry air. Determine the air/fuel ratio and the dew point of the products if the total pressure is 13.5 psia. Redo the problem if 200 percent excess air is used.

11·11 Redo the previous problem for the stoichiometric case if the air has an initial relative humidity of 60 percent and the inlet pressure is 14.7 psia at 77°F.

11·12 Determine the products of combustion and the air/fuel ratio by mass when a liquid fuel that is 84 percent carbon by weight and 16 percent hydrogen is burned with 50 percent excess air.

11·13 Propane (C_3H_8) reacts with air such that an Orsat analysis yields as products of combustion the following percentages: CO_2, 11.5; O_2, 2.7; CO, 0.7. Determine the reaction equation and the percent theoretical air for the reaction.

11·14 An unknown hydrocarbon fuel undergoes combustion such that the products have the following percentage composition as measured by an Orsat apparatus: CO_2,

9.0; CO, 1.1; O_2, 8.8; N_2, 81.1. Determine the reaction equation, air/fuel ratio, and the percent theoretical air on a mass basis.

11·15 How much energy is released when 1 mole of propane (C_3H_8) reacts with 80 percent theoretical air when both the inlet and exit states of the combustion are at the standard reference state? How much energy is released in this reaction if stoichiometric?

11·16 Determine the enthalpy of combustion at 77°F for ethane (C_2H_6) when gaseous ethane reacts, assuming liquid water in the products. Do the same for acetylene (C_2H_2). Determine your answer in Btu/lbm fuel.

11·17 Set up a table showing the enthalpy of combustion at standard reference conditions, assuming that water appears as a liquid in the product, for hydrogen, carbon monoxide, methane, ethane, and propane.

11·18 The enthalpy of combustion for gaseous propane is 955,000 Btu/lbmole at standard reference conditions if the water products are liquid. From this information, determine the lower heating value of this fuel at standard reference conditions.

11·19 There has been some interest in using hydrogen (H_2) and ammonia (NH_3) as possible automobile fuels. Calculate the air/fuel ratio for these two fuels and compare to the AFR for hydrocarbon fuels, assuming stoichiometric combustion.

11·20 Ammonia is being considered as a possible fuel for internal combustion engines. Calculate the HHV and LHV for ammonia at standard conditions if $\hat{h}_0 = -19{,}870$ Btu/lbmole for gaseous ammonia.

11·21 At standard conditions h_{fg} for ammonia is 503 Btu/lbm. Calculate the HHV and LHV for ammonia when used as fuel in the liquid phase (see Prob. 11·20).

11·22 Ammonia is burned in air in a test for an alternative automobile fuel. Stoichiometric combustion occurs at inlet conditions of 298 K at 1 atmosphere. Calculate the adiabatic flame temperature ($\hat{h}^\circ_{NH_3} = -19{,}870$ Btu/lbmole).

11·23 Calculate the HHV and LHV at standard conditions for the combustion of propane with oxygen assuming (*a*) gaseous propane and (*b*) liquid propane. The enthalpy of formation of water (liquid) is -286 kJ/gmole, and of water (gas) is -242 kJ/gmole; $\hat{h}^\circ_{fg}$ for propane is 16.3 kJ/gmole, and $\hat{h}^\circ_{C_3H_8}$ (gas) is -103.9 kJ/gmole.

11·24 Determine the enthalpy of combustion at 25°C for ethane (C_2H_6) when gaseous ethane reacts with air producing liquid water in the products. Do the same for acetylene (C_2H_2). Give the results in kJ/kg fuel.

11·25 Verify the results shown in Fig. 11·3 for the case of 100 percent excess air and an energy removal equal to 25 percent of the 25°C LHV.

11·26 Methane is burned in air. If 15 percent excess air is used with this reaction, calculate the volume percent oxygen in the flue gas on a wet (that is, including water vapor) and on a dry (neglecting the water vapor) basis.

11·27 The EPA standard for emissions of NO_x from gas-fired boilers is 0.2 lbm NO_x/million Btu. The heating value of natural gas is about 1000 Btu/standard ft^3. Determine the EPA standard in ppm if the exhaust gas measures 3 percent oxygen on a dry basis and the fuel is methane. Treat NO_x as having the molecular weight of NO_2.

11·28 Assume that gasoline is isooctane, C_8H_{18}. Calculate the CO_2 produced per day from the approximately 4×10^6 cars in Los Angeles County, assuming that each burns 2 gal gas/day.

11·29 Compare the energy released in burning 1 mole of CH_4 with stoichiometric air and burning 1 mole of C and 2 moles of H_2 with stoichiometric air. Assume combustion at the standard reference state. Why is the energy change different?

11·30 Calculate the enthalpy of combustion for the water-gas reaction, $CO(g) + H_2O(g) \rightarrow CO_2(g) + H_2(g)$, at standard reference conditions and at 1000 R.

11·31 Estimate the adiabatic flame temperature for a stoichiometric reaction of *n*-butane (C_4H_{10}) with air at an inlet temperature of 25°C.

11·32 Redo the previous problem for the adiabatic flame temperature if the reaction occurs with 400 percent theoretical air.

11·33 It is desired to burn liquid *n*-butane (C_4H_{10}) with sufficient excess air so that the inlet temperature to a gas turbine will not exceed 1750 R. Specify the required excess air to limit the temperature to this value, assuming that the standard reference condition exists at the inlet to the combustor.

11·34 Determine the adiabatic flame temperature that would be obtained from complete combustion of C_8H_{18} (octane) with theoretical air if (*a*) the inlet conditions are 1 atm at 77°F, and (*b*) the inlet conditions are 1000 R. (Note: an average $c_p = 7.90$ Btu/(lbmole·R) for C_8H_{18} in this temperature range; $\hat{h}^\circ_{C_8H_{18}} = -89{,}680$ Btu/lbmole.)

11·35 Estimate the adiabatic flame temperature when methane (CH_4) at 77°F reacts with 80 percent theoretical air also at 77°F. (Note that CO will be present in the products, since the reaction is one of incomplete combustion.)

11·36 Determine the adiabatic flame temperature for propane burning in air (*a*) stoichiometrically, (*b*) with twice the stoichiometric air, and (*c*) with four times the stoichiometric air. Assume complete combustion in each case.

11·37 Propane (C_3H_8) and oxygen in stoichiometric proportions react in a steady-flow, water-cooled burner. The reactants both enter at 90°F and 2 atm and emerge at 850°F. The products-flow rate is 250 lbm/h. What is the rate of energy transfer as heat to the cooling water? Assume that the products contain only CO_2 and H_2O.

11·38 Acetylene is burned in a constant-pressure, water-cooled burner at 30 psia. Twice the stoichiometric amount of air is provided, and the reactants enter at 40°F. The products emerge at 200°F, and the total flow rate is 10 lbm/min. Water enters the cooling jacket at 60°F and 10 psia and emerges at 1 atm and 180°F. What water-flow rate must be provided? The products contain only CO_2, H_2O, O_2, and N_2.

11·39 Gaseous acetylene at 77°F is burned with 400 percent theoretical air that enters the combustion chamber at 1000 R. The products of combustion leave at 2000 R. Calculate the heat loss from the combustion chamber per lbmole of acetylene burned.

11·40 A turboprop engine has the following characteristics:

Compressor pressure ratio, 6
Maximum cycle temperature, 2200 R
η_C compressor isentropic efficiency, 0.85
η_T turbine isentropic efficiency, 0.90
Fuel, *n*-octane (C_8H_{18})

The engine propels a plane at 400 mph at an altitude of 40,000 ft (4.4 psia, −48°F). Analyze the cycle on a per lbm basis, making suitable assumptions. Calculate all intermediate-state points, and the engine specific fuel consumption, lbm fuel/(hp·h).

11·41 A commercial heater must provide 1000 ft³/min of hot gas at 400°F, 1 atm pressure. Design an appropriate system, specifying the fuel, flow rates, and so on.

11·42 The burner for a large power station must provide boiler heating at the rate of 3×10^8 Btu/h. The maximum flame temperature is to be 1500°F, and the products can leave the boiler at no less than 1200°F. Design a burner system, specifying the fuel, flow rates, and so on.

11·43 The burner in a small engine must handle 10 lbm/s of air, heating it to 1600°F. Any liquid hydrocarbon fuel may be used. Design a burner system, specifying the fuel requirements.

11·44 Design a small air-breathing jet engine to provide 200 lbf of thrust for a flying belt operation. Specify the fuel requirements per min of flying time. What range might such a device obtain?

11·45 Orsat analyses are treated as though the gas sample contained no H_2O (dry gas analysis). Show that the presence of water vapor in the sample does not affect the results of the Orsat analysis.

CHAPTER TWELVE

EQUILIBRIUM

12·1 THE GENERAL APPROACH

In this chapter we shall discuss the problem of predicting the configuration of systems in thermodynamic equilibrium. For example, the theory will permit us to predict the equilibrium composition of a mixture of chemicals in an isolated vessel, will give us an upper bound on the voltage output of a fuel cell, and will tell us how many phases of an alloy can exist together at any particular temperature and pressure. It is clear that the concept and theory of thermodynamic equilibrium have very wide-ranging applications that are particularly important to physical chemists, metallurgists, solid-state physicists, and many engineers. Equilibrium phenomenon are particularly important for combustion and air pollution calculations, and we shall investigate some of the latter in this chapter.

The first and second laws provide the framework for the theory of thermodynamic equilibrium. In this theory the second law is used in a very quantitative manner. For isolated systems the equilibrium state is one of maximum entropy, and we have used this condition several times in the previous chapters. In this chapter we shall develop corresponding conditions for equilibrium of nonisolated systems with environments that maintain constant system temperature and pressure. The isolated system formed by the system under study and the environment will reach its maximum entropy when the system and environment are in equilibrium with one another. We shall show that in this case the *Gibbs function* of the *nonisolated* system is a *minimum*, hence the minimum Gibbs function gives the equilibrium state for a system exposed to an isothermal isobaric environment. Other similar criteria can be developed for systems in contact with other types of environments, using the first and second laws as the starting point.

In the first portion of the chapter we shall treat mixtures of nonreacting substances. This material culminates in the *Gibbs phase rule*, a consequence of the theory which permits one to predict the number of coexisting phases that may be present in such an equilibrium mixture. The remainder of the chapter is devoted to equilibrium in reacting mixtures, with particular emphasis on mixtures of perfect gases.

12·2 EQUILIBRIUM

Consider an isolated system consisting of a two-phase mixture, such as a gaseous water-vapor–nitrogen mixture in contact with liquid water containing nitrogen in solution. We seek the conditions of equilibrium between the two phases (see Fig. 12·1). Assuming that the allowable quantum states for the two phases are independent, the entropy of the system will be the sum of the entropies of the two phases. We imagine isolating the system; the equilibrium state will be the one of maximum system entropy, selected from all possible states having the same total internal energy, volume, and mass of each component. We denote the entropy of the combined system by S_C and have

$$S_C = S_A(U_A, V_A, \mathcal{N}_{1A}, \ldots, \mathcal{N}_{nA}) + S_B(U_B, V_B, \mathcal{N}_{1B}, \ldots, \mathcal{N}_{nB})$$

The isolation constraints require

$$U_A + U_B = \text{constant}$$

$$V_A + V_B = \text{constant}$$

$$\mathcal{N}_{iA} + \mathcal{N}_{iB} = \text{constant}$$

Thus we are free to vary only $U_A, V_A, \mathcal{N}_{1A}, \ldots, \mathcal{N}_{nA}$ in seeking the maximum of S_C. The condition

$$\left(\frac{\partial S_C}{\partial U_A}\right)_{V_A, \mathcal{N}_{1A}, \ldots, \mathcal{N}_{nA}} = 0$$

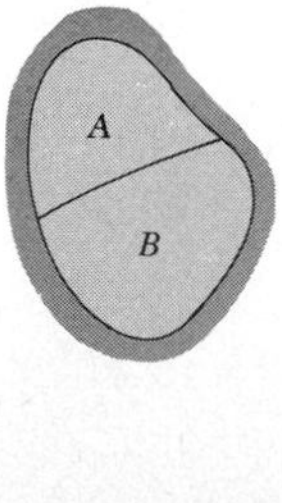

A

B

Figure 12·1. A and B represent two phases

leads directly to

$$\blacktriangleright \qquad \frac{1}{T_A} = \frac{1}{T_B} \qquad T_A = T_B \tag{12·1}$$

The condition

$$\left(\frac{\partial S_C}{\partial V_A}\right)_{U_A,\, \mathcal{N}_{1A},\, \ldots,\, \mathcal{N}_{nA}} = 0$$

requires that

$$\frac{P_A}{T_A} = \frac{P_B}{T_B}$$

and, since the temperatures are equal,

$$\blacktriangleright \qquad P_A = P_B \tag{12·2}$$

The condition

$$\left(\frac{\partial S_C}{\partial \mathcal{N}_{iA}}\right)_{U_A,\, V_A,\, \mathcal{N}_{1A},\, \ldots,\, \mathcal{N}_{(i-1)A},\, \mathcal{N}_{(i+1)A},\, \ldots,\, \mathcal{N}_{nA}} = 0$$

indicates that

$$\left(\frac{\hat{\mu}_i}{T}\right)_A = \left(\frac{\hat{\mu}_i}{T}\right)_B \qquad i = 1, 2, \ldots, n$$

But since the temperature must also be equal, a necessary condition for equilibrium between the phases is

$$\blacktriangleright \qquad \hat{\mu}_{iA} = \hat{\mu}_{iB} \tag{12·3}$$

These results may be extended to mixtures with more than two phases; necessary conditions for equilibrium are

The temperature of every phase must be the same.
The pressure of every phase must be the same.
The electrochemical potential of each component must have the same value in every phase.

Consider again the system of Fig. 12·1. Suppose that the two phases are in thermal and mechanical equilibrium, and that the electrochemical potentials of all but constituent i are equal in both phases. The entropy change of the combined system associated with any interaction will then be simply

$$dS_C = dS_A + dS_B = \left(\frac{\hat{\mu}_{iB}}{T} - \frac{\hat{\mu}_{iA}}{T}\right) d\mathcal{N}_{iA}$$

where $d\mathcal{N}_{iA}$ represents an infinitesimal number of moles of constituent i transferred from B to A. Since the combined system is isolated, the second law requires that $dS_C \geq 0$, and

$$(\hat{\mu}_{iB} - \hat{\mu}_{iA})\, d\mathcal{N}_{iA} \geq 0$$

The equality is associated with reversible transfer between the two phases, which can occur only if $\hat{\mu}_{iA} = \hat{\mu}_{iB}$. For the inequality to hold, if

$$\hat{\mu}_{iB} > \hat{\mu}_{iA} \qquad \text{then} \qquad d\mathcal{N}_{iA} > 0$$

and if

$$\hat{\mu}_{iB} < \hat{\mu}_{iA} \qquad \text{then} \qquad d\mathcal{N}_{iA} < 0$$

We conclude that the electrochemical potential acts as a driving force for mass transfer. Any species will try to move from the phase having the higher electrochemical potential for that species to the phase having lower electrochemical potential. This fact forms the basis for predictions and analysis of the solubility or insolubility of one species in another. It is also useful in studying the distribution of free electrons between dissimilar metals in contact.

12·3 EVALUATION OF THE ELECTROCHEMICAL POTENTIAL

Consider a single phase. Increasing the mass of the phase and keeping the temperature, pressure, and component proportions unchanged will not change the energy, volume, or entropy per unit of mass of the phase. In other words, because a phase is homogeneous, any sample taken from it will have the same intensive state. The condition may be expressed by

$$U(T, P, \lambda\mathcal{N}_1, \ldots, \lambda\mathcal{N}_n) = \lambda U(T, P, \mathcal{N}_1, \ldots, \mathcal{N}_n)$$
$$V(T, P, \lambda\mathcal{N}_1, \ldots, \lambda\mathcal{N}_n) = \lambda V(T, P, \mathcal{N}_1, \ldots, \mathcal{N}_n)$$
$$S(T, P, \lambda\mathcal{N}_1, \ldots, \lambda\mathcal{N}_n) = \lambda S(T, P, \mathcal{N}_1, \ldots, \mathcal{N}_n)$$

Here λ is any factor. Of particular interest in mixture thermodynamics is the Gibbs function,

$$\blacktriangleright \qquad G \equiv U + PV - TS \qquad (12\cdot4)$$

Because of the relations above, it also follows that the Gibbs function of a phase is such that

$$G(T, P, \lambda\mathcal{N}_1, \ldots, \lambda\mathcal{N}_n) = \lambda G(T, P, \mathcal{N}_1, \ldots, \mathcal{N}_n)$$

Differentiating with respect to λ,

$$\sum_i \frac{\partial G(T, P, \lambda\mathcal{N}_1, \ldots)}{\partial(\lambda\mathcal{N}_i)} \mathcal{N}_i = G(T, P, \mathcal{N}_1, \ldots, \mathcal{N}_n)$$

This must hold for all values of λ; setting λ equal to unity,

$$\sum_i \mathcal{N}_i \frac{\partial G(T, P, \mathcal{N}_1, \ldots, \mathcal{N}_n)}{\partial \mathcal{N}_i} = G(T, P, \mathcal{N}_1, \ldots, \mathcal{N}_n) \tag{12·5}$$

We shall return to this result in a moment.

Taking the differential of the Gibbs function,

$$dG = dU + P\,dV + V\,dP - T\,dS - S\,dT$$

Now, the Gibbs equation for the mixture is [see Eq. (10·7)]

$$T\,dS = dU + P\,dV - \sum_i \hat{\mu}_i\,d\mathcal{N}_i$$

Upon combining, we find

$$dG = V\,dP - S\,dT + \sum_i \hat{\mu}_i\,d\mathcal{N}_i \tag{12·6}$$

We therefore recognize the electrochemical potentials as

$$\blacktriangleright \qquad \hat{\mu}_i = \left(\frac{\partial G}{\partial \mathcal{N}_i}\right)_{P,\,T,\,\mathcal{N}_1,\,\ldots,\,\mathcal{N}_{i-1},\,\mathcal{N}_{i+1},\,\ldots,\,\mathcal{N}_n} \tag{12·7}$$

Equation (12·5) then simplifies to

$$\blacktriangleright \qquad G = \sum_i \hat{\mu}_i \mathcal{N}_i$$

A particularly important case is that of the *pure phase*, containing one constituent. The above equation then yields

$$\blacktriangleright \qquad \hat{\mu} = \frac{G}{\mathcal{N}} = \hat{g} \tag{12·8}$$

Hence, *the electrochemical potential for a pure phase is simply the Gibbs function of the phase.*

Let's illustrate these ideas by consideration of a mixture of H_2O liquid and vapor at 100°C. The conditions for phase equilibrium are

$$T_f = T_g \qquad P_f = P_g \qquad \mu_f = \mu_g$$

We have tacitly used the first two throughout this text. Let's test the third. Since $\mu = g = h - Ts$ for each pure phase, from Table B·1*a* we find

$$\mu_f = 419.0 - 373.15 \times 1.3071 = -68.7 \text{ kJ/kg}$$

$$\mu_g = 2676.0 - 373.15 \times 7.3557 = -68.7 \text{ kJ/kg}$$

The electrochemical potentials of the two phases in equilibrium are indeed identical.

We now introduce a special conceptual system analogous to the TER and MER used in Chap. 7. A *constituent reservoir* (CR) is conceived as a large region of constant volume which can be connected to a system under study by a rigid, semipermeable, diathermal membrane (Fig. 12·2). The membrane is conceived to be permeable only to one constituent, hence the CR becomes a warehouse for a pure phase of that constituent. Energy transfer as heat across the membrane permits the CR to reach thermal equilibrium with a system. The CR is assumed to have no internal irreversibility and to possess a uniform internal state. Since its volume is fixed, any entropy changes will be given by [see Eq. (10·7)]

$$dS_{\text{CR}} = \frac{dU_{\text{CR}}}{T} - \frac{\hat{\mu}_i}{T}\, d\mathscr{N}_i \tag{12·9}$$

Consider now a single-phase mixture in contact with a CR containing a pure phase of component 1. The combined system is isolated (Fig. 12·2), and its equilibrium state will be that which maximizes its entropy, subject to the constraints that

$$U_M + U_{\text{CR}} = \text{constant}$$

$$V_M = \text{constant}$$

$$V_{\text{CR}} = \text{constant}$$

$$\mathscr{N}_{iM} = \text{constant} \qquad i = 2, 3, \ldots, n$$

$$\mathscr{N}_{1M} + \mathscr{N}_{1\text{CR}} = \text{constant}$$

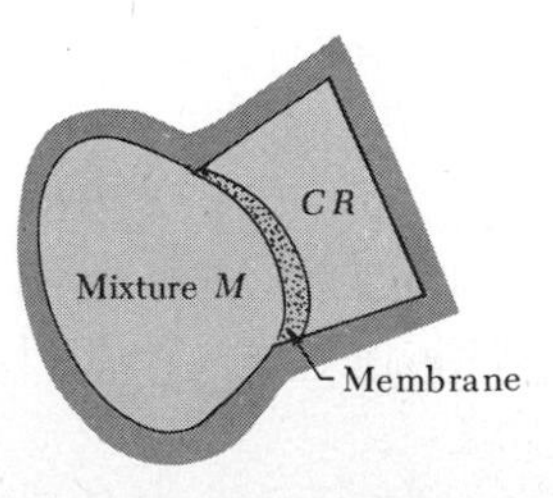

Figure 12·2. Evaluating the electrochemical potential with the help of a constituent reservoir

Here the subscript M denotes the mixture and CR denotes the reservoir. Because of the constraints, only U_M and $\mathcal{N}_{1M}$ can be varied in seeking the maximum of S_C, the entropy of the combined system. Necessary conditions for equilibrium are therefore

$$\left(\frac{\partial S_C}{\partial U_M}\right)_{\mathcal{N}_{1M}} = 0$$

$$\left(\frac{\partial S_C}{\partial \mathcal{N}_{1M}}\right)_{U_M} = 0$$

The first of these leads to

$$\left[\left(\frac{\partial S}{\partial U}\right)_{V,\,\mathcal{N}_1,\,\ldots,\,\mathcal{N}_n}\right]_M = \left[\left(\frac{\partial S}{\partial U}\right)_{V,\,\mathcal{N}}\right]_{CR}$$

which means that

$$T_M = T_{CR}$$

The second condition gives

$$\left[\left(\frac{\partial S}{\partial \mathcal{N}_1}\right)_{U,\,V,\,\mathcal{N}_2,\,\ldots,\,\mathcal{N}_n}\right]_M = \left[\left(\frac{\partial S}{\partial \mathcal{N}}\right)_{U,\,V}\right]_{CR}$$

Since $T_M = T_{CR}$, this requires

$$\hat{\mu}_{1M} = \hat{\mu}_{CR}$$

But the reservoir is a pure phase, so that $\hat{\mu}_{CR} = \hat{g}_{CR}$. This result may be shown to hold in general, and thus

▶ $$\hat{\mu}_i = \hat{g}_i(T, P^*) \tag{12·10}$$

where P^* is the pressure that would exist in a pure phase of the ith component in contact with the mixture through a rigid diathermal membrane permeable only to the ith component. In principle this provides a means for measuring the electrochemical potential of every component in a mixture.

12·4 ELECTROCHEMICAL POTENTIALS IN A MIXTURE OF PERFECT GASES

We can obtain a simple expression for the electrochemical potential of a constituent in a mixture of perfect gases. The Gibbs function for the mixture is

$$G \equiv U + PV - TS$$

$$= \sum_i \mathcal{N}_i \hat{u}_i + V \sum_i P_i - T \sum_i \mathcal{N}_i \hat{s}_i = \sum_i \mathcal{N}_i \hat{g}_i(T, P_i) \tag{12·11}$$

In particular, for a two-constituent mixture,

$$G = \mathcal{N}_1 \hat{g}_1(T, P_1) + \mathcal{N}_2 \hat{g}_2(T, P_2)$$

Then, using Eq. (12·7),

$$\begin{aligned}\hat{\mu}_1 &= \left(\frac{\partial G}{\partial \mathcal{N}_1}\right)_{T,\,P,\,\mathcal{N}_2} \\ &= \hat{g}_1(T, P_1) + \left[\mathcal{N}_1 \left(\frac{\partial \hat{g}_1}{\partial P_1}\right)_T \left(\frac{\partial P_1}{\partial \mathcal{N}_1}\right)_{\mathcal{N}_2,\,P,\,T} + \mathcal{N}_2 \left(\frac{\partial \hat{g}_2}{\partial P_2}\right)_T \left(\frac{\partial P_2}{\partial \mathcal{N}_1}\right)_{\mathcal{N}_2,\,P,\,T}\right]\end{aligned} \tag{12·12}$$

The partial pressures are related to the mole fractions by Eq. (10·19)

$$P_1 = P \frac{\mathcal{N}_1}{\mathcal{N}_1 + \mathcal{N}_2} \qquad P_2 = P \frac{\mathcal{N}_2}{\mathcal{N}_1 + \mathcal{N}_2}$$

Hence

$$\left(\frac{\partial P_1}{\partial \mathcal{N}_1}\right)_{\mathcal{N}_2,\,P} = \frac{P}{\mathcal{N}_1 + \mathcal{N}_2} - \frac{P\mathcal{N}_1}{(\mathcal{N}_1 + \mathcal{N}_2)^2} = \frac{\mathcal{N}_2 P}{(\mathcal{N}_1 + \mathcal{N}_2)^2}$$

$$\left(\frac{\partial P_2}{\partial \mathcal{N}_1}\right)_{\mathcal{N}_2,\,P} = -\frac{P\mathcal{N}_2}{(\mathcal{N}_1 + \mathcal{N}_2)^2}$$

From Eqs. (8·49*a*) and (8·3) it follows that

$$\left(\frac{\partial \hat{g}_1}{\partial P_1}\right)_T = \frac{\mathcal{R}T}{P_1} \qquad \left(\frac{\partial \hat{g}_2}{\partial P_2}\right)_T = \frac{\mathcal{R}T}{P_2}$$

Thus the term within brackets in Eq. (12·12) is

$$\mathcal{N}_1 \frac{\mathcal{R}T}{P_1}\left(\frac{\mathcal{N}_2 P}{\mathcal{N}^2}\right) - \mathcal{N}_2 \frac{\mathcal{R}T}{P_2}\left(\frac{\mathcal{N}_2 P}{\mathcal{N}^2}\right) = \mathcal{R}T \frac{\mathcal{N}_2}{\mathcal{N}}\left(\frac{\mathcal{N}_1}{\mathcal{N}}\frac{P}{P_1} - \frac{\mathcal{N}_2}{\mathcal{N}}\frac{P}{P_2}\right)$$

and vanishes by Eq. (10·19); hence Eq. (12·12) reduces to

$$\hat{\mu}_1 = \hat{g}_1(T, P_1)$$

Thus, *the molal electrochemical potential of a constituent is equal to its molal Gibbs function evaluated at its partial pressure and the mixture temperature.* This result may be shown to be true for every constituent in a mixture of perfect gases,

▶ $$\hat{\mu}_i = \hat{g}_i(T, P_i) \tag{12·13}$$

An alternative expression can be obtained using Eq. (8·18). We have

$$\hat{\mu}_i = \hat{g} = \hat{h}(T) - T\hat{s}(T, P_i)$$

Now, from Eq. (8·18),

$$\hat{s}(T, P_i) = \hat{s}(T, P) - \mathscr{R} \ln \frac{P_i}{P}$$

so

$$\hat{\mu}_i = \hat{h}(T) - T\hat{s}(T, P) + \mathscr{R}T \ln \frac{P_i}{P}$$

Hence, in a mixture of perfect gases,

$$\hat{\mu}_i = \hat{g}_i(T, P) + \mathscr{R}T \ln \chi_i \tag{12·14}$$

This latter form is particularly convenient in analysis of reacting perfect-gas mixtures.

As an example of calculating the chemical potential of a perfect gas, consider 1 lbmole of oxygen and 3.76 lbmoles of nitrogen at a total pressure of 20 psia at 600 R. To calculate the electrochemical potential of the oxygen, we start with

$$\hat{\mu}_{O_2} = \hat{g}_{O_2}(T, P) + \mathscr{R}T \ln \chi_{O_2}$$

Now

$$\chi_{O_2} = \frac{1}{4.76} = 0.21$$

$$\hat{g}_{O_2} = \hat{h}_{O_2} - T\hat{s}_{O_2}$$

but

$$\hat{s}_{O_2} = \hat{\phi}_{O_2} - \mathscr{R} \ln \frac{P}{P_0}$$

where

$$P_0 = 1 \text{ atm}$$

Then, using data from Table B·13,

$$\hat{g}_{O_2}(T, P) = 443.2 - 600 \times \left(49.762 - 1.986 \ln \frac{20}{14.7}\right)$$

$$= -29{,}048 \text{ Btu/lbmole}$$

and

$$\hat{\mu}_{O_2} = -29{,}048 + (1.986)(600) \ln (0.21)$$

$$= -30{,}903 \text{ Btu/lbmole}$$

12·5 THE GIBBS PHASE RULE

Since the molal Gibbs function is an intensive property, the electrochemical potential must also be intensive (this can be seen from the definition). Consequently it depends on the proportions of the mixture but not on the amounts of each component present. Within a given phase the temperature and pressure may be independently varied (by heating and compression) and, in addition, the proportions within the phase may be varied by altering the mole fractions. However, the mole fractions must add up to unity, and therefore only $n-1$ of these are independently variable. Thus within a given phase we may think of the electrochemical potential as being uniquely determined by specifications of T, P, and all but one of the mole fractions,

$$\hat{\mu}_i = \hat{\mu}_i(T, P, \chi_1, \chi_2, \ldots, \chi_{n-1})$$

With this functional relation, and with the condition that the electrochemical potential of a component must have the same value in all coexisting phases, we can now establish an important rule governing the number of possible coexisting phases.

Consider a mixture of phases of a single component. The electrochemical potential of each phase is a function of temperature and pressure. We denote these by

$$\hat{\mu}'(T, P), \hat{\mu}''(T, P), \hat{\mu}'''(T, P), \ldots$$

Henceforth the primes refer to phases. Now, suppose we have an equilibrium mixture of two phases of the substance. Then, from the condition of equilibrium,

$$\hat{\mu}'(T, P) = \hat{\mu}''(T, P)$$

This condition provides a relation between T and P, which must be satisfied for all equilibrium mixture states that contain two phases. It may be rewritten functionally as

$$T = T(P)$$

showing that the temperature and pressure of a mixture of two phases of a single substance are not independently variable.

Suppose the mixture is in an equilibrium state with three phases of the same substance present. The equilibrium condition is then

$$\hat{\mu}'(T, P) = \hat{\mu}''(T, P) = \hat{\mu}'''(T, P)$$

This amounts to two equations in the variables T and P, which suffice to determine the unique value of T and P at which the three phases can coexist. Thus, while it is possible to have a two-phase mixture of a substance over a range of temperatures, a mixture of three specified phases can exist only at a single temperature and pressure, called the *triple point*. A substance that can exist in several phases may have several triple points; six are evident in the phase diagram for water of Fig. 12·3.

While a single substance can exist in many phases, only three phases can coexist in equilibrium. Coexistence of four phases would require that three condi-

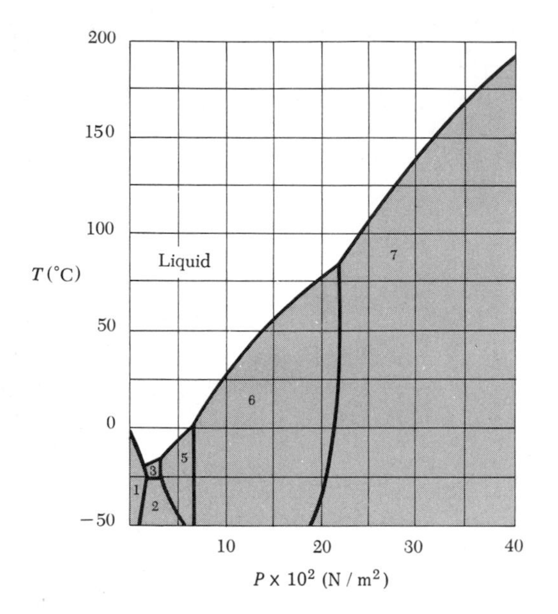

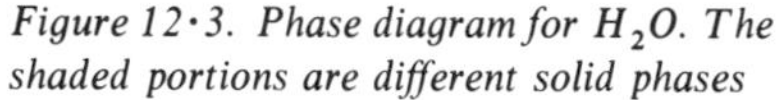
Figure 12·3. Phase diagram for H_2O. The shaded portions are different solid phases

tions of equilibrium in two variables (P and T) be satisfied, and no solution would be possible in general.

We might restate these observations as follows: given a mixture of a single component, the number of properties from the set (T, P) which may be independently fixed is equal to 3 minus the number of phases. This is a special case of the Gibbs phase rule.

Consider now a mixture of n components in p phases. For each component there will be a total of $p - 1$ equations of equilibrium of the form

$$\hat{\mu}_i'(T, P, \chi_1', \ldots, \chi_{n-1}') = \hat{\mu}_i''(T, P, \chi_1'', \ldots, \chi_{n-1}'')$$

Thus there will be a total of $n(p - 1)$ conditions of equilibrium which must be satisfied. The state of each phase is described by its temperature, pressure, and $n - 1$ mole fractions. The set of possible intensive variables which might conceivably be varied is therefore

▶ $$(T, P, \chi_1', \chi_2', \ldots, \chi_{n-1}', \chi_1'', \ldots, \chi_{n-1}^{p})$$

and is exactly $p(n - 1) + 2$ in number. But because of the $n(p - 1)$ equilibrium conditions, the number of these which are freely variable is only

$$f = p(n - 1) + 2 - n(p - 1)$$

or

▶ $$f = n - p + 2 \qquad (12 \cdot 15)$$

Here f denotes the number of independently variable intensive properties *from the list above*. This result is known as the *Gibbs phase rule*.

For the single-component mixture discussed earlier, we have the following:

$f = 2$ for one phase
$f = 1$ for two phases
$f = 0$ for three phases
No more than three phases permitted

It should be pointed out that the number of independently variable extensive properties from the set

$$(U, V, \mathcal{N}_1, \ldots, \mathcal{N}_n)$$

is always $n + 2$. The Gibbs rule does *not* pertain to this set, but rather to the set of *intensive* properties listed above.

To illustrate application of the phase rule, consider a two-component mixture. The largest number of coexisting phases possible is four ($n = 2$, $p = 4$, $f = 0$). This will occur only at particular values of pressure, temperature, and phase composition. At a given pressure three phases may coexist in states for which the following conditions are satisfied:

$$\hat{\mu}'_1(T, P, \chi'_1) = \hat{\mu}''_1(T, P, \chi''_1) = \hat{\mu}'''_1(T, P, \chi'''_1)$$
$$\hat{\mu}'_2(T, P, \chi'_1) = \hat{\mu}''_2(T, P, \chi''_1) = \hat{\mu}'''_2(T, P, \chi'''_1)$$

This set of four independent equations contains the five variables T, P, χ'_1, χ''_1, χ'''_1, and in principle could be reduced to the form

$$T = T(P)$$

Thus at a given pressure three phases could coexist only at one temperature and mixture composition ($n = 2$, $p = 3$, $f = 1$). With two phases ($n = 2$, $p = 2$, $f = 2$) the equilibrium conditions are functionally equivalent to

$$T = T(P, \chi'_1)$$

Hence the temperature at which two phases can coexist is determined by pressure and the mole fraction of one component in one of the phases. In a single-phase mixture ($n = 2$, $p = 1$, $f = 3$) the temperature, pressure, and one mole fraction can be arbitrarily prescribed.

A typical phase diagram for a liquid-vapor binary† mixture is shown in Fig. 12·4. The graph is made on a plane of constant pressure from the three-

† Having two components.

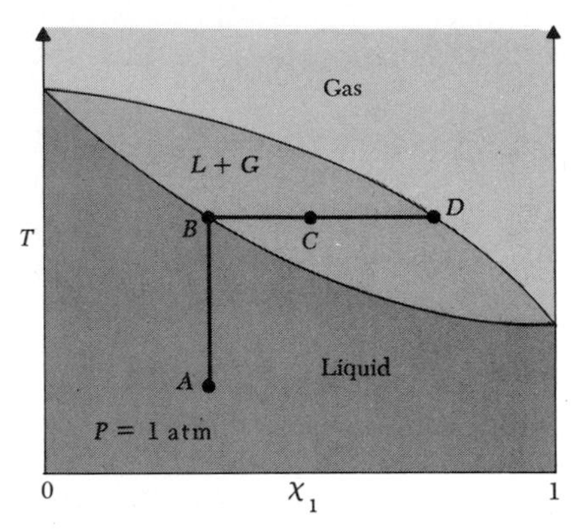

Figure 12·4. Typical phase diagram for a liquid-vapor binary mixture

dimensional T-P-χ_1 space. If a liquid mixture originally at A is heated, evaporation will occur when state B is reached. The first vapor to escape will have the composition of D. A mixture at state C will actually contain liquid at state B and vapor at state D, much as a two-phase, single-component mixture contains saturated liquid and saturated vapor.

Phase diagrams for solid-liquid binary solutions often appear as shown in Fig. 12·5. Again only a single constant-pressure plane is shown. Two solid phases, labeled α and β, exist. The curves ABC and $A'B'C'$ are called *solidus* curves, and the curve AEA' is the *liquidus* curve. Suppose we start cooling a liquid having the composition of state O. When F is reached, the β-phase solid will start to precipitate with the composition G. The only composition for which the condensed solid will have the identical composition as the liquid is the *eutectic* solution (E).

The line BEB' represents the locus of three phase states. The Gibbs phase rule requires that all such states have a common temperature for a given pressure, and consequently the line is an isotherm.

It is difficult to sketch the three-dimensional T-P-χ_1 phase space. If we could, the eutectic point would trace out a curved line, and at a particular pressure a vapor phase would be encountered. The phase rule tells us that there would be a

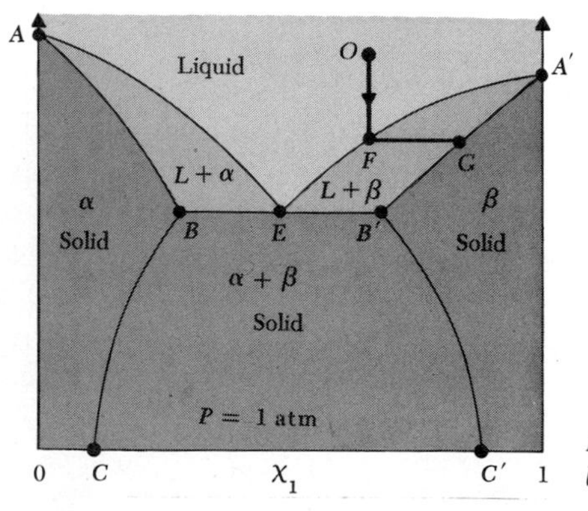

Figure 12·5. Typical phase diagram for a solid-liquid binary mixture

unique point in the phase space at which this would occur. No more than four coexisting phases could ever be found.

Phase diagrams for solid-liquid systems may be familiar from courses dealing with metallurgy. If so, you may recall that a eutectic solution makes good alloy castings, because it freezes very uniformly. You probably also wondered why certain multiple-phase lines were isotherms, while others were not. Now it can be seen that thermodynamics provides a sound theoretical basis for these things: the phase rule.

12·6 ALTERATION OF THE SATURATED-VAPOR PRESSURE BY AN INERT GAS

As a further application of the thermodynamic theory of equilibrium consider a vessel containing a liquid, above which is a mixture of its vapor and an inert gas (see Fig. 12·6). We treat the inert gas and vapor as a mixture of independent substances. Liquid will evaporate until the gas space is saturated; we wish to learn how much the presence of the inert gas alters the saturated-vapor pressure from that which would be obtained without the inert component.

The conditions of equilibrium between phases require that the chemical potential of each component have the same value in all phases. The liquid constitutes one phase and the gas mixture the other. We assume that no inert gas is present in the liquid, so it constitutes a pure phase. The electrochemical potential of the pure liquid phase [see Eq. (12·8)] is

$$\hat{\mu}_f = \hat{g}_f(T, P)$$

where $\hat{g}_f(T, P)$ denotes the Gibbs function of the liquid as a function of the temperature T and total pressure P. Under our perfect-gas idealizations the electrochemical potential of the vapor in the mixture [see Eq. (12·13)] is

$$\hat{\mu}_g = \hat{g}_g(T, P_g)$$

where $\hat{g}_g(T, P_g)$ denotes the molal Gibbs function of the vapor as a function of T and the vapor pressure P_g. Imagine increasing the amount of the inert gas present,

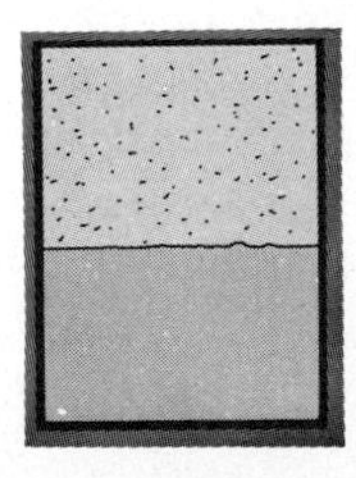

Figure 12·6. The system

keeping the temperature constant. Since the electrochemical potentials $\hat{\mu}_f$ and $\hat{\mu}_g$ remain equal, they must change in the same amount, and we have

$$\left(\frac{\partial \hat{g}_f}{\partial P}\right)_T dP = \left(\frac{\partial \hat{g}_g}{\partial P_g}\right)_T dP_g$$

In Chap. 8 we showed [Eq. (8·49*a*)] that

$$\left(\frac{\partial g}{\partial P}\right)_T = v(P, T)$$

Therefore the change in the partial pressure of the vapor is related to the change in the total (liquid) pressure by

$$dP_g = \frac{\hat{v}_f(P, T)}{\hat{v}_g(P_g, T)} dP$$

If the molal volume of the vapor is very large compared to that of the liquid, such as is the case with water at room temperature, the vapor-pressure change will be very small compared to the change in the total pressure, and in many engineering calculations it is satisfactory to neglect alteration of the vapor pressure by the presence of an inert gas.

If we idealize that the specific volume of the liquid is constant and bring in the perfect-gas equation of state to evaluate $\hat{v}_g$, we obtain

$$\mathscr{R}T \frac{dP_g}{P_g} = \hat{v}_f \, dP$$

Integrating between two configurations 1 and 2,

$$\mathscr{R}T \ln \frac{P_{g2}}{P_{g1}} = \hat{v}_f \cdot (P_2 - P_1)$$

or

$$\frac{P_{g2}}{P_{g1}} = \exp\left[\frac{\hat{v}_f(P_2 - P_1)}{\mathscr{R}T}\right]$$

For example, consider starting with a vessel containing nothing but water and water vapor at 20°C. The initial vapor pressure will be equal to the saturation pressure of pure water, which we find from Table B·1*a* as 0.002338 MPa. The liquid specific volume is 0.001002 m³/kg, hence $\hat{v}_f = 0.001002 \times 18.016 = 0.01805$ m³/kgmole. Imagine adding enough inert gas, such as air, to bring the total pressure up to 1 atm, cooling as necessary to keep the temperature constant. The final water vapor pressure P_{g2} will be given by

$$\frac{P_{g2}}{P_{g1}} = \exp\left[\frac{0.01805 \times (0.1013 - 0.0023) \times 10^6}{8314 \times 293}\right] = e^{0.00073}$$

$$= 1.00073$$

Thus the change in saturation pressure produced by the presence of the inert gas is very small for air–water-vapor mixtures at 1 atm and may be neglected in engineering calculations. This will be true whenever $\hat{v}_f/\hat{v}_g \ll 1$. However, if $\hat{v}_f$ is an appreciable fraction of $\hat{v}_g$, the presence of an inert constituent in the gas phase can markedly influence the saturation pressure.

12·7 GENERAL CONDITIONS FOR CHEMICAL EQUILIBRIUM OF A MIXTURE

An extremely important aspect of the theory of chemical reactions is determination of the equilibrium composition of a mixture of chemically reactive constituents. In this section we derive some general conditions for chemical equilibrium from basic thermodynamic considerations. These conditions will be expressed in terms of thermodynamic properties of the mixture; however, a system that is not in equilibrium is not in a thermodynamic state, and there is some difficulty in ascribing properties such as temperature and entropy in the absence of equilibrium. A similar problem arose when we discussed equilibrium between phases of a mixture. There we assumed that each phase was separately in a thermodynamic equilibrium state, though the combined system of phases was not. The same approach is adopted here. To discuss the properties of a mixture not in chemical equilibrium, that is, not having the composition of the equilibrium mixture, we imagine shutting off the reaction, and evaluate the properties of the mixture as if it were a mixture of nonreacting gases in a thermodynamic equilibrium state. By this procedure we shall be able to establish the conditions of chemical equilibrium, and thereby the composition in the true equilibrium state.

Restricting our consideration to a mixture of simple compressible substances, we apply the first and second laws of thermodynamics to the mixture, which we assume to be at temperature T and pressure P. The energy balance gives

$$dU = đQ + đW$$

and the entropy production is

$$đ\mathscr{P}_s = dS - \frac{đQ}{T} \geq 0$$

Since the pressure is uniform,

$$đW = -P\,dV$$

Combining the above three equations, we obtain

$$T\,đ\mathscr{P}_s = T\,dS - dU - P\,dV \geq 0 \tag{12·16}$$

Equation (12·16) tells us that any reactions that takes place must be such as

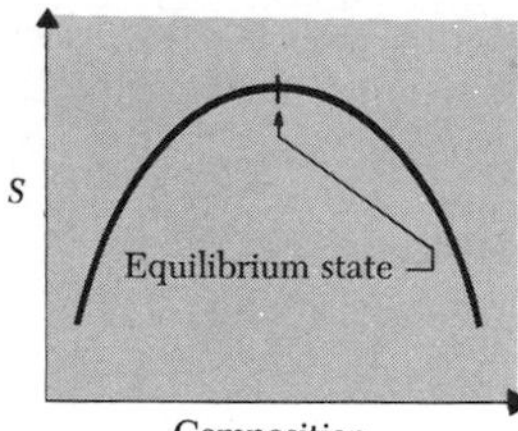

Figure 12·7. Equilibrium at constant volume and energy

to produce entropy. In particular, a reaction taking place in an insulated constant-volume vessel, where $dU = 0$ and $dV = 0$, must be such that

$$dS \geq 0 \tag{12·17}$$

If we want to determine the equilibrium composition of a mixture reacting under these conditions, we need only determine the mixture entropy as a function of composition; the equilibrium composition for an isolated reaction is then the one which maximizes the entropy (Fig. 12·7).

We are more often interested in finding equilibrium compositions under other conditions. The condition for equilibrium in a reaction taking place at constant volume and temperature can be expressed in terms of the *Helmholtz function*,

$$A \equiv U - TS \tag{12·18}$$

The differential of the Helmholtz function is

$$dA = dU - T\,dS - S\,dT$$

Upon combination with Eq. (12·16) we obtain

$$dA + S\,dT + P\,dV \leq 0 \tag{12·19}$$

Hence a reaction occurring at constant temperature and volume must be such as to decrease the Helmholtz function. The equilibrium composition of a mixture

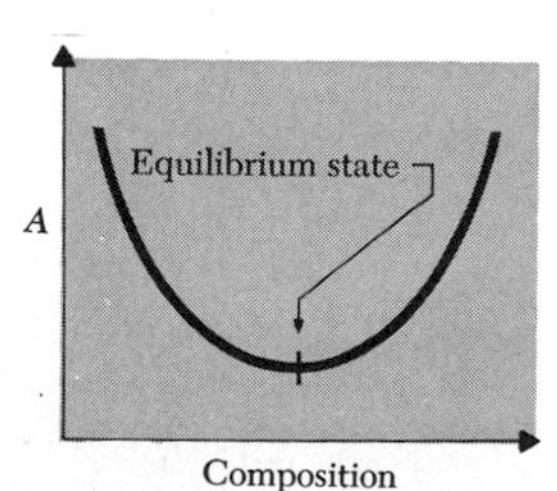

Figure 12·8. Equilibrium at constant volume and temperature

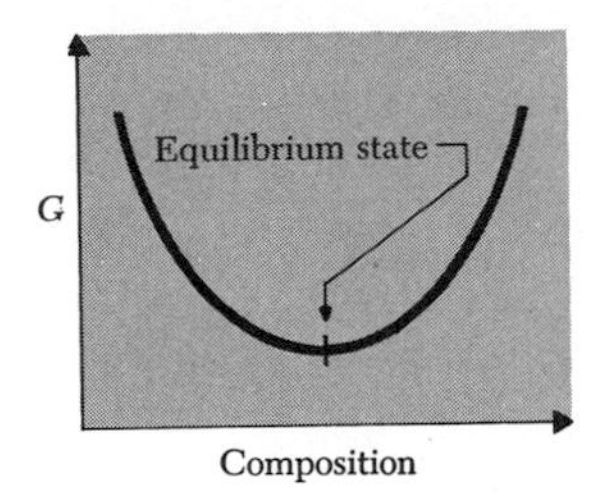

Figure 12·9. Equilibrium at constant pressure and temperature

reacting in an isothermal constant-volume vessel is therefore the one having the least value of the Helmholtz function for that volume and temperature (Fig. 12·8).

The most interesting situation is the reaction at constant pressure and temperature. To find the conditions for equilibrium under these conditions we need to bring in a function whose differential will involve the terms $T\,dS - dU - P\,dV$ in such a way as to cancel them out of Eq. (12·16) and replace them by differentials of pressure and temperature (which are then zero). The Gibbs function fills this need,

$$G \equiv U + PV - TS \tag{12·20}$$

Differentiating,

$$dG = dU + P\,dV + V\,dP - T\,dS - S\,dT$$

Combining with Eq. (12·16), we find

$$\blacktriangleright \qquad dG - V\,dP + S\,dT \leq 0 \tag{12·21}$$

Therefore any reaction proceeding at constant temperature and pressure will be such that the Gibbs function of the mixture will continually decrease until it reaches its minimum value at the final, equilibrium composition (Fig. 12·9).

The Gibbs function is instrumental in allowing us to determine the equilibrium composition of any reactive mixture of known pressure and temperature, regardless of whether or not these were kept constant during all the reaction. Since the electrochemical potential relates the change in the Gibbs function of a mixture (considered as nonreacting) to changes in the amounts of its constituents [see Eq. (12·7)], it too is very important in chemical thermodynamics.

12·8 DEGREES OF REACTION FREEDOM: THE SIMPLE REACTIVE MIXTURE

Any reaction that takes place in a reactive mixture will be limited by the fact that atoms are conserved (that is, by the chemical equation). For example, consider the mixture CH_4, O_2, H_2O, CO, and CO_2. If no material is added to the mixture, any change in the composition must result from chemical reaction. Denoting a change in mole number by $d\mathcal{N}$, the constraining conditions are,† for conservation of C,

$$d\mathcal{N}_{CH_4} + d\mathcal{N}_{CO} + d\mathcal{N}_{CO_2} = 0 \tag{12·22a}$$

for conservation of O,

$$2d\mathcal{N}_{O_2} + d\mathcal{N}_{H_2O} + d\mathcal{N}_{CO} + 2d\mathcal{N}_{CO_2} = 0 \tag{12·22b}$$

and for conservation of H,

$$4d\mathcal{N}_{CH_4} + 2d\mathcal{N}_{H_2O} = 0 \tag{12·22c}$$

We have three constraints and five constituents. Thus only two *degrees of reaction freedom* exist for the mixture. For instance, we could consider CH_4 and CO the two independent components; then, from Eqs. (12·22), we find

$$d\mathcal{N}_{H_2O} = -2d\mathcal{N}_{CH_4} \tag{12·23a}$$

$$d\mathcal{N}_{CO_2} = -d\mathcal{N}_{CH_4} - d\mathcal{N}_{CO} \tag{12·23b}$$

$$d\mathcal{N}_{O_2} = \tfrac{1}{2}d\mathcal{N}_{CO} + 2d\mathcal{N}_{CH_4} \tag{12·23c}$$

It is evident that the number of degrees of reaction freedom of a mixture is equal to the number of possible compounds minus the number of kinds of atoms represented. We shall call a mixture having one degree of reaction freedom a *simple reactive mixture*. For example, the mixture CH_4, O_2, H_2O, and CO_2 is a simple reactive mixture.

12·9 EQUATIONS OF REACTION EQUILIBRIUM

Consider a simple reactive mixture having a chemical equation of the form

$$\blacktriangleright \quad \nu_1 C_1 + \nu_2 C_2 \rightleftarrows \nu_3 C_3 + \nu_4 C_4 \tag{12·24}$$

where $C_1, \ldots, C_4$ denote the chemical constituents, and $\nu_1, \ldots, \nu_4$ are the stoichiometric coefficients. From Eq. (12·7) it follows that the difference in the Gibbs function of the mixture (considered as nonreacting) between any two states having the same temperature and pressure, but infinitesimally different compositions, is

$$dG_{T,P} = \hat{\mu}_1\, d\mathcal{N}_1 + \hat{\mu}_2\, d\mathcal{N}_2 + \hat{\mu}_3\, d\mathcal{N}_3 + \hat{\mu}_4\, d\mathcal{N}_4 \tag{12·25}$$

† As in calculus, $d\mathcal{N}$ always denotes an infinitesimal *increase* in $\mathcal{N}$. Each mole of CH_4 contains the same number of carbon atoms as 1 mole of CO, and twice as many hydrogen atoms as 1 mole of H_2O.

However, in our reacting mixture the changes in mole numbers are related through the chemical equation, and

$$d\mathcal{N}_2 = \frac{\nu_2}{\nu_1}\, d\mathcal{N}_1 \tag{12·26a}$$

$$d\mathcal{N}_3 = -\frac{\nu_3}{\nu_1}\, d\mathcal{N}_1 \tag{12·26b}$$

$$d\mathcal{N}_4 = -\frac{\nu_4}{\nu_1}\, d\mathcal{N}_1 \tag{12·26c}$$

From Eqs. (12·25) and (12·26), we see that for any reaction at constant temperature and pressure, since G must decrease,

▶ $$(\hat{\mu}_1 \nu_1 + \hat{\mu}_2 \nu_2 - \hat{\mu}_3 \nu_3 - \hat{\mu}_4 \nu_4)\, d\mathcal{N}_1 \le 0$$

It then follows that

If

$\hat{\mu}_1 \nu_1 + \hat{\mu}_2 \nu_2 > \hat{\mu}_3 \nu_3 + \hat{\mu}_4 \nu_4$

then $d\mathcal{N}_1 < 0$ *and the reaction proceeds to the right, and if*

$\hat{\mu}_1 \nu_1 + \hat{\mu}_2 \nu_2 < \hat{\mu}_3 \nu_3 + \hat{\mu}_4 \nu_4$

then $d\mathcal{N}_1 > 0$ *and the reaction proceeds to the left.*

The condition of chemical equilibrium for a simple reactive mixture is therefore

▶ $$\hat{\mu}_1 \nu_1 + \hat{\mu}_2 \nu_2 = \hat{\mu}_3 \nu_3 + \hat{\mu}_4 \nu_4 \tag{12·27}$$

This is called the *equation of reaction equilibrium;* it is a relation among the intensive properties of the products and reactants. We see that in order to determine the composition of a mixture after chemical equilibrium has been attained we must know the electrochemical potentials as functions of temperature, pressure, and the mole fractions.

The conditions for equilibrium of more complex mixtures are obtained in a similar manner. For example, for the first mixture considered in the previous section, CH_4, O_2, H_2O, CO, and CO_2, the change in the Gibbs function for any infinitesimal reaction at constant temperature and pressure is

$$dG_{T,P} = \hat{\mu}_{CH_4}\, d\mathcal{N}_{CH_4} + \hat{\mu}_{O_2}\, d\mathcal{N}_{O_2} + \hat{\mu}_{H_2O}\, d\mathcal{N}_{H_2O} + \hat{\mu}_{CO}\, d\mathcal{N}_{CO} + \hat{\mu}_{CO_2}\, d\mathcal{N}_{CO_2}$$

Using the constraining conditions, Eq. (12·23), we find

$$dG_{T,P} = (\hat{\mu}_{CH_4} + 2\hat{\mu}_{O_2} - 2\hat{\mu}_{H_2O} - \hat{\mu}_{CO_2})\, d\mathcal{N}_{CH_4} + (\tfrac{1}{2}\hat{\mu}_{O_2} + \hat{\mu}_{CO} - \hat{\mu}_{CO_2})\, d\mathcal{N}_{CO}$$

At equilibrium the Gibbs function must be a minimum with respect to each and every independent variation of the mixture composition. In other words, dG must be zero for any $d\mathcal{N}_{CH_4}$ and any $d\mathcal{N}_{CO}$. This requires the *two* equations of reaction equilibrium,

$$\hat{\mu}_{CH_4} + 2\hat{\mu}_{O_2} = 2\hat{\mu}_{H_2O} + \hat{\mu}_{CO_2} \tag{12·28a}$$

$$\hat{\mu}_{CO_2} = \tfrac{1}{2}\hat{\mu}_{O_2} + \hat{\mu}_{CO} \tag{12·28b}$$

These equations can be directly associated with the two chemical equations

$$CH_4 + 2O_2 \rightleftarrows 2H_2O + CO_2 \tag{12·29a}$$

$$CO_2 \rightleftarrows \tfrac{1}{2}O_2 + CO \tag{12·29b}$$

In general, one equation of reaction equilibrium will be obtained for each degree of reaction freedom. The coefficients of the electrochemical potentials for each equation of reaction equilibrium will be identical with the coefficients of the constituent in the associated simple chemical equation. It should be mentioned that the job of determining what constituents to put in the mixture list is not a simple one and generally requires considerable experience.

12·10 REACTIONS IN PERFECT-GAS MIXTURES

The equation of reaction equilibrium provides the means for determining equilibrium composition of reacting perfect-gas mixtures. Previously we found that the electrochemical potential of one constituent of a perfect-gas mixture is given by

$$\hat{\mu}_i = \hat{g}_i(T, P) + \mathcal{R}T \ln \chi_i \tag{12·30}$$

Since the equation of reaction equilibrium will involve the Gibbs functions of the constituents, it is convenient to define the Gibbs-function change for a complete unit reaction ΔG_r

▶
$$\Delta G_r = \sum_{\text{prod}} \nu_i \hat{g}_i - \sum_{\text{react}} \nu_i \hat{g}_i \tag{12·31}$$

For the special reaction of Eq. (12·27),

$$\Delta G_r \equiv \nu_3 \hat{g}_3(T, P) + \nu_4 \hat{g}_4(T, P) - \nu_1 \hat{g}_1(T, P) - \nu_2 \hat{g}_2(T, P) \tag{12·32}$$

Thus

$$\Delta G_r = \nu_3(\hat{\mu}_3 - \mathcal{R}T \ln \chi_3) + \nu_4(\hat{\mu}_4 - \mathcal{R}T \ln \chi_4) - \nu_1(\hat{\mu}_1 - \mathcal{R}T \ln \chi_1)$$
$$- \nu_2(\hat{\mu}_2 - \mathcal{R}T \ln \chi_2)$$

Using Eq. (12·27), the $\hat{\mu}$ terms cancel. Then, the ln terms combine to give

$$\Delta G_r = \mathscr{R}T \ln \left(\frac{\chi_1^{\nu_1}\chi_2^{\nu_2}}{\chi_3^{\nu_3}\chi_4^{\nu_4}} \right) \tag{12·33}$$

If ΔG_r is known, this equation plus equations indicating the relative abundances of the elements in the mixture may be solved simultaneously to obtain the constituent mole fractions in the equilibrium mixture.

The value of ΔG_r will depend on both temperature and pressure. However, the pressure effect is easily separated. Denoting P_0 as an arbitrarily chosen reference pressure, it follows from Eq. (8·18) and the definition of the Gibbs function that

$$\hat{g}(T, P) = \hat{g}(T, P_0) + \mathscr{R}T \ln \frac{P}{P_0} \tag{12·34}$$

Hence, substituting into Eq. (12·32),

$$\Delta G_r(T, P) = \Delta G_r(T, P_0) + \mathscr{R}T \ln \left(\frac{P}{P_0} \right)^{\nu_3 + \nu_4 - \nu_1 - \nu_2}$$

Note that $\Delta G_r(T, P_0)$ is a function only of T for a selected reference pressure. Equation (12·33) then may be written as

$$\left(\frac{\chi_3^{\nu_3}\chi_4^{\nu_4}}{\chi_1^{\nu_1}\chi_2^{\nu_2}} \right) \left(\frac{P}{P_0} \right)^{\nu_3 + \nu_4 - \nu_1 - \nu_2} = \exp \left[\frac{-\Delta G_r(T, P_0)}{\mathscr{R}T} \right]$$

Since the right-hand side is a function of temperature, the left-hand side is independent of pressure. We define the *equilibrium constant* for the reaction by

▶ $$K(T) \equiv \left(\frac{\chi_3^{\nu_3}\chi_4^{\nu_4}}{\chi_1^{\nu_1}\chi_2^{\nu_2}} \right) \left(\frac{P}{P_0} \right)^{\nu_3 + \nu_4 - \nu_1 - \nu_2} \tag{12·35}$$

Note that the equilibrium constant is a dimensionless property of the equilibrium mixture and is a function only of the mixture temperature.†

† It does, however, depend on the choice of P_0. In other texts $K(T)$ is normally defined as

$$\frac{\chi_3^{\nu_3}\chi_4^{\nu_4}}{\chi_1^{\nu_1}\chi_2^{\nu_2}} P^{\nu_3 + \nu_4 - \nu_1 - \nu_2}$$

which gives it the dimensions of pressure to the $\nu_3 + \nu_4 - \nu_1 - \nu_2$ power. Since tabulations in the literature always employ P in atm, the numerical values of these equilibrium constants will agree with Eq. (12·35) for $P_0 = 1$ atm.

The term ΔG_r is sometimes called the *free-energy change* for the reaction. The reference pressure P_0 is normally taken as 1 atm; values of ΔG_r (298 K, 1 atm) are tabulated in handbooks under the "standard free-energy change" for the reaction. Note that

$$\blacktriangleright \qquad K(T) = \exp\left[\frac{-\Delta G_r(T, P_0)}{\mathscr{R}T}\right] \tag{12·36}$$

The equilibrium constant can be calculated if ΔG_r is known. Values of K for several reactions are given in Table B·14 in the form $\log_{10} K$. Equation (12·35) is often called the *law of mass action;* it is not a law in the sense of the first and second laws of thermodynamics but is merely a definition of the equilibrium constant.

The equilibrium composition of a given simple reactive mixture of perfect gases can be determined from the equilibrium constant. Perfect-gas mixtures with more degrees of reaction freedom require one additional equilibrium constant for each additional degree of freedom.

Two other quantities appearing in the literature on chemical thermodynamics are the enthalpy and entropy changes for a complete unit reaction,

$$\blacktriangleright \qquad \Delta H_r \equiv v_3 \hat{h}_3 + v_4 \hat{h}_4 - v_1 \hat{h}_1 - v_2 \hat{h}_2$$

$$\blacktriangleright \qquad \Delta S_r \equiv v_3 \hat{s}_3 + v_4 \hat{s}_4 - v_1 \hat{s}_1 - v_2 \hat{s}_2$$

Note that ΔG_r may be calculated as:

$$\Delta G_r = \Delta H_r - T\,\Delta S_r$$

For a mixture of perfect gases ΔH_r will depend only on temperature, since the enthalpies are functions only of temperature. However, ΔS_r will depend on both temperature and pressure.

12·11 CALCULATION EXAMPLES

Equilibrium constant calculation. Consider the reaction

$$CO_2 \rightleftarrows CO + \tfrac{1}{2}O_2$$

at 1000 R. Let us calculate the equilibrium constant for this reaction. We must first calculate ΔG_r, using Eq. (12·31), and then we can calculate K (1000 R) from Eq. (12·36). Now, by definition

$$\hat{g} = \hat{h} - T\hat{s}$$

Using molal enthalpy and entropy data from Table B·13, at 1000 R, 1 atm,

$$\hat{g}_{O_2} = 3362.4 - 1000 \times 53.477 = -50{,}115 \text{ Btu/lbmole}$$

$$\hat{g}_{CO} = -44{,}254.3 - 1000 \times 51.646 = -95{,}900 \text{ Btu/lbmole}$$

$$\hat{g}_{CO_2} = -164{,}531 - 1000 \times 57.212 = -221{,}743 \text{ Btu/lbmole}$$

So

$$\Delta G_r = \hat{g}_{CO} + \tfrac{1}{2}\hat{g}_{O_2} - \hat{g}_{CO_2}$$

$$= -95{,}900 + \tfrac{1}{2} \times (-50{,}115) - (-221{,}743)$$

$$= +100{,}786 \text{ Btu/lbmole}$$

Then,

$$\ln K = \frac{-100{,}786}{1.986 \times 1000} = -50.7$$

$$\log_{10} K = \frac{1}{2.30} \ln K = \frac{-50.7}{2.30} = -22.0$$

Note this agrees well with the value interpolated from Table B·14.

Equilibrium composition. As a further example, let us determine the equilibrium composition of a mixture of CO, CO_2, and O_2 as a function of temperature at a pressure of 1 atm. We will suppose that the mixture contains 1 C atom to 3.125 O atoms. This happens to be the stoichiometric C/O ratio for the combustion of isooctane (C_8H_{18}) with air. While the equilibrium resulting from that combustion process is not the same as would occur for the single reaction we are considering, it is instructive to use the corresponding C/O ratio. We then have for this single reaction

$$CO_2 \rightleftarrows CO + \tfrac{1}{2}O_2 \tag{12·37}$$

Now using P_0 as 1 atm and χ as the mole fraction, we have for the equilibrium condition

$$K(T) = \frac{\chi_{CO}^{\nu_{CO}}\chi_{O_2}^{\nu_{O_2}}}{\chi_{CO_2}^{\nu_{CO_2}}}\left(\frac{P}{P_0}\right)^{\nu_{CO}+\nu_{O_2}-\nu_{CO_2}}$$

where $K(T)$ is the equilibrium constant and the ν's are the stoichiometric coefficients. Then for this reaction we have

$$K(T) = \frac{\chi_{CO}\chi_{O_2}^{1/2}}{\chi_{CO_2}}\left(\frac{P}{P_0}\right)^{1+1/2-1}$$

or, since $P = P_0$,

$$K(T) = \frac{\chi_{CO}\chi_{O_2}^{1/2}}{\chi_{CO_2}} \tag{12·38a}$$

As a side condition we know that the mole fractions must total unity:

$$\chi_{O_2} + \chi_{CO_2} + \chi_{CO} = 1 \tag{12·38b}$$

A third relation follows from the given initial atomic composition.

$$\frac{\chi_{CO_2} + \chi_{CO}}{2\chi_{CO_2} + \chi_{CO} + 2\chi_{O_2}} = \frac{1}{3.125} \tag{12·38c}$$

To solve these three simultaneous equations let us set $\chi_{CO_2} = a$, which shall then be our unknown. From the equations we find

$$\chi_{CO} + \chi_{O_2} = 1 - a$$

or

$$\chi_{O_2} = 1 - a - \chi_{CO}$$

Substituting

$$\frac{\chi_{CO} + a}{2a + \chi_{CO} + 2\chi_{O_2}} = \frac{1}{3.125}$$

$$\frac{\chi_{CO} + a}{\chi_{CO} + 2 - 2\chi_{CO}} = \frac{1}{3.125}$$

or

$$\chi_{CO} = 0.484 - 0.758a$$

$$\chi_{O_2} = 0.516 - 0.242a$$

Using the remaining relation,

$$K(T) = \frac{(0.484 - 0.758a)(0.516 - 0.242a)^{1/2}}{a} \tag{12·38d}$$

Let us now solve this equation for a for several different temperatures but at a pressure of 1 atm. The equilibrium constant is given usually as $\log_{10} K$. We have for K the information in Table 12·1.

Table 12·1 Equilibrium constant for the reaction $CO_2 \rightleftarrows CO + \frac{1}{2}O_2$

$\log_{10} K$	T, K	K
−45.043	298	9.05×10^{-46}
−10.199	1000	6.34×10^{-11}
− 2.863	2000	1.37×10^{-3}
− 0.469	3000	0.340
0.699	4000	5.00

Table 12·2 Equilibrium concentration for the single reaction $CO_2 \rightleftarrows CO + \frac{1}{2}O_2$

	T, K	χ_{CO}	χ_{CO_2}	χ_{O_2}
$P = 1$ atm				
C/O = 1 : 3.12	298	negligible	0.637	0.362
	1000	< 0.001	0.637	0.362
	2000	0.003	0.635	0.363
	3000	0.198	0.378	0.425
	4000	0.437	0.063	0.501
C/O = 1 : 2	3000	0.363	0.454	0.182
C/O = 1 : 5	3000	0.113	0.264	0.623
$P = 10$ psia				
C/O = 1 : 5	3000	0.128	0.247	0.625
$P = 30$ psia				
C/O = 1 : 3.12	3000	0.159	0.429	0.412
C/O = 1 : 5	3000	0.089	0.296	0.615
$P = 10$ psia				
C/O = 1 : 5	5000	0.326	0.009	0.665

Solving by trial and error we find the concentration *at equilibrium* to be as noted in Table 12·2. Some additional results are also found in the table. In particular, at 3000 K, $\chi_{CO_2} = 0.378$, $\chi_{CO} = 0.198$, and $\chi_{O_2} = 0.425$. What can we conclude from the exercise? First, at low temperatures only CO_2 and O_2 are present. The dissociation of CO_2 does not begin until about 2000 K. The reaction $CO_2 \rightleftarrows CO + \frac{1}{2}O_2$ is endothermic when going from left to right. It is a general conclusion that endothermic reactions are more complete at higher temperatures, and this is shown by the results given in the table. The equilibrium state is also affected by pressure.

Effect of pressure. Let us redo the previous example for the 3000 K condition but at a pressure of 30 psia. Since $K = K(T)$, the equilibrium constant has the same value as before. Applying Eq. (12·35) we have

$$\frac{\chi_{CO}\chi_{O_2}^{1/2}}{\chi_{CO_2}} = 0.340\left(\frac{30}{14.7}\right)^{-1/2} = 0.238$$

Solving with this value in Eq. (12·38d), we find

$$\chi_{CO_2} = 0.429$$

$$\chi_{CO} = 0.159$$

$$\chi_{O_2} = 0.412$$

Note that the equilibrium composition is indeed affected by the pressure. An increased pressure tends to reduce the number of moles (hence the number of molecules) in the equilibrium mixture. Consequently, the effect here of increased pressure is to increase the CO_2 concentration. The C/O ratio must of course also affect the equilibrium concentration. We see that an initial state with an excess of O atoms, that is, fuel lean, results in lower concentrations of CO.

Conversely, excess fuel or, as here, an increased ratio of C to O atoms, promotes the formation of CO and the concentration increases. At 3000 K we find for C/O ratios of 1 : 2, 1 : 3.12, and 1 : 5, CO mole fractions, respectively, of 0.363, 0.198, and 0.113. A hasty conclusion might be simply to run engines with excess air to reduce CO concentrations. Unfortunately, this causes the formation of other problem pollutants.

Effect of temperature. Again consider the reaction between CO_2, CO, and O_2. This time we shall seek the equilibrium for 5000 K and 10 psia. For variety let's write the reaction as

$$CO + \tfrac{1}{2}O_2 \rightleftarrows CO_2 \tag{12·39}$$

which interchanges the role of products and reactants from our previous approach. Note, however, that the equilibrium composition is independent of the direction of the reaction. Table B·14 does not explicitly give $K(T)$ for this reaction. However, ΔG_r for the reaction of Eq. (12·39) is simply the negative of that for the reaction of Eq. (12·37). Hence the $K(T)$ for one reaction is simply the *reciprocal* of the $K(T)$ for the other [see Eq. (12·36)], and the logarithms of K are simply the *negatives* of one another. Hence, for the reaction of Eq. (12·39), Table B·14 implies

$$\log_{10} K = -1.387$$

$$K = 0.0410$$

For an initial ratio now of one C atom to five O atoms we find that $\chi_{CO} = 0.326$, $\chi_{CO_2} = 0.009$, and $\chi_{O_2} = 0.665$. The reaction given in Eq. (12·39) is exothermic when proceeding from left to right. Note that in our higher-temperature example less CO_2 is formed and the reaction is thus less complete. As mentioned previously, it is a general rule that exothermic reactions are less complete at higher temperatures and endothermic reactions are more complete at higher temperatures.

The above example is a particularly simple one involving only three species. The more general combustion equation for gasoline, treated here as isooctane, is

$$C_8H_{18} + XO_2 + YN_2 \rightleftarrows aCO_2 + bH_2O + cO_2 + dH_2 + eCO$$
$$+ fH + gOH + hO + iN_2 + jN + kNO + \cdots$$

To work out the coefficients would require knowledge of all the possible reactions such as

$$H_2 + CO_2 \rightleftarrows CO + H_2O$$

$$N_2 \rightleftarrows 2N$$

$$O_2 \rightleftarrows 2O$$

$$NO \rightleftarrows \tfrac{1}{2}O_2 + \tfrac{1}{2}N_2$$

$$H_2O \rightleftarrows H + OH$$

$$H_2 \rightleftarrows 2H$$

$$2H_2O \rightleftarrows 2H_2 + O_2$$

To find the equilibrium composition with this large number of possible reactions clearly requires the use of a computer. However, the composition of the products of combustion can be predicted from equilibrium considerations, just as we have done for our single reaction. Knowing which reactions to consider is, however, not an easy task and requires experience. Since we can write one equation of conservation of atoms for each element, the number of independent equilibrium equations which must be written is equal to the number of species considered less the number of elements contained in these species. In this example we need seven equilibrium reactions and four equations for conservation of species, that is, C, O, N, and H.

In Chap. 11 we assumed that all the reactants are used up and no CO or NO is produced in the combustion process. The real fact is that the combustion process is usually incomplete and therefore noxious products can result. The unburned hydrocarbons, or partially burned ones, along with CO and the oxides of nitrogen, are the major pollutants from automobiles.

To understand why reactions do not go to completion requires knowledge of both the equilibrium conditions for reactions and the kinetics of reactions. Reaction rates are strongly dependent upon temperature. If the temperature drops before the reactions have had time to reach equilibrium, the mixture composition may be *frozen* at a nonequilibrium composition. This occurs for some of the critical pollutant-formation reactions in internal combustion engines.

12·12 VAN'T HOFF EQUATION

The dependence of the equilibrium constant on temperature can be related to the enthalpy change for a complete unit reaction. Differentiating Eq. (12·36)

$$\frac{dK}{dT} = \frac{1}{\mathscr{R}T^2}\left[\Delta G_r(T, P_0) - T\,\frac{d\Delta G_r(T, P_0)}{dT}\right]K$$

The fourth differential equation of state derived in Chap. 8 was

$$dg = -s\,dT + v\,dP$$

Applying this to the changes for a complete unit reaction at fixed pressure we find

$$\frac{d\Delta G_r(T, P_0)}{dT} = -\Delta S_r$$

From the equations above and the definition of the specific Gibbs function it follows that

$$\Delta G_r = \Delta H_r - T\,\Delta S_r$$

Hence the derivative of the equilibrium constant may be expressed as

$$\frac{1}{K}\frac{dK}{dT} = \frac{\Delta H_r(T)}{\mathscr{R}T^2}$$

or

$$\blacktriangleright \qquad \frac{d(\ln K)}{dT} = \frac{\Delta H_r(T)}{\mathscr{R}T^2} \qquad (12\cdot 40)$$

This is called the *van't Hoff equation*, and it is very important in chemical thermodynamics. It permits determination of the enthalpy change for a complete unit reaction solely from equilibrium-composition data. Conversely, it permits us to evaluate changes in the equilibrium constant (a second-law parameter) solely from energy balance (first-law) data.

Equation (12·40) shows that for endothermic reactions ($\Delta H_r > 0$), K will increase with increasing temperature, while for exothermic reactions ($\Delta H_r < 0$) K will decrease with increasing temperature. Since ΔH_r is only weakly dependent on temperature, we can integrate Eq. (12·40) from reference state 1, assuming that ΔH_r is constant, to obtain:

$$\ln\frac{K}{K_1} = -\frac{\Delta H_r}{\mathscr{R}}\left(\frac{1}{T} - \frac{1}{T_1}\right) \qquad (12\cdot 41)$$

Note that $\ln K$ is linear in $1/T$ in this approximation; consequently, plots of $\ln K$ versus $1/T$ are often used to determine $\Delta H_r/\mathscr{R}$ from experimental data.

As an example let us look at a reaction of interest in magnetohydrodynamics: the determination of ΔH_r for the ionization reaction of cesium. We have

$$\mathrm{Cs} \rightleftarrows \mathrm{Cs}^+ + e^-$$

From the data of Table B·14

$$T_1 = 4000 \text{ K} \qquad \log_{10} K_1 = 4.07 \qquad K_1 = 11{,}800$$
$$T = 4500 \text{ K} \qquad \log_{10} K = 4.43 \qquad K = 25{,}900$$

Assuming that ΔH_r is approximately constant, we have from Eq. (12·41)

$$\Delta H_r = \frac{-\mathscr{R} \ln (25{,}900/11{,}800)}{1/4500 - 1/4000}$$
$$\Delta H_r = 5.5 \times 10^4 \text{ cal/(gmole·K)}$$

If this result is converted to eV/molecule we have

$$\Delta H_r = 5.5 \times 10^4 \frac{\text{cal}}{\text{gmole·K}} \times \frac{\text{gmole}}{6.023 \times 10^{23} \text{ molecules}}$$
$$\times \frac{4.186 \text{ J}}{\text{cal}} \frac{\text{eV}}{1.06 \times 10^{-18} \text{ J}}$$
$$\Delta H_r = 3.6 \text{ eV/molecule}$$

Since ΔH_r is positive, the reaction is endothermic and requires 3.6 eV to ionize each molecule. This is in fact the *ionization potential* of cesium at these elevated temperatures.

12·13 APPLICATION TO THE FUEL CELL

Consider a fuel cell burning hydrogen and oxygen, as shown in Fig. 12·10. The maximum electric energy output per mole of water formed and the maximum cell voltage can be determined with the theory developed in this chapter.

We assume steady flow, steady state, and negligible kinetic- and potential-energy changes for the fluids and consider only the electrostatic potential energy of the electrons. The cell temperature and fluid temperatures are assumed to be equal. An energy balance, made over an infinitesimal period, then gives

$$(\hat{h}\, đ\mathscr{N})_{H_2O} - (\hat{h}\, đ\mathscr{N})_{H_2} - (\hat{h}\, đ\mathscr{N})_{O_2} + đQ + \mathsf{e}(\mathscr{E}_b - \mathscr{E}_a)\, đN_e = 0$$

where e denotes the charge on an electron, and $đN_e$ is the number of electrons passing through the control volume. Applying the second law, we obtain

$$đ\mathscr{P}_s = (\hat{s}\, đ\mathscr{N})_{H_2O} - (\hat{s}\, đ\mathscr{N})_{H_2} - (\hat{s}\, đ\mathscr{N})_{O_2} + \frac{đQ}{T} \geq 0$$

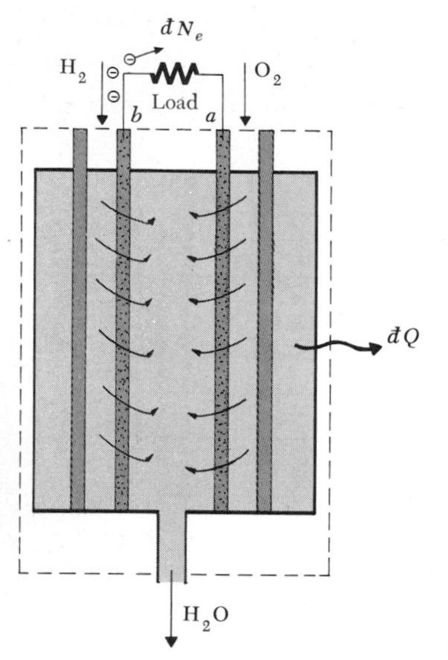

Figure 12·10. The fuel cell

The equality will hold if the processes within the cell are reversible. Otherwise the inequality must be used. Combining the first- and second-law equations, we obtain

$$\mathrm{e}(\mathscr{E}_b - \mathscr{E}_a)\, dN_e \leq -[(\hat{g}\, đ\mathscr{N})_{H_2O} - (\hat{g}\, đ\mathscr{N})_{H_2} - (\hat{g}\, đ\mathscr{N})_{O_2}] \qquad (12 \cdot 42)$$

The chemical equation is

$$H_2 + \tfrac{1}{2}O_2 \rightleftarrows H_2O$$

Since stoichiometric reaction is assumed, Eq. (12·42) may be written as

$$đE_{\text{elect}} = (\mathscr{E}_b - \mathscr{E}_a)\mathrm{e}\, dN_e \leq -\Delta G_r\, đ\mathscr{N}_{H_2O} \qquad (12 \cdot 43)$$

Here $đE_{\text{elect}}$ represents the useful electric energy output. The value of ΔG_r for this reaction is equivalent to $-238{,}000$ J/gmole of H_2O. Hence the maximum possible electric energy output of such a cell is 238,000 J for each gmole of water produced.

The maximum cell voltage can be obtained by considering the reactions at the electrodes. At the anode

$$H_2 + 2OH^- \rightarrow 2H_2O + 2e^-$$

and at the cathode

$$\tfrac{1}{2}O_2 + H_2O + 2e^- \rightarrow 2OH^-$$

The net amount of water production associated with the transfer of two electrons through the load is one molecule. The number of electrons passed for each mole of water formation is then $2N_0$, where N_0 is Avogadro's number. Dividing Eq. (12·43) by $-2eN_0$ (a positive quantity), it becomes

$$\mathscr{E}_a - \mathscr{E}_b \leq -\frac{\Delta G_r}{2N_0 e} = \frac{-238{,}000}{2 \times 6.023 \times 10^{23} \times (-1.602 \times 10^{-19})}$$

$$\mathscr{E}_a - \mathscr{E}_b \leq 1.23 \text{ V}$$

The efficiency of the fuel cell is defined as the ratio of the actual energy output per mole of water formation to the maximum possible output calculated above. Like the isentropic efficiency of a compressor, it compares the actual output to the output of an idealized device and is not an energy-conversion efficiency.

12·14 CURRENT METHODS AND APPLICATIONS OF THE THERMODYNAMIC THEORY OF COMBUSTION

The equilibrium composition of a perfect-gas mixture with more than one degree of reaction freedom required simultaneous solution of a large set of algebraic equations, with additional equations coming from the additional reaction equations and their associated equilibrium constants.

The standardized enthalpy and absolute entropy, together with the equilibrium constant, provide a sound theoretical basis for computation of states of reacting mixtures and energy analyses of control volumes in which reactions occur. Usually it is necessary to resort to digital-computer calculations, and rather elaborate computer programs for combustion analysis have been developed by groups engaged in serious combustion calculations. Hand calculations for a large class of hydrocarbon reactions are now possible with the aid of charts and tables.† Some of these charts employ slightly different datum states and methods, but the fundamental ideas are the same as we have presented here.

Some typical problems in present technology in which this sort of theory is employed include the reacting-gas effects in the high-temperature air around a reentry body, reactions in the boundary layers on rocket nozzle walls, air pollution control devices, waste utilization and disposal furnaces, new metal smelting procedures, new low-contaminant gas welding devices, and high-efficiency, clean combustors for central power stations. These are but a few of the challenging areas open to qualified thermodynamicists.

† Many tables use F for the Gibbs function. See for example the *JANAF Interim Tables of Thermochemical Data*, vols. 1–4, Dow Chemical Company, Midland, Mich., 1960.

SELECTED READING

Lee, J., and F. Sears, *Thermodynamics*, 2d ed., chap. 14, Addison-Wesley Publishing Co., Inc., Reading, Mass, 1963.

Van Wylen, G., and R. Sonntag, *Fundamentals of Classical Thermodynamics*, 2d ed., chap. 13, John Wiley & Sons, Inc., New York, 1973.

QUESTIONS

12•1 What is the chemical potential?
12•2 What is the phase rule?
12•3 Can two phases of ice plus liquid and vapor H_2O exist in equilibrium in any state?
12•4 Why is a triple point better than a two-phase point as the fixed point on a temperature scale?
12•5 What is the greatest number of coexisting phases for a three-component (tertiary) mixture?
12•6 What happens to the pressure of a vapor above its liquid when an inert nondissolving gas is introduced with the vapor? Does the vapor condense or does more evaporate when the inert gas is added at constant temperature?
12•7 What is the condition for equilibrium in a mixture of reacting substances held at constant temperature and pressure?
12•8 How is the chemical potential related to the Gibbs function of a mixture (considered nonreactive)?
12•9 What is meant by degree of reaction freedom?
12•10 What is a simple reactive mixture?
12•11 What is the enthalpy of formation?
12•12 What is the standard free energy of formation?
12•13 What is the enthalpy of reaction?
12•14 To what kind of reactive mixture does the equilibrium constant pertain?
12•15 How can you tell what will happen when some substances are put together in a constant-pressure constant-temperature chamber?
12•16 How would you go about calculating the composition of air, considered a mixture of O_2, O, N_2, N, and NO, at 3000 K?
12•17 Partial pressures can be measured by simple volumetric means. How, then, can the enthalpy of reaction be determined without any energy measurements?
12•18 How many equilibrium constants are needed for determination of the equilibrium composition of a mixture of CO_2, CO, O_2, O, H_2O, OH, N_2, NO, and NO_2?
12•19 Can thermodynamics be used to determine the output voltage of a battery?

PROBLEMS

12•1 Using data in Appendix B, verify that $g_g = g_f$ for mercury at 60 psia.
12•2 The Gibbs function of formation is defined as the difference between the Gibbs functions of the compound and its forming elements at standard reference conditions. Determine the Gibbs function of formation for CO_2 and H_2O (liquid).

12·3 Consider the alloy phase diagram shown.

(*a*) What states are also states on the phase diagrams of the single components?

(*b*) How many phases coexist in region *ACE*?

(*c*) How many phases coexist on the line *EB*?

(*d*) How many phases are present at point *E*? How many independent intensive properties does the system have at *E*?

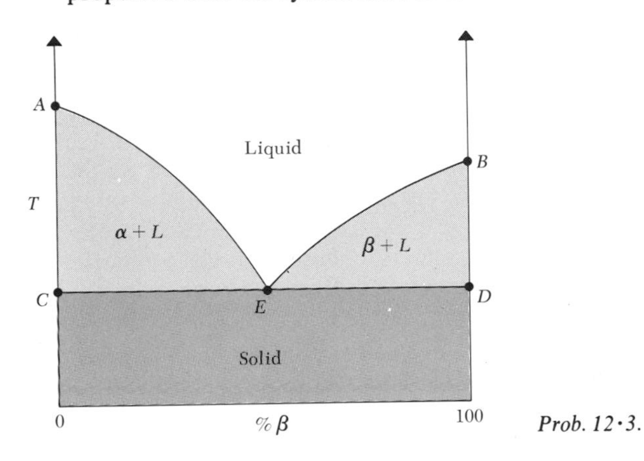

Prob. 12·3.

12·4 Consider the phase diagram for NaCl–H_2O. Explain why ice melts when salt is sprinkled on it. What is the lowest temperature that might be reached if ice, salt, water, and water vapor initially at 0°C are mixed at constant pressure in an insulated container?

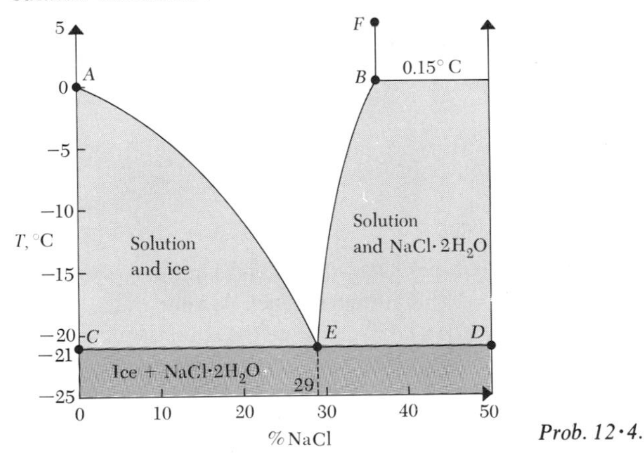

Prob. 12·4.

12·5 Air frequently becomes trapped in refrigeration systems. If the gas in a Freon-12 system contains 10 percent air (by mass), how much will the Freon-12 saturation pressure be changed when the system temperature is 80°F?

12·6 A laboratory analysis shows that CO_2 becomes 10 percent dissociated into CO and O_2 at 4300 R when the total pressure is 1 atm. Find the equilibrium constant from this information and compare to that given in Table B·14.

12·7 Determine the degree of dissociation of CO_2 at 5000 R, 1 atm, if 100 percent excess O_2 is supplied to the reaction.

12·8 Hydrogen is to be burned with oxygen to produce a 3000°F flame. Neglecting dissociation, determine the composition of the product gases and specify the ratios of the oxygen- and hydrogen-flow rates. $P = 1$ atm, and the burner is adiabatic.

12·9 Find the mole fractions of O present in equilibrium O_2 at 1000 K and at 5000 K ($P = 1$ atm).

12·10 Find the mole fractions of N present in equilibrium N_2 at 1000 K and 5000 K ($P = 1$ atm).

12·11 Assuming that air is composed of O_2, O, N_2, N, and NO, and that only O_2 and N_2 are present in significant amounts at room temperatures in the ratio of 3.76 moles of N_2 per mole of O_2, determine the composition of equilibrium air at 1000 K and 1 atm.

12·12 Using the data in Table B·13, calculate the equilibrium constant for the reaction $CO_2 \rightleftarrows CO + \frac{1}{2}O_2$ at 1000 K and compare your result with the value given in Table B·14.

12·13 Using the data in Table B·13, calculate the equilibrium constant for the dissociation of O_2 at 2500 K and compare your result with the value given in Table B·14.

12·14 Using the data in Table B·13, calculate the equilibrium constant for the gaseous reaction $CO_2 + H_2 \rightleftarrows CO + H_2O$ at 298 K. Derive an expression which would let you obtain this instead from Table B·14 in terms of the equilibrium constants for simpler reactions.

12·15 Using the data in the *Handbook of Chemistry and Physics* and the van't Hoff equation, determine the value of the equilibrium constant for the reaction $C + 2H_2 \rightleftarrows CH_4$ at 500°C. Assume that ΔH_r does not vary appreciably over the low-temperature range.

12·16 Experimental results give at 1 atm the following information for the reaction $\frac{1}{2}Cl_2(g) \rightarrow Cl(g)$:

T, K	ΔG, J/gmole Cl
100	115,244
1,000	65,079
3,000	−56,418

Find the equilibrium constant at each condition and estimate ΔH.

12·17 The water-gas reaction is $CO + H_2O \rightleftarrows CO_2 + H_2$. This reaction is under consideration as a means of providing H_2 as an automotive fuel. Assume that 1 mole of CO reacts with 1 mole of H_2O at 1800 K, 1 atm. Determine the equilibrium composition. Will lowering the temperature increase or decrease the yield of H_2? (Neglect species other than these four.)

$CO_2 + H_2 \rightleftarrows CO + H_2O$

T, K	$\log_{10} K$
298	−5.018
100	−0.159
1500	0.409
1800	0.577
2000	0.656

12·18 For the water-gas reaction (see Prob. 12·17) what effect does an increase in CO have on the H_2 yield at 1800 K?

12·19 Consider the formation of nitric oxide (NO) according to the reaction $\frac{1}{2}N_2 + \frac{1}{2}O_2 \rightleftarrows NO$. Calculate at 1 atm and for several temperatures the equilibrium composition of NO for an N/O ratio of 4 : 1 and 40 : 1. The latter corresponds roughly to the composition of flue gas. What measures can you then suggest for NO control? Since pressures in an automobile cylinder are not 1 atm, comment on the effect of pressure on the equilibrium composition.

12·20 Consider the reaction $N_2 \rightleftarrows 2N$. (*a*) At the peak combustion temperature in a car (2500 to 2800 K) will this reaction affect the production of NO? (Neglect pressure effects.) (*b*) Peak pressures are about 35 atm in an automobile cylinder. Does this high pressure increase or decrease the production of N via this reaction?

12·21 For the reaction $H_2O \rightleftarrows H_2 + \frac{1}{2}O_2$ the equilibrium constant is known to be $\log_{10} K = -3.531$ at 3600 R. Estimate the constant at $T = 4000$ R using the van't Hoff equation.

12·22 From the equilibrium constants given in Table B·14, estimate ΔH_r for the reaction $CO_2 \rightleftarrows CO + \frac{1}{2}O_2$ at $T = 2000$ K. Compare to a calculated value for ΔH_r based on Table B·13.

12·23 One mole of CO reacts with 1 mole of O_2 in a steady-flow process. Both inlet conditions are 77°F at 1 atm. The final products are a mixture of CO, CO_2, and O_2 at 1 atm. Determine the equilibrium compositions at 3000 and 2800 K.

12·24 For Prob. 12·23, determine the heat transfer from the system, for the cases where the temperature of the products is 3000 and 2800 K. At what exit temperature is the system adiabatic?

12·25 Calculate and plot the percentage of ionization of cesium at 10^{-2} and 10^{-4} atm over the range of ionization (the electrons may be treated as a monatomic perfect gas with $\hat{M} = 5.48 \times 10^{-4}$ g/gmole).

12·26 Calculate and plot the percentage of ionization in sodium at 10^{-2} and 10^{-4} atm over the range of ionization (the electrons may be treated as a monatomic perfect gas with $\hat{M} = 5.48 \times 10^{-4}$ g/gmole).

12·27 Determine the number of free electrons and ions per cm^3 in argon seeded with 1 percent cesium (by mass) at 1 atm and 1000 K (the electrons may be treated as a monatomic perfect gas with $\hat{M} = 5.48 \times 10^{-4}$ g/gmole). Neglect argon ionization.

12·28 Determine the number of free electrons and ions per cm^3 in argon seeded with 1% sodium at 1 atm and 1000 K (the electrons may be treated as a monatomic perfect gas with $\hat{M} = 5.48 \times 10^{-4}$ g/gmole). Neglect argon ionization.

12·29 Work the previous two problems. Which seed seems most desirable to attain a plasma with a high electric conductivity?

12·30 Determine the maximum energy output per gmole of CO_2 for a fuel cell operating on CO and O_2 at 298 K and 1 atm. Assume that the products contain only CO_2.

12·31 A particular plasma gas dynamics experiment requires a plasma at 800 K, 1 atm, having 10^{14} free electrons per cm^2. Specify an inert carrier gas, an easily ionizable seed, and the mass fractions that would produce the desired plasma.

CHAPTER THIRTEEN

AN INTRODUCTION TO COMPRESSIBLE FLOW

13·1 THE MOMENTUM PRINCIPLE

We have had experience to this point with analyzing various power systems using the first and second laws of thermodynamics. We can find the exit velocity from a nozzle just from an application of the first law if the entrance and exit states are fixed. But as yet we do not know either why supersonic nozzles are designed with a converging-diverging shape, such as shown in Fig. 13·1, or even what must be done to accelerate a flow past the speed of sound. Such design constitutes a problem in *compressible-fluid flow*. This topic is important in a variety of engineering applications. These include the design of gas and steam turbines, rocket nozzles, high-speed airplanes, steam ejectors, and combustion systems.

Thermodynamics is an important tool in working with compressible flows. To analyze such problems completely, momentum equations are required along with conservation of mass, energy, equations of state, and the second law. Consequently, the study of compressible flow provides an excellent opportunity to integrate thermodynamics with fluid mechanics.

The student is now familiar with all the basic principles used to solve compressible flow problems. These are:

1. Conservation of mass
2. Conservation of energy (the first law of thermodynamics)

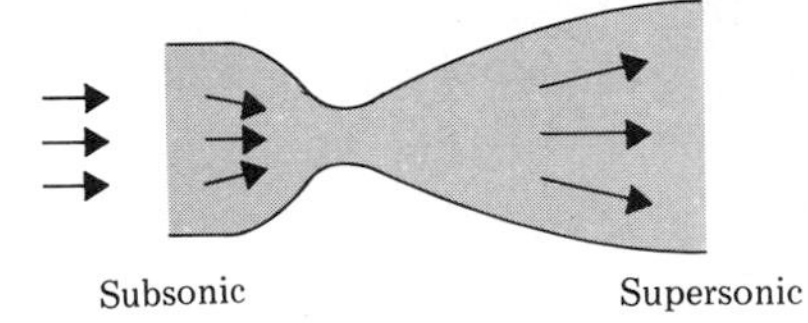

Figure 13·1. A supersonic nozzle. The Mach number is greater than 1 at the exit

3. The second law of thermodynamics
4. The momentum principle

The student is presumably well grounded in applications of the momentum principle (Newton's second law) to systems of *fixed mass* (any control mass). Before we can analyze compressible-flow systems it is necessary to extend Newton's second law to the control volume. Special forms of the control-volume momentum principle are usually used in elementary fluid mechanics courses, and the student should relate our development to the special cases previously encountered.

To derive the control-volume momentum principle we start with the momentum equation as applied to a control mass (Fig. 13·2). Let **F** denote the *net* force, that is, the sum of all surface and body forces, acting on the control mass at time t. Note that **F** is in general a *vector* with three components. **F** is related to the rate of change in the momentum of the control mass by Newton's law,

$$g_c\mathbf{F} = \frac{d}{dt}(\mathbf{M}_{CM}) \tag{13·1}$$

The momentum **M** is also a vector. For a continuous body the momentum may be expressed as an integral of the velocity **V** over the mass within the control volume,

$$\mathbf{M} = \int_{CM} \rho\mathbf{V}\, dV \tag{13·2}$$

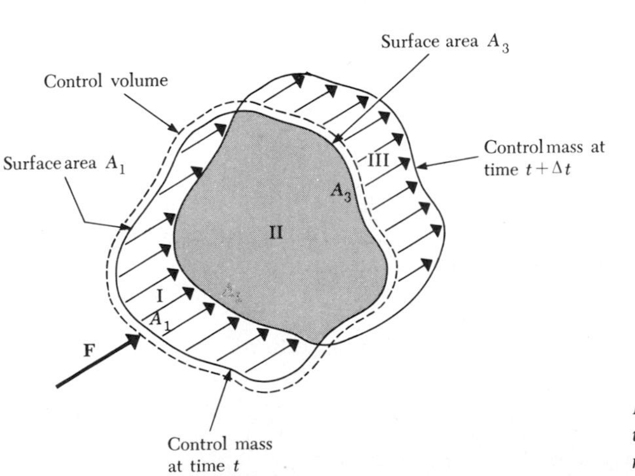

Figure 13·2. The control volume coincides with the control mass at time t

Note that $\rho\, dV$ represents an element of mass within the space occupied by the control mass (CM). Newton's law then says

$$g_c \mathbf{F} = \frac{d\mathbf{M}_{CM}}{dt} \tag{13·3}$$

The control-volume equation is obtained simply by expressing $d\mathbf{M}_{CM}/dt$ in terms of integrals related to the *control volume* (CV) rather than to the control mass (CM). From the basic definition of a derivative,

$$\frac{d\mathbf{M}_{CM}}{dt} = \lim_{\Delta t \to 0} \frac{\mathbf{M}_{CM}(t + \Delta t) - \mathbf{M}_{CM}(t)}{\Delta t} \tag{13·4}$$

Now, we choose the control mass at t as that matter within the control volume at time t. Then, since they contain the same matter at time t, their momenta are equal,

$$\mathbf{M}_{CM}(t) = \mathbf{M}_{CV}(t)$$

However, at time $t + \Delta t$ they do not contain the same matter, hence

$$\mathbf{M}_{CM}(t + \Delta t) \neq \mathbf{M}_{CV}(t + \Delta t)$$

But we can relate $\mathbf{M}_{CM}(t + \Delta t)$ to $\mathbf{M}_{CV}(t + \Delta t)$ and other terms with the help of Fig. 13·2. Note that

$$\mathbf{M}_{CM}(t + \Delta t) = \mathbf{M}_{II}(t + \Delta t) + \mathbf{M}_{III}(t + \Delta t)$$

$$\mathbf{M}_{CV}(t + \Delta t) = \mathbf{M}_{I}(t + \Delta t) + \mathbf{M}_{II}(t + \Delta t)$$

where I, II, and III refer to the three regions in space shown in Fig. 13·2. Hence

$$\mathbf{M}_{CM}(t + \Delta t) = \mathbf{M}_{CV}(t + \Delta t) + \mathbf{M}_{III}(t + \Delta t) - \mathbf{M}_{I}(t + \Delta t)$$

Substituting into Eq. (13·4), and regrouping terms

$$\frac{d\mathbf{M}_{CM}}{dt} = \lim_{\Delta t \to 0} \left(\frac{\mathbf{M}_{CV}(t + \Delta t) - \mathbf{M}_{CV}(t)}{\Delta t} \right) + \lim_{\Delta t \to 0} \frac{\mathbf{M}_{III}(t + \Delta t) - \mathbf{M}_{I}(t + \Delta t)}{\Delta t} \tag{13·5}$$

The first limit above is, by definition, the rate of change in momentum within the control *volume*,

$$\frac{d\mathbf{M}_{CV}}{dt} \equiv \lim_{\Delta t \to 0} \left(\frac{\mathbf{M}_{CV}(t + \Delta t) - \mathbf{M}_{CV}(t)}{\Delta t} \right)$$

The second limit in Eq. (13·5) is related to the *flows* of momentum across the *control-volume boundaries*. For small Δt an elemental volume of III can be represented by

$$dV_{\text{III}} = \mathbf{V}_{\text{rel}}\ \Delta t \cdot d\mathbf{A}$$

Here $d\mathbf{A}$ is an element of the *surface area* bounding regions II and III, and $\mathbf{V}_{\text{rel}}$ is the *outflow* velocity of the fluid relative to the control-volume boundaries; $\mathbf{V}_{\text{rel}}\ \Delta t$ is the *distance* outward from the boundaries which the fluid travels in time Δt. Similarly,

$$dV_{\text{I}} = -\mathbf{V}_{\text{rel}}\ \Delta t \cdot d\mathbf{A}$$

where the minus sign is needed because we retain the convention that $\mathbf{V}_{\text{rel}}$ is taken positive for flow *out* of the control volume. Then, the second limit in Eq. (13·5) becomes

$$\lim_{\Delta t \to 0} \frac{\int_{A3} \rho \mathbf{V}(\mathbf{V}_{\text{rel}}\ \Delta t \cdot d\mathbf{A}) + \int_{A1} \rho \mathbf{V}(\mathbf{V}_{\text{rel}}\ \Delta t \cdot d\mathbf{A})}{\Delta t}$$

Since $A_1 + A_3 = A_{CV}$, the surface area of the control *volume*, this limit is simply

$$\int_{A_{CV}} \rho \mathbf{V}(\mathbf{V}_{\text{rel}} \cdot d\mathbf{A}) = \int_{CS} \mathbf{V}(\rho \mathbf{V}_{\text{rel}} \cdot d\mathbf{A})$$

where we use CS to denote the *control surface*, that is, the surface area of the control volume. Note that $\rho \mathbf{V}_{\text{rel}} \cdot d\mathbf{A}$ represents an elemental mass-flow rate across the control-volume boundary, and $\mathbf{V}$ is the momentum per unit of mass at this point on the boundary. This term then represents the net *momentum-outflow rate* from the *control volume*.

We now have

$$\frac{d\mathbf{M}_{CM}}{dt} = \frac{d\mathbf{M}_{CV}}{dt} + \int_{CS} \mathbf{V}(\rho \mathbf{V}_{\text{rel}} \cdot d\mathbf{A}) \tag{13·6}$$

Since the right-hand side involves only control-*volume* quantities, Eq. (13·1) becomes

▶
$$g_c \mathbf{F} = \frac{d\mathbf{M}_{CV}}{dt} + \int_{CS} \mathbf{V}(\rho \mathbf{V}_{\text{rel}} \cdot d\mathbf{A}) \tag{13·7}$$

We could express $\mathbf{M}_{CV}$ as an integral over the control volume and write Eq. (13·7) as

▶
$$g_c \mathbf{F} = \frac{d}{dt} \int_{CV} \rho \mathbf{V}\ dV + \int_{CS} \mathbf{V}(\rho \mathbf{V}_{\text{rel}} \cdot d\mathbf{A}) \tag{13·8}$$

The right-hand side of Eq. (13·8) can be interpreted as the *rate of production of momentum* $\dot{\mathscr{P}}_{\mathbf{M}}$. The *momentum principle* then says

$$\dot{\mathscr{P}}_{\mathbf{M}} = g_c \mathbf{F} \tag{13·9}$$

Note that, if there is no outflow, Eq. (13·8) reduces to the control-mass equation. Equation (13·8) as written is also correct for a *moving* control volume with expanding boundaries, provided that $\mathbf{V}$ is measured in an inertial reference frame and $\mathbf{V}_{\text{rel}}$ is measured relative to the local control-volume surface.

Table 13·1 summarizes the basic principles in their appropriate control-volume form.

The methodology for momentum-principle analysis follows that laid down for energy balances in Chap. 5.

1. Define the control volume. Since it is quantities at the boundary that will enter the equation, the boundaries should be placed either where you know something or where you want to know something.
2. Define on the sketch the sign for *positive* values for each force and velocity, just as we did for heat and work in Chap. 5. If a force is to be calculated, define it positive in what seems a reasonable direction. If the calculation yields a negative value, it of course means that the force actually acts opposite the arrow. Be careful to show *all* forces acting *on* this control volume (these are *opposite* those forces that the control volume exerts on its environment).
3. Define the positive coordinate directions (on the system sketch). This is necessary because of the vector nature of forces and momentum.

Table 13·1 Summary of equations for a control volume†

Conservation of mass: $\dot{\mathscr{P}}_M = 0$

$$\dot{\mathscr{P}}_M \equiv \frac{d}{dt}\int_{CV} \rho \, dV + \int_{CS} \rho \mathbf{V}_{\text{rel}} \cdot d\mathbf{A}$$

Conservation of energy: $\dot{\mathscr{P}}_E = 0$

$$\dot{\mathscr{P}}_E \equiv \frac{d}{dt}\int_{CV} e\rho \, dV + \int_{CS} (e + Pv)\rho \mathbf{V}_{\text{rel}} \cdot d\mathbf{A} - \dot{Q}_{\text{in}} + \dot{W}_{\text{out}}$$

Second law: $\dot{\mathscr{P}}_S \geq 0$

$$\dot{\mathscr{P}}_S \equiv \frac{d}{dt}\int_{CV} s\rho \, dV + \int_{CS} s\rho \mathbf{V}_{\text{rel}} \cdot d\mathbf{A} - \int_{CS} \frac{q''}{T} dA$$

Momentum principle: $\dot{\mathscr{P}}_{\mathbf{M}} = g_c F$

$$\dot{\mathscr{P}}_{\mathbf{M}} \equiv \frac{d}{dt}\int_{CV} \mathbf{V}\rho \, dV + \int_{CS} \mathbf{V}(\rho \mathbf{V}_{\text{rel}} \cdot d\mathbf{A})$$

† $\mathbf{V}$ is the absolute velocity in an inertial frame, $\rho \mathbf{V}_{\text{rel}} \cdot d\mathbf{A}$ denotes *mass-outflow* rate, $e \equiv u + \mathsf{V}^2/2g_c + \cdots$, and q'' denotes the *input heat flux* (heat-transfer rate per unit of surface area).

4. Express the net forces in the positive directions in terms of the variables in the problem. The system sketch can often be used to define these terms.
5. Express the rates of momentum outflow and inflow, and the rate of momentum storage within the control volume (*zero in steady flow, steady state*), and then the total rate of production of momentum.
6. Apply the momentum principle.
7. Develop other equations from the other basic principles, bring in state information, and so on, as necessary to complete the problem.

Examples following illustrate this methodology.

13·2 SOME MOMENTUM-PRINCIPLE APPLICATIONS

Nozzle supporting force. Consider the nozzle flow of Fig. 13·3. Suppose we wish to calculate the force in the pipe. We first choose an appropriate control volume, placing the control surface where we either know or want to know something. So, let's use the control volume shown in the figure. Next, we calculate the net force on the control volume. We only need to work with one component here. Note the definition of F_p. Then, choosing the positive x direction to the right, the net force in this direction is

$$\mathsf{F}_x = P_1 A_1 - P_2 A_2 - P_a(A_1 - A_2) + \mathsf{F}_p$$

Next we compute the rate of production of momentum. Assuming steady flow, steady state, there is no change in the momentum storage within the control volume (though the momentum of the fluid passing through the control volume changes significantly). Assuming one-dimensional flow at 1 and 2, the rate of production of momentum is

$$\underset{\substack{\text{momentum}\\\text{production}\\\text{rate}}}{\dot{\mathscr{P}}_{\mathbf{M}}} = \underset{\substack{\text{momentum}\\\text{outflow}\\\text{rate}}}{\dot{M}_2 \mathsf{V}_2} - \underset{\substack{\text{momentum}\\\text{inflow}\\\text{rate}}}{\dot{M}_1 \mathsf{V}_1} + \underset{\substack{\text{momentum}\\\text{storage}\\\text{rate}}}{0}$$

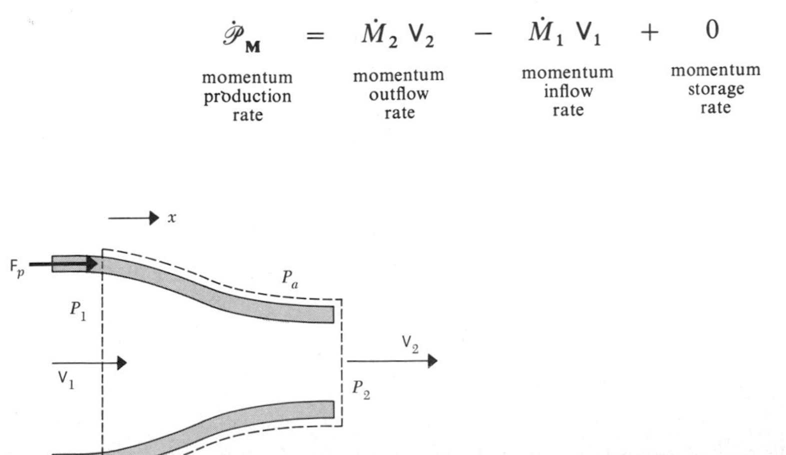

Figure 13·3. A force analysis on a nozzle

Then, the momentum principle says

$$g_c[P_1 A_1 - P_2 A_2 - P_a(A_1 - A_2) + \mathsf{F}_p] = \dot{M}_2 \mathsf{V}_2 - \dot{M}_1 \mathsf{V}_1$$

Solving for F_p,

$$\mathsf{F}_p = \frac{1}{g_c}(\dot{M}_2 \mathsf{V}_2 - \dot{M}_1 \mathsf{V}_1) + P_2 A_2 - P_1 A_1 + P_a(A_1 - A_2)$$

We must know everything on the right before we can calculate $\mathbf{F}_p$. This usually requires additional analysis. For example, by conservation of mass,

$$\dot{M}_2 = \dot{M}_1$$

A thermodynamic analysis will usually be required to relate P_2 and V_2 to P_1 and V_1. The *continuity equation* ($\dot{M} = A\rho\mathsf{V}$) would also be useful.

Steam-air jet ejector. High-pressure steam is used to suck air out of a large tank for low-vacuum experiments. The analysis of such an ejector is based in part on the momentum principle.

The control volume is shown in Fig. 13·4. Note that we assume one-dimensional flow at 1, 2, and 3, and neglect the friction force on the ejector wall. We also assume steady flow and steady state.

Choosing the positive x direction to the right, the net force on the control volume in this direction is

$$\mathsf{F}_x = P_1 A_1 + P_2 A_2 - P_3 A_3$$

The rate of production of momentum in the x direction is

$$\underset{\substack{\text{momentum}\\ \text{production}\\ \text{rate}}}{\dot{\mathscr{P}}_{\mathbf{M}}} = \underset{\substack{\text{momentum}\\ \text{outflow}\\ \text{rate}}}{\dot{M}_3 \mathsf{V}_3} - \underset{\substack{\text{momentum}\\ \text{inflow}\\ \text{rate}}}{(\dot{M}_1 \mathsf{V}_1 + \dot{M}_2 \mathsf{V}_2)}$$

Hence the momentum principle yields

$$g_c(P_1 A_1 + P_2 A_2 - P_3 A_3) = \dot{M}_3 \mathsf{V}_3 - (\dot{M}_1 \mathsf{V}_1 + \dot{M}_2 \mathsf{V}_2)$$

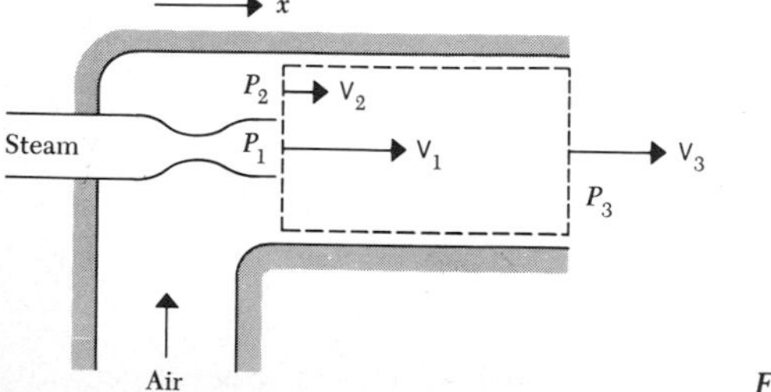

Figure 13·4. A steam jet ejector

If the mass-flow rates and inlet states 1 and 2 were known, the unknowns would be P_3, V_3, and $\dot{M}_3$. The mass balance, the energy balance, and the continuity equation ($\dot{M} = A\rho V$), plus appropriate state information (here for air, steam, and air-water mixtures) would suffice to permit calculation of the discharge state.

Force on a bend. Subcooled liquid water is pumped in a PWR system at a flow rate of 50×10^6 kg/h. Entering the reactor the pipe bends at a right angle (Fig. 13·5). Determine the force on the bend if the pipe diameter is 1.25 m and flow conditions inside the pipe are 290°C at 150 atm.

The control volume is shown on Fig. 13·5; note the positive x and y directions. We idealize as follows:

Steady state, steady flow
One-dimensional flow at sections 1 and 2
No pressure drop between 1 and 2
Negligible weight of water in the control volume
One-atm pressure on the outside of the pipe

In order to apply the momentum principle it will be necessary to determine V_1 and V_2. From continuity,

$$\dot{M} = \rho A V$$

or

$$V = \frac{\dot{M}}{\rho A} = \frac{4\dot{M}}{\rho \pi D^2}$$

At 290°C and 150 atm we can approximate the density of water from the incompressible-liquid equation of state and use the saturated liquid density,

$$\rho = 1/0.001366 = 732 \text{ kg/m}^3$$

So

$$V = 4 \times 50 \times 10^6/[732 \times \pi \times (1.25)^2]$$
$$= 55{,}600 \text{ m/h} = 15.5 \text{ m/s}$$

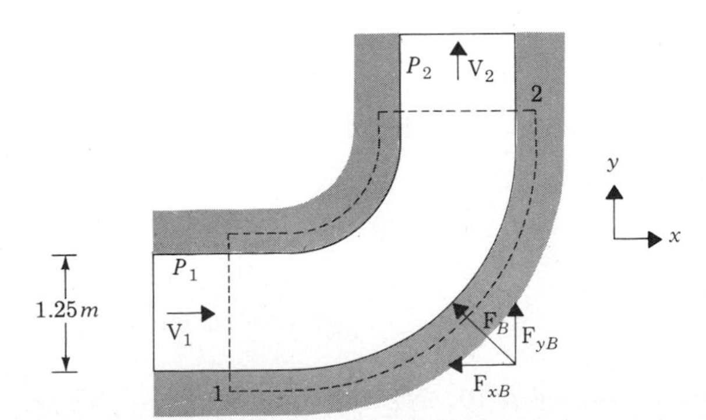

Figure 13·5. Force on a bend

We see that the x component of force on the bend F_{xB} is to the left or negative direction. So the net force in the x direction is

$$\mathsf{F}_x = -\mathsf{F}_{xB} + P_1 A_1 - P_{\text{atm}} A_1$$

For steady-state, steady-flow conditions the rate of production of momentum in the x direction is

$$\underset{\substack{\text{momentum}\\ \text{production}\\ \text{rate}}}{\dot{\mathscr{P}}_{\mathbf{M}}} = \underset{\substack{\text{momentum}\\ \text{outflow}\\ \text{rate}}}{0} - \underset{\substack{\text{momentum}\\ \text{inflow}\\ \text{rate}}}{\dot{M}_1 \mathsf{V}_1}$$

Applying the momentum principle,

$$-g_c \mathsf{F}_{xB} + g_c P_1 A_1 - g_c P_{\text{atm}} A_1 = -\dot{M}_1 \mathsf{V}_1$$

or, since $g_c = 1$ in SI and $P_1 - P_{\text{atm}} = 149$ atm,

$$\begin{aligned}\mathsf{F}_{xB} &= 50 \times 10^6 \times 15.5 \times (1/3600) + 149 \times 101325 \times (\pi/4) \times (1.25)^2\\ &= 2.15 \times 10^5 + 1.85 \times 10^7 = 18.7 \times 10^6 \text{ N}\end{aligned}$$

In the y direction we have a net force of

$$\mathsf{F}_y = \mathsf{F}_{yB} - P_2 A_2 + P_{\text{atm}} A_2$$

For the production of momentum,

$$\underset{\substack{\text{momentum}\\ \text{production}\\ \text{rate}}}{\dot{\mathscr{P}}_{\mathbf{M}}} = \underset{\substack{\text{momentum}\\ \text{outflow}\\ \text{rate}}}{\dot{M}_2 \mathsf{V}_2} - \underset{\substack{\text{momentum}\\ \text{inflow}\\ \text{rate}}}{0}$$

Hence

$$g_c \mathsf{F}_{yB} - g_c P_2 A_1 + g_c P_{\text{atm}} A_2 = \dot{M}_2 \mathsf{V}_2$$

$$\mathsf{F}_{yB} = \frac{1}{g_c} \dot{M}_2 \mathsf{V}_2 + (P_2 - P_{\text{atm}}) A_2$$

Since the pipe diameter remains the same and V_2 is the same as V_1, it follows that F_{yB} has the same magnitude as F_{xB},

$$\mathsf{F}_{yB} = 18.7 \times 10^6 \text{ N}$$

The total force on the bend is the vector sum of $\mathsf{F}_{xB} + \mathsf{F}_{yB}$ and is given by

$$\mathsf{F}_B = 18.7 \times 10^6/\cos 45^\circ = 26.4 \times 10^6 \text{ N } (\approx 6 \times 10^6 \text{ lbf})$$

In this example most of the force on the bend is due to the pressure difference and not to the change in momentum. We had to make the analysis, however, in order to find this out.

13·3 STAGNATION PROPERTIES IN COMPRESSIBLE FLOW

In treating compressible fluid flow it is convenient to work with a hypothetical reference state and its associated properties. The *isentropic stagnation state* is the state a flowing fluid would reach if it were brought to rest isentropically in a steady-flow, adiabatic, zero work output device. This reference state is commonly designated with the subscript zero. From an energy balance (Fig. 13·6) we find that the stagnation enthalpy h_0 is related to the enthalpy and velocity of the moving fluid by

$$h_0 = h + \frac{V^2}{2g_c} \tag{13·10}$$

The hypothetical isentropic stagnation process is shown in Fig. 13·6*b*.

For a perfect gas, since $h = h(T)$, the isentropic stagnation temperature T_0 can be related to the temperature and velocity of the flowing fluid. For a perfect gas with constant c_P,

$$h_0 - h = c_P(T_0 - T)$$

hence

$$\frac{T_0}{T} = 1 + \frac{V^2}{2g_c c_P T} \tag{13·11}$$

Using the perfect-gas equations relating entropy to temperature and pressure [Eq. (8·27)], a relationship for the stagnation pressure may be derived,

$$\frac{P_0}{P} = \left(\frac{T_0}{T}\right)^{k/(k-1)} = \left(1 + \frac{V^2}{2g_c c_P T}\right)^{k/(k-1)} \tag{13·12}$$

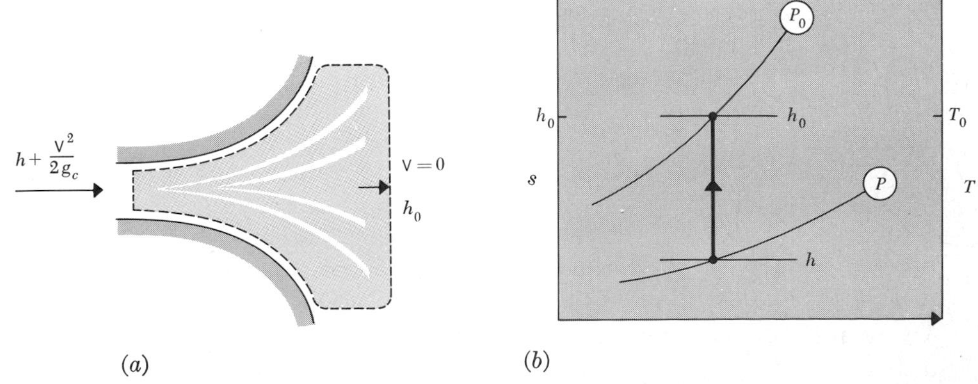

Figure 13·6. Definition of the isentropic stagnation state

As an example, let's calculate the isentropic stagnation conditions for air [$c_P = 1004$ J/(kg·K)] flowing at 400 m/s, at 20°C and 0.3 MPa.

$$T_0 = 293 + \frac{(400)^2}{2 \times 1 \times 1004} = 293 + 79.7 = 372.7 \text{ K}$$

$$P_0 = 0.3(372.7/293)^{3.5} = 0.696 \text{ MPa} = 6.86 \text{ atm}$$

13·4 MACH NUMBER, SPEED OF SOUND, REGIMES IN COMPRESSIBLE FLOW

The Mach number is defined as the ratio of the actual velocity V to the speed at which sound would propagate through the fluid. We shall denote the sound speed by c, and the Mach number by M. Then,

$$\mathsf{M} \equiv \frac{\mathsf{V}}{\mathsf{c}} \tag{13·13}$$

Three flow regimes are defined:

For $\mathsf{M} < 1$ the flow is subsonic
For $\mathsf{M} > 1$ the flow is supersonic
For $\mathsf{M} = 1$ the flow is exactly sonic

The speed of sound is an important parameter in compressible flow, both physically and as a reference speed. We can derive an equation for the speed of sound from an analysis of the motion of a small pressure wave through a compressible fluid, since sound is in fact carried by weak pressure waves. In place of a source of sound let us set up a piston-cylinder arrangement as shown in Fig. 13·7*a*. By putting the piston into motion to the right with a small velocity $d\mathsf{V}$, we can set up a pressure wave that travels down the tube at the sonic velocity c. The wave divides the fluid at rest, and unaware of the piston motion, from the fluid moving with the piston velocity $d\mathsf{V}$ to the right. The fluid near the piston will have a slightly increased pressure and will be slightly more dense.

It is easier to analyze this process if we follow the motion as an observer traveling with the wave front (Fig. 13·7*b*). In this case it appears to us that the fluid is moving toward us from the right with a velocity c, and moving away on the left at velocity $\mathsf{c} - d\mathsf{V}$. In this reference frame we can use a steady-state analysis. Applying the conservation of energy to the (adiabatic) control volume shown,

$$h + \frac{\mathsf{c}^2}{2g_c} = (h + dh) + \frac{(\mathsf{c} - d\mathsf{V})^2}{2g_c}$$

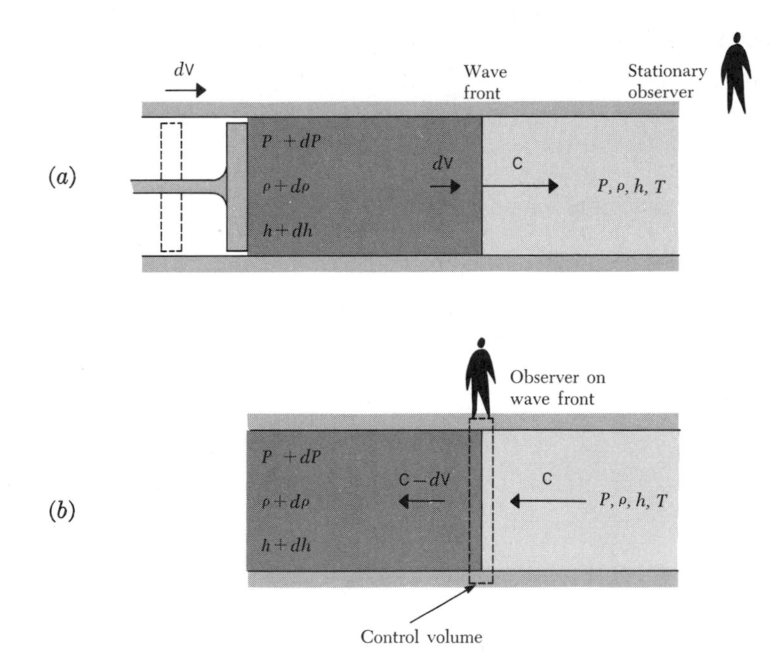

Figure 13·7. An analysis to determine the speed of sound

Neglecting higher-order differentials, we have

$$dh - \frac{\mathrm{c}\, d\mathrm{V}}{g_c} = 0 \tag{13·14}$$

Conservation of mass requires

$$\rho A\mathrm{c} = (\rho + d\rho)A(\mathrm{c} - d\mathrm{V})$$

Neglecting second-order differentials,

$$\mathrm{c}\, d\rho - \rho\, d\mathrm{V} = 0 \tag{13·15}$$

Since the wave is infinitesimal, we can consider the process reversible, and it is of course adiabatic. Consequently, the process can be assumed isentropic. Utilizing the Gibbs equation with $ds = 0$,

$$dh = v\, dP = \frac{dP}{\rho} \tag{13·16}$$

Substituting in the energy equation,

$$\frac{\mathrm{c}\,d\mathrm{V}}{g_c} = \frac{dP}{\rho} \tag{13·17}$$

and combining with the continuity equation,

$$\frac{\mathrm{c}^2\,d\rho}{\rho g_c} = \frac{dP}{\rho} \tag{13·18}$$

Since the process is isentropic, $dP/d\rho = (\partial P/\partial \rho)_s$, and thus

$$\mathrm{c}^2 = g_c\left(\frac{\partial P}{\partial \rho}\right)_s$$

so,

▶
$$\mathrm{c} = \sqrt{g_c\left(\frac{\partial P}{\partial \rho}\right)_s} \tag{13·19}$$

Note that the same relation can be derived using the momentum and continuity equations. Alternatively, the momentum equation may be used to show that the process is isentropic.

For a perfect gas with constant specific heats it may easily be shown that

$$\left(\frac{\partial P}{\partial \rho}\right)_s = kRT$$

hence

$$\mathrm{c} = \sqrt{g_c kRT} = \sqrt{g_c k\left(\frac{\mathscr{R}}{\hat{M}}\right)T} \tag{13·20}$$

Thus, for a given gas, the speed of sound depends only upon the *local* temperature. Note that c is a *thermodynamic property* of the fluid.

Let's calculate the speed of sound in air at 70°F and at 1200°F, and at an altitude of 50,000 feet (−70°F). At 70°F,

$$\mathrm{c} = \sqrt{g_c kRT} = \sqrt{32.2\,\frac{\text{lbm}\cdot\text{ft}}{\text{lbf}\cdot\text{s}^2} \times 1.4 \times 53.3\,\frac{\text{ft}\cdot\text{lbf}}{\text{lbm}\cdot\text{R}} \times 530\text{ R}}$$

$$= 1125 \text{ ft/s}$$

At 1200°F one finds c = 1996 ft/s, while at −70°F c = 968 ft/s. For hydrogen $R = 767$ ft·lbf/(lbm·R), and c is about 3.8 times that in air, while for a heavy gas such as Freon c is considerably less than that in air. Note that the speed of sound at the isentropic stagnation state is greater than that of the flowing fluid, since $T_0 > T$.

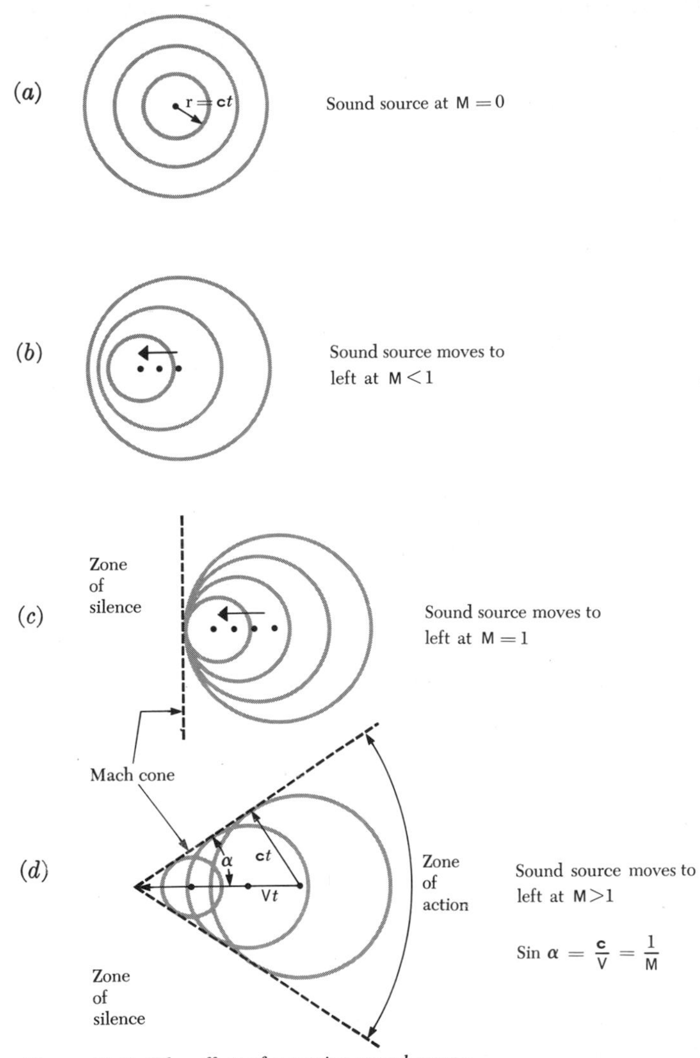

Figure 13·8. The effect of a moving sound source

Helium is much lighter than air, so the speed of sound is much greater. At room temperature, 293 K, we find the speed of sound in helium to be

$$c = \sqrt{g_c kRT} = \sqrt{1 \times 1.67 \times (8314/4)\ \text{J/(kg·K)} \times 293\ \text{K}}$$
$$= 1008\ \text{m/s} = 3307\ \text{ft/s}$$

For the speed of sound in air, using SI units, we have

$$c = 20.04\sqrt{T} \qquad \text{m/s}$$

where T is in K.

Some further helpful ideas follow. Let us imagine a source of sound at a given point as shown in Fig. 13·8*a*. In time *t* the sound waves traverse a distance *ct* relative to the fluid. If the source of sound is stationary or moves slowly ($\mathsf{M} < 1$), the disturbance is felt throughout the fluid (Fig. 13·8*a* and *b*). When the source moves at just the speed of sound, the disturbances created are not able to propagate ahead of the source, and zones are created within which the disturbance is felt (*zone of action*) and not felt (*zone of silence*). If $\mathsf{M} > 1$ the zone of action is a *cone* (the *Mach cone*). The passage of a supersonic aircraft overhead is announced to persons on the ground when the Mach cone passes their position. Note that the angle of the Mach cone is related to the Mach number.

The ratios T_0/T and P_0/P can be expressed in terms of the Mach number. Noting that

$$\frac{\mathsf{V}^2}{2g_c c_P T} = \frac{Rk}{2c_P} \frac{\mathsf{V}^2}{g_c kRT} = \frac{k-1}{2}\mathsf{M}^2$$

Eq. (13·11) gives

$$\blacktriangleright \qquad \frac{T_0}{T} = 1 + \frac{k-1}{2}\mathsf{M}^2 \qquad (13\cdot21)$$

and Eq. (13·12) gives

$$\blacktriangleright \qquad \frac{P_0}{P} = \left(1 + \frac{k-1}{2}\mathsf{M}^2\right)^{k/(k-1)} \qquad (13\cdot22)$$

The student should derive the corresponding density ratio.

13·5 A NOZZLE DESIGN PROBLEM

Let's now bring these ideas together in an example. Suppose we wish to design a nozzle to produce an air flow at a Mach number of 1.75. The air inlet state is to be 80 psia, 500°F. The mass-flow rate will be 100 lbm/s. We wish to calculate the flow area as a function of the local speed. We can then shape the nozzle to obtain a suitable distribution of acceleration along the flow.

Our flow model will be as follows:

One-dimensional flow in the nozzle core
Negligible effects of wall boundary†
Isentropic flow in the core
Steady flow, steady state
Velocity negligible at the inlet state
Perfect gas with constant c_P

† Boundary layers on the nozzle walls, produced by viscous effects, are normally very thin in these circumstances.

With these assumptions each point in the core flow will have the same isentropic stagnation state, with

$$T_0 = 960 \text{ R} \qquad P_0 = 80 \text{ psia}$$

Hence, at any point, Eqs. (13·21) and (13·22) apply. Then, choosing values for M at some point in the nozzle, the local T and P can be calculated. This fixes the thermodynamic state, hence the local density and speed of sound. Then the local velocity can be calculated from $\mathsf{V} = \mathsf{M}c$, and finally the flow area calculated from $A = \dot{M}/\rho\mathsf{V}$. Table 13·2 summarizes this calculation. Note that the speed of sound decreases as the Mach number increases along the nozzle. It may be surprising that the flow area first decreases, but then must be *increased* to obtain further acceleration of the flow. Note that the point of least area (the *nozzle throat*) appears to be near the point where M = 1. We see striking differences between the flow in the subsonic and supersonic portions of the nozzle. The reasons for this are evident when the density is considered. Note that ρ falls off slowly as the flow expands in the subsonic region, and then decreases very rapidly as the flow expands to supersonic speeds. This expansion is so rapid that the flow area must be *enlarged* as we move downstream past the throat.

Here we carried out calculations for a specific case. Considerable insight and convenience are provided by the general treatment and nondimensional results developed in the next section.

Suppose we want to increase the discharge velocity. There is a *maximum velocity* a flow can reach isentropically from its stagnation state. This occurs when $T \to 0$. From Eq. (13·21),

$$\mathsf{V}_{max} = \sqrt{2g_c c_P T_0} \tag{13·23}$$

This is important in rocket design. For our stagnation temperature of 960 R, $\mathsf{V}_{max} = 3360$ ft/s, and this is the maximum speed we could obtain by further increase in the nozzle exit area.

Table 13·2

P psia	T R	V ft/s	ρ lbm/ft^3	$c = \sqrt{g_c kRT}$ ft/s	A ft^2	M
80	960	0	0.225	1520	∞	0.000
70	924	655	0.205	1490	0.745	0.440
60	885	950	0.183	1454	0.575	0.652
50	840	1200	0.161	1419	0.519	0.845
42	798	1392	0.142	1380	0.505	1.010
40	787	1440	0.137	1372	0.506	1.050
30	725	1680	0.112	1319	0.532	1.270
20	645	1940	0.084	1245	0.612	1.560
15	595	2100	0.068	1195	0.700	1.750

13·6 ONE-DIMENSIONAL STEADY ISENTROPIC FLOW

The flow discussed in the previous section was modeled as frictionless and adiabatic or, equivalently, as *isentropic*. This section deals with a general treatment of one-dimensional steady isentropic flow, a model that is quite useful in engineering analysis.

The manner in which the different flow properties vary as a function of one another can be shown most simply from differential equations relating the variables. We shall develop these now. The solutions to these equations may be obtained by manipulating them into separable forms and integrating or, alternatively, one can work directly with algebraic equations like Eqs. (13·21) and (13·22). The latter method is easier, hence we shall use the differential equations merely for insight.

The appropriate energy equation is

$$h_0 = h + \frac{V^2}{2g_c} = \text{constant}$$

Differentiating,

$$\blacktriangleright \qquad dh + \frac{V\,dV}{g_c} = 0 \qquad (13\cdot24)$$

This equation relates enthalpy and velocity changes along the flow. As the velocity goes up, the enthalpy goes down, and vice versa. This represents the continual exchange of organized energy (bulk kinetic energy and flow work) and disorganized energy (the internal energy of the gas).

The continuity equation is

$$\rho A V = \text{constant}$$

In differential form,

$$AV\,d\rho + \rho A\,dV + \rho V\,dA = 0$$

or

$$\blacktriangleright \qquad \frac{d\rho}{\rho} + \frac{dV}{V} + \frac{dA}{A} = 0 \qquad (13\cdot25)$$

This equation reveals the complex coupling between area, density, and velocity changes. *If the density did not change*, A would decrease as V increases. At very high speeds $V \to V_{max}$, hence the velocity no longer changes as the area increases. Equation (13·25) then shows that, as the density decreases the area increases, as observed in the last section.

Now, the Gibbs equation requires

$$T\,ds = du + P\,dv = dh - \frac{1}{\rho}\,dP$$

Since $ds = 0$,

$$dh = \frac{1}{\rho}\,dP \tag{13·26}$$

This equation tells us that the enthalpy and pressure will vary in the same manner, with the pressure decreasing as the enthalpy decreases (and the velocity increases). Combining the energy and Gibbs equations, one has

$$\frac{\mathsf{V}\,d\mathsf{V}}{g_c} + \frac{1}{\rho}\,dP = 0 \tag{13·27}$$

This equation again tells us that as the velocity increases the pressure must decrease. Equation (13·27) is a differential form of *Bernoulli's equation.* Note that it cannot be integrated unless we know how ρ varies with P. For the *special case* of $\rho =$ constant, as encountered in elementary fluid mechanics courses, integration gives the Bernoulli equation, *but only for incompressible flows.*†

Now, from the equation for the sound speed c [Eq. (13·19)],

$$\frac{dP}{d\rho} = \left(\frac{\partial P}{\partial \rho}\right)_s = \frac{\mathsf{c}^2}{g_c}$$

so

$$dP = \frac{\mathsf{c}^2}{g_c}\,d\rho \tag{13·28}$$

This equation tells us that as the pressure increases the density will also increase, and vice versa.

Finally, we combine Eqs. (13·24) to (13·28) to obtain

$$\frac{dA}{A} = \frac{g_c\,dP}{\rho \mathsf{V}^2}\left(1 - \frac{\mathsf{V}^2}{\mathsf{c}^2}\right) = \frac{g_c\,dP}{\rho \mathsf{V}^2}(1 - \mathsf{M}^2) = \frac{d\mathsf{V}}{\mathsf{V}}(\mathsf{M}^2 - 1) \tag{13·29}$$

This equation tells us how A must vary with pressure. For *subsonic flow* ($\mathsf{M} < 1$), dP and dA will be of the *same sign,* hence a *reduction* in A is necessary to produce a reduction in P (flow acceleration). For *supersonic flow* ($\mathsf{M} > 1$), dP and dA will be of *opposite signs,* and an *increase* in A is required for further reduction in P

† However, gas flows in which the density varies only slightly are usually treated by the incompressible form of the Bernoulli equation.

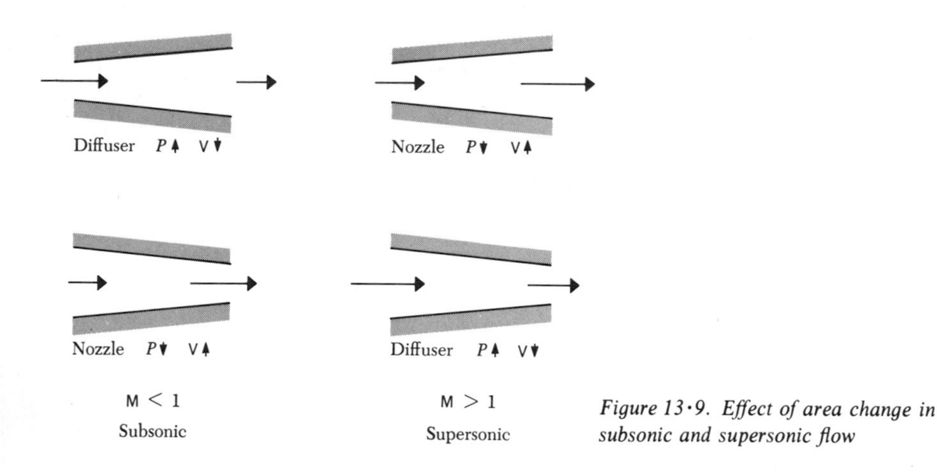

Figure 13·9. Effect of area change in subsonic and supersonic flow

(increase in V). This supports the suggestions made on the basis of the calculation example in the last section. We remark that the developments of this section have not yet required any equation-of-state assumptions, and therefore these conclusions hold for all gases and liquids. Figure 13·9 summarizes the variations in properties in subsonic and supersonic isentropic flow.

Now we turn to the generalized dimensionless equations representing the behavior for isentropic flow. To nondimensionalize we need suitable reference values for T, P, A, and so on. We have seen that the stagnation state provides useful reference values T_0, P_0, ρ_0, c_0. The area at stagnation is infinite, and we must look elsewhere for a suitable reference area. The *sonic* point, or throat, where $M = 1$ provides the reference area, which is smaller than the area at any other point in the flow. Conditions at $M = 1$ are customarily denoted by the superscript *. A^* is then the throat area.

Conditions at the sonic point follow immediately from Eqs. (13·21) and (13·22) (assuming perfect gas with constant c_P),

$$\frac{T_0}{T^*} = \left(1 + \frac{k-1}{2}\right) \tag{13·30}$$

$$\frac{P_0}{P^*} = \left(1 + \frac{k-1}{2}\right)^{k/(k-1)} \tag{13·31}$$

For a gas with $k = 1.4$,

$$\frac{T^*}{T_0} = 0.833 \qquad \frac{P^*}{P_0} = 0.528 \qquad \frac{\rho^*}{\rho_0} = 0.634$$

Expressions for A/A^*, ρ/ρ^*, T/T^*, and so on, as functions of M and k may be developed very easily from the algebraic forms of the continuity and state equations. We leave this as an exercise for the student. Extensive tables of these results are available,† and a short table for $k = 1.4$ is included as Table B·18.

To illustrate the use of the isentropic flow tables, let's redo the previous example for the design of a supersonic nozzle, but this time using the gas tables. Let's treat the exit pressure of 15 psia as specified, and the exit Mach number as a quantity to be determined.

State 1 $\quad P_0 = 80 \text{ psia} \quad$ *assume* $\mathsf{V}_1 = 0$

$T_0 = 960 \text{ R}$

State 2 $\quad P_2 = 15 \text{ psia}$

$s_2 = s_1$

Let us first find the final Mach number. Now,

$$\frac{P_2}{P_0} = \frac{15}{80} = 0.1875$$

Using this value for P/P_0, we find from Table B·18,

$$\mathsf{M}_2 = 1.75$$

We have thus immediately determined that the nozzle is supersonic, since $\mathsf{M}_2 > 1$. Since the ratio of P_2/P_0 is less than the critical pressure ratio ($P^*/P_0 = 0.528$ for $k = 1.4$), this had to be the case.

We must establish the throat conditions before we can calculate any flow areas. Using the ratios at $\mathsf{M} = 1$,

$$P^* = 0.528 \times 80 = 42.2 \text{ psia}$$

$$T^* = 0.833 \times 960 = 800 \text{ R}$$

Now,

$$\rho_0 = \frac{P_0}{RT_0} = \frac{80 \times 144}{53.3 \times 960} = 0.225 \text{ lbm/ft}^3$$

thus

$$\rho^* = 0.634 \times 0.225 = 0.143 \text{ lbm/ft}^3$$

Alternatively,

$$\rho^* = \frac{P^*}{RT^*} = \frac{42.2 \times 144}{53.3 \times 800} = 0.143 \text{ lbm/ft}^3$$

† See J. Keenan and J. Kaye, *Gas Tables*, John Wiley & Sons, Inc., New York, 1948.

At the throat†

$$\mathsf{M} = 1 = \frac{\mathsf{V}^*}{\mathsf{c}^*}$$

$$\mathsf{c}^* = \sqrt{g_c kRT^*} = 1385 \text{ ft/s}$$

$$\mathsf{V}^* = 1385 \text{ ft/s}$$

Thus
$$A^* = \frac{\dot{M}}{\rho^* \mathsf{V}^*} = \frac{100}{0.143 \times 1385} = 0.50 \text{ ft}^2$$

in agreement with the previous result. Finally, at the exit, we read (for $\mathsf{M} = 1.75$) the various ratios and then calculate as follows:

$$\frac{T}{T_0} = 0.620 \qquad T = 0.620 \times 960 = 595 \text{ R}$$

$$\mathsf{M}^* \equiv \frac{\mathsf{V}}{\mathsf{V}^*} = 1.509 \qquad \mathsf{V} = 1.509 \times 1385 = 2090 \text{ ft/s}$$

$$\frac{A}{A^*} = 1.386 \qquad A = 1.386 \times 0.50 = 0.69 \text{ ft}^2$$

$$\frac{\rho}{\rho_0} = 0.303 \qquad \rho = 0.303 \times 0.225 = 0.068 \text{ lbm/ft}^3$$

The answers are in good agreement with the previous example, although interpolation errors show up in some values.

A relationship between the flow rate, throat area, and stagnation state is useful. We start with

$$\dot{M} = A^* \rho^* \mathsf{V}^*$$

Manipulating and substituting for V^*

$$\dot{M} = A^* \frac{\rho^*}{\rho_0} \rho_0 \sqrt{g_c kRT^*} = \frac{\rho^*}{\rho_0} \sqrt{g_c kR\left(\frac{T^*}{T_0}\right)}\, A^* \rho_0 \sqrt{T_0}$$

$$= \frac{\rho^*}{\rho_0} \sqrt{g_c kR\left(\frac{T^*}{T_0}\right)}\, A^* \frac{P_0}{RT_0} \sqrt{T_0}$$

$$= \left(\frac{\rho^*}{\rho_0} \sqrt{\frac{g_c k}{R} \frac{T^*}{T_0}}\right) A^* \frac{P_0}{\sqrt{T_0}} \tag{13·32}$$

† Note that c^* is denoted by V^* in many tables. Also, $\mathsf{M}^* \equiv \mathsf{V}/\mathsf{V}^* = \mathsf{V}/\mathsf{c}^*$.

The term in parentheses is a constant, and a dimensional number for a given gas. For air, one has

$$\frac{\rho^*}{\rho_0} = 0.6339 \qquad k = 1.4$$

$$\frac{T^*}{T_0} = 0.8333 \qquad R = 286.9 \text{ J/(kg·K)}$$

Thus, in the SI,

$$\dot{M} = 0.0404 \frac{A^* P_0}{\sqrt{T_0}} \quad \textit{for air} \tag{13·33a}$$

where $\dot{M}$ is in kg/s, A^* is in m^2, P_0 is in N/m^2, and T_0 is in K.

In the old English system,

▶ $$\dot{M} = 0.532 A^* \frac{P_0}{\sqrt{T_0}} \quad \textit{for air} \tag{13·33b}$$

where $\dot{M}$ is in lbm/s, A^* is in ft^2, P_0 is in lbf/ft^2, and T_0 is in R. This is called Fliegner's formula.

We remark that the sonic state, like the stagnation state, is a *reference* state which may or may not occur in the flow. Hence the concept of A^* is useful even in purely subsonic compressible flows.

13·7 CHOKING IN ISENTROPIC FLOW

Suppose we have a converging duct fed (Fig. 13·10) at P_0, T_0, and discharging into a box in which the back pressure on the nozzle P_B can be varied. Imagine starting with $P_B = P_0$ (curve 1). Then, if we drop the back pressure a little (curve 2), there will be a little flow through the nozzle. The nozzle exit pressure P_E will be equal to P_B as long as no shock waves form in the flow. As we shall see in the next section, shock waves form only in supersonic flows, so long as the flow is subsonic at the exit $P_B = P_E$. As we drop P_B, the exit Mach number M_E will increase but, because of the duct geometry, $\mathsf{M}_E = 1$ forms an upper limit. The sonic condition will be reached at E (curve 3) when $P_B/P_0 = P^*/P_0$, which for $k = 1.4$ occurs when $P_B/P_0 = 0.528$. At this point the flow rate is given by Eq. (13·33). *This is the maximum flow that can be passed by this nozzle for the given stagnation condition.* The duct is said to be *choked* when $\mathsf{M}_E = 1$, and the value of P_B at this point is called the *critical back pressure.* If we drop P_B further, the news that P_B has decreased cannot propagate upstream against the sonic flow, hence the same maximum flow will be maintained for all $P_B < P^*$. For $P_B > P^*$ we can "yell" upstream, requesting more flow, and our call will be heard. But for $P_B < P^*$ our

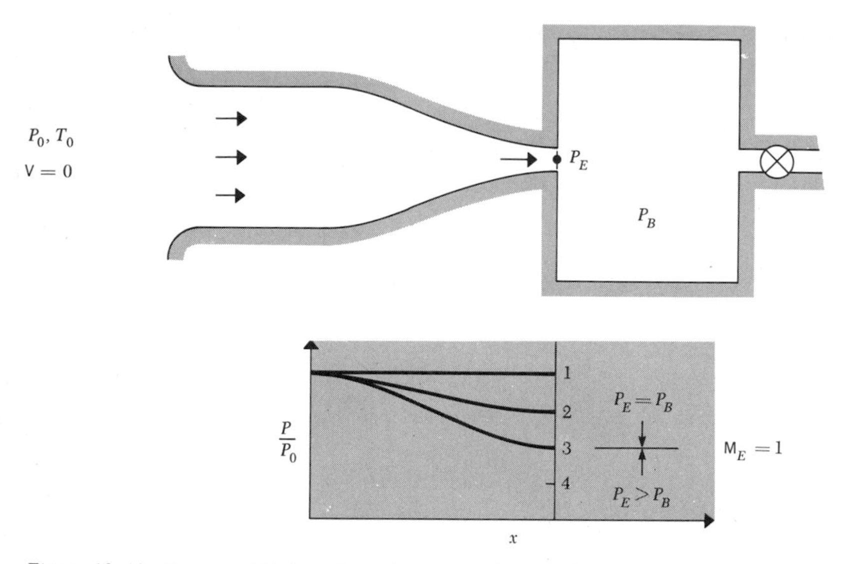

Figure 13·10. Compressible flow through a converging nozzle

shouts are swept downstream faster than they can propagate upstream, and our request goes unanswered. Figure 13·11 shows the behavior of $\dot{M}$ and P_E/P_0 with P_B/P_0.

We can of course increase the flow rate through the nozzle by altering the upstream stagnation conditions. An increase in stagnation pressure will increase the flow rate, but at the higher flow rate, if the flow remains choked, the Mach number will still be 1 at the throat.

Choked flow (or critical flow) nozzles are often used as flow-rate measuring devices. Since the flow rate is independent of back pressure under the choked conditions, measurement of P_0, T_0, and A^* suffice to fix $\dot{M}$. Choked nozzles also provide a means for maintaining a steady flow into a system, such as a chemical reactor, independent of fluctuations in the reaction chamber pressure.

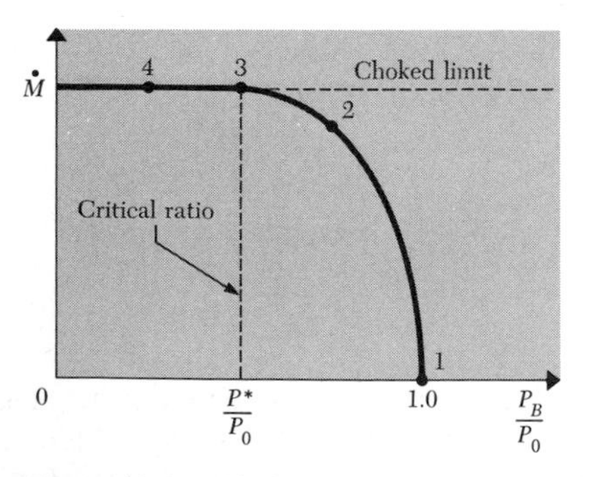

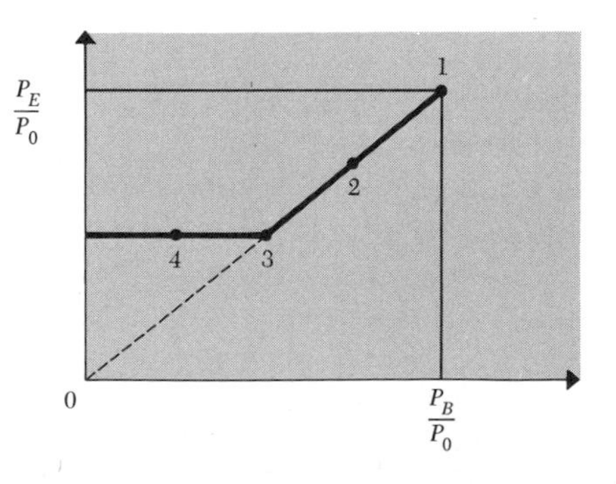

Figure 13·11. Choking in a converging nozzle

13·8 ISENTROPIC FLOW IN CONVERGENT-DIVERGENT PASSAGES

To obtain supersonic velocities it is necessary to have a *converging-diverging* nozzle. Figure 13·12 illustrates what happens in this case as the back pressure is lowered. Again, decreasing the back pressure from state 1 causes flow to start. The flow accelerates to the throat but, since $(P_T/P_0) > (P^*/P_0)$, the velocity is subsonic at the throat so that the diverging portion of the duct acts as a simple subsonic diffuser, and the pressure rises again (curve 2). At some back pressure the velocity in the throat reaches the sonic value, and $\mathsf{M} = 1$. The velocity can then either decelerate with a pressure rise to 3 or continue to accelerate. Point 4 represents a back pressure for which acceleration occurs *isentropically* such that the flow is supersonic at the nozzle exit and $P_{E4} = P_{B4}$. This condition of supersonic flow past the throat with isentropic conditions and the exit pressure equal to the back pressure is called the *design pressure ratio* of the nozzle. If the pressure is lowered to state 5, no further decrease in *exit* pressure occurs and the adjustment in pressure occurs outside the nozzle.

The conditions between 3 and 4 represent an exit pressure ratio P_E/P_0 which requires a shock to occur in the diverging section of the duct or an adjustment to occur outside the nozzle. A shock is a highly irreversible phenomenon and consequently represents a nonisentropic flow condition. We shall study shocks in the next section.

Example. Air with $k = 1.4$ flows through a convergent-divergent nozzle with a throat area of 0.0006 m². The stagnation conditions are $T_0 = 500$ K, $P_0 = 1.3 \times 10^6$ N/m². Let's first find the maximum possible flow rate.

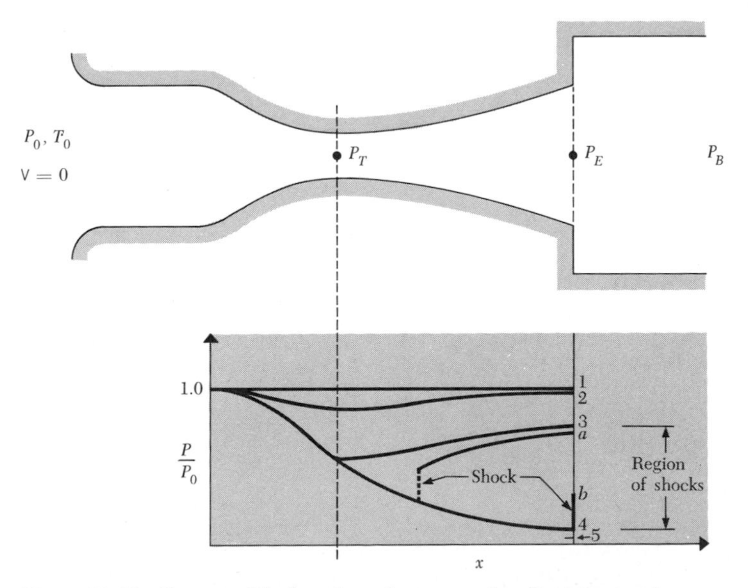

Figure 13·12. Compressible flow through a converging-diverging nozzle

We know that the maximum flow occurs when $\mathsf{M} = 1$ at the throat. Under these conditions Fliegner's formula for air [Eq. (13·33)] gives

$$\dot{M} = \frac{0.0404 \times 0.0006 \times 1.3 \times 10^6}{\sqrt{500}} = 1.409 \text{ kg/s}$$

The throat conditions are then

$$P^* = 0.528P_0 = 0.686 \times 10^6 \text{ N/m}^2$$

$$T^* = 0.833T_0 = 416 \text{ K}$$

Suppose now that the nozzle exit has an area of 0.0014 m²; let's determine the exit conditions assuming *isentropic* flow through the nozzle and $\mathsf{M} = 1$ at the throat. The question immediately arises as to whether the diverging section acts as a supersonic nozzle or a subsonic diffuser. We shall work out both cases. In Table B·18 we see that there are two conditions for $A/A^* = 0.0014/0.0006 = 2.33$. One corresponds to $\mathsf{M}_E = 2.37$, and one to $\mathsf{M}_E = 0.257$.

For $\mathsf{M} = 2.37$ we read values and then calculate the exit condition as follows:

$$\frac{P}{P_0} = 0.072 \qquad P_E = 9.36 \times 10^4 \text{ N/m}^2$$

$$\frac{T}{T_0} = 0.471 \qquad T_E = 235.5 \text{ K}$$

$$\frac{\rho}{\rho_0} = 0.152 \qquad \rho_E = 0.152\frac{P_0}{RT_0} = 1.377 \text{ kg/m}^3$$

$$\mathsf{M}^* \equiv \frac{\mathsf{V}}{\mathsf{V}^*} = 1.782$$

Now the exit velocity can be determined from either $\mathsf{V} = 1.782\mathsf{V}^*$, or from

$$\mathsf{V}_E = \mathsf{M}_E\, c_E = 2.37\sqrt{g_c kRT_E}$$

From the latter we find

$$\mathsf{V}_E = 728.9 \text{ m/s}$$

Let's check by determining $\dot{M}$ at the exit.

$$\dot{M} = \rho A\mathsf{V} = 1.377 \times 0.0014 \times 728.9 = 1.406 \text{ kg/s}$$

in good agreement with our value determined at the throat.

The subsonic exit condition is handled in the same manner. For $\mathsf{M} = 0.257$,

$$\frac{P}{P_0} = 0.955 \qquad P_E = 1.24 \times 10^6 \text{ N/m}^2$$

$$\frac{T}{T_0} = 0.987 \qquad T_E = 493.5 \text{ K}$$

$$\frac{\rho}{\rho_0} = 0.967 \qquad \rho_E = 8.763 \text{ kg/m}^3$$

$$\mathsf{V}_E = 0.257\sqrt{g_c kRT_E} = 114.4 \text{ m/s}$$

Which of these two solutions is correct? The answer is that both are correct. For isentropic flow with $\mathsf{M} = 1$ at the throat, the exit condition can be at either the calculated subsonic or supersonic condition. However, what would happen in an actual laboratory test would depend upon how the test was made. If the back pressure were gradually lowered from point 1 to point 3, the flow would be subsonic at the exit. Continued lowering of the pressure would result in a shock wave occurring in the divergent sections of the nozzle until the pressure was reduced to state 4, at which point the flow would be supersonic in the divergent section 1 (see Fig. 13·12).

The student probably knows that modern rocket motors use convergent-divergent nozzles, and may be surprised to learn that steam turbines also use such nozzles to produce high-velocity flow. Supersonic jet engine inlets use convergent-divergent passages in the opposite manner, taking an initially supersonic inlet flow down to a low subsonic speed suitable for compression and combustion (see the example ending the chapter).

13·9 SHOCK WAVES

We have seen that under certain conditions the flow in the divergent section of a supersonic nozzle appears to lie in between two isentropic paths, and in Fig. 13·12 we designated this region as one where shock waves occur.

Experiments reveal that shock waves are highly localized irreversibilities in the flow. In a layer of the order of a few molecular mean free paths thick the flow passes from a supersonic to a subsonic state, the velocity decreases markedly, and the pressure sharply rises. Enormous decelerations are imposed on a fluid particle passing through a shock, decelerations produced by very strong viscous effects within the shock layer.

Analysis of the details of the shock structure is a very complex problem in compressible viscous nonequilibrium flow. However, the overall changes that occur across the shock can be analyzed by control-volume methods if we place the boundaries just upstream and downstream of the shock. At the boundaries

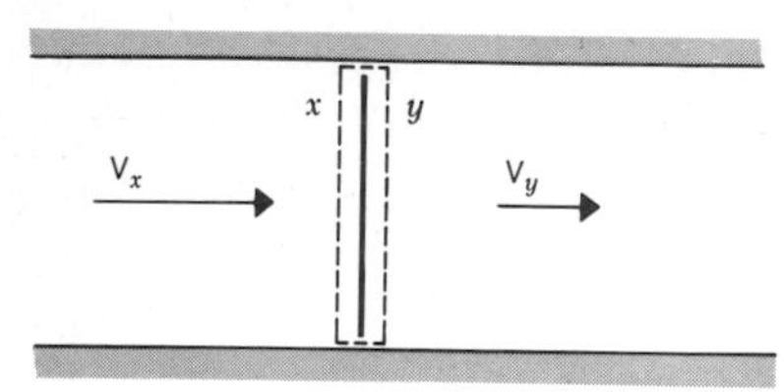

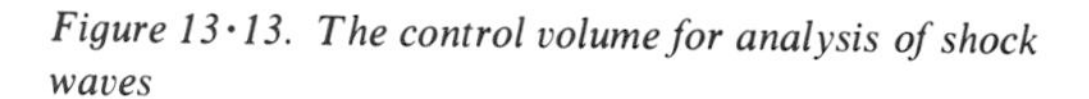
Figure 13·13. The control volume for analysis of shock waves

viscous effects can be neglected, and thermodynamic equilibrium assumed, which greatly simplifies the problem. We shall treat only *normal* shocks, that is, shock waves perpendicular to the flow.

Shocks provide an excellent example of an analysis that combines *all* the thermodynamic principles with the momentum principle. The control volume we use is shown in Fig. 13·13. Assuming steady flow and steady state and neglecting energy transfer as heat from the control volume, application of these basic principles leads to the following equations. Note that the subscripts x and y denote the upstream and downstream states, respectively.

Continuity

$$\frac{\dot{M}}{A} = \rho_x \mathsf{V}_x = \rho_y \mathsf{V}_y \tag{13·34}$$

Momentum

$$A(P_x - P_y) = \frac{\dot{M}}{g_c}(\mathsf{V}_y - \mathsf{V}_x) \tag{13·35}$$

Energy

$$h_x + \frac{\mathsf{V}_x^2}{2g_c} = h_y + \frac{\mathsf{V}_y^2}{2g_c} = h_0 \tag{13·36}$$

Second law

$$s_y - s_x \geq 0 \tag{13·37}$$

When combined with equations of state, Eqs. (13·34) to (13·36) will suffice to solve for the downstream conditions given the upstream conditions, and Eq. (13·37) will rule out solutions that are not possible.

Some insight can be obtained by studying two partial solutions of these equations. We introduce the *mass velocity* G,

$$G \equiv \rho \mathsf{V} \tag{13·38}$$

Note that G is the mass-flow rate per unit of area. The continuity equation tells us that G is constant in the flow. Combining the continuity and energy equations, and dropping the x and y subscripts,

$$h_0 = h + \frac{G^2}{2g_c\rho^2} = \text{constant} \tag{13·39}$$

Given the values of G and h_0, this equation relates h and ρ, hence will trace out a line on any thermodynamic plane. This line, the locus of points with the same mass velocity and stagnation enthalpy, is called a *Fanno line*. The end states of our shock wave must lie somewhere on this line.

The Fanno line is of some interest in another connection. Adiabatic flow in a constant-area duct with friction, when modeled in a one-dimensional manner, has both constant G and constant h_0, hence must follow a Fanno line. We shall discuss this process later in this chapter.

The second partial solution is obtained from a combination of the continuity and momentum equations. Dropping the subscripts, the momentum equation [Eq. (13·35)] may be written in terms of the *impulse pressure* I as

$$I \equiv P + \frac{\rho V^2}{g_c} = \text{constant}$$

Combining with continuity,

$$I = P + \frac{G^2}{\rho g_c} \tag{13·40}$$

Given the values for I and G, this equation relates P and ρ, hence will trace out another line on any thermodynamic plane. This line is the locus of states with the same impulse pressure and mass velocity and is called a *Rayleigh line*. The end states of our shock must lie somewhere on this line, since $I_x = I_y$ and $G_x = G_y$.

The Rayleigh line is also a model for flow in a constant-area duct with heat transfer, but *without* friction. While this is not a very realistic situation, certain qualitative features of the model process are relevant to real compressible flow heat-transfer processes, and we shall discuss them later in this chapter.

Since the end state of the shock must lie on both the Fanno and Rayleigh lines, the end states must be fixed by their intersection, as shown in Fig. 13·14. Only the intersection state of lower entropy can be the upstream state x, hence the shock process must pass from the lower to the upper intersection in Fig. 13·14. This graphical shock solution method is particularly useful for shocks in real gases, where the absence of simple algebraic equations of state precludes analytic solution of the shock equations.

If we use the perfect gas with constant c_P as a fluid model, closed-form

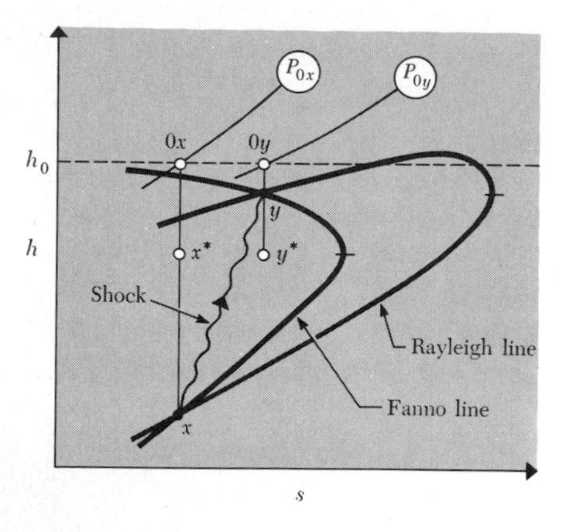

Figure 13·14. The shock process on the h-s plot

algebraic solutions can be obtained for the shock equations. For such a gas the impulse pressure is

$$P + \frac{\rho V^2}{g_c} = P + \frac{\rho M^2 c^2}{g_c} = P + \frac{\rho(g_c kRT)M^2}{g_c}$$

$$= P(1 + kM^2) \qquad (13 \cdot 41)$$

Since the impulse pressure has the same value on both sides of the shock waves,

▶ $$\frac{P_y}{P_x} = \frac{1 + kM_x^2}{1 + kM_y^2} \qquad (13 \cdot 42)$$

Since the stagnation enthalpy is constant, for the perfect gas the isentropic stagnation temperatures are identical on both sides of the shock,

$$T_{0x} = T_{0y} \qquad (13 \cdot 43)$$

hence, from Eq. (13·21)

▶ $$\frac{T_y}{T_x} = \frac{1 + [(k-1)M_x^2/2]}{1 + [(k-1)M_y^2/2]} \qquad (13 \cdot 44)$$

The continuity equation gives

▶ $$\frac{\rho_y}{\rho_x} = \frac{M_x c_x}{M_y c_y} = \sqrt{\frac{T_x}{T_y}} \frac{M_x}{M_y} \qquad (13 \cdot 45)$$

The perfect-gas equation may be put in a similar form,

$$\frac{P_x}{P_y} = \frac{\rho_x}{\rho_y} \frac{T_x}{T_y} \qquad (13 \cdot 46)$$

Equations (13·42) to (13·46) may be combined to obtain one equation relating M_x and M_y. One finds

$$\blacktriangleright \qquad \mathsf{M}_y^2 = \frac{\mathsf{M}_x^2 + [2/(k-1)]}{[2k\mathsf{M}_x^2/(k-1)] - 1} \qquad (13\cdot47)$$

Thus for a given M_x we can now solve directly for M_y. Table B·19 gives the results of this computation for several values of M_x. Once these are determined, Eq. (13·42) can be used to determine P_y/P_x as listed in the table. Equation (13·44) can then be used to determine T_y/T_x. Equation (13·45) is then used to fix the conditions on ρ_y/ρ_x and $\mathsf{V}_y/\mathsf{V}_x$. The last column listed in the table gives P_{0y}/P_{0x}, and the irreversibility of the shock shows up as a reduction in isentropic stagnation pressure (see Fig. 13·13). By suitable manipulations, one may show that

$$\blacktriangleright \qquad \frac{P_{0y}}{P_{0x}} = \left[\frac{(k+1)\mathsf{M}_x^2/2}{1+(k-1)\mathsf{M}_x^2/2}\right]^{k/(k-1)} \Bigg/ \left[\frac{2k}{k+1}\mathsf{M}_x^2 - \frac{k-1}{k+1}\right]^{1/(k-1)} \qquad (13\cdot48)$$

Calculations show that, if $\mathsf{M}_x > 1$, $\mathsf{M}_y < 1$, and the flow abruptly slows down as it passes through the shock. Much of the kinetic energy upstream is suddenly scrambled and converted to molecular (internal) energy by the shock, hence the flow leaves the shock slower and hotter than it entered. The second-law equation says that the reverse process, that is, a sudden *unscrambling* of the internal energy into organized kinetic energy, is impossible.

A bullet or an airplane moving at supersonic speed carries a *bow shock* (Fig. 13·15) with it. The shock suddenly announces the coming of the object and quickly moves the motionless air out of the way. The bow shock is curved, but at the nose of the object is perpendicular to the flow, and the theory of normal shocks developed here applies. The sudden increase in temperature as the flow passes through the shock makes it very hot near the nose, and reentry bodies must have adequate protection against this high-temperature gas. The bow shock is weak far from the body, hence approaches a Mach cone. Some shock pressure rise remains even at great distances from the body, and it is this pressure jump that is known as *sonic boom.*

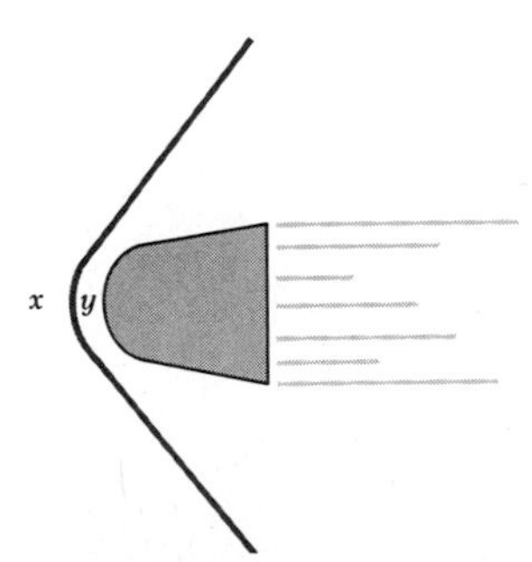

Figure 13·15. A bow shock

Example. To illustrate the use of normal shock tables and equations, suppose a stream of air with a Mach number of 3 passes through a normal shock. The upstream pressure and temperature are 150,000 N/m^2 and 0°C. Let's calculate the downstream conditions.

From Table B·19, for $\mathsf{M}_x = 3.0$,

$$\mathsf{M}_y = 0.475$$

$$\frac{P_y}{P_x} = 10.333$$

$$\frac{T_y}{T_x} = 2.697$$

$$\frac{\rho_y}{\rho_x} = \frac{\mathsf{V}_x}{\mathsf{V}_y} = 3.857$$

$$\frac{P_{0y}}{P_{0x}} = 0.328$$

Thus

$$P_y = 10.333 \times 150{,}000 = 1.549 \times 10^6 \text{ N/m}^2 = 15.29 \text{ atm}$$

$$T_y = 2.679 \times 273 = 731 \text{ K}$$

The magnitude of the pressure and temperature jumps should be impressive. What about the velocity change?

$$\mathsf{M}_x = \frac{\mathsf{V}_x}{\mathsf{c}_x} = 3.0 \qquad \mathsf{c}_x = \sqrt{g_c kRT} = 331 \text{ m/s}$$

$$\mathsf{V}_x = 993 \text{ m/s} \qquad \mathsf{V}_y = \frac{993}{3.857} = 257 \text{ m/s}$$

To calculate the isentropic stagnation pressures we first use Table B·18, at $\mathsf{M} = 3$,

$$\frac{P}{P_0} = 0.02722 = \frac{P_x}{P_{0x}}$$

so

$$P_{0x} = \frac{150{,}000}{0.02722} = 5.51 \times 10^6 \text{ N/m}^2 = 54.3 \text{ atm}$$

and

$$P_{0y} = 5.51 \times 10^6 \times 0.328 = 1.81 \times 10^6 \text{ N/m}^2 = 17.8 \text{ atm}$$

Let's verify that the entropy change across the shock is positive. Using Eq. (8·27),

$$s_y - s_x = c_P \ln\left(\frac{T_y}{T_x}\right) - R \ln\left(\frac{P_y}{P_x}\right)$$

$$= 1004.8 \ln (2.679) - 286.9 \ln (10.33)$$

$$= 320 \text{ J/(kg·K)}$$

A very useful result for normal shocks follows from Fliegner's formula [Eq. (13·33)]. Since both $\dot{M}$ and T_0 are the same on both sides of a normal shock, the product A^*P_0 is also preserved. Hence

$$\frac{A_y^*}{A_x^*} = \frac{P_{0x}}{P_{0y}} \tag{13·49}$$

13·10 COMPRESSIBILITY EFFECTS IN FLOWS WITH FRICTION AND HEAT TRANSFER

We mentioned in the last section that the Fanno line equation also holds for a one-dimensional model of adiabatic flow in a constant-area duct with friction. While the quantitative predictions of this simple model are not very accurate, the qualitative effects are correct, and we shall discuss these now.

For adiabatic flow the entropy must increase in the flow direction. Hence a Fanno process must follow its Fanno line to the right (Fig. 13·16). We know from our normal shock calculation that the upper branch of the Fanno curve corresponds to *subsonic* flow. The nose corresponds to the sonic point. Hence since friction will tend to move the fluid state to the right on the Fanno line, the Mach number of subsonic flows *increases* in the downstream direction, while in supersonic flows friction acts to decrease the Mach number. *Friction therefore tends to drive the flow toward the sonic point* (Fig. 13·16).

In the previous chapters we analyzed adiabatic friction flows by neglecting the flow kinetic energy and found that the flow enthalpy remained unchanged. Now we see that this is only a valid model at low Mach numbers, where the Fanno line approaches a line of constant h (Fig. 13·16).

Suppose we have a short duct with a given h_0 and G, that is, a given Fanno line, with a given subsonic exit Mach number. If we add more length to the duct, the new exit Mach number will be increased, but the exit state will lie on the same Fanno line. If we lengthen the duct sufficiently to make the exit state sonic, there is nowhere else to go on the Fanno line, and a further increase in length is impossible without a reduction in the mass-flow rate. Hence subsonic flows can become *choked* by friction. There is a *maximum flow rate* that can be passed by a pipe with given stagnation conditions. This has important consequences in engineering

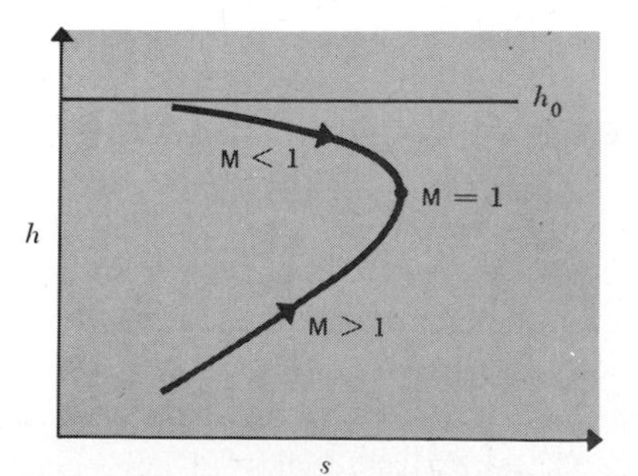

Figure 13·16. A Fanno line on the h-s plot

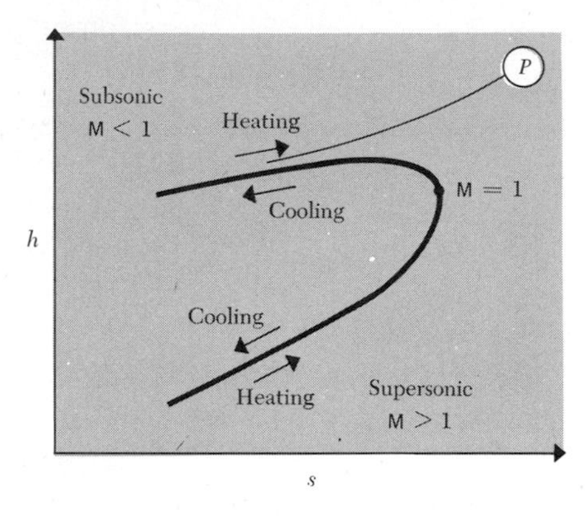

Figure 13·17. A Rayleigh line on the h-s plot

practice. Choking also occurs in supersonic flow with friction, usually in a very short length. It is this that makes it very difficult to use such flows in applications.

Flows with heating or cooling in a constant-area duct, *neglecting friction*, may be treated in a one-dimensional model by the *Rayleigh process*. This model is qualitatively correct but quantitatively accurate only at very low Mach numbers. The Rayleigh process diagram is shown in Fig. 13·17. Since this idealized process is reversible, the direction of entropy change is determined by the sign of the heat input, as shown in Fig. 13·17. We have also identified the subsonic and supersonic branches and the sonic point.

Heating a compressible flow has the same effect as friction, namely, the Mach number is driven toward unity. Hence, for a given flow rate, there is a *maximum heat input* for which the prescribed flow can be passed by the duct. Compressible flows therefore exhibit *choking due to heating*, which has very important applications, particularly in the design of very high-temperature gas heat exchangers or gas-cooled nuclear reactors.

In earlier chapters we often modeled steady-flow heating or cooling by a constant-pressure process. As Fig. 13·17 suggests, this is valid in frictionless flow at low Mach numbers, where the changes in the kinetic energy of the flow are negligible.

It is not obvious from the curves, but *cooling* increases the flow stagnation pressure. Hence, in principle, a nonmechanical pump can be made by *cooling* a compressible flow. This wonderful idea is defeated by friction, which rears its ugly head in all real devices and outweighs the pumping effects of cooling.

13·11 EXAMPLES OF COMPRESSIBLE-FLOW CALCULATIONS

Compressible-flow nozzle. A test on a compressible-flow nozzle requires using a gas with a k of 1.5. Further, the gas must be nonoxidizing and safe to use. Let's specify a gas mixture to meet these criteria and then determine the pressure, temperature, and density at the critical conditions, i.e., where $\mathsf{M} = 1$, if $P_0 = 0.5$ MPa and $T_0 = 450$ K. We are also asked to determine the required throat area to pass 1 kg/s at these conditions. Since we want a nonoxidizing atmosphere which is also safe, a nitrogen-helium mixture can be specified. The k for nitrogen is 1.4 and that for helium is 1.66. Table 13·3 summarizes the required calculations. Initially we choose to use as a trial mixture 2 moles of nitrogen and 1 mole of helium. The required mixture is $\chi_{N_2} = 0.49$ and $\chi_{He} = 0.51$ and yields a k of 1.502. From Eqs. (13·21) and (13·22) we can determine the critical conditions. We have for $\mathsf{M} = 1$,

$$\frac{T_0}{T^*} = \left(\frac{1+k}{2}\right) = 1.25 \qquad \text{so} \qquad T^* = 360 \text{ K}$$

$$\frac{P_0}{P^*} = \left(\frac{1+k}{2}\right)^{k/(k-1)} = 1.953 \qquad \text{so} \qquad P^* = 0.256 \text{ MPa}$$

$$\frac{\rho_0}{\rho^*} = \left(\frac{1+k}{2}\right)^{1/(k-1)} = 1.562 \qquad \text{and} \qquad \rho_0 = \frac{P_0}{RT_0}$$

so

$$\rho^* = \frac{0.5 \times 10^6}{R \times 450 \times 1.562} = 1.348 \text{ kg/m}^3$$

Here $R = \mathscr{R}/\hat{M}$ and $\hat{M} = \sum \chi_i \hat{M}_i$. For this mixture,

$$\hat{M} = 13.72 + 2.04 = 15.76 \text{ kg/kgmole}$$

To determine the area we can use the relationship for choked flow [Eq. (13·32)],

$$A^* = \frac{\dot{M}\sqrt{T_0}}{P_0\left(\dfrac{\rho^*}{\rho_0}\sqrt{\dfrac{g_c k T^*}{R T_0}}\right)}$$

$$A^* = \frac{(1)\sqrt{450}}{(0.5 \times 10^6)\left(\dfrac{1}{1.562} \times \sqrt{\dfrac{1 \times 1.5}{527 \times 1.25}}\right)}$$

$$A^* = 0.00139 \text{ m}^2$$

If it is assumed that we have an axisymmetric nozzle, the throat diameter is then 4.2 cm.

Table 13·3 Calculations for required gas mixture

Moles N_2	Moles He	χ_{N_2}	χ_{He}	$\hat{c}_{P,\text{mix}}$	$\hat{c}_{v,\text{mix}}$	k_{mix}
2	1	0.667	0.333	26.35	18.05	1.478
2	2	0.50	0.50	24.95	16.69	1.494
2	2.08	0.49	0.51	24.86	16.608	1.502

Jet engine analysis. Suppose a jet engine operates in level flight at a fixed altitude and is in a steady-state, steady-flow condition. We wish to determine the fuel-flow rate to the engine (*a*) in a frame of reference on the engine (i.e., from the point of view of an observer on the engine) and (*b*) in a frame of reference on the earth. Will the fuel-flow rates be equal? Should they be?

Let us start our analysis by working with the observer on the ground. Figure 13·18*a* notes the system we shall designate as the prime system. We shall first do the momentum analysis.

Assume one-dimensional flow at 1 and 2
Assume uniform pressure around the control volume

The net force in the x direction is, using the above assumptions,

$$\mathsf{F}_x = \mathsf{F}_E$$

The rate of production of momentum, as seen by the ground observer, is

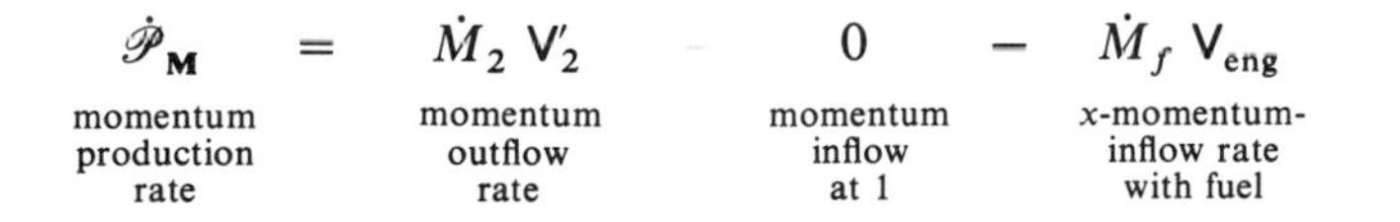

Applying the momentum principle

$$g_c \mathsf{F}_E = \dot{M}_2 \mathsf{V}'_2 - \dot{M}_f \mathsf{V}_{\text{eng}}$$

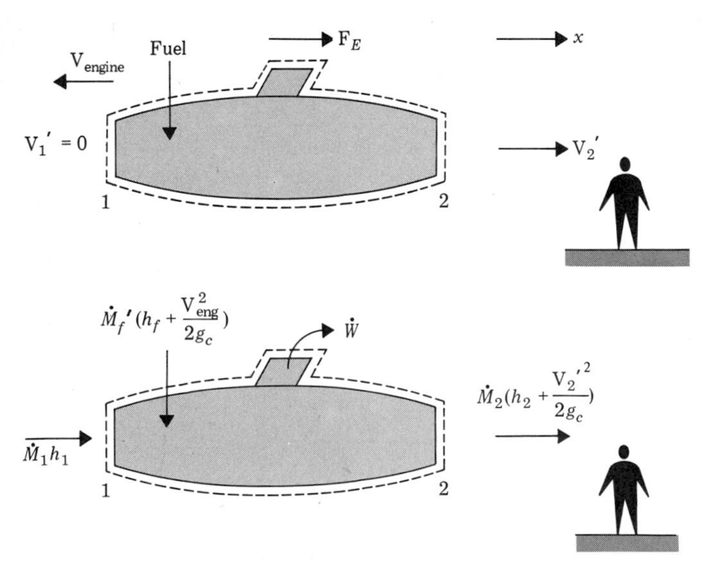

Figure 13·18(a). The control volume for the momentum analysis in the prime system. Observer on the ground; (b) the control volume for the energy analysis. Observer on the ground

We shall use this result to determine the work term in the energy balance. The energy balance terms are noted in Fig. 13·18*b*. We shall assume

Adiabatic control volume
Negligible kinetic energy associated with the fuel injection, i.e., $V_f \ll V_{eng}$

The observer on the ground sees no inlet velocity as the air enters the engine, hence no inlet kinetic energy. However, the observer sees the engine overcoming the force F_E as the engine (and vehicle) move through the air. The energy balance must include the power output of the engine. Following Fig. 13·18*b*,

$$\dot{M}_1 h_1 + \dot{M}'_f\left(h_f + \frac{V_{eng}^2}{2g_c}\right) = \dot{W} + \dot{M}_2\left(h_2 + \frac{V_2'^2}{2g_c}\right)$$

Now the power term is given by

$$\dot{W} = F_E \times V_{eng} = \frac{1}{g_c}(\dot{M}_2 V'_2 V_{eng} + \dot{M}'_f V_{eng} V_{eng})$$

Further, we know from continuity that

$$\dot{M}_2 = \dot{M}_1 + \dot{M}'_f$$

Substituting these relations we find

$$\dot{M}_1 h_1 + \dot{M}'_f\left(h_f + \frac{V_{eng}^2}{2g_c}\right) = (\dot{M}_1 + \dot{M}'_f)\frac{V'_2 V_{eng}}{g_c} + \frac{\dot{M}'_f V_{eng} V_{eng}}{g_c} + (\dot{M}_1 + \dot{M}'_f)\left(h_2 + \frac{V_2'^2}{2g_c}\right)$$

Solving for the ratio $\dot{M}'_f/\dot{M}_1$,

$$\frac{\dot{M}'_f}{\dot{M}_1} = \frac{(h_1 - h_2) - V'_2 V_{eng}/g_c - V_2'^2/(2g_c)}{(h_2 - h_f) + V'_2 V_{eng}/g_c + V_2'^2/(2g_c) + V_{eng}^2/(2g_c)}$$

We can now turn to the analysis as seen by the observer on the engine (Fig. 13·19). In this reference frame no work is done by the engine, since there is no distance through which the force acts. Thus we do not need to evaluate the force for use in the energy analysis. However, the student should do this and show that the result given above is again obtained.

Figure 13·19*b* notes the terms for the energy balance. We have

$$\dot{M}_1\left(h_1 + \frac{V_1^2}{2g_c}\right) + \dot{M}_f h_f = \dot{M}_2[h_2 + V_2^2/(2g_c)]$$

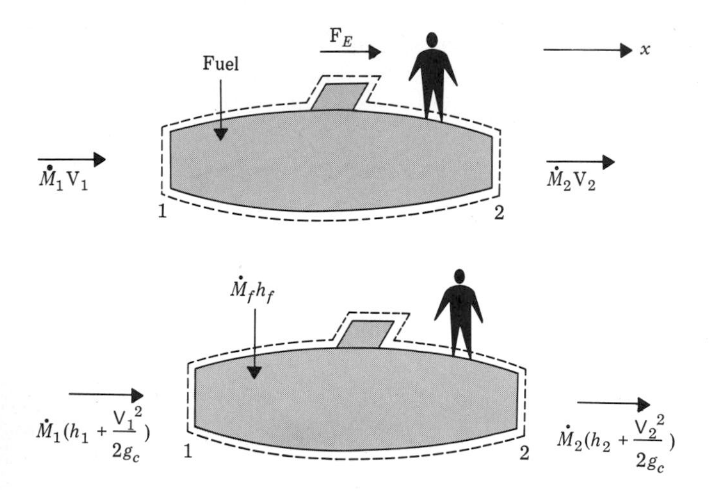

Figure 13·19(a). The control volume for the momentum analysis. Observer on the engine; (b) the control volume for the energy analysis. Observer on the engine

Note that the observer on the engine sees no inlet kinetic energy with the fuel, since in this frame of reference the fuel has no bulk motion and we are neglecting, as before, the kinetic energy associated with the velocity of the fuel. Again

$$\dot{M}_2 = \dot{M}_1 + \dot{M}_f$$

So

$$\dot{M}_1\left(h_1 + \frac{V_1^2}{2g_c}\right) + \dot{M}_f h_f = (\dot{M}_1 + \dot{M}_f)\left(h_2 + \frac{V_2^2}{2g_c}\right)$$

Solving for the ratio of the fuel-flow rate to the inlet flow rate, we have

$$\frac{\dot{M}_f}{\dot{M}_1} = \frac{(h_1 - h_2) + [V_1^2/(2g_c) - V_2^2/(2g_c)]}{(h_2 - h_f) + V_2^2/(2g_c)}$$

Now we must relate the velocities in the two reference frames so that we can compare our results. Figure 13·20 indicates the relationships. We have

$$V_2' = V_2 - V_{\text{eng}} \qquad V_1 = V_{\text{eng}}$$

Substituting these into the $\dot{M}_f/\dot{M}_1$ equation directly above we find

$$\frac{\dot{M}_f}{\dot{M}_1} = \frac{(h_1 - h_2) + V_{\text{eng}}^2/(2g_c) - V_2'^2/(2g_c) - 2V_2' V_{\text{eng}}/(2g_c) - V_{\text{eng}}^2/(2g_c)}{(h_2 - h_f) + V_{\text{eng}}^2/(2g_c) + V_{\text{eng}} V_2'/g_c + V_2'^2/(2g_c)}$$

Clearing terms we have

$$\frac{\dot{M}_f}{\dot{M}_1} = \frac{(h_1 - h_2) - V_2'^2/(2g_c) - V_2' V_{\text{eng}}/g_c}{(h_2 - h_f) + V_{\text{eng}}^2/(2g_c) + V_{\text{eng}} V_2'/g_c + V_2'/(2g_c)}$$

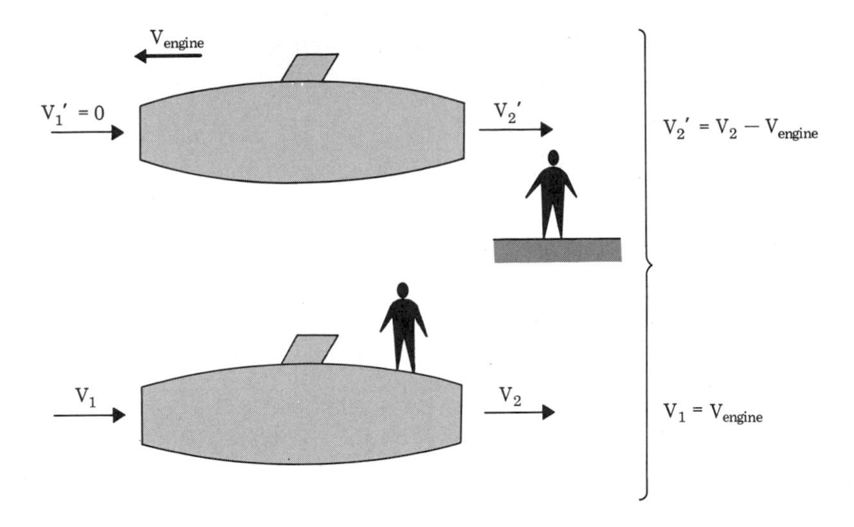

Figure 13·20. Relating the velocities

Comparing with the expression for $\dot{M}'_f/\dot{M}_1$ we see that the right-hand sides are identical. Thus, as anticipated, the fuel-flow rate is independent of the frame of reference used for the analysis. Note how this analysis required the use of both the energy and momentum equations, and how carefully one must identify the terms used in the equations.

Convergent-divergent duct flow. We can now see what happens in the convergent-divergent nozzle analyzed earlier in Sec. 13·8 (see Fig. 13·12). When the back pressure lies in between states 3 and 4, it is impossible for the fluid to follow an *isentropic path* down the nozzle, and it must undergo a shock in the divergent portion of the nozzle. As the back pressure is lowered, the shock moves down the nozzle until it stands just in the exit plane at condition *b*. As the pressure is lowered between *b* and 4, the pressure in the exit plane is lower than the back pressure. A compression outside the nozzle must occur. If the back pressure is lowered below state 4, there is an expansion outside the nozzle. Note that a lowering of the pressure below 3 does not increase the flow rate, since the conditions at the throat are already sonic.

Let's determine the exit conditions for which $\mathsf{M} = 1$ at the throat and the flow in the divergent section is subsonic throughout. Recall the design parameters (Sec. 13·5):

$$P_0 = 80 \text{ psia}$$

$$T_0 = 960 \text{ R}$$

$$A_{\text{throat}} = 0.505 \text{ ft}^2$$

$$A_{\text{exit}} = 0.70 \text{ ft}^2$$

Thus

$$\frac{A_E}{A^*} = \frac{0.70}{0.505} = 1.386$$

Corresponding to this value in Table B·18, we find a choice,

$$\mathsf{M} = 0.47 \qquad \text{or} \qquad \mathsf{M} = 1.75$$

We want the subsonic case, so for $\mathsf{M} = 0.47$,

$$\frac{P}{P_0} = 0.856 \qquad P_E = 0.856 \times 80 = 68.48 \text{ psia}$$

$$\frac{T}{T_0} = 0.956 \qquad T_E = 0.956 \times 960 = 918 \text{ R}$$

Now, let's determine the exit condition for which $\mathsf{M} = 1$ at the throat and $P_E = 60$ psia. This condition has a P_E less than the all-subsonic case of 68.48 psia, but greater than the design condition for all-supersonic flow with $P_E = 15$ psia. We therefore expect a shock in the divergent section of the nozzle. The *h-s* process representation for this case is shown in Fig. 13·21.

We have to find a solution technique which will lead to the given condition that $P_E = 60$ psia. We thus need to find a condition for M_x for which the rise in pressure across the shock to P_y plus the subsequent pressure rise in the subsonic diffuser section leads to $P_E = 60.0$ psia. Let us guess M_x values and find the solution by trial and error. Assume

$$\mathsf{M}_x = 1.50$$

Then, from Table B·19,

$$\mathsf{M}_y = 0.701$$

$$\frac{P_{0y}}{P_{0x}} = \frac{A_x^*}{A_y^*} = 0.930 \qquad \frac{P_y}{P_x} = 2.458$$

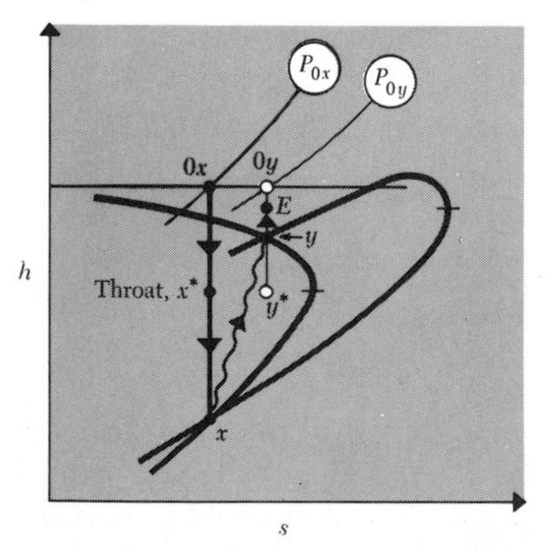

Figure 13·21. The process

At M = 1.5, from Table B·18,

$$\frac{P_x}{P_{0x}} = 0.272$$

So
$$P_x = 0.272 \times 80 = 21.8 \text{ psia}$$
$$P_y = 2.458 \times 21.8 = 53.58 \text{ psia}$$

The reference area A_y^* does not occur in the flow, but is nonetheless useful. Its value is

$$A_y^* = \frac{0.505}{0.930} = 0.543 \text{ ft}^2$$

Also
$$P_{0y} = 0.930 P_{0x} = 0.930 \times 80 = 74.4 \text{ psia}$$

Now, since the flow is isentropic from y to the exit,

$$A_y^* = A_E^* \qquad \text{and} \qquad P_{0y} = P_{0E}$$

Then
$$\left(\frac{A}{A^*}\right)_E = \frac{0.70}{0.543} = 1.29$$

This will establish the exit state in the isentropic flow table. Table B·18 gives, for $A/A^* = 1.29$ (subsonic branch),

$$\mathsf{M} = 0.527$$

$$\frac{P}{P_0} = 0.827$$

So,
$$P_E = 0.827 \times 74.4 = 61.5 \text{ psia}$$

But we wanted to come out with $P_E = 60$ psia. A repeat with $\mathsf{M}_x = 1.54$ does the job.

Finally, let's determine the back pressure for which M = 1 at the throat and the shock wave stands just in the exit. From Table B·19, for

$$\mathsf{M}_x = \mathsf{M}_E = 1.75 \qquad \frac{P_y}{P_x} = 3.406$$

We know that $P_E = P_x = 15$ psia in this case, hence

$$P_B = P_y = 3.406 \times 15 = 51 \text{ psia}$$

We have thus determined a complete range of operation for the nozzle. For $80 > P_b > 68.48$ the flow is subsonic in the nozzle; for $68.48 > P_b > 51$ psia a shock stands in the divergent section of the nozzle; for $51 > P_b > 15$ the adjustment from the exit pressure to the back pressure occurs as a compression outside the nozzle; for $P_b = 15$ psia the nozzle is at the design condition; for $P_b < 15$ psia the adjustment from the exit pressure to the back pressure occurs as an expansion outside the nozzle. The expansion and compression processes outside the nozzle are complicated and involve a diamond pattern of shock waves and expansions. This sort of pattern is observed in the engine discharge during launch of a rocket, which passes through all these regimes during startup and passage up through the atmosphere.

Steam nozzle design. Suppose we want to accelerate 2 kg/s steam flow from stagnation conditions at 10^6 N/m², 350°C, to a Mach number of 1.0. We must first decide how to model the fluid, that is, how much of the perfect-gas model can we employ with steam. The value of k is listed in Table B·6 as 1.33, and this value is all right for low-pressure steam [$P < 1.4 \times 10^6$ N/m² (200 psia)]. If we treat steam as a perfect gas,

$$\frac{P^*}{P_0} = \left(\frac{2}{k+1}\right)^{k/(k-1)} = 0.540$$

Note that this expression is not too sensitive to the value of k used. Let's presume that we want to expand down to

$$P_E = 0.54P_0 = 0.54 \times 10^6 \text{ N/m}^2$$

However, Fig. B·2 shows that $h = h(T, P)$ in the exit region of this nozzle, hence the real-gas effects on the velocity and density are possibly significant. We shall therefore use the real-gas enthalpy and density data in the calculation. The local velocity is calculated from the energy equation,

$$V = \sqrt{2g_c(h_0 - h)}$$

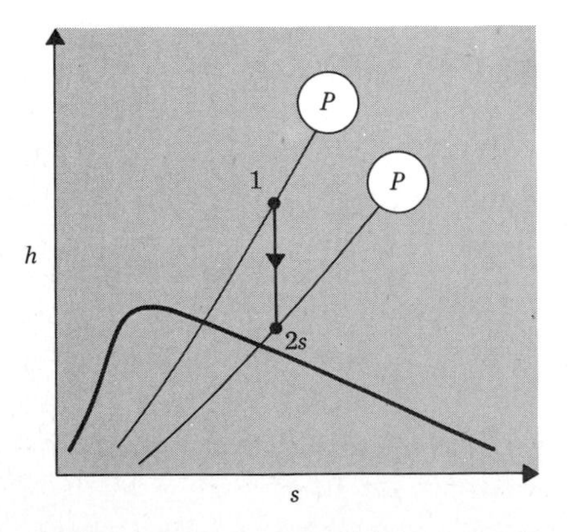

Figure 13·22. Process representation for isentropic flow of steam in a nozzle

Table 13·4 Area calculation

P N/m^2	h kJ/kg	V m/s	v m^3/kg	A m^2
10^6	3157.7	0	0.282	—
0.8×10^6	3100.0	339.7	0.336	0.00198
0.6×10^6	3022.0	520.9	0.416	0.00160
0.54×10^6	2995.0	570.4	0.447	0.00157

The calculation is made by following a line of constant s on the Mollier diagram (see Fig. 13·22) or by interpolating along a line of constant s in the tables. Continuity is used to calculate the local flow area. Mach 1 is reached when the area starts to increase again. The calculation is summarized in Table 13·4. Note that the exit density ratio is

$$\frac{\rho_E}{\rho_0} = \frac{0.282}{0.447} = 0.631$$

For a perfect gas with $k = 1.33$, $\rho_E/\rho_0 = 0.632$ at $\mathsf{M} = 1$. The agreement is obviously very good.

Nuclear ramjet design. Consider a nuclear ramjet engine as pictured in Fig. 13·23*a*. The engine is to be designed to fly at Mach 3 through air at $-60°$F and 2 psia. The reactor will heat the air to 1600°F and the pressure drop through the reactor is 5 psia. The flow velocity in the reactor is low, so states 4 and 5 are approximated as stagnation states. The areas are:

$$A_1 = 10 \text{ ft}^2$$

$$A_2 = A_3 = 2.8 \text{ ft}^2$$

$$A_7 = 15 \text{ ft}^2$$

A_6 is adjustable. This problem will entail a preliminary design of the ramjet engine.

We shall make various other modeling assumptions as we do the analysis. Since the inlet flow is supersonic, and we want to slow the flow to a low subsonic speed where we can heat it without chances of choking, we must somehow make the flow subsonic. While it is theoretically possible to design a convergent-divergent diffuser passage for steady flow, such passages of fixed geometry cannot be *started* because the bow shock must be swallowed and moved through the entire engine. We can, however, design the engine to permit the shock to stand in the throat, where the shock stagnation pressure drop loss is least. The divergent portion then acts as a subsonic diffuser but, because of boundary-layer separation

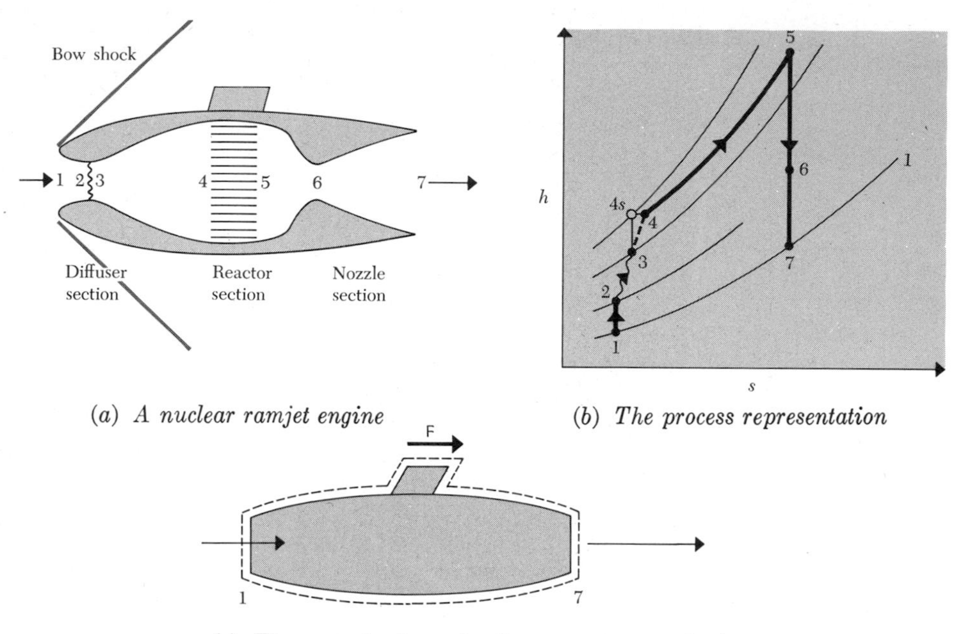

Figure 13·23.

produced by the shock, the subsonic diffusion process is far from isentropic. In order to account for separation effects in the subsonic diffuser, we can assume some typical value for the *diffuser pressure recovery factor*, defined by

$$C_P = \frac{\text{actual pressure rise}}{\text{isentropic pressure rise}}$$

A value of C_P of 0.6 is typical, and we shall use this in our analysis. We shall treat the supersonic diffuser and nozzle as isentropic flows, which is reasonable for a first analysis. Figure 13·23*b* shows the *h-s* process representation drawn considering these assumptions.

The analysis consists of working our way through the engine in the usual manner.

State 1

$$\mathsf{M} = 3 \qquad A = 10 \text{ ft}^2 \qquad P = 2 \text{ psia} \qquad T = 400 \text{ R specified}$$

From Table B·16, at $\mathsf{M} = 3$,

$$\frac{P}{P_0} = 0.0272 \qquad \frac{T}{T_0} = 0.357 \qquad \frac{A}{A^*} = 4.234$$

So,

$$P_0 = \frac{2}{0.0272} = 73.4 \text{ psia}$$

$$T_0 = \frac{400}{0.357} = 1120 \text{ R}$$

$$A^* = \frac{10}{4.234} = 2.36 \text{ ft}^2$$

The reference quantities above apply for the process 1-2. Also, at 1,

$$\mathsf{c}_1 = \sqrt{32.2 \times 1.4 \times 53.3 \times 400} = 982 \text{ ft/s}$$

$$\mathsf{V}_1 = 3 \times 982 = 2946 \text{ ft/s}$$

State 2

$$A = 2.8 \text{ ft}^2 \text{ specified}$$

$$\frac{A}{A^*} = \frac{2.8}{2.36} = 1.186$$

From Table B·18, at $A/A^* = 1.186$, $\mathsf{M}_2 = 1.51$,

$$\frac{P}{P_0} = 0.269 \qquad P_2 = 0.269 \times 73.4 = 19.6 \text{ psia}$$

State 2 corresponds to the upstream (x) state for the normal shock.

State 3

Shock at $\mathsf{M}_x = 1.51$. From Table B·19,

$$\mathsf{M}_y = \mathsf{M}_3 = 0.697$$

$$\frac{P_{0y}}{P_{0x}} = 0.926$$

$$\frac{P_y}{P_x} = 2.49$$

$$\frac{A_x^*}{A_y^*} = 0.926$$

So

$$P_{03} = 0.926 \times 73.4 = 67.8 \text{ psia}$$

$$P_3 = 2.49 \times 19.6 = 48.6 \text{ psia}$$

$$A_3^* = \frac{2.36}{0.927} = 2.55 \text{ ft}^2$$

State 4

We assume that the velocity is sufficiently low that the local pressure and isentropic stagnation pressures will be identical. For example, for $\mathsf{M}_4 = 0.2$, $P/P_0 = 0.493$ (Table B·18) and $A/A^* = 2.96$. Hence A_4 would have to be about $2.96 \times 2.55 = 7.5\ \text{ft}^2$, assuming isentropic diffusion. So, a sufficiently large area for nearly complete diffusion will be possible within the 10-ft² engine opening area. Then,

$$P_{4s} = P_{0y} = P_{03} = 67.8\ \text{psia}$$

If

$$C_P = 0.6,$$

$$P_4 = P_3 + C_P(P_{4s} - P_3) = 48.6 + 0.6 \times (67.8 - 48.6)$$
$$= 60.1\ \text{psia}$$

Since the flow is adiabatic from 1 to 4, h_0, hence T_0 is constant from 1 to 4. Then,

$$T_{04} = T_4 = T_{01} = 1120\ \text{R}$$

State 5

$$P_4 - P_5 = 5\ \text{psia specified, so}$$
$$P_5 = 60.1 - 5 = 55.1\ \text{psia} = P_{05}$$
$$T_5 = 2060\ \text{R (specified)} = T_{05}$$

States 6 and 7

The nozzle flow is modeled as isentropic. Any exit Mach number can be obtained by adjustment of A_6. Let's design to adjust A_6 so that P_7 matches the ambient pressure (2 psia) and the flow at 7 is supersonic.

$$P_7 = 2\ \text{psia} \qquad P_{07} = P_{05} = 55.1\ \text{psia}$$

So,

$$\left(\frac{P}{P_0}\right)_7 = \frac{2}{55.1} = 0.0363$$

In Table B·18, for $\dfrac{P}{P_0} = 0.0363$,

$$\mathsf{M} = 2.80 \qquad \frac{T}{T_0} = 0.389 \qquad \frac{A}{A^*} = 3.50$$

So, for state 7

$$T_7 = 0.389 \times 2060 = 800\ \text{R}$$
$$\mathsf{c}_7 = \sqrt{32.2 \times 1.4 \times 53.3 \times 800} = 1380\ \text{ft/s}$$
$$\mathsf{V}_7 = 2.80 \times 1380 = 3860\ \text{ft/s}$$

The required throat area is then

$$A_6 = \frac{15}{3.50} = 4.29 \text{ ft}^2$$

Now, the flow rate can be calculated from the data at any point. Let's take point 1:

$$\rho_1 = \frac{P_1}{RT_1} = \frac{2 \times 144}{53.3 \times 400} = 0.0135 \text{ lbm/ft}^3$$

$$\dot{M} = A_1 \rho_1 \mathsf{V}_1 = 10 \times 0.0135 \times 2946 = 400 \text{ lbm/s}$$

The reactor power requirement is then

$$\begin{aligned} \dot{Q} &= \dot{M}(h_5 - h_4) = \dot{M}c_P(T_5 - T_4) \\ &= 400 \times 0.24(2060 - 1120) = 90{,}500 \text{ Btu/s} \\ &= 95.5 \text{ MW} \end{aligned}$$

Finally, let's calculate the force delivered to the air frame by the engine. Applying the momentum principle to the control volume of Fig. 13·23*c*, neglecting any skin friction drag on the engine and assuming that there is no net pressure force on the control volume (recall $P_1 = P_7$),

$$\mathsf{F} = \frac{\dot{M}}{g_c}(\mathsf{V}_7 - \mathsf{V}_1) = \frac{400}{32.2} \times (3860 - 2946) = 11{,}360 \text{ lbf}$$

The specific impulse is

$$SI = \frac{\mathsf{F}}{\dot{M}} = \frac{11{,}360}{400} = 28.4 \text{ lbf/(lbm·s)}$$

Such propulsion systems have been suggested for use on airplanes such as the SST. The ramjet engines would be cut in after the plane has achieved supersonic speed and is at a high altitude and would run using conventional fuel sources, not nuclear energy. There are obvious problems with the safety of such a system.

This example deserves some discussion. Note that the exit Mach number was *less* than that at the inlet, and yet we derive thrust. This is because by heating the flow the stagnation temperature is increased, thus the velocity at the exit is substantially bigger than at the inlet. The specific impulse is *low* in comparison to subsonic jet engines (see Sec. 9·18). To achieve higher specific impulse we would have to increase the discharge velocity. If we wish to keep the exit pressure at 2 psia, the exit Mach number must be held at 2.80. The only hope is to increase the reactor outlet temperature, which is already pushing the metallurgical limit for an aircraft engine. If we could reach $T_5 = 2600°F$, then $V_7 = 4700$ ft/s, the thrust would be 22,300 lbf, and the specific impulse would be a respectable 56 lbf/(lbm·s), but the reactor power would be increased to about 190 MW. A thrust of 22,000 lbf is comparable to modern jet engines; a power level of 190 MW is comparable to the base load power requirements of a city of 200,000 people.

SELECTED READING

Obert, E. F., and R. L. Young, *Elements of Thermodynamics and Heat Transfer*, 2d ed., chap. 13, McGraw-Hill Book Company, New York, 1962.

Owczarek, J. A., *Fundamentals of Gas Dynamics*, chap. 1, sec. 6.2, International Textbook Company, Scranton, Pa., 1964.

Shapiro, A., *Dynamics and Thermodynamics of Compressible Fluid Flow*, vol. I, chaps. 1–7, The Ronald Press Company, New York, 1953.

QUESTIONS

13·1 Why is $\mathsf{F} \neq Ma$ for a control volume?
13·2 What is the momentum principle?
13·3 Sketch the pressure distribution on the wall of a (convergent) rocket nozzle. This force is clearly down, so why does the rocket move up?
13·4 What is the isentropic stagnation state?
13·5 How is the speed of sound defined?
13·6 How would you determine the sound speed in a *real* gas from thermodynamic data?
13·7 What is one-dimensional gas dynamics?
13·8 When is the isentropic-flow model reasonable?
13·9 What is choking?
13·10 What happens to the state of a gas as it passes through a shock wave?
13·11 How can a shock wave form?
13·12 What is the role of thermodynamics in gas dynamics theory?

PROBLEMS

13·1 A proposed electric jet car takes in atmospheric air (14.7 psia, 70°F) at a low speed through a vertical intake in its roof, and discharges hot gases to the rear at 1500 ft/s. The air-flow rate is 5 lbm/s, and the pressure in the jet exhaust is 20 psia. The engine uses 500 kW of electric power to compress and heat the air. Calculate the nozzle discharge area and the drag force on the body, neglecting the horizontal force between the wheels and the road.

13·2 The reaction control system for smaller space vehicles uses nitrogen, stored in a high-pressure bottle. When maneuvering is desired a valve is opened, allowing the gas to escape through a nozzle. The pressure upstream of the nozzle is typically about 250 psia, and the pressure in the jet at the nozzle exit plane is about 1 psia. Estimate the specific impulse (thrust force per unit of mass flow) of this system (the upstream temperature may be estimated as that of the vehicle, say, 80°F). Typically one set of nozzles is sized to produce about 40 lbf thrust, and another set produces about 200 lbf. Estimate the flow rate through each of these nozzles during operation and the exit area of the nozzles.

13·3 In a hydrogen rocket, hydrogen enters the nozzle at about 4000 R and 1000 psia; the pressure in the exit plane is about 1.5 psia. Estimate the specific impulse of this system and calculate the required hydrogen-flow rate for 2×10^6 lbf thrust. Calculate the nozzle exit area.

13·4 In an ion rocket engine cesium ions and electrons are accelerated by electric fields to about 0.0003 times the speed of light, mixed together to form a plasma, and emitted from the engine. Estimate the specific impulse of this system. Calculate the thrust force produced by expulsion of 100 lbm cesium/day.

13·5 Discuss the differences between the applications of the three thrusting systems of Probs. 13·2, 13·3, and 13·4.

13·6 Consider steady, frictionless, one-dimensional flow of a perfect gas in a constant-area duct. Energy as heat is added to the flow from the walls of the duct. Under these conditions show that

(*a*)
$$\frac{d\rho}{\rho} + \frac{d\mathsf{V}}{\mathsf{V}} = 0$$

(*b*)
$$\frac{d\mathsf{M}}{\mathsf{M}} = \frac{d\mathsf{V}}{\mathsf{V}} - \frac{1}{2}\frac{dT}{T}$$

(*c*)
$$dP = \frac{1}{g_c}\mathsf{V}^2\, d\rho$$

Which applies in other flow models?

13·7 Show that the model of Prob. 13·6 implies that the flow process is reversible.

13·8 Typical values for the specific impulse are: chemical rocket, 300; jet engine, 60; hydrogen used in a nuclear reactor rocket, 900; and an ion engine, 10,000. The units for these values are lbf·s/lbm. Determine these values in the SI system of units.

13·9 A mixing duct has two incoming streams of air. Stream A enters at 800 K and 300 m/s, and stream B enters at 300 K and 30 m/s. The inlet pressure is 1 atm. The inlet area for stream B is three times that for A. Find the velocity, pressure, and temperature downstream after mixing occurs. Assume one-dimensional flow and that the downstream area is the sum of the flow areas for A and B.

13·10 When Newton first tried to compute the propagation velocity of a pressure pulse (speed of sound), he assumed that the wave process is isothermal (rather than isentropic). Calculate the error in the sonic velocity using this assumption for a perfect gas. What is the numerical value of this error for air at room conditions?

13·11 Consider one-dimensional flow of a perfect gas in a constant-area adiabatic duct without friction. Changes in state come about as a result of changes in elevation in the earth's gravitational force field which acts in the negative z direction. Starting from first principles, derive the continuity, momentum, and energy equations in differential form. Then, discuss the direction of change (increase or decrease) in the Mach number, velocity, sound speed, density, pressure, and stagnation pressure for a positive increase in altitude z.

13·12 Starting with the steady-state energy equation, show that the velocity of a perfect gas leaving a frictionless adiabatic nozzle is given by

$$V = \sqrt{2g_c P_o v_o \frac{k}{k-1}\left[1 - \left(\frac{P_e}{P_o}\right)^{(k-1)/k}\right]}$$

The subscripts o and e correspond to the stagnation (zero velocity) and exit states, respectively.

13·13 A plane flies at a low altitude at 700 ft/s. Calculate the pressure rise that could be achieved if the air were brought nearly to rest by a diffuser in one of the engines. What is the pressure rise if the velocity is 2000 ft/s? Can you see a way to take advantage of the pressure rise in a turbojet engine?

13·14 Calculate the speed of sound at 530 R in air, helium, sulfur dioxide, steam, and isobutane.

13·15 Determine the error made in timing the 100-m dash if the timer uses the sound of the starter's gun as the initial time. Assume an ambient temperature of 27°C.

13·16 A hydrogen compressor operates at 20,000 rpm and has a rotor diameter of 10 cm. With the hydrogen at 400 K does the rotor tip speed exceed the speed of sound in hydrogen?

13·17 Freon 12 ($\hat{M} = 120.9$; $k = 1.14$) is compressed to a high pressure from 0.14 MPa at 15°C in a rotary compressor. The rotor diameter is 20 cm. What is the maximum rpm for the tip speed of the rotor to be less than Mach equal 1 in the Freon?

13·18 A cryogenic wind tunnel has been suggested as a means of reducing the power required to operate a normal-temperature wind tunnel. Assuming the air is cooled to the temperature of liquid nitrogen at 1 atm, calculate the air speed required to achieve Mach 3. At room temperature what air speed is required?

13·19 The inverse of the isentropic compressibility for water has a value of 2.27×10^9 N/m^2 at 90°C and 1 atm. Calculate the speed of sound in water at these conditions.

13·20 A diffuser is to be designed to reduce the Mach number of air from $M_1 = 0.75$ to $M_2 = 0.10$. Determine the diffuser area ratio A_1/A_2, assuming isentropic flow.

13·21 The standard atmosphere† is reproduced in short form in the table. Calculate the speed of sound at each of the given altitudes and discuss the implications on the flight of a subsonic airplane ($M = 0.9$).

U.S. Standard atmosphere

Altitude ft	Temperature R	Temperature °F	Pressure in Hg
0	518.67	59.0	29.92
10,000	483.03	23.3	20.58
20,000	447.41	−12.3	13.76
50,000	389.97	−69.7	3.44
100,000	408.57	−51.1	0.33
200,000	456.99	−2.67	0.58×10^{-2}

† U.S. Standard Atmosphere, 1962, U.S. Government Printing Office.

13·22 A vehicle moves at 2000 ft/s through the atmosphere. Calculate the stagnation temperature on its nose as a function of altitude (see Prob. 13·21).

13·23 The standard atmosphere is reproduced in short form in the table. Calculate the speed of sound at each of the given altitudes and discuss the implications on the flight of a subsonic airplane.

U.S. Standard atmosphere

Altitude m	Temperature K	Temperature °C	Pressure N/m^2
0	288.15	15.0	101,325
3,000	268.66	−4.49	70,121
6,000	249.19	−23.9	47,218
10,000	223.25	−49.9	26,499
20,000	216.65	−56.5	5,529
30,000	226.51	−46.6	1,197
60,000	255.77	−17.4	20.8

13·24 A vehicle moves at 60 m/s through the atmosphere. Calculate the stagnation temperature on its nose as a function of altitude (see Prob. 13·23).

13·25 Consider a flying saucer moving at a Mach number of 7 at an altitude of 60,000 m. What is the stagnation temperature at the nose of the saucer? (See Prob. 13·23.)

13·26 Steam flows at 1100 ft/s, 80 psia, 450°F. Determine the stagnation pressure and temperature.

13·27 Bernoulli's equation may be written as $P_0 = P + (1/2g_c)\rho V^2$. An equation that corrects for compressibility effects is

$$P_0 = P + \frac{1}{2g_c}\rho V^2\left(1 + \frac{M^2}{4} + \frac{2-k}{24}M^4 + \cdots\right)$$

Plot $C_P \equiv (P_0 - P)2g_c/\rho V^2$. At what Mach number is the error in Bernoulli's equation 1 percent, 2 percent, and 10 percent? How would you calculate C_P at $M = 3$?

13·28 Derive the two equations given in Prob. 13·27. List all the assumptions for each equation.

13·29 A pitot-static tube records a static pressure of 4 psig and a difference between the impact and static pressures of 10 in Hg. The barometer reads 29.9 in Hg at 70°F. Compute the air velocity assuming the air is (*a*) incompressible, (*b*) compressible.

13·30 Derive the equation $(\partial P/\partial \rho) = c^2/g_c$ from the continuity and momentum equations rather than the energy and continuity equations as was done in the text.

13·31 Using the continuity, momentum, and energy equations for frictionless, one-dimensional adiabatic flow, show that the flow process is isentropic.

13·32 A large diffuser with an intake area of 2 ft^2 and an exit of 6 ft^2 has air entering at 590 R at a pressure of 50 psia and a velocity of 750 ft/s. Find the thrust on the diffuser if its outside surface is at atmospheric pressure.

13·33 A proposed rocket has products of combustion with $k = 1.25$ and $\hat{M} = 20$. The combustion chamber operates at 3100 K and 6.8 MPa. Determine the specific impulse for expansion to a vacuum and find the thrust for a throat area of 0.01 m^2.

13·34 A rocket is designed to produce 4×10^6 N of thrust at sea level. The pressure in the combustion chamber is 4×10^6 N/m^2 at a stagnation temperature of 3000 K. Assuming that the propellant has approximately air properties, find the specific impulse, mass-flow rate, and exit area/throat area ratio. Assume the exit pressure is ambient at 1 atm.

13·35 For the above rocket, if the stagnation temperature can be increased by 20 percent, how much would this increase the thrust? If design changes can increase the stagnation pressure by 20 percent, how much would this increase the thrust?

13·36 The equation for maximum flow rate through a nozzle [Eq. (13·33)] can also be written as

$$\dot{M}_{\max} = \frac{A^* P_0}{\sqrt{RT_0}} \left[k \left(\frac{2}{k+1} \right)^{(k+1)/(k-1)} \right]^{1/2}$$

Show this relationship.

13·37 A converging nozzle is available with an inlet diameter of 15 cm and an exit diameter of 2 cm. It is desired to use this nozzle as a flowmeter for natural gas, treated here as methane. The inlet pressure is 6×10^6 N/m^2 at a temperature of 330 K. Calculate the maximum flow rate this nozzle could meter at these conditions.

13·38 Mixing helium with a heavier inert gas offers the possibility of reduced pumping power with only a modest decrease in the effectiveness of heat transfer. A test apparatus to investigate this has a critical flow nozzle installed. Determine the throat area (M = 1) required to pass 150 kg/h of a mixture that is 45 percent Ar and 55 percent He by volume if the upstream pressure is 0.6 MPa at 320 K.

13·39 A commercial jet flying at 12,000 m has a cabin temperature of 21°C and a pressure of 9×10^4 N/m^2. A meteorite suddenly makes a round hole 2 cm in diameter in the airplane cabin. At this altitude the outside temperature is 216 K and the outside pressure is 0.191 atm. If the cabin volume is 500 m^3 *estimate* how long it takes for the pressure to drop to 5×10^4 N/m^2. (*Hint:* Remember choked flow.)

13·40 A 1-m^3 rocket test chamber is filled with products of combustion at 3200 K and 7 MPa. The throat area for the exhaust is 0.1 m^2. Estimate the time for the pressure to drop to 0.7 MPa if $k = 1.25$ and $\hat{M} = 22$.

13·41 An experimental rocket for use at Great Desert University burns methane stoichiometrically with air. The air-methane mixture enters the rocket chamber vertically, is burned at 13.6 atm, and is exhausted through a nozzle with throat area of 0.000645 m^2. The gas is expanded to the room pressure of 1 atm. Calculate the flow rate, thrust, and specific impulse.

13·42 A converging nozzle has an efficiency of 0.94 and an exit pressure of 2 atm. If the entrance conditions are 2.5 atm, 330 K, compare the exit velocity and temperature to those that would occur if the nozzle were isentropic.

13·43 Dry steam flows through a converging nozzle such that the process is isentropic. The inlet state is 1.5 MPa at 400°C, and the exit area is 0.0006 m^2. Find the maximum flow rate possible. What is the flow rate if the back pressure is 0.5 MPa?

13·44 A converging-diverging nozzle is designed such that the inlet area is 10 in^2, the throat area is 2 in^2, and the exit area is 4 in^2. The inlet conditions are air at 100 psia, 540 R, with a flow rate of 3.5 lbm/s. Is the flow at the exit supersonic?

13·45 Air is accelerated by a nozzle in an isentropic process. The inlet velocity is 20 ft/s and the exit velocity is 750 ft/s. The inlet state is 68°F, 15 psia. Find the exit temperature, pressure, density, and stagnation temperature and pressure.

13·46 It is desired to accelerate air through a converging nozzle whose exit area is 2 in^2. If the upstream conditions are 50 psia, 200°F, what is the maximum flow that can pass through the nozzle? If the upstream pressure is doubled, how much is the maximum flow increased?

13·47 Determine the maximum flow for air, argon, carbon dioxide, and isobutane that can be passed through a converging nozzle if the upstream conditions are 100 psia at 500 R and the nozzle exit area is 1 in^2.

13·48 A convergent nozzle has an exit diameter of 1 in. Air enters the nozzle at 50 psia and 240°F. Assuming isentropic flow, calculate the mass-flow rates for an exit pressure of 40 psia, 26.4 psia, and 14.7 psia.

13·49 Low-velocity air enters the nozzles of a gas turbine at 2000 R and 60 psia. The nozzles are used to accelerate the air before it impinges upon the turbine blades. Expansion in the nozzles is such that the air leaves at 15 psia. If 15 lbm/s of air is flowing through each nozzle, find the exit velocity, the exit and throat areas, and the exit Mach number.

13·50 An air nozzle is designed to provide a flow rate of 1 kg/s at an exit velocity of 700 m/s. The stagnation pressure is 0.7 MPa; the stagnation temperature is 800 K. For isentropic flow determine the area of the nozzle at the throat and at the exit. What is the exit Mach number?

13·51 Air enters a converging-diverging nozzle at 100 psia and 500 R and a low velocity. It is desired to accelerate this flow to $M = 4$. What are the exit pressure, temperature, and density? What are the throat temperature, pressure, and density? What is the ratio of exit area to throat area?

13·52 Air enters a converging-diverging nozzle at 200 psia with a temperature of 240°F. The exit-to-throat area ratio is 1.55. If the exit velocity is subsonic, but the throat velocity is sonic, determine the exit temperature and pressure assuming isentropic flow. If the exit velocity is supersonic, determine the exit temperature and pressure, again assuming that the flow is isentropic. If the back pressure is 100 psia, what is the nature of the flow in the diverging section of the nozzle?

13·53 Steam at 100 psia and 200°F of superheat expands isentropically in a nozzle to a pressure of 20 psia. Find the mass-flow rate per ft^2 of throat area. How can you tell that this must be a converging-diverging nozzle ($k = 1.30$)?

13·54 Air is available at a pressure of 1×10^6 N/m^2 and a temperature of 400 K. It is desired to operate a converging-diverging nozzle isentropically with an exit Mach number of 1.75 and a nozzle throat area of 4×10^{-4} m^2 using this air. Find the range of back pressures over which the nozzle is choked, the mass-flow rate for a back pressure of 0 and for a back pressure of 0.9×10^6 N/m^2, and the required exit area.

13·55 During a blowdown accident methane undergoes a shock wave with upstream conditions of 270 K at 0.2 MPa. The upstream velocity is 500 m/s. Calculate the conditions downstream behind the shock.

13·56 The atmosphere of Venus is largely CO_2. A space probe designed to land on Venus travels at 12 km/s. At the altitude giving the largest heat transfer to the probe as it enters the Venus atmosphere the ambient temperature is 100 K at a pressure of 70 N/m^2. Determine the conditions behind the normal bow shock.

13·57 An Apollo-like spacecraft returns to earth from the moon. At an altitude of 48,000 m its reentry velocity is 4000 m/s. The ambient temperature is 270 K at a pressure of 97 N/m^2. Calculate the pressure and temperature behind the shock wave on the front of the spacecraft assuming a normal shock occurs.

13·58 A normal shock moves at 400 m/s through still air at 0°C and 90,000 N/m^2. After the wave passes find the velocity of the air.

13·59 A normal shock wave occurs in air such that the upstream conditions are Mach 2.4, 540 R, 14.7 psia. Calculate the conditions behind the shock.

13·60 A simple ramjet engine is shown below. At operating conditions the following data have been obtained: $M_1 = 2.0$, altitude 20,000 ft ($T_1 = 447$ R, $P_1 = 6.7$ psia), $A_1 = 2.5$ ft^2, $M_2 = 0.2$, $T_3 = 3200$ R. It is known that at 4 the flow is choked, the fuel has a heating value of 20,000 Btu/lbm, and a normal shock takes place at point 1. Assume that $M_3 \approx M_2$. Find the flow rate, the area A_2, the exhaust velocity V_4, the fuel consumption, and the thrust. List all idealizations employed.

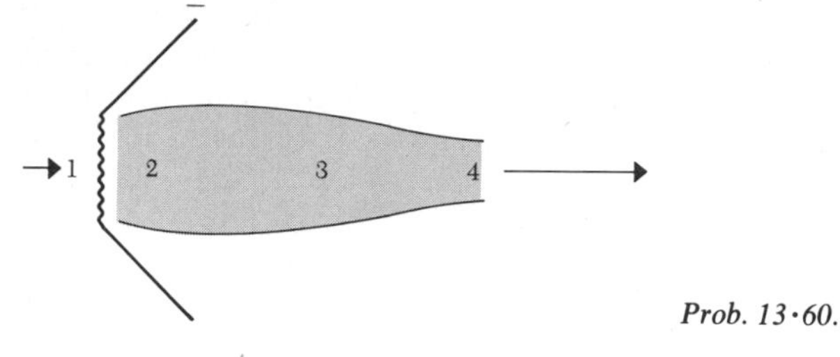

Prob. 13·60.

13·61 A blunt-shaped vehicle reenters the earth's atmosphere such that a Mach number of 10 is achieved. The shock wave stands slightly off of the body and is normal to the flow at the vehicle centerline. Estimate the conditions behind the normal shock ($T_x = -60$°F, $P_x = 2$ psia).

13·62 A blast wave caused by a nuclear explosion travels through still air (initially at 70°F and 14.7 psia) at a speed of 50,000 ft/s. Estimate the changes in pressure and temperature observed as the wave passes over an observer. Will he enjoy this experience?

13·63 Air flows through a convergent-divergent nozzle with an upstream stagnation pressure of 50 psia and a throat area of 1 in^2. If a normal shock occurs where $A/A^* =$ 2.0, what will be the Mach numbers on the upstream and downstream sides of the shock? If the Mach number in the exit plane is 0.5, what should be the back pressure to maintain the shock in the given location?

13·64 Consider a convergent-divergent nozzle operating with a shock wave in the divergent section of the nozzle. The upstream stagnation pressure is 200 psia at a temperature of 700 R. If the shock wave is located at the point where $M = 1.8$, find the exit pressure and Mach number, assuming that the exit area is three times the throat area.

13·65 Determine the exit velocity for water flowing through a nozzle if the upstream pressure is 100 psia at 70°F and the downstream pressure is 1 atm. Assume frictionless adiabatic flow.

13·66 A rocket is designed to operate with a stagnation pressure of 3 MPa at a stagnation temperature of 3400 K. The *isentropic* expansion will take place through a converging-diverging nozzle to an ambient pressure of 10^4 Pa. Determine the ambient pressure such that a normal shock stands at the exit plane. Describe the operation of the nozzle for pressures above and below this value. Calculate the thrust at the design condition for a throat diameter of 0.10 m. Assume $k = 1.4$.

13·67 A chemical processing plant requires 1 kg/s of O_2. This will be supplied from bottled O_2 at 20°C, 7 MPa, and must be delivered at 30 m/s at 90°C, 150,000 Pa.

The flow rate must be held constant during pressure variations of $\pm 30{,}000$ Pa. Design a system to provide this flow.

13•68 A chemical processing plant requires 2 lbm/s of O_2. This will be supplied from bottled O_2 at 70°F, 1000 psia, and must be delivered at 100 ft/s at 190°F, 20 psia. The flow rate must be constant during back-pressure variations of ± 5 psia. Design a system to provide this flow.

13•69 Discuss the phenomenon of choking. What difficulties can it cause and what advantages does it provide?

13•70 A gas dynamics experiment requires a gas with $c = 2000$ ft/s at 60°F. Specify an appropriate gas mixture.

13•71 A small thrusting backpack system capable of delivering a 250-N thrust for astronaut lunar mobility is required. Combustion systems cannot be used. Design a system. If the propellant storage volume is limited to 0.1 m^3, what is the flying time of your system?

13•72 A small wind tunnel for studies at $M = 2$ is desired. The test section area must be 6 by 6 in, and the stagnation conditions must be $T_0 = 1500$ R, $P_0 = 1$ atm. Design such a system, specifying all states, compressor and heater power requirements, nozzle parameters, and so on.

13•73 Do Prob. 13•72 for $M = 5$ with stagnation conditions appropriate for flight at 200,000 ft (see Prob. 13•21).

13•74 Reflecting on Chap. 12, criticize the idealizations in the normal shock theory presented in this chapter. Under what conditions is this theory least accurate? How could you formulate a better theory, assuming chemical equilibrium was maintained?

13•75 The states on either side of a shock wave are normally treated as equilibrium states. Consider a gas AB which dissociates by the reaction $AB \rightleftharpoons A + B$. Develop a theory for normal shocks in such a gas assuming (*a*) equilibrium downstream, (*b*) that the downstream composition is "frozen" at the upstream equilibrium composition. Compare your theories with single-component theory, quantitatively for typical gases at low and elevated temperatures.

CHAPTER FOURTEEN

AN INTRODUCTION TO HEAT TRANSFER

14·1 SOME BASIC CONCEPTS

By the time students have reached this final chapter they will have developed considerable skill in applying the principles of thermodynamics in engineering analysis. There is one important aspect of such analysis which we have yet to mention, and this is the evaluation of the *rate* at which energy is transferred as heat in a specified situation. The theory of heat-transfer rates rests on models and laws beyond the scope of thermodynamics, and heat transfer is usually treated as a separate and distinct subject. Heat transfer is of interest in many engineering disciplines. The thermal control of a space vehicle, the cooling of a component in an electronic system, the calculation of thermal expansion of a suspension bridge, and the problem of energy removal from a nuclear power reactor are all examples of engineering problems involving heat transfer. This wide range of important applications makes an elementary knowledge of heat-transfer theory of considerable importance to all engineers.

In this chapter we shall present a very limited introduction to heat transfer. The objectives of this chapter are (1) to make the student conversant with the terminology of heat transfer so he or she can work with thermal engineers, (2) to show how thermodynamic analysis can be further applied in a new area, and (3) to give the student some operational ability to solve simple heat-transfer problems.

Heat-transfer theory rests on certain fundamental equations called *rate equations*, which relate the rate of energy transfer as heat between two systems to the thermodynamic properties of these systems. These rate equations, when combined with energy balances and thermodynamic state equations, yield equations from which the temperature distribution and heat-transfer rates can be found. Heat-transfer theory is then essentially "thermodynamics with rate equations added."

Before we can write down any rate equations we must think carefully about the meaning of energy transfer as heat. We can often identify energy transfer as work between two systems; this is not heat. We can often identify energy convection from one system to another by mass which moves from one to the other; this also is not heat. Sometimes we can identify energy transfer due to organized charge flow; this is not heat either. Heat is the energy transfer which we *cannot* identify with work, or convection, or electric current; it is all the energy transfer we cannot see, all the disorganized transfer associated with the microscopic interchanges between the systems, which is not included in macroscopic representations of the work, charge flow, and so on.

We set up the concept of temperature as a "pointer" for heat transfer *in the absence of mass or current flows.* We talked qualitatively of temperature as the driving potential for energy transfer as heat. Indeed, in the absence of mass and current flows, the equations of heat transfer do relate the heat-transfer rate to temperature differences or gradients. However, temperature is not the sole potential for energy transfer as heat. There can be heat transfer, in the sense described, in the absence of any temperature differences but in the presence of voltage differences. Such effects are very important in thermoelectricity. Correspondingly, if substantial concentration gradients are present, there can be a heat flow driven by these gradients in an isothermal system. Hence the classical idea of heat as being driven by temperature differences is not quite correct, although it is indeed quite adequate for most situations. The case of *coupled flows* will not be treated in this chapter, and we shall restrict ourselves to situations in which temperature differences or gradients are the *only* driving force for heat transfer.†‡

There are *two* basic mechanisms for energy transfer as heat. The first mechanism is *conduction*, the process of energy transfer as heat through a stationary medium, such as copper, water, or air. In solids the energy transfer arises because atoms at the higher temperature vibrate more excitedly, hence they can transfer energy to more lackadaisical atoms nearby by microscopic work, that is, heat. In metals the free electrons also contribute to the heat-conduction process. In a liquid or gas the molecules are also mobile, and energy is also conducted by molecular collisions.

† See W. C. Reynolds, *Thermodynamics*, 2d ed., McGraw-Hill Book Company, New York, 1968, chap. 14, for a discussion of coupled irreversible flows.

‡ The statement "Heat is energy transfer by virtue of a temperature difference," is only true in the absence of coupled effects. This definition of heat is therefore appropriate for the material in this chapter.

The second heat-transfer mechanism is *radiation*, the transfer of energy by disorganized photon propagation. Any body continually emits photons randomly in direction and time, and the net energy transferred by these photons is counted as heat. In contrast, any *organized* photon energy, such as radio transmission, can be macroscopically identified and is not considered heat.

Often engineers speak of a third mechanism of heat transfer, *convection*. The term is applied to energy transfer between solids and moving fluids. However, it is not *heat* that is being convected but *internal energy*. There may be some energy transfer as heat by conduction within the moving fluid, and a convection analysis often involves a consideration of the effects of this conduction on the state of the fluid as determined by energy balances.

Heat-transfer analysts work both with the heat-transfer rate, which we will denote by

$$\dot{Q} \qquad \text{W or Btu/h}$$

and the *heat flux*, that is, the *heat-transfer rate per unit of area* normal to the direction of heat flow,

$$q'' \equiv \frac{\dot{Q}}{A} \qquad \text{W/m}^2 \text{ or Btu/(h}\cdot\text{ft}^2\text{)}$$

The double-prime notation is useful to remind the analyst that it is the flow per unit area that is considered. The rate equations are generally written first in terms of the heat flux, and then a simple multiplication by area or integration over the area gives the *total* heat-transfer rate. The concept of heat flux is somewhat analogous to that of current flux, the electric current flow per unit of conductor area. Since notation varies widely in the heat-transfer literature, the reader of a text or journal article must take care to identify the particular author's notations for heat-transfer rates and heat flux.

14·2 THE CONDUCTION RATE EQUATION

Consider two systems connected by a metal rod (Fig. 14·1). In the absence of coupled effects discussed in the last section, any energy transfer between the systems must be as heat. Our objective is to reason out a satisfactory equation for the rate of conduction heat transfer through the rod.

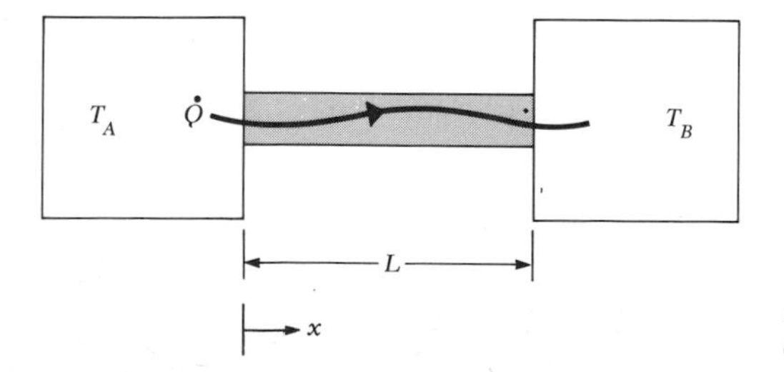

Figure 14·1. Transfer of energy as heat by conduction between two systems

We expect that the rate of energy transfer as heat must depend upon the thermodynamic states of the two systems, and on the physical and geometrical properties of the metal rod. However, in view of our concept of temperature, we might expect that the only properties of A and B relevant to the heat-transfer rate are their temperatures. Hence we might express the conduction heat-transfer rate (cal/s, Btu/h, W) as some function of T_A, T_B, and the bar,

$$\dot{Q} = f_1(T_A, T_B, \text{bar})$$

Alternatively, we could express $\dot{Q}$ as a function of one of the temperatures, the temperature difference, and the bar,

$$\dot{Q} = f_2(T_A - T_B, T_A, \text{bar})$$

If we can establish this function f_2 we shall have the needed rate equation.

One requirement on f_2 is that it must be zero when $T_A = T_B$. Then, for small $T_A - T_B$, a Taylor series expansion of f_2 about $T_A - T_B = 0$ gives†

$$\dot{Q} = \left[\frac{\partial f_2}{\partial(T_A - T_B)}\right]_{T_A - T_B = 0} (T_A - T_B) + \cdots \tag{14·1}$$

The derivative in Eq. (14·1) is evaluated when $T_A - T_B = 0$, that is, when there is equilibrium between A and B. It must therefore be a *property* of the bar, one that can be measured experimentally and tabulated as a function of the equilibrium state. Since $\dot{Q} > 0$ if $T_A > T_B$, this property must be positive. If the cross-sectional area of the bar is uniform, it seems likely that $\dot{Q}$ should be proportional to this area. However, the longer we make the bar, the less should be the energy flow for a given $T_A - T_B$. Hence the coefficient of $(T_A - T_B)$ in Eq. (14·1) most probably is proportional to the cross-sectional area A and inversely proportional to the length L. These considerations suggest that

$$\left[\frac{\partial f_2}{\partial(T_A - T_B)}\right]_{T_A - T_B = 0} = \frac{kA}{L} \tag{14·2}$$

where A is the bar area, L its length, and k is a property of the bar material.

Then,
$$\dot{Q} = \frac{kA}{L}(T_A - T_B) + \cdots$$

† To expand $f(x)$ in a Taylor series about a,

$$f(a + \Delta x) = f(a) + \left.\frac{df}{dx}\right|_{x=a} \Delta x + \left.\frac{d^2f}{dx^2}\right|_{x=a} \frac{\Delta x^2}{2!} + \cdots$$

Now let us shrink the bar and at the same time bring T_A very close to T_B. In the limit, as $L \rightarrow 0$ (note that x is in the direction of $\dot{Q}$)

$$\frac{T_A - T_B}{L} = -\frac{T_B - T_A}{L} \rightarrow -\frac{dT}{dx}$$

The higher-order terms are negligible in the limit, and Eq. (14·1) becomes a *rate equation for the heat flux* $q'' \equiv \dot{Q}/A$,

$$q'' = -k\frac{dT}{dx} \tag{14·3}$$

This is the *Fourier law*,† developed in 1822, the basic rate equation of conduction heat transfer. The property k is called the *thermal conductivity* of the bar material. The units of k are W/(m·K) or Btu/(h·ft·°F). Values of k vary widely from material to material. Appendix C gives some representative values. Note that k also depends on the state of the material, particularly upon its temperature. k can be determined from laboratory experiments based upon Eq. (14·3) or, in the case of gases and solids, predicted from molecular kinetic theory. The minus sign in Eq. (14·3) indicates that a negative temperature gradient produces heat transfer in the positive x direction.

For solids the conduction is due both to motions of free electrons within the solid and the action of the vibrations of the molecules within the solid. Since electric conductors have a large supply of free electrons, they are typically good heat conductors. The strength of these two mechanisms, hence k, is strongly temperature-dependent. For gases the conduction is due to molecular translational motions. These motions, hence k, increase with increasing temperature.

The thermal conductivity of a material is one of its *transport properties*. Others are the *viscosity* μ associated with the transport of momentum, and the *diffusion coefficient* associated with the transport of mass. As our Taylor series expansion [Eq. (14·1)] suggests, we can tabulate the transport properties as functions of the thermodynamic equilibrium state.

14·3 SOME SIMPLE CONDUCTION PROBLEMS

Conduction theory centers on the problem of solution of the differential equation for the temperature field in the conducting medium. The differential equation is obtained by applying an energy balance on an elemental volume of the system under study and bringing in the Fourier rate equation to relate the heat-transfer rates to and from the volume to the temperature field.

† For conduction in more than one direction in an isotropic media, Fourier's law relates the *heat-flux vector* $\mathbf{q}''$ to the temperature gradient, $\mathbf{q}'' = -k\,\mathbf{grad}\,T$.

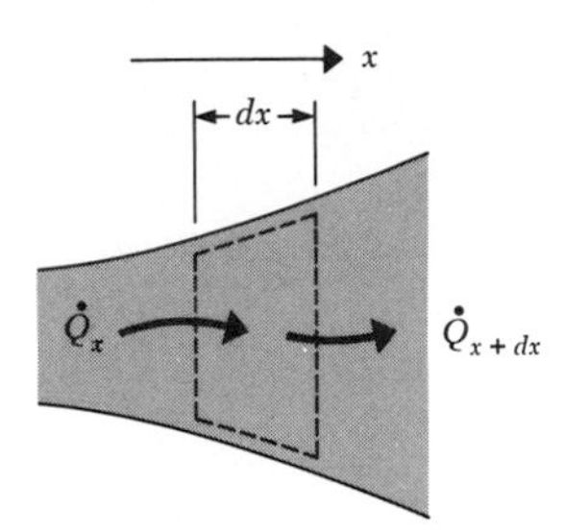

Figure 14·2. Conduction through a variable-area rod

Let's illustrate the method by development of the equation for steady-state one-dimensional conduction in a bar of varying area (Fig. 14·2). Assuming steady state, the energy balance is

$$\underset{\text{energy-input rate}}{\dot{Q}_x} = \underset{\text{energy-output rate}}{\dot{Q}_{x+dx}}$$

Now, from the Fourier rate equation,

$$\dot{Q}_x = -k\left(A\frac{dT}{dx}\right)_x \qquad \dot{Q}_{x+dx} = -k\left(A\frac{dT}{dx}\right)_{x+dx}$$

Note we are treating k as constant here. Now, from a Taylor series expansion of $\dot{Q}_{x+dx}$ about $\dot{Q}_x$,

$$\dot{Q}_{x+dx} = -k\left[\left(A\frac{dT}{dx}\right)_x + \frac{d}{dx}\left(A\frac{dT}{dx}\right)_x dx + \cdots\right]$$

Substituting in the energy balance,

$$-k\left(A\frac{dT}{dx}\right)_x = -k\left(A\frac{dT}{dx}\right)_x + \frac{d}{dx}\left(-kA\frac{dT}{dx}\right)_x dx + \cdots$$

The leading terms cancel, and our differential equation is obtained from the next-order terms,

$$\frac{d}{dx}\left(A\frac{dT}{dx}\right) = 0 \tag{14·4}$$

Equation (14·4) states that the gradient of energy flow through the slab is zero, that is, energy is conserved. Now, given an area profile $A(x)$ and some boundary conditions on T, we could solve Eq. (14·4) to find $T(x)$.

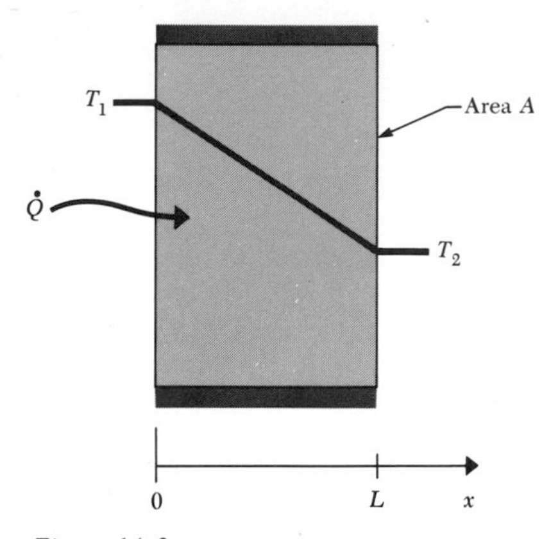

(a) Conduction through a slab. The temperature distribution through the slab is linear

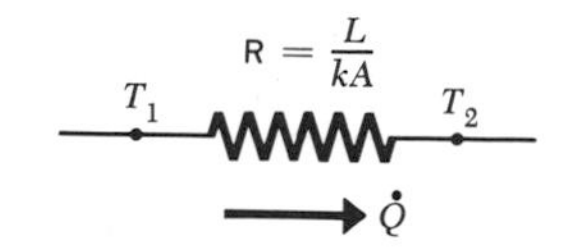

(b) The corresponding thermal circuit

Figure 14·3.

For example, consider the problem of heat transfer through a plane slab (Fig. 14·3). Here $A(x)$ is constant, and the heat flux is

$$q'' = -k \frac{dT}{dx} \tag{14·5}$$

The energy equation [Eq. (14·4)] tells us that the heat flux is *constant*, and that

$$\frac{d^2T}{dx^2} = 0 \tag{14·6}$$

Now, if we impose temperatures T_1 and T_2 at the two faces of the slab, the boundary conditions on $T(x)$ are

$$T(0) = T_1 \qquad T(L) = T_2 \tag{14·7}$$

The solution to Eq. (14·6) is

$$T = C_1 x + C_2$$

where C_1 and C_2 are constants of integration fixed by the boundary conditions. Applying Eqs. (14·7), one finds

$$C_2 = T_1 \qquad C_1 = \frac{T_2 - T_1}{L}$$

hence

$$T = T_1 + \frac{T_2 - T_1}{L} x \tag{14·8}$$

The temperature distribution is therefore linear across the slab. From Eq. (14·5), the heat flux is

$$q'' = k\left(\frac{T_1 - T_2}{L}\right)$$

so

$$\dot{Q} = Aq'' = \left(\frac{kA}{L}\right)(T_1 - T_2) \tag{14·9}$$

It is very often useful to think of heat flow in analogous electrical terms. Analogous to current is the heat flow $\dot{Q}$, and the temperature difference $T_1 - T_2$ is analogous to the voltage drop. The slab then is analogous to a *pure resistance*, and the factor analogous to the electric resistance is the *thermal resistance*,

▶ $$\mathsf{R} \equiv \frac{L}{kA} \tag{14·10}$$

Equation (14·9) can then be written as

▶ $$\dot{Q} = \frac{T_1 - T_2}{\mathsf{R}}$$

which is fully analogous to Ohm's law for an electric resistor. Figure 14·3 includes the heat-flow circuit diagram for this slab.

This network approach to heat flow is especially helpful in considering composite structures. If the heat flows in series first through one slab and then another, the thermal circuit is as shown in Fig. 14·4*a*. Note that the overall

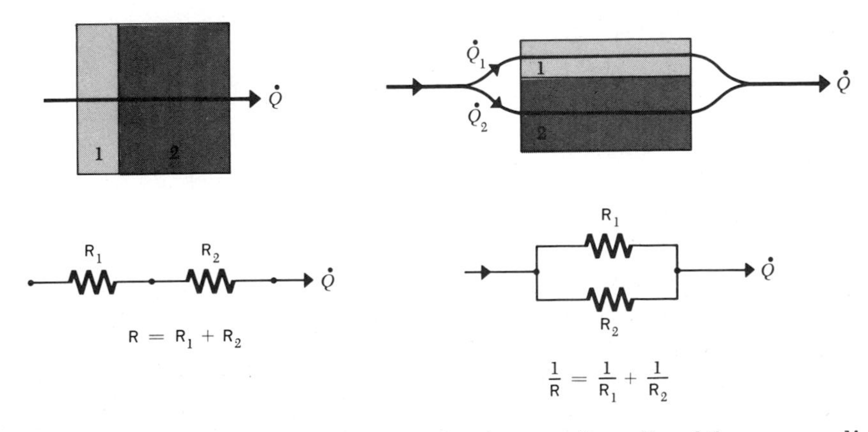

(*a*) *A composite wall and the corresponding series circuit*

(*b*) *A composite wall and the corresponding parallel circuit*

Figure 14·4.

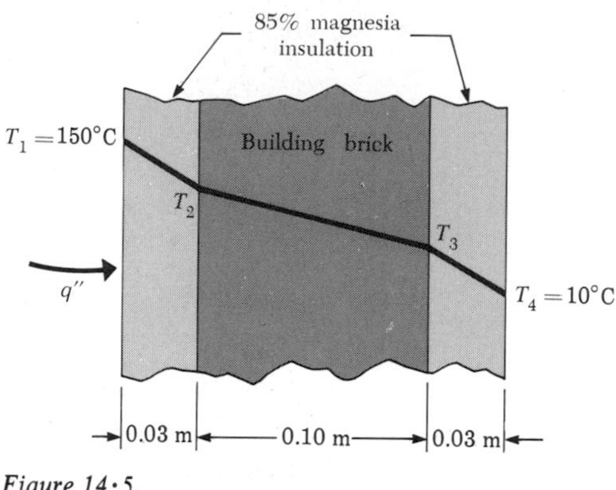

(a) *An insulated wall*

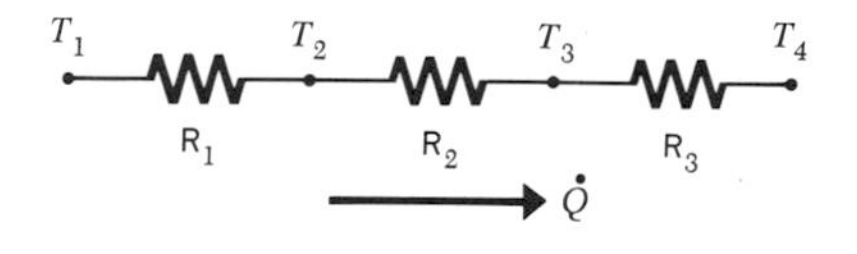

(b) *The corresponding thermal circuit*

Figure 14·5.

resistance is the sum of the component resistances, just as in an electric circuit. On the other hand, if the heat flow is carried in *parallel* by the heat conductors, the overall *thermal conductance* (1/R) is the sum of the thermal conductances, just as in a parallel electric circuit (Fig. 14·4*b*). The simplest types of heat-transfer analysis deal with series-parallel conduction circuits of this type, and once the network approach is understood such calculations are very simple.

For example, consider the composite slab of Fig. 14·5*a*. The solution (14·8) applies in each region, and the thermal circuit is shown in Fig. 14·5*b*. The total thermal resistance R is

$$\mathsf{R} = \mathsf{R}_1 + \mathsf{R}_2 + \mathsf{R}_3 = \frac{L_1}{k_1 A} + \frac{L_2}{k_2 A} + \frac{L_3}{k_3 A}$$

So, using thermal conductivity data from Table C·2,

$$\mathsf{R}A = \frac{0.03}{0.074} + \frac{0.10}{0.69} + \frac{0.03}{0.067}$$
$$= 0.405 + 0.144 + 0.447 = 0.996 \text{ m}^2\cdot\text{K/W}$$

Note that we treat the conductivity as constant through each slab, but use two different k values for the two magnesia slabs, employing estimates of their individual temperatures. The heat flux is then

$$q'' = \frac{T_1 - T_4}{\mathsf{R}A} = \frac{150 - 10}{0.996} = 140.5 \text{ W/m}^2$$

Then, since the same heat flux passes through each slab,

$$T_1 - T_2 = q''\mathsf{R}_1 A = 140.5 \times 0.405 = 56.9°\text{C}$$
$$T_2 - T_3 = q''\mathsf{R}_2 A = 140.5 \times 0.144 = 20.2°\text{C}$$
$$T_3 - T_4 = q''\mathsf{R}_3 A = 140.5 \times 0.447 = 62.8°\text{C}$$

Hence $\qquad T_2 = 93.1°\text{C} \qquad \text{and} \qquad T_3 = 72.8°\text{C}$

Converting to Btu/h·ft², we find,

$$q'' = 140.5 \text{ W/m}^2 \times 0.317 \text{ Btu}\cdot\text{m}^2/(\text{h}\cdot\text{ft}^2\cdot\text{W}) = 44.5 \text{ Btu/(h}\cdot\text{ft}^2)$$

According to Eq. (14·8), the temperature distribution is linear in each slab, as shown in Fig. 14·5*a*. Note that without the insulation the only thermal resistance would be due to the brick, and therefore the heat flux would be given by $(150 - 10)/0.144 = 972 \text{ W/m}^2$. As energy costs increase, the economic advantage of adding insulation also increases, and this example indicates the substantial drop in heat transfer that even small amounts of insulation can produce.

Suppose that the composite wall of the previous example has one 1-cm-diameter steel bolt passing through every 0.1 m² (Fig. 14·6*a*). The bolt will transfer energy as heat in *parallel* with the other materials, hence the thermal circuit is as shown in Fig. 14·6*b*. Now the bolt cross-sectional area per m² of wall is

$$\frac{A_b}{A} = \frac{\pi d^2}{4 \times 0.1} = 0.000785 \text{ m}^2/\text{m}^2$$

The change in the brick resistance caused by the bolt hole is negligible. However, the steel bolt has a relatively high thermal conductance, since $k = 40 \text{ W/(m}\cdot\text{K)}$, hence the effect of its parallel resistance is significant. Its resistance, per m² of wall area, is

$$\mathsf{R}A = \left(\frac{L}{k}\right)\frac{A}{A_b} = \frac{0.16}{40} \times \frac{1}{0.000785} = 5.095 \text{ m}^2\cdot°\text{C/W}$$

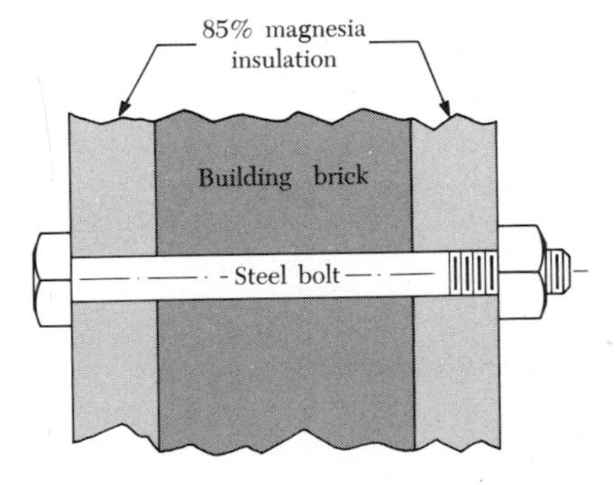

(*a*) *An insulated wall with a bolt. The bolt may increase the heat transfer rate significantly*

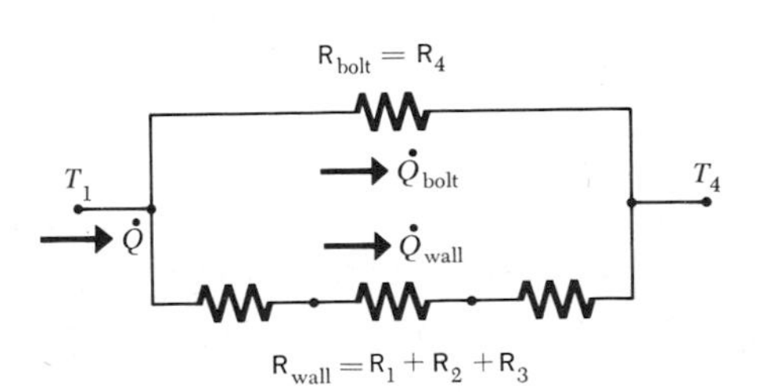

(*b*) *The corresponding thermal circuit*

Figure 14·6.

Then the overall resistance of the parallel circuit is given by the sum of the *thermal conductances* (reciprocal resistances). For $A = 1\ \text{m}^2$,

$$\frac{1}{\mathsf{R}} = \frac{1}{\mathsf{R}_1 + \mathsf{R}_2 + \mathsf{R}_3} + \frac{1}{\mathsf{R}_4}$$

$$= \frac{1}{0.996} + \frac{1}{5.095} = 1.20\ \text{W/°C}$$

so

$$\mathsf{R}A = \frac{1}{1.20} = 0.833\ \text{m}^2\cdot\text{°C/W}$$

Note that the bolts cause a substantial reduction in the thermal resistance of the wall. The total heat transfer rate per unit of surface area is then

$$q'' = \frac{T_1 - T_4}{\mathsf{R}} = \frac{140}{0.833} = 168\ \text{W/m}^2$$

The bolts increase the heat leak through the wall by almost 20 percent.

In the model above we assumed that the ends of the bolt were at the same temperatures as the faces of the slab. However, the bolt acts to depress the temperature on the hot face and increase the temperature on the cold face, and this will tend to reduce the heat flow through the bolt. Hence our analysis overpredicts the total heat-transfer rate. Refinement of the analysis would require a very complicated solution of the *three*-dimensional conduction problem in the composite slab; fortunately our simple analysis is normally quite adequate for engineering purposes.

Let's now turn to a second geometry. Consider the pipeline insulation of Fig. 14·7. We shall idealize the temperature distribution as *one-dimensional*, meaning that $T = T(r)$. Then, for a unit length of pipe, the conduction area at any point r is $2\pi r$. Hence the energy balance for steady heat flow gives [see (Eq. 14·4)]

$$\frac{d}{dr}\left(2\pi r\,\frac{dT}{dr}\right) = 0 \tag{14·11}$$

Now, if the temperatures are prescribed at two radii r_1 and r_2, the temperature distribution is obtained by solving Eq. (14·11). One finds

$$T = T_1 + \frac{T_2 - T_1}{\ln\,(r_2/r_1)}\ln\frac{r}{r_1} \tag{14·12}$$

Note that in this cylindrical geometry the temperature distribution is logarithmic, not linear as for the slab. Moreover, the heat flux is not constant with r, but

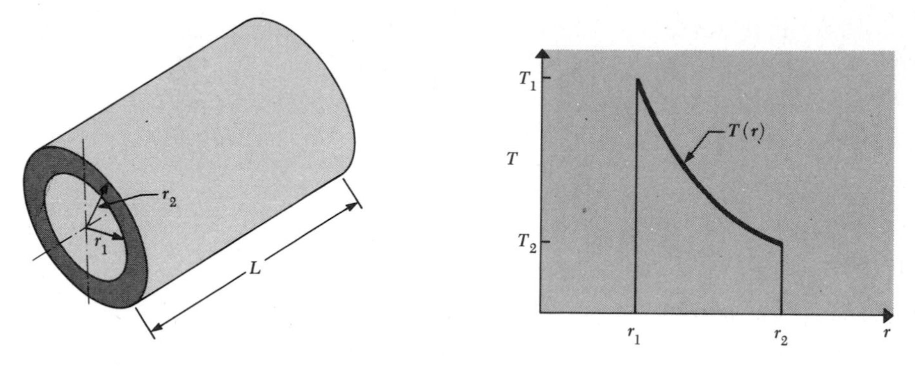

Figure 14·7. Conduction through a hollow cylinder. $T = T_1$ *at* $r = r_1$; $T = T_2$ *at* $r = r_2$

decreases as the area increases (q'' is proportional to $1/r$). The total heat-transfer rate may be calculated from the heat flux at any point. In particular, we may set

$$\dot{Q} = -kA \left.\frac{dT}{dr}\right|_{r=r_1} = -k2\pi r_1 L \frac{(T_2 - T_1)}{\ln (r_2/r_1)} \frac{1}{r_1}$$

So, for a cylinder of length L,

$$\dot{Q} = \frac{2\pi kL}{\ln (r_2/r_1)} (T_1 - T_2) \tag{14·13}$$

If we define the thermal resistance R of the annular insulation as

▶
$$\mathsf{R} = \frac{\ln (r_2/r_1)}{2\pi kL} \tag{14·14}$$

we again have a rate equation analogous to Ohm's law,

▶
$$\dot{Q} = \frac{T_1 - T_2}{\mathsf{R}}$$

Composite annular insulations, and insulations with parallel metallic penetrations, may be treated in the manner of the previous examples, using this definition of thermal resistance.

In nuclear reactors and electric resistance heaters there is a conversion of energy from some form to internal energy, and conduction is the mechanism by which this energy is removed from the material. We can extend our previous analyses to cover the case of such "heat sources." Let's denote by s the local volumetric rate of conversion of energy into internal energy; s will have dimensions of Btu/(h·ft³) or W/m³. The governing differential equations for plane and cylindrical geometries [see Eq. (14·4)] then become

Plane

$$\frac{d}{dx}\left(k\frac{dT}{dx}\right) + \mathsf{s} = 0 \tag{14·15}$$

Cylindrical

$$\frac{d}{dr}\left(kr\frac{dT}{dr}\right) + r\mathsf{s} = 0 \tag{14·16}$$

The development of these equations from first principles is left to the student. Note that we permit k to vary in Eqs. (14·15) and (14·16). We shall illustrate the method of development by consideration of a spherical geometry (Fig. 14·8). An energy balance on an elemental shell within the sphere gives, for a steady state,

$$(4\pi r^2\,dr)\mathsf{s} + \dot{Q}_r = \dot{Q}_{r+dr}$$

As in the development of Eq. (14·4), the heat-transfer rates are related to the temperature gradients by the Fourier rate equation

$$\dot{Q}_r = \left(-k4\pi r^2\frac{dT}{dr}\right)_r$$

$$\dot{Q}_{r+dr} = \left(-k4\pi r^2\frac{dT}{dr}\right)_r + \frac{d}{dr}\left(-k4\pi r^2\frac{dT}{dr}\right)_r dr + \cdots$$

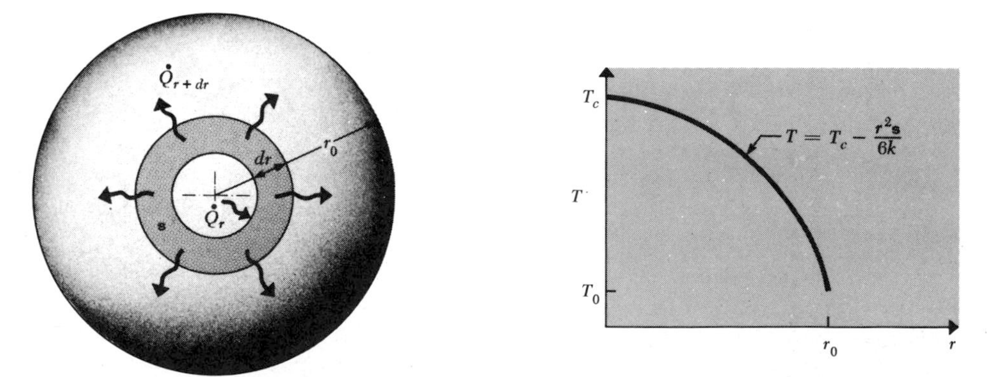

Figure 14·8. A sphere with a uniform heat source s *and the resulting temperature distribution*

Hence the governing equation for the temperature field for radial heat flow in the sphere is

Spherical

$$\frac{d}{dr}\left(kr^2\frac{dT}{dr}\right) + r^2\mathsf{s} = 0 \tag{14·17}$$

One important application of Eq. (14·17) is the calculation of the temperature at the center of a uranium fuel pellet (Fig. 14·8). Integrating Eq. (14·17) from $r = 0$, one has

$$kr^2\frac{dT}{dr} + \frac{r^3\mathsf{s}}{3} = C_1$$

So,

$$k\frac{dT}{dr} = \frac{C_1}{r^2} - \frac{r\mathsf{s}}{3}$$

Since the temperature gradient must be finite at $r = 0$, $C_1 = 0$. Then, we integrate again and obtain

$$T = \frac{-r^2\mathsf{s}}{6k} + C_2$$

The second integration constant C_2 is seen to represent the center temperature T_c. If the surface temperature at $r = r_0$ is T_0, we have

$$T_0 = T_c - \frac{r_0^2\mathsf{s}}{6k} \tag{14·18}$$

For example, suppose that the fuel pellet is a uranium alloy with $k = 13.8$ W/(m·°C), the pellet diameter is 12.7 mm, and the source strength s is 2.07×10^8 W/m³. Then

$$T_c - T_0 = \frac{r_0^2\mathsf{s}}{6k} = \frac{(0.0127)^2(2.07 \times 10^8)}{2^2 \times 6 \times 13.8}$$

$$T_c - T_0 = 100.8°\text{C}$$

Such a high temperature difference over such a small distance could cause serious thermal stresses that might lead to failure of the fuel element.

So far we have only discussed *steady, one-dimensional conduction* heat-transfer analysis. This simplified model breaks down when the temperature field becomes strongly two- or three-dimensional or time-dependent. The theory then involves solution of partial differential equations and is beyond the scope of this introductory treatment. However, a simplified model of transient heat conduction

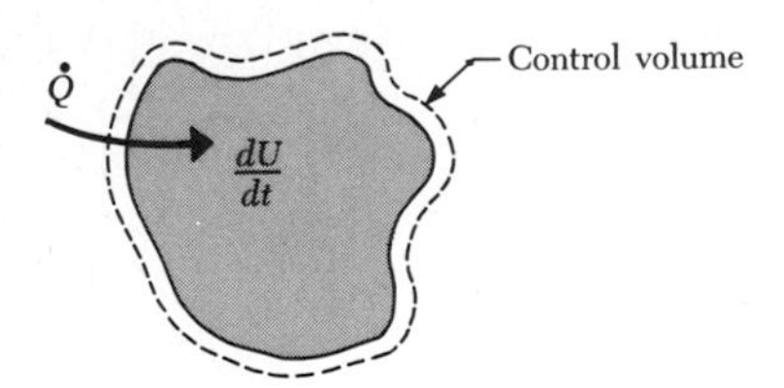

Figure 14·9. How can we compute the temperature-time history?

is often very useful, particularly if the main energy storage occurs in a good conductor and the main thermal resistance is caused by material with little energy storage capability. This model is known as the *lumped-parameter* model, and is a simple extension of the electrical analog already discussed.

Consider a piece of material receiving energy as heat at a rate $\dot{Q}$ (Fig. 14·9). An energy balance on the material gives

$$\dot{Q} = \frac{dU}{dt} \tag{14·19}$$

where U is the total internal energy of the material. If we can justifiably calculate U by the assumption that the temperature is uniform throughout the material at any instant, and if we can treat the material as a simple incompressible solid (Sec. 8·5),

$$dU \approx M\ du = Mc\ dT$$

and the energy balance gives

$$\blacktriangleright \qquad \dot{Q}_{\text{in}} = Mc \frac{dT}{dt} \tag{14·20}$$

If we again regard heat flow as analogous to current flow, and temperature as analogous to voltage, then if we define the *thermal capacitance* of the body by

$$\mathsf{C} \equiv Mc$$

Eq. (14·20) becomes fully analogous to the equation describing a simple electric capacitance,

$$\dot{Q} = \mathsf{C} \frac{dT}{dt} \tag{14·21}$$

To illustrate the use of this model consider the system of Fig. 14·10. Suppose at time $t = 0$ the system is at a uniform temperature T_0, and then the outside face of the asbestos insulation is suddenly heated to temperature T_1. We want to

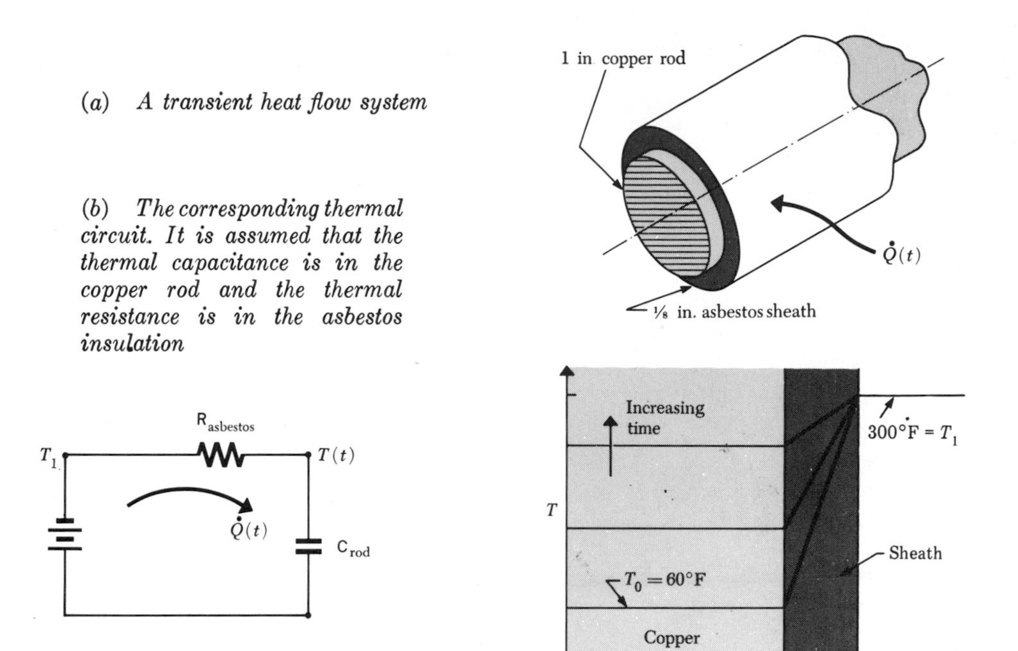

(a) *A transient heat flow system*

(b) *The corresponding thermal circuit. It is assumed that the thermal capacitance is in the copper rod and the thermal resistance is in the asbestos insulation*

(c) *Temperature as a function of time for the system*

Figure 14·10.

calculate the temperature-time history $T(t)$ for the copper rod. The thermal circuit is shown in Fig. 14·10*b*. Note that we model the copper rod as a lump of thermal capacitance and treat the insulation as purely resistive. The lumped-parameter equations are

Energy balance on rod

$$\dot{Q} = \mathsf{C}\,\frac{dT}{dt} \tag{14·22a}$$

Rate equation for insulation

$$\dot{Q} = \frac{1}{\mathsf{R}}(T_1 - T) \tag{14·22b}$$

Combining, we obtain the differential equation for $T(t)$

$$\frac{dT}{dt} = \frac{1}{\mathsf{RC}}(T_1 - T) \tag{14·23}$$

This is to be solved, subject to the *initial condition*

$$T(0) = T_0$$

The solution is

$$T = T_1 - (T_1 - T_0)e^{-t/(\mathsf{RC})} \tag{14·24}$$

which is identical to the analogous expression for the voltage rise in a suddenly energized RC electric circuit.

Let's complete the calculation for a 1-ft length of rod, taking $T_0 = 60°F$ and $T_1 = 300°F$. From Eq. (14·14),

$$\mathsf{R} = \frac{\ln (0.625/0.5)}{2\pi \times 0.096 \times 1} = 0.37 \text{ h·°F/Btu}$$

The rod thermal capacitance is

$$\mathsf{C} = Mc = \left(\frac{\pi D^2 L}{4}\right)\rho c = \frac{\pi(\frac{1}{12})^2}{4} \times 1 \times 559 \times 0.09$$

$$= 0.273 \text{ Btu/°F}$$

The RC product is called the *thermal time constant* of the circuit,

$$\mathsf{RC} = 0.37 \times 0.273 = 0.10 \text{ h} = 6 \text{ min}$$

Six min after the start of heating the rod temperature will be [Eq. (14·24)]

$$T = 300 - (300 - 60)e^{-6/6} = 212°F$$

Note that the temperature rise is about 67 percent completed in one time constant. After three time constants the response is 95 percent complete. Hence knowledge of the thermal time constant provides considerable insight into the transient thermal response.

The lumped-parameter technique provides a simple approach for transient heat conduction but is limited in application by the lumping assumptions. Numerical techniques are often used now to provide computer solutions to more difficult transient problems, and such general-purpose programs are available at modern computing centers.

14·4 CONVECTION

When a fluid passes over a hot solid surface, energy is transferred to the fluid from the wall by the process of conduction. This energy is then carried, or *convected*, downstream by the fluid, and diffused throughout the fluid by conduction within the fluid. This type of energy-transfer process is called *convection heat transfer*.

If the fluid flow process is induced by a pump or other circulating system, the term *forced convection* is used. In contrast, if the fluid flow arises because of fluid buoyancy caused by the heating, the process is called *free* or *natural* convection. Figure 14·11 qualitatively depicts several convection situations.

In order to understand convection it is helpful to have some idea about the temperature distribution in the fluid. If the surface is hotter than the fluid, the temperature distribution will be as in Fig. 14·11*a*. The convection of energy reduces the outward conduction in the fluid, and consequently the temperature gradient decreases away from the surface. If the fluid flows rapidly past the solid body, the region of fluid heated by the wall will be confined to a thin *boundary layer* near the surface. In contrast, in pipe flow (Fig. 14·11*b*) the boundary layer eventually fills the pipe, and thereafter all the flow feels the heating from the wall.

Until the early 1950s the subject of convection heat transfer was largely empirical. Experimenters found that the heat-transfer behavior of fluids in pipes, over wings, and so forth, could be correlated with the flow velocity, fluid properties, and the geometry (but not the material properties) of the solid surfaces. Since that time significant theoretical advances have been made, and it is now often possible to predict the heat-transfer performance in a new configuration with sufficient accuracy for engineering purposes. The theory of course substantiates the correlations, and in fact it is the correlations themselves (whether experimentally or theoretically determined) that the engineer uses in practical design calculations.

The first theoretical advances came in *laminar* flow, where the fluid moves in smooth layers sheared slowly by the action of fluid viscosity. More recent work has dealt with *turbulent* flow, which is much more complicated because of the stochastic nature of the turbulent motion and energy convection. Theories of convection are beyond the scope of this limited treatment, and the interested student is referred to the excellent heat-transfer texts listed at the chapter end.

In order to do simple engineering calculations involving convection, one needs a general understanding of the physical processes, and an understanding of (and the willingness to use) the correlations. There are many parameters affecting the convection heat transfer in a particular geometry. These include the system length scale L, the *fluid* thermal conductivity k, usually the fluid velocity V, density ρ, viscosity μ, specific heat c_P, and sometimes other factors relating to the manner of heating (uniform wall temperature or varying wall temperature). The heat flux from the solid surface will also depend on the surface temperature T_s and fluid temperature T_f, but it is usually assumed that only the *difference* in these temperatures, $\Delta T = T_s - T_f$, is significant. However, if the fluid properties vary markedly over the convecting region, then the absolute temperatures T_s and T_f may also be important factors in the correlation. It is obvious that with so many important variables any specific correlations will be unwieldy, and consequently the correlations are usually represented in terms of *dimensionless groupings* that permit much simpler representations. Also, the factors with less important influence, such as

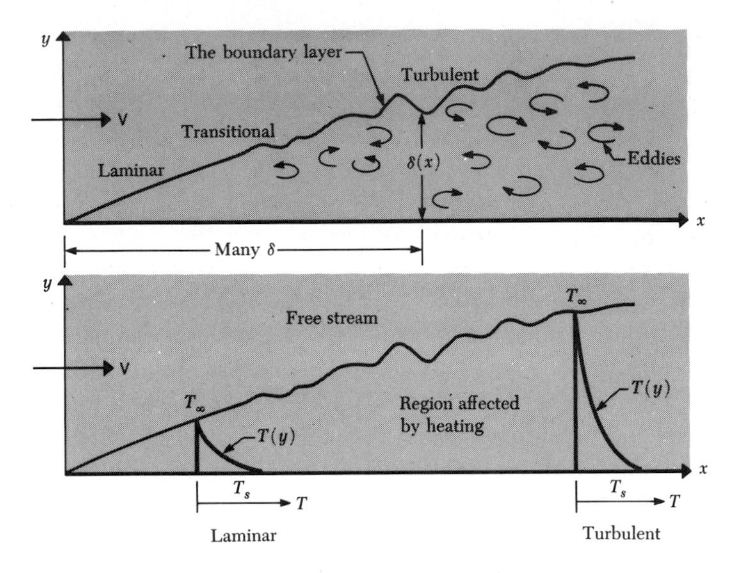

(a) The thermal boundary layer (thickness greatly enlarged) for flow over a surface at high Reynolds number. In turbulent flow the action of the eddies flattens the temperature profile. The boundary layer thickness $\delta(x) \ll x$

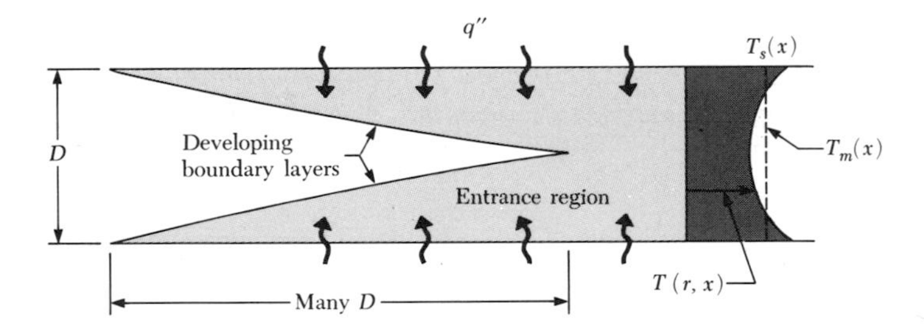

(b) Flow in the entrance region of a pipe. For laminar flow, $\mathrm{Re} = V_m D\rho/\mu < 2000$

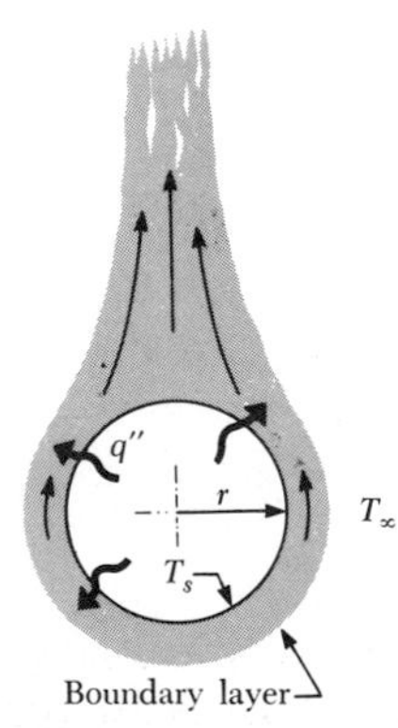

(c) Natural convection from a horizontal circular cylinder. The warm fluid rises because of buoyant effects

Figure 14·11.

fluid property variation and wall temperature distribution, are often ignored to simplify the correlations, and adequate estimations ($\pm$ 20 percent) of the heat-transfer rate can often be obtained from the simplified nondimensional correlations.

The heat-transfer rate is usually expressed through the *heat flux*, which may vary along the solid surface. The correlation for pipes and other heat-exchanger surfaces is generally expressed in terms of the *average* heat flux, while for external boundary layer flows, such as that over a wing or compressor blade, the *local* heat flux is usually used. The heat flux (local or average) is in turn related to the imposed temperature difference through the *convection heat-transfer coefficient* (*convective conductance*) h defined by

$$q'' = \mathsf{h}\,\Delta T \tag{14·25}$$

This may be interpreted as the *convection rate equation.* If h and ΔT are known, q'' can be calculated, and it is the correlations which give the engineer values for h. Note that h has the dimensions of *heat flux per unit of temperature difference*, for example, $W/(m^2 \cdot {}^\circ C)$, or $Btu/(h \cdot ft^2 \cdot {}^\circ F)$.

A thermal resistance for convection can be defined as

$$\mathsf{R} \equiv \frac{1}{\mathsf{h}A} \tag{14·26}$$

Then Eq. (14·25) can be written as

$$q'' = \frac{\Delta T}{\mathsf{R}} \tag{14·27}$$

A problem involving conduction through a wall with convection at the surfaces can now be treated by using a thermal circuit with series resistances as noted in Fig. 14·12. An *overall conductance* U per unit area can be defined for such a problem as

$$U \equiv \frac{1}{\sum_i \mathsf{R}_i} = \frac{1}{1/\mathsf{h}_0 + L_a/k_a + L_b/k_b + 1/\mathsf{h}_i}$$

The heat flux for this case is then given by

$$q'' = U(T_1 - T_5)$$

With boundary layer flow it is customary to use the free-stream temperature (the fluid temperature outside of the boundary layer) in defining ΔT. In pipe flow it is more convenient and sensible to use a suitable *average* temperature. The

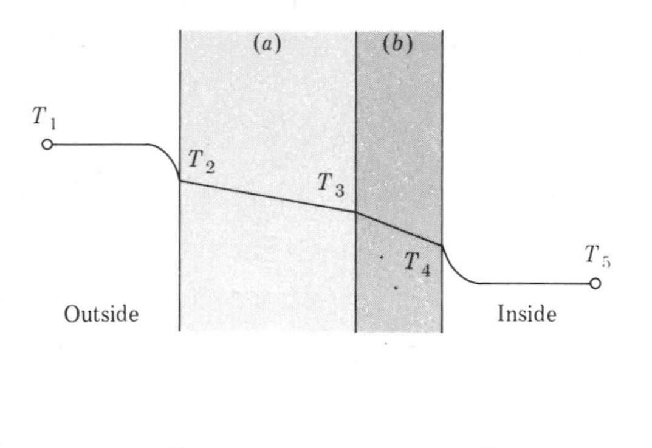

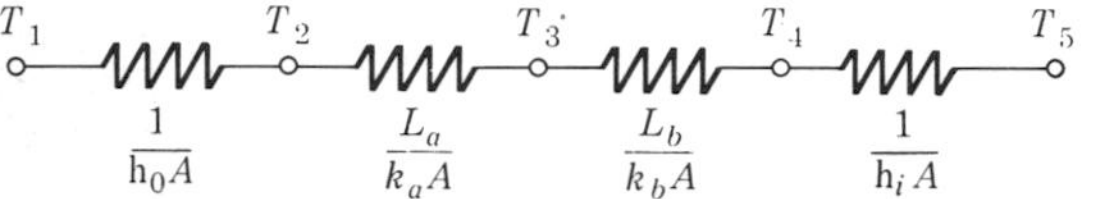

Figure 14·12. (a) Combined convection and conduction; (b) the thermal circuit

same idea is used in hydrodynamic analysis, where, for example, the mass flow in the tube is written as

$$\dot{M} = A\rho V_m$$

and V_m is the *average* velocity, defined by this equation. Similarly, we write the energy-flow rate in a pipe as

$$\dot{E} = \dot{M} h_m$$

where h_m is the *mixed mean enthalpy*. Note that if the heated flow were mixed adiabatically h_m would be the resulting enthalpy. The *mixed mean temperature* is that associated with h_m, that is, it is the temperature at which the fluid would emerge from an adiabatic zero pressure drop mixer. The concept of mixed mean properties permits us to treat the energy flows in boundary layers and pipes on a one-dimensional basis. The mixed mean temperature is especially important in internal flow, where it is used as the fluid temperature in the calculation of ΔT.

Two dimensionless groupings involving h are commonly used. These are the *Nusselt number*,

▶ $$\mathrm{Nu} \equiv \frac{hL}{k} \tag{14·28a}$$

and the *Stanton number*,

▶ $$\mathrm{St} \equiv \frac{h}{V \rho c_P} \tag{14·28b}$$

Once either of these numbers has been determined (by correlation, theory, or a good engineering guess), h, and then q'', and then finally $\dot{Q}$, can be calculated.

Nusselt and Stanton numbers are usually correlated in terms of the *Prandtl number*,

$$\text{Pr} \equiv \frac{\mu c_P}{k} \tag{14·29}$$

and the *Reynolds number*,

$$\text{Re} \equiv \frac{\mathsf{V} L \rho}{\mu} \tag{14·30}$$

The Prandtl number is indicative of the relative ability of the fluid to diffuse momentum and internal energy by molecular mechanisms. For oils, which are highly viscous, Pr is typically large (100 to 10,000), indicating rapid diffusion of momentum (by viscous action) compared to the slow diffusion of internal energy. In contrast, liquid metals typically have Pr = 0.003 to 0.01, indicating more rapid diffusion of internal energy. For almost all gases Pr ≈ 0.7, and for water Pr ≈ 2 to 8, depending upon the temperature.

The Reynolds number is indicative of the relative importance of inertial and viscous effects in the fluid motion. At low Re the viscous effects dominate, and the motion is laminar. At high Re the inertial effects lead to turbulence and dominate the momentum and energy-transfer processes in turbulent flow. Hence the nature of the correlation is usually quite different at low and high Re. It is also often difficult to predict the Nusselt number accurately in the *transition region* between laminar and turbulent flow.

In natural convection the Nusselt number is usually related to the Prandtl number and the *Grashof number*. The Grashof number is defined by

$$\text{Gr} = \frac{g\beta(\Delta T)L^3\rho^2}{\mu^2} \tag{14·31}$$

where β is the isobaric compressibility (Sec. 4·7). The Grashof number can be viewed as a measure of the relative strength of buoyancy and viscous forces. Natural convection is usually suppressed at sufficiently small Gr, begins at some *critical* value of Gr, depending upon the system, and then becomes more and more effective as Gr increases.

To summarize, convection heat transfer correlations are usually expressed in the form

$$\text{Nu} = f(\text{Pr, Re}) \qquad \text{or} \qquad \text{St} = f(\text{Pr, Re})$$

or, for natural convection,

$$\text{Nu} = f(\text{Gr, Pr})$$

and as such apply to a particular flow geometry. The surface temperature distribution and fluid property variations usually have minor influence on these correlations and often can be ignored for engineering calculations. For accurate prediction (± 10 percent) these secondary effects should be considered, and the beginning heat-transfer practitioner should at least be aware of their existence. The correlations in laminar flow are usually developed analytically and confirmed by experiment. Those for turbulent flow are usually determined experimentally but can sometimes be generated by an experienced heat-transfer analyst if needed for a new situation. Appendix C includes several graphs and tables giving representative correlations for convection heat transfer.

We remark that this discussion is limited to *single-phase* flow. Convection heat transfer in two-phase flow is more complicated and beyond the scope of our brief introduction.

The computing procedure for convection analysis is as follows:

1. Establish the values of Re (or Gr) and Pr, and determine Nu or St from the appropriate correlation.
2. Calculate h from the Nu or St value.

3*a*. If Δt is specified, calculate q'' or $\dot{Q}$ from $q'' = \mathsf{h}\,\Delta t$ or $\dot{Q} = \mathsf{h}A\,\Delta T$.

3*b*. Or, if $\dot{Q}$ or q'' is specified, calculate ΔT.

Examples of such calculations constitute the next section. The physical aspects of convection are discussed further in these examples, and they should be studied for this content.

14·5 CONVECTION CALCULATION EXAMPLES

Electrically heated pipe. Air flows at 5 kg/h through an electrically heated 0.5-cm-diameter tube, 0.5 m long, entering at 100°C. The electric power dissipation is 200 W. Calculate the outlet temperature and the maximum tube-wall temperature (Fig. 14·13).

We shall first use thermodynamics (an energy balance) on the air inside the tube to determine the mixed mean exit temperature T_2. From an energy balance

$$\dot{M}(h_2 - h_1) = \dot{Q}$$

Treating the air as a perfect gas, $h_2 - h_1 = c_P(T_2 - T_1)$. Then,

$$T_2 = T_1 + \frac{\dot{Q}}{\dot{M}c_P} = 100 + \frac{200 \times 3600}{5 \times 1.004 \times 10^3} = 243.4°\text{C}$$

We next use the heat-transfer data to determine the wall temperature. Now, at any location along the tube the heat flux is

$$q'' = \mathsf{h}(T_s - T_m)$$

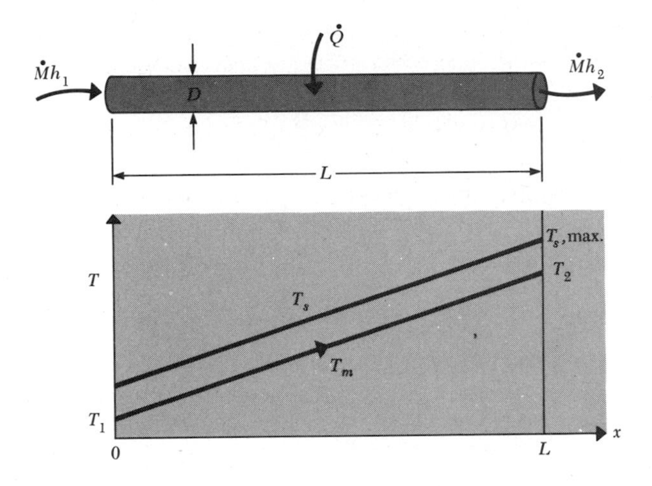

Figure 14·13. The wall and mixed mean temperatures along an electrically heated tube

where T_s and T_m are the local wall and mixed mean fluid temperatures, respectively. If we assume h is constant along the tube, the maximum wall temperature will occur at the exit. To choose the proper correlation from Table C·8, we need to know the Reynolds and Prandtl numbers. These contain the fluid properties, which vary somewhat with temperature. Let's base the Reynolds and Prandtl numbers on an average temperature of 170°C. Then, from Table C·5,

$$\mu = 24.5 \times 10^{-6} \text{ N·s/m}^2$$

$$k = 0.037 \text{ W/(m·K)}$$

$$\text{Pr} = 0.68$$

Then
$$\text{Re} = \frac{VD\rho}{\mu} = \frac{\dot{M}D}{A\mu} = \frac{4\dot{M}}{\pi\mu D}$$

$$= \frac{4 \times 5}{3600 \times \pi \times 24.5 \times 10^{-6} \times 0.005} = 14{,}435$$

Since the Reynolds number is well over the transition value of 2300, the flow is turbulent, and the appropriate correlation is (Table C·8)

$$\text{Nu} = 0.022 \text{ Re}^{0.8}\text{Pr}^{0.6}$$

$$= 0.022 \times (14{,}435)^{0.8} \times (0.68)^{0.6} = 37.1$$

Then, since Nu = hD/k,

$$\text{h} = \text{Nu}\frac{k}{D} = \frac{37.1 \times 0.037}{0.005} = 274.5 \text{ W/(m}^2\text{·°C)}$$

or
$$\text{h} = 48.3 \text{ Btu/(h·ft}^2\text{·°F)}$$

Now the heat flux is given by

$$q'' = \frac{Q}{L\pi D} = \frac{200}{\pi \times 0.5 \times 0.005} = 25{,}464 \text{ W/m}^2$$

Then, the local wall-to-fluid temperature difference is

$$T_s - T_m = \frac{q''}{\mathsf{h}} = \frac{25{,}464}{274.5} = 92.8°\text{C}$$

The maximum tube-wall temperature is therefore

$$243.4 + 92.8 = 336.2°\text{C}$$

To assess the accuracy of this prediction, we note that the fluid at the exit will have a temperature distribution across the tube ranging from about 335°C down to less than 243°C. Table C·5 reveals that k varies by about 15 percent over this range. We actually used a k appropriate for 170°C, which is about 20 percent below the k at 335°C. These considerations suggest that our prediction for h, hence for $T_s - T_m$, might be off by as much as 20 percent, which is about 20°C. Since the conductivity we used was probably low, the tube temperature we predict is probably high by 0 to 20°C. We often accept this accuracy in an engineering situation. The alternative is to use a more elaborate correlation which corrects for fluid property variations; such a correlation can be found in a heat-transfer textbook or reference manual.

Turbulent flow in a pipe consists of randomly moving globs of fluid, or "eddies," that serve as carriers of momentum and energy within the flow (Fig. 14·14). Since the eddies are very effective at energy transport, the heat-transfer coefficients in turbulent flow are typically much greater than for laminar flow. For example, we calculated Nu = 37 in this turbulent flow and would have found Nu = 4.36 in laminar flow. Much smaller temperature differences are therefore required to sustain a given heat flux in turbulent flow, hence high-performance heat-transfer devices are usually designed for turbulent flow. In

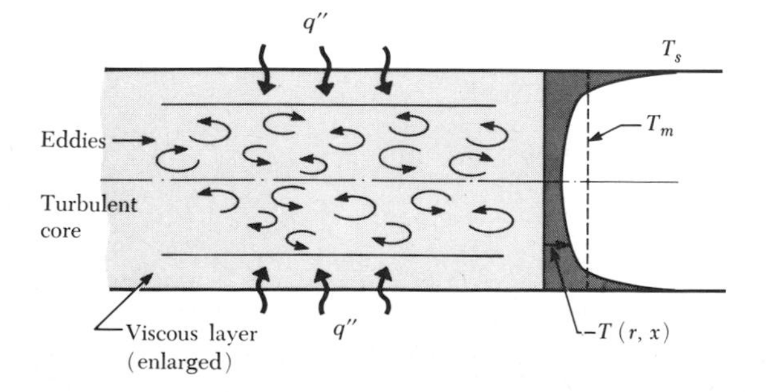

Figure 14·14. Turbulent flow in a heated circular tube. The eddies provide an effective means for energy transport

contrast, where thermal protection from hot gases is required, such as on the nose of a reentry body, laminar flow gives lower heat input, hence designers (and astronauts) strive to maintain laminar flow.

Window convection. A large building has glass walls 10 ft high and 20 ft wide. Calculate the convection heat transfer from the glass when the glass surface temperature is 60°F and the outdoor temperature is 20°F.

This is a problem in natural convection (Fig. 14·15). We can model the window as a *vertical plate*, for which correlations are given in Table C·8. First we must calculate the *Grashof* number; using data in Table C·5, at an average temperature of 40°F, for air

$$\mu = 0.042 \text{ lbm/(h·ft)} \qquad k = 0.014 \text{ Btu/(h·ft·°F)} \qquad \text{Pr} = 0.71$$

For ideal gases the isobaric compressibility β is simply $1/T$, so

$$\beta = \frac{1}{500 \text{ R}} = \frac{0.002}{\text{R}}$$

The air density is also needed. It can be calculated from the perfect-gas equation of state,

$$\rho = \frac{P}{RT} = \frac{14.7 \times 144}{53.3 \times 500} = 0.0794 \text{ lbm/ft}^3$$

Then,
$$\text{Gr} = \frac{\rho^2 g\beta(T_s - T_f)L^3}{\mu^2}$$

$$= \frac{(0.0794)^2 \times 32.2 \times (3600)^2 \times 0.002 \times (60 - 20) \times 10^3}{0.042^2}$$

$$= 1.2 \times 10^{11}$$

So,
$$\text{GrPr} = 1.2 \times 10^{11} \times 0.71 = 8.5 \times 10^{10}$$

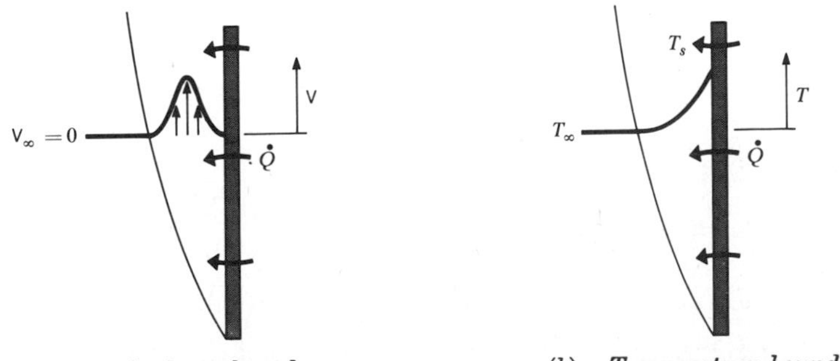

(*a*) *Velocity boundary layer* (*b*) *Temperature boundary layer*

Figure 14·15. Velocity and temperature profiles for natural convection from a vertical wall

According to Table C·8, in this range the natural convection will be turbulent, and the appropriate correlation is

$$\mathrm{Nu}_{av} = 0.021\ (\mathrm{GrPr})^{0.4}$$

Note that this gives the *average* Nusselt number over the height of the plate. This is just what we need to calculate the total heat-transfer rate.

$$\mathrm{Nu}_{av} = 0.021 \times (8.5 \times 10^{10})^{0.4} = 495$$

Then $$\mathsf{h}_{av} = \mathrm{Nu}_{av} \frac{k}{L} = \frac{495 \times 0.014}{10} = 0.7\ \mathrm{Btu/(h \cdot ft^2 \cdot {}^\circ F)}$$

So $$\dot{Q} = \mathsf{h}_{av} A \cdot (T_s - T_f) = 0.7 \times (10 \times 20) \times (60 - 20)$$
$$= 5600\ \mathrm{Btu/h}$$

This corresponds to an average heat flux of

$$q''_{av} = \frac{\dot{Q}}{A} = \frac{5600}{10 \times 20} = 28\ \mathrm{Btu/(h \cdot ft^2)}$$

The window heat loss $\dot{Q}$ is equivalent to 1650 W and, at 3 cents/kW·h the luxury of the glass could cost the user about 5 cents/h, or a few hundred dollars per year, on his or her heating bill, per window this size.

Boundary layer flow. Air at 60°F, 2 atm, flows at 25 ft/s between the two metal sheets shown in Fig. 14·16. Calculate the heat-transfer rate from each sheet when the sheet temperature is 100°F.

At first thought it might seem reasonable to model the space between the sheets by an "equivalent pipe," and the approach would be quite reasonable if the plates were much longer in comparison to their spacing. However, the boundary layers that form on the plate are relatively thin in this situation and do not fill the flow passage. Hence a boundary layer treatment is appropriate, and we shall use this example to illustrate several aspects of boundary layer convection.

The *flat plate* (Table C·8) forms a satisfactory model for this problem. Let's first calculate the Reynolds number at the end of the 6-in plates. The air density is

$$\rho = \frac{P}{RT} = \frac{2 \times 14.7 \times 144}{53.3 \times 540} = 0.146\ \mathrm{lbm/ft^3}$$

From Table C·5, at an average temperature, for air

$$\mu = 0.044\ \mathrm{lbm/(h \cdot ft)} = 1.22 \times 10^{-5}\ \mathrm{lbm/(s \cdot ft)}$$
$$k = 0.015\ \mathrm{Btu/(h \cdot ft \cdot {}^\circ F)}$$
$$\mathrm{Pr} = 0.7$$

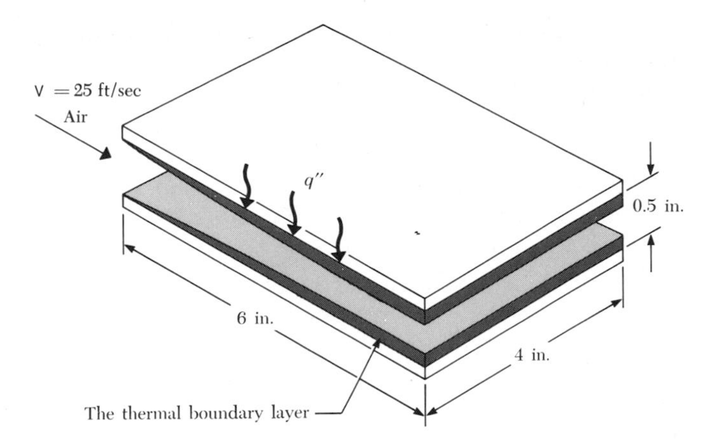

Figure 14·16. Flow between parallel plates. The thermal boundary layers do not meet, so an external flow analysis is used

So, with $L = 6$ in $= 0.5$ ft,

$$\text{Re}_L = \frac{\mathsf{V}\rho L}{\mu} = \frac{25 \times 0.146 \times 0.5}{1.22 \times 10^{-5}}$$
$$= 149{,}500$$

Hence the entire boundary layer is laminar, and the *average* Nusselt number is given by (Table C·8)

$$\text{Nu}_{\text{av}} = 0.664\ \text{Re}_L^{0.5}\text{Pr}^{0.33}$$
$$= 0.664 \times 149{,}500^{0.5} \times 0.7^{0.33} = 228$$

Then $$\mathsf{h}_{\text{av}} = \text{Nu}\frac{k}{L} = \frac{228 \times 0.015}{0.5} = 6.8\ \text{Btu/(h·ft}^2\text{·°F)}$$

So $$\dot{Q} = \mathsf{h}A(T_s - T_f) = 6.8 \times (0.5 \times 0.33) \times (100 - 60)$$
$$= 45\ \text{Btu/h per plate}$$

To check on the boundary layer nature of the flow, let's determine the thermal boundary layer thickness at the end of the plate. The hydrodynamic laminar boundary layer thickness δ can be estimated from†

$$\delta = 5\sqrt{\frac{x\mu}{\mathsf{V}\rho}} = 5\sqrt{\frac{0.5 \times 1.22 \times 10^{-5}}{25 \times 0.146}} = 6.4 \times 10^{-3}\ \text{ft} = 0.078\ \text{in}$$

† See your fluid mechanics text for a derivation.

So, the hydrodynamic boundary layers clearly do not fill the gap between the plates. Note that the boundary layer is extremely thin in comparison to its length.

The thermal boundary layer (the region affected by the heating) will in general be of different thickness than the hydrodynamic boundary layer. Since the Pr number is a measure of the relative ability of the fluid to diffuse momentum in comparison to internal energy, low-Prandtl-number fluids have thin hydrodynamic boundary layers and thick thermal boundary layers. In contrast, high-Prandtl-number fluids have thick hydrodynamic boundary layers and thin thermal layers. For gases, with $\mathrm{Pr} \approx 1$, the two layers will be about the same thickness, hence we can take 0.078 in as an estimate of the thermal boundary layer thickness for this example.

House cooling load. A house is constructed of double brick walls as noted in Fig. 14·17*a*. One wall is 3 m high and 6 m long. The wind is 4.5 m/s along the outside surface, and the outside ambient temperature is 40°C. It is desired to maintain the inside air at 25°C. Calculate the heat transfer through this wall.

To determine the heat transfer we will need to determine an h for the inside wall and for the outside wall surface. Also, we will need the conductivities of the wall materials. From Appendix C we find

$$k_b = 0.69 \text{ W/(m·K)}$$

$$k_m = 1.73 \text{ W/(m·K)}$$

$$\mu_{\text{air}} = 19 \times 10^{-6} \text{ N·s/m}^2$$

For the outside h_0 we can treat the wall as a flat plate with air flowing along it at 4.5 m/s. To calculate the Reynolds number we will need the air density. At 1 atm,

$$\rho = \frac{P}{RT} = \frac{101{,}325}{286.9 \times 313} = 1.128 \text{ kg/m}^3$$

Thus

$$\mathrm{Re} = \frac{VL\rho}{\mu} = \frac{4.5 \times 6 \times 1.128}{19 \times 10^{-6}}$$

$$= 1.6 \times 10^6$$

At this Re value the boundary layer is turbulent, since Re > 300,000. We shall neglect the fact that the first part of the wall sees a laminar boundary layer and treat the whole wall as if it had a turbulent boundary layer. Table C·8 gives for $\mathrm{Re}_L > 300{,}000$,

$$\mathrm{Nu} = 0.0295 \, \mathrm{Re}^{0.8} \mathrm{Pr}^{0.6}$$

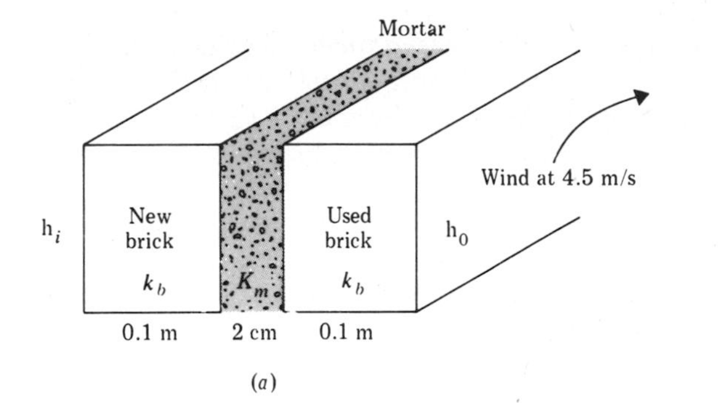

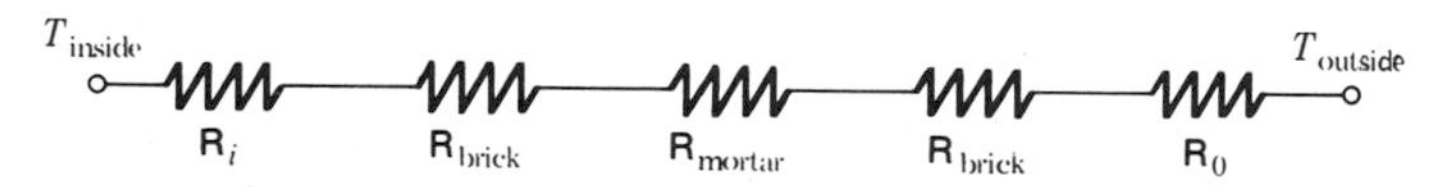

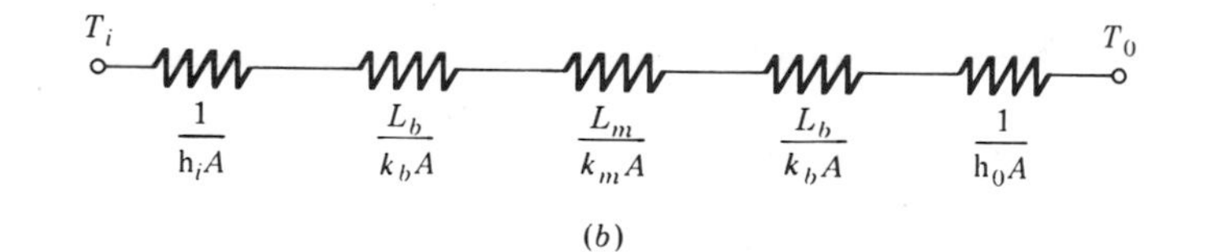

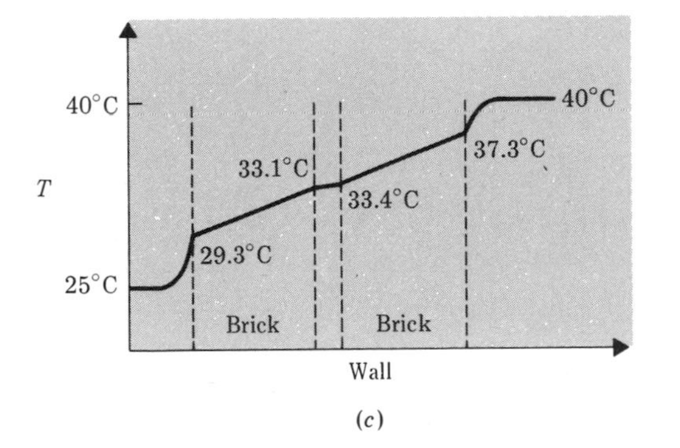

Figure 14·17. (a) Double brick wall construction; (b) the thermal circuit; (c) the temperature distribution

for $0.5 < \mathrm{Pr} < 10$. For air $\mathrm{Pr} = 0.70$, so this correlation is the appropriate one. We find for the *local* Nusselt number at the end of the wall

$$\mathrm{Nu}_x = 0.0295 \times (0.70)^{0.6} \times (1.6 \times 10^6)^{0.8}$$

$$\mathrm{Nu} = 2200$$

and

$$h_0 = \mathrm{Nu}\frac{k}{L} = \frac{2200 \times 0.0271}{6} = 9.9 \ \mathrm{W/(m^2 \cdot {}^\circ C)}$$

where we have determined the k for air from Appendix C.

An average h can be determined by integrating the local value from $x = 0$ to $x = L$. The resulting coefficient on the correlation is 0.036 in place of 0.0295. However, our analysis neglected the initial laminar boundary layer and the effect of surface roughness of the bricks, so we can approximate h_0 by using the value 9.9 W/(m$^2 \cdot$°C).

The inside wall surface h_i is due to natural convection if we assume that the air inside the house is still. We have just seen in the example on window convection that such h values are about 1 Btu/(h·ft^2·°F). While our temperature difference is smaller than in that example, the effect of an air conditioning blower is to create some forced convection inside the house. Thus we can obtain a good answer with

$$\mathsf{h}_i = 1 \text{ Btu/(h}\cdot\text{ft}^2\cdot{}^\circ\text{F)} \cong 6 \text{ W/(m}^2\cdot{}^\circ\text{C)}.$$

The overall resistance to heat transfer per m^2 of wall area can then be determined by using the series resistance noted in Fig. 14·17*b*. We have

$$\begin{aligned} \mathsf{R} &= \frac{1}{\mathsf{h}_i} + \frac{L_b}{k_b} + \frac{L_m}{k_m} + \frac{L_b}{k_b} + \frac{1}{\mathsf{h}_0} \\ &= \frac{1}{6} + \frac{0.1}{0.69} + \frac{0.02}{1.73} + \frac{0.1}{0.69} + \frac{1}{9.9} \\ &= 0.57 \text{ m}^2\cdot{}^\circ\text{C/W} \end{aligned}$$

The overall conductance is given by $U = 1/\mathsf{R} = 1.75$ W/m$^2\cdot$°C.

The heat transfer per m^2 through the wall is given by

$$q'' = \frac{\Delta T}{\mathsf{R}} = U\,\Delta T = 1.75 \times (40 - 25) = 26.3 \text{ W/m}^2$$

For the total wall area we have

$$\dot{Q} = 26.3 \times 3 \times 6 = 474 \text{ W or } 1617 \text{ Btu/h}$$

Knowing the heat flux we can establish the temperature distribution through the wall. This is noted in Fig. 14·17*c*. To obtain the overall thermal load on the house one would also need the energy input through the windows, ceiling, and other walls. If the wall is directly in the sun, it will reach a surface temperature higher than the ambient, and this effect must also be included in the analysis. The actual response of the wall temperature to the ambient surroundings is of course also a transient problem. A complete analysis of the cooling load for the house would have to take into account all these phenomena.

14·6 FINS

Fins are often used in heat-exchange devices to increase the convection heat-transfer area, thereby improving the heat-transfer performance. Figure 14·18 shows some typical examples of finned surfaces. Automobile radiators, household refrigerator condensers, and transistor heat sinks are familiar examples.

The theory of fin heat transfer is based on a combination of an energy balance and rate equations. These provide one with the differential equation for the temperature distribution in the fin. We shall illustrate the essential ideas by a special case. Figure 14·19 shows a simple fin of uniform rectangular cross section. In setting up the governing equation we must include the effects of convective heat transfer between the fin and the surrounding fluid and the conduction heat transfer within the fin. If the fin is thin relative to its length, we can treat the conduction as being one-dimensional and assume that $T = T(x)$. Further simplifications are obtained if we assume a steady state, treat the fin conductivity as constant, neglect radiation, and assume that the convection coefficient h and fluid temperature T_∞ are independent of x. Satisfactory solutions to engineering problems are usually obtained with this simple model.

With these simplifying assumptions the energy balance on the elemental control volume of Fig. 14·19 gives

$$\underbrace{\dot{Q}_x}_{\text{energy-input rate}} = \underbrace{\dot{Q}_{x+dx} + \dot{Q}_c}_{\text{energy-output rate}}$$

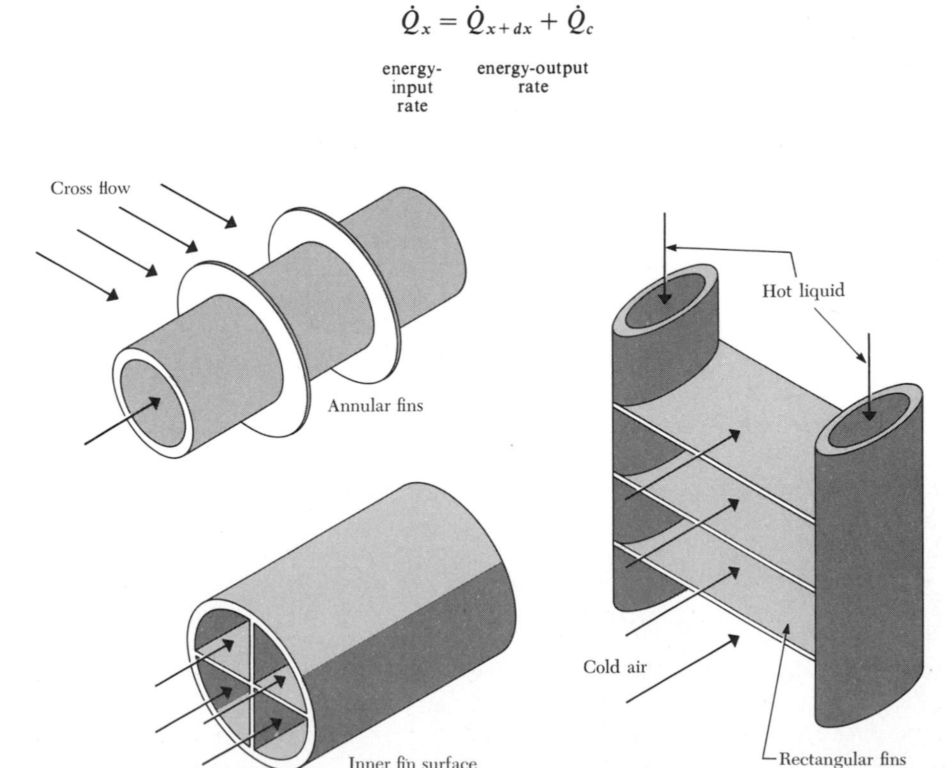

Figure 14·18. Types of fins

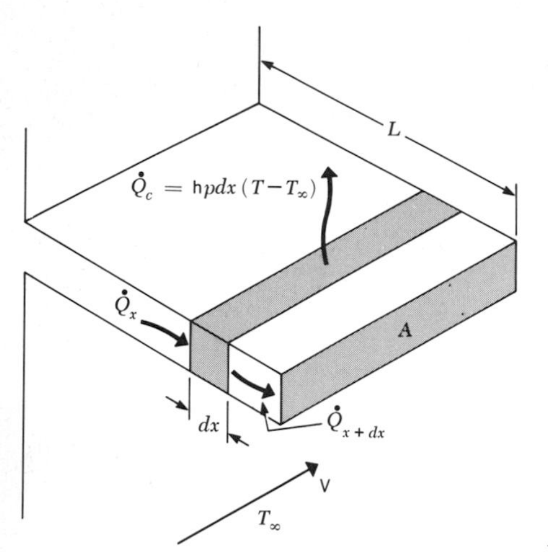

Figure 14·19. The control volume for a fin analysis

The conduction rate equation gives

$$\dot{Q}_{x+dx} - \dot{Q}_x = -kA \frac{d^2T}{dx^2} dx$$

where A is the conduction *cross-sectional area* (not surface area) of the fin. The convection rate equation gives

$$\dot{Q}_c = \mathsf{h}p\, dx(T - T_\infty)$$

where p is the convection perimeter of the fin (twice the width for a wide flat fin). Combining, the differential equation for $T(x)$ is found as

▶ $$\frac{d^2T}{dx^2} - \frac{\mathsf{h}p}{kA}(T - T_\infty) = 0 \qquad (14\cdot32)$$

This equation must be solved, subject to appropriate boundary conditions on $T(x)$. Three commonly used sets are

(*a*) *Infinite fin*

$$T(0) = T_0, \qquad T(\infty) \to T_\infty \qquad (14\cdot33a)$$

(*b*) *End temperatures specified*

$$T(0) = T_0, \qquad T(L) = T_e \qquad (14\cdot33b)$$

(*c*) *End leak negligible*

$$T(0) = T_0, \quad \left(\frac{dT}{dx}\right)_{x=L} = 0 \qquad (14\cdot33c)$$

Case (*a*) is useful for long fins, (*b*) is useful when the fin is stretched between two surfaces of known temperatures, and (*c*) is useful for thin fins of finite length or in configurations where $x = L$ is a point of symmetry in $T(x)$. For the infinite-fin model the solution may be written as

$$\frac{T - T_\infty}{T_0 - T_\infty} = e^{-mx} \qquad (14\cdot34)$$

where $$m^2 \equiv \frac{\mathsf{h}p}{kA}$$

The solutions for the other two cases are left as problems for the student.

The local convective heat flux $q''(x)$ for the infinite-length fin is

$$q''(x) = \mathsf{h}(T - T_\infty) = \mathsf{h}(T_0 - T_\infty)e^{-mx}$$

Note that the heat flux becomes less and less as x increases. The total heat-transfer rate can be found either by integrating the heat flux over the fin surface, or by considering the conduction into the fin at its root. Since the energy that leaves the fin by convection must have entered by conduction at the base, or root, of the fin, the two methods are equivalent. The latter is more convenient and gives

$$\dot{Q}_0 = -kA \left.\frac{dT}{dx}\right|_{x=0}$$

which for our model is

$$\dot{Q}_0 = kA \cdot m(T_0 - T_\infty) = \sqrt{\mathsf{h}pkA}\,(T_0 - T_\infty) \qquad (14\cdot35)$$

The form of this expression clearly shows that it is a combination of convection and conduction that acts to remove the energy conducted in at the fin root.

For a fin of finite length L the end-leakage-negligible model gives

$$\dot{Q}_0 = \sqrt{\mathsf{h}pkA}\,(T_0 - T_\infty)\tanh(mL) \qquad (14\cdot36)$$

If all this fin were at the root temperature T_0, more heat would be removed from the surface; let's call this maximum possible heat-transfer rate $\dot{Q}_{max}$.

Then $$\dot{Q}_{max} = \mathsf{h}pL(T_0 - T_\infty)$$

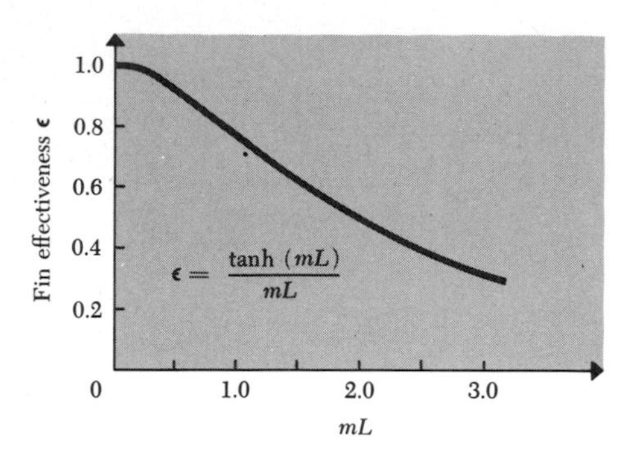

Figure 14·20. Effectiveness of rectangular fins

The *fin effectiveness* ε is defined by

$$\varepsilon = \frac{\dot{Q}_0}{\dot{Q}_{max}} \tag{14·37}$$

and is the ratio of the actual fin heat transfer to that which could be convected with all the surface at the fin root temperature. For the end-leakage-negligible model Eqs. (14·36) and (14·37) yield

$$\varepsilon = \frac{1}{mL} \tanh(mL) \tag{14·38}$$

Figure 14·20 shows $\varepsilon(mL)$ as given by Eq. (14·38). Note that increasing the length of a fin provides diminishing returns, and seldom does one find heat-exchanger equipment with effectiveness below 35 percent ($mL > 3$).

To illustrate a fin calculation, suppose we wish to transfer 50,000 Btu/h from a high-temperature surface at 800°F by blowing ambient air at 100°F between aluminum fins attached to the surface. The fins will be 6 in long, $\frac{1}{16}$ in thick, and 1 ft wide. Suppose a convection calculation reveals that $\mathsf{h} = 20$ Btu/(h·ft²·°F) in this situation, and we wish to determine how many fins must be used. The parameters are

$$\mathsf{h} = 20 \text{ Btu/(h·ft}^2\text{·°F)} \qquad p = 2 \times 1 = 2 \text{ ft}$$

$$k = 120 \text{ Btu/(h·ft·°F)} \qquad A = \frac{\frac{1}{16} \times 12}{144} = 0.0052 \text{ ft}^2$$

So

$$m = \sqrt{\frac{\mathsf{h}p}{kA}} = \sqrt{\frac{20 \times 2}{120 \times 0.0052}} = 8.0/\text{ft}$$

Using the infinite-fin model,

$$\dot{Q}_0 = 120 \times 0.0052 \times 8.0 \times (800 - 100) = 3500 \text{ Btu/h}$$

Let's check to see if the finite-fin model is needed. Since

$$mL = 8.0 \times 0.5 = 4.0$$

Eq. (14·36) gives

$$\dot{Q} = 3500 \times \tanh(4.0) \approx 3500$$

The fins are clearly long enough to be considered infinite (they are probably too long for good design). Then, to obtain 50,000 Btu/h, a total of 50,000/3500, or about 14, fins are required. A good engineer would recommend that the fins be shortened. For 3-in fins $mL = 2.0$, and the heat transfer rate per fin is

$$\dot{Q} = 3500 \times \tanh(2.0) = 3380 \text{ Btu/h}$$

and 14 or 15 fins will still remove the required 50,000 Btu/h but will use only half the material of the 6-in fins.

14·7 HEAT EXCHANGERS

A heat exchanger is typically a device in which energy is transferred from one fluid to another across a solid surface. Exchanger analysis and design therefore involve both convection and conduction. The *thermal circuit* concept can be used to simplify the analysis, and Fig. 14·21 shows the circuit at some point in the exchanger. Note that resistances have been assigned to the fluid. The *convection resistances* (unit area basis) are defined by

$$\mathsf{R}_C \equiv \frac{1}{\mathsf{h}}$$

The convection rate equation then becomes

$$q'' = \frac{\Delta T}{\mathsf{R}_C}$$

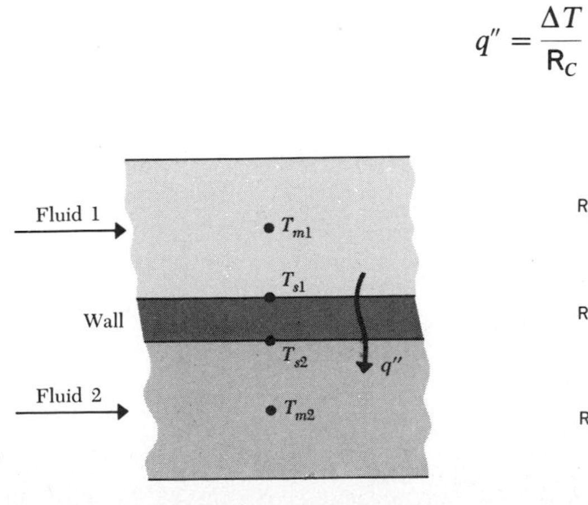

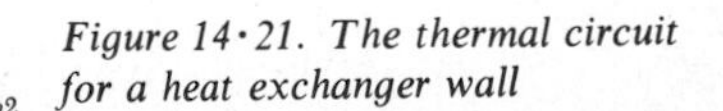

Figure 14·21. The thermal circuit for a heat exchanger wall

analogous again to Ohm's law. Treating the wall as a plane slab of thickness δ, the wall resistance is [unit area basis; see Eq. (14·10)]

$$\mathsf{R}_w = \frac{\delta}{k_w}$$

and we then have

$$q'' = \frac{(T_{s1} - T_{s2})}{\mathsf{R}_w}$$

The *overall* resistance is then (per unit area)

$$\mathsf{R} = \frac{1}{\mathsf{h}_1} + \frac{\delta}{k_w} + \frac{1}{\mathsf{h}_2}$$

It is customary to work with the *overall* conductance U defined by

$$\mathsf{U} = \frac{1}{\mathsf{R}} = \frac{1}{1/\mathsf{h}_1 + \delta/k_w + 1/\mathsf{h}_2}$$

The heat-transfer rate equation relating the local heat flux from fluid 1 to fluid 2 to their local mixed mean temperatures is then

$$q'' = \mathsf{U}(T_{m1} - T_{m2}) \tag{14·39}$$

Equation (14·39), together with differential equations for the local fluid mixed mean temperatures, forms the basis for heat-exchanger analysis.

For example, let's consider an exchanger in which the temperature of the hot fluid is uniform along the exchanger. This could be realized by condensing the hot fluid at a fixed pressure, or approximated by having a high hot fluid flow in comparison to the cold fluid flow. The exchanger and general shape of the cold fluid temperature distribution are shown in Fig. 14·22.

We could calculate the rate of heat transfer between the two fluids if we just knew the temperature rise of the cold fluid. But in order to find this temperature rise we need to calculate the distribution $T_{m2}(x)$ throughout the exchanger. To do this we need the differential equation for $T_{m2}(x)$ which is obtained from an *energy balance* on an elemental length of the exchanger (Fig. 14·22*c*). If we neglect axial heat conduction in the exchanger and flow kinetic energy, the energy balance is

$$\dot{M}\left(h_{m2} + \frac{dh_{m2}}{dx}\,dx + \cdots\right) - \dot{M}h_{m2} = q''A'\,dx$$

(*a*) *A parallel-flow heat exchanger*

Hot fluid

Cold fluid

$\dot{Q}$

$T_{m1} = T_{hot}$

T_{m2o}

T

T_{m2i}

x

(*c*) *The control volume to analyze the heat exchanger*

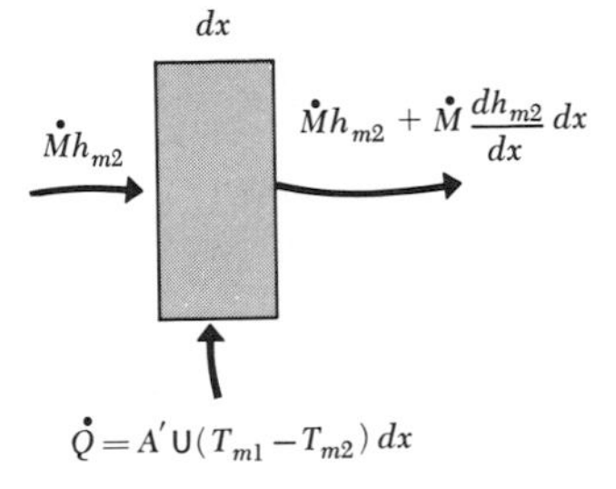

(*b*) *The mean temperature distributions in a parallel-flow heat exchanger with an isothermal hot side*

Figure 14·22.

Here h_{m2} is the mixed mean fluid enthalpy, and A' is the exchanger surface area per unit of length. Neglecting the effect of pressure drop on the fluid enthalpy, we may write

$$\frac{dh_{m2}}{dx} = c_P \frac{dT_{m2}}{dx}$$

Combining and simplifying, we have

$$\frac{dT_{m2}}{dx} = \frac{q''A'}{\dot{M}c_P} \tag{14·40}$$

The product of $\dot{M}c_P$ is sometimes called the *capacity rate* of the flow, and denoted by C,

$$C \equiv \dot{M}c_P$$

Now, we bring in the *overall rate equation* for the heat flux [Eq. (14·39)]. Then, Eq. (14·40) becomes

$$\blacktriangleright \quad \frac{dT_{m2}}{dx} = \frac{A'U}{C}(T_{m1} - T_{m2}) \tag{14·41}$$

This equation must be solved subject to the initial *condition*

$$T_{m2} = T_{m2i} \qquad \text{at} \qquad x = 0$$

The solution is

$$T_{m2} = T_{m2i} + (T_{m1} - T_{m2i})(1 - e^{-A'\mathsf{U}x/C}) \tag{14·42}$$

Of particular interest is the mixed mean outlet temperature T_{m2o}, obtained from Eq. (14·42) by setting $x = L$. The result may be expressed as

▶ $$\frac{T_{m2o} - T_{m2i}}{T_{m1} - T_{m2i}} = 1 - e^{-\text{NTU}} \tag{14·43}$$

where the single parameter NTU (*number of transfer units*) is defined as

▶ $$\text{NTU} \equiv \frac{A\mathsf{U}}{C} \tag{14·44}$$

and A is $A'L$, the *total* heat-transfer area. Note that the exchanger NTU depends both upon the exchanger geometry and the flow rate and properties of the two fluids. Given these quantities, one can calculate the NTU and then calculate the outlet temperature from Eq. (14·43). Or, given the required outlet temperature, the exchanger NTU can be calculated, and the information used to design an exchanger of appropriate size. The NTU concept is therefore central to heat-exchanger analysis.

We note that the left-hand side of Eq. (14·43) is the ratio of the actual temperature rise of fluid 2 to the *maximum possible* temperature rise. This is a special definition of the *heat exchanger effectiveness* ε, the general definition of which is

▶ $$\varepsilon = \frac{\text{actual heat-transfer rate}}{\text{maximum possible heat-transfer rate}} \tag{14·45}$$

The maximum possible heat-transfer rate is

$$\dot{Q}_{\max} = C_{\min}(T_{\text{hot in}} - T_{\text{cold in}}) \tag{14·46}$$

We see that for this particular heat exchanger the effectiveness is a function only of the NTU. In general, theory reveals that the effectiveness also depends upon the *capacity rate ratio* of the two fluids and upon the exchanger flow configuration,

$$\varepsilon = f\left(\text{NTU}, \frac{C_{\min}}{C_{\max}}, \text{flow arrangement}\right)$$

$$\text{NTU} \equiv \frac{A\mathsf{U}}{C_{\min}}$$

The functional dependence can be determined by solution of the appropriate differential equations, and some tables summarizing the solutions are included in Appendix C.

As an example, suppose we have a counterflow heat exchanger with 150 ft^2 of heat-transfer surface. Suppose that the hot and cold side convection coefficients, capacity ratios, and inlet temperatures are as given below and we wish to determine the outlet temperatures.

Cold side

$$C_c = 3000 \text{ Btu/(h}\cdot{}^\circ\text{F)}$$

$$\mathsf{h}_c = 45 \text{ Btu/(h}\cdot\text{ft}^2\cdot{}^\circ\text{F)}$$

$$T_{c_{in}} = 60^\circ\text{F}$$

Hot side

$$C_h = 6000 \text{ Btu/(h}\cdot{}^\circ\text{F)}$$

$$\mathsf{h}_h = 70 \text{ Btu/(h}\cdot\text{ft-}^\circ\text{F)}$$

$$T_{h_{in}} = 250^\circ\text{F}$$

To calculate the outlet temperatures we need the effectiveness. Since $C_c < C_h$, $C_{min} = C_c = 3000$ Btu/(h·°F) and

$$\dot{Q}_{max} = C_{min}(T_{h_{in}} - T_{c_{in}}) = 3000(250 - 60)$$

$$= 570{,}000 \text{ Btu/h}$$

$$\frac{C_{min}}{C_{max}} = \frac{3000}{6000} = 0.5$$

Neglecting the wall resistance, which is usually small compared to the convection resistances, the overall conductance is

$$\mathsf{U} = \frac{1}{(1/\mathsf{h}_h) + (1/\mathsf{h}_c)} = \frac{1}{(1/45) + (1/70)} = 27.4 \text{ Btu/(h}\cdot\text{ft}^2\cdot{}^\circ\text{F)}$$

so $$\text{NTU} = \frac{A\mathsf{U}}{C_{min}} = \frac{150 \times 27.4}{3000} = 1.37$$

For a counterflow exchanger with NTU = 1.37 and $C_{min}/C_{max} = 0.5$, Table C·9*a* gives $\varepsilon = 0.67$. Hence the actual heat-transfer rate is

$$\dot{Q} = \varepsilon\dot{Q}_{max} = 0.67 \times 570{,}000 = 380{,}000 \text{ Btu/h}$$

Finally, the cold and hot fluid exit temperatures can be calculated from energy balances,

$$\dot{Q} = C_h(T_{h_{in}} - T_{h_{out}}) \qquad \dot{Q} = C_c(T_{c_{out}} - T_{c_{in}})$$

So

$$T_{h_{out}} = 250 - \frac{380{,}000}{6000} = 186°F$$

$$T_{c_{out}} = 60 + \frac{380{,}000}{3000} = 187°F$$

Note that the hot fluid emerges cooler than the cold fluid emerges. This is possible with the counterflow arrangement but is not possible with parallel-flow exchangers.

14·8 RADIATION

We have already mentioned that all bodies radiate energy and that this energy consists of photons moving with random direction, phase, and frequency. When these photons are in the wavelength range 0.38 to 0.76 μm they affect our eyes as visible light. At wavelengths just longer than the visible limit the radiation is called *infrared* radiation, and at wavelengths just shorter than the visible limit the radiation is said to be in the *ultraviolet* region. These regions are a part of the electromagnetic-wave spectrum that ranges continuously from long-wavelength radio waves to the very short-wavelength cosmic rays.

When radiated photons reach another surface, they are either absorbed, reflected, or transmitted through the surface. Three surface properties which measure these quantities are

α absorptivity, the fraction of incident radiation absorbed
ρ reflectivity, the fraction of incident radiation reflected
τ transmissivity, the fraction of incident radiation transmitted

From energy considerations

$$\alpha + \rho + \tau = 1 \tag{14·47}$$

The reflected energy may be either *diffuse*, where the reflections are independent of the incident radiation angle, or *specular*, where the angle of reflection equals the angle of incidence. Most engineering surfaces exhibit a combination of the two types of reflection.

The energy radiation flux [Btu/(h·ft²)] from a surface is defined as the *emissive power E*. Thermodynamic reasoning shows that E is proportional to the fourth power of the absolute temperature.†

† See for example, W. C. Reynolds, *Thermodynamics*, 2d ed., McGraw-Hill Book Company, New York, 1968.

For a body with $\alpha = 1$, $\rho = \tau = 0$ (a *blackbody*)

$$E_b = \sigma T^4 \tag{14·48}$$

where σ is the *Stefan-Boltzmann constant,*

$$\sigma = 5.669 \times 10^{-8}\ \mathrm{W/(m^2 \cdot K^4)}$$
$$= 0.1714 \times 10^{-8}\ \mathrm{Btu/(h \cdot ft^2 \cdot R^4)}$$

Real bodies are not "black" and radiate less energy than the blackbody. To account for this one can define an *emissivity* ε in terms of the emissive powers of the real body and the blackbody, both evaluated at the same temperature,

▶ $$\varepsilon \equiv \frac{E}{E_b} \tag{14·49}$$

The absorptivity is related to the emissivity through *Kirchhoff's law* which states that, at *thermal equilibrium,*

$$\varepsilon = \alpha \tag{14·50}$$

Note that the absorptivity equals the emissivity only under conditions of thermal equilibrium, that is, when the incoming radiation to be absorbed comes from a body at the same temperature as the emitting body.

Radiation of all frequencies is emitted from a hot surface. The emissive power E therefore has contributions at all frequencies, and we write

$$E = \int_0^\infty E_\lambda \, d\lambda$$

where $E_\lambda \, d\lambda$ is the energy emitted in the frequency band between λ and $\lambda + d\lambda$. E_λ is a function of the surface and its temperature. For a blackbody, which is a surface that absorbs all incident radiation, E_λ takes the form shown in Fig. 14·23. Figure 14·23 also shows E_λ for a typical real surface. Simplified approximate radiation calculations can often be made by the *gray-body approximation,* in which E_λ is taken as $\varepsilon E_{\lambda b}$ at all wavelengths (see Fig. 14·23). More accurate calculations require consideration of variation in the emission with wavelength, variations in absorptivity and transmissivity with wavelength, and wavelength-dependent combinations of specular and diffuse radiation. One must usually rely extensively on computer analysis to solve a radiation problem including all these effects, and this area of heat transfer is currently the center of much research activity. Such calculations are beyond the scope of this brief introduction, and we shall restrict ourselves henceforth to the simple gray-body models.

All surfaces radiate energy, but there will be a net energy exchange only if

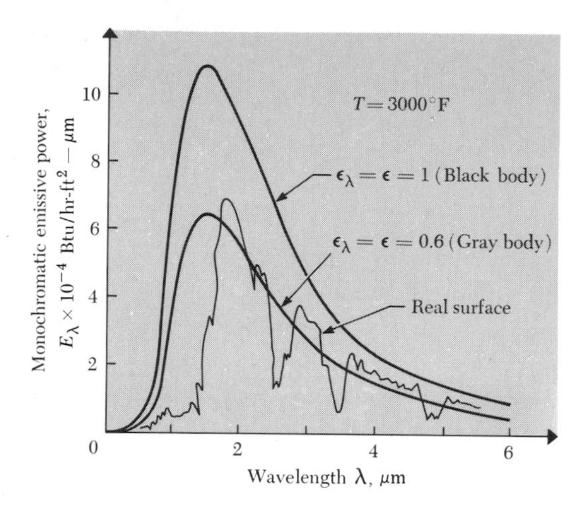

Figure 14·23. Blackbody and real surface emissive power as functions of wavelength and temperature

the bodies are at different temperatures. Two facing parallel plates will radiate to each other, but if their temperatures are equal there is no net energy transfer. To evaluate radiation between bodies when their temperatures are different the geometry must be considered. Two facing parallel plates radiating only from their facing sides see only each other, but two rectangular plates meeting at right angles do not radiate solely to each other, and their *shape factor* must be determined. The shape factor F_{12} represents the fraction of the radiation emitted from surface 1 that is intercepted by surface 2. A case of particular interest is that of a surface 1 that sees only a surface 2, such as one of two closely spaced plates or the inner of two concentric spheres. In this case

$$F_{12} = 1$$

$$F_{21} = \frac{A_1}{A_2} \tag{14·51}$$

For blackbody radiation between surfaces 1 and 2 the rate equations for the *net* energy transfer are

$$\dot{Q} = A_1 F_{12}(E_{b1} - E_{b2})$$
$$\dot{Q} = \sigma A_1 F_{12}(T_1^4 - T_2^4) \tag{14·52}$$

Since all the incident radiation is absorbed by a blackbody, there are no reflections to consider and thus the rate equation is relatively simple.

As an example, let's use the blackbody model to determine the radiation heat flux between two closely spaced parallel plates radiating only to each other if their temperatures are 1000 and 500°F. For parallel plates, $F_{12} = 1$. Then,

$$q'' = \sigma(T_1^4 - T_2^4) = (0.1714 \times 10^{-8})(1460^4 - 960^4)$$
$$= 6340 \text{ Btu/(h·ft}^2) = 20{,}000 \text{ W/m}^2$$

For gray bodies a factor is required in the rate equation to account for the reflections. The rate equation for radiation between two gray bodies *which see each other and nothing else* is given by

$$\dot{Q} = A_1 \mathscr{F}_{12}(E_{b1} - E_{b2}) \tag{14·53}$$

where

$$A_1 \mathscr{F}_{12} = \frac{1}{\dfrac{1-\varepsilon_1}{\varepsilon_1 A_1} + \dfrac{1}{A_1 F_{12}} + \dfrac{1-\varepsilon_2}{\varepsilon_2 A_2}} \tag{14·54}$$

Let's assume that the two parallel plates of the previous example each have emissivities of 0.5 and recalculate the heat flux using the gray-body model.

$$q'' = \frac{1}{A}\left(\frac{E_{b_1} - E_{b_2}}{\dfrac{1-\varepsilon_1}{\varepsilon_1 A_1} + \dfrac{1}{A_1 F_{12}} + \dfrac{1-\varepsilon_2}{\varepsilon_2 A_2}}\right)$$

Here

$$A = A_1 = A_2 \qquad \text{and} \qquad F_{12} = 1$$

so

$$q'' = \frac{\sigma(T_1^4 - T_2^4)}{\dfrac{1-\varepsilon_1}{\varepsilon_1} + 1 + \dfrac{1-\varepsilon_2}{\varepsilon_2}} = 2113 \text{ Btu/h} = 6666 \text{ W/m}^2$$

Note that, if the emissivity of each plate is only one-half that of a blackbody, heat flux is reduced by a factor of 3.

As a further example, consider a deep-space probe constructed as a 1-m-diameter polished aluminum sphere. Let's estimate the equilibrium temperature that the probe reaches if the solar energy received† is 300 W/m². Our energy balance gives

$$\underset{\text{input rate from sun}}{\dot{Q}} = \underset{\text{radiation-output rate}}{\dot{Q}}$$

For solar radiation the absorptivity of aluminum is about 0.3. The heat input is determined by the projected area of the sphere, and so

$$\dot{Q}_{\text{in}} = \alpha q'' A = 0.3 \times 300 \times \frac{\pi(1)^2}{4} = 70.7 \text{ W}$$

† The *solar constant* at the *outside* of the earth's atmosphere is about 440 Btu/(h·ft²) or 1388 W/m². Because of absorption, scattering, and so on, less radiation reaches the earth's surface, and, as a rule of thumb, about 300 Btu/(h·ft²) reaches the surface at midday at the latitudes of the United States. Since 300 Btu/(h·ft²) is equivalent to 946 W/m², the peak solar input is roughly 1 kW/m².

The outgoing radiation goes to space where the temperature is low enough to be neglected. Therefore

$$\dot{Q}_{\text{out}} = \varepsilon\sigma A T^4$$

Equating these, using $\varepsilon = 0.04$ (Table C·7), and using the total surface area of the sphere, we find

$$70.7 = \varepsilon\sigma A T^4$$

$$T = \left[\frac{70.7}{0.04 \times 5.669 \times 10^{-8} \times 4\pi \times (0.5)^2}\right]^{1/4}$$

$$T = 315 \text{ K}$$

Note that $\alpha \neq \varepsilon$ in our calculation. We evaluate α at solar frequencies and use an average ε appropriate for aluminum at lower temperatures. The low emissivity of aluminum at these temperatures compared to its relative high absorptivity for solar radiation explains this surprisingly high equilibrium temperature.

Many heat-transfer problems involve simultaneous convection, conduction, and radiation transfer, for example, determining the energy load (heat balance) on a residence during the daytime. To include radiation in the thermal network it is convenient to define a radiation heat-transfer coefficient h_r where

$$q''_{\text{rad}} = h_r \, \Delta T \tag{14·55}$$

In effect we have linearized the fourth-power temperature dependence that radiation has. This approach works well for small temperature differences. Combined convection and radiation from a surface can be treated as two modes of heat transfer acting in parallel, and a resistance per unit area for radiation can be defined as $1/h_r$, just as in the case of convective heat transfer.

Radiation heat transfer is important to understanding the energy balance for the earth-atmosphere system. Our energy input comes from the sun, and an equilibrium is maintained by having an equal energy output by radiation from the earth and atmosphere to space. The difference in wavelength between the solar radiation, much of which is in the visible region, and the longer infrared wavelength radiation emitted from the earth is important. Carbon dioxide and water vapor are absorbers of radiation, but primarily in the infrared region, not in the visible. This behavior is the cause of the greenhouse effect.

Radiation from the earth to space also is responsible for the formation of one form of temperature inversion in the atmosphere. Under clear skies and with low wind speed conditions the earth will cool faster at night than does the air above and a temperature inversion will result. A stable layer of air close to the ground forms and any pollutants emitted into that layer tend to remain there. Fortunately radiative inversions are usually removed the next morning as the solar input warms the surface of the earth.

14·9 SUMMARY

This brief chapter has only introduced the ideas of heat transfer. The important point is that heat transfer is founded on the concepts of thermodynamics, but because it is an irreversible process more information than thermodynamics can provide is required. The rate equations provide this additional information. The combination of the energy balance and the rate equations is sufficient to derive the differential equations that govern the energy transfer. Their solution provides knowledge of the temperature as a function of position and time. The theory of irreversible thermodynamics, of which heat transfer forms only a small part, provides us with a general framework within which heat transfer, diffusive mass transfer, electric conduction, and many other phenomena can be analyzed.

SELECTED READING

Holman, J. P., *Heat Transfer*, 3d ed., chaps. 1, 2, sec. 4.2, chaps. 5–8, 10, McGraw-Hill Book Company, New York, 1972.

Kays, W. M., *Convective Heat and Mass Transfer*, pp. ix–xi, chaps. 6, 8, 10, McGraw-Hill Book Company, New York, 1966.

Kreith, F., *Principles of Heat Transfer*, 3d ed., chaps. 1, 2, secs. 4.1–4.3, chaps. 5, 8, 9, 11, International Textbook Company, Scranton, Pa., 1973.

Obert, E. F., and R. L. Young, *Elements of Thermodynamics and Heat Transfer*, 2d ed., chaps. 19, 20, McGraw-Hill Book Company, New York, 1962.

QUESTIONS

14·1 How does heat-transfer theory differ from thermodynamics?
14·2 What is the objective of a heat-transfer analysis?
14·3 What are the two mechanisms for heat transfer?
14·4 What is convection?
14·5 What is the difference between heat, heat flux, heat-transfer rate, and the heat-transfer coefficient?
14·6 What are the fundamental conduction and convection rate equations?
14·7 How would you describe a thermal circuit to an electrical engineer?
14·8 When do thermal resistances act in parallel? When in series?
14·9 Why are convection correlations expressed nondimensionally?
14·10 What is a Nusselt number?
14·11 What is a Reynolds number?
14·12 What is the Grashof number?
14·13 What is the Prandtl number?
14·14 What is a telephone number?
14·15 Why do fins increase the rate of heat transfer from a surface?
14·16 Would fins work in space?
14·17 What is the NTU of a heat exchanger?

14·18 What happens to the NTU of a parallel-flow exchanger when its length is cut in half? What happens to its effectiveness?
14·19 What is the blackbody model?
14·20 Can you sketch the emission spectra of a blackbody?
14·21 How do real surfaces differ from blackbodies?
14·22 What heat-transfer processes are involved in atmospheric reentry?

PROBLEMS

14·1 A brick wall 8 in thick is finished on the inside with 1 in of plaster. Calculate the total thermal resistance of the wall and the heat-transfer rate per ft² when the overall temperature difference is 80°F. Is the sign of the temperature difference important?

14·2 A steam turbine casing of $\frac{1}{4}$-in-thick steel has an inside surface temperature of 212°F. The resistances for radiation and free convection on the outside are each about 0.67 h·ft²·°F/Btu. The outside ambient air temperature is 70°F. What is the outer surface temperature? Suppose the outer surface is covered with a 1-in layer of insulation [$k = 0.05$ Btu/(h·ft·°F)]; will the insulation be safe to touch?

14·3 The $\frac{1}{8}$-in-thick wall of a gas turbine casing is made from a nickel alloy [$k = 10$ Btu/(h·ft·°F)]. If the convective heat-transfer resistances on the inside and outside are $\frac{1}{35}$ and $\frac{1}{6}$ h·ft²·°F/Btu, respectively, find the heat-transfer rate per ft² if the overall temperature difference is 1200°F.

14·4 Prescribe sufficient insulation for the turbine casing of Prob. 14·3 to reduce the heat flux to 300 Btu/(h·ft²).

14·5 The wall for a cold storage is constructed of standard 2 by 4 fir studs ($1\frac{3}{4}$ in × $3\frac{3}{4}$ in) on 16-in centers. The walls are sheathed on both sides with $\frac{1}{2}$-in fir plywood and both external surfaces lined with $\frac{1}{16}$-in galvanized steel. The space between the studs is filled with ground cork. The inner surface of the inner liner is to be maintained at 15°F. The outside ambient air temperature is 80°F. The thermal conductivities of the various materials are:

Ground cork, 0.025 Btu/(h·ft·°F)
Fir, 0.087 Btu/(h·ft·°F)
Steel, 26 Btu/(h·ft·°F)

The outside radiation resistance will be about 1 h·ft²·°F/Btu, and the outside resistance for convection will be about the same.

What is the heat leak into the room per *average* ft² of wall? Plot the temperature through the wall at a point through a stud, and through the insulation. What percentage goes through the studs? How critical is the assumption of convection and radiation resistances in a problem of this type? Draw the complete thermal circuit showing clearly each of the various resistances involved. What idealization do you have to make to construct this circuit?

14·6 Calculate the conduction heat-transfer rate per ft of pipe through a single hollow cylinder made from 1.5 percent C steel 4 in thick on a 3-in inner radius. The inside temperature is 750°F, and the outside is 400°F.

14·7 Add 2 in of 85 percent magnesia to the outside of the cylinder in Prob. 14·6 and recalculate the heat-transfer rate for the same overall temperature difference. Calculate the temperature at the steel-insulation interface.

14•8 Calculate the reduction in heat transfer through a wall made up of jumbo bricks of width = 0.15 m [$k = 1$ W/(m·°C)] created by adding 5 cm of fiber insulating board to the inside surface of the bricks. The outside wall temperature is −10°C, and the inside wall temperature is 20°C.

14•9 A guarded plate heater is a device to determine thermal conductivity. Experimental results for a firebrick sample are the following: frontal area, 0.0004 m^2; thickness, 0.075 m; top face temperature, 20°C; bottom face temperature, 0°C. The energy input to maintain the top-face temperature is 0.01 W. Find the thermal conductivity from these data.

14•10 A double-glazed window has two sheets of glass separated by a thin layer of air. The air gap is thin enough so that convection between the glass panes may be neglected. Determine the reduction in heat transfer when a single-glazed window is replaced by a double-glazed window. The window glass is 0.4 cm thick, the outside h is 12 W/(m^2·°C), the inside h is 6 W/(m^2·°C), and the air gap is 0.5 cm thick.

14•11 Draw the thermal circuit for a wall constructed of 4.5 × 9.5 cm pine studs on 40-cm centers. The studs are sheathed on each side with 1-cm-thick plywood. On the outside surface 1.5-cm-thick wood shingles are used, while the inside wall is 0.5 cm of insulating board over which 1 cm of plaster has been applied. The area between the studs is filled in with glass wool.

14•12 For the above wall (Prob. 14·11) calculate the heat flux (W/m^2) through an *average* m^2 of wall if the outside temperature is −20°C and the inside temperature is 18°C. The inside combined heat-transfer coefficient for radiation and convection is about 10 W/(m^2·°C), while the outside convection coefficient is 20 W/(m^2·°C) and the outside radiation coefficient is 5 W/(m^2·°C). Calculate the required input from a furnace for 1 day if the wall area is 100 m^2.

14•13 An aluminum conductor is annular in shape and insulated on the outside surface. It is cooled internally. The outside diameter is 0.035 m, and the inner diameter is 0.01 m. A current flux of 3000 A/cm^2 is carried through the conductor. The coolant temperature is 50°C, and the inner surface temperature is 70°C. Find the temperature distribution through the annulus and the maximum temperature. Determine the heat-transfer coefficient between the coolant and the inner surface. If the coolant is water, find the Reynolds number required to achieve this coefficient. (The resistivity of aluminum is 2.8×10^{-8} Ohm·m.)

14•14 Derive an expression for the temperature distribution and heat transfer through a hollow sphere of inner radius r_i, outer radius r_0, and conductivity k. The inner and outer temperatures, T_i and T_0, are known.

14•15 Design an experiment to measure k for plastics.

14•16 Derive Eq. (14·15). Give an expression for the difference between the centerline and surface temperature in a slab with a uniform heat source [analogous to Eq. (14·18)].

14•17 A nuclear reactor has fuel elements that are slabs of uranium [$k = 18$ Btu/(h·ft·°F)] covered with a thin coating of aluminum. The uranium is $\frac{1}{4}$ in thick and the fission process in the reactor generates 96×10^5 Btu/(h·ft^3). Estimate the difference between the uranium centerline and surface temperature. (See Prob. 14·16.)

14•18 A thin slab of thickness L has a high temperature T_1 imposed on one face and a lower temperature T_2 imposed on the other face. In addition, a uniform heat source s Btu/(h·ft^3) exists within the slab. To analyze the thermal stress distribution it is required to determine the temperature distribution through the slab. Derive an expression for this temperature distribution and an expression for the point of maximum temperature.

14·19 A particular nuclear reactor operates with a source strength of 10^7 Btu/(h·ft^3). The fuel element is made up of 1-in-thick plates made from a uranium alloy [$k = 18$ Btu/(h·ft·°F)]. At one point in the reactor a temperature of 1000°F is expected on the left side of the plate and 800°F on the right side. Determine the location of maximum temperature and specify whether the element will fail. (Failure occurs if any temperature is greater than 1350°F.) (See Prob. 14·18.)

14·20 In a particular nuclear reactor the uranium fuel is in the form of solid circular rods, 1 in in diameter. The thermal conductivity of uranium is about 18 Btu/(h·ft^2·°F)/ft. High-pressure water at 2000 psia flows normal to the rods at a temperature of 575°F. The convection conductance (reciprocal resistance) at the rod surface is about 2000 Btu/(h·ft^2·°F), and it is specified that the uranium rod centerline temperature should not exceed 800°F. At this condition, what will be the reactor power level, expressed in Btu/ft^3 of uranium and in Btu/ft^2 of cooling surface? If the surface temperature exceeds the water saturation temperature, local boiling (which happens to be undesirable here) will occur. Is this a problem here?

14·21 A farmer produces hay for market. His hay baler forms *long* cylindrical hay bales about 12 in in diameter. As the hay cures, heat is uniformly generated at a measured rate of 800 Btu/(h·ft^3). If the outside surface temperature of the bale is 90°F and $k = 0.25$ Btu/(h·ft·°F), what is the maximum temperature within the bale?

14·22 Show that Laplace's equation, $(\partial^2 T/\partial x^2) + (\partial^2 T/\partial y^2) = 0$ in two dimensions, is the governing differential equation for the temperature field in a steady-state system with constant k and no sources.

14·23 Suppose the thermal conductivity of a material can be approximated by $k = k_0(1 + aT)$, where a is a constant. Derive an expression for the temperature distribution through a slab of this material if the inner and outer temperatures are T_i and T_0, respectively.

14·24 One brand of cookware has stainless-steel pots with copper bases. Discuss the engineering reasons for this design.

14·25 A copper wire 0.1 in in diameter initially at 200°F is cooled by 60°F air blowing over its surface. The convection conductance (reciprocal resistance) is 6 Btu/(h·ft^2·°F). Calculate the time constant of the system and the time required to cool to 80°F.

14·26 The wire of Prob. 14·25 carries 100 A. Derive a general expression for the temperature-time history of the wire, and plot for this specific case.

14·27 Consider the thermal circuit of Fig. 14·10*b*. Derive an expression for the capacitance temperature $T(t)$ for the case where the driving temperature is oscillated such that $T_1 = T_0 + A \cos \omega t$, where ω and A are constants. Plot the amplitude of the rod temperature oscillation (normalized on A) versus RC. For the specific rod of Fig. 14·10, calculate the frequency for which the rod temperature oscillation is $\frac{1}{2}A$.

14·28 A container with a surface area of 0.13 m^2 contains 4 kg of water boiling at 100°C. An average heat-transfer coefficient from the container to room air is estimated to be 25 W/(m^2·°C). Calculate the time required for the water to cool to 60°C when placed in an ambient temperature of 20°C. Neglect the effect of the thermal capacitance of the container.

14·29 A large mild-steel pipe of 30 cm diameter and 1 cm wall thickness carries hot water at 150°C. Assuming natural convection from the outside of the pipe, calculate the heat loss per m of pipe for a 20°C ambient temperature (spring conditions) and for a −20°C ambient temperature (winter conditions). The Reynolds number for the water flow is 10^4.

14·30 For the conditions of Prob. 14·29 add on 0.03 m of "stop-heat" insulation with a

$k = 0.04$ W/(m·°C) and calculate the reduction in heat transfer that occurs. With the insulation what is the heat loss from the pipe per m/year?

14·31 Data taken on a steel pipe [$k = 34.6$ W/(m·°C)] indicate that for water flowing at 400°C heat transfer per m of pipe occurs at 1200 W/m to an ambient temperature of 25°C. The outside h is estimated to be 16 W/(m^2·°C). Find the inside heat-transfer coefficient if the pipe has a 0.5-cm wall thickness and an O.D. of 0.075 m. After several months of service the heat loss has dropped to 1000 W/m. Calculate the thermal resistance due to scale buildup on the inner wall.

14·32 A swirl-cooled, radial efflux waste-heat unit consists of a long pipe with a helical ribbon down the middle. The helix forces the fluid to turn and mix as it moves down the pipe. Laboratory tests indicate that for turbulent flow the unit gives heat-transfer coefficients about three times those in a straight pipe; for laminar flow an increase of 2.5 occurs. Determine the heat-transfer coefficient in this unit for:

(*a*) Water flowing in a pipe of 0.02 m I.D. at 451 kg/h at 40°C

(*b*) Mercury flowing in a pipe of 0.01 m I.D. at 0.8 m/s at 200°C, assuming a uniform wall heat flux

(*c*) Ethylene glycol flowing in a pipe of 2.5 cm I.D. at 1.3 m/s and 60°C

(*d*) Air flowing in a pipe of 0.07 m I.D. at 400°C with a flow rate of 6.4 kg/h and a uniform wall temperature

14·33 Mercury flows at 200°C through a 0.02-m tube at a velocity of 1 m/s. At a particular point in the uniformly heated tube the wall temperature is 230°C. Find the heat flux at this point.

14·34 Using the correlations given in Appendix C, determine values of h for

(*a*) Air flow at 30 ft/s, 1 atm, $T = 100$°F in a 1-in-diameter pipe

(*b*) Air flow as above except at 10 atm

(*c*) Water flow inside a 1-in-diameter pipe at 5 ft/s, 100°F

Discuss the physical reasons for the differences in these values.

14·35 Estimate the rate of heat transfer per ft^2 on a thin, nearly flat airfoil 6 in downstream from the leading edge. The fluid velocity is 50 ft/s, the airfoil is maintained at 100°F, the fluid temperature is 70°F, and the pressure 1 atm.

14·36 In the Sea Lab experiments it is necessary to maintain the atmosphere at a temperature of 31°C because of the large heat loss from the skin of the participants. The atmosphere was largely helium. Verify the need to maintain a high temperature.

14·37 Air at 60°F, 1 atm, is blown across an electrically heated copper wire 0.460 in in diameter. The wire carries 325 A and has a resistance of 0.26 Ω/mi. Calculate the air velocity required to keep the wire surface at 125°F.

14·38 A semiconductor device has to dissipate 1 W. The device is a $\frac{1}{2}$-in-diameter cylinder $1\frac{1}{4}$ in high. If the ambient air is at 70°F, find the equilibrium temperature of the device.

14·39 Steam at 250°F, 1.0 atm, flows through an uninsulated 6-in-diameter pipe between two buildings 100 ft apart. If the air is at 40°F and there is no wind, find the heat loss from the pipe.

14·40 Steam flows through a 4-in-diameter pipe at 8 ft/s, 500°F, 100 psia. The pipe is $\frac{1}{2}$-in-thick iron and has a 1-in-thick covering of glass wool. If the ambient temperature is 60°F, estimate the heat loss per ft of pipe.

14·41 A 1-in-diameter copper pipe running under the house carries hot water at 120°F from the utility room to the kitchen, a distance of 10 ft. The flow rate is 1 gal/min, and the outside air temperature is 30°F. If the outside h is 3 Btu/(h·ft^2·°F), is insulation warranted?

14·42 Air is to be heated at 1 atm from 60°F to 300°F. The flow rate is 10 lb/min. A thin electrically heated tube (or tubes) will be used, and the tube temperature cannot exceed 550°F. Design a system to meet these specifications. Give the tube diameter and length, the total power input, and the maximum tube temperature.

14·43 Repeat Prob. 14·42 for helium and discuss the physical reasons for the differences in the designs.

14·44 Optimize the design of Prob. 14·42 minimizing the *volume* of the heater system.

14·45 A 4-in-diameter copper sphere is heated to 500°F during a chemical reaction. Estimate the time required for the sphere to cool down to 150°F if air at a velocity of 30 ft/s and a temperature of 60°F, 1 atm, is available as the cooling medium.

14·46 A long, thin nickel-chrome fin has a temperature of 400°F at its base. If the ambient temperature is 70°F and the fin is of rectangular shape having a 2-in perimeter and 0.10-in^2 area, calculate the rate of heat transfer from the fin assuming that h = 7 Btu/(h·ft^2·°F).

14·47 A recent heat-exchanger design called for a plastic piece (failure point 220°F) at the end of a rectangular metal fin. The fin is stainless steel, $\frac{1}{8}$ in thick and 4 in long. The convective conductance is 3 Btu/(h·ft^2·°F), the base temperature is 400°F, and the gas temperature is 70°F. Will the plastic fail?

14·48 A rectangular steel fin on a compressor case is 0.3 m wide, 0.004 m thick, and 0.08 m long. The base temperature is 160°C, and the ambient temperature is 15°C. An average h is determined to be 50 W/(m^2·°C). Find the temperature at the fin tip and the heat transfer per fin. How many fins would be required to dissipate 1 kW?

14·49 An aluminum cooking pot has an aluminum handle that is roughly shaped as a 3 × 0.3 cm rectangle. If the pot contains boiling water, how long should the handle be before you would use it to pick up the pot? (Burn temperature is about 60°C.)

14·50 A stainless-steel-clad thermometer is placed 1 in into a superheated steam duct at a right angle to the flow direction. The thermometer reads 247°F when the steam velocity is 15 ft/s, the pressure 14.7 psia, and the base of the thermometer is at the wall temperature of 150°F. Assuming the thermometer is a $\frac{1}{4}$-in-diameter stainless rod, find the true steam temperature. What is the steam temperature if the thermometer reads the same but is inserted 3 in into the steam?

14·51 The barrel of an engine cylinder consists of a $\frac{3}{8}$-in-thick steel liner on the outside of which is bonded a $\frac{1}{4}$-in aluminum muff. Circumferential cooling fins are then bonded to the outside of the muff. The fins are 1.5 in long, 0.030 in thick, with a space between fins of 0.085 in, and the muff diameter is 7 in. During a complete engine cycle, the average temperature of the gases in the cylinder is 1500°F. The cooling air temperature averages 70°F.

Making suitable simplifying assumptions and educated guesses, calculate the temperatures at the inner cylinder wall, the inner surface of the aluminum muff, the base of the fins, and the tip of the fins. Calculate the heat-transfer rate per ft^2 of muff area.

14·52 Equation (14·32) assumes that A is constant. For the case of circular fins on a rod this is not true, since the cross-sectional area is a function of the radius. Derive the analogous "fin equation," including the effect of changing area; give the fin equation for circular fins of uniform thickness.

14·53 Consider a thin plate with a uniform heat source s [Btu/(h·ft^3)]. The plate is exposed to a flow for which the average convective conductance over each side of the plate is h. The plate is suspended in the flow between two side plates, as shown in the figure. The side-plate temperature is T_0, and the flow temperature is T_∞. Derive

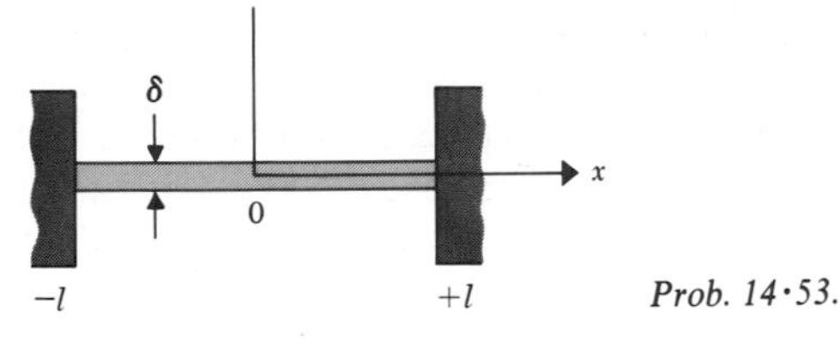

Prob. 14·53.

an expression for the temperature distribution $T(x)$ along the plate using the coordinate system shown. Discuss engineering situations in which this theory would be useful.

14·54 Air at the rate of 15,000 lbm/h enters a heat exchanger at 200°F and 50 psia. The air is cooled in a single-pass counterflow arrangement. If the water coolant enters at 60°F at 20,000 lbm/h, find the air and water outlet temperatures. The overall heat-transfer coefficient is 30 Btu/(h·ft²·°F) and is based on a tube area of 200 ft².

14·55 Repeat Prob. 14·54 for parallel-flow and cross-flow configurations. Discuss the advantages of the various flow arrangements.

14·56 Air is to be cooled by passing it through a tube immersed in still water at 10°C. The air enters at 90°C and must be cooled to 15°C. The air velocity should not exceed 25 m/s in the tube, and the flow rate is 2 kg/min. The nominal air density is 6 kg/m³. Specify a suitable tube diameter and length for this problem.

14·57 It is required to specify the size of a heat exchanger to meet the following conditions: water inlet temperature, 40°C; water exit temperature, 85°C; oil inlet temperature, 150°C. The water-flow rate is 1000 kg/h, the oil-flow rate is 3500 kg/h, and the oil specific heat is 1 kJ/(kg·°C). Determine the number of transfer units and the oil outlet temperature for a parallel-flow exchanger.

14·58 Liquid carbon dioxide at −40°C is pumped through 10 m of high-conductivity tubing exposed to gaseous carbon dioxide at 27°C and 1 atm pressure. The flow rate is 1 kg/min, the tubing is 2.5 cm O.D., and the wall thickness is 2.5 mm. Determine the overall U based on the outside area. Treating the system as a heat exchanger find the effectiveness.

14·59 A simple tube-within-a-tube heat exchanger is to heat an organic fluid $[c_P = 0.6$ Btu/(lbm·°F)$]$ from 50 to 150°F. The fluid-flow rate is 100 lbm/h. The heating fluid is water which cools from 200 to 100°F as it passes through the exchanger. The overall heat-transfer coefficient based on the area of the organic side is 50 Btu/(h·ft²·°F). Sketch to scale the hot and cold fluid temperature distributions along the exchanger, including flow directions. What water-flow rate and what surface area are required to meet this design? If the direction of the water flow were reversed, what would be the outlet temperature of the organic fluid assuming the same flow rates, *inlet* temperature, and heat exchanger?

14·60 It is required to specify the size of a heat exchanger to meet the following conditions: water inlet temperature, 100°F; water exit temperature, 185°F; oil inlet temperature, 300°F. The water-flow rate is 4500 lbm/h, the oil-flow rate is 8000 lbm/h, and the oil specific heat is 0.45 Btu/(lbm·°F). How many transfer units are required to meet these conditions and what is the oil outlet temperature? Assume that because of piping problems a parallel-flow arrangement must be used.

14·61 Hot water at 180°F passes through 50 ft of copper tubing underneath a house. The tube is completely exposed to still air at a temperature of 50°F. The flow rate is to be 1 gal/min. The tube O.D. is 1 in and the I.D. is 0.83 in. Determine an overall U Btu/(h·ft²·°F) based on the outside area. Treating the system as a heat exchanger, find the exchanger effectiveness. Calculate the exit water temperature and then make a recommendation as to whether insulation is warranted.

14·62 A counterflow heat exchanger operates using river water as the coolant. During the winter water is available at 45°F. The exchanger is to be designed to cool 4000 lbm/h of oil [$c_P = 0.50$ Btu/(lbm·°F)] entering at 400°F to 210°F. Specify a water-flow rate in lbm/h and the NTU required to achieve the design. Will your design work in the summer when the river inlet is at 65°F?

14·63 Air is to be cooled by passing it through a tube immersed in still water at 60°F. The air enters at 180°F and must be cooled to 65°F. The air velocity should not exceed 100 ft/s in the tube, and the flow rate is 5 lbm/min. The nominal air density is 0.4 lbm/ft^3. Specify a suitable tube diameter and length for this operation.

14·64 Design a concentric tube heat exchanger to cool 10 gal/min of water from 180°F to 80°F. Water entering at 60°F will be used as the coolant. Neither stream should exceed 15 ft/s. Specify the tube geometry and the required coolant flow rate.

14·65 Optimize the design of Prob. 14·64, minimizing the exchanger volume.

14·66 Calculate the radiant energy emission flux (W/m^2) from a blackbody at 20°C, 500°C, and 1000°C.

14·67 Two parallel plates are at 1000 and 500°C. Calculate the radiation heat transfer per m^2 if (*a*) the plates are black, (*b*) the plates are gray with an emissivity = 0.6, (*c*) the plates are black and a third black plate is inserted between them.

14·68 A black asphalt street ($\varepsilon = 1$) absorbs radiant energy at the rate of 1000 W/m^2 during the summer. The ambient temperature is 35°C, and the convective coefficient between the air and the street is estimated to be 10 W/(m^2·°C). Estimate the equilibrium street temperature.

14·69 If the average earth-surface temperature is 288 K with our present solar input, and we have an increase in energy to be radiated from the earth of 5 percent, how much must the temperature of the earth rise to accommodate this increase?

14·70 The sun can be treated as a blackbody radiating at 6000 K. The energy received at the surface of our atmosphere from the sun is about 2 cal/(cm^2·min). Calculate the energy intercepted by the earth-atmosphere system in J/day and Btu/day.

14·71 A solar collector 1 by 2 m consists of a flat plate of glass directly in contact with water flowing under it. The collector is insulated on the lower surface. Water flows in at 13,300 kg/h at 35°C and leaves at 65°C. Find the heat losses by radiation and convection from the top face if the solar input at this latitude and time of day is 940 W/m^2 and the glass transmits 82 percent of this input. What is the overall collection efficiency for this device?

14·72 A pan of water is set outside on a clear, cloudless, still night. The water is at 20°C initially, the ambient at 4°C. The sky radiating temperature is about −10°C. Sketch the water temperature versus time, and explain qualitatively the shape that you draw. What is the mechanism by which smudge pots protect fruit from freezing on cold nights?

14·73 Estimate the heat loss through a 1.5-in-diameter "peephole" in a scrap-metal furnace. The inside temperature is 2540°F, and the outside temperature seen by the hole is 140°F.

14·74 A flat piece of oxidized steel receives radiant energy from the sun at the rate of 300 Btu/(h·ft^2). If the ambient temperature is 70°F and a convective heat-transfer coefficient of 1.8 Btu/(h·ft^2·°F) is appropriate, determine the equilibrium temperature of the steel, assuming no heat loss from the back side of the plate.

14·75 A cylindrical container 20 cm long and 5 cm in diameter contains liquid helium

(7 R). There is an evacuated space between it and an outer annular container, which in turn contains liquid nitrogen at 140 R. Both containers are of a polished nickel alloy. Estimate the rate of heat leak into the helium and the percent evaporative loss per h.

14·76 A black asphalt street ($\varepsilon = 1$) absorbs radiant energy at the rate of 350 Btu/(h·ft^2) during the summer. If a mild breeze is blowing such that the ambient temperature is 90°F and the heat-transfer coefficient between the air and street is 2.8 Btu/(h·ft^2·°F), estimate the street temperature.

14·77 Liquid oxygen is to be stored in a spherical container 1 ft in diameter at 1 atm. The system is insulated by an evacuated space between the inner sphere and a surrounding concentric sphere of 1.5 ft I.D. Both spheres are made of aluminum ($\varepsilon = 0.03$), and the temperature of the outer sphere is 40°F. Estimate the rate of heat flow by radiation to the oxygen in the container, and the rate of evaporation of the oxygen, assuming it is vented to keep the pressure constant.

14·78 In a steelmaking apparatus two parallel plates, large compared to the distance between them, are at 1000 and 500 R.

(*a*) Calculate the radiation exchange per ft^2 of area if $\alpha_1 = \alpha_2 = 1$.

(*b*) A third parallel black plate is located between the two plates. What is its steady-state temperature? How is the heat transfer altered?

(*c*) Suppose the hotter of the two plates is oxidized steel, while the cooler is oxidized aluminum. What is the rate of heat transfer and what is the temperature of the black plate between?

(*d*) Discuss briefly the significance of this exercise.

14·79 A large duct conveys a high-temperature stream of air at 15 ft/s. The duct wall is at 200°F and a thermocouple ($\varepsilon = 0.6$) is placed in the middle of the duct. Neglecting conduction up the wire, what is the gas temperature when the thermocouple reads 1000°F? Treat the thermocouple as a $\frac{1}{4}$-in cylinder in cross-flow.

14·80 A thermocouple is to be designed to measure accurately the temperature of hot atmospheric air (500°C) in a cool duct (80°C wall). The thermocouple wire is 20 gauge and is strung across the flow. Design suitable radiation shielding to permit a gas-temperature measurement to within 1°C. The air velocity is 5 m/s.

APPENDIX A

UNIT SYSTEMS AND DIMENSIONAL EQUIVALENTS

The conceptual basis of unit systems was discussed in Chap. 1. Table A·1 shows the dimensions of mechanical quantities in four commonly used unit systems. Note that the constant in Newton's second law, g_c, is arbitrarily taken as unity, except in the old English system where its value of 32.17 $\text{ft}\cdot\text{lbm}/(\text{lbf}\cdot\text{s}^2)$ is conceptually experimental. Three common electromagnetic unit systems are shown in Table A·2. Note that the differences are in the manner that charge is handled; some differences in the definitions are also involved. Table A·3 contains a selected list of dimensional equivalents, and Table A·4 gives values for several important physical constants.

Dimensional equivalents are most easily obtained by considering a given magnitude of the quantity in the two unit systems. Two examples of this procedure follow.

1. Find the equivalent of 1 N of force in the old English system. To determine this dimensional equivalent, consider a body having a mass of 1 kg being accelerated by a force at the rate of 1 m/s². The force acting on the body is, in the mks system,

$$\mathsf{F} = ma = 1\ \text{kg}\cdot\text{m/s}^2$$

The mass of the body in the old English system is (Table A·3) 2.2046 lbm. Its acceleration is

$$1 \text{ m} \times 3.280 \text{ (ft/m)/s}^2 = 3.280 \text{ ft/s}^2$$

In the old English system the force is

$$\mathsf{F} = \frac{1}{g_c} ma = \frac{2.2046 \text{ lbm} \times 3.280 \text{ ft/s}^2}{32.17 \text{ ft} \cdot \text{lbm/(lbf} \cdot \text{s}^2)} = 0.2248 \text{ lbf}$$

Therefore

$$1 \text{ N} \equiv 1 \text{ kg} \cdot \text{m/s}^2 = 0.2248 \text{ lbf}$$

2. Determine the dimensional equivalents of the magnetic induction B and the magnetization M in the rationalized practical mksc and absolute magnetostatic cgs unit systems. This calculation is complicated by the difference in definition. We have

$$\mathsf{B} = \mathsf{H} + 4\pi\mathsf{M} \qquad \textit{magnetostatic cgs system} \qquad (\text{A}\cdot 1a)$$

$$\mathsf{B} = \mu_0(\mathsf{H} + \mathsf{M}) \qquad \textit{mksc system} \qquad (\text{A}\cdot 1b)$$

Consider two charges, each of 1 C, separated by a distance of 1 m, moving in

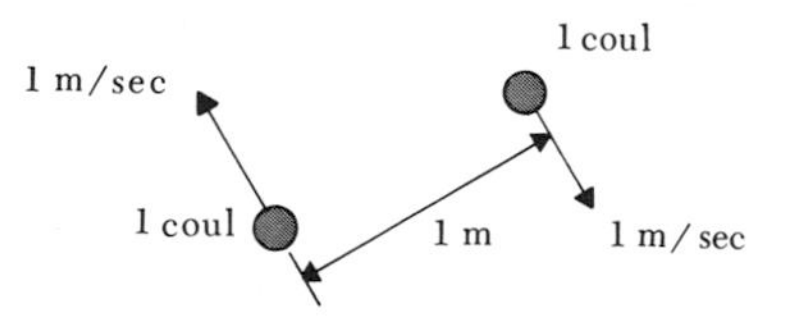

opposite directions perpendicular to the line between them at 1 m/s. From the Biot-Savart law the force in the mksc system is

$$\mathsf{F} = \frac{1.256 \times 10^{-6}}{4\pi} \text{ kg} \cdot \text{m/C}^2 \times 1 \text{ C}^2 \times 1 \text{ m}^2/\text{s}^2 \times \frac{1 \text{ m}}{1 \text{ m}^3}$$

$$= \frac{1.256 \times 10^{-6}}{4\pi} \text{ kg} \cdot \text{m/s}^2 = 10^{-7} \text{ kg} \cdot \text{m/s}^2$$

In the absolute magnetostatic unit system the charge, distance, and speed have the magnitudes

$$\mathcal{Q} = 0.1 \text{ g}^{1/2} \cdot \text{cm}^{1/2}$$

$$R = 100 \text{ cm}$$

$$\mathsf{V} = 100 \text{ cm/s}$$

The force in the magnetostatic cgs system is therefore

$$\mathsf{F} = 0.1^2 \text{ g}\cdot\text{cm} \times 100^2 \text{ cm}^2/\text{s}^2 \frac{100 \text{ cm}}{100^3 \text{ cm}^3} = 0.01 \text{ g}\cdot\text{cm/s}^2$$

The magnetic induction B, in either system, is

$$\mathsf{B} = \frac{\mathsf{F}}{\mathsf{V}\mathcal{Q}} \tag{A·2}$$

In the rationalized practical mksc system

$$\begin{aligned}\mathsf{B} &= \frac{10^{-7} \text{ kg}\cdot\text{m/s}^2}{1 \text{ C} \times 1 \text{ m/s}} \\ &= 10^{-7} \text{ kg/(C}\cdot\text{s)}\end{aligned}$$

In the absolute magnetostatic cgs system

$$\begin{aligned}\mathsf{B} &= \frac{0.01 \text{ g}\cdot\text{cm/s}^2}{0.1 \text{ g}^{1/2}\cdot\text{cm}^{1/2} \times 100 \text{ cm/s}} \\ &= 0.001 \text{ g}^{1/2}/(\text{cm}^{1/2}\cdot\text{s})\end{aligned}$$

In terms of the aliases

$$1 \text{ T} \equiv 1 \text{ kg/(C}\cdot\text{s)} \qquad \text{(Tesla)}$$

$$1 \text{ G} \equiv 1 \text{ g}^{1/2}/(\text{cm}^{1/2}\cdot\text{s}) \qquad \text{(Gauss)}$$

we have

$$10^{-7} \text{ T} = 0.001 \text{ G}$$

So, for B,

$$1 \text{ T} = 10^4 \text{ G}$$

We hasten to point out that the correspondence between T and G does not apply to H and M. To find the dimensional equivalents for H we consider a field with $\mathsf{B} = 1$ T and $\mathsf{M} = 0$. Then, in the rationalized practical mksc system,

$$\begin{aligned}\mathsf{H} &= \frac{\mathsf{B}}{\mu_0} \\ &= \frac{1 \text{ kg/(C}\cdot\text{s)}}{1.256 \times 10^{-6} \text{ kg}\cdot\text{m/C}^2} = \frac{1}{1{:}256 \times 10^{-6}} \text{ C/(m}\cdot\text{s)}\end{aligned}$$

In the absolute magnetostatic cgs system $\mathsf{B} = \mathsf{H} = 10^4$ G. Equating the two values for H, we find

$$1 \text{ G of } \mathsf{H} = \frac{1}{1.256 \times 10^{-2}} \text{ C/(m·s)}$$

To obtain the equivalents for M we imagine a field of 1 T with $\mathsf{H} = 0$. Then, in the rationalized practical mksc unit system,

$$\mathsf{M} = \frac{\mathsf{B}}{\mu_0} = \frac{1}{1.256 \times 10^{-6}} \text{ C/(m·s)}$$

In the absolute magnetostatic cgs system

$$\mathsf{M} = \frac{\mathsf{B}}{4\pi} = \frac{10^4}{4\pi} \text{ G}$$

Equating the two magnetizations, we find

$$1 \text{ G of } \mathsf{M} = 10^3 \text{ C/(m·s)}$$

The dimensional equivalence for B might be put in the form

$$1 \text{ G of } \mathsf{B} = 10^{-4} \text{ kg/(C·s)}$$

This example illustrates the importance of keeping straight just what the dimensional equivalent is for. One G is not always equivalent to the same number of C/(m·s); sometimes it is equivalent to a number of kg/(C·s).

Converting units from one system to another is a confusing task which can be made simple by noting that multiplication of anything by unity leaves it unchanged. The number 1 can be written in many useful ways, two of which are

$$1 = \frac{12 \text{ in}}{1 \text{ ft}}$$

$$1 = \frac{778.16 \text{ ft·lbf}}{1 \text{ Btu}}$$

Then, to express an energy density of 50 Btu/in^3 in terms of ft·lbf/ft^3,

$$\frac{E}{V} = 50 \text{ Btu/in}^3 \times 778.16 \text{ ft·lbf/Btu} \times (12 \text{ in/ft})^3$$
$$= 6.72 \times 10^7 \text{ ft·lbf/ft}^3$$

You are encouraged to acquire the habit of always writing down units in a numerical problem. Dimensions provide an easy check on equations and are an essential part of the answer to any engineering problem.

Table A·1 Mechanical unit systems

	SI	Cgs	Absolute old English	Old English†
Primary quantities and their units				
Length	meter, m	centimeter, cm	foot, ft	foot, ft
Mass	kilogram, kg	gram, g	...	pound mass, lbm
Time	second, s	second, s	second, s	second, s
Force	...	...	pound force, lbf	pound force, lbf
Newton's second law, $\mathsf{F} = \dfrac{m\,dV}{g_c\,dt}$	$g_c \equiv 1$ (selected)	$g_c \equiv 1$ (selected)	$g_c \equiv 1$ (selected)	$g_c = 32.17$ ft·lbm/(lbf·s^2) (experimental)
Secondary quantities and their units				
Force	kg·m/s^2	g·cm/s^2	...	...
Mass	...	...	lbf·s^2/ft	...
Energy, $đW = \mathsf{F}\,dX$	kg·m^2/s^2	g·cm^2/s^2	ft·lbf	ft·lbf
Power, $\dot{W}$	kg·m^2/s^3	g·cm^2/s^3	ft·lbf/s	ft·lbf/s
Aliases				
Force	1 N ≡ 1 kg·m/s^2	1 dyn ≡ 1 g·cm/s^2		
Mass	...	...	1 slug ≡ 1 lbf·s^2/ft	
Energy	1 J ≡ 1 kg·m^2/s^2 = 1 N·m	1 erg ≡ 1 g·cm^2/s^2 = 1 dyn·cm		
Power	1 W ≡ 1 kg·m^2/s^3 = 1 J/s			

† A body having a weight of 1 lbf on the surface of the earth will have a mass of approximately 1 lbm.

Table A·2 Electromagnetic unit systems

	Mksc (SI)†	Absolute cgs electrostatic (esu)	Absolute cgs magnetostatic (emu)
Primary quantities and their units			
Length	meter, m	centimeter, cm	centimeter, cm
Mass	kilogram, kg	gram, g	gram, g
Time	second, s	second, s	second, s
Charge	coulomb, C	...	...
Coulomb's law,	$k_C = \dfrac{1}{4\pi\varepsilon_0}$	$k_C \equiv 1$ (selected)	$k_C = 0.8992 \times 10^{17}$ cm^2·s^{-2} (experimental)
$\mathsf{F}_{12} = k_C \dfrac{\mathcal{Q}_1\mathcal{Q}_2}{R_{12}^2}$	$\varepsilon_0 = 8.854 \times 10^{-12}$ C^2·s^2· kg^{-1}·m^{-3} (experimental)		
Biot-Savart law,			
$\mathsf{F}_{12} = k_B \dfrac{\mathcal{Q}_1\mathcal{Q}_2 \mathbf{V}_1 \times (\mathbf{V}_2 \times \mathbf{R})}{\lvert\mathbf{R}\rvert^3}$	$k_B = \dfrac{\mu_0}{4\pi}$	$k_B = 1.112 \times 10^{-17}$ s^2·cm^{-2} (experimental)	$k_B \equiv 1$ (selected)
	$\mu_0 = 1.256 \times 10^{-6}$ kg·m·C^{-2} (experimental)		
Secondary quantities and their units			
Charge	...	g$^{1/2}$·cm$^{3/2}$·s^{-1}	g$^{1/2}$·cm$^{1/2}$
Current density, $J = \dot{\mathcal{Q}}/A$	C·m^{-2}·s^{-1}	g$^{1/2}$·cm$^{-1/2}$·s^{-2}	g$^{1/2}$·cm$^{-3/2}$·s^{-1}
Current, $I = \int \mathbf{J} \cdot d\mathbf{A}$	C·s^{-1}	g$^{1/2}$·cm$^{3/2}$·s^{-2}	g$^{1/2}$·cm$^{1/2}$·s^{-1}
Electric and magnetic fields, $\mathbf{F} = \mathcal{Q}(\mathbf{E} + \mathbf{V} \times \mathbf{B})$			
Electric field strength **E**	kg·m·C^{-1}·s^{-2}	g$^{1/2}$·cm$^{-1/2}$·s^{-1}	g$^{1/2}$·cm$^{-1/2}$·s^{-2}
Magnetic induction **B**	kg·s^{-1}·C^{-1}	g$^{1/2}$·cm$^{-3/2}$	g$^{1/2}$·cm$^{-1/2}$·s^{-1}

Electrical potential $\mathscr{E}$	$kg \cdot m^2 \cdot C^{-1} \cdot s^{-2}$	$g^{1/2} \cdot cm^{1/2} \cdot s^{-1}$	$g^{1/2} \cdot cm^{3/2} \cdot s^{-2}$
$d\mathscr{E} = \mathbf{E} \cdot d\mathbf{L}$			
Electric displacement, polarization	$\mathbf{D} = \varepsilon_0 \mathbf{E} + \mathbf{P}$	$\mathbf{D} = \mathbf{E} + 4\pi\mathbf{P}$	
Displacement **D**	$C \cdot m^{-2}$	$g^{1/2} \cdot cm^{-1/2} \cdot s^{-1}$	
Polarization **P**	$C \cdot m^{-2}$	$g^{1/2} \cdot cm^{-1/2} \cdot s^{-1}$	
Magnetic field strength, magnetization	$\mathbf{B} = \mu_0(\mathbf{H} + \mathbf{M})$		$\mathbf{B} = \mathbf{H} + 4\pi\mathbf{M}$
Magnetic field strength **H**	$C \cdot m^{-1} \cdot s^{-1}$		$g^{1/2} \cdot cm^{-1/2} \cdot s^{-1}$
Magnetization **M**	$C \cdot m^{-1} \cdot s^{-1}$		$g^{1/2} \cdot cm^{-1/2} \cdot s^{-1}$
Electric flux, $\Phi_E = \int \mathbf{E} \cdot d\mathbf{A}$	$kg \cdot m^3 \cdot C^{-1} \cdot s^{-2}$	$g^{1/2} \cdot cm^{3/2} \cdot s^{-1}$	
Magnetic flux, $\Phi_B = \int \mathbf{B} \cdot d\mathbf{A}$	$kg \cdot m^2 \cdot C^{-1} \cdot s^{-1}$		$g^{1/2} \cdot cm^{3/2} \cdot s^{-1}$
Electrical conductivity, J/E	$C^2 \cdot s \cdot kg^{-1} \cdot m^{-3}$	s^{-1}	$s \cdot cm^{-2}$
Resistance, $R = \Delta\mathscr{E}/I$	$kg \cdot m^2 \cdot C^{-2} \cdot s^{-1}$	$s \cdot cm^{-1}$	$cm \cdot s^{-1}$
Capacitance, $C = I/(d\mathscr{E}/dt)$	$C^2 \cdot s^2 \cdot kg^{-1} \cdot m^{-2}$	cm	$s^2 \cdot cm^{-1}$
Inductance, $L = \Delta\mathscr{E}/(dI/dt)$	$kg \cdot m^2 \cdot C^{-2}$	$s^2 \cdot cm^{-1}$	cm
Aliases			
Charge	...	$1\ statC \equiv 1\ g^{1/2} \cdot cm^{3/2} \cdot s^{-1}$	$1\ aC \equiv 1\ g^{1/2} \cdot cm^{1/2}$
Current	$1\ A \equiv 1\ C \cdot s^{-1}$	$1\ statA \equiv 1\ g^{1/2} \cdot cm^{3/2} \cdot s^{-2}$	$1\ aA \equiv 1\ g^{1/2} \cdot cm^{1/2} \cdot s^{-1}$
Magnetic induction	$1\ T \equiv 1\ kg \cdot s^{-1} \cdot C^{-1}$		$1\ G \equiv 1\ g^{1/2} \cdot cm^{-1/2} \cdot s^{-1}$
Electric potential	$1\ V \equiv 1\ kg \cdot m^2 \cdot C^{-1} \cdot s^{-2}$	$1\ statV \equiv 1\ g^{1/2} \cdot cm^{1/2} \cdot s^{-1}$	$1\ aV \equiv 1\ g^{1/2} \cdot cm^{3/2} \cdot s^{-2}$
Magnetic flux	$1\ Wb \equiv 1\ kg \cdot m^2 \cdot C^{-1} \cdot s^{-1}$		
Resistance	$1\ \Omega \equiv 1\ kg \cdot m^2 \cdot C^{-2} \cdot s^{-1}$	$1\ stat\Omega \equiv 1\ s \cdot cm^{-1}$	$1\ a\Omega \equiv 1\ cm \cdot s^{-1}$
Capacitance	$1\ F \equiv 1\ C^2 \cdot s^2 \cdot kg^{-1} \cdot m^{-2}$	$1\ statF \equiv 1\ cm$	$1\ aF \equiv 1\ s^2 \cdot cm^{-1}$
Inductance	$1\ H \equiv 1\ kg \cdot m^2 \cdot C^{-2}$	$1\ statH \equiv 1\ s^2 \cdot cm^{-1}$	$1\ aH \equiv 1\ cm$

† In SI current replaces charge as a primary quantity.

Table A·3 Selected dimensional equivalents

Length	$1\ \text{m} = 3.2808\ \text{ft} = 39.37\ \text{in}$
	$1\ \text{cm} \equiv 10^{-2}\ \text{m} = 0.394\ \text{in} = 0.0328\ \text{ft}$
	$1\ \text{mm} \equiv 10^{-3}\ \text{m}$
	$1\ \mu\text{m} \equiv 10^{-6}\ \text{m}$
	$1\ \text{Å} \equiv 10^{-10}\ \text{m}$
	$1\ \text{km} = 0.621\ \text{mi}$
	$1\ \text{mi} = 5280\ \text{ft}$
Area	$1\ \text{m}^2 = 10.76\ \text{ft}^2$
	$1\ \text{cm}^2 = 10^{-4}\ \text{m}^2 = 0.155\ \text{in}^2$
Volume	$1\ \text{gal} \equiv 0.13368\ \text{ft}^3 = 3.785\ \text{liters}$
	$1\ \text{liter} \equiv 10^{-3}\ \text{m}^3$
Time	$1\ \text{h} \equiv 3600\ \text{s} = 60\ \text{min}$
	$1\ \text{ms} \equiv 10^{-3}\ \text{s}$
	$1\ \mu\text{s} \equiv 10^{-6}\ \text{s}$
	$1\ \text{ns} = 10^{-9}\ \text{s}$
Mass	$1\ \text{kg} \equiv 1000\ \text{g} = 2.2046\ \text{lbm} = 6.8521 \times 10^{-2}\ \text{slug}$
	$1\ \text{slug} \equiv 1\ \text{lbf}\cdot\text{s}^2/\text{ft} = 32.174\ \text{lbm}$
Force	$1\ \text{N} \equiv 1\ \text{kg}\cdot\text{m/s}^2$
	$1\ \text{dyn} \equiv 1\ \text{g}\cdot\text{cm/s}^2$
	$1\ \text{lbf} = 4.448 \times 10^5\ \text{dyn} = 4.448\ \text{N}$
Energy	$1\ \text{J} \equiv 1\ \text{kg}\cdot\text{m}^2/\text{s}^2$
	$1\ \text{Btu} \equiv 778.16\ \text{ft}\cdot\text{lbf} = 1.055 \times 10^{10}\ \text{ergs} = 252\ \text{cal} = 1055.0\ \text{J}$
	$1\ \text{cal} \equiv 4.186\ \text{J}$
	$1\ \text{kcal} \equiv 4186\ \text{J} = 1000\ \text{cal}$
	$1\ \text{erg} \equiv 1\ \text{g}\cdot\text{cm}^2/\text{s}^2 = 10^{-7}\ \text{J}$
	$1\ \text{eV} \equiv 1.602 \times 10^{-19}\ \text{J}$
	$1\ \text{Q} \equiv 10^{18}\ \text{Btu} = 1.055 \times 10^{21}\ \text{J}$
	$1\ \text{Quad} = 10^{15}\ \text{Btu}$
	$1\ \text{kJ} = 0.947813\ \text{Btu} = 0.23884\ \text{kcal}$
Power	$1\ \text{W} \equiv 1\ \text{kg}\cdot\text{m}^2/\text{s}^3 = 1\ \text{J/s}$
	$1\ \text{hp} \equiv 550\ \text{ft}\cdot\text{lbf/s}$
	$1\ \text{hp} = 2545\ \text{Btu/h} = 746\ \text{W}$
	$1\ \text{kW} \equiv 1000\ \text{W} = 3412\ \text{Btu/h}$
Pressure	$1\ \text{atm} \equiv 14.696\ \text{lbf/in}^2 = 760\ \text{torr} = 101325\ \text{N/m}^2$
	$1\ \text{mm Hg} = 0.01934\ \text{lbf/in}^2 \equiv 1\ \text{torr}$
	$1\ \text{dyn/cm}^2 = 145.04 \times 10^{-7}\ \text{lbf/in}^2$
	$1\ \text{bar} \equiv 10^5\ \text{N/m}^2 = 14.504\ \text{lbf/in}^2 \equiv 10^6\ \text{dyn/cm}^2$
	$1\ \mu \equiv 10^{-6}\ \text{m Hg} = 10^{-3}\ \text{mm Hg}$
	$1\ \text{Pa} \equiv 1\ \text{N/m}^2 = 1.4504 \times 10^{-4}\ \text{lbf/in}^2$
	$1\ \text{in Hg} \equiv 3376.8\ \text{N/m}^2$
	$1\ \text{in H}_2\text{O} \equiv 248.8\ \text{N/m}^2$
Power per unit area	$1\ \text{W/m}^2 = 0.3170\ \text{Btu/(h}\cdot\text{ft}^2) = 0.85984\ \text{kcal/(h}\cdot\text{m}^2)$
Heat-transfer coefficient	$1\ \text{W/(m}^2\cdot{}^\circ\text{C}) = 0.1761\ \text{Btu/(h}\cdot\text{ft}^2\cdot{}^\circ\text{F}) = 0.85984\ \text{kcal/(h}\cdot\text{m}^2\cdot{}^\circ\text{C})$
Energy per unit mass	$1\ \text{kJ/kg} = 0.4299\ \text{Btu/lbm} = 0.23884\ \text{kcal/kg}$
Specific heat	$1\ \text{kJ/(kg}\cdot{}^\circ\text{C}) = 0.23884\ \text{Btu/(lbm}\cdot{}^\circ\text{F}) = 0.23884\ \text{kcal/(kg}\cdot{}^\circ\text{C})$
Thermal conductivity	$1\ \text{W/(m}\cdot{}^\circ\text{C}) = 0.5778\ \text{Btu/(h}\cdot\text{ft}\cdot{}^\circ\text{F}) = 0.85984\ \text{kcal/(h}\cdot\text{m}\cdot{}^\circ\text{C})$

Table A·3 (*continued*)

Viscosity	$1\ kg/(m\cdot s) = 1\ N\cdot s/m^2 = 0.6720\ lbm/(ft\cdot s) = 10$ Poise
Temperature	$C^\circ = 1.8\ F^\circ$ 0°C corresponds to 32°F, 273.16 K, and 491.69 R
Magnetic quantities	$1\ G \equiv 1\ g^{1/2}/(cm^{1/2}\cdot s)$ $1\ G = 10^3\ C/(m\cdot s)$ for M $1\ G = (1/4\pi) \times 10^3\ C/(m\cdot s)$ for H $1\ G = 10^{-4}\ T$ for B $1\ T \equiv 1\ kg/(C\cdot s)$

Table A·4 Physical constants

Avogadro's number	$N_0 = 6.022 \times 10^{23}$/gmole
Boltzmann's constant	$k = 1.380 \times 10^{-23}$ J/K
Gas constant	$\mathscr{R} = 1545.33$ ft·lbf/(lbmole·R) $= 8.3143$ J/(gmole·K) $= 8314.3$ J/(kgmole·K) $= 1.9858$ Btu/(lbmole·R) $= 1.9858$ cal/(gmole·K)
Planck's constant	$h = 6.626 \times 10^{-34}$ J·s
Coulomb constant	$1/4\pi\varepsilon_0 = 8.987 \times 10^9\ kg\cdot m^3/(C^2\cdot s^2)$
Biot-Savart constant	$\mu_0/4\pi = 1.0000 \times 10^{-7}\ kg\cdot m/C^2$
Electronic charge	$e = -1.6021 \times 10^{-19}$ C
Speed of light	$c = 2.998 \times 10^8$ m/s
Newton constant	$g_c = 32.174$ ft·lbm/(lbf·s²)
Gravitational constant	$k_G = 6.67 \times 10^{-11}\ m^3/(kg\cdot s^2)$

APPENDIX B

THERMODYNAMIC PROPERTIES OF SUBSTANCES

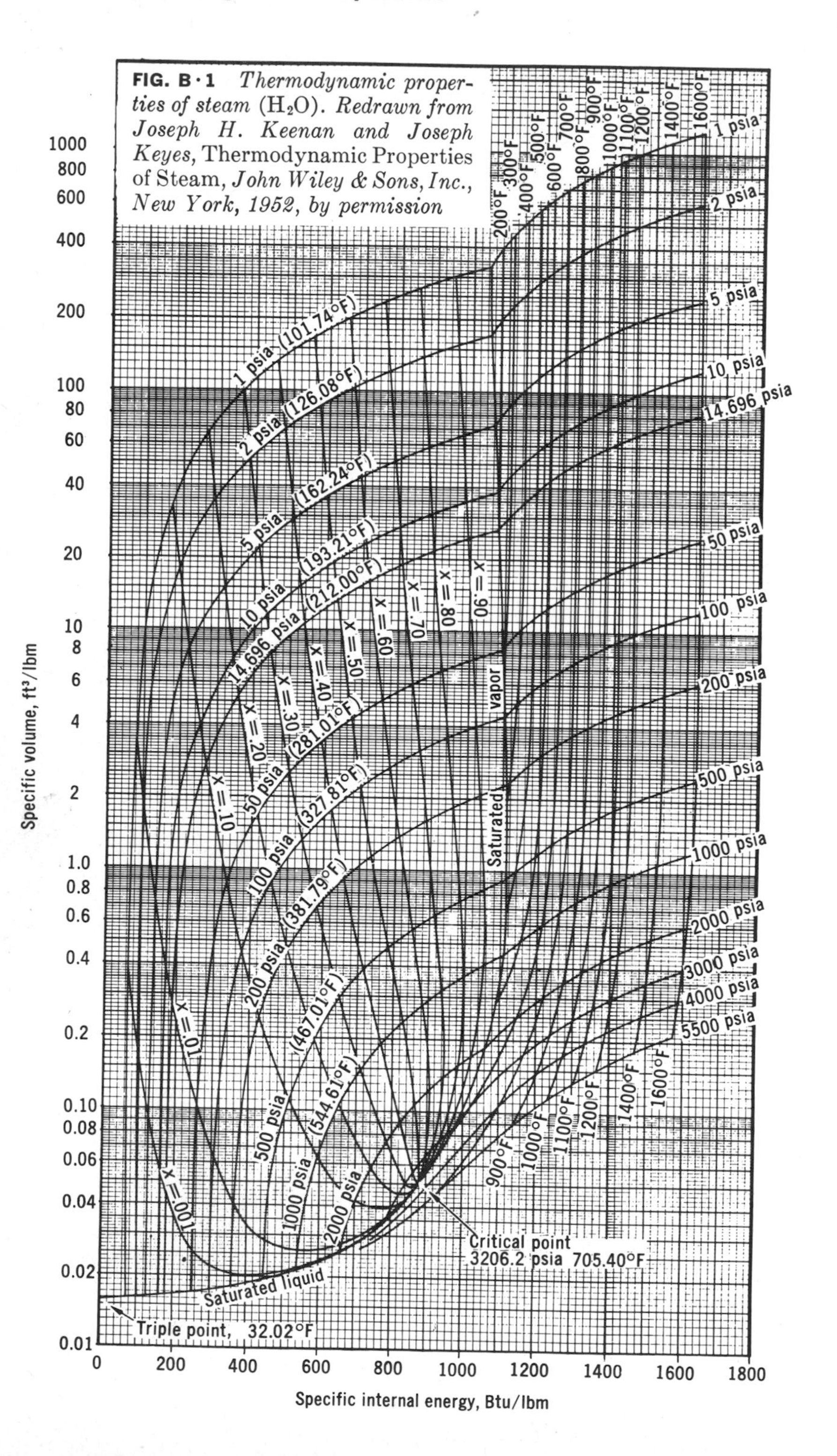

FIG. B·1 *Thermodynamic properties of steam* (H_2O). *Redrawn from Joseph H. Keenan and Joseph Keyes,* Thermodynamic Properties of Steam, *John Wiley & Sons, Inc., New York, 1952, by permission*

FIG. B·2 *Mollier diagram for steam* (H_2O). *From Joseph H. Keenan and Joseph Keyes*, Thermodynamic Properties of Steam, *John Wiley & Sons, Inc., New York, 1936, by permission*

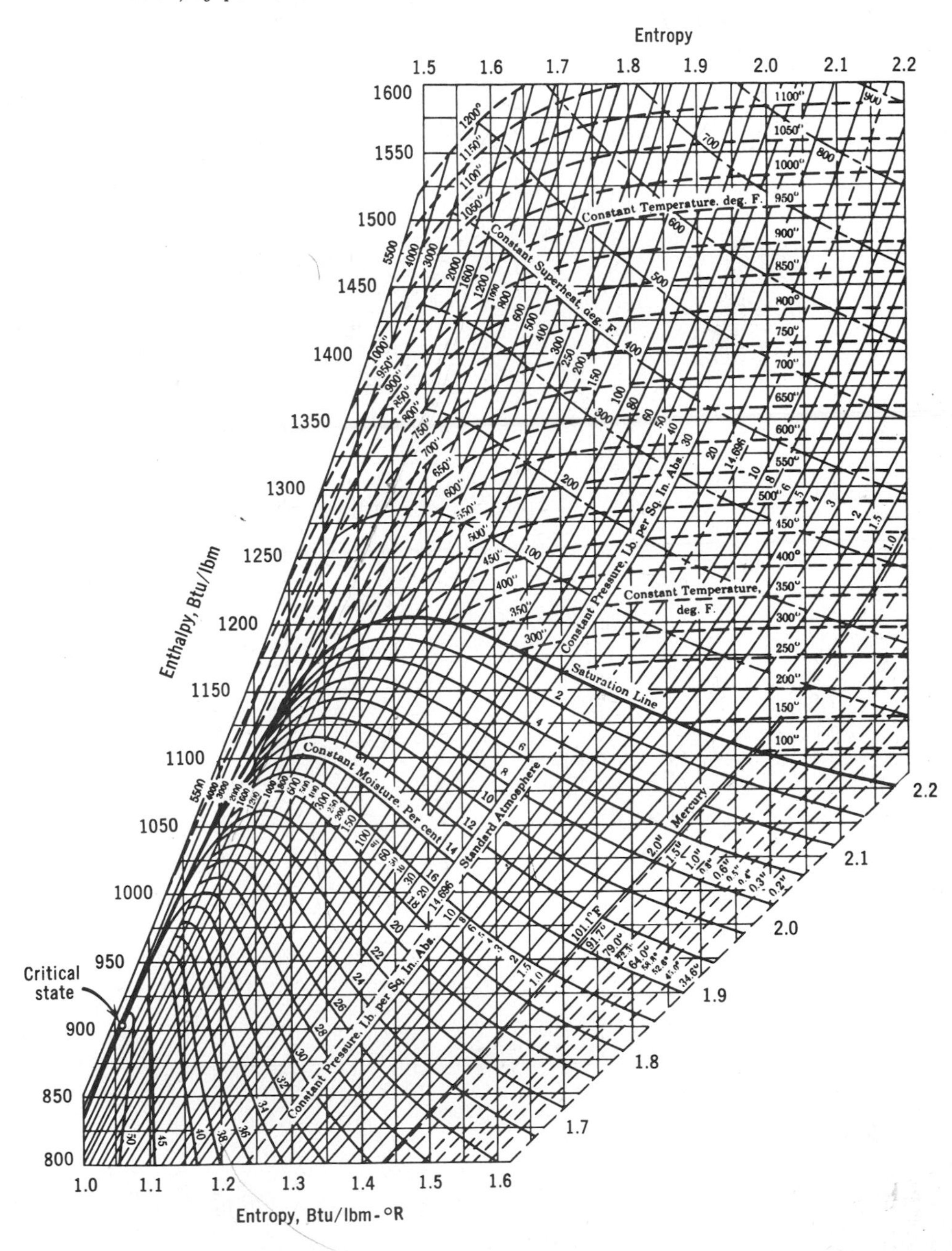

FIG. B·3 *Temperature-entropy diagram for steam* (H_2O). *From Joseph H. Keenan and Joseph Keyes,* Thermodynamic Properties of Steam, *John Wiley & Sons, Inc., New York, as adapted by Lay,* Thermodynamics, *Charles E. Merrill, Inc., Englewood Cliffs, N.J., 1963, by permission*

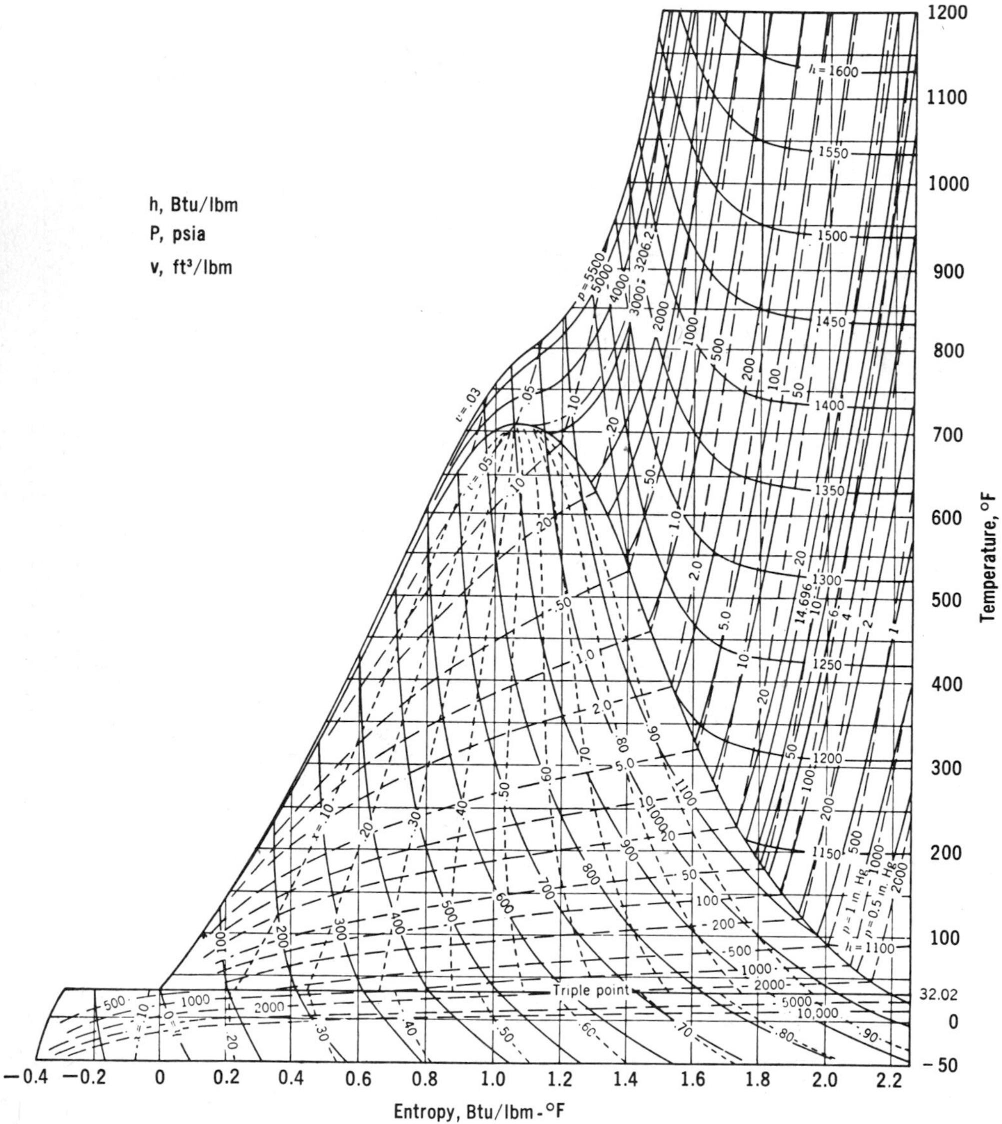

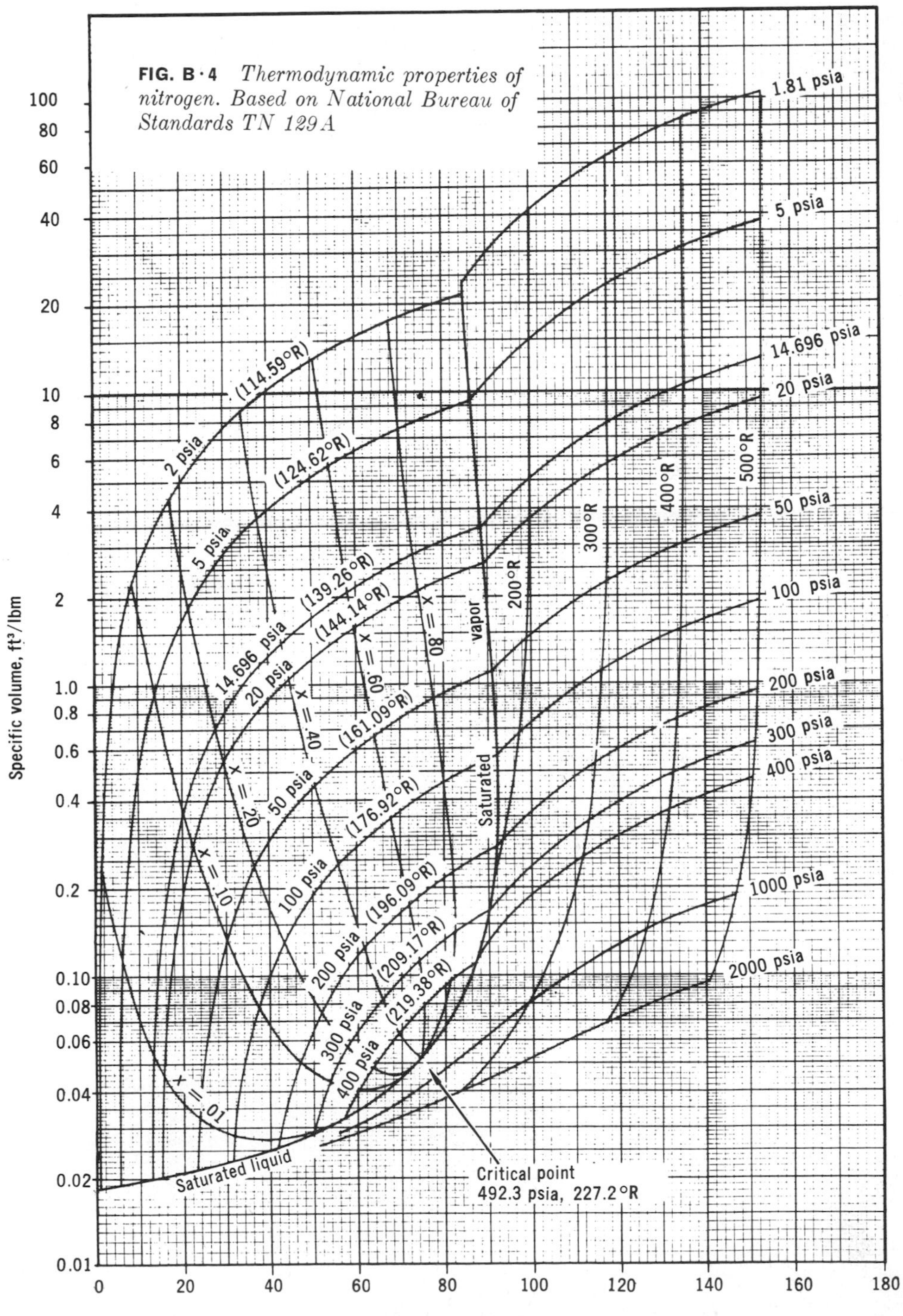

FIG. B·4 *Thermodynamic properties of nitrogen. Based on National Bureau of Standards TN 129A*

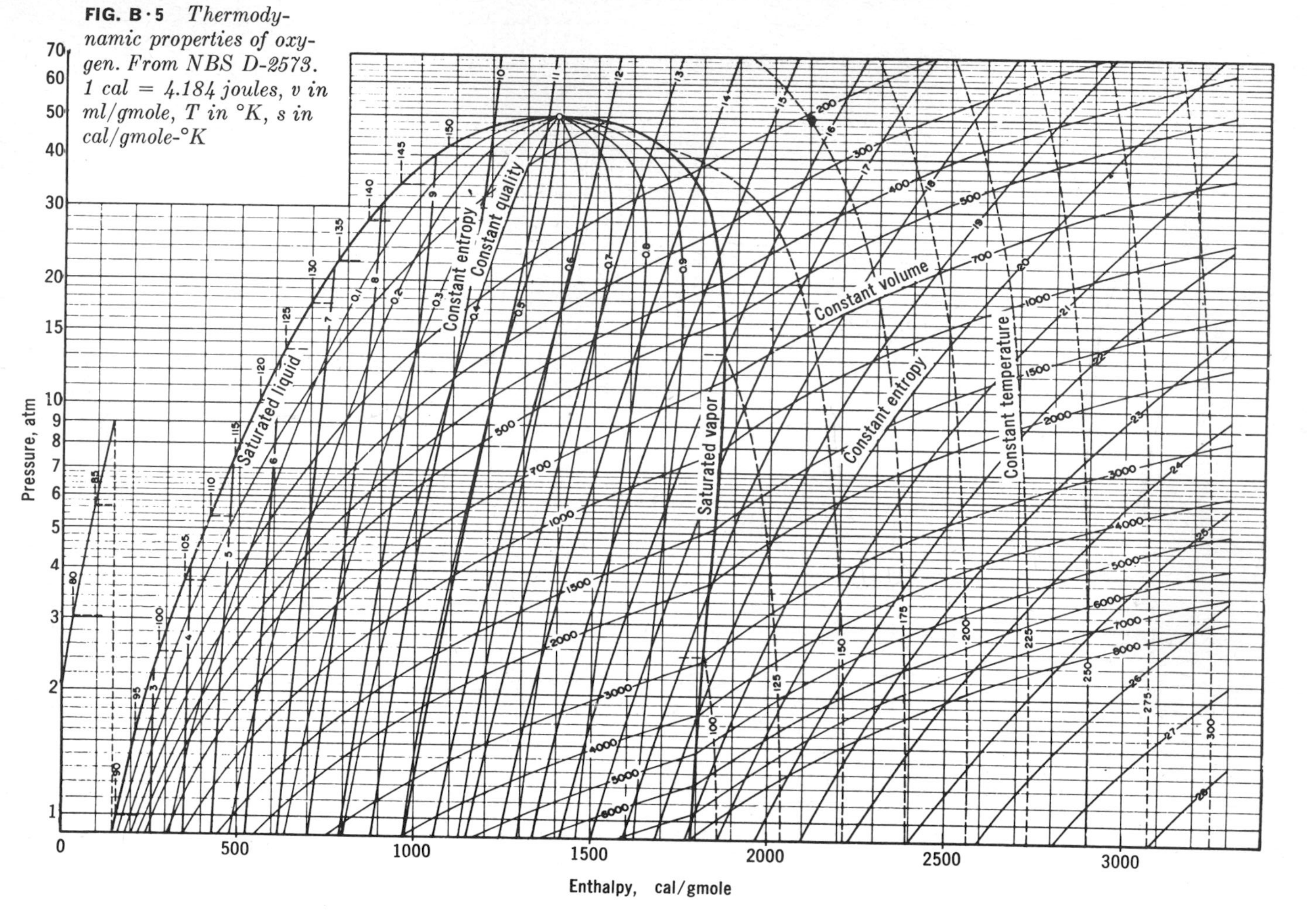

FIG. B·5 *Thermodynamic properties of oxygen. From NBS D-2573. 1 cal = 4.184 joules, v in ml/gmole, T in °K, s in cal/gmole-°K*

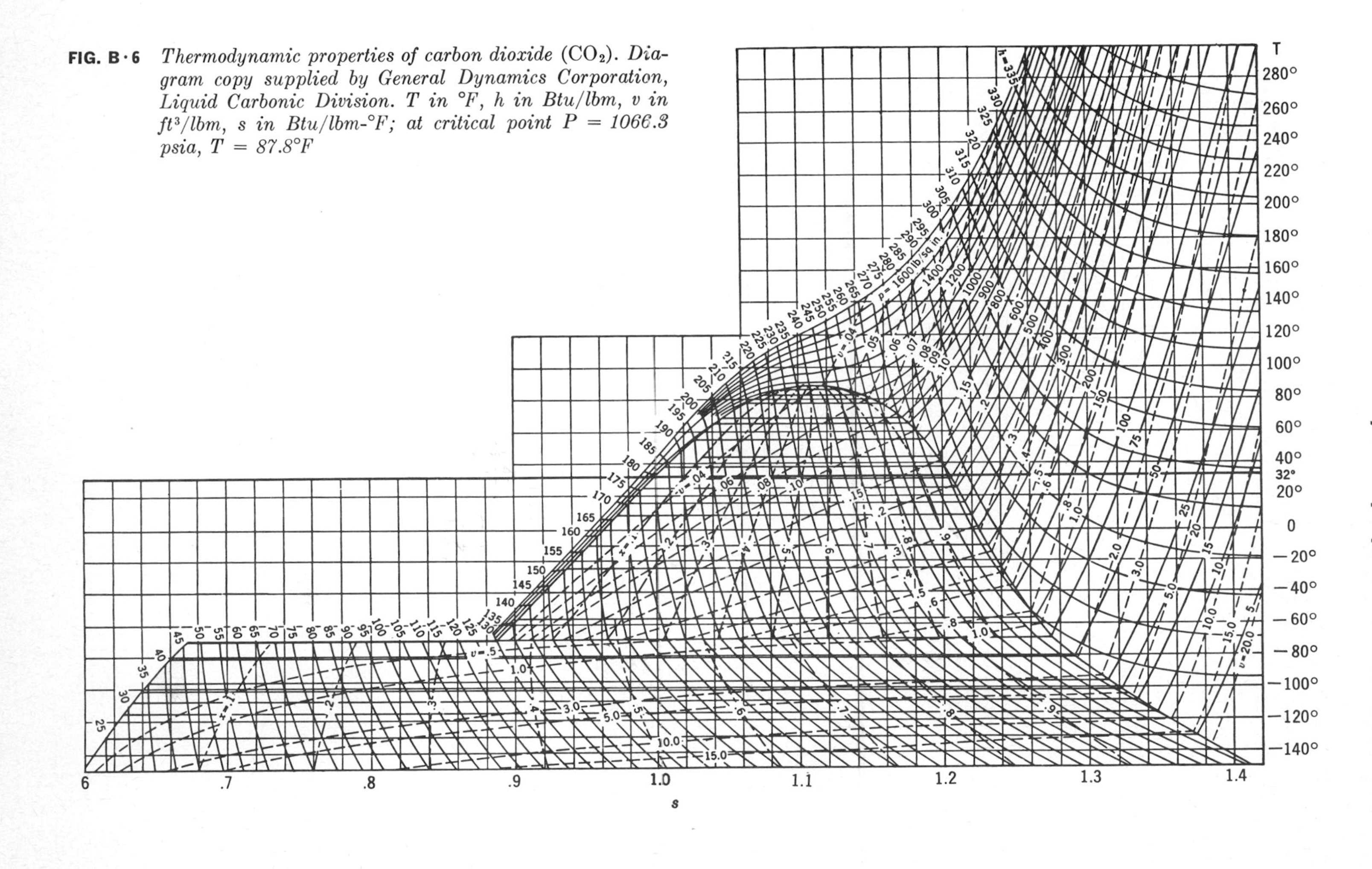

FIG. B·6 *Thermodynamic properties of carbon dioxide* (CO_2). *Diagram copy supplied by General Dynamics Corporation, Liquid Carbonic Division. T in °F, h in Btu/lbm, v in ft*^{3}*/lbm, s in Btu/lbm-°F; at critical point P = 1066.3 psia, T = 87.8°F*

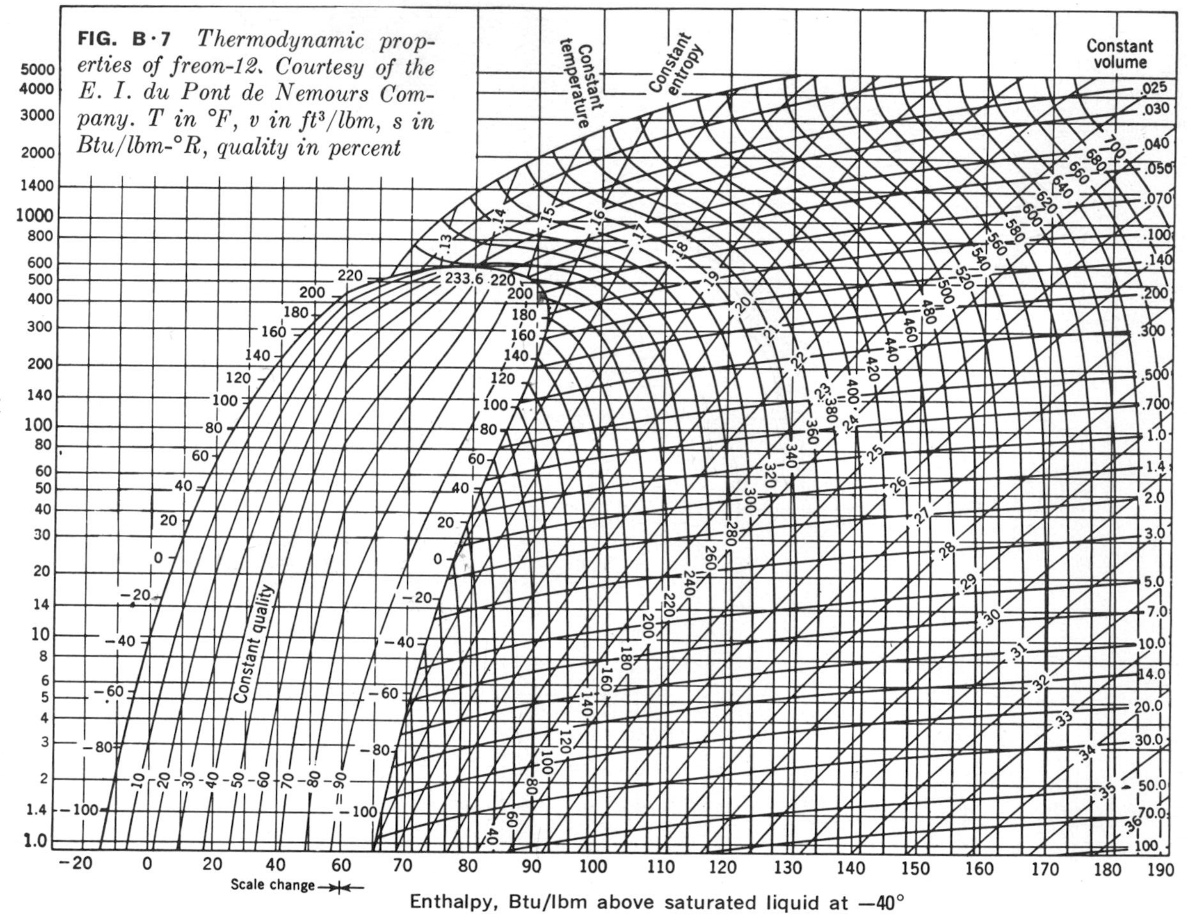

FIG. B·7 *Thermodynamic properties of freon-12. Courtesy of the E. I. du Pont de Nemours Company. T in °F, v in ft³/lbm, s in Btu/lbm-°R, quality in percent*

FIG. B·8 *Thermodynamic properties of air. From the National Bureau of Standards*

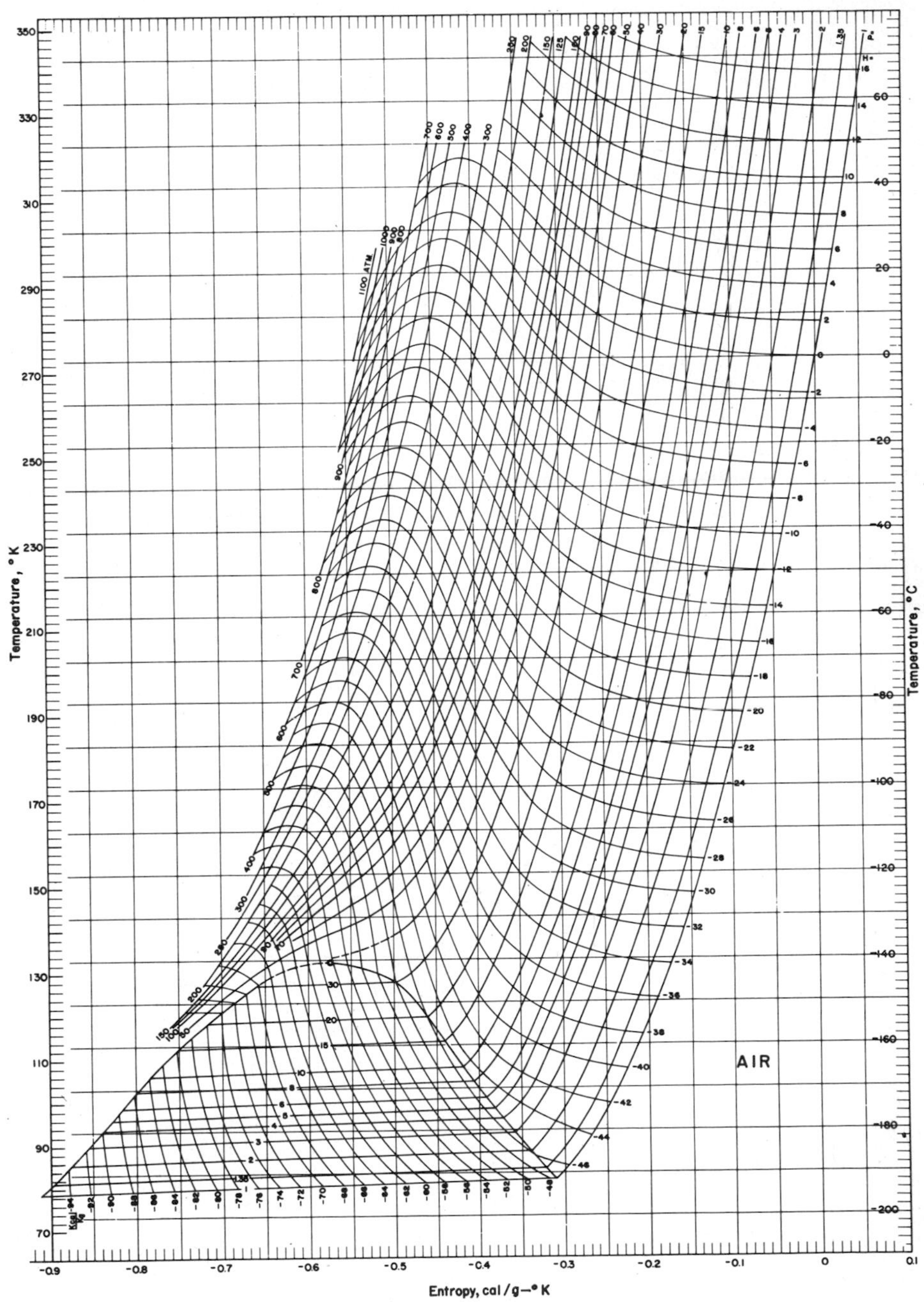

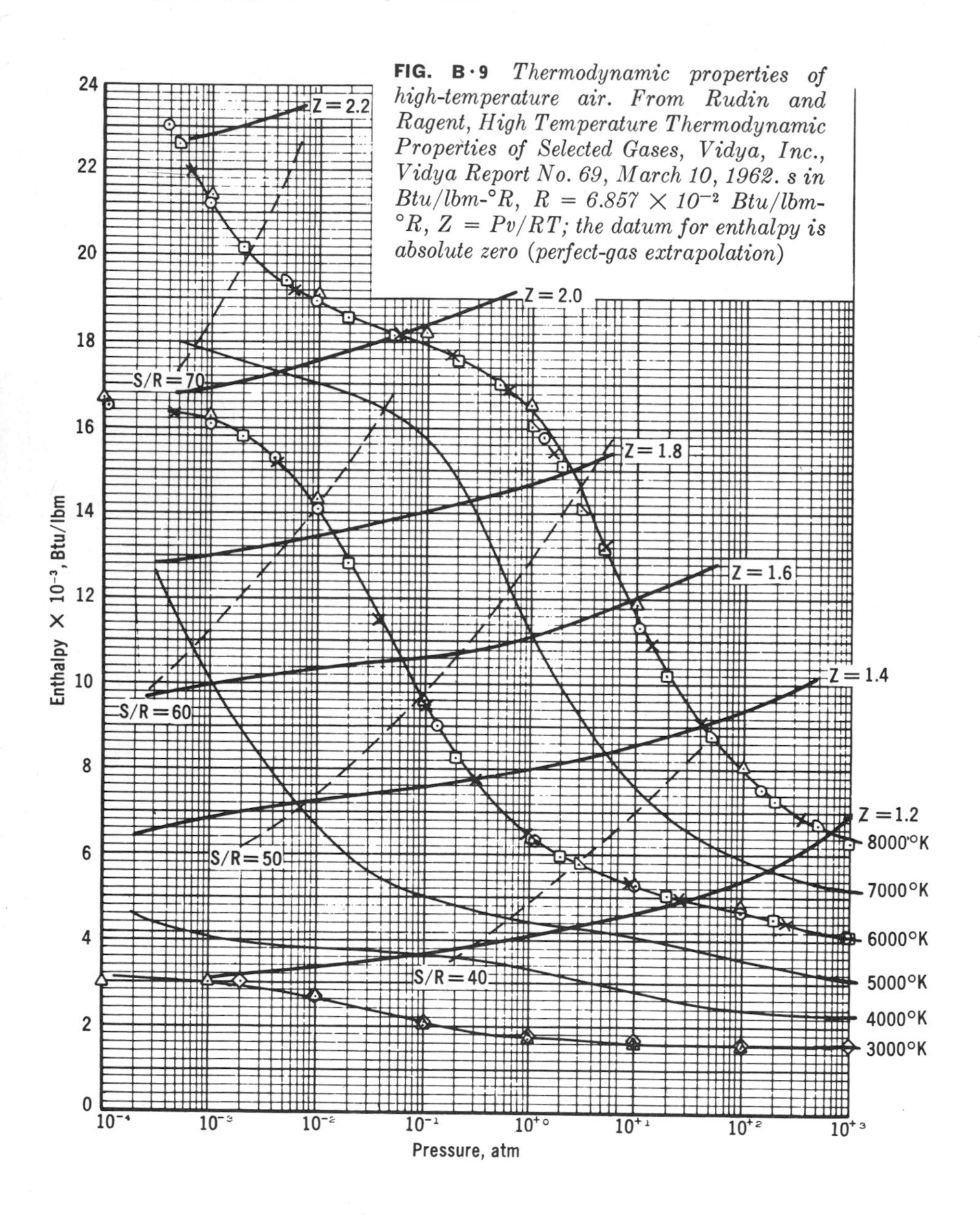

FIG. B·9 *Thermodynamic properties of high-temperature air. From Rudin and Ragent, High Temperature Thermodynamic Properties of Selected Gases, Vidya, Inc., Vidya Report No. 69, March 10, 1962. s in Btu/lbm-°R, $R = 6.857 \times 10^{-2}$ Btu/lbm-°R, $Z = Pv/RT$; the datum for enthalpy is absolute zero (perfect-gas extrapolation)*

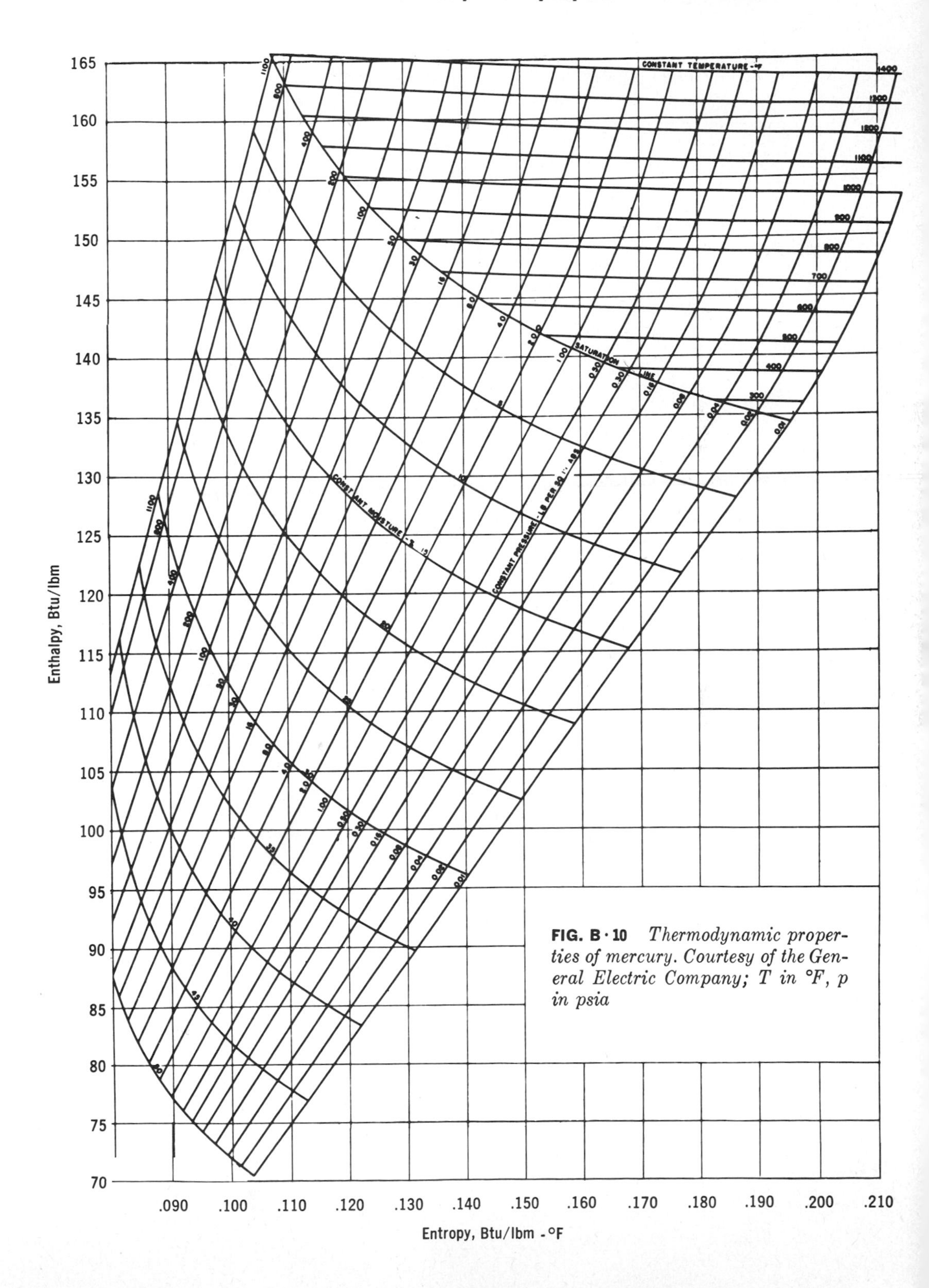

FIG. B · 10 *Thermodynamic properties of mercury. Courtesy of the General Electric Company; T in °F, p in psia*

FIG. B·11 *Thermodynamic properties of cesium. Redrawn from WADD Report 61-96*

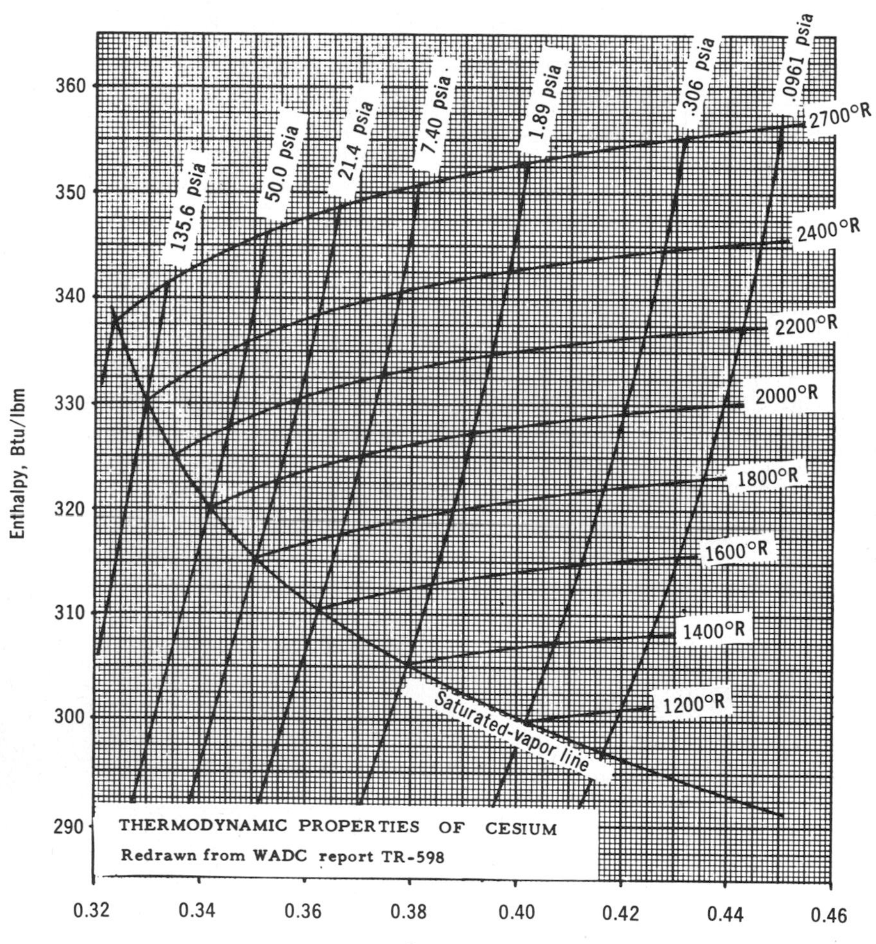

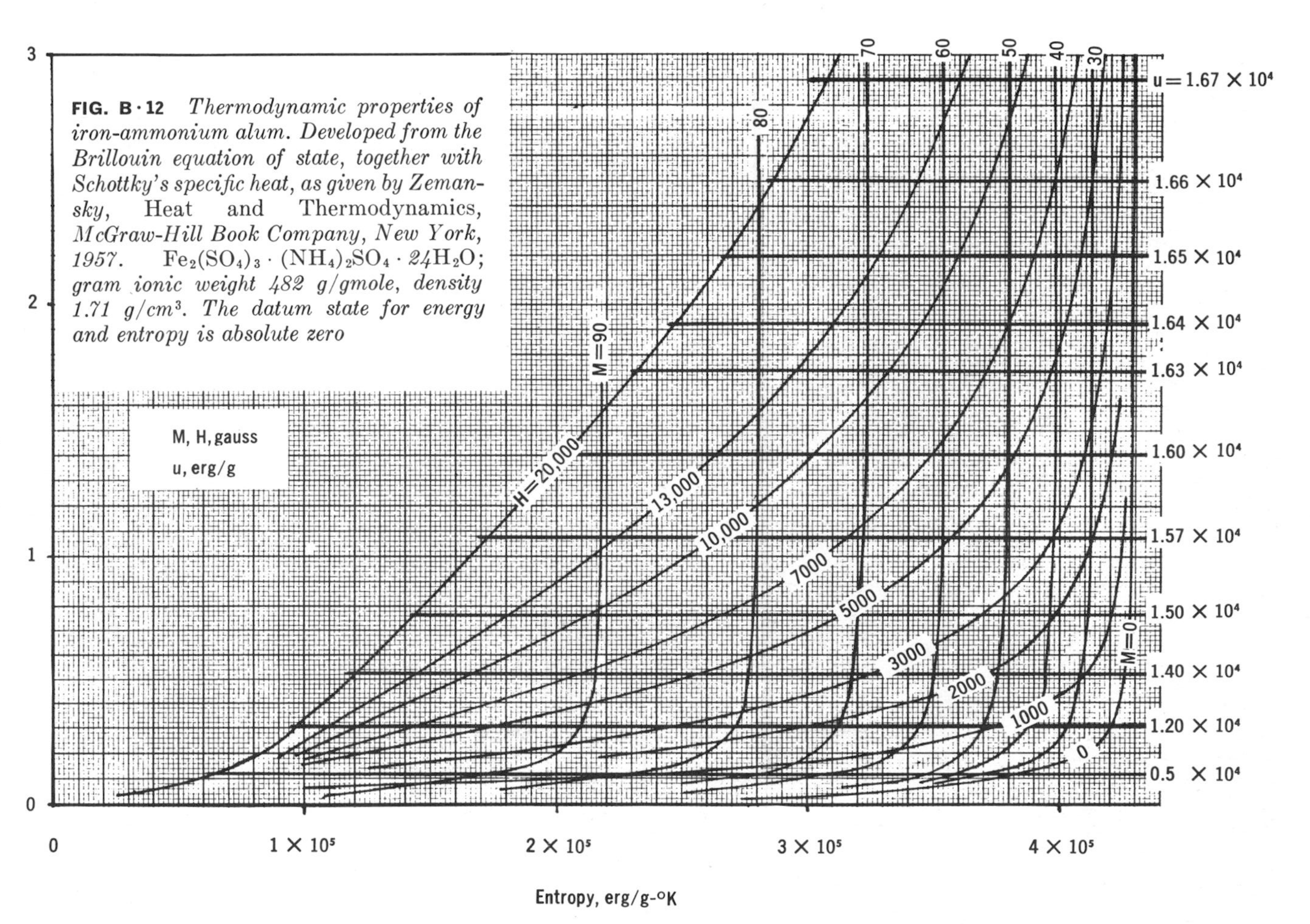

FIG. B·12 *Thermodynamic properties of iron-ammonium alum. Developed from the Brillouin equation of state, together with Schottky's specific heat, as given by Zemansky,* Heat and Thermodynamics, *McGraw-Hill Book Company, New York, 1957.* $Fe_2(SO_4)_3 \cdot (NH_4)_2SO_4 \cdot 24H_2O$; *gram ionic weight 482 g/gmole, density 1.71 g/cm³. The datum state for energy and entropy is absolute zero*

FIG. B·13a *Psychrometric chart (for 1 atm pressure). Courtesy of the General Electric Company. Barometric pressure, 14.696 lbf/in.² (1 lbm = 7000 grains)*

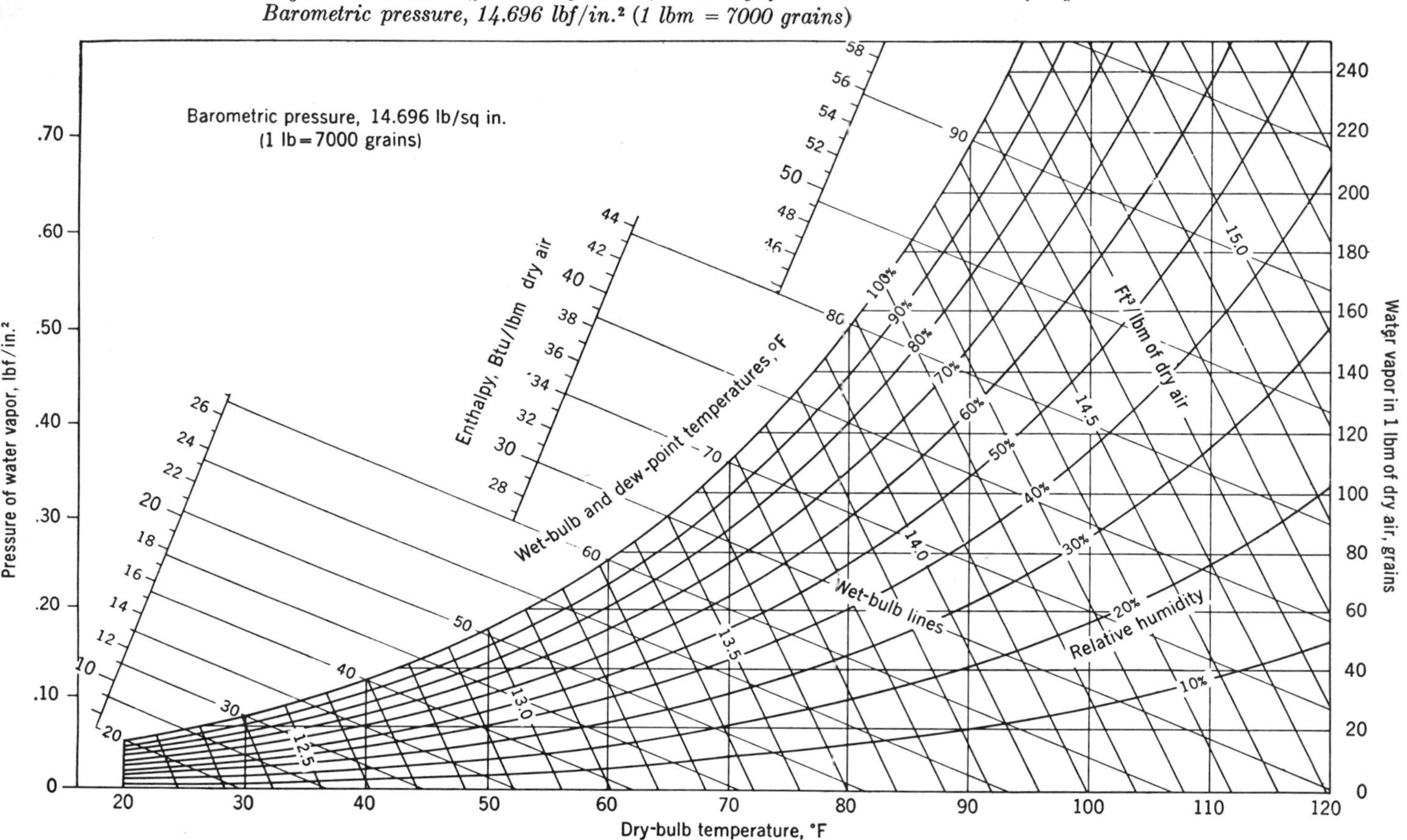

FIG. B·13b *Psychrometric chart (SI).*

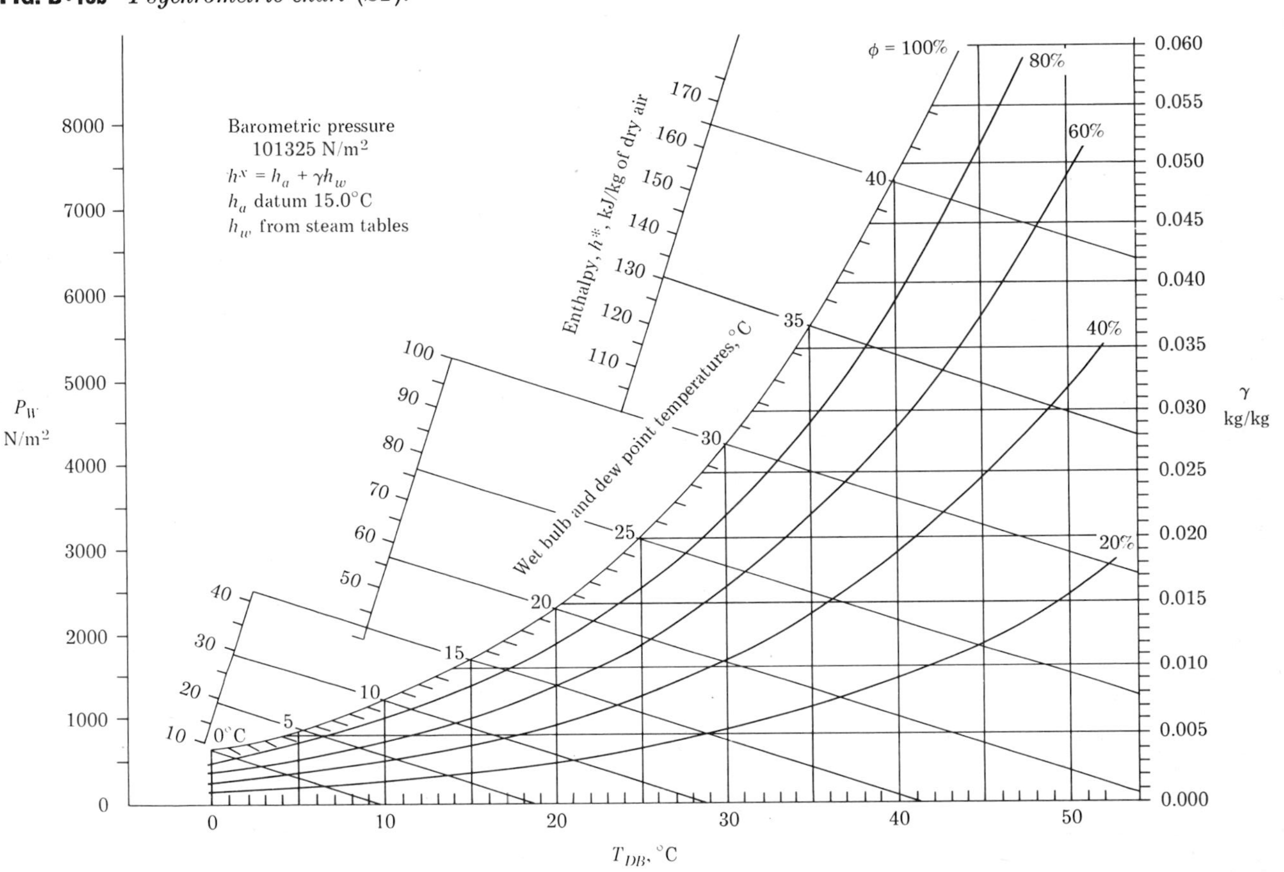

FIG. B · 14a *Generalized compressibility chart—low-pressure range. Data from L. C. Nelson and E. F. Obert, Generalized Compressibility Charts*, Chem. En., *vol. 61, p. 203, 1954.* Note: $v/v_c \equiv P_c v/RT_c$

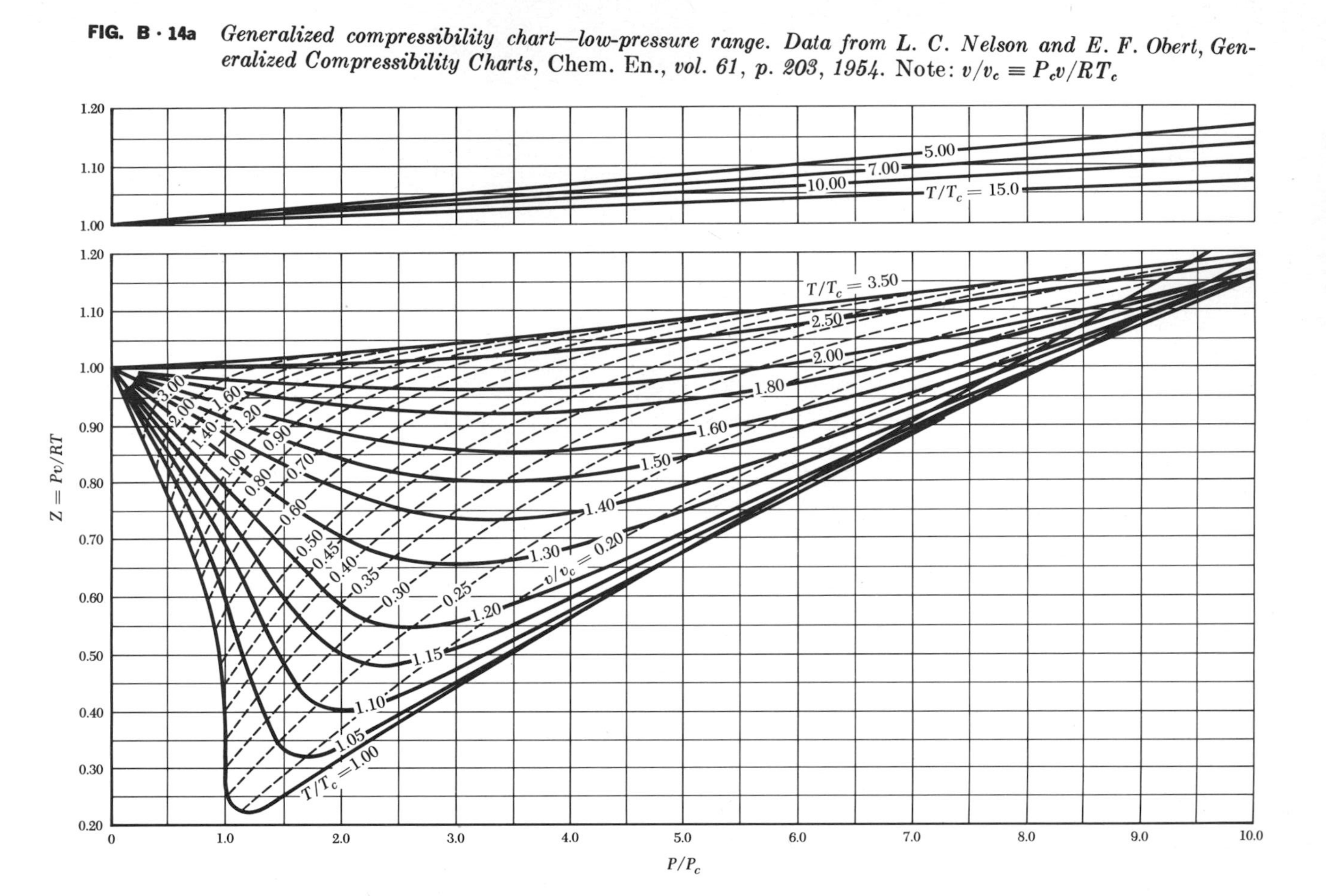

FIG. B · 14b *Generalized compressibility chart—high-pressure range. Adapted from E. F. Obert,* Concepts of Thermodynamics, *McGraw-Hill Book Company, New York, 1960.* Note: $v/v_c \equiv P_c v/RT_c$

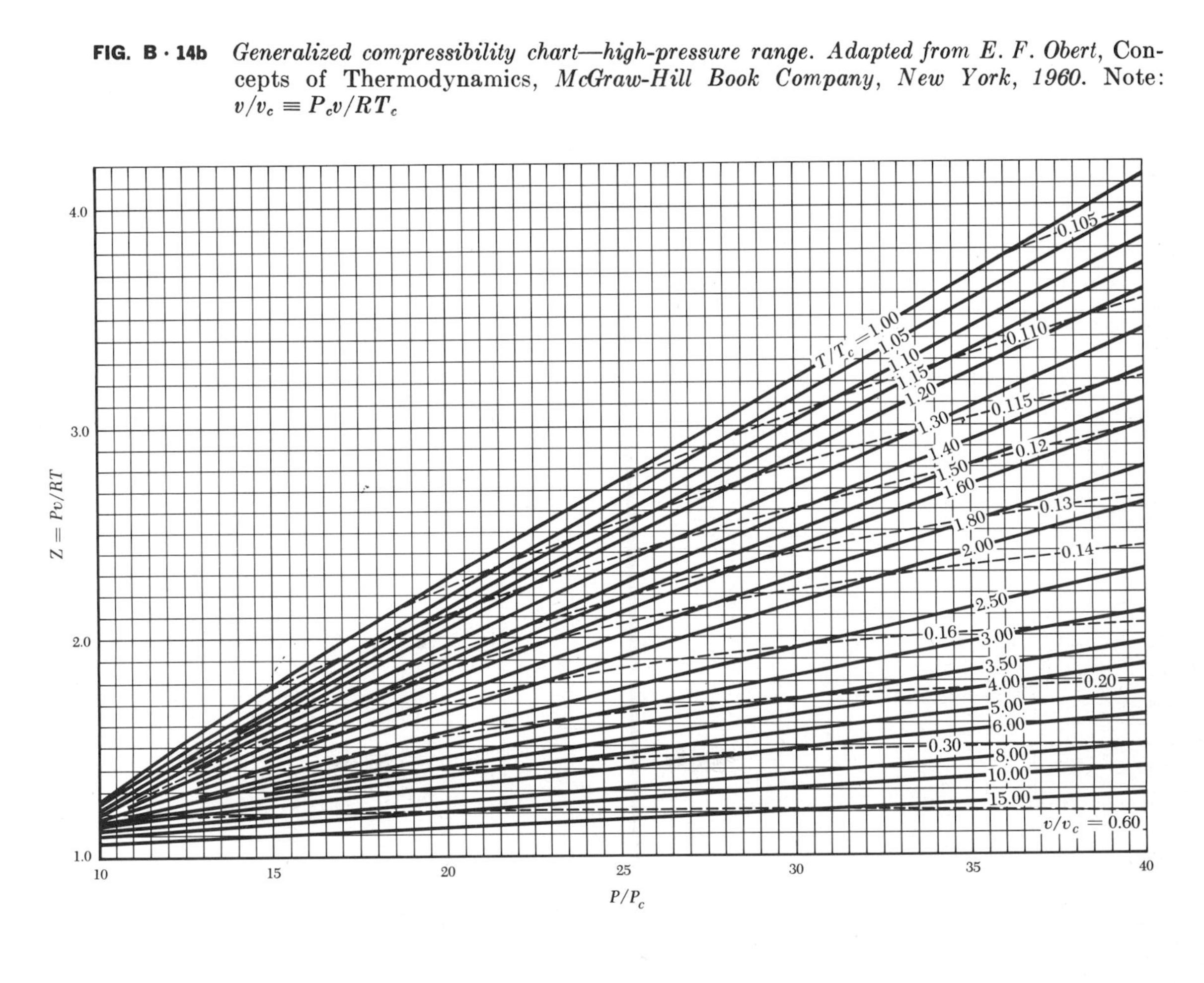

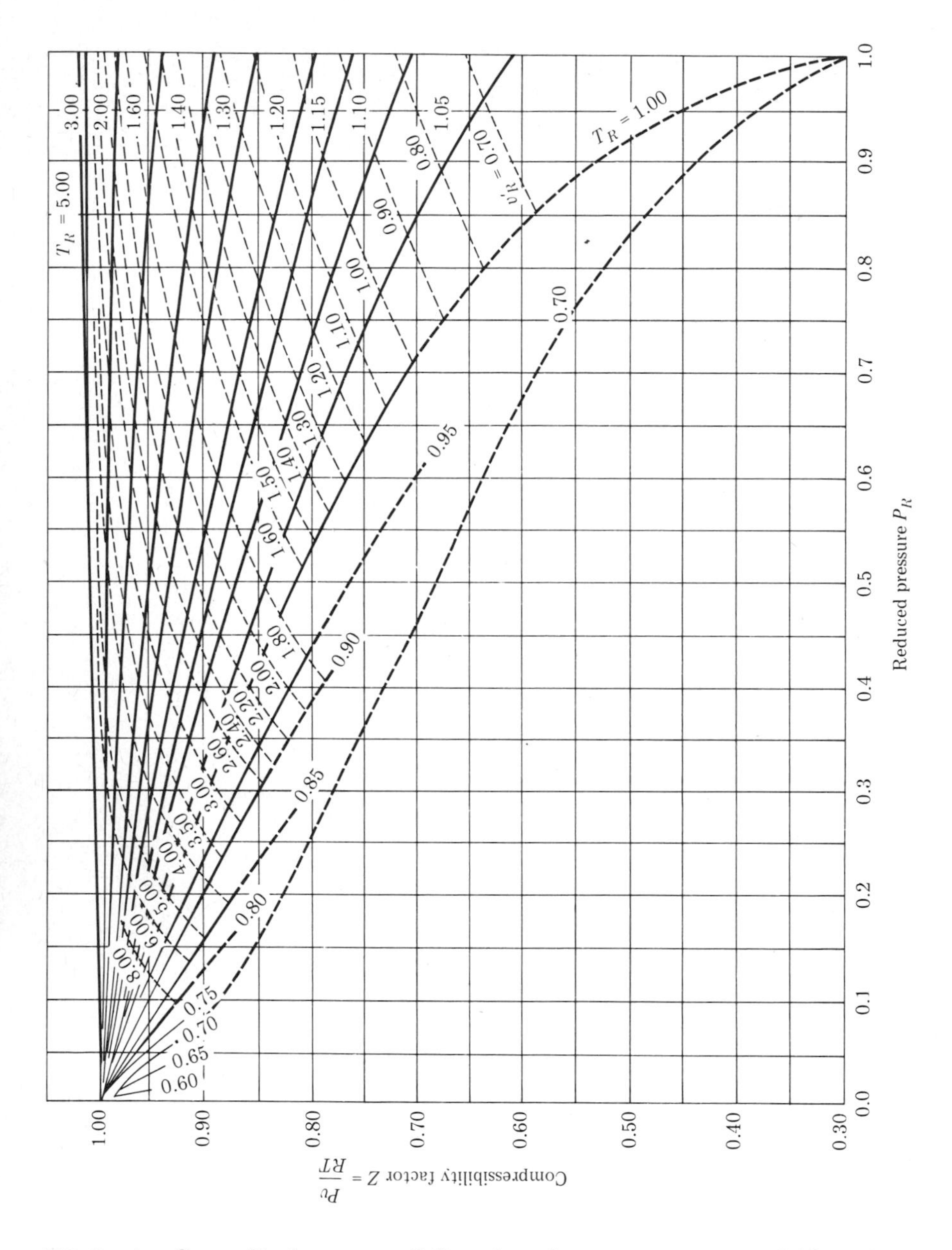

FIG. B·14c *Generalized compressibility chart–low-pressure range. (Adapted from E. F. Obert,* Concepts of Thermodynamics, *McGraw-Hill, 1960.)*

FIG. B·15 *Generalized enthalpy chart. Adapted from K. Wark,* Thermodynamics, *McGraw-Hill Book Company, New York, 1966*

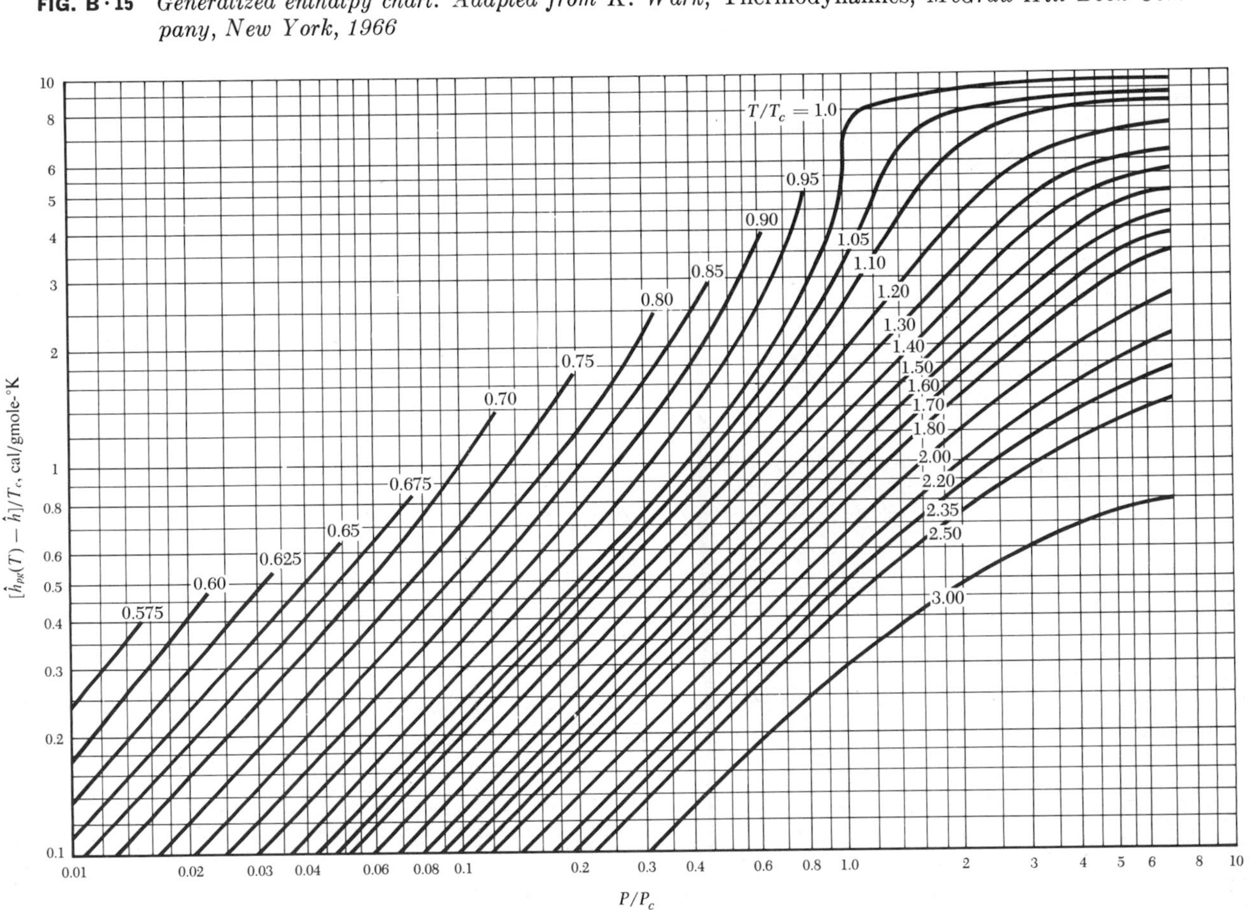

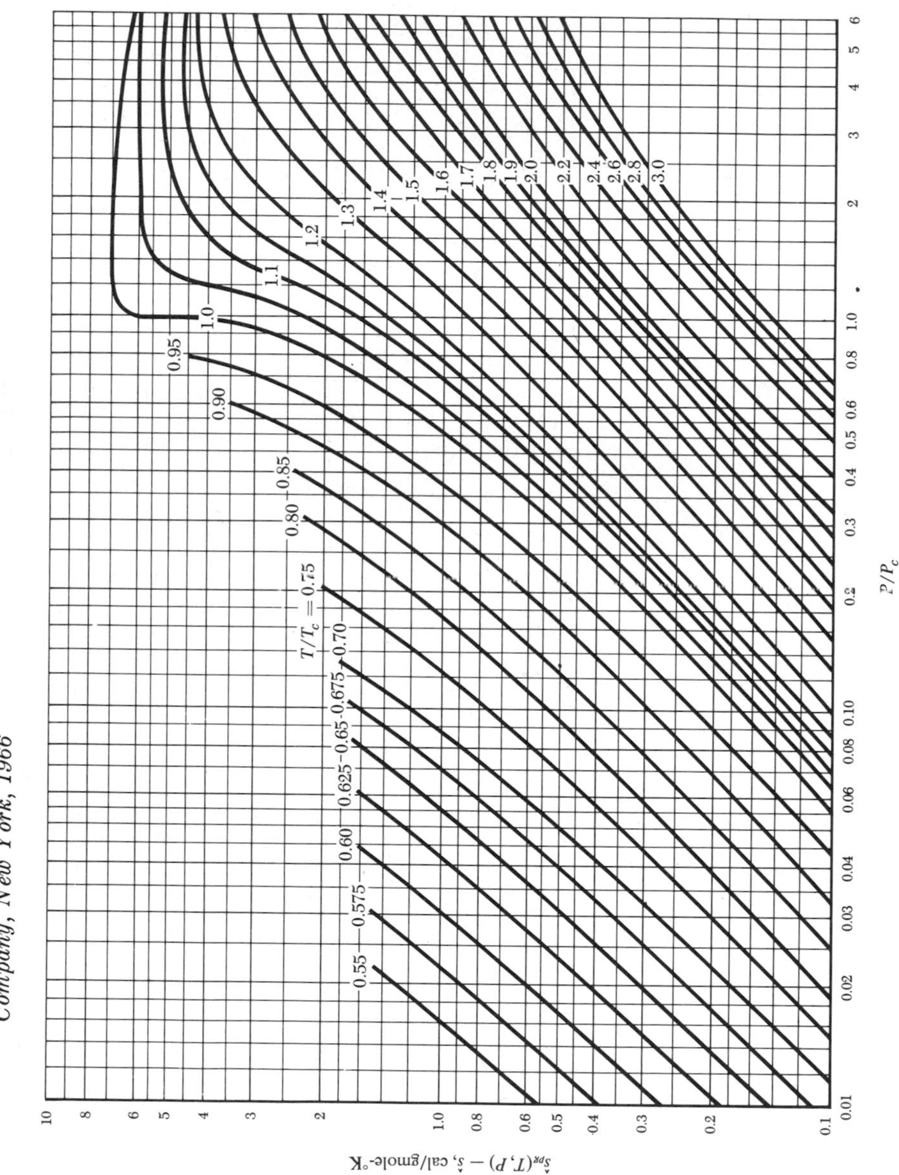

FIG. B·16 *Generalized entropy chart. Adapted from K. Wark,* Thermodynamics, *McGraw-Hill Book Company, New York, 1966*

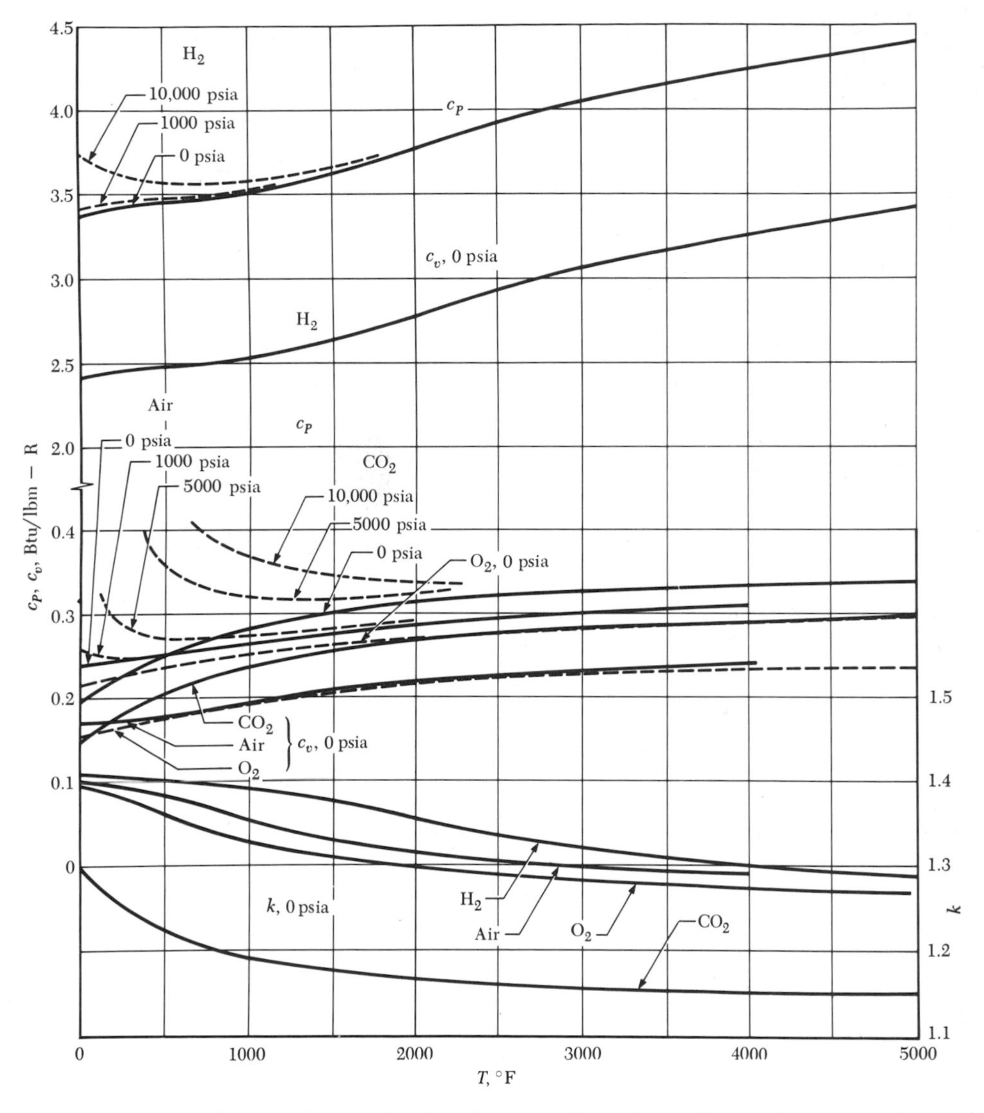

FIG. B·17 *Specific heats of selected gases. Data from National Bureau of Standards*

Table B·1*a* Properties of saturated H_2O—temperature table (SI units)†

T, °C	P, MPa	Volume, m³/kg		Energy, kJ/kg		Enthalpy, kJ/kg			Entropy, kJ/(kg · K)		
		v_f	v_g	u_f	u_g	h_f	h_{fg}	h_g	s_f	s_{fg}	s_g
0.010	0.0006113	0.001000	206.1	0.0	2375.3	0.0	2501.3	2501.3	0.0000	9.1571	9.1571
2	0.0007056	0.001000	179.9	8.4	2378.1	8.4	2496.6	2505.0	0.0305	9.0738	9.1043
5	0.0008721	0.001000	147.1	21.0	2382.2	21.0	2489.5	2510.5	0.0761	8.9505	9.0266
10	0.001228	0.001000	106.4	42.0	2389.2	42.0	2477.7	2519.7	0.1510	8.7506	8.9016
15	0.001705	0.001001	77.93	63.0	2396.0	63.0	2465.9	2528.9	0.2244	8.5578	8.7822
20	0.002338	0.001002	57.79	83.9	2402.9	83.9	2454.2	2538.1	0.2965	8.3715	8.6680
25	0.003169	0.001003	43.36	104.9	2409.8	104.9	2442.3	2547.2	0.3672	8.1916	8.5588
30	0.004246	0.001004	32.90	125.8	2416.6	125.8	2430.4	2556.2	0.4367	8.0174	8.4541
35	0.005628	0.001006	25.22	146.7	2423.4	146.7	2418.6	2565.3	0.5051	7.8488	8.3539
40	0.007383	0.001008	19.52	167.5	2430.1	167.5	2406.8	2574.3	0.5723	7.6855	8.2578
45	0.009593	0.001010	15.26	188.4	2436.8	188.4	2394.8	2583.2	0.6385	7.5271	8.1656
50	0.01235	0.001012	12.03	209.3	2443.5	209.3	2382.8	2592.1	0.7036	7.3735	8.0771
55	0.01576	0.001015	9.569	230.2	2450.1	230.2	2370.7	2600.9	0.7678	7.2243	7.9921
60	0.01994	0.001017	7.671	251.1	2456.6	251.1	2358.5	2609.6	0.8310	7.0794	7.9104
65	0.02503	0.001020	6.197	272.0	2463.1	272.0	2346.2	2618.2	0.8934	6.9384	7.8318
70	0.03119	0.001023	5.042	292.9	2469.5	293.0	2333.8	2626.8	0.9549	6.8012	7.7561
75	0.03858	0.001026	4.131	313.9	2475.9	313.9	2321.4	2635.3	1.0155	6.6678	7.6833
80	0.04739	0.001029	3.407	334.8	2482.2	334.9	2308.8	2643.7	1.0754	6.5376	7.6130
85	0.05783	0.001032	2.828	355.8	2488.4	355.9	2296.0	2651.9	1.1344	6.4109	7.5453
90	0.07013	0.001036	2.361	376.8	2494.5	376.9	2283.2	2660.1	1.1927	6.2872	7.4799
95	0.08455	0.001040	1.982	397.9	2500.6	397.9	2270.2	2668.1	1.2503	6.1664	7.4167
100	0.1013	0.001044	1.673	418.9	2506.5	419.0	2257.0	2676.0	1.3071	6.0486	7.3557
110	0.1433	0.001052	1.210	461.1	2518.1	461.3	2230.2	2691.5	1.4188	5.8207	7.2395
120	0.1985	0.001060	0.8919	503.5	2529.2	503.7	2202.6	2706.3	1.5280	5.6024	7.1304
130	0.2701	0.001070	0.6685	546.0	2539.9	546.3	2174.2	2720.5	1.6348	5.3929	7.0277

Table B·1*a* Properties of saturated H_2O—temperature table (SI units) (*continued*)

		Volume, m^3/kg		Energy, kJ/kg		Enthalpy, kJ/kg			Entropy, kJ/(kg · K)		
T, °C	P, MPa	v_f	v_g	u_f	u_g	h_f	h_{fg}	h_g	s_f	s_{fg}	s_g
140	0.3613	0.001080	0.5089	588.7	2550.0	589.1	2144.8	2733.9	1.7395	5.1912	6.9307
150	0.4758	0.001090	0.3928	631.7	2559.5	632.2	2114.2	2746.4	1.8422	4.9965	6.8387
160	0.6178	0.001102	0.3071	674.9	2568.4	675.5	2082.6	2758.1	1.9431	4.8079	6.7510
170	0.7916	0.001114	0.2428	718.3	2576.5	719.2	2049.5	2768.7	2.0423	4.6249	6.6672
180	1.002	0.001127	0.1941	762.1	2583.7	763.2	2015.0	2778.2	2.1400	4.4466	6.5866
190	1.254	0.001141	0.1565	806.2	2590.0	807.5	1978.8	2786.4	2.2363	4.2724	6.5087
200	1.554	0.001156	0.1274	850.6	2595.3	852.4	1940.8	2793.2	2.3313	4.1018	6.4331
210	1.906	0.001173	0.1044	895.5	2599.4	897.7	1900.8	2798.5	2.4253	3.9340	6.3593
220	2.318	0.001190	0.08620	940.9	2602.4	943.6	1858.5	2802.1	2.5183	3.7686	6.2869
230	2.795	0.001209	0.07159	986.7	2603.9	990.1	1813.9	2804.0	2.6105	3.6050	6.2155
240	3.344	0.001229	0.05977	1033.2	2604.0	1037.3	1766.5	2803.8	2.7021	3.4425	6.1446
250	3.973	0.001251	0.05013	1080.4	2602.4	1085.3	1716.2	2801.5	2.7933	3.2805	6.0738
260	4.688	0.001276	0.04221	1128.4	2599.0	1134.4	1662.5	2796.9	2.8844	3.1184	6.0028
270	5.498	0.001302	0.03565	1177.3	2593.7	1184.5	1605.2	2789.7	2.9757	2.9553	5.9310
280	6.411	0.001332	0.03017	1227.4	2586.1	1236.0	1543.6	2779.6	3.0674	2.7905	5.8579
290	7.436	0.001366	0.02557	1278.9	2576.0	1289.0	1477.2	2766.2	3.1600	2.6230	5.7830
300	8.580	0.001404	0.02168	1332.0	2563.0	1344.0	1405.0	2749.0	3.2540	2.4513	5.7053
310	9.856	0.001447	0.01835	1387.0	2546.4	1401.3	1326.0	2727.3	3.3500	2.2739	5.6239
320	11.27	0.001499	0.01549	1444.6	2525.5	1461.4	1238.7	2700.1	3.4487	2.0883	5.5370
330	12.84	0.001561	0.01300	1505.2	2499.0	1525.3	1140.6	2665.9	3.5514	1.8911	5.4425
340	14.59	0.001638	0.01080	1570.3	2464.6	1594.2	1027.9	2622.1	3.6601	1.6765	5.3366
350	16.51	0.001740	0.008815	1641.8	2418.5	1670.6	893.4	2564.0	3.7784	1.4338	5.2122
360	18.65	0.001892	0.006947	1725.2	2351.6	1760.5	720.7	2481.2	3.9154	1.1382	5.0536
370	21.03	0.002213	0.004931	1844.0	2229.0	1890.5	442.2	2332.7	4.1114	0.6876	4.7990
374.136	22.088	0.003155	0.003155	2029.6	2029.6	2099.3	0.0	2099.3	4.4305	0.0000	4.4305

† Saturated liquid entropies have been adjusted to make the Gibbs functions of the liquid and vapor phases exactly equal. For this reason there are some small differences between values presented here and the original tables.

Source: Recalculated from equations given in *Steam Tables*, by Keenan, Keyes, Hill, and Moore (Wiley, 1969, by permission).

Table B·1*b* Properties of saturated H_2O—pressure table (SI units)†

		Volume, m³/kg		Energy, kJ/kg		Enthalpy, kJ/kg			Entropy, kJ/(kg · K)		
P, MPa	T, °C	v_f	v_g	u_f	u_g	h_f	h_{fg}	h_g	s_f	s_{fg}	s_g
0.000611	0.01	0.001000	206.1	0.0	2375.3	0.0	2501.3	2501.3	0.0000	9.1571	9.1571
0.0008	3.8	0.001000	159.7	15.8	2380.5	15.8	2492.5	2508.3	0.0575	9.0007	9.0582
0.001	7.0	0.001000	129.2	29.3	2385.0	29.3	2484.9	2514.2	0.1059	8.8706	8.9765
0.0012	9.7	0.001000	108.7	40.6	2388.7	40.6	2478.5	2519.1	0.1460	8.7639	8.9099
0.0014	12.0	0.001001	93.92	50.3	2391.9	50.3	2473.1	2523.4	0.1802	8.6736	8.8538
0.0016	14.0	0.001001	82.76	58.9	2394.7	58.9	2468.2	2527.1	0.2101	8.5952	8.8053
0.0018	15.8	0.001001	74.03	66.5	2397.2	66.5	2464.0	2530.5	0.2367	8.5259	8.7626
0.002	17.5	0.001001	67.00	73.5	2399.5	73.5	2460.0	2533.5	0.2606	8.4639	8.7245
0.003	24.1	0.001003	45.67	101.0	2408.5	101.0	2444.5	2545.5	0.3544	8.2240	8.5784
0.004	29.0	0.001004	34.80	121.4	2415.2	121.4	2433.0	2554.4	0.4225	8.0529	8.4754
0.006	36.2	0.001006	23.74	151.5	2424.9	151.5	2415.9	2567.4	0.5208	7.8104	8.3312
0.008	41.5	0.001008	18.10	173.9	2432.1	173.9	2403.1	2577.0	0.5924	7.6371	8.2295
0.01	45.8	0.001010	14.67	191.8	2437.9	191.8	2392.8	2584.6	0.6491	7.5019	8.1510
0.012	49.4	0.001012	12.36	206.9	2442.7	206.9	2384.1	2591.0	0.6961	7.3910	8.0871
0.014	52.6	0.001013	10.69	220.0	2446.9	220.0	2376.6	2596.6	0.7365	7.2968	8.0333
0.016	55.3	0.001015	9.433	231.5	2450.5	231.5	2369.9	2601.4	0.7719	7.2149	7.9868
0.018	57.8	0.001016	8.445	241.9	2453.8	241.9	2363.9	2605.8	0.8034	7.1425	7.9459
0.02	60.1	0.001017	7.649	251.4	2456.7	251.4	2358.3	2609.7	0.8319	7.0774	7.9093
0.03	69.1	0.001022	5.229	289.2	2468.4	289.2	2336.1	2625.3	0.9439	6.8256	7.7695
0.04	75.9	0.001026	3.993	317.5	2477.0	317.6	2319.1	2636.7	1.0260	6.6449	7.6709
0.06	85.9	0.001033	2.732	359.8	2489.6	359.8	2293.7	2653.5	1.1455	6.3873	7.5328
0.08	93.5	0.001039	2.087	391.6	2498.8	391.6	2274.1	2665.7	1.2331	6.2023	7.4354
0.1	99.6	0.001043	1.694	417.3	2506.1	417.4	2258.1	2675.5	1.3029	6.0573	7.3602
0.12	104.8	0.001047	1.428	439.2	2512.1	439.3	2244.2	2683.5	1.3611	5.9378	7.2989
0.14	109.3	0.001051	1.237	458.2	2517.3	458.4	2232.0	2690.4	1.4112	5.8360	7.2472

Table B·1*b* Properties of saturated H_2O—pressure table (SI units) *(continued)*

		Volume, m^3/kg		Energy, kJ/kg		Enthalpy, kJ/kg			Entropy, kJ/(kg · K)		
P, MPa	T, °C	v_f	v_g	u_f	u_g	h_f	h_{fg}	h_g	s_f	s_{fg}	s_g
0.16	113.3	0.001054	1.091	475.2	2521.8	475.3	2221.2	2696.5	1.4553	5.7472	7.2025
0.18	116.9	0.001058	0.9775	490.5	2525.9	490.7	2211.1	2701.8	1.4948	5.6683	7.1631
0.2	120.2	0.001061	0.8857	504.5	2529.5	504.7	2201.9	2706.6	1.5305	5.5975	7.1280
0.3	133.5	0.001073	0.6058	561.1	2543.6	561.5	2163.8	2725.3	1.6722	5.3205	6.9927
0.4	143.6	0.001084	0.4625	604.3	2553.6	604.7	2133.8	2738.5	1.7770	5.1197	6.8967
0.6	158.9	0.001101	0.3157	669.9	2567.4	670.6	2086.2	2756.8	1.9316	4.8293	6.7609
0.8	170.4	0.001115	0.2404	720.2	2576.8	721.1	2048.0	2769.1	2.0466	4.6170	6.6636
1	179.9	0.001127	0.1944	761.7	2583.6	762.8	2015.3	2778.1	2.1391	4.4482	6.5873
1.2	188.0	0.001139	0.1633	797.3	2588.8	798.6	1986.2	2784.8	2.2170	4.3072	6.5242
1.4	195.1	0.001149	0.1408	828.7	2592.8	830.3	1959.7	2790.0	2.2847	4.1854	6.4701
1.6	201.4	0.001159	0.1238	856.9	2596.0	858.8	1935.2	2794.0	2.3446	4.0780	6.4226
1.8	207.2	0.001168	0.1104	882.7	2598.4	884.8	1912.3	2797.1	2.3986	3.9816	6.3802
2	212.4	0.001177	0.09963	906.4	2600.3	908.8	1890.7	2799.5	2.4478	3.8939	6.3417
3	233.9	0.001216	0.06668	1004.8	2604.1	1008.4	1795.7	2804.1	2.6462	3.5416	6.1878
4	250.4	0.001252	0.04978	1082.3	2602.3	1087.3	1714.1	2801.4	2.7970	3.2739	6.0709
6	275.6	0.001319	0.03244	1205.4	2589.7	1213.3	1571.0	2784.3	3.0273	2.8627	5.8900
8	295.1	0.001384	0.02352	1305.6	2569.8	1316.6	1441.4	2758.0	3.2075	2.5365	5.7440
9	303.4	0.001418	0.02048	1350.5	2557.8	1363.3	1378.8	2742.1	3.2865	2.3916	5.6781
10	311.1	0.001452	0.01803	1393.0	2544.4	1407.6	1317.1	2724.7	3.3603	2.2546	5.6149
12	324.8	0.001527	0.01426	1472.9	2513.7	1491.3	1193.6	2684.9	3.4970	1.9963	5.4933
14	336.8	0.001611	0.01149	1548.6	2476.8	1571.1	1066.5	2637.6	3.6240	1.7486	5.3726
16	347.4	0.001711	0.009307	1622.7	2431.8	1650.0	930.7	2580.7	3.7468	1.4996	5.2464
18	357.1	0.001840	0.007491	1698.9	2374.4	1732.0	777.2	2509.2	3.8722	1.2332	5.1054
20	365.8	0.002036	0.005836	1785.6	2293.2	1826.3	583.7	2410.0	4.0146	0.9135	4.9281
22.088	374.136	0.003155	0.003155	2029.6	2029.6	2099.3	0.0	2099.3	4.4305	0.0000	4.4305

† Saturated liquid entropies have been adjusted to make the Gibbs functions of the liquid and vapor phases exactly equal. For this reason there are some small differences between values presented here and the original tables.

Source: Recalculated from equations given in *Steam Tables*, by Keenan, Keyes, Hill, and Moore (Wiley, 1969, by permission).

Table B·2 Thermodynamic properties of superheated steam—SI units†

P, MPa (T_{sat}, °C)		Temperature, °C												
		50	100	150	200	250	300	350	400	500	600	700	800	900
0.002	v, m³/kg	74.52	86.08	97.63	109.2	120.7	132.3	143.8	155.3	178.4	201.5	224.6	247.6	270.7
(17.5)	u, kJ/kg	2445.2	2516.3	2588.3	2661.6	2736.2	2812.2	2889.8	2969.0	3132.3	3302.5	3479.7	3663.9	3855.1
	h, kJ/kg	2594.3	2688.4	2783.6	2879.9	2977.6	3076.7	3177.4	3279.6	3489.1	3705.5	3928.8	4159.1	4396.5
	s, kJ/(kg · K)	8.9227	9.1936	9.4328	9.6479	9.8442	10.0251	10.1935	10.3513	10.6414	10.9044	11.1465	11.3718	11.5832
0.005	v, m³/kg	29.78	34.42	39.04	43.66	48.28	52.90	57.51	62.13	71.36	80.59	89.82	99.05	108.3
(32.9)	u, kJ/kg	2444.7	2516.0	2588.1	2661.4	2736.1	2812.2	2889.8	2968.9	3132.3	3302.5	3479.6	3663.9	3855.0
	h, kJ/kg	2593.6	2688.1	2783.3	2879.8	2977.5	3076.6	3177.3	3279.6	3489.1	3705.4	3928.8	4159.1	4396.5
	s, kJ/(kg · K)	8.4982	8.7699	9.0095	9.2248	9.4212	9.6022	9.7706	9.9284	10.2185	10.4815	10.7236	10.9489	11.1603
0.01	v, m³/kg	14.87	17.20	19.51	21.83	24.14	26.45	28.75	31.06	35.68	40.29	44.91	49.53	54.14
(45.8)	u, kJ/kg	2443.9	2515.5	2587.9	2661.3	2736.0	2812.1	2889.7	2968.9	3132.3	3302.5	3479.6	3663.8	3855.0
	h, kJ/kg	2592.6	2687.5	2783.0	2879.5	2977.3	3076.5	3177.2	3279.5	3489.0	3705.4	3928.7	4159.1	4396.4
	s, kJ/(kg · K)	8.1757	8.4487	8.6890	8.9046	9.1010	9.2821	9.4506	9.6084	9.8985	10.1616	10.4037	10.6290	10.8404
0.02	v, m³/kg		8.585	9.748	10.91	12.06	13.22	14.37	15.53	17.84	20.15	22.45	24.76	27.07
(60.1)	u, kJ/kg		2514.5	2587.3	2660.9	2735.7	2811.9	2889.5	2968.8	3132.2	3302.4	3479.6	3663.8	3855.0
	h, kJ/kg		2686.2	2782.3	2879.1	2977.0	3076.3	3177.0	3279.4	3488.9	3705.3	3928.7	4159.1	4396.4
	s, kJ/(kg · K)		8.1263	8.3678	8.5839	8.7807	8.9619	9.1304	9.2884	9.5785	9.8417	10.0838	10.3091	10.5205
0.05	v, m³/kg		3.418	3.889	4.356	4.820	5.284	5.747	6.209	7.134	8.057	8.981	9.904	10.83
(81.3)	u, kJ/kg		2511.6	2585.6	2659.8	2735.0	2811.3	2889.1	2968.4	3131.9	3302.2	3479.5	3663.7	3854.9
	h, kJ/kg		2682.5	2780.1	2877.6	2976.0	3075.5	3176.4	3278.9	3488.6	3705.1	3928.5	4158.9	4396.3
	s, kJ/(kg · K)		7.6955	7.9409	8.1588	8.3564	8.5380	8.7069	8.8650	9.1554	9.4186	9.6608	9.8861	10.0975
0.07	v, m³/kg		2.434	2.773	3.108	3.441	3.772	4.103	4.434	5.095	5.755	6.415	7.074	7.734
(89.9)	u, kJ/kg		2509.6	2584.5	2659.1	2734.5	2811.0	2888.8	2968.2	3131.8	3302.1	3479.4	3663.6	3854.9
	h, kJ/kg		2680.0	2778.6	2876.7	2975.3	3075.0	3176.1	3278.6	3488.4	3704.9	3928.4	4158.8	4396.2
	s, kJ/(kg · K)		7.5349	7.7829	8.0020	8.2001	8.3821	8.5511	8.7094	8.9999	9.2632	9.5054	9.7307	9.9422
0.1	v, m³/kg		1.696	1.936	2.172	2.406	2.639	2.871	3.103	3.565	4.028	4.490	4.952	5.414
(99.6)	u, kJ/kg		2506.6	2582.7	2658.0	2733.7	2810.4	2888.4	2967.8	3131.5	3301.9	3479.2	3663.5	3854.8
	h, kJ/kg		2676.2	2776.4	2875.3	2974.3	3074.3	3175.5	3278.1	3488.1	3704.7	3928.2	4158.7	4396.1
	s, kJ/(kg · K)		7.3622	7.6142	7.8351	8.0341	8.2165	8.3858	8.5442	8.8350	9.0984	9.3406	9.5660	9.7775

P, MPa (T_{sat}, °C)		150	200	250	300	350	400	450	500	550	600	700	800	900
0.15	v, m³/kg	1.285	1.444	1.601	1.757	1.912	2.067	2.222	2.376	2.530	2.685	2.993	3.301	3.609
(111.4)	u, kJ/kg	2579.8	2656.2	2732.5	2809.5	2887.7	2967.3	3048.4	3131.1	3215.6	3301.6	3479.0	3663.4	3854.6
	h, kJ/kg	2772.6	2872.9	2972.7	3073.0	3174.5	3277.3	3381.7	3487.6	3595.1	3704.3	3927.9	4158.5	4395.9
	s, kJ/(kg · K)	7.4201	7.6441	7.8446	8.0278	8.1975	8.3562	8.5057	8.6473	8.7821	8.9109	9.1533	9.3787	9.5903
0.2	v, m³/kg	0.9596	1.080	1.199	1.316	1.433	1.549	1.665	1.781	1.897	2.013	2.244	2.475	2.706
(120.2)	u, kJ/kg	2576.9	2654.4	2731.2	2808.6	2886.9	2966.7	3047.9	3130.7	3215.2	3301.4	3478.8	3663.2	3854.5
	h, kJ/kg	2768.8	2870.5	2971.0	3071.8	3173.5	3276.5	3381.0	3487.0	3594.7	3704.0	3927.7	4158.3	4395.8
	s, kJ/(kg · K)	7.2803	7.5074	7.7094	7.8934	8.0636	8.2226	8.3723	8.5140	8.6489	8.7778	9.0203	9.2458	9.4574
0.4	v, m³/kg	0.4708	0.5342	0.5951	0.6548	0.7139	0.7726	0.8311	0.8893	0.9475	1.006	1.121	1.237	1.353
(143.6)	u, kJ/kg	2564.5	2646.8	2726.1	2804.8	2884.0	2964.4	3046.0	3129.2	3213.9	3300.2	3477.9	3662.5	3853.9
	h, kJ/kg	2752.8	2860.5	2964.2	3066.7	3169.6	3273.4	3378.4	3484.9	3592.9	3702.4	3926.5	4157.4	4395.1
	s, kJ/(kg · K)	6.9307	7.1714	7.3797	7.5670	7.7390	7.8992	8.0497	8.1921	8.3274	8.4566	8.6995	8.9253	9.1370
0.6	v, m³/kg		0.3520	0.3938	0.4344	0.4742	0.5137	0.5529	0.5920	0.6309	0.6697	0.7472	0.8245	0.9017
(158.9)	u, kJ/kg		2638.9	2720.9	2801.0	2881.1	2962.0	3044.1	3127.6	3212.5	3299.1	3477.1	3661.8	3853.3
	h, kJ/kg		2850.1	2957.2	3061.6	3165.7	3270.2	3375.9	3482.7	3591.1	3700.9	3925.4	4156.5	4394.4
	s, kJ/(kg · K)		6.9673	7.1824	7.3732	7.5472	7.7086	7.8600	8.0029	8.1386	8.2682	8.5115	8.7375	8.9494
0.8	v, m³/kg		0.2608	0.2931	0.3241	0.3544	0.3843	0.4139	0.4433	0.4726	0.5018	0.5601	0.6181	0.6761
(170.4)	u, kJ/kg		2630.6	2715.5	2797.1	2878.2	2959.7	3042.2	3125.9	3211.2	3297.9	3476.2	3661.1	3852.8
	h, kJ/kg		2839.2	2950.0	3056.4	3161.7	3267.1	3373.3	3480.6	3589.3	3699.4	3924.3	4155.7	4393.6
	s, kJ/(kg · K)		6.8167	7.0392	7.2336	7.4097	7.5723	7.7245	7.8680	8.0042	8.1341	8.3779	8.6041	8.8161
1	v, m³/kg		0.2060	0.2327	0.2579	0.2825	0.3066	0.3304	0.3541	0.3776	0.4011	0.4478	0.4943	0.5407
(179.9)	u, kJ/kg		2621.9	2709.9	2793.2	2875.2	2957.3	3040.2	3124.3	3209.8	3296.8	3475.4	3660.5	3852.2
	h, kJ/kg		2827.9	2942.6	3051.2	3157.7	3263.9	3370.7	3478.4	3587.5	3697.9	3923.1	4154.8	4392.9
	s, kJ/(kg · K)		6.6948	6.9255	7.1237	7.3019	7.4658	7.6188	7.7630	7.8996	8.0298	8.2740	8.5005	8.7127
1.5	v, m³/kg		0.1325	0.1520	0.1697	0.1866	0.2030	0.2192	0.2352	0.2510	0.2668	0.2981	0.3292	0.3603
(198.3)	u, kJ/kg		2598.1	2695.3	2783.1	2867.6	2951.3	3035.3	3120.3	3206.4	3293.9	3473.2	3658.7	3850.8
	h, kJ/kg		2796.8	2923.2	3037.6	3147.4	3255.8	3364.1	3473.0	3582.9	3694.0	3920.3	4152.6	4391.2
	s, kJ/(kg · K)		6.4554	6.7098	6.9187	7.1025	7.2697	7.4249	7.5706	7.7083	7.8393	8.0846	8.3118	8.5243

† Saturated liquid entropies have been adjusted to make the Gibbs functions of the liquid and vapor phases exactly equal. For this reason there are some small differences between values presented here and the original tables.

Source: Recalculated from equations given in *Steam Tables*, by Keenan, Keyes, Hill, and Moore (Wiley, 1969, by permission).

Table B·2 Thermodynamic properties of superheated steam—SI units† (*continued*)

P, MPa (T_{sat}, °C)		Temperature, °C												
		250	300	350	400	450	500	550	600	650	700	750	800	900
2	v, m³/kg	0.1114	0.1255	0.1386	0.1512	0.1635	0.1757	0.1877	0.1996	0.2114	0.2232	0.2350	0.2467	0.2700
(212.4)	u, kJ/kg	2679.6	2772.6	2859.8	2945.2	3030.4	3116.2	3203.0	3290.9	3380.2	3471.0	3563.2	3657.0	3849.3
	h, kJ/kg	2902.5	3023.5	3137.0	3247.6	3357.5	3467.6	3578.3	3690.1	3803.1	3917.5	4033.2	4150.4	4389.4
	s, kJ/(kg · K)	6.5461	6.7672	6.9571	7.1279	7.2853	7.4325	7.5713	7.7032	7.8290	7.9496	8.0656	8.1774	8.3903
3	v, m³/kg	0.07058	0.08114	0.09053	0.09936	0.1079	0.1162	0.1244	0.1324	0.1404	0.1484	0.1563	0.1641	0.1798
(233.9)	u, kJ/kg	2644.0	2750.0	2843.7	2932.7	3020.4	3107.9	3196.0	3285.0	3375.2	3466.6	3559.4	3653.6	3846.5
	h, kJ/kg	2855.8	2993.5	3115.3	3230.8	3344.0	3456.5	3569.1	3682.3	3796.5	3911.7	4028.2	4146.0	4385.9
	s, kJ/(kg · K)	6.2880	6.5398	6.7436	6.9220	7.0842	7.2346	7.3757	7.5093	7.6364	7.7580	7.8747	7.9871	8.2008
4	v, m³/kg		0.05884	0.06645	0.07341	0.08003	0.08643	0.09269	0.09885	0.1049	0.1109	0.1169	0.1229	0.1347
(250.4)	u, kJ/kg		2725.3	2826.6	2919.9	3010.1	3099.5	3189.0	3279.1	3370.1	3462.1	3555.5	3650.1	3843.6
	h, kJ/kg		2960.7	3092.4	3213.5	3330.2	3445.2	3559.7	3674.4	3789.8	3905.9	4023.2	4141.6	4382.3
	s, kJ/(kg · K)		6.3622	6.5828	6.7698	6.9371	7.0908	7.2343	7.3696	7.4981	7.6206	7.7381	7.8511	8.0655
6	v, m³/kg		0.03616	0.04223	0.04739	0.05214	0.05665	0.06101	0.06525	0.06942	0.07352	0.07758	0.08160	0.08958
(275.6)	u, kJ/kg		2667.2	2789.6	2892.8	2988.9	3082.2	3174.6	3266.9	3359.6	3453.2	3547.6	3643.1	3837.8
	h, kJ/kg		2884.2	3043.0	3177.2	3301.8	3422.1	3540.6	3658.4	3776.2	3894.3	4013.1	4132.7	4375.3
	s, kJ/(kg · K)		6.0682	6.3342	6.5415	6.7201	6.8811	7.0296	7.1685	7.2996	7.4242	7.5433	7.6575	7.8735
8	v, m³/kg		0.02426	0.02995	0.03432	0.03817	0.04175	0.04516	0.04845	0.05166	0.05481	0.05791	0.06097	0.06702
(295.1)	u, kJ/kg		2590.9	2747.7	2863.8	2966.7	3064.3	3159.8	3254.4	3349.0	3444.0	3539.6	3636.1	3832.1
	h, kJ/kg		2785.0	2987.3	3138.3	3272.0	3398.3	3521.0	3642.0	3762.3	3882.5	4002.9	4123.8	4368.3
	s, kJ/(kg · K)		5.7914	6.1309	6.3642	6.5559	6.7248	6.8786	7.0214	7.1553	7.2821	7.4027	7.5182	7.7359
10	v, m³/kg			0.02242	0.02641	0.02975	0.03279	0.03564	0.03837	0.04101	0.04358	0.04611	0.04859	0.05349
(311.1)	u, kJ/kg			2699.2	2832.4	2943.3	3045.8	3144.5	3241.7	3338.2	3434.7	3531.5	3629.0	3826.3
	h, kJ/kg			2923.4	3096.5	3240.8	3373.6	3500.9	3625.3	3748.3	3870.5	3992.6	4114.9	4361.2
	s, kJ/(kg · K)			5.9451	6.2127	6.4197	6.5974	6.7569	6.9037	7.0406	7.1696	7.2919	7.4086	7.6280
12	v, m³/kg			0.01721	0.02108	0.02412	0.02680	0.02929	0.03164	0.03390	0.03610	0.03824	0.04034	0.04447
(324.8)	u, kJ/kg			2641.1	2798.3	2918.8	3026.6	3128.9	3228.7	3327.2	3425.3	3523.4	3621.8	3820.6
	h, kJ/kg			2847.6	3051.2	3208.2	3348.2	3480.3	3608.3	3734.0	3858.4	3982.3	4105.9	4354.2
	s, kJ/(kg · K)			5.7604	6.0754	6.3006	6.4879	6.6535	6.8045	6.9445	7.0757	7.1998	7.3178	7.5390

P, MPa (T_{sat}, °C)		400	450	500	550	600	650	700	750	800	850	900	950	1000
15	v, m³/kg	0.01565	0.01845	0.02080	0.02293	0.02491	0.02680	0.02861	0.03037	0.03210	0.03379	0.03546	0.03711	0.03875
(342.2)	u, kJ/kg	2740.7	2879.5	2996.5	3104.7	3208.6	3310.4	3410.9	3511.0	3611.0	3711.2	3811.9	3913.2	4015.4
	h, kJ/kg	2975.4	3156.2	3308.5	3448.6	3582.3	3712.3	3840.1	3966.6	4092.4	4218.0	4343.8	4469.9	4596.6
	s, kJ/(kg · K)	5.8819	6.1412	6.3451	6.5207	6.6784	6.8232	6.9580	7.0848	7.2048	7.3192	7.4288	7.5340	7.6356
20	v, m³/kg	0.00994	0.01270	0.01477	0.01656	0.01818	0.1969	0.02113	0.02251	0.02385	0.02516	0.02645	0.02771	0.02897
(365.8)	u, kJ/kg	2619.2	2806.2	2942.8	3062.3	3174.0	3281.5	3386.5	3490.0	3592.7	3695.1	3797.4	3900.0	4003.1
	h, kJ/kg	2818.1	3060.1	3238.2	3393.4	3537.6	3675.3	3809.1	3940.3	4069.8	4198.3	4326.4	4454.3	4582.5
	s, kJ/(kg · K)	5.5548	5.9025	6.1409	6.3356	6.5056	6.6591	6.8002	6.9317	7.0553	7.1723	7.2839	7.3907	7.4933
22.088	v, m³/kg	0.00818	0.01104	0.01305	0.01475	0.01627	0.01768	0.01901	0.02029	0.02152	0.02272	0.02389	0.02505	0.02619
(374.136)	u, kJ/kg	2552.9	2772.1	2919.0	3043.9	3159.1	3269.1	3376.1	3481.1	3585.0	3688.3	3791.4	3894.5	3998.0
	h, kJ/kg	2733.7	3015.9	3207.2	3369.6	3518.4	3659.6	3796.0	3929.2	4060.3	4190.1	4319.1	4447.9	4576.6
	s, kJ/(kg · K)	5.4013	5.8072	6.0634	6.2670	6.4426	6.5998	6.7437	6.8772	7.0024	7.1206	7.2330	7.3404	7.4436
30	v, m³/kg	0.00279	0.00674	0.00868	0.01017	0.01145	0.01260	0.01366	0.01466	0.01562	0.01655	0.01745	0.01833	0.01920
	u, kJ/kg	2067.3	2619.3	2820.7	2970.3	3100.5	3221.0	3335.8	3447.0	3555.6	3662.6	3768.5	3873.8	3978.8
	h, kJ/kg	2151.0	2821.4	3081.0	3275.4	3443.9	3598.9	3745.7	3886.9	4024.3	4159.0	4291.9	4423.6	4554.7
	s, kJ/(kg · K)	4.4736	5.4432	5.7912	6.0350	6.2339	6.4066	6.5614	6.7030	6.8341	6.9568	7.0726	7.1825	7.2875
40	v, m³/kg	0.00191	0.00369	0.00562	0.00698	0.00809	0.00906	0.00994	0.01076	0.01152	0.01226	0.01296	0.01365	0.01432
	u, kJ/kg	1854.5	2365.1	2678.4	2869.7	3022.6	3158.0	3283.6	3402.9	3517.9	3629.8	3739.4	3847.5	3954.6
	h, kJ/kg	1930.8	2512.8	2903.3	3149.1	3346.4	3520.6	3681.3	3833.1	3978.8	4120.0	4257.9	4393.6	4527.6
	s, kJ/(kg · K)	4.1143	4.9467	5.4707	5.7793	6.0122	6.2063	6.3759	6.5281	6.6671	6.7957	6.9158	7.0291	7.1365
60	v, m³/kg	0.00163	0.00208	0.00296	0.00396	0.00483	0.00560	0.00627	0.00689	0.00746	0.00800	0.00851	0.00900	0.00948
	u, kJ/kg	1745.3	2053.9	2390.5	2658.8	2861.1	3028.8	3177.2	3313.6	3441.6	3563.6	3681.0	3795.0	3906.4
	h, kJ/kg	1843.4	2179.0	2567.9	2896.2	3151.2	3364.5	3553.6	3726.8	3889.1	4043.3	4191.5	4335.0	4475.2
	s, kJ/(kg · K)	3.9325	4.4128	4.9329	5.3449	5.6460	5.8838	6.0832	6.2569	6.4118	6.5523	6.6814	6.8012	6.9135
80	v, m³/kg	0.00152	0.00177	0.00219	0.00276	0.00339	0.00398	0.00452	0.00502	0.00548	0.00591	0.00632	0.00671	0.00709
	u, kJ/kg	1687.0	1944.9	2218.9	2483.9	2711.8	2904.7	3073.2	3225.3	3365.7	3497.3	3622.3	3742.1	3857.8
	h, kJ/kg	1808.3	2086.9	2393.9	2704.9	2982.7	3222.8	3434.7	3626.6	3803.8	3970.1	4127.9	4279.1	4425.2
	s, kJ/(kg · K)	3.8338	4.2328	4.6432	5.0331	5.3609	5.6284	5.8521	6.0445	6.2137	6.3652	6.5026	6.6289	6.7459

† Saturated liquid entropies have been adjusted to make the Gibbs functions of the liquid and vapor phases exactly equal. For this reason there are some small differences between values presented here and the original tables.

Source: Recalculated from equations given in *Steam Tables*, by Keenan, Keyes, Hill, and Moore (Wiley, 1969, by permission).

Table B·1a Properties of saturated H_2O—temperature table (old English units)†

		Volume, ft³/lbm		Energy, Btu/lbm		Enthalpy, Btu/lbm			Entropy, Btu/(lbm · R)		
T, °F	P, psia	v_f	v_g	u_f	u_g	h_f	h_{fg}	h_g	s_f	s_{fg}	s_g
32.018	0.08866	0.01602	3302.	0.0	1021.2	0.0	1075.4	10.75.4	0.0000	2.1871	2.1871
35	0.09992	0.01602	2948.	3.0	1022.2	3.0	1073.7	1076.7	0.0061	2.1705	2.1766
40	0.1217	0.01602	2445.	8.0	1023.8	8.0	1070.9	1078.9	0.0162	2.1432	2.1594
45	0.1475	0.01602	2037.	13.0	1025.5	13.0	1068.1	1081.1	0.0262	2.1163	2.1425
50	0.1780	0.01602	1704.	18.1	1027.2	18.1	1065.2	1083.3	0.0361	2.0900	2.1261
55	0.2140	0.01603	1431.	23.1	1028.8	23.1	1062.4	1085.5	0.0458	2.0643	2.1101
60	0.2563	0.01603	1207.	28.1	1030.4	28.1	1059.6	1087.7	0.0555	2.0390	2.0945
65	0.3057	0.01604	1021.	33.1	1032.1	33.1	1056.8	1089.9	0.0651	2.0142	2.0793
70	0.3632	0.01605	867.6	38.1	1033.7	38.1	1053.9	1092.0	0.0746	1.9898	2.0644
75	0.4300	0.01606	739.7	43.1	1035.4	43.1	1051.1	1094.2	0.0840	1.9659	2.0499
80	0.5073	0.01607	632.7	48.1	1037.0	48.1	1048.3	1096.4	0.0933	1.9425	2.0358
85	0.5964	0.01609	543.1	53.1	1038.6	53.1	1045.5	1098.6	0.1025	1.9195	2.0220
90	0.6989	0.01610	467.6	58.1	1040.2	58.1	1042.6	1100.7	0.1116	1.8969	2.0085
95	0.8162	0.01611	403.9	63.0	1041.9	63.1	1039.8	1102.9	0.1206	1.8747	1.9953
100	0.9503	0.01613	350.0	68.0	1043.5	68.0	1037.0	1105.0	0.1296	1.8528	1.9824
110	1.276	0.01617	265.1	78.0	1046.7	78.0	1031.3	1109.3	0.1473	1.8103	1.9576
120	1.695	0.01621	203.0	88.0	1049.9	88.0	1025.5	1113.5	0.1646	1.7692	1.9338
130	2.225	0.01625	157.2	98.0	1053.0	98.0	1019.7	1117.7	0.1817	1.7294	1.9111
140	2.892	0.01629	122.9	107.9	1056.2	108.0	1013.9	1121.9	0.1985	1.6909	1.8894
150	3.722	0.01634	96.98	117.9	1059.3	117.9	1008.2	1126.1	0.2150	1.6535	1.8685
160	4.745	0.01640	77.23	127.9	1062.3	128.0	1002.1	1130.1	0.2313	1.6173	1.8486
180	7.515	0.01651	50.20	148.0	1068.3	148.0	990.2	1138.2	0.2631	1.5480	1.8111
200	11.53	0.01663	33.63	168.0	1074.2	168.1	977.8	1145.9	0.2941	1.4823	1.7764
212	14.696	0.01672	26.80	180.1	1077.6	180.1	970.4	1150.5	0.3122	1.4447	1.7569
220	17.19	0.01677	23.15	188.2	1079.8	188.2	965.3	1153.5	0.3241	1.4202	1.7443

Table B·1*a* Properties of saturated H_2O—temperature table (old English units) (*continued*)

		Volume, ft³/lbm		Energy, Btu/lbm		Enthalpy, Btu/lbm			Entropy, Btu/(lbm · R)		
T, °F	P, psia	v_f	v_g	u_f	u_g	h_f	h_{fg}	h_g	s_f	s_{fg}	s_g
240	24.97	0.01692	16.33	208.4	1085.3	208.4	952.3	1160.7	0.3534	1.3611	1.7145
260	35.42	0.01708	11.77	228.6	1090.5	228.7	938.9	1167.6	0.3820	1.3046	1.6866
280	49.19	0.01726	8.650	249.0	1095.4	249.2	924.9	1174.1	0.4100	1.2504	1.6604
300	66.98	0.01745	6.472	269.5	1100.0	269.7	910.5	1180.2	0.4373	1.1985	1.6358
320	89.60	0.01765	4.919	290.1	1104.2	290.4	895.4	1185.8	0.4641	1.1484	1.6125
340	117.9	0.01787	3.792	310.9	1108.0	311.3	879.5	1190.8	0.4904	1.0998	1.5902
360	152.9	0.01811	2.961	331.8	1111.4	332.3	862.9	1195.2	0.5163	1.0527	1.5690
380	195.6	0.01836	2.339	352.9	1114.3	353.6	845.4	1199.0	0.5417	1.0068	1.5485
400	247.1	0.01864	1.866	374.3	1116.6	375.1	826.9	1202.0	0.5668	0.9618	1.5286
420	308.5	0.01894	1.502	395.8	1118.3	396.9	807.2	1204.1	0.5916	0.9177	1.5093
440	381.2	0.01926	1.219	417.6	1119.3	419.0	786.3	1205.3	0.6162	0.8740	1.4902
460	466.3	0.01961	0.9961	439.7	1119.6	441.4	764.1	1205.5	0.6405	0.8309	1.4714
480	565.5	0.02000	0.8187	462.2	1118.9	464.3	740.3	1204.6	0.6647	0.7879	1.4526
500	680.0	0.02043	0.6761	485.1	1117.4	487.7	714.8	1202.5	0.6889	0.7448	1.4337
520	811.4	0.02091	0.5605	508.5	1114.8	511.7	687.2	1198.9	0.7132	0.7015	1.4147
540	961.4	0.02145	0.4658	532.6	1111.0	536.4	657.4	1193.8	0.7375	0.6577	1.3952
560	1132.	0.02207	0.3877	557.3	1105.8	562.0	625.0	1187.0	0.7622	0.6129	1.3751
580	1324.	0.02278	0.3225	583.0	1098.9	588.6	589.4	1178.0	0.7873	0.5669	1.3542
600	1541.	0.02363	0.2677	609.9	1090.0	616.6	549.8	1166.4	0.8131	0.5188	1.3319
620	1784.	0.02465	0.2209	638.3	1078.5	646.4	505.0	1151.4	0.8399	0.4678	1.3077
640	2057.	0.02593	0.1805	668.7	1063.2	678.6	453.3	1131.9	0.8683	0.4122	1.2805
660	2362.	0.02767	0.1446	702.0	1042.3	714.3	391.2	1105.5	0.8992	0.3493	1.2485
680	2705.	0.03032	0.1113	741.7	1011.0	756.9	309.8	1066.7	0.9352	0.2718	1.2070
700	3090.	0.03666	0.07444	801.7	947.7	822.7	167.7	990.4	0.9903	0.1447	1.1350
705.445	3203.8	0.05053	0.05053	872.6	872.6	902.5	0.0	902.5	1.0582	0.0000	1.0582

† Saturated liquid entropies have been adjusted to make the Gibbs functions of the liquid and vapor phases exactly equal. For this reason there are some small differences between values presented here and the original tables.

Source: Recalculated from equations given in *Steam Tables*, by Keenan, Keyes, Hill, and Moore (Wiley, 1969, by permission).

Table B·1*b* Properties of saturated H_2O—pressure table (old English units)†

		Volume, ft³/lbm		Energy, Btu/lbm		Enthalpy, Btu/lbm			Entropy, Btu/(lbm·R)		
P, psia	T, °F	v_f	v_g	u_f	u_g	h_f	h_{fg}	h_g	s_f	s_{fg}	s_g
0.0887	32.018	0.01602	3302.0	0.0	1021.2	0.0	1075.4	1075.4	0.0000	2.1871	2.1871
0.1	35.0	0.01602	2946.0	3.0	1022.2	3.0	1073.7	1076.7	0.0061	2.1705	2.1768
0.12	39.6	0.01602	2477.0	7.7	1023.7	7.7	1071.0	1078.7	0.0155	2.1451	2.1606
0.14	43.6	0.01602	2140.0	11.7	1025.0	11.7	1068.8	1080.5	0.0234	2.1237	2.1471
0.16	47.1	0.01602	1886.0	15.2	1026.2	15.2	1066.8	1082.0	0.0304	2.1050	2.1354
0.18	50.3	0.01602	1686.0	18.3	1027.2	18.3	1065.1	1083.4	0.0366	2.0886	2.1252
0.2	53.1	0.01603	1526.0	21.2	1028.2	21.2	1063.5	1084.7	0.0422	2.0738	2.1160
0.25	59.3	0.01603	1235.0	27.4	1030.2	27.4	1060.0	1087.4	0.0542	2.0425	2.0967
0.3	64.5	0.01604	1040.0	32.5	1031.9	32.5	1057.1	1089.6	0.0641	2.0168	2.0809
0.4	72.8	0.01606	792.0	40.9	1034.7	40.9	1052.4	1093.3	0.0799	1.9762	2.0561
0.6	85.2	0.01609	540.0	53.3	1038.7	53.3	1045.3	1098.6	0.1028	1.9187	2.0215
0.8	94.3	0.01611	411.7	62.4	1041.6	62.4	1040.2	1102.6	0.1195	1.8775	1.9970
1	101.7	0.01614	333.6	69.7	1044.0	69.7	1036.0	1105.7	0.1326	1.8455	1.9781
1.2	107.9	0.01616	280.9	75.9	1046.0	75.9	1032.5	1108.4	0.1435	1.8193	1.9628
1.4	113.2	0.01618	243.0	81.2	1047.7	81.2	1029.5	1110.7	0.1529	1.7969	1.9498
1.6	117.9	0.01620	214.3	85.9	1049.2	85.9	1026.8	1112.7	0.1611	1.7775	1.9386
1.8	122.2	0.01621	191.8	90.2	1050.6	90.2	1024.3	1114.5	0.1684	1.7604	1.9288
2	126.0	0.01623	173.8	94.0	1051.8	94.0	1022.1	1116.1	0.1750	1.7450	1.9200
3	141.4	0.01630	118.7	109.4	1056.6	109.4	1013.1	1122.5	0.2009	1.6854	1.8863
4	152.9	0.01636	90.64	120.9	1060.2	120.9	1006.4	1127.3	0.2198	1.6428	1.8626
6	170.0	0.01645	61.98	138.0	1065.4	138.0	996.2	1134.2	0.2474	1.5820	1.8294
8	182.8	0.01653	47.35	150.8	1069.2	150.8	988.5	1139.3	0.2676	1.5384	1.8060
10	193.2	0.01659	38.42	161.2	1072.2	161.2	982.1	1143.3	0.2836	1.5043	1.7879
12	201.9	0.01665	32.40	170.0	1074.7	170.0	976.7	1146.7	0.2970	1.4762	1.7732
14	209.6	0.01670	28.05	177.6	1076.9	177.7	971.9	1149.6	0.3085	1.4523	1.7608

Table B·1*b* Properties of saturated H_2O—pressure table (old English units) (*continued*)

		Volume, ft³/lbm		Energy, Btu/lbm		Enthalpy, Btu/lbm			Entropy, Btu/(lbm·R)		
P, psia	T, °F	v_f	v_g	u_f	u_g	h_f	h_{fg}	h_g	s_f	s_{fg}	s_g
14.696	212.0	0.01672	26.80	180.1	1077.6	180.1	970.4	1150.5	0.3122	1.4447	1.7569
16	216.3	0.01675	24.75	184.4	1078.8	184.5	967.6	1152.1	0.3186	1.4315	1.7501
18	222.4	0.01679	22.17	190.6	1080.5	190.6	963.8	1154.4	0.3277	1.4129	1.7406
20	228.0	0.01683	20.09	196.2	1082.0	196.2	960.2	1156.4	0.3359	1.3963	1.7322
30	250.3	0.01700	13.75	218.8	1088.0	218.9	945.4	1164.3	0.3683	1.3315	1.6998
40	267.3	0.01715	10.50	236.0	1092.3	236.2	933.8	1170.0	0.3922	1.2847	1.6769
60	292.7	0.01738	7.177	262.0	1098.3	262.2	915.8	1178.0	0.4274	1.2172	1.6446
80	312.1	0.01757	5.474	281.9	1102.6	282.2	901.4	1183.6	0.4535	1.1681	1.6216
100	327.9	0.01774	4.434	298.3	1105.8	298.6	889.2	1187.8	0.4745	1.1291	1.6036
120	341.3	0.01789	3.730	312.3	1108.3	312.7	878.4	1191.1	0.4921	1.0967	1.5888
140	353.1	0.01802	3.221	324.6	1110.3	325.0	868.8	1193.8	0.5074	1.0688	1.5762
160	363.6	0.01815	2.836	335.6	1112.0	336.2	859.8	1196.0	0.5209	1.0443	1.5652
180	373.1	0.01827	2.533	345.7	1113.4	346.3	851.5	1197.8	0.5330	1.0225	1.5555
200	381.9	0.01839	2.289	354.9	1114.6	355.6	843.7	1199.3	0.5441	1.0025	1.5466
300	417.4	0.01890	1.544	393.0	1118.1	394.1	809.8	1203.9	0.5885	0.9232	1.5117
400	444.7	0.01934	1.162	422.8	1119.4	424.2	781.3	1205.5	0.6219	0.8639	1.4858
600	486.3	0.02013	0.7702	469.4	1118.5	471.6	732.5	1204.1	0.6724	0.7742	1.4466
800	518.4	0.02087	0.5691	506.6	1115.0	509.7	689.6	1199.3	0.7112	0.7050	1.4162
1000	544.8	0.02159	0.4459	538.4	1109.9	542.4	650.0	1192.4	0.7434	0.6471	1.3905
1200	567.4	0.02232	0.3623	566.7	1103.5	571.7	612.2	1183.9	0.7714	0.5961	1.3675
1400	587.3	0.02307	0.3016	592.6	1096.0	598.6	575.5	1174.1	0.7966	0.5497	1.3463
1600	605.1	0.02386	0.2552	616.9	1087.4	624.0	538.9	1162.9	0.8198	0.5062	1.3260
2000	636.0	0.02565	0.1881	662.4	1066.6	671.9	464.4	1136.3	0.8624	0.4239	1.2863
2600	674.1	0.02938	0.1210	729.2	1021.8	743.3	336.8	1080.1	0.9237	0.2971	1.2208
3203.8	705.445	0.05053	0.05053	872.6	872.6	902.5	0.0	902.5	1.0582	0.0000	1.0582

† Saturated liquid entropies have been adjusted to make the Gibbs functions of the liquid and vapor phases exactly equal. For this reason there are some small differences between values presented here and the original tables.

Source: Recalculated from equations given in *Steam Tables*, by Keenan, Keyes, Hill, and Moore (Wiley, 1969, by permission).

Table B·2 Thermodynamic properties of superheated steam—old English units†

P, psia (T_{sat}, °F)		Temperature, °F 100	200	300	400	500	600	700	800	900	1000	1100	1200	1300
0.2	v, ft³/lbm	1666.0	1964.0	2262.0	2560.0	2858.0	3156.0	3454.0	3752.0	4050.0	4347.0	4645.0	4943.0	5241.0
(53.1)	u, Btu/lbm	1043.9	1077.7	1112.1	1147.1	1182.8	1219.3	1256.7	1294.9	1334.0	1373.9	1414.8	1456.7	1499.4
	h, Btu/lbm	1105.6	1150.4	1195.8	1241.9	1288.6	1336.1	1384.5	1433.7	1483.8	1534.8	1586.8	1639.6	1693.4
	s, Btu/(lbm · R)	2.1550	2.2287	2.2928	2.3497	2.4011	2.4483	2.4919	2.5326	2.5708	2.6070	2.6414	2.6743	2.7058
0.5	v, ft³/lbm	665.9	785.5	904.8	1024.0	1143.0	1262.0	1382.0	1501.0	1620.0	1739.0	1858.0	1977.0	2096.0
(79.5)	u, Btu/lbm	1043.7	1077.6	1112.0	1147.1	1182.8	1219.3	1256.7	1294.9	1334.0	1373.9	1414.8	1456.7	1499.4
	h, Btu/lbm	1105.3	1150.3	1195.8	1241.8	1288.6	1336.1	1384.5	1433.7	1483.8	1534.8	1586.8	1639.6	1693.4
	s, Btu/(lbm · R)	2.0537	2.1276	2.1917	2.2487	2.3001	2.3472	2.3909	2.4316	2.4698	2.5060	2.5404	2.5733	2.6048
1	v, ft³/lbm		392.5	452.3	511.9	571.5	631.1	690.7	750.3	809.9	869.5	929.0	988.6	1048.0
(101.7)	u, Btu/lbm		1077.5	1112.0	1147.0	1182.8	1219.3	1256.7	1294.9	1333.9	1373.9	1414.8	1456.7	1499.4
	h, Btu/lbm		1150.1	1195.6	1241.7	1288.5	1336.1	1384.5	1433.7	1483.8	1534.8	1586.8	1639.6	1693.4
	s, Btu/(lbm · R)		2.0510	2.1152	2.1722	2.2237	2.2708	2.3144	2.3551	2.3934	2.4296	2.4640	2.4969	2.5283
2	v, ft³/lbm		196.0	226.0	255.9	285.7	315.5	345.3	375.1	404.9	434.7	464.5	494.3	524.1
(126.0)	u, Btu/lbm		1077.2	1111.8	1146.9	1182.7	1219.3	1256.6	1294.8	1333.9	1373.9	1414.8	1456.7	1499.4
	h, Btu/lbm		1149.7	1195.4	1241.6	1288.4	1336.0	1384.4	1433.7	1483.8	1534.8	1586.7	1639.6	1693.4
	s, Btu/(lbm · R)		1.9741	2.0386	2.0957	2.1472	2.1944	2.2380	2.2787	2.3170	2.3532	2.3876	2.4205	2.4519
5	v, ft³/lbm		78.15	90.24	102.2	114.2	126.1	138.1	150.0	161.9	173.9	185.8	197.7	209.6
(162.2)	u, Btu/lbm		1076.2	1111.3	1146.6	1182.5	1219.1	1256.5	1294.7	1333.8	1373.8	1414.8	1456.6	1499.4
	h, Btu/lbm		1148.6	1194.8	1241.2	1288.2	1335.8	1384.3	1433.5	1483.7	1534.7	1586.7	1639.5	1693.3
	s, Btu/(lbm · R)		1.8717	1.9369	1.9943	2.0460	2.0932	2.1369	2.1776	2.2159	2.2522	2.2866	2.3194	2.3509
10	v, ft³/lbm		38.85	44.99	51.03	57.04	63.03	69.01	74.98	80.95	86.91	92.88	98.84	104.8
(193.2)	u, Btu/lbm		1074.7	1110.4	1146.1	1182.2	1218.8	1256.3	1294.6	1333.7	1373.7	1414.7	1456.5	1499.3
	h, Btu/lbm		1146.6	1193.7	1240.5	1287.7	1335.5	1384.0	1433.3	1483.5	1534.6	1586.5	1639.4	1693.2
	s, Btu/(lbm · R)		1.7929	1.8594	1.9173	1.9692	2.0166	2.0603	2.1011	2.1394	2.1757	2.2101	2.2430	2.2745
14.7	v, ft³/lbm			30.52	34.67	38.77	42.86	46.93	51.00	55.07	59.13	63.19	67.25	71.30
(212.0)	u, Btu/lbm			1109.6	1145.6	1181.8	1218.6	1256.1	1294.4	1333.6	1373.6	1414.6	1456.5	1499.3
	h, Btu/lbm			1192.6	1239.9	1287.3	1335.2	1383.8	1433.1	1483.3	1534.4	1586.4	1639.3	1693.2
	s, Btu/(lbm · R)			1.8159	1.8743	1.9264	1.9739	2.0177	2.0586	2.0969	2.1332	2.1676	2.2005	2.2320

P, psia (T_{sat}, °F)		300	400	500	600	700	800	900	1000	1100	1200	1300	1400	1500
20	v, ft³/lbm	22.36	25.43	28.46	31.47	34.47	37.46	40.45	43.44	46.42	49.41	52.39	55.37	58.35
(227.9)	u, Btu/lbm	1108.7	1145.1	1181.5	1218.3	1255.9	1294.3	1333.5	1373.5	1414.5	1456.4	1499.2	1542.9	1587.6
	h, Btu/lbm	1191.4	1239.2	1286.8	1334.8	1383.5	1432.2	1483.2	1534.3	1586.3	1639.2	1693.1	1747.8	1803.5
	s, Btu/(lbm · R)	1.7807	1.8397	1.8921	1.9397	1.9836	2.0245	2.0629	2.0991	2.1336	2.1665	2.1980	2.2283	2.2574
40	v, ft³/lbm	11.04	12.62	14.16	15.69	17.20	18.70	20.20	21.70	23.20	24.69	26.18	27.68	29.17
(267.2)	u, Btu/lbm	1105.1	1143.0	1180.1	1217.3	1255.1	1293.7	1333.0	1373.1	1414.2	1456.1	1498.9	1542.7	1587.4
	h, Btu/lbm	1186.8	1236.4	1284.9	1333.4	1382.4	1432.1	1482.5	1533.7	1585.9	1638.9	1692.8	1747.6	1803.3
	s, Btu/(lbm · R)	1.6995	1.7608	1.8142	1.8623	1.9065	1.9476	1.9861	2.0224	2.0570	2.0899	2.1214	2.1517	2.1809
60	v, ft³/lbm	7.260	8.353	9.399	10.42	11.44	12.45	13.45	14.45	15.45	16.45	17.45	18.45	19.44
(292.7)	u, Btu/lbm	1101.3	1140.8	1178.6	1216.3	1254.4	1293.0	1332.5	1372.7	1413.8	1455.8	1498.7	1542.5	1587.2
	h, Btu/lbm	1181.9	1233.5	1283.0	1332.1	1381.4	1431.2	1481.8	1533.2	1585.4	1638.5	1692.4	1747.3	1803.0
	s, Btu/(lbm · R)	1.6497	1.7136	1.7680	1.8167	1.8611	1.9024	1.9410	1.9775	2.0121	2.0450	2.0766	2.1069	2.1361
80	v, ft³/lbm		6.217	7.017	7.794	8.561	9.321	10.08	10.83	11.58	12.33	13.08	13.83	14.58
(312.0)	u, Btu/lbm		1138.5	1177.2	1215.3	1253.6	1292.4	1332.0	1372.3	1413.5	1455.5	1498.4	1542.3	1587.0
	h, Btu/lbm		1230.6	1281.1	1330.7	1380.3	1430.4	1481.1	1532.6	1584.9	1638.1	1692.1	1747.0	1802.8
	s, Btu/(lbm · R)		1.6792	1.7348	1.7840	1.8287	1.8702	1.9089	1.9455	1.9801	2.0131	2.0447	2.0751	2.1043
100	v, ft³/lbm		4.934	5.587	6.216	6.834	7.445	8.053	8.657	9.260	9.861	10.46	11.06	11.66
(327.8)	u, Btu/lbm		1136.2	1175.7	1214.2	1252.8	1291.8	1331.4	1371.9	1413.1	1455.2	1498.2	1542.0	1586.8
	h, Btu/lbm		1227.5	1279.1	1329.3	1379.2	1429.6	1480.5	1532.1	1584.5	1637.7	1691.8	1746.7	1802.5
	s, Btu/(lbm · R)		1.6519	1.7087	1.7584	1.8035	1.8451	1.8840	1.9206	1.9553	1.9884	2.0200	2.0504	2.0796
140	v, ft³/lbm		3.466	3.952	4.412	4.860	5.301	5.739	6.173	6.605	7.036	7.466	7.895	8.324
(353.1)	u, Btu/lbm		1131.4	1172.7	1212.1	1251.2	1290.5	1330.4	1371.0	1412.4	1454.6	1497.7	1541.6	1586.4
	h, Btu/lbm		1221.2	1275.1	1326.4	1377.1	1427.9	1479.1	1531.0	1583.5	1636.9	1691.1	1746.1	1802.0
	s, Btu/(lbm · R)		1.6090	1.6684	1.7193	1.7650	1.8070	1.8461	1.8829	1.9178	1.9509	1.9826	2.0130	2.0423
180	v, ft³/lbm		2.648	3.042	3.409	3.763	4.110	4.453	4.793	5.131	5.467	5.802	6.137	6.471
(373.1)	u, Btu/lbm		1126.2	1169.6	1210.0	1249.6	1289.3	1329.4	1370.2	1411.7	1454.0	1497.2	1541.2	1586.0
	h, Btu/lbm		1214.4	1270.9	1323.5	1374.9	1426.2	1477.7	1529.8	1582.6	1636.1	1690.4	1745.6	1801.5
	s, Btu/(lbm · R)		1.5751	1.6374	1.6895	1.7359	1.7783	1.8177	1.8546	1.8896	1.9229	1.9546	1.9851	2.0144

† Saturated liquid entropies have been adjusted to make the Gibbs functions of the liquid and vapor phases exactly equal. For this reason there are some small differences between values presented here and the original tables.

Source: Recalculated from equations given in *Steam Tables*, by Keenan, Keyes, Hill, and Moore (Wiley, 1969, by permission).

Table B·2 Thermodynamic properties of superheated steam—old English units† (continued)

P, psia (T_{sat}, °F)		Temperature, °F												
		400	500	600	700	800	900	1000	1100	1200	1300	1400	1500	1600
200	v, ft³/lbm	2.361	2.724	3.058	3.379	3.693	4.003	4.310	4.615	4.918	5.220	5.521	5.822	6.123
(381.8)	u, Btu/lbm	1123.5	1168.0	1208.9	1248.8	1288.6	1328.9	1369.8	1411.4	1453.7	1496.9	1540.9	1585.8	1631.6
	h, Btu/lbm	1210.8	1268.8	1322.0	1373.8	1425.3	1477.0	1529.3	1582.1	1635.7	1690.1	1745.3	1801.3	1858.1
	s, Btu/(lbm · R)	1.5602	1.6240	1.6769	1.7236	1.7662	1.8057	1.8427	1.8778	1.9111	1.9429	1.9734	2.0027	2.0310
250	v, ft³/lbm		2.150	2.426	2.688	2.943	3.193	3.440	3.685	3.929	4.172	4.414	4.655	4.896
(401.0)	u, Btu/lbm		1163.8	1206.1	1246.7	1287.0	1327.6	1368.7	1410.5	1453.0	1496.3	1540.4	1585.3	1631.1
	h, Btu/lbm		1263.3	1318.3	1371.1	1423.2	1475.3	1527.9	1581.0	1634.8	1689.3	1744.6	1800.7	1857.6
	s, Btu/(lbm · R)		1.5950	1.6496	1.6972	1.7403	1.7801	1.8174	1.8526	1.8860	1.9179	1.9485	1.9779	2.0062
300	v, ft³/lbm		1.766	2.004	2.227	2.442	2.653	2.860	3.066	3.270	3.473	3.675	3.877	4.078
(417.4)	u, Btu/lbm		1159.5	1203.2	1244.6	1285.4	1326.3	1367.7	1409.6	1452.2	1495.6	1539.8	1584.8	1630.7
	h, Btu/lbm		1257.5	1314.5	1368.3	1421.0	1473.6	1526.4	1579.8	1633.8	1688.4	1743.8	1800.0	1857.0
	s, Btu/(lbm · R)		1.5703	1.6268	1.6753	1.7189	1.7591	1.7966	1.8319	1.8655	1.8975	1.9281	1.9575	1.9859
400	v, ft³/lbm		1.284	1.476	1.650	1.816	1.978	2.136	2.292	2.446	2.599	2.752	2.904	3.055
(444.7)	u, Btu/lbm		1150.1	1197.3	1240.4	1282.1	1323.7	1365.5	1407.8	1450.7	1494.3	1538.7	1583.8	1629.8
	h, Btu/lbm		1245.2	1306.6	1362.5	1416.6	1470.1	1523.6	1577.4	1631.8	1686.8	1742.4	1798.8	1855.9
	s, Btu/(lbm · R)		1.5284	1.5894	1.6398	1.6846	1.7254	1.7634	1.7991	1.8329	1.8650	1.8958	1.9253	1.9537
600	v, ft³/lbm		0.7947	0.9456	1.073	1.190	1.302	1.411	1.517	1.622	1.726	1.829	1.931	2.033
(486.3)	u, Btu/lbm		1128.0	118.45	1231.5	1275.4	1318.4	1361.2	1404.2	1447.7	1491.7	1536.4	1581.8	1628.0
	h, Btu/lbm		1216.2	1289.5	1350.6	1407.6	1462.9	1517.8	1572.7	1627.8	1683.4	1739.5	1796.3	1853.7
	s, Btu/(lbm · R)		1.4594	1.5322	1.5874	1.6345	1.6768	1.7157	1.7521	1.7863	1.8188	1.8499	1.8796	1.9082
800	v, ft³/lbm			0.6776	0.7829	0.8764	0.9640	1.048	1.130	1.210	1.289	1.367	1.445	1.522
(518.3)	u, Btu/lbm			1170.1	1222.1	1268.4	1312.9	1356.7	1400.5	1444.6	1489.1	1534.2	1579.8	1626.2
	h, Btu/lbm			1270.4	1338.0	1398.2	1455.6	1511.9	1567.8	1623.8	1680.0	1736.6	1793.7	1851.5
	s, Btu/(lbm · R)			1.4863	1.5473	1.5971	1.6410	1.6809	1.7180	1.7527	1.7856	1.8169	1.8469	1.8756
1000	v, ft³/lbm			0.5140	0.6080	0.6878	0.7610	0.8305	0.8976	0.9630	1.027	1.090	1.153	1.215
(544.7)	u, Btu/lbm			1153.7	1212.0	1261.2	1307.3	1352.2	1396.8	1441.5	1486.4	1531.9	1577.8	1624.4
	h, Btu/lbm			1248.8	1324.5	1388.5	1448.1	1505.9	1562.9	1619.7	1676.5	1733.7	1791.2	1849.3
	s, Btu/(lbm · R)			1.4452	1.5137	1.5666	1.6122	1.6532	1.6910	1.7263	1.7595	1.7911	1.8212	1.8501

P, psia (T_{sat}, °F)		800	900	1000	1100	1200	1300	1400	1500	1600	1700	1800	1900	2000
2000	v, ft³/lbm	0.3071	0.3534	0.3945	0.4325	0.4685	0.5031	0.5368	0.5697	0.6020	0.6340	0.6656	0.6971	0.7284
(636.0)	u, Btu/lbm	1220.1	1276.8	1328.1	1377.2	1425.2	1472.7	1520.2	1567.6	1615.4	1663.5	1712.0	1761.0	1810.6
	h, Btu/lbm	1333.8	1407.6	1474.1	1537.2	1598.6	1659.0	1718.8	1778.5	1838.2	1898.1	1958.3	2019.0	2080.1
	s, Btu/(lbm · R)	1.4564	1.5128	1.5600	1.6019	1.6400	1.6753	1.7084	1.7397	1.7694	1.7978	1.8251	1.8513	1.8767
3000	v, ft³/lbm	0.1757	0.2160	0.2485	0.2772	0.3036	0.3285	0.3524	0.3754	0.3978	0.4198	0.4416	0.4631	0.4844
(695.5)	u, Btu/lbm	1167.6	1241.8	1301.7	1356.2	1408.0	1458.5	1508.1	1557.3	1606.3	1655.3	1704.5	1754.0	1803.9
	h, Btu/lbm	1265.2	1361.7	1439.6	1510.0	1576.6	1640.8	1703.7	1765.6	1827.1	1888.4	1949.6	2011.1	2072.8
	s, Btu/(lbm · R)	1.3677	1.4416	1.4969	1.5436	1.5850	1.6226	1.6573	1.6897	1.7203	1.7494	1.7771	1.8037	1.8293
4000	v, ft³/lbm	0.1052	0.1462	0.1752	0.1995	0.2213	0.2414	0.2603	0.2784	0.2959	0.3129	0.3296	0.3462	0.3625
	u, Btu/lbm	1095.0	1201.5	1272.9	1333.9	1390.1	1443.7	1495.7	1546.7	1597.1	1647.2	1697.1	1747.1	1797.3
	h, Btu/lbm	1172.9	1309.7	1402.6	1481.6	1553.9	1622.4	1688.4	1752.8	1816.1	1878.8	1941.1	2003.3	2065.6
	s, Btu/(lbm · R)	1.2742	1.3791	1.4451	1.4975	1.5425	1.5825	1.6190	1.6528	1.6843	1.7140	1.7422	1.7691	1.7950
5000	v, ft³/lbm	0.05933	0.1038	0.1312	0.1530	0.1720	0.1892	0.2052	0.2203	0.2348	0.2489	0.2626	0.2761	0.2895
	u, Btu/lbm	987.2	1155.1	1242.0	1310.6	1371.6	1428.6	1483.2	1536.1	1587.9	1639.0	1689.7	1740.3	1790.8
	h, Btu/lbm	1042.1	1251.1	1363.4	1452.2	1530.8	1603.7	1673.0	1739.9	1805.2	1869.3	1932.7	1995.7	2058.6
	s, Btu/(lbm · R)	1.1586	1.3192	1.3990	1.4579	1.5068	1.5495	1.5878	1.6228	1.6553	1.6857	1.7144	1.7417	1.7678
6000	v, ft³/lbm	0.03942	0.07588	0.1021	0.1222	0.1393	0.1545	0.1685	0.1817	0.1942	0.2063	0.2180	0.2295	0.2409
	u, Btu/lbm	896.9	1102.9	1209.1	1286.4	1352.7	1413.3	1470.5	1525.4	1578.7	1630.9	1682.4	1733.4	1784.3
	h, Btu/lbm	940.6	1187.2	1322.4	1422.1	1507.3	1584.9	1657.6	1727.1	1794.3	1859.9	1924.5	1988.3	2051.7
	s, Btu/(lbm · R)	1.0710	1.2601	1.3563	1.4224	1.4754	1.5208	1.5610	1.5974	1.6309	1.6620	1.6912	1.7189	1.7452
7000	v, ft³/lbm	0.03341	0.05760	0.08172	0.1004	0.1161	0.1299	0.1425	0.1542	0.1653	0.1759	0.1862	0.1963	0.2062
	u, Btu/lbm	855.0	1049.7	1175.0	1261.7	1333.5	1397.8	1457.7	1514.6	1569.4	1622.8	1675.0	1726.7	1777.8
	h, Btu/lbm	898.3	1124.3	1280.9	1391.8	1483.9	1566.1	1642.3	1714.4	1783.5	1850.7	1916.3	1981.0	2045.0
	s, Btu/(lbm · R)	1.0321	1.2049	1.3163	1.3899	1.4471	1.4953	1.5374	1.5751	1.6096	1.6414	1.6711	1.6991	1.7257
8000	v, ft³/lbm	0.03061	0.04657	0.06722	0.08445	0.09892	0.1116	0.1231	0.1337	0.1437	0.1533	0.1625	0.1715	0.1803
	u, Btu/lbm	830.7	1003.7	1141.0	1236.8	1314.2	1382.3	1444.9	1503.8	1560.1	1614.6	1667.7	1719.9	1771.4
	h, Btu/lbm	876.0	1072.6	1240.5	1361.9	1460.6	1547.5	1627.1	1701.7	1772.9	1841.5	1908.3	1973.7	2038.4
	s, Btu/(lbm · R)	1.0098	1.1598	1.2793	1.3598	1.4212	1.4720	1.5160	1.5552	1.5906	1.6231	1.6533	1.6817	1.7085

† Saturated liquid entropies have been adjusted to make the Gibbs functions of the liquid and vapor phases exactly equal. For this reason there are some small differences between values presented here and the original tables.

Source: Recalculated from equations given in *Steam Tables*, by Keenan, Keyes, Hill, and Moore (Wiley, 1969, by permission).

Table B·3 Thermodynamic properties of saturated mercury

		Enthalpy, Btu/lbm			Entropy, Btu/(lbm·R)			Volume, ft³/lbm	
P, psia	T, °F	h_f	h_{fg}	h_g	s_f	s_{fg}	s_g	v_f	v_g
0.01	233.57	6.668	127.732	134.400	0.01137	0.18428	0.19565	1.21×10^{-3}	3637
0.02	259.88	7.532	127.614	135.146	0.01259	0.17735	0.18994	1.21	1893
0.03	276.22	8.068	127.540	135.608	0.01332	0.17332	0.18664	1.21	1292
0.05	297.97	8.778	127.442	136.220	0.01427	0.16821	0.18248	1.21	799
0.1	329.73	9.814	127.300	137.114	0.01561	0.16126	0.17687	1.22	416
0.2	364.25	10.936	127.144	138.080	0.01699	0.15432	0.17131	1.22×10^{-3}	217.3
0.3	385.92	11.639	127.047	138.686	0.01783	0.15024	0.16807	1.22	148.6
0.4	401.98	12.159	126.975	139.134	0.01844	0.14736	0.16580	1.22	113.7
0.5	415.00	12.568	126.916	139.484	0.01892	0.14511	0.16403	1.22	92.18
0.6	425.82	12.929	126.868	139.797	0.01932	0.14328	0.16260	1.23	77.84
0.8	443.50	13.500	126.788	140.288	0.01994	0.14038	0.16032	1.23×10^{-3}	59.58
1	457.72	13.959	126.724	140.683	0.02045	0.13814	0.15859	1.24	48.42
2	504.93	15.476	126.512	141.988	0.02205	0.13116	0.15321	1.24	25.39
3	535.25	16.439	126.377	142.816	0.02302	0.12706	0.15008	1.24	17.50
5	575.7	17.741	126.193	143.934	0.02430	0.12188	0.14618	1.24	10.90
7	604.7	18.657	126.065	144.722	0.02516	0.11846	0.14362	1.25×10^{-3}	8.04
10	637.0	19.685	125.919	145.604	0.02610	0.11483	0.14093	1.25	5.81
20	706.0	21.864	125.609	147.473	0.02800	0.10779	0.13579	1.26	3.09
40	784.4	24.345	125.255	149.600	0.03004	0.10068	0.13072	1.27	1.648
60	835.7	25.940	125.024	150.964	0.03127	0.09652	0.12779	1.28	1.144
80	874.8	27.149	124.849	152.008	0.03218	0.09356	0.12574	1.29×10^{-3}	0.885
100	906.8	28.152	124.706	152.858	0.03290	0.09127	0.12417	1.29	0.725
150	969.4	30.090	124.424	154.514	0.03425	0.08707	0.12132	1.30	0.507
200	1017.2	31.560	124.209	155.769	0.03523	0.08411	0.11934	1.31	0.392
250	1057.2	32.784	124.029	156.813	0.03603	0.08178	0.11781	1.31	0.322
300	1091.2	33.824	123.876	157.700	0.03669	0.07989	0.11658	1.32×10^{-3}	0.276
400	1148.4	35.565	123.620	159.185	0.03775	0.07688	0.11463	1.32	0.215
450	1173.2	36.315	123.509	159.824	0.03820	0.07566	0.11386	1.32	0.194
500	1196.0	37.006	123.406	160.412	0.03861	0.07455	0.11316	1.33	0.177
600	1236.8	38.245	123.221	161.466	0.03932	0.07264	0.11196	1.34	0.151
700	1273.3	39.339	123.058	162.397	0.03993	0.07102	0.11095	1.34×10^{-3}	0.132
800	1306.1	40.324	122.910	163.234	0.04047	0.06961	0.11008	1.34	0.118
900	1336.2	41.226	122.775	164.001	0.04095	0.06837	0.10932	1.35	0.106
1000	1364.0	42.056	122.649	164.705	0.04139	0.06726	0.10865	1.35	0.098
1100	1390.0	42.828	122.533	165.361	0.04179	0.06625	0.10804	1.36	0.090

From Lucian A. Sheldon, *Thermodynamic Properties of Mercury Vapor*, General Electric Company. Liquid densities from WADC TR-59-598.

Table B·4 Thermodynamic properties of saturated cesium

T, R	P, psia	Equilibrium molal mass, lbm/lbmole	Enthalpy, Btu/lbm		Entropy, Btu/(lbm·R)		Volume, ft^3/lbm	
			h_f	h_g	s_f	s_g	v_f	v_g
700	3.41×10^{-5}	133.0	4.14×10^{1}	2.84×10^{2}	1.796×10^{-1}	5.262×10^{-1}	8.765×10^{-3}	1.654×10^{6}
800	5.24×10^{-4}	133.2	4.71×10^{1}	2.87×10^{2}	1.873×10^{-1}	4.878×10^{-1}	8.841×10^{-3}	1.231×10^{5}
900	4.38×10^{-3}	133.5	5.29×10^{1}	2.90×10^{2}	1.940×10^{-1}	4.585×10^{-1}	8.918×10^{-3}	1.653×10^{4}
1000	2.39×10^{-2}	133.9	5.86×10^{1}	2.94×10^{2}	2.000×10^{-1}	4.356×10^{-1}	8.994×10^{-3}	3.347×10^{3}
1100	9.61×10^{-2}	134.6	6.43×10^{1}	2.97×10^{2}	2.055×10^{-1}	4.172×10^{-1}	9.070×10^{-3}	9.128×10^{2}
1200	3.06×10^{-1}	135.3	7.00×10^{1}	3.00×10^{2}	2.105×10^{-1}	4.022×10^{-1}	9.147×10^{-3}	3.109×10^{2}
1300	8.15×10^{-1}	136.3	7.57×10^{1}	3.02×10^{2}	2.150×10^{-1}	3.897×10^{-1}	9.223×10^{-3}	1.255×10^{2}
1400	1.89×10^{0}	137.3	8.15×10^{1}	3.054×10^{2}	2.193×10^{-1}	3.793×10^{-1}	9.300×10^{-3}	5.791×10^{1}
1500	3.91×10^{0}	138.5	8.72×10^{1}	3.080×10^{2}	2.232×10^{-1}	3.704×10^{-1}	9.376×10^{-3}	2.971×10^{1}
1600	7.40×10^{0}	139.7	9.29×10^{1}	3.104×10^{2}	2.269×10^{-1}	3.629×10^{-1}	9.452×10^{-3}	1.661×10^{1}
1700	1.29×10^{1}	141.0	9.86×10^{1}	3.128×10^{2}	2.304×10^{-1}	3.564×10^{-1}	9.529×10^{-3}	9.970×10^{0}
1800	2.14×10^{1}	142.3	1.04×10^{2}	3.152×10^{2}	2.337×10^{-1}	3.508×10^{-1}	9.605×10^{-3}	6.346×10^{0}
1900	3.34×10^{1}	143.6	1.10×10^{2}	3.177×10^{2}	2.367×10^{-1}	3.460×10^{-1}	9.682×10^{-3}	4.245×10^{0}
2000	5.00×10^{1}	144.9	1.15×10^{2}	3.201×10^{2}	2.397×10^{-1}	3.418×10^{-1}	9.758×10^{-3}	2.961×10^{0}
2100	7.20×10^{1}	146.2	1.21×10^{2}	3.226×10^{2}	2.425×10^{-1}	3.382×10^{-1}	9.834×10^{-3}	2.141×10^{0}
2200	1.00×10^{2}	147.5	1.27×10^{2}	3.251×10^{2}	2.451×10^{-1}	3.351×10^{-1}	9.911×10^{-3}	1.597×10^{0}
2300	1.35×10^{2}	148.8	1.32×10^{2}	3.276×10^{2}	2.477×10^{-1}	3.323×10^{-1}	9.987×10^{-3}	1.224×10^{0}
2400	1.78×10^{2}	150.0	1.38×10^{2}	3.302×10^{2}	2.501×10^{-1}	3.299×10^{-1}	1.006×10^{-2}	9.604×10^{-1}
2500	2.30×10^{2}	151.1	1.44×10^{2}	3.328×10^{2}	2.524×10^{-1}	3.278×10^{-1}	1.014×10^{-2}	7.694×10^{-1}
2600	2.92×10^{2}	152.3	1.50×10^{2}	3.355×10^{2}	2.547×10^{-1}	3.260×10^{-1}	1.022×10^{-2}	6.277×10^{-1}
2700	3.63×10^{2}	153.4	1.55×10^{2}	3.382×10^{2}	2.568×10^{-1}	3.244×10^{-1}	1.029×10^{-2}	5.206×10^{-1}

W. D. Weatherford et al., *Properties of Inorganic Energy-conversion and Heat-transfer Fluids for Space Applications*, WADD Report 61-96, November 1961.

Table B·5 Thermodynamic properties of saturated Freon 12 (Dichlorodifluoromethane)

		Volume, ft^3/lbm			Enthalpy, Btu/lbm			Entropy, Btu/(lbm·R)		
T, °F	P, psia	v_f	v_{fg}	v_g	h_f	h_{fg}	h_g	s_f	s_{fg}	s_g
−130	0.41224	0.009736	70.7203	70.730	−18.609	81.577	62.968	−0.04983	0.24743	0.19760
−120	0.64190	0.009816	46.7312	46.741	−16.565	80.617	64.052	−0.04372	0.23731	0.19359
−110	0.97034	0.009899	31.7671	31.777	−14.518	79.663	65.145	−0.03779	0.22780	0.19002
−100	1.4280	0.009985	21.1541	22.164	−12.466	78.714	66.248	−0.03200	0.21883	0.18683
−90	2.0509	0.010073	15.8109	15.821	−10.409	77.764	67.355	−0.02637	0.21034	0.18398
−80	2.8807	0.010164	11.5228	11.533	−8.3451	76.812	68.467	−0.02086	0.20229	0.18143
−70	3.9651	0.010259	8.5584	8.5687	−6.2730	75.853	69.580	−0.01548	0.19464	0.17916
−60	5.3575	0.010357	6.4670	6.4774	−4.1919	74.885	70.693	−0.01021	0.18716	0.17714
−50	7.1168	0.010459	4.9637	4.9742	−2.1011	73.906	71.805	−0.00506	0.18038	0.17533
−40	9.3076	0.010564	3.8644	3.8750	0	72.913	72.913	0	0.17373	0.17373
−30	11.999	0.010674	3.0478	3.0585	2.1120	71.903	74.015	0.00496	0.16733	0.17229
−20	15.267	0.010788	2.4321	2.4429	4.2357	70.874	75.110	0.00983	0.16119	0.17102
−10	19.189	0.010906	1.9628	1.9727	6.3716	69.824	76.196	0.01462	0.15527	0.16989
0	23.849	0.011030	1.5979	1.6089	8.5207	68.750	77.271	0.01932	0.14956	0.16888
10	29.335	0.011160	1.3129	1.3241	10.684	67.651	78.335	0.02395	0.14403	0.16798
20	35.736	0.011296	1.0875	1.0988	12.863	66.522	79.385	0.02852	0.13867	0.16719
30	43.148	0.011438	0.90736	0.91880	15.058	65.361	80.419	0.03301	0.13347	0.16648
40	51.667	0.011588	0.76198	0.77357	17.273	64.163	81.436	0.03745	0.12841	0.16586
50	61.394	0.011746	0.64362	0.65537	19.507	62.926	82.433	0.04184	0.12346	0.16530
60	72.433	0.011913	0.54648	0.55839	21.766	61.643	83.409	0.04618	0.11861	0.16479
70	84.888	0.012089	0.46609	0.47818	24.050	60.309	84.359	0.05048	0.11386	0.16434
80	98.870	0.012277	0.39907	0.41135	26.365	58.917	85.282	0.05475	0.10917	0.16392
90	114.49	0.012478	0.34281	0.35529	28.713	57.461	86.174	0.05900	0.10453	0.16353
100	131.86	0.012693	0.29525	0.30794	31.100	55.929	87.029	0.06323	0.09992	0.16315
110	151.11	0.012924	0.25577	0.26769	33.531	54.313	87.844	0.06745	0.09534	0.16279
120	172.35	0.013174	0.22019	0.23326	36.013	52.597	88.610	0.07168	0.09073	0.16241
130	195.71	0.013447	0.19019	0.20364	38.553	50.768	89.321	0.07583	0.08609	0.16202
140	221.32	0.013746	0.16424	0.17799	41.162	48.805	89.967	0.08021	0.08138	0.16159
150	249.31	0.014078	0.14156	0.15564	43.850	46.684	90.534	0.08453	0.07657	0.16110
160	279.82	0.014449	0.12159	0.13604	46.633	44.373	91.006	0.08893	0.07260	0.16053
170	313.00	0.014871	0.10386	0.11873	49.529	41.830	91.359	0.09342	0.06643	0.15985
180	349.00	0.015360	0.08794	0.10330	52.562	38.999	91.561	0.09804	0.06096	0.15900
190	387.98	0.015942	0.073476	0.089418	55.769	35.792	91.561	0.10284	0.05511	0.15793
200	430.09	0.016659	0.060069	0.076728	59.203	32.075	91.278	0.10789	0.04862	0.15651
210	475.52	0.017601	0.047242	0.064843	62.959	27.599	90.558	0.11332	0.03921	0.15453
220	524.43	0.018986	0.035154	0.053140	67.246	21.790	89.036	0.11943	0.03206	0.15149
230	577.03	0.021854	0.017581	0.039435	72.893	12.229	85.122	0.12739	0.01773	0.14512
233.6	596.9	0.02870	0	0.02870	78.86	0	78.86	0.1359	0	0.1359

Table B·6a Nominal thermodynamic properties of gases at low pressures (old English units)†

Substance	$\hat{M}$ lbm/lbmole, g/gmole	c_P Btu/(lbm·R)	$\hat{c}_P$ Btu/(lbmole·R)	c_v Btu/(lbm·R)	$\hat{c}_v$ Btu/(lbmole·R)	R ft·lbf/(lbm·R)	R Btu/(lbm·R)	$k = c_P/c_v$
Argon, Ar	39.94	0.125	4.99	0.075	3.00	38.69	0.0497	1.67
Helium, He	4.003	1.24	4.96	0.744	2.97	386.0	0.4961	1.67
Hydrogen, H_2	2.016	3.42	6.89	2.435	4.90	766.5	0.9850	1.40
Nitrogen, N_2	28.02	0.248	6.95	0.177	4.96	55.15	0.0709	1.40
Oxygen, O_2	32.00	0.219	7.01	0.157	5.02	48.29	0.0621	1.39
Carbon monoxide, CO	28.01	0.249	6.97	0.178	4.98	55.17	0.0709	1.40
Air	28.97	0.240	6.95	0.172	4.96	53.34	0.0685	1.40
Water vapor, H_2O	18.016	0.446	8.04	0.336	6.05	85.78	0.1102	1.33
Methane, CH_4	16.04	0.532	8.53	0.408	6.54	96.3	0.1238	1.30
Carbon dioxide, CO_2	44.01	0.202	8.89	0.157	6.90	35.1	0.0451	1.29
Sulfur dioxide, SO_2	64.07	0.154	9.87	0.123	7.88	24.1	0.0310	1.25
Acetylene, C_2H_2	26.04	0.409	10.65	0.333	8.66	59.3	0.0763	1.23
Ethylene, C_2H_4	28.05	0.410	11.50	0.339	9.51	55.1	0.0708	1.21
Ethane, C_2H_6	30.07	0.422	12.69	0.356	10.70	51.4	0.0660	1.19
Propane, C_3H_8	44.09	0.404	17.81	0.358	15.82	35.0	0.0450	1.13
Isobutane, C_4H_{10}	58.12	0.420	24.41	0.386	22.42	26.0	0.0342	1.09

† $R = \mathscr{R}/\hat{M}$ and $c_v = c_P - R$.

Table B·6b **Nominal thermodynamic properties of gases at low pressures (SI units)†**

Substance	$\hat{M}$ kg/kgmole	c_P kJ/(kg·K)	$\hat{c}_P$ kJ/(kgmole·K)	c_v kJ/(kg·K)	$\hat{c}_v$ kJ/(kgmole·K)	R kJ/(kg·K)	k c_P/c_v
Argon, Ar	39.94	0.523	20.89	0.315	12.57	0.208	1.67
Helium, He	4.003	5.200	20.81	3.123	12.50	2.077	1.67
Hydrogen, H_2	2.016	14.32	28.86	10.19	20.55	4.124	1.40
Nitrogen, N_2	28.02	1.038	29.08	0.742	20.77	0.296	1.40
Oxygen, O_2	32.00	0.917	29.34	0.657	21.03	0.260	1.39
Carbon monoxide, CO	28.01	1.042	29.19	0.745	20.88	0.297	1.40
Air	28.97	1.004	29.09	0.718	20.78	0.286	1.40
Water vapor, H_2O	18.016	1.867	33.64	1.406	25.33	0.461	1.33
Carbon dioxide, CO_2	44.01	0.845	37.19	0.656	28.88	0.189	1.29
Sulfur dioxide, SO_2	64.07	0.644	41.26	0.514	32.94	0.130	1.25
Methane, CH_4	16.04	2.227	35.72	1.709	27.41	0.518	1.30
Propane, C_3H_8	44.09	1.691	74.56	1.502	66.25	0.189	1.13

† The R values in this table were determined from $R = \mathscr{R}/\hat{M}$, and the c_v values from $c_v = c_P - R$. For purposes of internal consistency more digits have been reported than justified by the experimental data. In calculations of entropy changes using the perfect-gas equation of state, it is recommended that the value of k be computed to calculator accuracy.

Table B·7 Properties of copper at 1 atm†

T K	$\hat{v}$ cm^3/gmole	β, 1/K	κ, cm^2/dyn	$\hat{c}_P$ cal/(gmole·K)	$\hat{c}_v$ cal/(gmole·K)	$k = \hat{c}_P/\hat{c}_v$
0	(7.0)	(0)	$(0.710) \times 10^{-12}$	(0)	(0)	1.00
50	(7.002)	$(11.5) \times 10^{-6}$	(0.712)	1.38	1.38	1.00
100	(7.008)	31.5	0.721	3.88	3.86	1.00
150	(7.018)	41.0	0.733	5.01	4.95	1.01
200	(7.029)	45.6	0.748	5.41	5.32	1.02
250	7.043	48.0	0.762	5.65	5.52	1.02
300	7.062	49.2	0.776	5.87	5.71	1.03
500	7.115	54.2	0.837	6.25	5.95	1.05
800	7.256	60.7	0.922	6.70	6.14	1.09
1200	7.452	69.7	1.030	7.34	6.34	1.16

From M. Zemansky, *Thermodynamics*, McGraw-Hill Book Company, New York, 1957.
† Values in parentheses are extrapolated.

Table B·8 Thermodynamic properties at the critical point

Substance	T, K	P, atm	$\hat{v}$, cm^3/gmole	$Z = Pv/RT$
Air	132.41	37.25	92.35	
Argon, Ar	150.72	47.99	75	0.291
Helium, He	5.19	2.26	58	0.308
Carbon monoxide, CO	132.91	34.529	93	0.294
Hydrogen, H_2	33.24	12.797	65	0.304
Nitrogen, N_2	126.2	33.54	90	0.291
Oxygen, O_2	154.78	50.14	74	0.292
Carbon dioxide, CO_2	304.20	72.90	94	0.275
Sulfur dioxide, SO_2	430.7	77.8	122	0.269
Water, H_2O	647.27	218.167	56	0.230
Acetylene, C_2H_2	309.5	61.6	113	0.274
Ethane, C_2H_6	305.48	48.20	148	0.285
Ethylene, C_2H_4	283.06	50.50	124	0.270
n-Butane, C_4H_{10}	425.17	37.47	255	0.274
Methane, CH_4	190.7	45.8	99	0.290
Propane, C_3H_8	370.01	42.1	200	0.277

From E. F. Obert, *Concepts of Thermodynamics*, McGraw-Hill Book Company, New York, 1960. By permission.

Table B·9 Thermodynamic properties of air at low pressures

T R	T °F	h Btu/lbm	p_r	u Btu/lbm	v_r	ϕ Btu/(lbm·R)
100	−360	23.7	0.00384	16.9	9640	0.1971
120	−340	28.5	0.00726	20.3	6120	0.2408
140	−320	33.3	0.01244	23.7	4170	0.2777
160	−300	38.1	0.01982	27.1	2990	0.3096
180	−280	42.9	0.0299	30.6	2230	0.3378
200	−260	47.7	0.0432	34.0	1715	0.3630
220	−240	52.5	0.0603	37.4	1352	0.3858
240	−220	57.2	0.0816	40.8	1089	0.4067
260	−200	62.0	0.1080	44.2	892	0.4258
280	−180	66.8	0.1399	47.6	742	0.4436
300	−160	71.6	0.1780	51.0	624	0.4601
320	−140	76.4	0.2229	54.5	532	0.4755
340	−120	81.2	0.2754	57.9	457	0.4900
360	−100	86.0	0.336	61.3	397	0.5037
380	−80	90.8	0.406	64.7	347	0.5166
400	−60	95.5	0.486	68.1	305	0.5289
420	−40	100.3	0.576	71.5	270	0.5406
440	−20	105.1	0.678	74.9	241	0.5517
460	0	109.9	0.791	78.4	215.3	0.5624
480	+20	114.7	0.918	81.8	193.6	0.5726
500	40	119.5	1.059	85.2	174.9	0.5823
520	60	124.3	1.215	88.6	158.6	0.5917
540	80	129.1	1.386	92.0	144.3	0.6008
560	100	133.9	1.574	95.5	131.8	0.6095
580	120	138.7	1.780	98.9	120.7	0.6179
600	140	143.5	2.00	102.3	110.9	0.6261
620	160	148.3	2.25	105.8	102.1	0.6340
640	180	153.1	2.51	109.2	94.3	0.6416
660	200	157.9	2.80	112.7	87.3	0.6490
680	220	162.7	3.11	116.1	81.0	0.6562
700	240	167.6	3.45	119.6	75.2	0.6632
720	260	172.4	3.81	123.0	70.1	0.6700
740	280	177.2	4.19	126.5	65.4	0.6766
760	300	182.1	4.61	130.0	61.1	0.6831
780	320	186.9	5.05	133.5	57.2	0.6894
800	340	191.8	5.53	137.0	53.6	0.6956
820	360	196.7	6.03	140.5	50.4	0.7016
840	380	201.6	6.67	144.0	47.3	0.7075
860	400	206.5	7.15	147.5	44.6	0.7132
880	420	211.4	7.76	151.0	42.0	0.7189
900	440	216.3	8.41	154.6	39.6	0.7244
920	460	221.2	9.10	158.1	37.4	0.7298
940	480	226.1	9.83	161.7	35.4	0.7351
960	500	231.1	10.61	165.3	33.5	0.7403
980	520	236.0	11.43	168.8	31.8	0.7454

Table B·9 (*continued*)

T R	T °F	h Btu/lbm	p_r	u Btu/lbm	v_r	ϕ Btu/(lbm·R)
1000	540	241.0	12.30	172.4	30.1	0.7504
1020	560	246.0	13.22	176.0	28.6	0.7554
1040	580	251.0	14.18	179.7	27.2	0.7602
1060	600	256.0	15.20	183.3	25.8	0.7650
1080	620	261.0	16.28	186.9	24.6	0.7696
1100	640	266.0	17.41	190.6	23.4	0.7743
1120	660	271.0	18.60	194.2	22.3	0.7788
1140	680	276.1	19.86	197.9	21.3	0.7833
1160	700	281.1	21.2	201.6	20.29	0.7877
1180	720	286.2	22.6	205.3	19.38	0.7920
1200	740	291.3	24.0	209.0	18.51	0.7963
1220	760	296.4	25.2	212.8	17.70	0.8005
1240	780	301.5	27.1	216.5	16.93	0.8047
1260	800	306.6	28.8	220.3	16.20	0.8088
1280	820	311.8	30.6	224.0	15.52	0.8128
1300	840	316.9	32.4	227.8	14.87	0.8168
1320	860	322.1	34.3	231.6	14.25	0.8208
1340	880	327.3	36.3	235.4	13.67	0.8246
1360	900	332.5	38.4	239.2	13.12	0.8285
1380	920	337.7	40.6	243.1	12.59	0.8323
1400	940	342.9	42.9	246.9	12.10	0.8360
1420	960	348.1	45.3	250.8	11.62	0.8398
1440	980	353.4	47.8	254.7	11.17	0.8434
1460	1000	358.6	50.3	258.5	10.74	0.8470
1480	1020	363.9	53.0	262.4	10.34	0.8506
1500	1040	369.2	55.9	266.3	9.95	0.8542
1520	1060	374.5	58.8	270.3	9.58	0.8568
1540	1080	379.8	61.8	274.2	9.23	0.8611
1560	1100	385.1	65.0	278.1	8.89	0.8646
1580	1120	390.4	68.3	282.1	8.57	0.8679
1600	1140	395.7	71.7	286.1	8.26	0.8713
1620	1160	401.1	75.3	290.0	7.97	0.8746
1640	1180	406.4	79.0	294.0	7.69	0.8779
1660	1200	411.8	82.8	298.0	7.42	0.8812
1680	1220	417.2	86.8	302.0	7.17	0.8844
1700	1240	422.6	91.0	306.1	6.92	0.8876
1720	1260	428.0	95.2	310.1	6.69	0.8907
1740	1280	433.4	99.7	314.1	6.46	0.8939
1760	1300	438.8	104.3	318.2	6.25	0.8970
1780	1320	444.3	109.1	322.2	6.04	0.9000
1800	1340	449.7	114.0	326.3	5.85	0.9031
1820	1360	455.2	119.2	330.4	5.66	0.9061
1840	1380	460.6	124.5	334.5	5.48	0.9091
1860	1400	466.1	130.0	338.6	5.30	0.9120
1880	1420	471.6	135.6	342.7	5.13	0.9150

Table B·9 (*continued*)

T R	T °F	h Btu/lbm	p_r	u Btu/lbm	v_r	ϕ Btu/(lbm·R)
1900	1440	477.1	141.5	346.8	4.97	0.9179
1920	1460	482.6	147.6	351.0	4.82	0.9208
1940	1480	488.1	153.9	355.1	4.67	0.9236
1960	1500	493.6	160.4	359.3	4.53	0.9264
1980	1520	499.1	167.1	363.4	4.39	0.9293
2000	1540	504.7	174.0	367.6	4.26	0.9320
2020	1560	510.3	181.2	371.8	4.13	0.9348
2040	1580	515.8	188.5	376.0	4.01	0.9376
2060	1600	521.4	196.2	380.2	3.89	0.9403
2080	1620	527.0	204.0	384.4	3.78	0.9430
2100	1640	532.6	212	388.6	3.67	0.9456
2120	1660	538.2	220	392.8	3.56	0.9483
2140	1680	543.7	229	397.0	3.46	0.9509
2160	1700	549.4	238	401.3	3.36	0.9535
2180	1720	555.0	247	405.5	3.27	0.9561
2200	1740	560.6	257	409.8	3.18	0.9587
2220	1760	566.2	266	414.0	3.09	0.9612
2240	1780	571.9	276	418.3	3.00	0.9638
2260	1800	577.5	287	422.6	2.92	0.9663
2280	1820	583.2	297	426.9	2.84	0.9688
2300	1840	588.8	308	431.2	2.76	0.9712
2320	1860	594.5	319	435.5	2.69	0.9737
2340	1880	600.2	331	439.8	2.62	0.9761
2360	1900	605.8	343	444.1	2.55	0.9785
2380	1920	611.5	355	448.4	2.48	0.9809
2400	1940	617.2	368	452.7	2.42	0.9833
2420	1960	622.9	380	457.0	2.36	0.9857
2440	1980	628.6	394	461.4	2.30	0.9880
2460	2000	634.3	407	465.7	2.24	0.9904
2480	2020	640.0	421	470.0	2.18	0.9927
2500	2040	645.8	436	474.4	2.12	0.9950
2520	2060	651.5	450	478.8	2.07	0.9972
2540	2080	657.2	466	483.1	2.02	0.9995
2560	2100	663.0	481	487.5	1.971	1.0018
2580	2120	668.7	497	491.9	1.922	1.0040
2600	2140	674.5	514	496.3	1.876	1.0062
2620	2160	680.2	530	500.6	1.830	1.0084
2640	2180	686.0	548	505.0	1.786	1.0106
2660	2200	691.8	565	509.4	1.743	1.0128
2680	2220	697.6	583	513.8	1.702	1.0150
2700	2240	703.4	602	518.3	1.662	1.0171
2720	2260	709.1	621	522.7	1.623	1.0193
2740	2280	714.9	640	527.1	1.585	1.0214
2760	2300	720.7	660	531.5	1.548	1.0235
2780	2320	726.5	681	536.0	1.512	1.0256
2800	2340	733.3	702	540.4	1.478	1.0277
2820	2360	738.2	724	544.8	1.444	1.0297

Table B·9 (*continued*)

T R	T °F	h Btu/lbm	p_r	u Btu/lbm	v_r	ϕ Btu/(lbm·R)
2840	2380	744.0	746	549.3	1.411	1.0318
2860	2400	749.8	768	553.7	1.379	1.0338
2880	2420	755.6	791	558.2	1.348	1.0359
2900	2440	761.4	815	562.7	1.318	1.0379
2920	2460	767.3	839	567.1	1.289	1.0399
2940	2480	773.1	864	571.6	1.261	1.0419
2960	2500	779.0	889	576.1	1.233	1.0439
2980	2520	784.8	915	580.6	1.206	1.0458
3000	2540	790.7	941	585.0	1.180	1.0478
3020	2560	796.5	969	589.5	1.155	1.0497
3040	2580	802.4	996	594.0	1.130	1.0517
3060	2600	808.3	1025	598.5	1.106	1.0536
3080	2620	814.2	1054	603.0	1.083	1.0555
3100	2640	820.0	1083	607.5	1.060	1.0574
3120	2660	825.9	1114	612.0	1.038	1.0593
3140	2680	831.8	1145	616.6	1.016	1.0612
3160	2700	837.7	1176	621.1	0.995	1.0630
3180	2720	843.6	1209	625.6	0.975	1.0649
3200	2740	849.5	1242	630.1	0.955	1.0668
3220	2760	855.4	1276	634.6	0.935	1.0686
3240	2780	861.3	1310	639.2	0.916	1.0704
3260	2800	867.2	1345	643.7	0.898	1.0722
3280	2820	873.1	1381	648.3	0.880	1.0740
3300	2840	879.0	1418	652.8	0.862	1.0758
3320	2860	884.9	1455	657.4	0.845	1.0776
3340	2880	890.9	1494	661.9	0.828	1.0794
3360	2900	896.8	1533	666.5	0.812	1.0812
3380	2920	902.7	1573	671.0	0.796	1.0830
3400	2940	908.7	1613	675.6	0.781	1.0847
3420	2960	914.6	1655	680.2	0.766	1.0864
3440	2980	920.6	1697	684.8	0.751	1.0882
3460	3000	926.5	1740	689.3	0.736	1.0899
3480	3020	932.4	1784	693.9	0.722	1.0916
3500	3040	938.4	1829	698.5	0.709	1.0933
3520	3060	944.4	1875	703.1	0.695	1.0950
3540	3080	950.3	1922	707.6	0.682	1.0967
3560	3100	956.3	1970	712.2	0.670	1.0984
3580	3120	962.2	2018	716.8	0.637	1.1000
3600	3140	968.2	2068	721.4	0.645	1.1017
3620	3160	974.2	2118	726.0	0.633	1.1034
3640	3180	980.2	2170	730.6	0.621	1.1050
3660	3200	986.1	2222	735.3	0.610	1.1066
3680	3220	992.1	2276	739.9	0.599	1.1083

Abridged from J. H. Keenan and J. Kaye, *Gas Tables*, John Wiley & Sons, Inc., New York, 1948. By permission of the authors.

Table B·10 Specific heats of selected liquids

Substance	State	c_P, Btu/(lbm·R)	Substance	State	c_P, Btu/(lbm·R)
Water	1 atm, 32°F	1.007	Glycerin	1 atm, 50°F	0.554
	1 atm, 77°F	0.998		1 atm, 120°F	0.617
	1 atm, 212°F	1.007	Bismuth	1 atm, 800°F	0.0345
Ammonia	sat., 0°F	1.08		1 atm, 1000°F	0.0369
	sat., 120°F	1.22		1 atm, 1400°F	0.0393
Freon 12	sat., −40°F	0.211	Mercury	1 atm, 50°F	0.033
	sat., 0°F	0.217		1 atm, 600°F	0.032
	sat., 120°F	0.244	Sodium	1 atm, 200°F	0.33
Benzene	1 atm, 60°F	0.43		1 atm, 1000°F	0.30
	1 atm, 150°F	0.46	*n*-Butane	1 atm, 32°F	0.550
Light oil	1 atm, 60°F	0.43	Propane	1 atm, 32°F	0.576
	1 atm, 300°F	0.54			

Based on values from *Handbook of Chemistry and Physics*, American Rubber Company.

Table B·11 Specific heats of selected solids ($P = 1$ atm)

Substance	T, °C	c_P, cal/(g·K)	Substance	T, °C	c_P, cal/(g·K)
Ice	−200	0.168	Lead	−270	0.00001
	−140	0.262		−259	0.0073
	−60	0.392		−100	0.0283
	−11	0.468		0	0.0297
	−2.6	0.500		+100	0.0320
Aluminum	−250	0.0039		300	0.0356
	−200	0.076	Iron	20	0.107
	−100	0.167	Silver	20	0.0558
	0	0.208	Sodium	20	0.295
	+100	0.225	Tungsten	20	0.034
	300	0.248	Graphite	20	0.17
	600	0.277	Wood	20	0.42
Platinum	−256	0.00123	Rubber	20	0.44
	−152	0.0261	Mica	20	0.21
	0	0.0316			
	+500	0.0349			
	1000	0.0381			

Based on values from *Handbook of Chemistry and Physics*, American Rubber Company.

Table B·12 Standardized enthalpies and absolute entropies of substances in equilibrium states at 77°F (25°C) and 1 atm

Substance	$\hat{h}^\circ$ Btu/lbmole	$\hat{h}^\circ$ J/gmole	$\hat{s}^\circ$ Btu/(lbmole·R)	$\hat{s}^\circ$ J/(gmole·K)
Hydrogen, H_2	0	0	31.194	130.60
Carbon, C	0	0	1.360	5.694
Oxygen, O_2	0	0	48.986	205.09
Nitrogen, N_2	0	0	45.755	191.56
Water (liquid), H_2O	−122,976	−286,022	16.72	70.00
Carbon dioxide, CO_2	−169,183	−393,492	51.032	213.66
Carbon monoxide, CO	−47,517	−110,516	47.272	197.92
Methane, CH_4	−32,179	−74,843	44.47	186.19
Ethane, C_2H_6	−36,401	−84,662	54.81	229.48
Propane, C_3H_8	−44,647	−103,848	64.47	269.92
n-Butane, C_4H_{10}	−53,627	−124,727	74.05	310.03
Acetylene, C_2H_2	97,495	226,757	47.966	200.82
Benzene, liquid, C_6H_6	35,653	82,923	64.30	269.21

Table B·13 Standardized enthalpies and absolute entropies for gases at low pressure†

T	Oxygen		Nitrogen		Hydrogen		Air‡	
R	$\hat{h}$	$\hat{\phi}$	$\hat{h}$	$\hat{\phi}$	$\hat{h}$	$\hat{\phi}$	$\hat{h}$	$\hat{\phi}$
537	0	48.986	0	45.755	0	31.194	0	46.336
600	443.2	49.762	438.4	46.514	435.3	31.959	438.3	47.107
700	1154.2	50.858	1135.4	47.588	1129.9	33.031	1136.2	48.183
800	1876.9	51.821	1834.9	48.522	1826.8	33.961	1838.7	49.121
900	2612.8	52.688	2538.6	49.352	2525.0	34.784	2547.0	49.955
1000	3362.4	53.477	3248.4	50.099	3224.2	35.520	3263.2	50.710
1100	4125.3	54.204	3965.5	50.783	4014.3	36.188	3987.7	51.400
1200	4900.7	54.879	4690.5	51.413	4625.5	36.798	4721.0	52.038
1300	5687.8	55.508	5424.4	52.001	5328.4	37.360	5463.7	52.633
1400	6485.3	56.099	6167.4	52.551	6033.5	37.883	6215.8	53.190
1500	7292.0	56.656	6919.4	53.071	6741.2	38.372	6976.8	53.715
1600	8107.4	57.182	7680.2	53.561	7452.2	38.830	7746.6	54.212
1700	8930.5	57.680	8449.4	54.028	8167.1	39.264	8524.4	54.683
1800	9760.7	58.155	9226.8	54.472	8886.5	39.675	9310.1	55.132
1900	10597.0	58.607	10012.1	54.896	9610.6	40.067	10103.3	55.561
2000	11438.9	59.039	10804.9	55.303	10339.8	40.441	10903.4	55.971
2100	12285.8	59.451	11604.5	55.694	11074.2	40.799	11710.0	56.365
2200	13137.5	59.848	12410.3	56.068	11814.1	41.143	12522.3	56.743
2300	13993.7	60.228	13221.7	56.429	12559.5	41.475	13340.1	57.106
2400	14854.1	60.594	14038.4	56.777	13310.3	41.794	14162.9	57.456
2500	15718.3	60.946	14860.0	57.112	14067.0	42.104	14990.2	57.794
2600	16586.3	61.287	15686.3	57.436	14829.4	42.403	15822.0	58.120
2700	17457.8	61.616	16516.9	57.750	15597.5	42.692	16658.0	58.436
2800	18332.7	61.934	17351.6	58.053	16371.5	42.973	17497.6	58.742
2900	19211.0	62.242	18190.0	58.348	17151.2	43.247	18341.2	59.037
3000	20092.6	62.540	19032.0	58.632	17936.6	43.514	19188.0	59.324
3100	20977.4	62.831	19877.3	58.910	18727.4	43.773	20038.3	59.603
3200	21865.4	63.113	20725.5	59.179	19523.8	44.026	20891.4	59.874
3300	22756.5	63.386	21576.5	59.442	20325.2	44.273	21747.2	60.137
3400	23650.8	63.654	22430.2	59.697	21131.6	44.513	22605.9	60.394
3500	24548.2	63.914	23286.4	59.944	21942.6	44.748	23467.4	60.643
3600	25448.8	64.168	24144.9	60.186	22758.2	44.978	24331.0	60.887
3700	26352.4	64.415	25005.6	60.422	23578.2	45.203	25197.2	61.124
3800	27259.0	64.657	25868.4	60.652	24402.5	45.423	26065.8	61.356
3900	28168.5	64.893	26733.3	60.877	25230.8	45.638	26936.3	61.582
4000	29081.0	65.123	27599.9	61.097	26063.2	45.849	27808.9	61.803
4100	29996.5	65.350	28468.5	61.310	26899.5	46.056	28683.8	62.019
4200	30914.8	65.571	29338.6	61.520	27739.5	46.257	29560.4	62.230
4300	31836.0	65.788	30210.4	61.726	28583.2	46.456	30438.8	62.437
4400	32759.9	66.000	31083.6	61.927	29430.6	46.651	31318.9	62.639
4500	33686.7	66.208	31958.3	62.123	30281.3	46.842	32200.7	62.837
4600	34616.3	66.413	32834.3	62.316	31135.4	47.030	33084.3	63.031
4700	35548.5	66.613	33711.6	62.504	32022.7	47.215	33969.6	63.223
4800	36483.5	66.809	34590.0	62.689	32853.1	47.396	34855.8	63.408
4900	37421.0	67.003	35469.6	62.870	33716.6	47.574	35744.1	63.591
5000	38361.2	67.193	36350.3	63.049	34583.0	47.749	36633.4	63.771

Table B·13 (*continued*)

T	Water vapor		Carbon monoxide		Carbon dioxide		Monatomic gases	
R	$\hat{h}$	$\hat{\phi}$	$\hat{h}$	$\hat{\phi}$	$\hat{h}$	$\hat{\phi}$	$\hat{h}$	$\hat{\phi}$
537	−103968.0	45.079	−47517.0	47.272	−169183.0	51.032	0	36.942
600	−103471.6	45.970	−47078.5	48.044	−168612.3	52.038	212.7	37.512
700	−102660.9	47.219	−46380.5	49.120	−167661.2	53.503	809.2	38.278
800	−101839.4	48.316	−45678.3	50.058	−166660.3	54.839	1305.7	38.941
900	−101005.4	49.298	−44970.1	50.892	−165615.6	56.070	1802.1	39.525
1000	−100157.4	50.191	−44254.3	51.646	−164531.1	57.212	2298.6	40.048
1100	−99294.3	51.013	−43529.7	52.337	−163410.6	58.281	2795.1	40.522
1200	−98415.9	51.777	−42795.7	52.976	−162257.9	59.283	3291.5	40.954
1300	−97521.8	52.494	−42051.9	53.571	−161076.3	60.229	3788.0	41.351
1400	−96611.5	53.168	−41298.5	54.129	−159868.5	61.124	4284.5	41.719
1500	−95684.9	53.808	−40535.4	54.655	−158636.5	61.974	4781.0	42.061
1600	−94741.4	54.418	−39763.1	55.154	−157383.5	62.783	5277.4	42.382
1700	−93780.9	54.999	−38982.2	55.628	−156111.8	63.555	5773.9	42.683
1800	−92803.3	55.559	−38193.3	56.078	−154821.0	64.292	6270.4	42.967
1900	−91808.8	56.097	−37396.7	56.509	−153514.7	64.999	6766.8	43.235
2000	−90797.3	56.617	−36593.3	56.922	−152193.8	65.676	7263.3	43.490
2100	−89769.4	57.119	−35783.2	57.317	−150859.8	66.327	7759.8	43.732
2200	−88725.5	57.605	−34967.1	57.696	−149513.5	66.953	8256.2	43.963
2300	−87665.7	58.077	−34145.5	58.062	−148156.2	67.557	8752.7	44.183
2400	−86590.6	58.535	−33319.1	58.414	−146788.5	68.139	9249.2	44.395
2500	−85500.9	58.980	−32487.7	58.754	−145411.3	68.702	9745.7	44.597
2600	−84396.8	59.414	−31652.2	59.081	−144025.4	69.245	10242.1	44.792
2700	−83279.1	59.837	−30812.5	59.398	−142631.3	69.771	10738.6	44.980
2800	−82148.3	60.248	−29969.3	59.705	−141229.7	70.282	11235.1	45.160
2900	−81005.1	60.650	−29122.7	60.002	−139821.0	70.776	11731.5	45.334
3000	−79850.0	61.043	−28273.1	60.290	−138405.9	71.255	12228.0	45.503
3100	−78683.5	61.426	−27420.5	60.569	−136984.6	71.722	12724.5	45.663
3200	−77506.1	61.801	−26565.3	60.841	−135557.8	72.175	13220.9	45.823
3300	−76318.1	62.167	−25707.5	61.105	−134125.8	72.616	13717.4	45.973
3400	−75120.3	62.526	−24847.2	61.362	−132688.9	73.045	14213.9	46.124
3500	−73912.8	62.876	−23984.7	61.612	−131247.3	73.462	14710.5	46.266
3600	−72696.2	63.221	−23119.9	61.855	−129801.5	73.870	15206.8	46.408
3700	−71470.9	63.557	−22253.0	62.093	−128351.9	74.267	15703.3	46.542
3800	−70237.4	63.887	−21384.2	62.325	−126898.4	74.655	16199.8	46.676
3900	−68996.1	64.210	−20513.6	62.551	−125441.5	75.033	16696.3	46.803
4000	−67747.2	64.528	−19641.3	62.772	−123981.1	75.404	17192.7	46.931
4100	−66490.9	64.839	−18767.4	62.988	−122517.4	75.765	17689.2	47.052
4200	−65227.9	65.144	−17892.1	63.198	−121050.5	76.119	18185.6	47.173
4300	−63958.3	65.444	−17015.3	63.405	−119580.4	76.464	18682.1	47.289
4400	−62682.4	65.738	−16137.3	63.607	−118107.4	76.803	19178.6	47.404
4500	−61400.4	66.028	−15257.9	63.805	−116631.5	77.135	19675.0	47.515
4600	−60112.7	66.312	−14377.2	63.998	−115152.8	77.460	20171.4	47.625
4700	−58819.4	66.591	−13495.5	64.188	−113671.4	77.779	20668.0	47.731
4800	−57520.8	66.866	−12612.6	64.374	−112187.6	78.091	21164.5	47.836
4900	−56217.3	67.135	−11728.7	64.556	−110701.2	78.398	21661.0	47.938
5000	−54908.9	67.401	−10843.8	64.735	−109212.5	78.698	22157.4	48.039

Derived from J. H. Keenan and J. Kaye, *Gas Tables*, John Wiley & Sons, Inc., 1948, using standardized enthalpies based on NBS Circulars 467 and 500.

† $\hat{s} = \hat{\phi}(T) - \mathscr{R} \ln (P/P_0)$; $P_0 = 1$ atm; $\hat{h}$ in Btu/lbmole; $\hat{s}$ in Btu/(lbmole·R).

‡ 78.03 percent N_2, 20.99 percent O_2, 0.98 percent A.

Table B·14 Logarithms to the base 10 of the equilibrium constant K for the reaction

$$\nu_1 C_1 + \nu_2 C_2 \rightleftarrows \nu_3 C_3 + \nu_4 C_4 \qquad K(T) = \frac{\chi_3^{\nu_3}\chi_4^{\nu_4}}{\chi_1^{\nu_1}\chi_2^{\nu_2}}\left(\frac{P}{P_0}\right)^{\nu_3+\nu_4-\nu_1-\nu_2} \qquad (P_0 = 1\ \text{atm})$$

T, K	$H_2 \rightleftarrows 2H$	$O_2 \rightleftarrows 2O$	$H_2O \rightleftarrows H_2 + \frac{1}{2}O_2$	$H_2O \rightleftarrows OH + \frac{1}{2}H_2$	$CO_2 \rightleftarrows CO + \frac{1}{2}O_2$	$N_2 \rightleftarrows 2N$	$\frac{1}{2}O_2 + \frac{1}{2}N_2 \rightleftarrows NO$	$Na \rightleftarrows Na^+ + e^-$	$Cs \rightleftarrows Cs^+ + e^-$
298	−71.210	−80.620	−40.047	−46.593	−45.043	−119.434	−15.187	−32.3	−25.1
400	−51.742	−58.513	−29.241	−33.910	−32.41	−87.473	−11.156	−24.3	−17.5
600	−32.667	−36.859	−18.663	−21.470	−20.07	−56.206	−7.219	−14.6	−10.0
800	−23.074	−25.985	−13.288	−15.214	−13.90	−40.521	−5.250	−9.58	−6.15
1000	−17.288	−19.440	−10.060	−11.444	−10.199	−31.084	−4.068	−6.54	−3.79
1200	−13.410	−15.062	−7.896	−8.922	−7.742	−24.619	−3.279	−4.47	−2.18
1400	−10.627	−11.932	−6.334	−7.116	−5.992	−20.262	−2.717	−2.97	−1.010
1600	−8.530	−9.575	−5.175	−5.758	−4.684	−16.869	−2.294	−1.819	−0.108
1800	−6.893	−7.740	−4.263	−4.700	−3.672	−14.225	−1.966	−0.913	+0.609
2000	−5.579	−6.269	−3.531	−3.852	−2.863	−12.016	−1.703	−0.175	+1.194
2200	−4.500	−5.064	−2.931	−3.158	−2.206	−10.370	−1.488	+0.438	+1.682
2400	−3.598	−4.055	−2.429	−2.578	−1.662	−8.992	−1.309	+0.956	+2.098
2600	−2.833	−3.206	−2.003	−2.087	−1.203	−7.694	−1.157	+1.404	+2.46
2800	−2.176	−2.475	−1.638	−1.670	−0.807	−6.640	−1.028	+1.792	+2.77
3000	−1.604	−1.840	−1.322	−1.302	−0.469	−5.726	−0.915	+2.13	+3.05
3200	−1.104	−1.285	−1.046	−0.983	−0.175	−4.925	−0.817	+2.44	+3.29
3500	−0.458	−0.571	−0.693	−0.557	+0.201	−3.893	−0.692	+2.84	+3.62
4000	+0.406	+0.382	−0.221	−0.035	+0.699	−2.514	−0.526	+3.38	+4.07
4500	+1.078	+1.125	+0.153	+0.392	+1.081	−1.437	−0.345	+3.82	+4.43
5000	+1.619	+1.719	+0.450	+0.799	+1.387	−0.570	−0.298	+4.18	+4.73

Based on information provided by the National Bureau of Standards; ionization of Cs and Na from the Saha equation.

Table B·15 The elements

Name	Symbol	Atomic number	International atomic weight, 1966
Actinium	Ac	89	——
Aluminum	Al	13	26.9815
Americium	Am	95	——
Antimony,			——
stibium	Sb	51	121.75
Argon	Ar	18	39.948
Arsenic	As	33	74.9216
Astatine	At	85	——
Barium	Ba	56	137.34
Berkelium	Bk	97	——
Beryllium	Be	4	9.0122
Bismuth	Bi	83	208.980
Boron	B	5	10.811
Bromine	Br	35	79.904
Cadmium	Cd	48	112.40
Calcium	Ca	20	40.08
Californium	Cf	98	——
Carbon	C	6	12.01115
Cerium	Ce	58	140.12
Cesium	Cs	55	132.905
Chlorine	Cl	17	35.453
Chromium	Cr	24	51.996
Cobalt	Co	27	58.9332
Columbium, see			——
niobium			
Copper	Cu	29	63.546
Curium	Cm	96	——
Dysprosium	Dy	66	162.50
Einsteinium	Es	99	——
Erbium	Er	68	167.26
Europium	Eu	63	151.96
Fermium	Fm	100	——
Fluorine	F	9	18.9984
Francium	Fr	87	——
Gadolinium	Gd	64	157.25
Gallium	Ga	31	69.72
Germanium	Ge	32	72.59
Gold, aurum	Au	79	196.967
Hafnium	Hf	72	178.49
Helium	He	2	4.0026
Holmium	Ho	67	164.930
Hydrogen	H	1	1.00797
Indium	In	49	114.82
Iodine	I	53	126.9044
Iridium	Ir	77	192.2
Iron, ferrum	Fe	26	55.847
Krypton	Kr	36	83.80
Lanthanum	La	57	138.91
Lawrencium	Lr	103	(257)
Lead, plumbum	Pb	82	207.19
Lithium	Li	3	6.939
Lutetium	Lu	71	174.97

Table B·15 *(continued)*

Name	Symbol	Atomic number	International atomic weight, 1966	Name	Symbol	Atomic number	International atomic weight, 1966
Magnesium	Mg	12	24.312	Rhodium	Rh	45	102.905
Manganese	Mn	25	54.9380	Rubidium	Rb	37	85.47
Mendelevium	Md	101		Ruthenium	Ru	44	101.07
Mercury,				Samarium	Sm	62	150.35
hydrargyrum	Hg	80	200.59	Scandium	Sc	21	44.956
Molybdenum	Mo	42	95.94	Selenium	Se	34	78.96
Neodymium	Nd	60	144.24	Silicon	Si	14	28.086
Neon	Ne	10	20.183	Silver, argentum	Ag	47	107.868
Neptunium	Np	93		Sodium, natrium	Na	11	22.9898
Nickel	Ni	28	58.71	Strontium	Sr	38	87.62
Niobium,				Sulfur	S	16	32.064
columbium	Nb	41	92.906	Tantalum	Ta	73	180.948
Nitrogen	N	7	14.0067	Technetium	Tc	43	
Nobelium	No	102		Tellurium	Te	52	127.60
Osmium	Os	76	190.2	Terbium	Tb	65	158.924
Oxygen	O	8	15.9994	Thallium	Tl	81	204.37
Palladium	Pd	46	106.4	Thorium	Th	90	232.038
Phosphorus	P	15	30.9738	Thulium	Tm	69	168.934
Platinum	Pt	78	195.09	Tin, stannum	Sn	50	118.69
Plutonium	Pu	94		Titanium	Ti	22	47.90
Polonium	Po	84		Tungsten,			
Potassium,				wolfram	W	74	183.85
kalium	K	19	39.102	Uranium	U	92	238.03
Praseodymium	Pr	59	140.907	Vanadium	V	23	50.942
Promethium	Pm	61		Xenon	Xe	54	131.30
Protactinium	Pa	91		Ytterbium	Yb	70	173.04
Radium	Ra	88		Yttrium	Y	39	88.905
Radon	Rn	86		Zinc	Zn	30	65.37
Rhenium	Re	75	186.2	Zirconium	Zr	40	91.22

Table B·16 Molar specific heats at constant pressure for the ideal-gas state (300 to 1500 K) $\hat{c}_P = a + bT + cT^2$

Gas	Formula	a cal/(gmole·K)	$b \times 10^3$ cal/(gmole·K^2)	$c \times 10^6$ cal/(gmole·K^3)
Acetylene	C_2H_2	7.331	12.622	−3.889
Ammonia	NH_3	6.086	8.812	−1.506
Benzene	C_6H_6	−0.409	77.621	−26.429
Carbon dioxide	CO_2	6.214	10.396	−3.545
Carbon monoxide	CO	6.420	1.665	−0.196
Chlorine	Cl_2	7.576	2.424	−0.965
Ethyl alcohol	C_2H_6O	6.990	39.741	−11.926
Hydrogen	H_2	6.947	−0.200	0.481
Hydrogen chloride	HCl	6.732	0.433	0.370
Methane	CH_4	3.381	18.044	−4.300
Nitrogen	N_2	6.524	1.250	−0.001
Oxygen	O_2	6.148	3.102	−0.923
Sulfur dioxide	SO_2	7.116	9.512	3.511
Water	H_2O	7.256	2.298	0.283

From M. W. Zemansky and H. C. Van Ness, *Basic Engineering Thermodynamics*, McGraw-Hill Book Company, New York, 1966. By permission.

Table B·17 Constants for equations of state

	Van der Waals gas		Beattie-Bridgeman gas				
	a	b	A_0	a	B_0	b	c
Substance	$\frac{atm \cdot ft^6}{lbmole^2}$	$\frac{ft^3}{lbmole}$	$\frac{atm \cdot ft^6}{lbmole^2}$	$\frac{ft^3}{lbmole}$	$\frac{ft^3}{lbmole}$	$\frac{ft^3}{lbmole}$	$\frac{ft^3 \cdot R^3}{lbmole}$
Air	343.8	0.585	334.1	0.309	0.739	−0.0176	4.05×10^6
Hydrogen, H_2	63.02	0.427	50.57	−0.0811	0.336	−0.698	0.0471×10^6
Nitrogen, N_2	346.0	0.618	344.92	0.419	0.808	−0.111	3.92×10^6
Oxygen, O_2	349.5	0.510	382.53	0.410	0.741	0.0674	4.48×10^6
Carbon monoxide, CO	374.7	0.630	344.9	0.419	0.808	−0.111	3.92×10^6
Carbon dioxide, CO_2	924.2	0.685	1284.9	1.143	1.678	1.159	61.65×10^6
Methane, CH_4	578.9	0.684	584.6	0.297	0.895	−0.254	11.98×10^6
Propane, C_3H_8	2374.0	1.446	305.8	1.173	2.90	0.688	112.2×10^6
n-Butane, C_4H_{10}	3675.0	1.944	456.5	1.948	3.944	1.51	327.02×10^6

From E. F. Obert, *Concepts of Thermodynamics*, McGraw-Hill Book Company, New York, 1960. By permission. Evaluated from critical data and from Beattie and Bridgeman, *Proceedings of the American Academy of Arts and Sciences*, vol. 63, pp. 229–308 (1928); *Journal of the American Chemical Society*, vol. 50, p. 3133 (1928).

Table B·18 One-dimensional isentropic compressible-flow functions for an ideal gas with constant specific heat and $k = 1.4$

M	$M^* = V/c^*$	$\frac{A}{A^*}$	$\frac{P}{P_0}$	$\frac{\rho}{\rho_0}$	$\frac{T}{T_0}$
0	0	∞	1.00000	1.00000	1.00000
0.10	0.10943	5.8218	0.99303	0.99502	0.99800
0.20	0.21822	2.9635	0.97250	0.98027	0.99206
0.30	0.32572	2.0351	0.93947	0.95638	0.98232
0.40	0.43133	1.5901	0.89562	0.92428	0.96899
0.50	0.53452	1.3398	0.84302	0.88517	0.95238
0.60	0.63480	1.1882	0.78400	0.84045	0.93284
0.70	0.73179	1.09437	0.72092	0.79158	0.91075
0.80	0.82514	1.03823	0.65602	0.74000	0.88652
0.90	0.91460	1.00886	0.59126	0.68704	0.86058
1.00	1.00000	1.00000	0.52828	0.63394	0.83333
1.10	1.08124	1.00793	0.46835	0.58169	0.80515
1.20	1.1583	1.03044	0.41238	0.53114	0.77640
1.30	1.2311	1.06631	0.36092	0.48291	0.74738
1.40	1.2999	1.1149	0.31424	0.43742	0.71839
1.50	1.3646	1.1762	0.27240	0.39498	0.68965
1.60	1.4254	1.2502	0.23527	0.35573	0.66138
1.70	1.4825	1.3376	0.20259	0.31969	0.63372
1.80	1.5360	1.4390	0.17404	0.28682	0.60680
1.90	1.5861	1.5552	0.14924	0.25699	0.58072
2.00	1.6330	1.6875	0.12780	0.23005	0.55556
2.10	1.6769	1.8369	0.10935	0.20580	0.53135
2.20	1.7179	2.0050	0.09352	0.18405	0.50813
2.30	1.7563	2.1931	0.07997	0.16458	0.48591
2.40	1.7922	2.4031	0.06840	0.14720	0.46468
2.50	1.8258	2.6367	0.05853	0.13169	0.44444
2.60	1.8572	2.8960	0.05012	0.11787	0.42517
2.70	1.8865	3.1830	0.04295	0.10557	0.40684
2.80	1.9140	3.5001	0.03685	0.09462	0.38941
2.90	1.9398	3.8498	0.03165	0.08489	0.37286
3.00	1.9640	4.2346	0.02722	0.07623	0.35714
3.50	2.0642	6.7896	0.01311	0.04523	0.28986
4.00	2.1381	10.719	0.00658	0.02766	0.23810
4.50	2.1936	16.562	0.00346	0.01745	0.19802
5.00	2.2361	25.000	$189(10)^{-5}$	0.01134	0.16667
6.00	2.2953	53.180	$633(10)^{-6}$	0.00519	0.12195
7.00	2.3333	104.143	$242(10)^{-6}$	0.00261	0.09259
8.00	2.3591	190.109	$102(10)^{-6}$	0.00141	0.07246
9.00	2.3772	327.189	$474(10)^{-7}$	0.000815	0.05814
10.00	2.3904	535.938	$236(10)^{-7}$	0.000495	0.04762
∞	2.4495	∞	0	0	0

Abridged from J. H. Keenan and J. Kaye, *Gas Tables*, John Wiley & Sons, Inc., New York, 1948. By permission of the authors.

Table B·19 One-dimensional normal-shock functions for an ideal gas with constant specific heat and $k = 1.4$

M_x	M_y	$\frac{P_y}{P_x}$	$\frac{\rho_y}{\rho_x}$	$\frac{T_y}{T_x}$	$\frac{P_{oy}}{P_{ox}} = \frac{A_x^*}{A_y^*}$	$\frac{P_{oy}}{P_x}$
1.00	1.00000	1.00000	1.00000	1.00000	1.00000	1.8929
1.10	0.91177	1.2450	1.1691	1.06494	0.99892	2.1328
1.20	0.84217	1.5133	1.3416	1.1280	0.99280	2.4075
1.30	0.78596	1.8050	1.5157	1.1909	0.97935	2.7135
1.40	0.73971	2.1200	1.6896	1.2547	0.95819	3.0493
1.50	0.70109	2.4583	1.8621	1.3202	0.92978	3.4133
1.60	0.66844	2.8201	2.0317	1.3880	0.89520	3.8049
1.70	0.64055	3.2050	2.1977	1.4583	0.85573	4.2238
1.80	0.61650	3.6133	2.3592	1.5316	0.81268	4.6695
1.90	0.59562	4.0450	2.5157	1.6079	0.76735	5.1417
2.00	0.57735	4.5000	2.6666	1.6875	0.72088	5.6405
2.10	0.56128	4.9784	2.8119	1.7704	0.67422	6.1655
2.20	0.54706	5.4800	2.9512	1.8569	0.62812	6.7163
2.30	0.53441	6.0050	3.0846	1.9968	0.58331	7.2937
2.40	0.52312	6.5533	3.2119	2.0403	0.54015	7.8969
2.50	0.51299	7.1250	3.3333	2.1375	0.49902	8.5262
2.60	0.50387	7.7200	3.4489	2.2383	0.46012	9.1813
2.70	0.49563	8.3383	3.5590	2.3429	0.42359	9.8625
2.80	0.48817	8.9800	3.6635	2.4512	0.38946	10.569
2.90	0.48138	9.6450	3.7629	2.5632	0.35773	11.302
3.00	0.47519	10.333	3.8571	2.6790	0.32834	12.061
4.00	0.43496	18.500	4.5714	4.0469	0.13876	21.068
5.00	0.41523	29.000	5.0000	5.8000	0.06172	32.654
10.00	0.38757	116.50	5.7143	20.388	0.00304	129.217
∞	0.37796	∞	6.000	∞	0	∞

Abridged from J. H. Keenan and J. Kaye, *Gas Tables*, John Wiley & Sons, Inc., New York, 1948. By permission of the authors.

APPENDIX C

INFORMATION FOR HEAT TRANSFER ANALYSIS

Table C·1 Properties of metals

Metal	T °C	T °F	ρ kg/m³	ρ lbm/ft³	c_P kJ/(kg·K)	c_P Btu/(lbm·°F)	k W/(m·K)	k Btu/(h·ft·°F)
Aluminum, pure	20	68	2,707	169	0.896	0.214	204	118
	200	392	···	···	···	···	215	124
	400	752	···	···	···	···	249	144
Lead	20	68	11,373	710	0.130	0.031	35	20
	300	572	···	···	···	···	29.8	17.2
Iron								
Pure	20	68	7,897	493	0.452	0.108	73	42
	300	572	···	···	···	···	55	32
	1,000	1,832	···	···	···	···	35	20
Wrought	20	68	7,849	490	0.46	0.11	59	34
Carbon steel (max. 0.5% C)	20	68	7,833	489	0.465	0.111	54	31
Carbon steel (1.5%)	20	68	7,753	484	0.486	0.116	36	21
	400	752	···	···	···	···	33	19
	1,200	2,192	···	···	···	···	29	17
Stainless steel	20	68	···	···	···	···	12–45	7–26
Copper								
Pure	20	68	8,954	559	0.383	0.0915	386	223
	300	572	···	···	···	···	369	213
	600	1,112	···	···	···	···	353	204
Bronze (75% Cu, 25% Zn)	20	68	8,666	541	0.343	0.082	26	15
Brass, (70% Cu, 30% Zn)	20	68	8,522	532	0.385	0.092	111	64
Silver, pure	20	68	10,524	657	0.234	0.0559	407	235
Tungsten	20	68	19,350	1,208	0.134	0.0321	163	94

Adapted from E. R. G. Eckert and R. M. Drake, *Analysis of Heat and Mass Transfer*, 3d ed., McGraw-Hill Book Company, New York, 1972. By permission.

Table C·2 Thermal conductivity of nonmetals

Substance	T, °C	k, W/(m·K)	T, °F	k, Btu/(h·ft·°F)
Structural and heat-resistant materials				
Asphalt	20–55	0.75	68–132	0.43–0.44
Brick				
Building brick, common	20	0.69	68	0.40
Building brick, face	···	1.31	···	0.76
Diatomaceous earth, molded and fired	200	0.24	400	0.14
	870	0.31	1600	0.18
Fireclay brick, burnt at 2426°F	500	1.04	932	0.60
	800	1.07	1472	0.62
	1100	1.09	2012	0.63
Cement, portland	···	0.29	···	0.17
Concrete, stone 1–2–4 mix	20	1.37	69	0.79
Glass, window	20	0.78	68	0.45 (av.)
Glass, borosilicate	30–70	1.09	86–167	0.63
Plaster, gypsum	21	0.48	70	0.28
Plaster, metal lath	21	0.47	70	0.27
Plaster, wood lath	21	0.28	70	0.16
Stone				
Granite	···	1.7–4.0	···	1.0–2.3
Limestone	100–300	1.26–1.33	210–570	0.73–0.77
Wood, across the grain				
Balsa, 8.8 lb/ft^3	30	0.055	86	0.032
Yellow pine	24	0.147	75	0.085
White pine	30	0.112	86	0.065
Insulating material				
Asbestos				
Asbestos-cement boards	20	0.744	68	0.43
Asbestos sheets	50	0.166	124	0.096
Asbestos cement	···	2.07	···	1.2
Asbestos, loosely packed	−45	0.148	−50	0.086
	0	0.154	32	0.089
	100	0.160	210	0.093
Corkboard, 10 lb/ft^3	30	0.043	86	0.025
Cork, regranulated	32	0.044	90	0.026
Cork, ground	32	0.043	90	0.025
Diatomaceous earth (Sil-o-cel)	0	0.060	32	0.035
Fiber insulating board	21	0.048	70	0.028
Glass wool, 1.5 lb/ft^3	24	0.038	75	0.022
Kapok	30	0.035	86	0.020
Magnesia, 85%	38	0.067	100	0.039
	93	0.071	200	0.041
	148	0.074	300	0.043
	204	0.080	400	0.046
Sawdust	24	0.059	75	0.034
Ice	0	2.22	32	1.28
Rock wool	32	0.039	90	0.0225
Cement mortar	···	1.73	···	1.0

Adapted from A. I. Brown and S. M. Marco, *Introduction to Heat Transfer*, 3d ed., McGraw-Hill Book Company, New York, 1958. By permission.

Table C·3 Properties of saturated liquids (SI units)

T, °C	ρ, kg/m^3	c_p, J/(kg·K)	ν, m^2/s	k, W/(m·K)	α, m^2/s	Pr	β, K^{-1}
Water, H_2O							
0	1,002.28	4.2178×10^3	1.788×10^{-5}	0.552	1.308×10^{-7}	13.6	
20	1,000.52	4.1818	1.006	0.597	1.430	7.02	0.18×10^{-3}
40	994.59	4.1784	0.658	0.628	1.512	4.34	
60	985.46	4.1843	0.478	0.651	1.554	3.02	
80	974.08	4.1964	0.364	0.668	1.636	2.22	
100	960.63	4.2161	0.294	0.680	1.680	1.74	
120	945.25	4.250	0.247	0.685	1.708	1.446	
140	928.27	4.283	0.214	0.684	1.724	1.241	
160	909.69	4.342	0.190	0.680	1.729	1.099	
180	889.03	4.417	0.173	0.675	1.724	1.004	
200	866.76	4.505	0.160	0.665	1.706	0.937	
220	842.41	4.610	0.150	0.652	1.680	0.891	
240	815.66	4.756	0.143	0.635	1.639	0.871	
260	785.87	4.949	0.137	0.611	1.577	0.874	
280	752.55	5.208	0.135	0.580	1.481	0.910	
300	714.26	5.728	0.135	0.540	1.324	1.019	
Carbon dioxide, CO_2							
−50	1,156.34	1.84×10^3	0.119×10^{-6}	0.0855	0.4021×10^{-7}	2.96	
−40	1,117.77	1.88	0.118	0.1011	0.4810	2.46	
−30	1,076.76	1.97	0.117	0.1116	0.5272	2.22	
−20	1,032.39	2.05	0.115	0.1151	0.5445	2.12	
−10	983.38	2.18	0.113	0.1099	0.5133	2.20	
0	926.99	2.47	0.108	0.1045	0.4578	2.38	
10	860.03	3.14	0.101	0.0971	0.3608	2.80	
20	772.57	5.0	0.091	0.0872	0.2219	4.10	14.00×10^{-3}
30	597.81	36.4	0.080	0.0703	0.0279	28.7	
Dichlorodifluoromethane (Freon), CCl_2F_2							
−50	1,546.75	0.8750×10^3	0.310×10^{-6}	0.067	0.501×10^{-7}	6.2	2.63×10^{-3}
−40	1,518.71	0.8847	0.279	0.069	0.514	5.4	
−30	1,489.56	0.8956	0.253	0.069	0.526	4.8	
−20	1,460.57	0.9073	0.235	0.071	0.539	4.4	
−10	1,429.49	0.9203	0.221	0.073	0.550	4.0	
0	1,397.45	0.9345	0.214	0.073	0.557	3.8	
10	1,364.30	0.9496	0.203	0.073	0.560	3.6	
20	1,330.18	0.9659	0.198	0.073	0.560	3.5	
30	1,295.10	0.9835	0.194	0.071	0.560	3.5	
40	1,257.13	1.0019	0.191	0.069	0.555	3.5	
50	1,215.96	1.0216	0.190	0.067	0.545	3.5	
Ethylene glycol, $[C_2H_4(OH)_2]$							
0	1,130.75	2.294×10^3	57.53×10^{-6}	0.242	0.934×10^{-7}	615	
20	1,116.65	2.382	19.18	0.249	0.939	204	0.65×10^{-3}
40	1,101.43	2.474	8.69	0.256	0.939	93	
60	1,087.66	2.562	4.75	0.260	0.932	51	
80	1,077.56	2.650	2.98	0.261	0.921	32.4	
100	1,058.50	2.742	2.03	0.263	0.908	22.4	

Table C·3 (*continued*)

T, °C	ρ, kg/m³	c_p, J/(kg·K)	ν, m²/s	k, W/(m·K)	α, m²/s	Pr	β, K⁻¹
Mercury, Hg							
0	13,628.22	0.1403×10^3	0.124×10^{-6}	8.20	42.99×10^{-7}	0.0288	
20	13,579.04	0.1394	0.114	8.69	46.06	0.0249	1.82×10^{-4}
50	13,505.84	0.1386	0.104	9.40	50.22	0.0207	
100	13,384.58	0.1373	0.0928	10.51	57.16	0.0162	
150	13,264.28	0.1365	0.0853	11.49	63.54	0.0134	
200	13,144.94	0.1570	0.0802	12.34	69.08	0.0116	
250	13,025.60	0.1357	0.0765	13.07	74.06	0.0103	
315.5	12,847	0.134	0.0673	14.02	81.5	0.0083	

Adapted from E. R. G. Eckert and R. M. Drake, *Analysis of Heat and Mass Transfer*, McGraw-Hill Book Company, New York, 1972. By permission.

Table C·3 Properties of saturated liquids (old English units)

T, °F	ρ, lbm/ft^3	c_p, Btu/(lbm·°F)	$\nu = \mu/\rho$, ft^2/sec	k, Btu/(h·ft·°F)	Pr	β, 1/R
Carbon dioxide (CO_2)						
−58	72.19	0.44	0.128×10^{-5}	0.0494	2.96	
−40	69.78	0.45	0.127	0.0584	2.46	
−22	67.22	0.47	0.126	0.0645	2.22	
−4	64.45	0.49	0.124	0.0665	2.12	
14	61.39	0.52	0.122	0.0635	2.20	
32	57.87	0.59	0.117	0.0604	2.38	
50	53.69	0.75	0.109	0.0561	2.80	
68	48.23	1.2	0.098	0.0504	4.10	3.67×10^{-3}
86	37.32	8.7	0.086	0.0406	28.7	
Dichlorodifluoromethane (Freon) (CCl_2F_2)						
−58	96.56	0.2090	0.334×10^{-5}	0.039	6.2	1.4×10^{-4}
−22	92.99	0.2139	0.272	0.040	4.8	
14	89.24	0.2198	0.238	0.042	4.0	
32	87.24	0.2232	0.230	0.042	3.8	
68	83.04	0.2307	0.213	0.042	3.5	
104	78.48	0.2393	0.206	0.040	3.5	
122	75.91	0.2440	0.204	0.039	3.5	
Ethylene glycol [$C_2H_4(OH)_2$]						
32	70.59	0.548	61.92×10^{-5}	0.140	615	
68	69.71	0.569	20.64	0.144	204	0.36×10^{-3}
104	68.76	0.591	9.35	0.148	93	
140	67.90	0.612	5.11	0.150	51	
176	67.27	0.633	3.21	0.151	32.4	
212	66.08	0.655	2.18	0.152	22.4	
Mercury (Hg)						
32	850.78	0.0335	0.133×10^{-5}	4.74	0.0288	
68	847.71	0.0333	0.123	5.02	0.0249	1.01×10^{-4}
122	843.14	0.0331	0.112	5.43	0.0207	
212	835.57	0.0328	0.0999	6.07	0.0162	
392	820.61	0.0325	0.0863	7.13	0.0116	
482	813.16	0.0324	0.0823	7.55	0.0103	
600	802	0.032	0.0724	8.10	0.0083	

Adapted from E. R. G. Eckert and R. M. Drake, *Heat and Mass Transfer*, 2d ed., McGraw-Hill Book Company, New York, 1959. By permission.

Table C·4 Properties of water, saturated liquid (old English units)

T, °F	c_p, $\frac{\text{Btu}}{\text{lbm}\cdot°\text{F}}$	ρ, $\frac{\text{lbm}}{\text{ft}^3}$	μ, $\frac{\text{lbm}}{\text{ft}\cdot\text{h}}$	k, $\frac{\text{Btu}}{\text{h}\cdot\text{ft}\cdot°\text{F}}$	Pr, $\frac{c_p\mu}{k}$	$\frac{g\beta\rho^2 c_p}{\mu k}$, $\frac{1}{\text{ft}^3\cdot°\text{F}}$
32	1.009	62.42	4.33	0.327	13.35	
40	1.005	62.42	3.75	0.332	11.35	0.3×10^8
50	1.002	62.38	3.17	0.338	9.40	1.0×10^8
60	1.000	62.34	2.71	0.344	7.88	1.7×10^8
70	0.998	62.27	2.37	0.349	6.78	2.3×10^8
80	0.998	62.17	2.08	0.355	5.85	3.0×10^8
90	0.997	62.11	1.85	0.360	5.12	3.9×10^8
100	0.997	61.99	1.65	0.364	4.53	5.2×10^8
110	0.997	61.84	1.49	0.368	4.04	6.6×10^8
120	0.997	61.73	1.36	0.372	3.64	7.7×10^8
130	0.998	61.54	1.24	0.375	3.30	8.9×10^8
140	0.998	61.39	1.14	0.378	3.01	10.2×10^8
150	0.999	61.20	1.04	0.381	2.73	12.0×10^8
160	1.000	61.01	0.97	0.384	2.53	13.9×10^8
170	1.001	60.79	0.90	0.386	2.33	15.5×10^8
180	1.002	60.57	0.84	0.389	2.16	17.1×10^8
190	1.003	60.35	0.79	0.390	2.03	
200	1.004	60.13	0.74	0.392	1.90	
220	1.007	59.63	0.65	0.395	1.66	
240	1.010	59.10	0.59	0.396	1.51	
260	1.015	58.51	0.53	0.396	1.36	
280	1.020	57.94	0.48	0.396	1.24	
300	1.026	57.31	0.45	0.395	1.17	
350	1.044	55.59	0.38	0.391	1.02	
400	1.067	53.65	0.33	0.384	1.00	
450	1.095	51.55	0.29	0.373	0.85	
500	1.130	49.02	0.26	0.356	0.83	
550	1.200	45.92	0.23			
600	1.362	42.37	0.21			

A. I. Brown and S. M. Marco, *Introduction to Heat Transfer*, 3d ed., McGraw-Hill Book Company, New York, 1958. By permission.

Table C·5*a* Properties of air at moderate pressure (old English units)

T, °F	μ, lbm/(hr·ft)	k, Btu/(h·ft·°F)	c_P, Btu/(lbm·°F)	Pr
−100	0.0319	0.0104	0.239	0.739
0	0.0394	0.0131	0.240	0.718
100	0.0459	0.0157	0.240	0.706
200	0.0519	0.0181	0.241	0.693
300	0.0574	0.0203	0.243	0.686
400	0.0626	0.0225	0.245	0.681
500	0.0675	0.0246	0.248	0.680
600	0.0721	0.0265	0.250	0.680
700	0.0765	0.0284	0.254	0.682
800	0.0806	0.0303	0.257	0.684
900	0.0846	0.0320	0.260	0.687
1000	0.0884	0.0337	0.263	0.690

From *National Bureau of Standards Circular* 564, 1955.

Table C·5*b* Properties of air (SI units)

T, K	c_p, J/(kg·K)	μ, kg/(m·s)	ν, m²/s (1 atm)	k, W/(m·K)	Pr
100	1.0266×10^3	0.6924×10^{-5}	1.923×10^{-6}	0.009246	0.770
150	1.0099	1.0283	4.343	0.013735	0.753
200	1.0061	1.3289	7.490	0.01809	0.739
250	1.0053	1.488	9.49	0.02227	0.722
300	1.0057	1.983	15.68	0.02624	0.708
350	1.0090	2.075	20.76	0.03003	0.697
400	1.0140	2.286	25.90	0.03365	0.689
450	1.0207	2.484	28.86	0.03707	0.683
500	1.0295	2.671	37.90	0.04038	0.680
550	1.0392	2.848	44.34	0.04360	0.680
600	1.0551	3.018	51.34	0.04659	0.680
650	1.0635	3.177	58.51	0.04953	0.682
700	1.0752	3.332	66.25	0.05230	0.684
750	1.0856	3.481	73.91	0.05509	0.686
800	1.0978	3.625	82.29	0.05779	0.689
850	1.1095	3.765	90.75	0.06028	0.692
900	1.1212	3.899	99.3	0.06279	0.696
950	1.1321	4.023	108.2	0.06525	0.699
1000	1.1417	4.152	117.8	0.06752	0.702
1100	1.160	4.44	138.6	0.0732	0.704
1200	1.179	4.69	159.1	0.0782	0.707
1300	1.197	4.93	182.1	0.0837	0.705
1400	1.214	5.17	205.5	0.0891	0.705
1500	1.230	5.40	229.1	0.0946	0.705
1600	1.248	5.63	254.5	0.100	0.705

Table C·6a Properties of gases at moderate pressures (old English units)

T, °F	c_P, Btu/(lbm·°F)	μ, lbm/(s·ft)	k, Btu/(h·ft·°F)	Pr
Helium				
−200	1.242	84.3×10^{-7}	0.0536	0.70
−100	1.242	105.2	0.0680	0.694
0	1.242	122.1	0.0784	0.70
200	1.242	154.9	0.0977	0.71
400	1.242	184.8	0.114	0.72
600	1.242	209.2	0.130	0.72
1000	1.242	256.5	0.159	0.72
Hydrogen				
−190	3.010	3.760×10^{-6}	0.0567	0.718
−100	3.234	4.578	0.0741	0.719
−10	3.358	5.321	0.0902	0.713
80	3.419	6.023	0.105	0.706
170	3.448	6.689	0.119	0.697
260	3.461	7.300	0.132	0.690
440	3.465	8.491	0.157	0.675
620	3.472	9.599	0.182	0.664
800	3.481	10.68	0.203	0.659
980	3.505	11.69	0.222	0.664
1160	3.540	12.62	0.238	0.676
Oxygen				
−190	0.2192	7.721×10^{-6}	0.00790	0.773
−100	0.2181	9.979	0.01054	0.745
−10	0.2187	12.01	0.01305	0.725
80	0.2198	13.86	0.01546	0.709
170	0.2219	15.56	0.01774	0.702
260	0.2250	17.16	0.02000	0.695
440	0.2322	20.10	0.02411	0.697
620	0.2399	22.79	0.02792	0.704
Nitrogen				
−100	0.2491	8.700×10^{-6}	0.01054	0.747
80	0.2486	11.99	0.01514	0.713
260	0.2498	14.77	0.01927	0.691
440	0.2521	17.27	0.02302	0.684
800	0.2620	21.59	0.02960	0.691
1160	0.2738	25.19	0.03507	0.711
1340	0.2789	26.88	0.03741	0.724
1700	0.2875	29.90	0.04151	0.748

Table C·6*a* (*continued*)

T, °F	c_p, Btu/(lbm·°F)	μ, lbm/(s·ft)	k, Btu/(h·ft·°F)	Pr
Carbon dioxide				
−64	0.187	7.462×10^{-6}	0.006243	0.818
−10	0.192	8.460	0.007444	0.793
80	0.208	10.051	0.009575	0.770
170	0.215	11.561	0.01183	0.755
260	0.225	12.98	0.01422	0.738
350	0.234	14.34	0.01674	0.721
440	0.242	15.63	0.01937	0.702
530	0.250	16.85	0.02208	0.685
620	0.257	18.03	0.02491	0.668

Adapted from E. R. G. Eckert and R. M. Drake, *Heat and Mass Transfer*, 2d ed., McGraw-Hill Book Company, New York, 1959. By permission.

Table C·6b Properties of gases at moderate pressures (SI units)

T, K	c_p, J/(kg·K)	μ, kg/ms	ν, m²/s	k, W/(m·K)	Pr
Helium					
3	5.200×10^3	8.42×10^{-7}		0.0106	
33	5.200	50.2	3.42×10^{-6}	0.0353	0.74
144	5.200	125.5	37.11	0.0928	0.70
200	5.200	156.6	64.38	0.1177	0.694
255	5.200	181.7	95.50	0.1357	0.70
366	5.200	230.5	173.6	0.1691	0.71
477	5.200	275.0	269.3	0.197	0.72
589	5.200	311.3	375.8	0.225	0.72
700	5.200	347.5	494.2	0.251	0.72
800	5.200	381.7	634.1	0.275	0.72
900	5.200	413.6	781.3	0.298	0.72
Hydrogen					
30	10.840×10^3	1.606×10^{-6}	1.895×10^{-6}	0.0228	0.759
50	10.501	2.516	4.880	0.0362	0.721
100	11.229	4.212	17.14	0.0665	0.712
200	13.540	6.813	55.53	0.1282	0.719
300	14.314	8.963	109.5	0.182	0.706
400	14.491	10.864	177.1	0.228	0.690
500	14.507	12.636	257.0	0.272	0.675
600	14.537	14.285	349.7	0.315	0.664
800	14.675	17.40	569	0.384	0.664
1000	14.968	20.16	822	0.440	0.686
1200	15.366	22.75	1107	0.488	0.715
Oxygen					
100	0.9479×10^3	7.768×10^{-6}	1.946×10^{-6}	0.00903	0.815
150	0.9178	11.490	4.387	0.01367	0.773
200	0.9131	14.850	7.593	0.01824	0.745
250	0.9157	17.87	11.45	0.02259	0.725
300	0.9203	20.63	15.86	0.02676	0.709
350	0.9291	23.16	20.80	0.03070	0.702
400	0.9420	25.54	26.18	0.03461	0.695
450	0.9567	27.77	31.99	0.03828	0.694
500	0.9722	29.91	38.34	0.04173	0.697
550	0.9881	31.97	45.05	0.04517	0.700
600	1.0044	33.92	52.15	0.04832	0.704

Table C·6*b* *(continued)*

T, K	c_p, J/(kg·K)	μ, kg/ms	ν, m²/s	k, W/(m·K)	Pr
Nitrogen					
100	1.0722×10^3	6.862×10^{-6}	1.971×10^{-6}	0.009450	0.786
200	1.0429	12.947	7.568	0.01824	0.747
300	1.0408	17.84	15.63	0.02620	0.713
400	1.0459	21.98	25.74	0.03335	0.691
500	1.0555	25.70	37.66	0.03984	0.684
600	1.0756	29.11	51.19	0.04580	0.686
700	1.0969	32.13	65.13	0.05123	0.691
800	1.1225	34.84	81.46	0.05609	0.700
900	1.1464	37.49	91.06	0.06070	0.711
1000	1.1677	40.00	117.2	0.06475	0.724
1100	1.1857	42.28	136.0	0.06850	0.736
1200	1.2037	44.50	156.1	0.07184	0.748
Carbon dioxide					
220	0.783×10^3	11.105×10^{-6}	4.490×10^{-6}	0.010805	0.818
250	0.804	12.590	5.813	0.012884	0.793
300	0.871	14.958	8.321	0.016572	0.770
350	0.900	17.205	11.19	0.02047	0.755
400	0.942	19.32	14.39	0.02461	0.738
450	0.980	21.34	17.90	0.02897	0.721
500	1.013	23.26	21.67	0.03352	0.702
550	1.047	25.08	25.74	0.03821	0.685
600	1.076	26.83	30.02	0.04311	0.668

Adapted from E. R. G. Eckert and R. M. Drake, *Analysis of Heat and Mass Transfer*, McGraw-Hill Book Company, New York, 1972. By permission.

Table C·7 Normal total emissivity of various surfaces

Surface	T, °F	Emissivity, ε	T, °C
A. Metals and their oxides			
Aluminum			
Highly polished plate, 98.3% pure	440–1070	0.039–0.057	226–576
Commercial sheet	212	0.09	100
Heavily oxidized	299–940	0.20–0.31	148–504
Al-surfaced roofing	100	0.216	38
Chromium (see nickel alloys for Ni–Cr steels), polished	100–2000	0.08–0.36	38–1093
Copper			
Polished	242	0.023	117
Polished	212	0.052	100
Plate, heated long time, covered with thick oxide layer	77	0.78	25
Gold, pure, highly polished	440–1160	0.018–0.035	226–627
Iron and steel (not including stainless)			
Steel, polished	212	0.066	100
Iron, polished	800–1880	0.14–0.38	427–1027
Cast iron, newly turned	72	0.44	22
Cast iron, turned and heated	1620–1810	0.60–0.70	882–988
Mild steel	450–1950	0.20–0.32	232–1065
Oxidized surfaces			
Iron plate, pickled, then rusted red	68	0.61	20
Iron, dark-gray surface	212	0.31	100
Rough ingot iron	1700–2040	0.87–0.95	927–1115
Sheet steel with strong, rough oxide layer	75	0.80	24
Nickel			
Polished	212	0.072	100
Nickel oxide	1200–2290	0.59–0.86	649–1254
Nickel alloys			
Copper nickel, polished	212	0.059	100
Nichrome wire, bright	120–1830	0.65–0.79	49–1000
Nichrome wire, oxidized	120–930	0.95–0.98	49–500
Platinum, polished plate, pure	440–1160	0.054–0.104	226–627
Silver			
Polished, pure	440–1160	0.020–0.032	226–627
Polished	100–700	0.022–0.031	38–371
Stainless steels			
Polished	212	0.074	100
Type 301	450–1725	0.54–0.63	232–940
Tin, bright tinned iron	76	0.043 and 0.064	24
Tungsten, filament	6000	0.39	3315

Table C·7 (*continued*)

Surface	T, °F	Emissivity, ε	T, °C
B. Refractories, building materials, paints, and miscellaneous			
Alumina (85–99.5% Al_2O_3, 0–12% SiO_2, 0–1% Fe_2O_3); effect of mean grain size (μm)			
10 μm	...	0.30–0.18	
50 μm	...	0.39–0.28	
100 μm	...	0.50–0.40	
Asbestos, board	74	0.96	23
Brick			
Red, rough, but no gross irregularities	70	0.93	21
Fireclay	1832	0.75	1000
Carbon			
T carbon (Gebruder Siemens) 0.9% ash, started with emissivity at 260°F of 0.72, but on heating changed to values given	260–1160	0.81–0.79	127–627
Filament	1900–2560	0.526	1038–1404
Rough plate	212–608	0.77	100–320
Lampblack, rough deposit	212–932	0.84–0.78	100–500
Concrete tiles	1832	0.63	1000
Enamel, white fused, on iron	66	0.90	19
Glass			
Smooth	72	0.94	22
Pyrex, lead, and soda	500–1000	0.95–0.85	260–538
Paints, lacquers, varnishes			
Snow-white enamel varnish on rough iron plate	73	0.906	23
Black shiny lacquer, sprayed on iron	76	0.875	24
Black shiny shellac on tinned iron sheet	70	0.821	21
Black matte shellac	170–295	0.91	77–146
Black or white lacquer	100–200	0.80–0.95	38–93
Flat black lacquer	100–200	0.96–0.98	38–93
Aluminum paints and lacquers			
10% Al, 22% lacquer body, on rough or smooth surface	212	0.52	100
Other Al paints, varying age and Al content	212	0.27–0.67	100
Porcelain, glazed	72	0.92	22
Quartz, rough, fused	70	0.93	21
Roofing paper	69	0.91	21
Rubber, hard, glossy plate	74	0.94	23
Water	32–212	0.95–0.963	0–100

Adapted from W. H. McAdams, *Heat Transmission*, 3d ed., McGraw-Hill Book Company, New York, 1954. By permission.

Table C·8 Convection heat transfer correlations

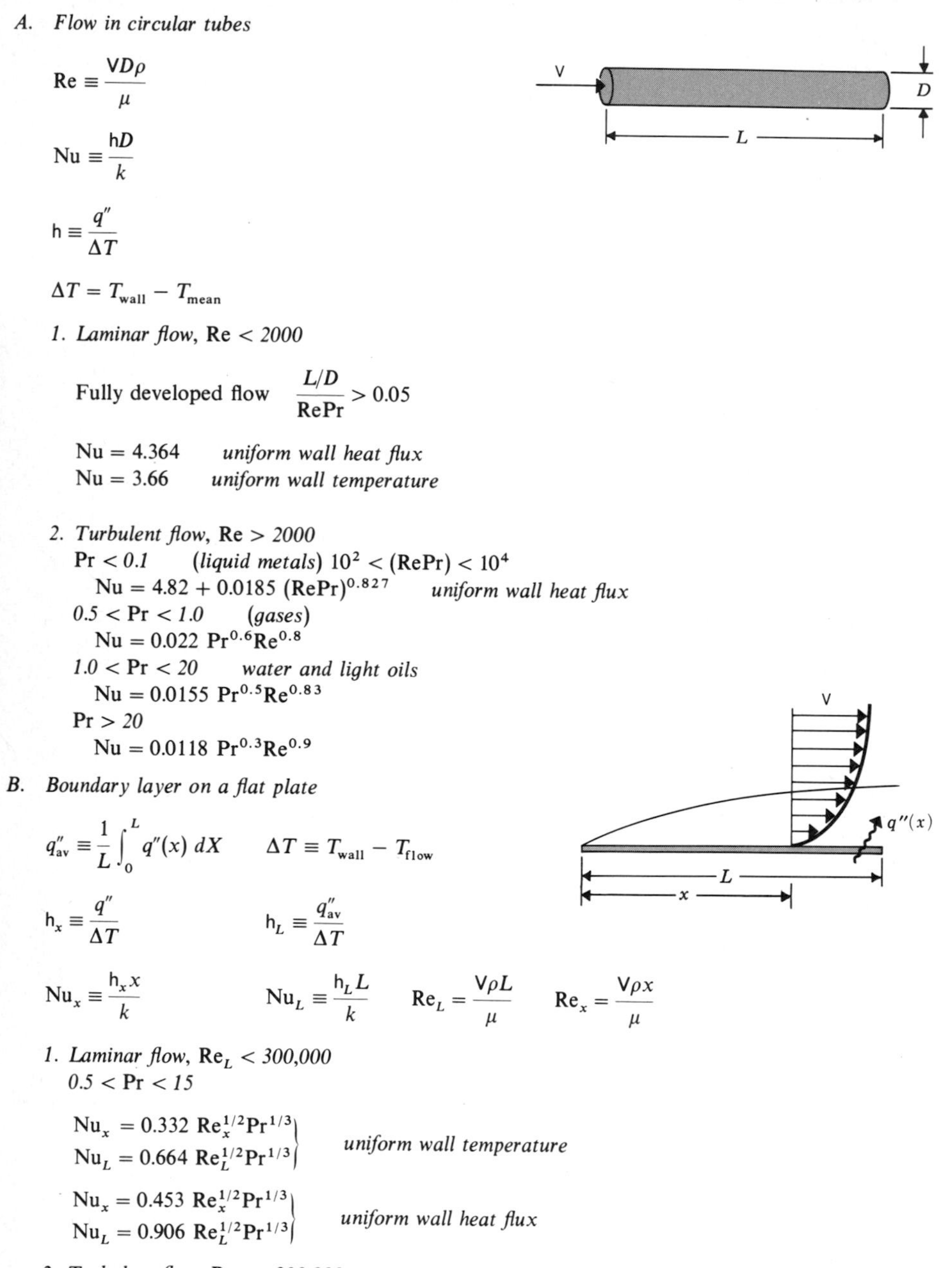

A. Flow in circular tubes

$$\mathrm{Re} \equiv \frac{\mathsf{V}D\rho}{\mu}$$

$$\mathrm{Nu} \equiv \frac{\mathsf{h}D}{k}$$

$$\mathsf{h} \equiv \frac{q''}{\Delta T}$$

$$\Delta T = T_{\text{wall}} - T_{\text{mean}}$$

1. *Laminar flow,* Re < *2000*

Fully developed flow $\dfrac{L/D}{\mathrm{RePr}} > 0.05$

Nu = 4.364 *uniform wall heat flux*
Nu = 3.66 *uniform wall temperature*

2. *Turbulent flow,* Re > *2000*

Pr < *0.1* (*liquid metals*) $10^2 < (\mathrm{RePr}) < 10^4$
$\mathrm{Nu} = 4.82 + 0.0185\,(\mathrm{RePr})^{0.827}$ *uniform wall heat flux*
0.5 < Pr < *1.0* (*gases*)
$\mathrm{Nu} = 0.022\ \mathrm{Pr}^{0.6}\mathrm{Re}^{0.8}$
1.0 < Pr < *20* *water and light oils*
$\mathrm{Nu} = 0.0155\ \mathrm{Pr}^{0.5}\mathrm{Re}^{0.83}$
Pr > *20*
$\mathrm{Nu} = 0.0118\ \mathrm{Pr}^{0.3}\mathrm{Re}^{0.9}$

B. Boundary layer on a flat plate

$$q''_{\text{av}} \equiv \frac{1}{L}\int_0^L q''(x)\,dX \qquad \Delta T \equiv T_{\text{wall}} - T_{\text{flow}}$$

$$\mathsf{h}_x \equiv \frac{q''}{\Delta T} \qquad \mathsf{h}_L \equiv \frac{q''_{\text{av}}}{\Delta T}$$

$$\mathrm{Nu}_x \equiv \frac{\mathsf{h}_x x}{k} \qquad \mathrm{Nu}_L \equiv \frac{\mathsf{h}_L L}{k} \qquad \mathrm{Re}_L = \frac{\mathsf{V}\rho L}{\mu} \qquad \mathrm{Re}_x = \frac{\mathsf{V}\rho x}{\mu}$$

1. *Laminar flow,* Re_L < *300,000*
0.5 < Pr < *15*

$\mathrm{Nu}_x = 0.332\ \mathrm{Re}_x^{1/2}\mathrm{Pr}^{1/3}$
$\mathrm{Nu}_L = 0.664\ \mathrm{Re}_L^{1/2}\mathrm{Pr}^{1/3}$
} *uniform wall temperature*

$\mathrm{Nu}_x = 0.453\ \mathrm{Re}_x^{1/2}\mathrm{Pr}^{1/3}$
$\mathrm{Nu}_L = 0.906\ \mathrm{Re}_L^{1/2}\mathrm{Pr}^{1/3}$
} *uniform wall heat flux*

2. *Turbulent flow,* Re_L > *300,000*
0.5 < Pr < *10*
$\mathrm{Nu}_x = 0.0295\ \mathrm{Pr}^{0.6}\mathrm{Re}^{0.8}$

Table C·8 *(continued)*

C. Single circular cylinder or sphere in cross flow

$$\text{Re} \equiv \frac{VD\rho}{\mu} \qquad \text{Nu} \equiv \frac{hD}{k}$$

$$h \equiv \frac{q''_{av}}{\Delta T} \qquad \Delta T = T_{wall} - T_{fluid}$$

1. *Cylinder*
   $\text{Nu} = C(\text{Re})^n$

Re	C, gases	C, liquids	n
4–40	0.821	0.911 $\text{Pr}^{1/3}$	0.385
40–4000	0.615	0.683 $\text{Pr}^{1/3}$	0.466
4000–40,000	0.174	0.193 $\text{Pr}^{1/3}$	0.618
40,000–400,000	0.0239	0.0266 $\text{Pr}^{1/3}$	0.805

2. *Sphere*
   $\text{Nu} = 0.37\ \text{Re}^{0.6}$
   $17 < \text{Re} < 70{,}000$, gases
   $\text{Nu} = (1.2 + 0.53\ \text{Re}_D^{0.54})\ \text{Pr}^{0.3}$
   $1 < \text{Re} < 200{,}000$, $\text{Pr} > 3$

D. Natural convection from a horizontal cylinder (Nu, Gr *based on diameter D*)
$\text{Nu} = C(\text{GrPr})^n$

GrPr	C	n
10^3–10^9	0.53	$\frac{1}{4}$
10^9–10^{12}	0.13	$\frac{1}{3}$

E. Natural convection from vertical surfaces (Nu, Gr *based on height L*)
$\text{Nu} = C(\text{GrPr})^n$

GrPr	C	n
10^5–10^9	0.555	0.25
$> 10^9$	0.021	0.4

Compiled from W. M. Kays, *Convective Heat and Mass Transfer*, McGraw-Hill Book Company, New York, 1966; and W. H. McAdams, *Heat Transmission*, 3d ed., McGraw-Hill Book Company, New York, 1954.

Table C·9*a* Counterflow heat-exchanger performance

NTU	ε for indicated capacity rate ratios, C_{min}/C_{max}				
	0	0.25	0.50	0.75	1.00
0	0	0	0	0	0
0.25	0.221	0.216	0.209	0.205	0.200
0.50	0.393	0.378	0.362	0.347	0.333
0.75	0.528	0.502	0.476	0.452	0.429
1.00	0.632	0.598	0.565	0.532	0.500
1.25	0.713	0.674	0.635	0.598	0.556
1.50	0.777	0.735	0.691	0.645	0.600
1.75	0.826	0.784	0.737	0.687	0.636
2.00	0.865	0.823	0.774	0.722	0.667
2.50	0.918	0.880	0.832	0.776	0.714
3.00	0.950	0.919	0.874	0.817	0.750
3.50	0.970	0.945	0.905	0.848	0.778
4.00	0.982	0.962	0.927	0.873	0.800
4.50	0.989	0.974	0.944	0.893	0.818
5.00	0.993	0.982	0.957	0.909	0.833
∞	1	1	1	1	1

From W. M. Kays and A. L. London, *Compact Heat Exchangers*, 2d ed., McGraw-Hill Book Company, New York, 1964. By permission.

Table C·9*b* Parallel-flow heat-exchanger performance

NTU	ε for indicated capacity rate ratios, C_{min}/C_{max}				
	0	0.25	0.50	0.75	1.00
0	0	0	0	0	0
0.25	0.221	0.215	0.208	0.202	0.197
0.50	0.393	0.372	0.352	0.333	0.316
0.75	0.528	0.487	0.450	0.418	0.388
1.00	0.632	0.571	0.518	0.472	0.432
1.25	0.713	0.632	0.564	0.507	0.459
1.50	0.777	0.677	0.596	0.530	0.475
1.75	0.826	0.710	0.618	0.544	0.489
2.00	0.865	0.734	0.633	0.554	0.491
2.50	0.918	0.765	0.651	0.564	0.497
3.00	0.950	0.781	0.659	0.568	0.498
3.50	0.970	0.790	0.663	0.570	0.499
4.00	0.982	0.795	0.665	0.571	0.500
4.50	0.989	0.797	0.666	0.571	0.500
5.00	0.993	0.799	0.666	0.571	0.500
∞	1	0.800	0.6667	0.5714	0.500

From W. M. Kays and A. L. London, *Compact Heat Exchangers*, 2d ed., McGraw-Hill Book Company, New York, 1964. By permission.

Table C·9c Cross-flow heat exchanger with both fluids unmixed

NTU	ε for the indicated capacity rate ratios, C_{min}/C_{max}: 0	0.25	0.50	0.75	1.00
0	0	0	0	0	0
0.25	0.221	0.215	0.208	0.204	0.199
0.50	0.393	0.376	0.357	0.341	0.327
0.75	0.528	0.495	0.467	0.437	0.412
1.00	0.632	0.587	0.546	0.511	0.477
1.25	0.713	0.659	0.610	0.565	0.522
1.50	0.777	0.715	0.657	0.606	0.560
1.75	0.826	0.760	0.700	0.643	0.590
2.00	0.865	0.797	0.732	0.671	0.613
2.50	0.918	0.851	0.782	0.715	0.652
3.00	0.950	0.888	0.819	0.750	0.680
3.50	0.970	0.920	0.848	0.775	0.703
4.00	0.982	0.930	0.870	0.800	0.722
4.50	0.989	0.946	0.888	0.815	0.737
5.00	0.993	0.959	0.900	0.830	0.750
∞	1	1	1	1	1

From W. M. Kays and A. L. London, *Compact Heat Exchangers*, 2d ed., McGraw-Hill Book Company, New York, 1964. By permission.

NOMENCLATURE

A	Area
	Helmholtz function, $U - TS$
a	Acceleration
	Helmholtz function per unit of mass, $u - Ts$
$\hat{a}$	Helmholtz function per mole, $\hat{u} - T\hat{s}$
B, $\mathbf{B}$	Magnetic induction
C	Constant, defined where used
	Curie constant of a substance
	Thermal capacitance
CM	Control mass
CV	Control volume
cop	Coefficient of performance
c	Speed of light
	Speed of sound
c_P	Specific heat at constant pressure
$\hat{c}_P$	Molal specific heat at constant pressure
c_v	Specific heat at constant volume
$\hat{c}_v$	Molal specific heat at constant volume
c	$c = c_P = c_v$ for an incompressible substance
c_H	Specific heat in a constant applied magnetic field
c_M	Specific heat at constant magnetization
D, $\mathbf{D}$	Electric displacement
$\mathcal{E}$	Electrostatic potential
e	Charge of an electron
E	Energy
	Emissive power
e	Energy per unit of mass
E, $\mathbf{E}$	Electric field strength
F	Shape factor
$\mathfrak{F}$	Gray body radiation factor
F, $\mathbf{F}$	Force

$f(x)$	Function of x
f	Fraction
g_c	Constant in Newton's law, $\mathsf{F} = (1/g_c)Ma$
g	Local acceleration of gravity
g	Gibbs function per unit of mass, $h - Ts$
$\hat{g}$	Gibbs function per mole, $\hat{h} - T\hat{s}$
G	Gibbs function, $U + PV - TS$
	Mass velocity, $\rho\mathsf{V}$
ΔG_r	Gibbs-function change for a complete unit reaction
G, $\mathbf{G}$	Gravitational field strength
$\mathfrak{g}$	Magnetic Gibbs function per unit of mass, $\mathfrak{h} - Ts$
h	Enthalpy per unit of mass, $u + Pv$
$\hat{h}$	Enthalpy per mole, $u + Pv$
$\mathfrak{h}$	Magnetic enthalpy, $u - \mu_0 v\mathbf{H} \cdot \mathbf{M}$
h	Planck's constant
	Heat transfer coefficient
H, $\mathbf{H}$	Magnetic field strength
Δh_f°	Enthalpy of formation of a mole of compound from its elements at the standard reference state
ΔH_r	Enthalpy change for a complete unit reaction
i	Electric current
I	Moment of inertia
	Impulse function
k	Ratio of specific heats, c_P/c_v
	Thermal conductivity
k	Boltzmann constant
k_N	Constant in Newton's law, $k_N = 1/g_c$
k_C	Constant in Coulomb's law, $k_C = 1/(4\pi\epsilon_0)$
k_B	Constant in Biot-Savart's law, $k_B = \mu_0/(4\pi)$
k_G	Constant in the gravitational law
K	Equilibrium constant for ideal-gas reactions
KE	Kinetic energy
$\mathcal{K}$	An amount of entropy transfer with heat
$\dot{\mathcal{K}}$	Rate of entropy transfer with heat
L	Length
M	Mass
m	Molecular weight (dimensionless)
	Fin parameter, $m^2 = \mathsf{h}p/kA$
$\hat{M}$	Molal mass
m	Mass of a particle
$\dot{M}$	Mass flow rate
M, $\mathbf{M}$	Magnetic dipole moment per unit of volume
	Momentum

M	Mach number
n, N	Number of particles
$\mathfrak{N}$	Number of moles
N_0	Avogadro's number
NTU	Number of transfer units
Nu	Nusselt number, $\mathsf{h}L/k$
Pr	Prandtl number, $\mu c_P/k$
PE	Potential energy
P	Pressure
P, $\mathbf{P}$	Electric-dipole moment per unit of volume
$\mathcal{P}_E$, $\mathcal{P}_S$, $\mathcal{P}_{\mathbf{M}}$	Amounts of energy, entropy, momentum production
$\dot{\mathcal{P}}_E$, $\dot{\mathcal{P}}_S$, $\dot{\mathcal{P}}_{\mathbf{M}}$	Rates of energy, entropy, momentum production
p	Probability
	Perimeter of fin
p	Momentum of a particle
p_r	Reduced pressure
P^*	Pressure ratio
	Pressure in a pure phase
Q	An amount of energy transfer as heat
$\dot{Q}$	Rate of energy transfer as heat
$\mathcal{Q}$	Charge
q''	Heat flux
$\mathbf{q}''$	Heat-flux vector
R	Thermal resistance
Re	Reynolds number, $\mathsf{V}L\rho/\mu$
R	Gas constant for a particular gas, $R = \mathcal{R}/\hat{M}$
$\mathcal{R}$	Universal gas constant
r	Radius
S	Entropy
s	Entropy per unit of mass
$\hat{s}$	Entropy per mole
$\hat{s}^\circ$	Absolute entropy of a substance, per mole, at the standard reference state
ΔS_r	Entropy change for a complete unit reaction
s	Heat source strength
St	Stanton number, $\mathsf{h}/(\mathsf{V}\rho c_P)$
T	Absolute temperature
T^*	Temperature ratio
t	Time
U	Internal energy
u	Internal energy per unit of volume
u	Internal energy per unit of mass
$\hat{u}$	Internal energy per mole

U	Overall heat transfer conductance
V	Volume
V, $\mathbf{V}$	Velocity
V_m	Mean velocity
v	Volume per unit of mass
$\hat{v}$	Volume per mole
W	An amount of energy transfer as work
$\dot{W}$	Rate of energy transfer as work
x	Quality of a two-phase mixture
X	Directional coordinate
x_i	Generalized intensified property (specific volume, magnetization, etc.)
X_i	Gradient of a natural intensified property
Z	Compressibility, $Z = Pv/RT$
x, y, z	Coordinates
α	Isentropic compressibility
	Absorptivity
β	Isobaric compressibility
δ	Boundary layer thickness
ϵ	Emissivity
	Fin effectiveness
ϵ_i	Energy of a system in quantum state i
ε	Energy of a particle
ϵ_0	Permittivity of a vacuum
ϕ	Function of temperature for an ideal gas
	Relative humidity
	Scalar potential
γ	Specific humidity
κ	Isothermal compressibility
λ	Wavelength
μ	Electrochemical potential per unit of mass
	Viscosity
$\hat{\mu}$	Electrochemical potential per mole
μ_0	Permeability of a vacuum
Ω	Number of quantum states
ω	Angular frequency
Ψ	Volume fraction
Φ	Mass fraction
η	Cycle energy-conversion efficiency
η_s	Isentropic efficiency
σ	Stefan-Boltzmann constant
	Entropy production per unit of volume
	Interfacial tension
θ	Angle

ν	Frequency
	Kinematic viscosity
ν_i	Stoichiometric coefficients in a chemical equation
τ	Torque
	Transmissivity
ρ	Density
	Reflectivity
χ	Mole fraction

Special notations

d	An infinitesimal increase in a property of matter
đ	An infinitesimal amount of transfer by some mechanism
Δ	A finite increase in a property of matter
	$\Delta \equiv$ final − initial
$(\partial y/\partial x)_z$	The partial derivative of y with respect to x, obtained from the function $y(x,z)$
$f(x)$, $f(x,y)$	Functional relations
$f \cdot (x)$	f times x, used to avoid misreading as the function $f(x)$
x	The time-average value of a quantity of x
$\mathbf{B}$	The vector $\mathbf{B}$
B	The magnitude of vector $\mathbf{B}$
$\hat{u}$	A molal quantity
$\equiv$	Identity symbol, used when the equation defines the quantity on the left
$\overset{d}{=}$	Means "has the dimensions of"
$\sum_i x_i$	The sum $x_1 + x_2 + \cdots + x_n$
$\prod_i x_i$	The product $x_1 x_2 x_3 \cdots x_n$
$\dot{W}$, $\dot{Q}$, $\dot{M}$	Rates of transfer or flow; *not* to be interpreted as time derivatives
SI	System Internationale

Frequently used subscripts

W_{12}, Q_{12}	Amounts of energy transfer as work and heat corresponding to a charge from state 1 to state 2
W_1, Q_1	Amounts of energy transfer as work and heat for process 1; *not* to be interpreted as the "work and heat at state 1"
h_f, h_g, h_s	Saturated-liquid, saturated-vapor, and saturated-solid states
h_{fg}, s_{fg}	$h_g - h_f$, $s_g - s_f$, etc.
h_{crit}	The critical state
χ_i	"Dummy indices" which could take on any of the possible integer values
T_0	Stagnation conditions
T_{sat}	Saturation conditions
T_x	Upstream of a shock wave
T_y	Downstream of a shock wave

SELECTED ANSWERS

1·8	Asteroid, 5.1×10^{24} J
1·9	1 kg = 6.67×10^{-1} m^3/s^2; 1 C = 0.774 m^3/s^2
1·15	Coal, 4.3×10^6 tons/year; ash, 12.8×10^6 tons
2·11	150 ft·lbf; 15,150 ft·lbf; 30,150 ft·lbf
2.12	28,800 ft·lbf; 19,950 ft·lbf; 43,200 ft·lbf; 67,200 ft·lbf
2·17	1.27×10^5 J, 517 m/s
2·19	72.5 Btu input
2·22	$W = P_1V_1$, $Q = W$
3·4	43.3 Btu/lbm
3·7	11.67 Btu/lbm
3·8	21 kJ/kg
3·9	50.7 Btu/lbm
4·1	0.173 ft^3/lbm, h = 80.7 Btu/lbm, u = 74.9 Btu/lbm, 0.0045, 0.9955
4·9	$x = 23\%$
4·14	0.292 Btu/(lbm·R)
4·21	394°F
4·24	$c_v \simeq 4$ Btu/lbm·°F, $c_P = \infty$
4·27	$\mathbf{M}/C$; C/T; $C/\mathbf{M}$; $-C\mathbf{H}/\mathbf{M}^2$
4·33	$c_v = 0.738$ kJ/(kg·K); $c_p = 1.034$ kJ/(kg·K)
5·2	$h_{fg} = 78$ Btu/lbm
5·5	81.3 Btu

5·9 19 Btu

5·13 1-2: $Q_{in} = c_p(T_2 - T_1)$; $W_{out} = R(T_2 - T_1)$
2-3: $Q_{out} = c_v(T_2 - T_3)$; $W = 0$
3-1: $Q_{out} = W_{in} = RT_1 \ln(P_1/P_3)$
eff. = 0.0884

5·20 2740 Btu/min

5·23 3.7 hp; 2.9 hp

5·31 $\simeq$960°F

5·37 385°F; 0.184

6·2 $1/N$ if molecules are indistinguishable
$1/2^N$ if distinguishable

6·10 1/6; 1/6; 10/36;
H: 1, 5/6, 26/36, ..., 1/2
T: 0, 1/6, 10/36, ..., 1/2
S = 0, 0.450, 0.591, ..., 0.693

6·15 $e^{10^{25.2}}$; $10^{(0.378 \times 10^{25.2} - 22)}$

7·4 $(\partial T/\partial u)_{\mathbf{M}} > 0$

7·17 769 Btu/(h·R)

7·20 6580 kW

7·22 0.06 K

7·27 $(1 - x) \cong 0.09$

8·2 420 ft/s

8·5 844 R, 61.5 Btu/lbm, 7.8 Btu/lbm

8·13 $u_2 - u_1 = c_p(T_2 - T_1)$
$s_2 - s_1 = c_p \ln(T_2 - T_1) - (vA/2)(P_2^2 - P_1^2)$

8·18 632.2 kJ/kg, 633.5 kJ/kg, 657.8 kJ/kg

8.24 184 kJ/kg

8·35 $v = 0.137$ ft^3/lbm, $V = 343$ ft^3

8·41 68 atm, 47.2 atm, 47.4 atm

8·51 2164 psia

8·63 $u_2 - u_1 = c_p(T_2 - T_1)$
$s_2 - s_1 = c_p\ ln(T_2/T_1) - (vA/2)(P_2^2 - P_1^2)$

9·1 0.73

9·6 3135 ft/s

9·8 0.21

9·10 81.2%

9·22 26.5%, $W = 705.6$ kJ/kg

9·29 0.39, 8.6×10^8 Btu/h, 6.6×10^5 lbm/h

9·46 2.56 hp

9·53 Isentropic, 170 Btu/lbm, isothermal 107 Btu/lbm

9·76 No bypass, compressor 48 Btu/lbm, turbine 65 Btu/lbm, reactor 90.5 Btu/lbm

9·83 $T_{max} = 4080$R, 50%

9·86 212 lbf/(lbm/s), 1030 lbf/(lbm/s)

9·89 500(N·s)/kg

10·6	$k = 1.42$, $\hat{M} = 2.31$
10·10	445 K, 3360 kJ/kg mole-mix
10·17	$\chi_{He} = 0.342$; $\chi_{O_2} = 0.658$, $M = 22.4$ lbm/lbmole $c_P = 0.282$ Btu/(lbm·R)
10·24	575 R
10·28	0.0057 kg/kg dry air, 0.96 kg/min
10·34	65°F, 43%
10·44	66°F, 66%, 0.0092 lbm H_2O/lbm-dry-air
10·48	0.42
10·56	$\chi_{CO_2} = \chi_{He} = 0.5$
11·4	15.2 lbm/lb-fuel
11·10	104°F, 13.2 lbm-air/lbm-fuel, 71°F
11·16	−22,350 Btu/lbm-fuel, −21,500 Btu/lbm-fuel
11·26	Wet, 2.5%; dry, 3%
11·31	4330 R
11·37	1.017×10^6 Btu/h
11·39	156,000 Btu/lb mole-fuel
12·5	$\Delta P \simeq 0.05$ psia
12·6	$K = 0.0242$; $\log_{10} K = -1.62$
12·14	$\log_{10} K = -4.98$
12·23	$0.65\ CO_2 + 0.35\ CO + 0.675\ O_2$(3000 K)
12·27	4.7×10^{15} elec/cm^3
13·2	S.I. = 70.5 lbf/(lbm/sec), (40 lbf) 1×10^{-2}ft^2
13·10	Error $\simeq$ 15%
13·14	1128, 3306, 720, 1393, 704 ft/s
13·17	238 ft/s
13·29	692 ft/s, 665 ft/s
13·37	3.05 kg/s
13·48	0.64 lbm/s, 0.791 lbm/s, 0.791 lbm/s
13·51	Exit; 119 R, 0.658 psia, $A = 10.7A^*$
13·58	106 m/s
13·64	155 psia, M = 0.25
14·3	5775 Btu/(h·ft^2)
14·10	44% reduction
14·20	$q'' = 1.36 \times 10^5$ Btu/(h·ft^2), $T = 642$°F
14·32	7370 W/(m^2·K), H_2O
14·34	7, 47, 1280 Btu/(h·ft^2·°F)
14·38	$T \simeq 220$°F
14·47	Plastic ok
14·57	Ntu = 1.25, $T = 96.3$°C
14·68	63°C
14·75	2.3 cal/h

INDEX